An Introduction to Analysis

An Introduction to Analysis

Fourth Edition

Pearson Modern Classic

William R. Wade
University of Tennessee

Editor-in-Chief: Deirdre Lynch
Senior Acquisitions Editor: William Hoffman
Associate Editor: Caroline Celano
Project Manager, Production: Robert S. Merenoff
Associate Managing Editor: Bayani Mendoza de Leon
Senior Managing Editor: Linda Mihatov Behrens
Operations Specialists: Ilene Kahn/Carol Melville
Executive Marketing Manager: Jeff Weidenaar
Marketing Assistant: Kendra Bassi
Senior Design Supervisor: Andrea Nix
Art Director / Cover Designer: Barbara Atkinson
Compositor / Art studio: Integra Software Services, Pvt. Ltd, Pondicherry, India
Photographer: ©Matt C. Jones
Cover image: 360 Bridge in Austin, TX

Library of Congress Cataloging-in-Publication Data

Names: Wade, W. R.
Title: An introduction to analysis / William R. Wade, University of Tennessee.
Description: Fourth edition [2018 reissue]. | New York, NY : Pearson, [2018]
  | Series: Pearson modern classic | Originally published in 2010, reissued
  as part of Pearson's modern classic series. | Includes bibliographical
  references and index.
Identifiers: LCCN 2016055242| ISBN 9780134707624 (pbk.) | ISBN 0134707621
  (pbk.)
Subjects: LCSH: Mathematical analysis--Textbooks.
Classification: LCC QA300 .W25 2018 | DDC 515--dc23
LC record available at https://lccn.loc.gov/2016055242

4   18

ISBN 13: 978-0-13-470762-4
ISBN 10:   0-13-470762-1

To Cherri, Peter, and Benjamin

Contents

Preface

This text provides a bridge from "sophomore" calculus to graduate courses which use analytic ideas (e.g., real and complex analysis, partial and ordinary differential equations, numerical analysis, fluid mechanics, and differential geometry). For a two-semester course, the first semester should end with Chapter 8. For a three-quarter course, the second quarter should begin in Chapter 6 and end somewhere in the middle of Chapter 11.

Our presentation is divided into two parts. The first half, Chapters 1 through 7 together with Appendices A and B, gradually introduces the central ideas of analysis in a one-dimensional setting. The second half, Chapters 8 through 14 together with Appendices C through F, covers multidimensional theory.

More specifically, Chapter 1 introduces the real number system as a complete, ordered field; Chapters 2 through 5 cover calculus on the real line; and Chapters 6 and 7 discuss infinite series, including uniform and absolute convergence. The first two sections of Chapter 8 give a short introduction to the algebraic structure of $\mathbf{R}^n$, including the connection between linear functions and matrices.

At that point instructors have two options. They can continue covering Chapters 8 and 9 to explore topology and convergence in the concrete Euclidean space setting, or they can cover these same concepts in the abstract metric space setting (Chapter 10). Since either of these options provides the necessary foundation for the rest of the book, instructors are free to choose the approach they feel best suits their aims.

With this background material out of the way, Chapters 11 through 13 develop the machinery and theory of vector calculus. Chapter 14 gives a short introduction to Fourier series, including summability and convergence of Fourier series, growth of Fourier coefficients, and uniqueness of trigonometric series.

Separating the one-dimensional from the n-dimensional material is not the most efficient way to present the material, but it does have two advantages. The more abstract, geometric concepts can be postponed until students have been given a thorough introduction to analysis on the real line. And, students have two chances to master some of the deeper ideas of analysis (e.g., convergence of sequences, limits of functions, and uniform continuity).

We have made this text flexible in another way by including core material and enrichment material. The core material provides a foundation for the typical one-year course in analysis. Besides making the book a better reference, the enrichment material has been included for two other reasons: curious students can use it to delve deeper into the core material or as a jumping off place to pursue more general topics, and instructors can use it to supplement their course or to add variety from year to year.

Enrichment and optional materials are marked with an asterisk. Exercises which use enrichment material are also marked with an asterisk, and the material needed to solve them is cited in the Answers and Hints section. To make course planning easier, each enrichment section begins with a statement which indicates whether that section uses material from any other enrichment section. Since no core material depends on enrichment material, any of the latter can be skipped without loss in the integrity of the course.

Most enrichment sections (5.5, 5.6, 6.5, 6.6, 7.5, 9.5, 11.6, 12.6, 14.1) are independent and can be covered in any order after the core material which precedes them has been dealt with. Sections 9.6 and 12.5 require 9.5, Section 14.3 requires 5.5 only to establish Lemma 14.24. This result can easily be proved for continuously differentiable functions, thereby avoiding mention of functions of bounded variation. In particular, the key ideas in Section 14.3 can be covered without the background material from Section 5.5 anytime after finishing Chapter 7.

Since for many students this is the last (for some the only) place to see a rigorous development of vector calculus, we focus our attention on classical, nitty-gritty analysis. By avoiding abstract concepts such as vector spaces and the Lebesgue integral, we have room for a thorough, comprehensive introduction. We include sections on improper integration, the gamma function, Lagrange multipliers, the Inverse and Implicit Function Theorem, Green's Theorem, Gauss's Theorem, Stokes's Theorem, and a full account of the change of variables formula for multiple Riemann integrals.

We assume the reader has completed a three-semester or four-quarter sequence in elementary calculus. Because many of our students now take their elementary calculus in high school (where theory may be almost nonexistent), we assume that the reader is familiar only with the mechanics of calculus, i.e., can differentiate, integrate, and graph simple functions of the form $y = f(x)$ or $z = f(x, y)$. We also assume the reader has had an introductory course in linear algebra, i.e., can add, multiply, and take determinants of matrices with real entries, and are familiar with Cramer's Rule. (Appendix C, which contains an exposition of all definitions and theorems from linear algebra used in the text, can be used as review if the instructor deems it necessary.)

Always we emphasize the fact that the concepts and results of analysis are based on simple geometric considerations and on analogies with material already known to the student. The aim is to keep the course from looking like a collection of tricks and to share enough of the motivation behind the mathematics so that students are prepared to construct their own proofs when asked. We begin complicated proofs with a short paragraph (marked STRATEGY:) which shows why the proof works; for example, the Archimedean Principle (Theorem 1.16), Density of Rationals (Theorem 1.18), Cauchy's Theorem (Theorem 2.29), Change of Variables in **R** (Theorem 5.34), Riemann's Theorem about rearrangements (Theorem 6.29), the Implicit Function Theorem (Theorem 11.47), the Borel Covering Lemma (Lemma 9.9), and the fact that a curve is smooth when $\phi' \neq \mathbf{0}$ (Remark 13.10). We precede abstruse definitions or theorems with a short paragraph which describes, in simple terms, what behavior we are examining, and why (e.g., Cauchy sequences, one-sided limits, upper and lower Riemann sums, the Integral Test, Abel's Formula, uniform convergence, the total derivative, compact sets, differentiable curves and surfaces, smooth curves, and orientation equivalence). And we include examples to show why each hypothesis of a major theorem is necessary (e.g., the Nested Interval Property, the Bolzano–Weierstrass Theorem, the Mean Value Theorem, the Heine–Borel Theorem, the Inverse Function Theorem, the existence of exact differentials, and Fubini's Theorem).

Each section contains a collection of exercises which range from very elementary (to be sure the student understands the concepts introduced in that section) to more challenging (to give the student practice in using these concepts to expand the theory). To minimize frustration, some of the more difficult exercises have several parts which serve as an outline to a solution of the problem. To keep from producing students who know theory but cannot apply it, each set of exercises contains a mix of computational and theoretical assignments. (Exercises which play a prominent role later in the text are marked with a box. These exercises are an integral part of the course and all of them should be assigned.)

Since many students have difficulty reading and understanding mathematics, we have paid close attention to style and organization. We have consciously limited the vocabulary, kept notation consistent from chapter to chapter, and presented the proofs in a unified style. Individual sections are determined by subject matter, not by length of lecture, so that students can comprehend related results in a larger context. Examples and important remarks are numbered and labeled so that students can read the text in small chunks. (Many of these, included for the student's benefit, need not be covered in class.) Paragraphs are short and focused so that students are not overwhelmed by long-winded explanations. To help students discern between central results and peripheral ones, the word *Theorem* is used relatively sparingly; preliminary results and results which are only used in one section are called Remarks, Lemmas, and Examples. And we have broken with tradition by stating definitions explicitly with an "if and only if." (How can we chide our students for using the converse of a result when it appears that we do so about half the time we apply a definition?)

NEW TO THIS EDITION

We have changed many of the computational exercises so that the answers are simpler and easier to obtain. We have replaced most of the beginning calculus–style exercises with slightly more conceptual exercises that emphasize the same ideas, but at a higher level. We have added many theoretical exercises of medium difficulty. We have scattered true-false questions throughout the first six chapters. These are designed to confront common misconceptions that some students tend to acquire at this level. We have gathered introductory material that was scattered over several sections into a new section entitled *Introduction*. This section includes two accessible examples about why proof is necessary and why we cannot always trust what we see. We have reduced the number of axioms from four to three by introducing the Completeness Axiom first, and using it to prove the Well-Ordering Principle and the Principle of Mathematical Induction. We have moved Taylor's Formula from Chapter 7 to Chapter 4 to offer another example of the utility of the Mean Value Theorem. We have given the Heine–Borel Theorem its own section and included several exercises designed to give students practice in making a local condition on a compact set into a global one. We have reorganized Section 12.1 (Jordan regions) to simplify the presentation and make it easier to teach. We have omited Chapter 15, And we have corrected a number of misprints.

We wish to thank Mr. P. W. Wade and Professors S. Fridli, G. S. Jordan, Mefharet Kocatepe, J. Long, M. E. Mays, M. S. Osborne, P. W. Schaefer, F. E. Schroeck, and Ali Sinan Sertoz, who carefully read parts of the first edition and made many valuable suggestions and corrections. Also, I wish to express my gratitude to Ms. C. K. Wade for several lively discussions of a pedagogical nature, which helped shape the organization and presentation of this material. I wish to thank Der-Chen Chang (Georgetown University), Wen D. Chang (Alabama State), Patrick N. Dowling (Miami University), Jeffery Ehme (Spelman College), Dana S. Fine (University of Massachusetts-Dartmouth), Stephen Fisher (Northwestern University), Scott Fulton (Clarkson University), Kevin Knudson (Mississippi State University), Maria Nogin (California State University- Fresno), Gary Weiss (University of Cincinnati), Peter Wolfe (University of Maryland), and Mohammed Yahdi (Ursinus College) who looked at the fourth edition while it was in manuscript form and did some pre-revision reviews. I wish to thank Professor Stan Perrine (Charleston Southern University) for checking the penultimate version of the manuscript for accuracy and typographical errors. Finally, I wish to make special mention of Professor Lewis Lum, (Portland State University) who continues to make many careful and perspicuous comments about style, elegance of presentation, and level of rigor which have found their way into this fourth edition.

If there remain any typographical errors, I plan to keep an up-to-date list at my Web site (http://www.math.utk.edu/~wade). If you find errors which are not listed at that site, please contact me via e-mail at iwade@utk.edu.

William R. Wade
Mathematics Department
University of Tennessee
Knoxville, TN 37996-1300

CHAPTER 1

The Real Number System

You have already had several calculus courses in which you evaluated limits, differentiated functions, and computed integrals. You may even remember some of the major results of calculus, such as the Chain Rule, the Mean Value Theorem, and the Fundamental Theorem of Calculus. Although you are probably less familiar with multivariable calculus, you have taken partial derivatives, computed gradients, and evaluated certain line and surface integrals.

In view of all this, you must be asking: Why another course in calculus? The answer to this question is twofold. Although some proofs may have been presented in earlier courses, it is unlikely that the subtler points (e.g., completeness of the real numbers, uniform continuity, and uniform convergence) were covered. Moreover, the skills you acquired were mostly computational; you were rarely asked to prove anything yourself. This course develops the theory of calculus carefully and rigorously from basic principles and gives you a chance to learn how to construct your own proofs. It also serves as an introduction to analysis, an important branch of mathematics which provides a foundation for numerical analysis, functional analysis, harmonic analysis, differential equations, differential geometry, real analysis, complex analysis, and many other areas of specialization within mathematics.

1.1 INTRODUCTION

Every rigorous study of mathematics begins with certain undefined concepts, primitive notions on which the theory is based, and certain postulates, properties which are assumed to be true and given no proof. Our study will be based on the primitive notions of real numbers and sets, which will be discussed in this section.

We shall use standard notation for sets and real numbers. For example, $\mathbf{R}$ or $(-\infty, \infty)$ represents the set of *real numbers*, $\emptyset$ represents the *empty set* (the set with no elements), $a \in A$ means that a is an *element of* A, and $a \notin A$ means that a is *not* an element of A. We can represent a given finite set in two ways. We can list its elements directly, or we can describe it using sentences or equations. For example, the set of solutions to the equation $x^2 = 1$ can be written as

$$\{1, -1\} \quad \text{or} \quad \{x : x^2 = 1\}.$$

A set A is said to be a *subset* of a set B (notation: $A \subseteq B$) if and only if every element of A is also an element of B. If A is a subset of B but there is at least one element $b \in B$ that does not belong to A, we shall call A a *proper subset* of B (notation: $A \subset B$). Two sets A and B are said to be *equal* (notation: $A = B$)

1

if and only if $A \subseteq B$ and $B \subseteq A$. If A and B are not equal, we write $A \neq B$. A set A is said to be *nonempty* if and only if $A \neq \emptyset$.

The *union* of two sets A and B (notation: $A \cup B$) is the set of elements x such that x belongs to A or B or both. The *intersection* of two sets A and B (notation: $A \cap B$) is the set of elements x such that x belongs to both A and B. The *complement* of B relative to A (notation: $A \setminus B$, sometimes B^c if A is understood) is the set of elements x such that x belongs to A but does not belong to B. For example,

$$\{-1, 0, 1\} \cup \{1, 2\} = \{-1, 0, 1, 2\}, \qquad \{-1, 0, 1\} \cap \{1, 2\} = \{1\},$$
$$\{1, 2\} \setminus \{-1, 0, 1\} = \{2\} \quad \text{and} \quad \{-1, 0, 1\} \setminus \{1, 2\} = \{-1, 0\}.$$

Let X and Y be sets. The *Cartesian product* of X and Y is the set of *ordered pairs* defined by

$$X \times Y := \{(x, y) : x \in X, y \in Y\}.$$

(The symbol $:=$ means "equal by definition" or "is defined to be.") Two points (x, y), $(z, w) \in X \times Y$ are said to be *equal* if and only if $x = z$ and $y = w$.

Let X and Y be sets. A *relation* on $X \times Y$ is any subset of $X \times Y$. Let $\mathcal{R}$ be a relation on $X \times Y$. The *domain* of $\mathcal{R}$ is the collection of $x \in X$ such that (x, y) belongs to $\mathcal{R}$ for some $y \in Y$. The *range* of $\mathcal{R}$ is the collection of $y \in Y$ such that (x, y) belongs to $\mathcal{R}$ for some $x \in X$. When $(x, y) \in \mathcal{R}$, we shall frequently write $x\mathcal{R}y$.

A *function* f from X into Y (notation: $f : X \to Y$) is a relation on $X \times Y$ whose domain is X (notation: $\mathrm{Dom}(f) := X$) such that for each $x \in X$ there is a *unique* (one and only one) $y \in Y$ that satisfies $(x, y) \in f$. If $(x, y) \in f$, we shall call y the *value* of f at x (notation: $y = f(x)$ or $f : x \longmapsto y$) and call x a *preimage* of y under f. We said *a* preimage because, in general, a point in the range of f might have more than one preimage. For example, since $\sin(k\pi) = 0$ for $k = 0, \pm 1, \pm 2, \ldots$, the value 0 has infinitely many preimages under $f(x) = \sin x$.

If f is a function from X into Y, we will say that f is *defined* on X and call Y the *codomain* of f. The *range* of f is the collection of all values of f; that is, the set $\mathrm{Ran}(f) := \{y \in Y : f(x) = y \text{ for some } x \in X\}$. In general, then, the range of a function is a subset of its codomain and each $y \in \mathrm{Ran}(f)$ has one or more preimages. If $\mathrm{Ran}(f) = Y$ and each $y \in Y$ has exactly one preimage, $x \in X$, under f, then we shall say that $f : X \to Y$ *has an inverse*, and shall define the *inverse function* $f^{-1} : Y \to X$ by $f^{-1}(y) := x$, where $x \in X$ satisfies $f(x) = y$.

At this point it is important to notice a consequence of defining a function to be a set of ordered pairs. By the definition of equality of ordered pairs, two functions f and g are equal if and only if they have the same domain, and same values; that is, $f, g : X \to Y$, and $f(x) = g(x)$ for all $x \in X$. If they have different domains, they are different functions.

For example, let $f(x) = g(x) = x^2$. Then $f : [0, 1) \to [0, 1)$ and $g : (-1, 1) \to [0, 1)$ are two different functions. They both have the same range, $[0, 1)$, but each $y \in (0, 1)$ has exactly one preimage under f, namely $\sqrt{y}$, and two preimages under g, namely $\pm\sqrt{y}$. In particular, f has an inverse function, $f^{-1}(x) = \sqrt{x}$, but g does not. Making distinctions like this will actually make our life easier later in the course.

For the first half of this course, most of the concrete functions we consider will be *real-valued functions of a real variable* (i.e., functions whose domains and ranges are subsets of **R**). We shall often call such functions simply *real functions*.

You are already familiar with many real functions.

1) The *exponential function* $e^x : \mathbf{R} \to (0, \infty)$ and its inverse function, the *natural logarithm*

$$\log x := \int_1^x \frac{dt}{t},$$

defined and real-valued for each $x \in (0, \infty)$. (Although this last function is denoted by $\ln x$ in elementary calculus texts, most analysts denote it, as we did just now, by $\log x$. We will follow this practice throughout this text. For a more constructive definition, see Exercise 4.5.5.)

2) The *trigonometric functions* (whose formulas are) represented by $\sin x$, $\cos x$, $\tan x$, $\cot x$, $\sec x$, and $\csc x$, and the inverse trigonometric functions $\arcsin x$, $\arccos x$, and $\arctan x$ whose ranges are, respectively, $[-\pi/2, \pi/2]$, $[0, \pi]$, and $(-\pi/2, \pi/2)$.

3) The *power functions* x^α, which can be defined constructively (see Appendix A.10 and Exercise 3.3.11) or by using the exponential function:

$$x^\alpha := e^{\alpha \log x}, \qquad x > 0, \quad \alpha \in \mathbf{R}.$$

We assume that you are familiar with the various algebraic laws and identities that these functions satisfy. A list of the most widely used trigonometric identities can be found in Appendix B. The most widely used properties of the power functions are $x^0 = 1$ for all $x \neq 0$; $x^n = x \cdot \ldots \cdot x$ (there are n factors here) when $n = 1, 2, \ldots$ and $x \in \mathbf{R}$; $x^\alpha > 0$, $x^\alpha \cdot x^\beta = x^{\alpha+\beta}$, and $(x^\alpha)^\beta = x^{\alpha \cdot \beta}$ for all $x > 0$ and $\alpha, \beta \in \mathbf{R}$; $x^\alpha = \sqrt[m]{x}$ when $\alpha = 1/m$ for some $m \in = 1, 2, \ldots$ and the indicated root exists and is real; and $0^\alpha := 0$ for all $\alpha > 0$. (The symbol 0^0 is left undefined because it is indeterminate [see Example 4.31].)

We also assume that you can differentiate algebraic combinations of these functions using the basic formulas $(\sin x)' = \cos x$, $(\cos x)' = -\sin x$, and $(e^x)' = e^x$, for $x \in \mathbf{R}$; $(\log x)' = 1/x$ and $(x^\alpha)' = \alpha x^{\alpha-1}$, for $x > 0$ and $\alpha \in \mathbf{R}$; and

$$(\tan x)' = \sec^2 x \quad \text{for } x \neq \frac{(2n + 1)\pi}{2}, \quad n \in \mathbf{Z}.$$

(You will have an opportunity to develop some of these rules in the exercises, e.g., see Exercises 4.2.9, 4.4.6, 4.5.3, 5.3.7, and 5.3.8.) Even with these assumptions, we shall repeat some material from elementary calculus.

We mentioned postulates in the opening paragraph. In the next two sections, we will introduce three postulates (containing a total of 13 different properties) which characterize the set of real numbers. Although you are probably already familiar with all but the last of these properties, we will use them to prove other equally familiar properties (e.g., in Example 1.4 we will prove that if $a \neq 0$, then $a^2 > 0$).

Why would we assume some properties and prove others? At one point, mathematicians thought that all laws about real numbers were of equal weight. Gradually, during the late 1800s, we discovered that many of the well-known laws satisfied by **R** are in fact consequences of others. The net result of this research is that the 13 properties listed below are considered to be fundamental properties describing **R**. All other laws satisfied by real numbers are secondary in the sense that they can be proved using these fundamental properties.

Why would we prove a law that is well known, perhaps even "obvious"? Why not just assume all known properties about **R** and proceed from there? We want this book to be reasonably self-contained, because this will make it easier for you to begin to construct your own proofs. We want the first proofs you see to be easily understood, because they deal with familiar properties that are unobscured by new concepts. But most importantly, we want to form a habit of proving all statements, even seemingly "obvious" statements.

The reason for this hard-headed approach is that some "obvious" statements are false. For example, divide an 8×8-inch square into triangles and trapezoids as shown on the left side of Figure 1.1. Since the 3-inch sides of the triangles perfectly match the 3-inch sides of the trapezoids, it is "obvious" that these triangles and trapezoids can be reassembled into a rectangle (see the right side of Figure 1.1). Or is it? The area of the square is $8 \times 8 = 64$ square inches but the area of the rectangle is $5 \times 13 = 65$ square inches. Since you cannot increase area by reassembling pieces, what looked right was in fact wrong. By computing slopes, you can verify that the rising diagonal on the right side of Figure 1.1 is, in fact, four distinct line segments that form a long narrow diamond which conceals that extra one square inch.

NOTE: *Reading a mathematics book is different from reading any other kind of book. When you see phrases like "you can verify" or "it is easy to see," you should use pencil and paper to do the calculations to be sure what we've said is correct.*

Here is another example. Grab a calculator and graph the functions $y = \log x$ and $y = \sqrt[100]{x}$. It is easy to see, using calculus, that $\log x$ and $\sqrt[100]{x}$ are both increasing and concave downward on $[0, \infty)$. Looking at the graphs (see Figure 1.2), it's "obvious" that $\log x$ is much larger than $\sqrt[100]{x}$ no matter how big x is. Or is it? Let's evaluate each function at e^{1000}: $\log(e^{1000}) = 1000 \log e = 1000$ is much smaller than $\sqrt[100]{e^{1000}} = e^{10} \approx 22,000$. Evidently, the graph of $y = \sqrt[100]{x}$

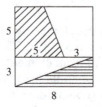
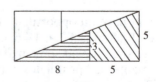

FIGURE 1.1

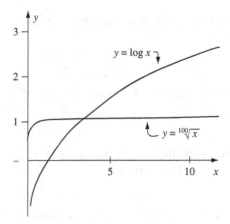

FIGURE 1.2

eventually crosses that of $y = \log x$. With a little calculus, you can prove that $\log x < \sqrt[100]{x}$ forever after that (see Exercise 4.4.6a).

What can be learned from these examples? We cannot always trust what we think we see. We must, as above, find some mathematical way of testing our perception, either verifying that it is correct, or rejecting it as wrong. This type of phenomenon is not a rare occurrence. You will soon encounter several other plausible statements that are, in fact, false. In particular, you must harbor a skepticism that demands proofs of all statements not assumed in postulates, even the "obvious" ones.

What, then, are you allowed to use when solving the exercises? You may use any property of real numbers (e.g., $2+3 = 5$, $2 < 7$, or $\sqrt{2}$ is irrational) without reference or proof. You may use any algebraic property of real numbers involving equal signs [e.g., $(x + y)^2 = x^2 + 2xy + y^2$ or $(x + y)(x - y) = x^2 - y^2$] and the techniques of calculus to find local maxima or minima of a given function without reference or proof. After completing the exercises in Section 1.2, you may also use any algebraic property of real numbers involving inequalities (e.g., $0 < a < b$ implies $0 < a^x < b^x$ for all $x > 0$) without reference or proof.

1.2 ORDERED FIELD AXIOMS

In this section we explore the algebraic structure of the real number system. We shall assume that the set of real numbers, **R**, is a field (i.e., that **R** satisfies the following postulate).

Postulate 1.   [FIELD AXIOMS]. There are functions $+$ and $\cdot$, defined on $\mathbf{R}^2 := \mathbf{R} \times \mathbf{R}$, which satisfy the following properties for every $a, b, c \in \mathbf{R}$:

Closure Properties. $a + b$ and $a \cdot b$ belong to **R**.

Associative Properties. $a + (b + c) = (a + b) + c$ and $a \cdot (b \cdot c) = (a \cdot b) \cdot c$.

Commutative Properties. $a + b = b + a$ and $a \cdot b = b \cdot a$.

Distributive Law. $a \cdot (b + c) = a \cdot b + a \cdot c$.

Existence of the Additive Identity. There is a unique element $0 \in \mathbf{R}$ such that $0 + a = a$ for all $a \in \mathbf{R}$.

Existence of the Multiplicative Identity. There is a unique element $1 \in \mathbf{R}$ such that $1 \neq 0$ and $1 \cdot a = a$ for all $a \in \mathbf{R}$.

Existence of Additive Inverses. For every $x \in \mathbf{R}$ there is a unique element $-x \in \mathbf{R}$ such that

$$x + (-x) = 0.$$

Existence of Multiplicative Inverses. For every $x \in \mathbf{R} \setminus \{0\}$ there is a unique element $x^{-1} \in \mathbf{R}$ such that

$$x \cdot (x^{-1}) = 1.$$

We note in passing that the word *unique* can be dropped from the statements in Postulate 1 (see Appendix A).

We shall usually denote $a + (-b)$ by $a - b$, $a \cdot b$ by ab, a^{-1} by $\frac{1}{a}$ or $1/a$, and $a \cdot b^{-1}$ by $\frac{a}{b}$ or a/b. Notice that by the existence of additive and multiplicative inverses, the equation $x + a = 0$ can be solved for each $a \in \mathbf{R}$, and the equation $ax = 1$ can be solved for each $a \in \mathbf{R}$ provided that $a \neq 0$.

From these few properties (i.e., from Postulate 1), we can derive all the usual algebraic laws of real numbers, including the following:

$$(-1)^2 = 1, \tag{1}$$

$$0 \cdot a = 0, \quad -a = (-1) \cdot a, \quad -(-a) = a, \quad a \in \mathbf{R}, \tag{2}$$

$$-(a - b) = b - a, \quad a, b \in \mathbf{R}, \tag{3}$$

and

$$a, b \in \mathbf{R} \quad \text{and} \quad ab = 0 \quad \text{imply} \quad a = 0 \quad \text{or} \quad b = 0. \tag{4}$$

We want to keep our attention sharply focused on analysis. Since the proofs of algebraic laws like these lie more in algebra than analysis (see Appendix A), we will not present them here. In fact, with the exception of the absolute value and the Binomial Formula, we will assume all material usually presented in a high school algebra course (including the quadratic formula and graphs of the conic sections).

Postulate 1 is sufficient to derive all algebraic laws of $\mathbf{R}$, but it does not completely describe the real number system. The set of real numbers also has an order relation (i.e., a concept of "less than").

Postulate 2. [ORDER AXIOMS]. There is a relation $<$ on $\mathbf{R} \times \mathbf{R}$ that has the following properties:

Trichotomy Property. Given $a, b \in \mathbf{R}$, one and only one of the following statements holds:

$$a < b, \quad b < a, \quad \text{or} \quad a = b.$$

Transitive Property. For $a, b, c \in \mathbf{R}$,

$$a < b \quad \text{and} \quad b < c \quad \text{imply} \quad a < c.$$

The Additive Property. For $a, b, c \in \mathbf{R}$,

$$a < b \quad \text{and} \quad c \in \mathbf{R} \quad \text{imply} \quad a + c < b + c.$$

The Multiplicative Properties. For $a, b, c \in \mathbf{R}$,

$$a < b \quad \text{and} \quad c > 0 \quad \text{imply} \quad ac < bc$$

and

$$a < b \quad \text{and} \quad c < 0 \quad \text{imply} \quad bc < ac.$$

By $b > a$ we shall mean $a < b$. By $a \le b$ and $b \ge a$ we shall mean $a < b$ or $a = b$. By $a < b < c$ we shall mean $a < b$ and $b < c$. In particular, $2 < x < 1$ makes no sense at all.

WARNING. *There are two Multiplicative Properties, so every time you multiply an inequality by an expression, you must carefully note the sign of that expression and adjust the inequality accordingly. For example, $x < 1$ does NOT imply that $x^2 < x$ unless $x > 0$. If $x < 0$, then by the Second Multiplicative Property, $x < 1$ implies $x^2 > x$.*

We shall call a number $a \in \mathbf{R}$ *nonnegative* if $a \ge 0$ and *positive* if $a > 0$. Postulate 2 has a slightly simpler formulation using the set of positive elements as a primitive concept (see Exercise 1.2.11). We have introduced Postulate 2 as above because these are the properties we use most often.

The real number system $\mathbf{R}$ contains certain special subsets: the set of *natural numbers*

$$\mathbf{N} := \{1, 2, \ldots\},$$

obtained by beginning with 1 and successively adding 1s to form $2 := 1 + 1$, $3 := 2 + 1$, and so on; the set of *integers*

$$\mathbf{Z} := \{\ldots, -2, -1, 0, 1, 2, \ldots\}$$

(*Zahl* is German for number); the set of *rationals* (or fractions or quotients)

$$\mathbf{Q} := \left\{ \frac{m}{n} : m, n \in \mathbf{Z} \text{ and } n \ne 0 \right\};$$

and the set of *irrationals*

$$\mathbf{Q}^c = \mathbf{R} \setminus \mathbf{Q}.$$

Equality in $\mathbf{Q}$ is defined by

$$\frac{m}{n} = \frac{p}{q} \quad \text{if and only if} \quad mq = np.$$

Recall that each of the sets $\mathbf{N}, \mathbf{Z}, \mathbf{Q}$, and $\mathbf{R}$ is a proper subset of the next; that is,

$$\mathbf{N} \subset \mathbf{Z} \subset \mathbf{Q} \subset \mathbf{R}.$$

For example, every rational is a real number (because $m/n := mn^{-1}$ is a real number by Postulate 1), but $\sqrt{2}$ is an irrational.

Since we did not really define $\mathbf{N}$ and $\mathbf{Z}$, we must make certain assumptions about them. If you are interested in the definitions and proofs, see Appendix A.

1.1 Remark. We will assume that the sets $\mathbf{N}$ and $\mathbf{Z}$ satisfy the following properties.

i) If $n, m \in \mathbf{Z}$, then $n + m$, $n - m$, and mn belong to $\mathbf{Z}$.
ii) If $n \in \mathbf{Z}$, then $n \in \mathbf{N}$ if and only if $n \geq 1$.
iii) There is no $n \in \mathbf{Z}$ that satisfies $0 < n < 1$.

Using these properties, we can prove that $\mathbf{Q}$ satisfies Postulate 1 (see Exercise 1.2.9).

We notice in passing that none of the other special subsets of $\mathbf{R}$ satisfies Postulate 1. $\mathbf{N}$ satisfies all but three of the properties in Postulate 1: $\mathbf{N}$ has no additive identity (since $0 \notin \mathbf{N}$), $\mathbf{N}$ has no additive inverses (e.g., $-1 \notin \mathbf{N}$), and only one of the nonzero elements of $\mathbf{N}$ (namely, 1) has a multiplicative inverse. $\mathbf{Z}$ satisfies all but one of the properties in Postulate 1: Only two nonzero elements of $\mathbf{Z}$ have multiplicative inverses (namely, 1 and -1). $\mathbf{Q}^c$ satisfies all but four of the properties in Postulate 1: $\mathbf{Q}^c$ does not have an additive identity (since $0 \notin \mathbf{R} \setminus \mathbf{Q}$), does not have a multiplicative identity (since $1 \notin \mathbf{R} \setminus \mathbf{Q}$), and does not satisfy either closure property. Indeed, since $\sqrt{2}$ is irrational, the sum of irrationals may be rational ($\sqrt{2} + (-\sqrt{2}) = 0$) and the product of irrationals may be rational ($\sqrt{2} \cdot \sqrt{2} = 2$).

Notice that any subset of $\mathbf{R}$ satisfies Postulate 2. Thus $\mathbf{Q}$ satisfies both Postulates 1 and 2. The remaining postulate, introduced in Section 1.3, identifies a property that $\mathbf{Q}$ does not possess. In particular, Postulates 1 through 3 distinguish $\mathbf{R}$ from each of its special subsets $\mathbf{N}, \mathbf{Z}, \mathbf{Q}$, and $\mathbf{Q}^c$. These postulates actually characterize $\mathbf{R}$; that is, $\mathbf{R}$ is the only set that satisfies Postulates 1 through 3. (Such a set is called a *complete Archimedean ordered field*. We may as well admit a certain arbitrariness in choosing this approach. $\mathbf{R}$ has been developed axiomatically in at least five other ways [e.g., as a one-dimensional continuum or as a set of binary decimals with certain arithmetic operations]. The decision to present $\mathbf{R}$ using Postulates 1 through 3 is based partly on economy and partly on personal taste.)

Postulates 1 and 2 can be used to derive all identities and inequalities which are true for real numbers [e.g., see implications (5) through (9) below]. Since arguments based on inequalities are of fundamental importance to analysis, we begin to supply details of proofs at this stage.

What is a proof? Every mathematical result (for us this includes examples, remarks, lemmas, and theorems) has hypotheses and a conclusion. There are three main methods of proof: mathematical induction, direct deduction, and contradiction.

Mathematical induction, a special method for proving statements that depend on positive integers, will be covered in Section 1.4.

To construct a *deductive proof*, we assume the hypotheses to be true and proceed step by step to the conclusion. Each step is justified by a hypothesis, a definition, a postulate, or a mathematical result that has already been proved. (Actually, this is usually the way we write a proof. When constructing your own proofs, you may find it helpful to work forward from the hypotheses as far as you can and then work backward from the conclusion, trying to meet in the middle.)

To construct a *proof by contradiction*, we assume the hypotheses to be true, the conclusion to be false, and work step by step deductively until a *contradiction* occurs; that is, a statement that is obviously false or that is contrary to the assumptions made. At this point the proof by contradiction is complete. The phrase "suppose to the contrary" always indicates a proof by contradiction (e.g., see the proof of Theorem 1.9).

What about false statements? How do we "prove" that a statement is false? We can show that a statement is false by producing a single, concrete example (called a *counterexample*) that satisfies the hypotheses but not the conclusion of that statement. For example, to show that the statement "$x > 1$ implies $x^2 - x - 2 \neq 0$" is false, we need only observe that $x = 2$ is greater than 1 but $2^2 - 2 - 2 = 0$.

Here are some examples of deductive proofs. (*Note*: The symbol ■ indicates that the proof or solution is complete.)

1.2 EXAMPLE.

If $a \in \mathbf{R}$, prove that

$$a \neq 0 \quad \text{implies} \quad a^2 > 0. \tag{5}$$

In particular, $-1 < 0 < 1$.

Proof. Suppose that $a \neq 0$. By the Trichotomy Property, either $a > 0$ or $a < 0$.

Case 1. $a > 0$. Multiply both sides of this inequality by a, using the First Multiplicative Property. We obtain $a^2 = a \cdot a > 0 \cdot a$. Since (by (2)), $0 \cdot a = 0$ we conclude that $a^2 > 0$.

Case 2. $a < 0$. Multiply both sides of this inequality by a. Since $a < 0$, it follows from the Second Multiplicative Property that $a^2 = a \cdot a > 0 \cdot a = 0$. This proves that $a^2 > 0$ when $a \neq 0$.

Since $1 \neq 0$, it follows that $1 = 1^2 > 0$. Adding -1 to both sides of this inequality, we conclude that $0 = 1 - 1 > 0 - 1 = -1$.   ■

1.3 EXAMPLE.

If $a \in \mathbf{R}$, prove that

$$0 < a < 1 \quad \text{implies} \quad 0 < a^2 < a \quad \text{and} \quad a > 1 \quad \text{implies} \quad a^2 > a. \tag{6}$$

Proof. Suppose that $0 < a < 1$. Multiply both sides of this inequality by a using the First Multiplicative Property. We obtain $0 = 0 \cdot a < a^2 < 1 \cdot a = a$. In particular, $0 < a^2 < a$.

On the other hand, if $a > 1$, then $a > 0$ by Example 1.2 and the Transitive Property. Multiplying $a > 1$ by a, we conclude that $a^2 = a \cdot a > 1 \cdot a = a$. ∎

Similarly (see Exercise 1.2.2), we can prove that

$$0 \leq a < b \quad \text{and} \quad 0 \leq c < d \quad \text{imply} \quad ac < bd, \tag{7}$$

$$0 \leq a < b \quad \text{implies} \quad 0 \leq a^2 < b^2 \quad \text{and} \quad 0 \leq \sqrt{a} < \sqrt{b}, \tag{8}$$

and

$$0 < a < b \quad \text{implies} \quad \frac{1}{a} > \frac{1}{b} > 0. \tag{9}$$

Much of analysis deals with estimation (of error, of growth, of volume, etc.) in which these inequalities and the following concept play a central role.

1.4 Definition.

The *absolute value* of a number $a \in \mathbf{R}$ is the number

$$|a| := \begin{cases} a & a \geq 0 \\ -a & a < 0. \end{cases}$$

When proving results about the absolute value, we can always break the proof up into several cases, depending on when the parameters are positive, negative, or zero. Here is a typical example.

1.5 Remark. *The absolute value is multiplicative; that is,* $|ab| = |a|\,|b|$ *for all* $a, b \in \mathbf{R}$.

Proof. We consider four cases.

Case 1. $a = 0$ or $b = 0$. Then $ab = 0$, so by definition, $|ab| = 0 = |a|\,|b|$.

Case 2. $a > 0$ and $b > 0$. By the First Multiplicative Property, $ab > 0 \cdot b = 0$. Hence by definition, $|ab| = ab = |a|\,|b|$.

Case 3. $a > 0$ and $b < 0$, or, $b > 0$ and $a < 0$. By symmetry, we may suppose that $a > 0$ and $b < 0$. (That is, if we can prove it for $a > 0$ and $b < 0$, then by reversing the roles of a and b, we can prove it for $a < 0$ and $b > 0$.) By the Second Multiplicative Property, $ab < 0$. Hence by Definition 1.4, (2), associativity, and commutativity,

$$|ab| = -(ab) = (-1)(ab) = a((-1)b) = a(-b) = |a|\,|b|.$$

Case 4. $a < 0$ and $b < 0$. By the Second Multiplicative Property, $ab > 0$. Hence by Definition 1.4,

$$|ab| = ab = (-1)^2(ab) = (-a)(-b) = |a|\,|b|. \qquad \blacksquare$$

We shall soon see that there are more efficient ways to prove results about absolute values than breaking the argument into cases.

The following result is useful when solving inequalities involving absolute value signs.

1.6 Theorem.   [FUNDAMENTAL THEOREM OF ABSOLUTE VALUES].
Let $a \in \mathbf{R}$ and $M \geq 0$. Then $|a| \leq M$ if and only if $-M \leq a \leq M$.

Proof. Suppose first that $|a| \leq M$. Multiplying by -1, we also have $-|a| \geq -M$.
Case 1. $a \geq 0$. By Definition 1.4, $|a| = a$. Thus by hypothesis,

$$-M \leq 0 \leq a = |a| \leq M.$$

Case 2. $a < 0$. By Definition 1.4, $|a| = -a$. Thus by hypothesis,

$$-M \leq -|a| = a < 0 \leq M.$$

This proves that $-M \leq a \leq M$ in either case.

Conversely, if $-M \leq a \leq M$, then $a \leq M$ and $-M \leq a$. Multiplying the second inequality by -1, we have $-a \leq M$. Consequently, $|a| = a \leq M$ if $a \geq 0$, and $|a| = -a \leq M$ if $a < 0$. $\qquad \blacksquare$

NOTE: In a similar way we can prove that $|a| < M$ if and only if $-M < a < M$.

Here is another useful result about absolute values.

1.7 Theorem.   *The absolute value satisfies the following three properties.*

  i) [POSITIVE DEFINITE] *For all $a \in \mathbf{R}$, $|a| \geq 0$ with $|a| = 0$ if and only if $a = 0$.*
 ii) [SYMMETRIC] *For all $a, b \in \mathbf{R}$, $|a - b| = |b - a|$.*
iii) [TRIANGLE INEQUALITIES] *For all $a, b \in \mathbf{R}$,*

$$|a + b| \leq |a| + |b| \quad and \quad \big|\,|a| - |b|\,\big| \leq |a - b|.$$

Proof. i) If $a \geq 0$, then $|a| = a \geq 0$. If $a < 0$, then by Definition 1.4 and the Second Multiplicative Property, $|a| = -a = (-1)a > 0$. Thus $|a| \geq 0$ for all $a \in \mathbf{R}$.

If $|a| = 0$, then by definition $a = |a| = 0$ when $a \geq 0$ and $a = -|a| = 0$ when $a < 0$. Thus $|a| = 0$ implies that $a = 0$. Conversely, $|0| = 0$ by definition.

 ii) By Remark 1.5, $|a - b| = |-1|\,|b - a| = |b - a|$.

iii) To prove the first inequality, notice that $|x| \leq |x|$ holds for any $x \in \mathbf{R}$. Thus Theorem 1.6 implies that $-|a| \leq a \leq |a|$ and $-|b| \leq b \leq |b|$. Adding these inequalities (see Exercise 1.2.1), we obtain

$$-(|a| + |b|) \leq a + b \leq |a| + |b|.$$

Hence by Theorem 1.6 again, $|a + b| \leq |a| + |b|$.

To prove the second inequality, apply the first inequality to $(a - b) + b$. We obtain

$$|a| - |b| = |a - b + b| - |b| \leq |a - b| + |b| - |b| = |a - b|.$$

By reversing the roles of a and b and applying part ii), we also obtain

$$|b| - |a| \leq |b - a| = |a - b|.$$

Multiplying this last inequality by -1 and combining it with the preceding one verifies

$$-|a - b| \leq |a| - |b| \leq |a - b|.$$

We conclude by Theorem 1.6 that $\big| \, |a| - |b| \, \big| \leq |a - b|$. ∎

Notice once and for all that this last inequality implies that $|a| - |b| \leq |a - b|$ for all $a, b \in \mathbf{R}$. We will use this inequality several times.

WARNING. *Some students mistakenly mix absolute values and the Additive Property to conclude that $b < c$ implies $|a + b| < |a + c|$. It is important from the beginning to recognize that this implication is false unless both $a + b$ and $a + c$ are nonnegative. For example, if $a = 1$, $b = -5$, and $c = -1$, then $b < c$ but $|a + b| = 4$ is not less than $|a + c| = 0$.*

A correct way to estimate using absolute value signs usually involves one of the triangle inequalities.

1.8 EXAMPLE.

Prove that if $-2 < x < 1$, then $|x^2 - x| < 6$.

Proof. By hypothesis, $|x| < 2$. Hence by the triangle inequality and Remark 1.5,

$$|x^2 - x| \leq |x|^2 + |x| < 4 + 2 = 6.$$ ∎

The following result (which is equivalent to the Trichotomy Property) will be used many times in this and subsequent chapters.

1.9 Theorem. *Let $x, y, a \in \mathbf{R}$.*

i) $x < y + \varepsilon$ *for all $\varepsilon > 0$ if and only if $x \leq y$.*
ii) $x > y - \varepsilon$ *for all $\varepsilon > 0$ if and only if $x \geq y$.*
iii) $|a| < \varepsilon$ *for all $\varepsilon > 0$ if and only if $a = 0$.*

Proof. i) Suppose to the contrary that $x < y + \varepsilon$ for all $\varepsilon > 0$ but $x > y$. Set $\varepsilon_0 = x - y > 0$ and observe that $y + \varepsilon_0 = x$. Hence by the Trichotomy Property, $y + \varepsilon_0$ cannot be greater than x. This contradicts the hypothesis for $\varepsilon = \varepsilon_0$. Thus $x \leq y$.

Conversely, suppose that $x \leq y$ and $\varepsilon > 0$ is given. Either $x < y$ or $x = y$. If $x < y$, then $x + 0 < y + 0 < y + \varepsilon$ by the Additive and Transitive Properties. If $x = y$, then $x < y + \varepsilon$ by the Additive Property. Thus $x < y + \varepsilon$ for all $\varepsilon > 0$ in either case. This completes the proof of part i).

ii) Suppose that $x > y - \varepsilon$ for all $\varepsilon > 0$. By the Second Multiplicative Property, this is equivalent to $-x < -y + \varepsilon$, hence by part i), equivalent to $-x \leq -y$. By the Second Multiplicative Property, this is equivalent to $x \geq y$.

iii) Suppose that $|a| < \varepsilon = 0 + \varepsilon$ for all $\varepsilon > 0$. By part i), this is equivalent to $|a| \leq 0$. Since it is always the case that $|a| \geq 0$, we conclude by the Trichotomy Property that $|a| = 0$. Therefore, $a = 0$ by Theorem 1.7i. ∎

Let a and b be real numbers. A *closed interval* is a set of the form

$$[a, b] := \{x \in \mathbf{R} : a \leq x \leq b\}, \qquad [a, \infty) := \{x \in \mathbf{R} : a \leq x\},$$
$$(-\infty, b] := \{x \in \mathbf{R} : x \leq b\}, \quad \text{or} \quad (-\infty, \infty) := \mathbf{R},$$

and an *open interval* is a set of the form

$$(a, b) := \{x \in \mathbf{R} : a < x < b\}, \quad (a, \infty) := \{x \in \mathbf{R} : a < x\},$$
$$(-\infty, b) := \{x \in \mathbf{R} : x < b\}, \quad \text{or} \quad (-\infty, \infty) := \mathbf{R}.$$

By an *interval* we mean a closed interval, an open interval, or a set of the form

$$[a, b) := \{x \in \mathbf{R} : a \leq x < b\} \quad \text{or} \quad (a, b] := \{x \in \mathbf{R} : a < x \leq b\}.$$

Notice, then, that when $a < b$, the intervals $[a, b]$, $[a, b)$, $(a, b]$, and (a, b) correspond to line segments on the real line, but when $b < a$, these "intervals" are all the empty set.

An interval I is said to be *bounded* if and only if it has the form $[a, b]$, (a, b), $[a, b)$, or $(a, b]$ for some $-\infty < a \leq b < \infty$, in which case the numbers a, b will be called the *endpoints* of I. All other intervals will be called *unbounded*. An interval with endpoints a, b is called *degenerate* if $a = b$ and *nondegenerate* if $a < b$. Thus a degenerate open interval is the empty set, and a degenerate closed interval is a point.

Analysis has a strong geometric flavor. Geometry enters the picture because the real number system can be identified with the real line in such a way that $a < b$ if and only if a lies to the left of b (see Figures 1.2, 2.1, and 2.2). This gives us a way of translating analytic results on $\mathbf{R}$ into geometric results on the number line, and vice versa. We close with several examples.

The absolute value is closely linked to the idea of length. The *length* of a bounded interval I with endpoints a, b is defined to be $|I| := |b - a|$, and the *distance* between any two points $a, b \in \mathbf{R}$ is defined by $|a - b|$.

Inequalities can be interpreted as statements about intervals. By Theorem 1.6, $|a| \le M$ if and only if a belongs to the closed interval $[-M, M]$; and by Theorem 1.9, a belongs to the open interval $(-\varepsilon, \varepsilon)$ for all $\varepsilon > 0$ if and only if $a = 0$.

We will use this point of view in Chapters 2 through 5 to give geometric interpretations to the calculus of functions defined on **R**, and in Chapters 11 through 13 to extend this calculus to functions defined on the Euclidean spaces $\mathbf{R}^n$.

EXERCISES

In each of the following exercises, verify the given statement carefully, proceeding step by step. Validate each step that involves an inequality by using some statement found in this section.

1.2.0 Let $a, b, c, d \in \mathbf{R}$ and consider each of the following statements. Decide which are true and which are false. Prove the true ones and give counterexamples to the false ones.

a) If $a < b$ and $c < d < 0$, then $ac > bd$.
b) If $a \le b$ and $c > 1$, then $|a + c| \le |b + c|$.
c) If $a \le b$ and $b \le a + c$, then $|a - b| \le c$.
d) If $a < b - \varepsilon$ for all $\varepsilon > 0$, then $a < 0$.

1.2.1. Suppose that $a, b, c \in \mathbf{R}$ and $a \le b$.

a) Prove that $a + c \le b + c$.
b) If $c \ge 0$, prove that $a \cdot c \le b \cdot c$.

1.2.2. Prove (7), (8), and (9). Show that each of these statements is false if the hypothesis $a \ge 0$ or $a > 0$ is removed.

1.2.3. **This exercise is used in Section 6.3.** The *positive part* of an $a \in \mathbf{R}$ is defined by

$$a^+ := \frac{|a| + a}{2}$$

and the *negative part* by

$$a^- := \frac{|a| - a}{2}.$$

a) Prove that $a = a^+ - a^-$ and $|a| = a^+ + a^-$.
b) Prove that

$$a^+ = \begin{cases} a & a \ge 0 \\ 0 & a \le 0 \end{cases} \quad \text{and} \quad a^- = \begin{cases} 0 & a \ge 0 \\ -a & a \le 0. \end{cases}$$

1.2.4. Solve each of the following inequalities for $x \in \mathbf{R}$.

a) $|2x + 1| < 7$
b) $|2 - x| < 2$

c) $|x^3 - 3x + 1| < x^3$

d) $\dfrac{x}{x-1} < 1$

e) $\dfrac{x^2}{4x^2 - 1} < \dfrac{1}{4}$

1.2.5. Let $a, b \in \mathbf{R}$.

a) Prove that if $a > 2$ and $b = 1 + \sqrt{a-1}$, then $2 < b < a$.
b) Prove that if $2 < a < 3$ and $b = 2 + \sqrt{a-2}$, then $0 < a < b$.
c) Prove that if $0 < a < 1$ and $b = 1 - \sqrt{1-a}$, then $0 < b < a$.
d) Prove that if $3 < a < 5$ and $b = 2 + \sqrt{a-2}$, then $3 < b < a$.

1.2.6. The *arithmetic mean* of $a, b \in \mathbf{R}$ is $A(a, b) = (a+b)/2$, and the *geometric mean* of $a, b \in [0, \infty)$ is $G(a, b) = \sqrt{ab}$. If $0 \le a \le b$, prove that $a \le G(a, b) \le A(a, b) \le b$. Prove that $G(a, b) = A(a, b)$ if and only if $a = b$.

1.2.7. Let $x \in \mathbf{R}$.

a) Prove that $|x| \le 2$ implies $|x^2 - 4| \le 4|x - 2|$.
b) Prove that $|x| \le 1$ implies $|x^2 + 2x - 3| \le 4|x - 1|$.
c) Prove that $-3 \le x \le 2$ implies $|x^2 + x - 6| \le 6|x - 2|$.
d) Prove that $-1 < x < 0$ implies $|x^3 - 2x + 1| < 1.26|x - 1|$.

1.2.8. For each of the following, find all values of $n \in \mathbf{N}$ that satisfy the given inequality.

a) $\dfrac{1-n}{1-n^2} < 0.01$

b) $\dfrac{n^2 + 2n + 3}{2n^3 + 5n^2 + 8n + 3} < 0.025$

c) $\dfrac{n-1}{n^3 - n^2 + n - 1} < 0.002$

1.2.9. a) Interpreting a rational m/n as $m \cdot n^{-1} \in \mathbf{R}$, use Postulate 1 to prove that

$$\frac{m}{n} + \frac{p}{q} = \frac{mq + np}{nq}, \quad \frac{m}{n} \cdot \frac{p}{q} = \frac{mp}{nq}, \quad -\frac{m}{n} = \frac{-m}{n}, \text{ and } \left(\frac{\ell}{n}\right)^{-1} = \frac{n}{\ell}$$

for $m, n, p, q, \ell \in \mathbf{Z}$ and $n, q, \ell \ne 0$.

b) Using Remark 1.1, Prove that Postulate 1 holds with $\mathbf{Q}$ in place of $\mathbf{R}$.

c) Prove that the sum of a rational and an irrational is always irrational. What can you say about the product of a rational and an irrational?

d) Let $m/n, p/q \in \mathbf{R}$ with $n, q > 0$. Prove that

$$\frac{m}{n} < \frac{p}{q} \quad \text{if and only if} \quad mq < np.$$

(Restricting this observation to $\mathbf{Q}$ gives a definition of "$<$" on $\mathbf{Q}$.)

1.2.10. Prove that

$$(ab + cd)^2 \leq (a^2 + c^2)(b^2 + d^2)$$

for all $a, b, c, d \in \mathbf{R}$.

1.2.11. a) Let $\mathbf{R}^+$ represent the collection of positive real numbers. Prove that $\mathbf{R}^+$ satisfies the following two properties.

 i) For each $x \in \mathbf{R}$, one and only one of the following holds:

$$x \in \mathbf{R}^+, \quad -x \in \mathbf{R}^+, \quad \text{or} \quad x = 0.$$

 ii) Given $x, y \in \mathbf{R}^+$, both $x + y$ and $x \cdot y$ belong to $\mathbf{R}^+$.

b) Suppose that $\mathbf{R}$ contains a subset $\mathbf{R}^+$ (not necessarily the set of positive numbers) which satisfies properties i) and ii). Define $x \prec y$ by $y - x \in \mathbf{R}^+$. Prove that Postulate 2 holds with $\prec$ in place of $<$.

1.3 COMPLETENESS AXIOM

In this section we introduce the last of three postulates that describe $\mathbf{R}$. To formulate this postulate, which distinguishes $\mathbf{Q}$ from $\mathbf{R}$, we need the following concepts.

1.10 Definition.

Let $E \subset \mathbf{R}$ be nonempty.

i) The set E is said to be *bounded above* if and only if there is an $M \in \mathbf{R}$ such that $a \leq M$ for all $a \in E$, in which case M is called an *upper bound* of E.
ii) A number s is called a *supremum* of the set E if and only if s is an upper bound of E and $s \leq M$ for all upper bounds M of E. (In this case we shall say that *E has a finite supremum s* and write $s = \sup E$.)

NOTE: Because French mathematicians (e.g., Borel, Jordan, and Lebesgue) did fundamental work on the connection between analysis and set theory, and *ensemble* is French for *set*, analysts frequently use E to represent a general set.

By Definition 1.10ii, a supremum of a set E (when it exists) is the smallest (or least) upper bound of E. By definition, then, in order to prove that $s = \sup E$ for some set $E \subset \mathbf{R}$, we must show two things: s is an upper bound, AND s is the smallest upper bound. Here is a typical example.

1.11 *EXAMPLE.*

If $E = [0, 1]$, prove that $\sup E = 1$.

Proof. By the definition of interval, 1 is an upper bound of E. Let M be any upper bound of E; that is, $M \geq x$ for all $x \in E$. Since $1 \in E$, it follows that $M \geq 1$. Thus 1 is the smallest upper bound of E. ∎

The following two remarks answer the question: How many upper bounds and suprema can a given set have?

1.12 Remark. *If a set has one upper bound, it has infinitely many upper bounds.*

Proof. If M_0 is an upper bound for a set E, then so is M for any $M > M_0$. ■

1.13 Remark. *If a set has a supremum, then it has only one supremum.*

Proof. Let s_1 and s_2 be suprema of the same set E. Then both s_1 and s_2 are upper bounds of E, whence by Definition 1.10ii, $s_1 \le s_2$ and $s_2 \le s_1$. We conclude by the Trichotomy Property that $s_1 = s_2$. ■

NOTE: *This proof illustrates a general principle. When asked to prove $a = b$, it is often easier to verify that $a \le b$ and $b \le a$ separately.*

The next result, a fundamental property of suprema, shows that the supremum of a set E can be approximated by a point in E (see Figure 1.3 for an illustration).

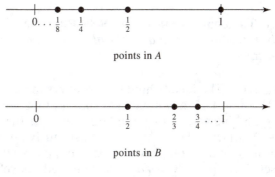

points in A

points in B

FIGURE 1.3

1.14 Theorem.   [APPROXIMATION PROPERTY FOR SUPREMA].
If E has a finite supremum and $\varepsilon > 0$ is any positive number, then there is a point $a \in E$ such that

$$\sup E - \varepsilon < a \le \sup E.$$

Proof. Suppose that the theorem is false. Then there is an $\varepsilon_0 > 0$ such that no element of E lies between $s_0 := \sup E - \varepsilon_0$ and $\sup E$. Since $\sup E$ is an upper bound for E, it follows that $a \le s_0$ for all $a \in E$; that is, s_0 is an upper bound of E. Thus, by Definition 1.10ii, $\sup E \le s_0 = \sup E - \varepsilon_0$. Adding $\varepsilon_0 - \sup E$ to both sides of this inequality, we conclude that $\varepsilon_0 \le 0$, a contradiction. ■

The Approximation Property can be used to show that the supremum of any subset of integers is itself an integer.

1.15 Theorem. *If $E \subset \mathbf{Z}$ has a supremum, then* $\sup E \in E$. *In particular, if the supremum of a set, which contains only integers, exists, that supremum must be an integer.*

Proof. Suppose that $s := \sup E$ and apply the Approximation Property to choose an $x_0 \in E$ such that $s - 1 < x_0 \le s$. If $s = x_0$, then $s \in E$, as promised. Otherwise, $s - 1 < x_0 < s$ and we can apply the Approximation Property again to choose $x_1 \in E$ such that $x_0 < x_1 < s$.

Subtract x_0 from this last inequality to obtain $0 < x_1 - x_0 < s - x_0$. Since $-x_0 < 1 - s$, it follows that $0 < x_1 - x_0 < s + (1 - s) = 1$. Thus $x_1 - x_0 \in \mathbf{Z} \cap (0, 1)$, a contradiction by Remark 1.1iii. We conclude that $s \in E$. ∎

The existence of suprema is the last assumption about **R** we make.

Postulate 3. [COMPLETENESS AXIOM]. If E is a nonempty subset of **R** that is bounded above, then E has a finite supremum.

We shall use Completeness Axiom many times. Our first two applications deal with the distribution of integers (Theorem 1.16) and rationals (Theorem 1.18) among real numbers.

1.16 Theorem. [ARCHIMEDEAN PRINCIPLE].
Given real numbers a and b, with $a > 0$, there is an integer $n \in \mathbf{N}$ such that $b < na$.

STRATEGY: The idea behind the proof is simple. By the Completeness Axiom and Theorem 1.15, any nonempty subset of integers that is bounded above has a "largest" integer. If k_0 is the largest integer that satisfies $k_0 a \le b$, then $n = (k_0 + 1)$ (which is larger than k_0) must satisfy $na > b$. In order to justify this application of the Completeness Axiom, we have two details to attend to: (1) Is the set $E := \{k \in \mathbf{N} : ka \le b\}$ bounded above? (2) Is E nonempty? The answer to the second question depends on whether $b < a$ or not. Here are the details.

Proof. If $b < a$, set $n = 1$. If $a \le b$, consider the set $E = \{k \in \mathbf{N} : ka \le b\}$. E is nonempty since $1 \in E$. Let $k \in E$ (i.e., $ka \le b$). Since $a > 0$, it follows from the First Multiplicative Property that $k \le b/a$. This proves that E is bounded above by b/a. Thus, by the Completeness Axiom and Theorem 1.15, E has a finite supremum s that belongs to E, in particular, $s \in \mathbf{N}$.

Set $n = s + 1$. Then $n \in \mathbf{N}$ and (since n is larger than s), n cannot belong to E. Thus $na > b$. ∎

Notice in Example 1.11 and Theorem 1.15 that the supremum of E belonged to E. The following result shows that this is not always the case.

1.17 EXAMPLE.
Let $A = \{1, \frac{1}{2}, \frac{1}{4}, \frac{1}{8}, \ldots\}$ and $B = \{\frac{1}{2}, \frac{2}{3}, \frac{3}{4}, \ldots\}$. Prove that $\sup A = \sup B = 1$.

Proof. It is clear that 1 is an upper bound of both sets. It remains to see that 1 is the smallest upper bound of both sets. For A, this is trivial. Indeed, if M is any upper bound of A, then $M \geq 1$ (since $1 \in A$). On the other hand, if M is an upper bound for B, but $M < 1$, then $1 - M > 0$. In particular, $1/(1 - M) \in \mathbf{R}$.

Choose, by the Archimedean Principle, an $n \in \mathbf{N}$ such that $n > 1/(1-M)$. It follows (do the algebra) that $x_0 := 1 - 1/n > M$. Since $x_0 \in B$, this contradicts the assumption that M is an upper bound of B (see Figure 1.3). ∎

The next proof shows how the Archimedean Principle is used to establish scale.

1.18 Theorem.   [DENSITY OF RATIONALS].
If $a, b \in \mathbf{R}$ satisfy $a < b$, then there is a $q \in \mathbf{Q}$ such that $a < q < b$.

STRATEGY: To find a fraction $q = m/n$ such that $a < q < b$, we must specify both numerator m and denominator n. Let's suppose first that $a > 0$ and that the set $E := \{k \in \mathbf{N} : k/n \leq a\}$ has a supremum, k_0. Then $m := k_0 + 1$, being greater than the supremum of E, cannot belong to E. Thus $m/n > a$. Is this the fraction we look for? Is $m/n < b$? Not unless n is large enough. To see this, look at a concrete example: $a = 2/3$ and $b = 1$. If $n = 1$, then E has no supremum, When $n = 2$, $k_0 = 1$ and when $n = 3$, $k_0 = 2$. In both cases $(k_0 + 1)/n = 1$ is too big. However, when $n = 4$, $k_0 = 2$ so $(k_0 + 1)/4 = 3/4$ is smaller than b, as required.

How can we prove that for each fixed $a < b$ there always is an n large enough so that if k_0 is chosen as above, then $(k_0 + 1)/n < b$? By the choice of k_0, $k_0/n \leq a$. Let's look at the worst case scenario: $a = k_0/n$. Then $b > (k_0 + 1)/n$ means

$$b > \frac{k_0 + 1}{n} = \frac{k_0}{n} + \frac{1}{n} = a + \frac{1}{n}$$

(i.e., $b - a > 1/n$). Such an n can always be chosen by the Archimedean Principle.

What about the assumption that $\sup E$ exists? This requires that E be nonempty and bounded above. Once n is fixed, E will be bounded above by na. But the only way that E is nonempty is that at the very least, $1 \in E$ (i.e., that $1/n \leq a$). This requires a second restriction on n. We begin our formal proof at this point.

Proof. Suppose first that $a > 0$. Since $b - a > 0$, use the Archimedean Principle to choose an $n \in \mathbf{N}$ that satisfies

$$n > \max \left\{ \frac{1}{a}, \frac{1}{b - a} \right\},$$

and observe that both $1/n < a$ and $1/n < b - a$.

Consider the set $E = \{k \in \mathbf{N} : k/n \leq a\}$. Since $1 \in E$, E is nonempty. Since $n > 0$, E is bounded above by na. Hence, by Theorem 1.15, $k_0 := \sup E$ exists and belongs to E, in particular, to **N**. Set $m = k_0 + 1$ and $q = m/n$. Since k_0 is the supremum of E, $m \notin E$. Thus $q > a$. On the other hand, since $k_0 \in E$, it follows from the choice of n that

$$b = a + (b - a) \geq \frac{k_0}{n} + (b - a) > \frac{k_0}{n} + \frac{1}{n} = \frac{m}{n} = q.$$

Now suppose that $a \leq 0$. Choose, by the Archimedean Principle, an integer $k \in \mathbf{N}$ such that $k > -a$. Then $0 < k + a < k + b$, and by the case already proved, there is an $r \in \mathbf{Q}$ such that $k + a < r < k + b$. Therefore, $q := r - k$ belongs to **Q** and satisfies the inequality $a < q < b$. ∎

For some applications, we also need the following concepts.

1.19 Definition.

Let $E \subset \mathbf{R}$ be nonempty.

i) The set E is said to be *bounded below* if and only if there is an $m \in \mathbf{R}$ such that $a \geq m$ for all $a \in E$, in which case m is called a *lower bound* of the set E.

ii) A number t is called an *infimum* of the set E if and only if t is a lower bound of E and $t \geq m$ for all lower bounds m of E. In this case we shall say that E has an infimum t and write $t = \inf E$.

iii) E is said to be *bounded* if and only if it is bounded both above and below.

When a set E contains its supremum (respectively, its infimum) we shall frequently write $\max E$ for $\sup E$ (respectively, $\min E$ for $\inf E$).

(Some authors call the supremum the *least upper bound* and the infimum the *greatest lower bound*. We will not use this terminology because it is somewhat old fashioned and because it confuses some students, since the **least** upper bound of given set is always greater than or equal to the **greatest** lower bound.)

To relate suprema to infima, we define the *reflection* of a set $E \subseteq \mathbf{R}$ by

$$-E := \{x : x = -a \text{ for some } a \in E\}.$$

For example, $-(1, 2] = [-2, -1)$.

The following result shows that the supremum of a set is the same as the negative of its reflection's infimum. This can be used to prove an Approximation Property and a Completeness Property for Infima (see Exercise 1.3.6).

1.20 Theorem. [REFLECTION PRINCIPLE].
Let $E \subseteq \mathbf{R}$ be nonempty.

i) *E has a supremum if and only if $-E$ has an infimum, in which case*

$$\inf(-E) = -\sup E.$$

ii) *E has an infimum if and only if* $-E$ *has a supremum, in which case*

$$\sup(-E) = -\inf E.$$

Proof. The proofs of these statements are similar. We prove only the first statement.

Suppose that E has a supremum s and set $t = -s$. Since s is an upper bound for E, $s \geq a$ for all $a \in E$, so $-s \leq -a$ for all $a \in E$. Therefore, t is a lower bound of $-E$. Suppose that m is any lower bound of $-E$. Then $m \leq -a$ for all $a \in E$, so $-m$ is an upper bound of E. Since s is the supremum of E, it follows that $s \leq -m$ (i.e., $t = -s \geq m$). Thus t is the infimum of $-E$ and $\sup E = s = -t = -\inf(-E)$.

Conversely, suppose that $-E$ has an infimum t. By definition, $t \leq -a$ for all $a \in E$. Thus $-t$ is an upper bound for E. Since E is nonempty, E has a supremum by the Completeness Axiom. ∎

Theorem 1.20 allows us to obtain information about infima from results about suprema, and vice versa (see the proof of the next theorem).

We shall use the following result many times.

1.21 Theorem. [MONOTONE PROPERTY].
Suppose that $A \subseteq B$ are nonempty subsets of **R**.

i) *If B has a supremum, then* $\sup A \leq \sup B$.
ii) *If B has an infimum, then* $\inf A \geq \inf B$.

Proof. i) Since $A \subseteq B$, any upper bound of B is an upper bound of A. Therefore, $\sup B$ is an upper bound of A. It follows from the Completeness Axiom that $\sup A$ exists, and from Definition 1.10ii that $\sup A \leq \sup B$.

ii) Clearly, $-A \subseteq -B$. Thus by part i), Theorem 1.20, and the Second Multiplicative Property,

$$\inf A = -\sup(-A) \geq -\sup(-B) = \inf B. \qquad \blacksquare$$

It is convenient to extend the definition of suprema and infima to all subsets of **R**. To do this we expand the definition of **R** as follows. The set of *extended real numbers* is defined to be $\overline{\mathbf{R}} := \mathbf{R} \cup \{\pm\infty\}$. Thus x is an extended real number if and only if either $x \in \mathbf{R}$, $x = +\infty$, or $x = -\infty$.

Let $E \subseteq \mathbf{R}$ be nonempty. We shall define $\sup E = +\infty$ if E is unbounded above and $\inf E = -\infty$ if E is unbounded below. Finally, we define $\sup \emptyset = -\infty$ and $\inf \emptyset = +\infty$. Notice, then, that the supremum of a subset E of **R** (respectively, the infimum of E) is finite if and only if E is nonempty and bounded above (respectively, nonempty and bounded below). Moreover, under the convention $-\infty < a$ and $a < \infty$ for all $a \in \mathbf{R}$, the Monotone Property still holds for this extended definition; that is, if A and B are subsets of **R** and $A \subseteq B$, then $\sup A \leq \sup B$ and $\inf A \geq \inf B$, provided we use the convention that $-\infty < \infty$.

EXERCISES

1.3.0. Decide which of the following statements are true and which are false. Prove the true ones and give counterexamples to the false ones.

a) If A and B are nonempty, bounded subsets of $\mathbf{R}$, then $\sup(A \cap B) \le \sup A$.

b) Let ε be a positive real number. If A is a nonempty, bounded subset of $\mathbf{R}$ and $B = \{\varepsilon x : x \in A\}$, then $\sup(B) = \varepsilon \sup(A)$.

c) If $A + B := \{a + b : a \in A \text{ and } b \in B\}$, where A and B are nonempty, bounded subsets of $\mathbf{R}$, then $\sup(A + B) = \sup(A) + \sup(B)$.

d) If $A - B := \{a - b : a \in A \text{ and } b \in B\}$, where A and B are nonempty, bounded subsets of $\mathbf{R}$, then $\sup(A - B) = \sup(A) - \sup(B)$

1.3.1. Find the infimum and supremum of each of the following sets.

a) $E = \{x \in \mathbf{R} : x^2 + 2x = 3\}$

b) $E = \{x \in \mathbf{R} : x^2 - 2x + 3 > x^2 \text{ and } x > 0\}$

c) $E = \{p/q \in \mathbf{Q} : p^2 < 5q^2 \text{ and } p, q > 0\}$

d) $E = \{x \in \mathbf{R} : x = 1 + (-1)^n/n \text{ for } n \in \mathbf{N}\}$

e) $E = \{x \in \mathbf{R} : x = 1/n + (-1)^n \text{ for } n \in \mathbf{N}\}$

f) $E = \{2 - (-1)^n/n^2 : n \in \mathbf{N}\}$

1.3.2. Prove that for each $a \in \mathbf{R}$ and each $n \in \mathbf{N}$ there exists a rational r_n such that $|a - r_n| < 1/n$.

1.3.3 . [DENSITY OF IRRATIONALS] **This exercise is used in Section 3.3**. Prove that if $a < b$ are real numbers, then there is an irrational $\xi \in \mathbf{R}$ such that $a < \xi < b$.

1.3.4. Prove that a lower bound of a set need not be unique but the infimum of a given set E is unique.

1.3.5. Show that if E is a nonempty bounded subset of $\mathbf{Z}$, then $\inf E$ exists and belongs to E.

1.3.6 . **This exercise is used in many sections, including 2.2 and 5.1**. Use the Reflection Principle and analogous results about suprema to prove the following results.

a) [APPROXIMATION PROPERTY FOR INFIMA] Prove that if a set $E \subset \mathbf{R}$ has a finite infimum and $\varepsilon > 0$ is any positive number, then there is a point $a \in E$ such that $\inf E + \varepsilon > a \ge \inf E$.

b) [COMPLETENESS PROPERTY FOR INFIMA] If $E \subseteq \mathbf{R}$ is nonempty and bounded below, then E has a (finite) infimum.

1.3.7. a) Prove that if x is an upper bound of a set $E \subseteq \mathbf{R}$ and $x \in E$, then x is the supremum of E.

b) Make and prove an analogous statement for the infimum of E.

c) Show by example that the converse of each of these statements is false.

1.3.8. Suppose that $E, A, B \subset \mathbf{R}$ and $E = A \cup B$. Prove that if E has a supremum and both A and B are nonempty, then $\sup A$ and $\sup B$ both exist, and $\sup E$ is one of the numbers $\sup A$ or $\sup B$.

1.3.9. A *dyadic rational* is a number of the form $k/2^n$ for some $k, n \in \mathbf{Z}$. Prove that if a and b are real numbers and $a < b$, then there exists a dyadic rational q such that $a < q < b$.

1.3.10. Let $x_n \in \mathbf{R}$ and suppose that there is an $M \in \mathbf{R}$ such that $|x_n| \leq M$ for $n \in \mathbf{N}$. Prove that $s_n = \sup\{x_n, x_{n+1}, \ldots\}$ defines a real number for each $n \in \mathbf{N}$ and that $s_1 \geq s_2 \geq \cdots$. Prove an analogous result about $t_n = \inf\{x_n, x_{n+1}, \ldots\}$.

1.3.11. If $a, b \in \mathbf{R}$ and $b - a > 1$, then there is at least one $k \in \mathbf{Z}$ such that $a < k < b$.

1.4 MATHEMATICAL INDUCTION

In this section we introduce the method of Mathematical Induction and use it to prove the Binomial Formula, a result that shows how to expand powers of a binomial expression (i.e., an expression of the form $a + b$).

We begin by obtaining another consequence of the Completeness Axiom, the Well-Ordering Principle, which is a statement about the existence of least elements of subsets of $\mathbf{N}$.

1.22 Theorem.   [WELL-ORDERING PRINCIPLE].
If E is a nonempty subset of $\mathbf{N}$, then E has a least element (i.e., E has a finite infimum and $\inf E \in E$).

Proof. Suppose that $E \subseteq \mathbf{N}$ is nonempty. Then $-E$ is bounded above, by -1, so by the Completeness Axiom $\sup(-E)$ exists, and by Theorem 1.15, $\sup(-E) \in -E$. Hence by Theorem 1.20, $\inf E = -\sup(-E)$ exists, and $\inf E \in -(-E) = E$. ∎

Our first application of the Well-Ordering Principle is called the *Principle of Mathematical Induction* or the *Axiom of Induction* (which, under mild assumptions, is equivalent to the Well-Ordering Principle—see Appendix A).

1.23 Theorem. *Suppose for each $n \in \mathbf{N}$ that $A(n)$ is a proposition (i.e., a verbal statement or formula) which satisfies the following two properties:*

i) *$A(1)$ is true.*
ii) *For every $n \in \mathbf{N}$ for which $A(n)$ is true, $A(n + 1)$ is also true.*

Then $A(n)$ is true for all $n \in \mathbf{N}$.

Proof. Suppose that the theorem is false. Then the set $E = \{n \in \mathbf{N} : A(n)$ is false$\}$ is nonempty. Hence by the Well-Ordering Principle, E has a least element, say x.

Since $x \in E \subseteq \mathbf{N} \subset \mathbf{Z}$, we have by Remark 1.1ii that $x \geq 1$. Since $x \in E$, we have by hypothesis i) that $x \neq 1$. In particular, $x - 1 > 0$. Hence, by Remark 1.1i and iii, $x - 1 \geq 1$ and $x - 1 \in \mathbf{N}$.

Since $x - 1 < x$ and x is a least element of E, the statement $A(x - 1)$ must be true. Applying hypothesis ii) to $n = x - 1$, we see that $A(x) = A(n + 1)$ must also be true; that is, $x \notin E$, a contradiction. ∎

Recall that if $x_0, x_1, \ldots, x_n$ are real numbers and $0 \le j \le n$, then

$$\sum_{k=j}^{n} x_k := x_j + x_{j+1} + \cdots + x_n$$

denotes the sum of the x_k's as k ranges from j to n. The following examples illustrate the fact that the Principle of Mathematical Induction can be used to prove a variety of statements involving integers.

1.24 EXAMPLE.

Prove that

$$\sum_{k=1}^{n} (3k - 1)(3k + 2) = 3n^3 + 6n^2 + n$$

for $n \in \mathbf{N}$.

Proof. Let $A(n)$ represent the statement

$$\sum_{k=1}^{n} (3k - 1)(3k + 2) = 3n^3 + 6n^2 + n.$$

For $n = 1$ the left side of this equation is $2 \cdot 5$ and the right side is $3 + 6 + 1$. Therefore, $A(1)$ is true. Suppose that $A(n)$ is true for some $n \ge 1$. Then

$$\sum_{k=1}^{n+1} (3k - 1)(3k + 2) = (3n + 2)(3n + 5) + \sum_{k=1}^{n} (3k - 1)(3k + 2)$$

$$= (3n + 2)(3n + 5) + 3n^3 + 6n^2 + n$$

$$= 3n^3 + 15n^2 + 22n + 10.$$

On the other hand, a direct calculation reveals that

$$3(n + 1)^3 + 6(n + 1)^2 + (n + 1) = 3n^3 + 15n^2 + 22n + 10.$$

Therefore, $A(n + 1)$ is true when $A(n)$ is. We conclude by induction that $A(n)$ holds for all $n \in \mathbf{N}$. ∎

Two formulas encountered early in an algebra course are the perfect square and cube formulas:

$$(a + b)^2 = a^2 + 2ab + b^2 \quad \text{and} \quad (a + b)^3 = a^3 + 3a^2b + 3ab^2 + b^3.$$

Our next application of the Principle of Mathematical Induction generalizes these formulas from $n = 2$ and 3 to arbitrary $n \in \mathbf{N}$.

Recall that Pascal's triangle is the triangular array of integers whose rows begin and end with 1s with the property that an interior entry on any row is obtained by adding the two numbers in the preceding row immediately above that entry. Thus the first few rows of Pascal's triangle are as below.

$$
\begin{array}{ccccccccccccc}
 & & & & & & 1 & & & & & & \\
 & & & & & 1 & & 1 & & & & & \\
 & & & & 1 & & 2 & & 1 & & & & \\
 & & & 1 & & 3 & & 3 & & 1 & & & \\
 & & 1 & & 4 & & 6 & & 4 & & 1 & & \\
 & 1 & & 5 & & 10 & & 10 & & 5 & & 1 & \\
1 & & 6 & & 15 & & 20 & & 15 & & 6 & & 1 \\
\end{array}
$$

Notice that the third and fourth rows are precisely the coefficients that appeared in the perfect square and cube formulas above.

We can write down a formula for each entry in each row of the Pascal triangle. The first (and only) entry in the first row is

$$
\binom{0}{0} := 1.
$$

Using the notation $0! := 1$ and $n! := 1 \cdot 2 \cdots (n-1) \cdot n$ for $n \in \mathbf{N}$, define the *binomial coefficient n choose k* by

$$
\binom{n}{k} := \frac{n!}{(n-k)!k!}
$$

for $0 \le k \le n$ and $n = 0, 1, \ldots$.

Since $\binom{n}{0} = \binom{n}{n} = 1$ for all $n \in \mathbf{N}$, the following result shows that the binomial coefficient n over k does produce the $(k+1)$st entry in the $(n+1)$st row of Pascal's triangle.

1.25 Lemma.
If $n, k \in \mathbf{N}$ and $1 \le k \le n$, then

$$
\binom{n+1}{k} = \binom{n}{k-1} + \binom{n}{k}.
$$

Proof. By definition,

$$
\binom{n}{k-1} + \binom{n}{k} = \frac{n!\,k}{(n-k+1)!k!} + \frac{n!(n-k+1)}{(n-k+1)!k!}
$$

$$
= \frac{n!(n+1)}{(n-k+1)!k!} = \binom{n+1}{k}. \qquad \blacksquare
$$

Binomial coefficients can be used to expand the nth power of a sum of two terms.

1.26 Theorem.   [BINOMIAL FORMULA].
If $a, b \in \mathbf{R}$, $n \in \mathbf{N}$, and 0^0 is interpreted to be 1, then

$$(a+b)^n = \sum_{k=0}^{n} \binom{n}{k} a^{n-k} b^k.$$

Proof. The proof is by induction on n. The formula is obvious for $n = 1$. Suppose that the formula is true for some $n \in \mathbf{N}$. Then by the inductive hypothesis and Postulate 1,

$$(a+b)^{n+1} = (a+b)(a+b)^n$$

$$= (a+b)\left(\sum_{k=0}^{n} \binom{n}{k} a^{n-k} b^k\right)$$

$$= \left(\sum_{k=0}^{n} \binom{n}{k} a^{n-k+1} b^k\right) + \left(\sum_{k=0}^{n} \binom{n}{k} a^{n-k} b^{k+1}\right)$$

$$= \left(a^{n+1} + \sum_{k=1}^{n} \binom{n}{k} a^{n-k+1} b^k\right) + \left(b^{n+1} + \sum_{k=0}^{n-1} \binom{n}{k} a^{n-k} b^{k+1}\right)$$

$$= a^{n+1} + \sum_{k=1}^{n} \left(\binom{n}{k} + \binom{n}{k-1}\right) a^{n-k+1} b^k + b^{n+1}.$$

Hence it follows from Lemma 1.25 that

$$(a+b)^{n+1} = a^{n+1} + \sum_{k=1}^{n} \binom{n+1}{k} a^{n+1-k} b^k + b^{n+1} = \sum_{k=0}^{n+1} \binom{n+1}{k} a^{n+1-k} b^k;$$

that is, the formula is true for $n+1$. We conclude by induction that the formula holds for all $n \in \mathbf{N}$. ∎

We close this section with two optional, well-known results that further demonstrate the power of the Completeness Axiom and its consequences.

*1.27 Remark. *If $x > 1$ and $x \notin \mathbf{N}$, then there is an $n \in \mathbf{N}$ such that $n < x < n+1$.*

Proof. By the Archimedean Principle, the set $E = \{m \in \mathbf{N} : x < m\}$ is nonempty. Hence by the Well-Ordering Principle, E has a least element, say m_0.

Set $n = m_0 - 1$. Since $m_0 \in E$, $n + 1 = m_0 > x$. Since m_0 is least, $n = m_0 - 1 \leq x$. Since $x \notin \mathbf{N}$, we also have $n \neq x$. Therefore, $n < x < n+1$. ∎

Using this last result, we can prove that the set of irrationals is nonempty.

*1.28 Remark. *If $n \in \mathbf{N}$ is not a perfect square (i.e., if there is no $m \in \mathbf{N}$ such that $n = m^2$), then $\sqrt{n}$ is irrational.*

Proof. Suppose to the contrary that $n \in \mathbf{N}$ is not a perfect square but $\sqrt{n} \in \mathbf{Q}$; that is, $\sqrt{n} = p/q$ for some $p, q \in \mathbf{N}$. Choose by Remark 1.27 an integer $m_0 \in \mathbf{N}$ such that

$$m_0 < \sqrt{n} < m_0 + 1. \tag{10}$$

Consider the set $E := \{k \in \mathbf{N} : k\sqrt{n} \in \mathbf{Z}\}$. Since $q\sqrt{n} = p$, we know that E is nonempty. Thus by the Well-Ordering Principle, E has a least element, say n_0.

Set $x = n_0(\sqrt{n} - m_0)$. By (10), $0 < \sqrt{n} - m_0 < 1$. Multiplying this inequality by n_0, we find that

$$0 < x < n_0. \tag{11}$$

Since n_0 is a least element of E, it follows from (11) that $x \notin E$. On the other hand,

$$x\sqrt{n} = n_0(\sqrt{n} - m_0)\sqrt{n} = n_0 n - m_0 n_0 \sqrt{n} \in \mathbf{Z}$$

since $n_0 \in E$. Moreover, since $x > 0$ and $x = n_0\sqrt{n} - n_0 m_0$ is the difference of two integers, $x \in \mathbf{N}$. Thus $x \in E$, a contradiction. ∎

EXERCISES

1.4.0. Decide which of the following statements are true and which are false. Prove the true ones and give counterexamples to the false ones.

   a) If $a \geq 0$ and $b \neq 0$, then $(a + b)^n \geq b^n$ for all $n \in \mathbf{N}$.

   b) If $a < 0 < b$, then $(a + b)^n \leq b^n$ for all $n \in \mathbf{N}$.

   c) If $n \in \mathbf{N}$ is even and both a and b are negative, then $(a + b)^n > a^n + na^{n-1}b$.

   d) If $a \neq 0$, then

$$\frac{1}{2^n} = \sum_{k=0}^{n} \binom{n}{k} \frac{(a - 2)^{n-k}}{a^n 2^{n-k}}$$

   for all $n \in \mathbf{N}$.

1.4.1. a) Prove that if $x_1 > 2$ and $x_{n+1} = 1 + \sqrt{x_n - 1}$ for $n \in \mathbf{N}$, then $2 < x_{n+1} < x_n$ holds for all $n \in \mathbf{N}$.

   b) Prove that if $2 < x_1 < 3$ and $x_{n+1} = 2 + \sqrt{x_n - 2}$ for $n \in \mathbf{N}$, then $0 < x_n < x_{n+1}$ holds for all $n \in \mathbf{N}$.

   c) Prove that if $0 < x_1 < 1$ and $x_{n+1} = 1 - \sqrt{1 - x_n}$ for $n \in \mathbf{N}$, then $0 < x_{n+1} < x_n$ holds for all $n \in \mathbf{N}$.

   d) Prove that if $3 < x_1 < 5$ and $x_{n+1} = 2 + \sqrt{x_n - 2}$, then $3 < x_{n+1} < x_n$ holds for all $n \in \mathbf{N}$.

1.4.2. Use the Binomial Formula or the Principle of Induction to prove each of the following.

a) $\sum_{k=0}^{n}(-1)^k \binom{n}{k} = 0$ for all $n \in \mathbf{N}$.

b) $(a+b)^n \geq a^n + b^n$ for all $n \in \mathbf{N}$ and $a, b \geq 0$.

c) $(1 + 1/n)^n \geq 2$ for all $n \in \mathbf{N}$.

d) $\sum_{k=1}^{n}\binom{n}{k} = \sum_{k=0}^{n-1} 2^k$ for all $n \in \mathbf{N}$.

1.4.3. Prove each of the following statements.

a) $2n + 1 < 2^n$ for $n = 3, 4, \ldots$.

b) $n < 2^n$ for $n = 1, 2, \ldots$.

c) $n^2 \leq 2^n + 1$ for $n = 1, 2, \ldots$.

d) $n^3 \leq 3^n$ for $n = 1, 2, \ldots$.

1.4.4 . **Parts a) and c) of this exercise are used in Sections 2.4 and 5.1.** Prove that the following formulas hold for all $n \in \mathbf{N}$.

a) $\sum_{k=1}^{n} k = \dfrac{n(n+1)}{2}$

b) $\sum_{k=1}^{n} k^2 = \dfrac{n(n+1)(2n+1)}{6}$

c) $\sum_{k=1}^{n} \dfrac{a-1}{a^k} = 1 - \dfrac{1}{a^n}$, $a \neq 0$

d) $\sum_{k=1}^{n} (2k-1)^2 = \dfrac{n(4n^2-1)}{3}$

1.4.5 . **This exercise is used in Section 2.3.** Prove that $0 \leq a < b$ implies $0 \leq a^n < b^n$ and $0 \leq \sqrt[n]{a} < \sqrt[n]{b}$ for all $n \in \mathbf{N}$.

1.4.6. Prove that $2^n + 3^n$ is a multiple of 5 for all odd $n \in \mathbf{N}$.

1.4.7. Prove that $2^n \leq n! + 2$ for $n \in \mathbf{N}$.

1.4.8. Prove that

$$2^n > \frac{n(n-1)(n-2)}{6}$$

for $n \in \mathbf{N}$.

1.4.9. a) Using Remark 1.28, prove that the square root of an integer m is rational if and only if $m = k^2$ for some $k \in \mathbf{N}$.

b) Prove that $\sqrt{n+3} + \sqrt{n}$ is rational for some $n \in \mathbf{N}$ if and only if $n = 1$.

c) Find all $n \in \mathbf{N}$ such that $\sqrt{n+7} + \sqrt{n}$ is rational.

1.4.10. Let $a_0 = 3$, $b_0 = 4$, and $c_0 = 5$.

a) Let $a_k = a_{k-1} + 2$, $b_k = 2a_{k-1} + b_{k-1} + 2$, and $c_k = 2a_{k-1} + c_{k-1} + 2$ for $k \in \mathbf{N}$. Prove that $c_k - b_k$ is constant for all $k \in \mathbf{N}$.

b) Prove that the numbers defined in part a) satisfy

$$a_k^2 + b_k^2 = c_k^2$$

for all $k \in \mathbf{N}$.

1.5 INVERSE FUNCTIONS AND IMAGES

Let $f : X \to Y$ (i.e., suppose that f is a function from one set X to another set Y). In this section, we obtain simple conditions for when f has an inverse, introduce images and inverse images induced by f, and explore how they interact with the algebra of sets.

First, recall from Section 1.1 that a function $f : X \to Y$ has an inverse function if and only if $\operatorname{Ran}(f) = Y$ and each $y \in Y$ has a unique preimage $x \in X$, in which case we define the *inverse function* f^{-1} by $f^{-1}(y) := x$. In particular, if $f : X \to Y$ has an inverse function, then

$$f^{-1}(f(x)) = x \quad \text{and} \quad f(f^{-1}(y)) = y \qquad (12)$$

for all $x \in X$ and $y \in Y$.

We introduce the following concepts in order to answer the question, "Is there an easy way to recognize when f has an inverse?"

1.29 Definition.

Let X and Y be sets and $f : X \to Y$.

i) f is said to be 1–1 (*one-to-one* or an *injection*) if and only if

$$x_1, x_2 \in X \quad \text{and} \quad f(x_1) = f(x_2) \quad \text{imply} \quad x_1 = x_2.$$

ii) f is said to be *onto* (or a *surjection*) if and only if for each $y \in Y$ there is an $x \in X$ such that $y = f(x)$.

iii) f is called a *bijection* if and only if it is both 1–1 and onto.

Sometimes, to emphasize the domain and range of f, we shall say that a bijection $f : X \to Y$ is 1–1 from X onto Y.

For example, the function $f(x) = x^2$ is 1–1 from $[0, \infty)$ onto $[0, \infty)$ but not 1–1 on any open interval containing 0.

We shall now prove that bijections always have inverse functions and that (12) characterizes those inverses.

1.30 Theorem. *Let X and Y be sets and $f : X \to Y$. Then the following three statements are equivalent.*

i) *f has an inverse;*
ii) *f is 1–1 from X onto Y;*

iii) *There is a function g : Y → X such that*

$$g(f(x)) = x \quad \text{for all } x \in X \tag{13}$$

and

$$f(g(y)) = y \quad \text{for all } y \in Y. \tag{14}$$

Moreover, for each f : X → Y, there is only one function g that satisfies (13) and (14). It is the inverse function f^{-1}.

Proof. *i) implies ii).* By definition, if f has an inverse, then $\text{Ran}(f) = Y$ (so f takes X onto Y) and each $y \in Y$ has a unique preimage in X [so, if $f(y_1) = f(y_2)$, then $y_1 = y_2$, i.e., f is 1–1 on X].

ii) implies iii). The proof that i) implies ii) also shows that if $f : X \to Y$ is 1–1 and onto, then f has an inverse. In particular, $g(y) := f^{-1}(y)$ satisfies (13) and (14) by (12).

iii) implies i). Suppose that there is a function $g : Y \to X$ which satisfies (13) and (14). If some $y \in Y$ has two preimages, say $x_1 \neq x_2$ in X, then $f(x_1) = y = f(x_2)$. It follows from (13) that $x_1 = g(f(x_1)) = g(f(x_2)) = x_2$, a contradiction. On the other hand, given $y \in Y$, set $x = g(y)$. Then $f(x) = f(g(y)) = y$ by (14), so $\text{Ran}(f) = Y$.

Finally, suppose that h is another function which satisfies (13) and (14), and fix $y \in Y$. By ii), there is an $x \in X$ such that $f(x) = y$. Hence by (13),

$$h(y) = h(f(x)) = x = g(f(x)) = g(y);$$

that is, $h = g$ on Y. It follows that the function g is unique. ∎

There are two ways to show that a given function f is 1–1 on a set X. We can suppose that $f(x_1) = f(x_2)$ for some $x_1, x_2 \in X$, and prove (using algebra, for example) that $x_1 = x_2$. If X is an interval in **R** and f is differentiable, there is an easier way to prove that f is 1–1 on X.

1.31 Remark. *Let I be an interval and let f : I → R. If the derivative of f is either always positive on I, or always negative on I, then f is 1–1 on I.*

Proof. By symmetry, we may suppose that the derivative f' of f satisfies $f'(x) > 0$ for all $x \in I$. We will use a result that almost everyone who has studied one variable calculus remembers (for a proof, see Theorem 4.17): If $f' > 0$ on an interval I, then f is strictly increasing on I; that is, $x_1, x_2 \in I$ and $x_1 < x_2$ imply that $f(x_1) < f(x_2)$.

To see why this implies that f is 1–1, suppose that $f(x_1) = f(x_2)$ for some x_1, x_2 in X. If $x_1 \neq x_2$, then it follows from the trichotomy property that either $x_1 < x_2$ or $x_2 < x_1$. Since f is strictly increasing on I, either $f(x_1) < f(x_2)$ or $f(x_2) < f(x_1)$. Both of these conclusions contradict the assumption that $f(x_1) = f(x_2)$. ∎

By Theorem 1.30, $f : X \to Y$ has an inverse function f^{-1} if and only if $f^{-1}(f(x)) = x$ for all $x \in X$ and $f(f^{-1}(y)) = y$ for all $y \in Y$. This suggests that we can find a formula for f^{-1} if $y = f(x)$ can be solved for x.

*1.32 EXAMPLE.

Prove that $f(x) = e^x - e^{-x}$ is 1–1 on **R** and find a formula for f^{-1} on Ran(f).

Solution. Since $f'(x) = e^x + e^{-x} > 0$ for all $x \in$ **R**, f is 1–1 on **R** by Remark 1.31.

Let $y = e^x - e^{-x}$. Multiplying this equation by e^x and collecting all nonzero terms on one side of the equation, we have

$$e^{2x} - ye^x - 1 = 0,$$

a quadratic in e^x. By the quadratic formula,

$$e^x = \frac{y \pm \sqrt{y^2 + 4}}{2}. \tag{15}$$

Since e^x is always positive, the minus sign must be discarded. Taking the logarithm of this last identity, we obtain $x = \log(y + \sqrt{y^2 + 4}) - \log 2$. Therefore,

$$f^{-1}(x) = \log(x + \sqrt{x^2 + 4}) - \log 2. \qquad \blacksquare$$

The following concepts greatly simplify the general theory of continuity (see Theorem 9.26, for example).

1.33 Definition.

Let X and Y be sets and $f : X \to Y$. The *image* of a set $E \subseteq X$ under f is the set

$$f(E) := \{y \in Y : y = f(x) \text{ for some } x \in E\}.$$

The *inverse image* of a set $E \subseteq Y$ under f is the set

$$f^{-1}(E) := \{x \in X : f(x) = y \text{ for some } y \in E\}. \tag{16}$$

When E is an interval, we will sometimes drop the extra parentheses; for example, write $f(a, b]$ for $f((a, b])$ and $f^{-1}(a, b]$ for $f^{-1}((a, b])$.

1.34 EXAMPLE.

Find the images and inverse images of the sets $I = (-1, 0)$ and $J = (0, 1]$ under the function $f(x) = x^2 + x$.

Solution. Since "find" doesn't mean "prove," we look at the graph $y = x^2 + x$. By definition, $f(I)$ consists of the y-values of $f(x)$ as x ranges over $I = (-1, 0)$.

Since f has roots at $x = 0, -1$ and has a minimum of -0.25 at $x = -0.5$, it is clear by looking at the graph that $f(I) = [-0.25, 0)$. Since $f^{-1}(I)$ consist of the x-values whose images belong to $I = (-1, 0)$, and the graph of f lies below the x-axis only when $-1 < x < 0$, it is also clear that $f^{-1}(I) = (-1, 0)$. Similarly, $f(J) = (0, 2]$ and

$$f^{-1}(J) = \left[\frac{-1-\sqrt{5}}{2}, -1 \right) \cup \left(0, \frac{-1+\sqrt{5}}{2} \right].$$

(Be sure to look at the graph of $y = x^2 + x$ and understand how these numbers were obtained.) ∎

WARNING. Unfortunately, there are now three meanings to f^{-1}: (1) $f^{-1}(x) = 1/f(x)$, the reciprocal of f which exists when f is real-valued and $f(x) \neq 0$; (2) $f^{-1}(x)$, the inverse function of f which exists when f is 1–1 and onto; (3) $f^{-1}(E)$, the inverse image of E under f, which always exists. Context will usually indicate which meaning we are using.

Notice that Definition 1.33 contains an asymmetry: $y \in f(E)$ means that $y = f(x)$ for some $x \in E$, but $x \in f^{-1}(E)$ does NOT mean that $x = f^{-1}(y)$ for some $y \in E$. For example, let $f(x) = \sin x$. Since $\sin(k\pi) = 0$ for all $k \in \mathbf{Z}$, the inverse image of $\{0\}$ under f is $f^{-1}(\{0\}) = \{k\pi : k \in \mathbf{Z}\}$, but since the range of $\arcsin x$ is $[-\pi/2, \pi/2]$, the image of $\{0\}$ under f^{-1} is $\arcsin\{0\} = \{0\}$.

Before we give an account of how images and inverse images interact with set algebra (specifically, what the image and inverse image of a union, an intersection, and a complement of sets are), we need to expand the algebra of sets to include unions and intersections of infinitely many sets. We need these concepts for some of the deeper results in the second half of this book because many of the proofs involve associating a set E_α with each α in a set A. With this end in mind, we introduce the following terminology.

A collection of sets $\mathcal{E}$ is said to be *indexed* by a set A if and only if there is a function F from A onto $\mathcal{E}$ (i.e., each $\alpha \in A$ is associated with one and only one set in $\mathcal{E}$). In this case we shall call A the *index set* of $\mathcal{E}$, say that $\mathcal{E}$ is *indexed* by A, and represent $F(\alpha)$ by E_α. In particular, $\mathcal{E}$ is indexed by A means $\mathcal{E} = \{E_\alpha\}_{\alpha \in A}$.

1.35 Definition.

Let $\mathcal{E} = \{E_\alpha\}_{\alpha \in A}$ be a collection of sets.

i) The *union* of the collection $\mathcal{E}$ is the set

$$\bigcup_{\alpha \in A} E_\alpha := \{x : x \in E_\alpha \text{ for some } \alpha \in A\}.$$

ii) The *intersection* of the collection $\mathcal{E}$ is the set

$$\bigcap_{\alpha \in A} E_\alpha := \{x : x \in E_\alpha \quad \text{for all } \alpha \in A\}.$$

For example,

$$\bigcup_{x \in (0,1]} [0, x) = [0, 1) \quad \text{and} \quad \bigcap_{x \in (0,1]} [0, x) = \{0\}.$$

The following important, often used result shows that there is an easy way to get from unions to intersections, and vice versa.

1.36 Theorem.   [DEMORGAN'S LAWS].
Let X be a set and $\{E_\alpha\}_{\alpha \in A}$ be a collection of subsets of X. If for each $E \subseteq X$ the symbol E^c represents the set $X \backslash E$, then

$$\left(\bigcup_{\alpha \in A} E_\alpha \right)^c = \bigcap_{\alpha \in A} E_\alpha^c \tag{17}$$

and

$$\left(\bigcap_{\alpha \in A} E_\alpha \right)^c = \bigcup_{\alpha \in A} E_\alpha^c. \tag{18}$$

Proof. Suppose that x belongs to the left side of (17); that is, $x \in X$ and $x \notin \bigcup_{\alpha \in A} E_\alpha$. By definition, $x \in X$ and $x \notin E_\alpha$ for all $\alpha \in A$. Hence, $x \in E_\alpha^c$ for all $\alpha \in A$; that is, x belongs to the right side of (17). These steps are reversible. This verifies (17). A similar argument verifies (18). ∎

The following result, which plays a prominent role in Chapters 9 and 12, describes images and inverse images of unions and intersections of sets.

1.37 Theorem. *Let X and Y be sets and $f : X \to Y$.*

i) *If $\{E_\alpha\}_{\alpha \in A}$ is a collection of subsets of X, then*

$$f\left(\bigcup_{\alpha \in A} E_\alpha \right) = \bigcup_{\alpha \in A} f(E_\alpha) \quad \text{and} \quad f\left(\bigcap_{\alpha \in A} E_\alpha \right) \subseteq \bigcap_{\alpha \in A} f(E_\alpha).$$

ii) *If B and C are subsets of X, then $f(C \backslash B) \supseteq f(C) \backslash f(B)$.*

iii) If $\{E_\alpha\}_{\alpha \in A}$ is a collection of subsets of Y, then

$$f^{-1}\left(\bigcup_{\alpha \in A} E_\alpha\right) = \bigcup_{\alpha \in A} f^{-1}(E_\alpha) \quad and \quad f^{-1}\left(\bigcap_{\alpha \in A} E_\alpha\right) = \bigcap_{\alpha \in A} f^{-1}(E_\alpha).$$

iv) If B and C are subsets of Y, then $f^{-1}(C \backslash B) = f^{-1}(C) \backslash f^{-1}(B)$.
v) If $E \subseteq f(X)$, then $f(f^{-1}(E)) = E$, but if $E \subseteq X$, then $f^{-1}(f(E)) \supseteq E$.

Proof. i) By definition, $y \in f(\bigcup_{\alpha \in A} E_\alpha)$ if and only if $y = f(x)$ for some $x \in E_\alpha$ and $\alpha \in A$. This is equivalent to $y \in \bigcup_{\alpha \in A} f(E_\alpha)$. Similarly, $y \in f(\bigcap_{\alpha \in A} E_\alpha)$ if and only if $y = f(x)$ for some $x \in \bigcap_{\alpha \in A} E_\alpha$. This implies that for all $\alpha \in A$ there is an $x_\alpha \in E_\alpha$ such that $y = f(x_\alpha)$. Therefore, $y \in \bigcap_{\alpha \in A} f(E_\alpha)$.

ii) If $y \in f(C) \backslash f(B)$, then $y = f(c)$ for some $c \in C$ but $y \neq f(b)$ for any $b \in B$. It follows that $y \in f(C \backslash B)$. Similar arguments prove parts iii), iv), and v). ∎

It is important to recognize that the set inequalities in parts i), ii), and v) can be strict unless f is 1–1 (see Exercises 1.5.6 and 1.5.7). For example, if $f(x) = x^2$, $E_1 = \{1\}$, and $E_2 = \{-1\}$, then $f(E_1 \cap E_2) = \emptyset$ is a proper subset of $f(E_1) \cap f(E_2) = \{1\}$.

EXERCISES

1.5.0. Decide which of the following statements are true and which are false. Prove the true ones and give counterexamples to the false ones.

a) Let $f(x) = \sin x$. Then the function

$$f : \left[\frac{\pi}{2}, \frac{3\pi}{2}\right] \to [-1, 1]$$

is a bijection, and its inverse function is arcsin x.
b) Suppose that A, B, and C are subsets of some set X and that $f : X \to X$. If $A \cap B \neq \emptyset$, then $f(A) \cap f(B \cup C) \neq \emptyset$.
c) Suppose that A and B are subsets of some set X. Then $(A \backslash B)^c = B \backslash A$.
d) If f takes $[-1, 1]$ onto $[-1, 1]$, then $f^{-1}(f(\{0\})) = \{0\}$.

1.5.1. a) For each of the following, prove that f is 1–1 on E and find $f(E)$.

α) $f(x) = 3x - 7$, $E = \mathbf{R}$
β) $f(x) = e^{1/x}$, $E = (0, \infty)$
γ) $f(x) = \tan x$, $E = (\pi/2, 3\pi/2)$
δ) $f(x) = x^2 + 2x - 5$, $E = (-\infty, -6]$
ε) $f(x) = 3x - |x| + |x - 2|$, $E = \mathbf{R}$
ζ) $f(x) = x/(x^2 + 1)$, $E = [-1, 1]$

*b) Find an explicit formula for f^{-1} on $f(E)$.

1.5.2. Find $f(E)$ and $f^{-1}(E)$ for each of the following.

   a) $f(x) = 2 - 3x, \ E = (-1, 2)$
   b) $f(x) = x^2 + 1, \ E = (-1, 2]$
   c) $f(x) = 2x - x^2, \ E = [-2, 2)$
   d) $f(x) = \log(x^2 - 2x + 2), \ E = (0, 3]$
   e) $f(x) = \cos x, \ E = [0, \infty)$

1.5.3. Give a simple description of each of the following sets.

   a) $\displaystyle\bigcup_{x \in [0,1]} [x - 2, x + 1]$

   b) $\displaystyle\bigcap_{x \in [0,1]} (x - 1, x + 1]$

   c) $\displaystyle\bigcap_{k \in \mathbf{N}} \left[-\tfrac{1}{k}, \tfrac{1}{k}\right]$

   d) $\displaystyle\bigcup_{k \in \mathbf{N}} \left[-\tfrac{1}{k}, 0\right]$

   e) $\displaystyle\bigcup_{k \in \mathbf{N}} \left[-k, \tfrac{1}{k}\right)$

   f) $\displaystyle\bigcap_{k \in \mathbf{N}} \left[\tfrac{k-1}{k}, \tfrac{k+1}{k}\right]$

1.5.4. Prove (18).

1.5.5. Prove Theorem 1.37iii, iv, and v.

1.5.6. Let $f(x) = x^2$.

   a) Find subsets B and C of $\mathbf{R}$ such that $f(C \backslash B) \neq f(C) \backslash f(B)$.
   b) Find a subset E of $\mathbf{R}$ such that $f^{-1}(f(E)) \neq E$.

$\boxed{\textbf{1.5.7}}$. **This exercise is used several times in Chapter 12.** Let X, Y be sets and $f : X \to Y$. Prove that the following are equivalent.

   a) f is 1–1 on X.
   b) $f(A \backslash B) = f(A) \backslash f(B)$ for all subsets A and B of X.
   c) $f^{-1}(f(E)) = E$ for all subsets E of X.
   d) $f(A \cap B) = f(A) \cap f(B)$ for all subsets A and B of X.

1.6 COUNTABLE AND UNCOUNTABLE SETS

In this section we will show how to use bijections to "count" infinite sets. We begin by examining what it means to count a finite set. When we count a finite set E, we assign consecutive numbers in $\mathbf{N}$ to the elements of E; that is, we construct a function f from $\{1, 2, \ldots, n\}$ to E, where n is the number of elements in E. For example, if E has three objects, then the "counting" function, f, takes $\{1, 2, 3\}$ to E. Now in order to count E properly, we must be careful to avoid two pitfalls. We must not count any element of E more than once (i.e., f must be 1–1), and we cannot miss any element of E (i.e., f must take $\{1, 2, 3\}$ onto E). Accordingly, we make the following definition.

1.38 Definition.

Let E be a set.

  i) E is said to be *finite* if and only if either $E = \emptyset$ or there exists a 1–1 function which takes $\{1, 2, \ldots, n\}$ onto E, for some $n \in \mathbf{N}$.

 ii) E is said to be *countable* if and only if there exists a 1–1 function which takes $\mathbf{N}$ onto E.

iii) E is said to be *at most countable* if and only if E is either finite or countable.

iv) E is said to be *uncountable* if and only if E is neither finite nor countable.

Loosely speaking, a set is countable if it has the same number of elements as $\mathbf{N}$, finite if it has less, and uncountable if it has more.

To show that a set E is countable, it suffices to exhibit a 1–1 function f from $\mathbf{N}$ onto E. For example, the set of even integers $E = \{2, 4, \ldots\}$ is countable because $f(k) := 2k$ is 1–1 from $\mathbf{N}$ onto E. Thus, two infinite sets can have the same number of elements even though one is a proper subset of the other. (In fact, this property can be used as a definition of "infinite set.")

The following result shows that not every infinite set is countable.

1.39 Remark. [CANTOR'S DIAGONALIZATION ARGUMENT]. *The open interval $(0, 1)$ is uncountable.*

STRATEGY: Suppose to the contrary that $(0, 1)$ is countable. Then by definition, there is a function f on $\mathbf{N}$ such that $f(1), f(2), \ldots$ exhausts the elements of $(0, 1)$. We could reach a contradiction if we could find a new number $x \in (0, 1)$ that is different from all the $f(k)$'s. How can we determine whether two numbers are different? One easy way is to look at their decimal expansions. For example, $0.1234 \neq 0.1254$ because they have different decimal expansions. Thus, we could find an x that has no preimage under f by making the decimal expansion of x different by at least one digit from the decimal expansion of EVERY $f(k)$.

There is a flaw in this approach that we must fix. Decimal expansions are unique except for finite decimals, which always have an alternative expansion that terminates in 9s (e.g., $0.5 = 0.4999\ldots$ and $0.24 = 0.23999\ldots$) (see Exercise 2.2.10). Hence, when specifying the decimal expansion of x, we must avoid decimals that terminate in 9s.

Proof. Suppose that there is a 1–1 function f that takes $\mathbf{N}$ onto the interval $(0, 1)$. Write the numbers $f(j)$, $j \in \mathbf{N}$, in decimal notation, using the finite expansion when possible, that is,

$$f(1) = 0.\alpha_{11}\alpha_{12}\ldots,$$
$$f(2) = 0.\alpha_{21}\alpha_{22}\ldots,$$
$$f(3) = 0.\alpha_{31}\alpha_{32}\ldots,$$
$$\ldots,$$

where α_{ij} represents the jth digit in the decimal expansion of $f(i)$ and none of these expansions terminates in 9s. Let x be the number whose decimal expansion is given by $0.\beta_1\beta_2\ldots$, where

$$\beta_k := \begin{cases} \alpha_{kk} + 1 & \text{if } \alpha_{kk} \leq 5 \\ \alpha_{kk} - 1 & \text{if } \alpha_{kk} > 5. \end{cases}$$

Clearly, x is a number in $(0, 1)$ whose decimal expansion does not contain one 9, much less terminate in 9s. Since f is onto, there is a $j \in \mathbf{N}$ such that $f(j) = x$. Since we have avoided 9s, the decimal expansions of $f(j)$ and x must be identical (e.g., $\alpha_{jj} = \beta_j := \alpha_{jj} \pm 1$). It follows that $0 = \pm 1$, a contradiction. ∎

It is natural to ask about the countability of the sets $\mathbf{Z}$, $\mathbf{Q}$, and $\mathbf{R}$. To answer these questions, we prove several preliminary results. First, to show that a set E is at most countable, we do not need to construct a ONE-TO-ONE function which takes $\mathbf{N}$ onto E.

1.40 Lemma.
A nonempty set E is at most countable if and only if there is a function g from $\mathbf{N}$ onto E.

Proof. If E is countable, then by Definition 1.38ii there is a (1–1) function f from $\mathbf{N}$ onto E, so $g := f$ takes $\mathbf{N}$ onto E. If E is finite, then there is an $n \in \mathbf{N}$ and a 1–1 function f that takes $\{1, 2, \ldots, n\}$ onto E. Hence

$$g(j) := \begin{cases} f(j) & j \leq n \\ f(1) & j > n \end{cases}$$

takes $\mathbf{N}$ onto E.

Conversely, suppose that g takes $\mathbf{N}$ onto E. We need to construct a function f that is 1–1 from some subset of $\mathbf{N}$ onto E. We will do this by eliminating the duplication in g. To this end, let $k_1 = 1$. If the set $E_1 := \{k \in \mathbf{N} : g(k) \neq g(k_1)\}$ is empty, then $E = \{g(k_1)\}$, thus evidently at most countable. Otherwise, let k_2 be the least element in E_1 and notice that $k_2 > k_1$.

Set $E_2 := \{k \in \mathbf{N} : g(k) \in E \backslash \{g(k_1), g(k_2)\}\}$. If E_2 is empty, then $E = \{g(k_1), g(k_2)\}$ is finite, hence at most countable. Otherwise, let k_3 be the least element in E_2. Since $g(k_3) \in E \backslash \{g(k_1), g(k_2)\}$, we have $g(k_3) \neq g(k_2)$ and $g(k_3) \neq g(k_1)$. Since g is a function, the first condition implies $k_3 \neq k_2$. Since k_2 is least in E_1, the second condition implies $k_2 < k_3$. Hence, $k_1 < k_2 < k_3$.

Continue this process. If it ever terminates, then some

$$E_j := \{k \in \mathbf{N} : g(k) \in E \setminus \{g(k_1), \ldots, g(k_j)\}\}$$

is empty, so E is finite, hence at most countable. If this process never terminates, then we generate integers $k_1 < k_2 < \cdots$ such that k_{j+1} is the least element of E_j for $j = 1, 2, \ldots$.

Set $f(j) = g(k_j)$, $j \in \mathbf{N}$. To show that f is 1–1, notice that $j \neq \ell$ implies that $k_j \neq k_\ell$, say $k_j < k_\ell$. Then $k_j \leq k_{\ell-1}$, so by construction

$$g(k_\ell) \in E \setminus \{g(k_1), \ldots, g(k_j), \ldots, g(k_{\ell-1})\} \subseteq E \setminus \{g(k_1), \ldots, g(k_j)\}.$$

In particular, $g(k_\ell) \neq g(k_j)$; that is, $f(\ell) \neq f(j)$.

To show that f is onto, let $x \in E$. Since g is onto, choose $\ell \in \mathbf{N}$ such that $g(\ell) = x$. Since by construction $j \leq k_j$, use the Archimedean Principle to choose a $j \in \mathbf{N}$ such that $k_j > \ell$. Since k_j is the least element in E_{j-1}, it follows that $g(\ell)$ cannot belong to $E \setminus \{g(k_1), \ldots, g(k_{j-1})\}$; that is, $g(\ell) = g(k_n)$ for some $n \in [1, j-1]$. In particular, $f(n) = g(k_n) = x$. ∎

Next, we show how set containment affects countability and use it to answer the question about countability of $\mathbf{R}$.

1.41 Theorem. *Suppose that A and B are sets.*

i) *If $A \subseteq B$ and B is at most countable, then A is at most countable.*
ii) *If $A \subseteq B$ and A is uncountable, then B is uncountable.*
iii) *$\mathbf{R}$ is uncountable.*

Proof. i) Since B is at most countable, choose by Lemma 1.40 a function g which takes $\mathbf{N}$ onto B. We may suppose that A is nonempty, hence fix an $a_0 \in A$. Then

$$f(n) := \begin{cases} g(n) & g(n) \in A \\ a_0 & g(n) \notin A \end{cases}$$

takes $\mathbf{N}$ onto A. Hence by Lemma 1.40, A is at most countable.

ii) If B were at most countable, then by part i), A would also be at most countable, a contradiction.

iii) By Remark 1.39, the interval $(0, 1)$ is an uncountable subset of $\mathbf{R}$. Thus, by part ii), $\mathbf{R}$ is uncountable. ∎

The following result shows that the Cartesian product of two countable sets is countable, and that a countable union of countable sets is countable.

1.42 Theorem. *Let $A_1, A_2, \ldots$ be at most countable sets.*

i) *Then $A_1 \times A_2$ is at most countable.*
ii) *If*

$$E = \bigcup_{j=1}^{\infty} A_j := \bigcup_{j \in \mathbf{N}} A_j := \{x : x \in A_j \quad \text{for some } j \in \mathbf{N}\},$$

then E is at most countable.

Proof. i) By Lemma 1.40, there exist functions ϕ (respectively, ψ) which take $\mathbf{N}$ onto A_1 (respectively, onto A_2). Hence $f(n, m) := (\phi(n), \psi(m))$ takes $\mathbf{N} \times \mathbf{N}$

onto $A_1 \times A_2$. If we can construct a function g which takes $\mathbf{N}$ onto $\mathbf{N} \times \mathbf{N}$, then by Exercise 1.6.5a, $f \circ g$ takes $\mathbf{N}$ onto $A_1 \times A_2$. Hence by Lemma 1.40, $A_1 \times A_2$ is at most countable.

To construct the function g, plot the points of $\mathbf{N} \times \mathbf{N}$ in the plane. Notice that we can connect these lattice points with a series of parallel backward-slanted lines; for example, the first line passes through $(1, 1)$, the second line passes through $(1, 2)$ and $(2, 1)$, and the third line passes through $(1, 3)$, $(2, 2)$, and $(3, 1)$. This suggests a method for constructing g. Set $g(1) = (1, 1)$, $g(2) = (1, 2)$, $g(3) = (2, 1)$, $g(4) = (3, 1), \ldots$.

If you wish to see an explicit formula for g, observe that the nth line passes through the set of lattice points

$$(1, n), (2, n - 1), (3, n - 2), \ldots, (n - 1, 2), (n, 1);$$

that is, through the set of lattice points (k, j) which satisfy $k + j = n + 1$. Since the sum of integers $1 + 2 + \cdots + (n - 1)$ is given by $(n - 1)n/2$ (see Exercise 1.4.4a), there are $(n - 1)n/2$ elements in the first $n - 1$ slanted lines. Hence a function which takes $\mathbf{N}$ onto the nth slanted line is given by

$$g(j) = (\ell, n + 1 - \ell), \tag{19}$$

where $j = \ell + (n - 1)n/2$. This function is defined on all of $\mathbf{N}$ because given $j \in \mathbf{N}$, we can use the Archimedean Principle and the Well-Ordering Principle to choose n least such that $j \le n(n + 1)/2$; that is, such that $j = \ell + (n - 1)n/2$ for some $\ell \in [1, n]$. Thus g takes $\mathbf{N}$ onto $\mathbf{N} \times \mathbf{N}$.

ii) By Lemma 1.40, choose functions f_j that take $\mathbf{N}$ onto A_j, $j \in \mathbf{N}$. Clearly, the function $h(k, j) := f_k(j)$ takes $\mathbf{N} \times \mathbf{N}$ onto E. Hence the function $h \circ g$, where g is defined by (19), takes $\mathbf{N}$ onto E. We conclude by Lemma 1.40 that E is at most countable. ∎

1.43 Remark. *The sets $\mathbf{Z}$ and $\mathbf{Q}$ are countable, but the set of irrationals is uncountable.*

Proof. $\mathbf{Z} = \mathbf{N} \cup (-\mathbf{N}) \cup \{0\}$ and $\mathbf{Q} = \bigcup_{n=1}^{\infty} \{p/n : p \in \mathbf{Z}\}$ are both countable by Theorem 1.42ii.

If $\mathbf{R} \backslash \mathbf{Q}$ were countable, then $\mathbf{R} = (\mathbf{R} \backslash \mathbf{Q}) \cup \mathbf{Q}$ would also be countable, a contradiction of Theorem 1.41iii. ∎

EXERCISES

1.6.0. Decide which of the following statements are true and which are false. Prove the true ones and give counterexamples to the false ones.

a) Suppose that E is a set. If there exists a function f from E onto $\mathbf{N}$, then E is at most countable.

b) A dyadic rational is a point $x \in \mathbf{R}$ such that $x = n/2^m$ for some $n \in \mathbf{Z}$ and $m \in \mathbf{N}$. The set of dyadic rationals is uncountable.

c) Suppose that A and B are sets and that $f : A \to B$ is 1–1. If A is uncountable, then B is uncountable.

d) If E_1, E_2, ... are finite sets, and

$$E := E_1 \times E_2 \times \cdots := \{(x_1, x_2, \ldots) : x_j \in E_j \quad \text{for all } j \in \mathbf{N}\},$$

then E is countable.

1.6.1. Prove that the set of odd integers $\{1, 3, \ldots\}$ is countable.

1.6.2. Prove that set of rational lattice points in space—that is, the set $\mathbf{Q}^3 := \{(x, y, z) : x, y, z \in \mathbf{Q}\}$—is countable.

1.6.3. Suppose that A and B are sets and that B is uncountable. If there exists a function which takes A onto B, prove that A is uncountable.

1.6.4. Suppose that A is finite and f is 1–1 from A onto B. Prove that B is finite.

1.6.5. Let $f : A \to B$ and $g : B \to C$ and define $g \circ f : A \to C$ by $(g \circ f)(x) := g(f(x))$.

a) Show that if f, g are 1–1 (respectively, onto), then $g \circ f$ is 1–1 (respectively, onto).

b) Prove that if f is 1–1 from A into B and $B_0 := \{y : y = f(x) \text{ for some } x \in A\}$, then f^{-1} is 1–1 from B_0 onto A.

c) Suppose that g is 1–1 from B onto C. Prove that f is 1–1 on A (respectively, onto B) if and only if $g \circ f$ is 1–1 on A (respectively, onto C).

1.6.6. Suppose that $n \in \mathbf{N}$ and $\phi : \{1, 2, \ldots, n\} \to \{1, 2, \ldots, n\}$.

a) Prove that ϕ is 1–1 if and only if ϕ is onto.

b) [PIGEONHOLE PRINCIPLE] Suppose that E is a finite set and that $f : E \to E$. Prove that f is 1–1 on E if and only if f takes E onto E.

1.6.7. A number $x_0 \in \mathbf{R}$ is called *algebraic of degree n* if it is the root of a polynomial $P(x) = a_n x^n + \cdots + a_1 x + a_0$, where $a_j \in \mathbf{Z}$, $a_n \neq 0$, and n is minimal. A number x_0 that is not algebraic is called *transcendental*.

a) Prove that if $n \in \mathbf{N}$ and $q \in \mathbf{Q}$, then n^q is algebraic.

b) Prove that for each $n \in \mathbf{N}$ the collection of algebraic numbers of degree n is countable.

c) Prove that the collection of transcendental numbers is uncountable. (Two famous transcendental numbers are π and e. For more information on transcendental numbers and their history, see Kline [5].)

CHAPTER 2

Sequences in R

2.1 LIMITS OF SEQUENCES

An *infinite sequence* (more briefly, a *sequence*) is a function whose domain is **N**. A sequence f whose *terms* are $x_n := f(n)$ will be denoted by $x_1, x_2, \ldots$ or $\{x_n\}_{n\in\mathbf{N}}$ or $\{x_n\}_{n=1}^{\infty}$, or $\{x_n\}$. Thus $1, 1/2, 1/4, 1/8, \ldots$ represents the sequence $\{1/2^{n-1}\}_{n\in\mathbf{N}}$; $-1, 1, -1, 1, \ldots$ represents the sequence $\{(-1)^n\}_{n\in\mathbf{N}}$; and $1, 2, 3, 4, \ldots$ represents the sequence $\{n\}_{n\in\mathbf{N}}$.

It is important not to confuse a sequence $\{x_n\}_{n\in\mathbf{N}}$ with the set $\{x_n : n \in \mathbf{N}\}$; these are two entirely different concepts. For example, as sequences, $1, 2, 3, 4, \ldots$ is different from $2, 1, 3, 4, \ldots$, but as sets, $\{1, 2, 3, 4, \ldots\}$ is identical with $\{2, 1, 3, 4, \ldots\}$. Again, the sequence $1, -1, 1, -1, \ldots$ is infinite, but the set $\{(-1)^n : n \in \mathbf{N}\}$ has only two points.

The limit concept is one of the fundamental building blocks of analysis. Recall from elementary calculus that a sequence of real numbers $\{x_n\}$ converges to a number a if x_n gets near a (i.e., the distance between a and x_n gets small) as n gets large. Thus, given $\varepsilon > 0$ (no matter how small), if n is large enough, $|x_n - a|$ is smaller than ε. This leads us to a formal definition of the limit of a sequence.

2.1 Definition.

A sequence of real numbers $\{x_n\}$ is said to *converge* to a real number $a \in \mathbf{R}$ if and only if for every $\varepsilon > 0$ there is an $N \in \mathbf{N}$ (which in general depends on ε) such that

$$n \geq N \quad \text{implies} \quad |x_n - a| < \varepsilon.$$

We shall use the following phrases and notation interchangeably:

a) $\{x_n\}$ converges to a; b) x_n converges to a; c) $a = \lim_{n\to\infty} x_n$; d) $x_n \to a$ as $n \to \infty$; e) the *limit* of $\{x_n\}$ exists and equals a.

When $x_n \to a$ as $n \to \infty$, you can think of x_n as a sequence of approximations to a, and ε as an upper bound for the error of these approximations. The number N in Definition 2.1 is chosen so that the error is less than ε when $n \geq N$. In general, the smaller ε gets, the larger N must be. (See, for example, Figure 2.1.)

Notice by definition that x_n converges to a if and only if $|x_n - a| \to 0$ as $n \to \infty$. In particular, $x_n \to 0$ if and only if $|x_n| \to 0$ as $n \to \infty$.

According to Definition 2.1, to prove that a particular limit exists, given an arbitrary $\varepsilon > 0$, no matter how small, we must describe how to choose an N such that $n \geq N$ implies $|x_n - a| < \varepsilon$. In particular, ε is usually introduced BEFORE

FIGURE 2.1

N is specified, and N often is defined to depend on ε. Since $|x_n - a| < \varepsilon$ for all $n \geq N$, N CANNOT depend on n.

Before we actually prove that some concrete limits exist, we introduce additional terminology. Let $\mathcal{P}_n$ be a property indexed by **N**. We shall say that $\mathcal{P}_n$ *holds for large n* if there is an $N \in \mathbf{N}$ such that $\mathcal{P}_n$ is true for all $n \geq N$. Hence a loose summary of Definition 2.1 is that x_n converges to a if and only if $|x_n - a|$ is small for large n. What we mean by this is that given any prescribed positive quantity ε (no matter how small), we can choose N large enough so that $|x_n - a|$ is less than ε for all $n \geq N$.

2.2 EXAMPLE.

i) Prove that $1/n \to 0$ as $n \to \infty$.
ii) If $x_n \to 2$, prove that $(2x_n + 1)/x_n \to 5/2$ as $n \to \infty$.

> **Proof.** i) Let $\varepsilon > 0$. Use the Archimedean Principle to choose $N \in \mathbf{N}$ such that $N > 1/\varepsilon$. By taking the reciprocal of this inequality, we see that $n \geq N$ implies $1/n \leq 1/N < \varepsilon$. Since $1/n$ are all positive, it follows that $|1/n| < \varepsilon$ for all $n \geq N$.
>
> STRATEGY for ii): By definition, we must show that
>
> $$\frac{2x_n + 1}{x_n} - \frac{5}{2} = \frac{2 - x_n}{2x_n}$$
>
> is small for large n. The numerator of this last fraction will be small for large n since $x_n \to 2$, as $n \to \infty$. What about the denominator? Since $x_n \to 2$, x_n will be greater than 1 for large n, so $2x_n$ will be greater than 2 for large n. Since we made n large twice, we will make two restrictions to determine the N that corresponds to ε in Definition 2.1. Let's try to write all this down carefully to be sure that it works out.
>
> ii) Let $\varepsilon > 0$. Since $x_n \to 2$, apply Definition 2.1 to this $\varepsilon > 0$ to choose $N_1 \in \mathbf{N}$ such that $n \geq N_1$ implies $|x_n - 2| < \varepsilon$. Next, apply Definition 2.1 with $\varepsilon = 1$ to choose N_2 such that $n \geq N_2$ implies $|x_n - 2| < 1$. By the Fundamental Theorem of Absolute Values, we have $n \geq N_2$ implies $x_n > 1$ (i.e., $2x_n > 2$).
>
> Set $N = \max\{N_1, N_2\}$ and suppose that $n \geq N$. Since $n \geq N_1$, we have $|2 - x_n| = |x_n - 2| < \varepsilon$. Since $n \geq N_2$, we have $0 < 1/(2x_n) < 1/2 < 1$. It follows that
>
> $$\left| \frac{2x_n + 1}{x_n} - \frac{5}{2} \right| = \frac{|2 - x_n|}{2x_n} < \frac{\varepsilon}{2x_n} < \varepsilon$$
>
> for all $n \geq N$. ∎

Notice that in the proof of Remark 2.1 we forced two properties that held for $n \geq N_j$, $j = 1, 2$, to hold for $n \geq N$ by setting N equal to the maximum of N_1 and N_2. It is clear that by this same process, if $N_1, \ldots, N_q$ have been chosen so that for each j a property $\mathcal{P}_j$ holds when $n \geq N_j$ and if $N = \max\{N_1, \ldots, N_q\}$, then all q properties $\mathcal{P}_1, \ldots, \mathcal{P}_q$ hold simultaneously when $n \geq N$. We shall use this device frequently below, but rarely write N explicitly as a maximum of integers N_j again.

The following two results show that a given sequence can have no limits or one limit, but no more.

2.3 EXAMPLE.

The sequence $\{(-1)^n\}_{n \in \mathbf{N}}$ has no limit.

Proof. Suppose that $(-1)^n \to a$ as $n \to \infty$ for some $a \in \mathbf{R}$. Given $\varepsilon = 1$, there is an $N \in \mathbf{N}$ such that $n \geq N$ implies $|(-1)^n - a| < \varepsilon$. For n odd this implies $|1 + a| = |-1 - a| < 1$, and for n even this implies $|1 - a| < 1$. Hence,

$$2 = |1 + 1| \leq |1 - a| + |1 + a| < 1 + 1 = 2;$$

that is, $2 < 2$, a contradiction. ∎

2.4 Remark. *A sequence can have at most one limit.*

Proof. Suppose that $\{x_n\}$ converges to both a and b. By definition, given $\varepsilon > 0$, there is an integer N such that $n \geq N$ implies $|x_n - a| < \varepsilon/2$ and $|x_n - b| < \varepsilon/2$. Thus it follows from the triangle inequality that

$$|a - b| \leq |a - x_n| + |x_n - b| < \varepsilon;$$

that is, $|a - b| < \varepsilon$ for all $\varepsilon > 0$. We conclude, by Theorem 1.9, that $a = b$. ∎

We shall use the following concept many times.

2.5 Definition.

By a *subsequence* of a sequence $\{x_n\}_{n \in \mathbf{N}}$, we shall mean a sequence of the form $\{x_{n_k}\}_{k \in \mathbf{N}}$, where each $n_k \in \mathbf{N}$ and $n_1 < n_2 < \cdots$.

Thus a subsequence $x_{n_1}, x_{n_2}, \ldots$ of $x_1, x_2, \ldots$ is obtained by "deleting" from $x_1, x_2, \ldots$ all x_n's except those such that $n = n_k$ for some k. For example, $1, 1, \ldots$ is a subsequence of $(-1)^n$ obtained by deleting every other term (set $n_k = 2k$), and $1/2, 1/4, \ldots$ is a subsequence of $1/n$ obtained by deleting all nondyadic fractions; that is, deleting $1/3, 1/5, 1/6, 1/7, \ldots$ (set $n_k = 2^k$).

Subsequences are sometimes used to correct a sequence that behaves badly or to speed up convergence of another that converges slowly. For example, $\{1/n\}$ converges much more slowly to zero than its subsequence $\{1/2^n\}$, and $\{(-1)^n\}$ does not converge at all (see Example 2.3 above), but its subsequence $1, 1, \ldots$ converges to 1 immediately.

If $x_n \to a$ as $n \to \infty$, then the x_n's get near a as n gets large. Since n_k gets large as k does, it comes as no surprise that any subsequence of a convergent sequence also converges.

2.6 Remark. *If $\{x_n\}_{n\in\mathbf{N}}$ converges to a and $\{x_{n_k}\}_{k\in\mathbf{N}}$ is any subsequence of $\{x_n\}_{n\in\mathbf{N}}$, then x_{n_k} converges to a as $k \to \infty$.*

Proof. Let $\varepsilon > 0$ and choose $N \in \mathbf{N}$ such that $n \geq N$ implies $|x_n - a| < \varepsilon$. Since $n_k \in \mathbf{N}$ and $n_1 < n_2 < \cdots$, it is easy to see by induction that $n_k \geq k$ for all $k \in \mathbf{N}$. Hence, $k \geq N$ implies $|x_{n_k} - a| < \varepsilon$; that is, $x_{n_k} \to a$ as $k \to \infty$. ∎

The following concepts also play an important role for the theory of sequences.

2.7 Definition.

Let $\{x_n\}$ be a sequence of real numbers.

i) The sequence $\{x_n\}$ is said to be *bounded above* if and only if the set $\{x_n : n \in \mathbf{N}\}$ is bounded above.

ii) The sequence $\{x_n\}$ is said to be *bounded below* if and only if the set $\{x_n : n \in \mathbf{N}\}$ is bounded below.

iii) $\{x_n\}$ is said to be *bounded* if and only if it is bounded both above and below.

Combining Definitions 2.7 and 1.10, we see that $\{x_n\}$ is bounded above (respectively, below) if and only if there is an $M \in \mathbf{R}$ such that $x_n \leq M$ for all $n \in \mathbf{N}$ (respectively, if and only if there is an $m \in \mathbf{R}$ such that $x_n \geq m$ for all $n \in \mathbf{N}$). It is easy to check (see Exercise 2.1.4) that $\{x_n\}$ is bounded if and only if there is a $C > 0$ such that $|x_n| \leq C$ for all $n \in \mathbf{N}$. In this case we shall say that $\{x_n\}$ is *bounded*, or *dominated*, by C.

Is there a relationship between convergent sequences and bounded sequences?

2.8 Theorem. *Every convergent sequence is bounded.*

STRATEGY: The idea behind the proof is simple (see Figure 2.1). Suppose that $x_n \to a$ as $n \to \infty$. By definition, for large N the sequence $x_N, x_{N+1}, \ldots$ must be close to a, hence bounded. Since the finite sequence $x_1, \ldots, x_{N-1}$ is also bounded, it should follow that the whole sequence is bounded. We now make this precise.

Proof. Given $\varepsilon = 1$, there is an $N \in \mathbf{N}$ such that $n \geq N$ implies $|x_n - a| < 1$. Hence by the triangle inequality, $|x_n| < 1 + |a|$ for all $n \geq N$. On the other hand, if $1 \leq n \leq N$, then

$$|x_n| \leq M := \max\{|x_1|, |x_2|, \ldots, |x_N|\}.$$

Therefore, $\{x_n\}$ is dominated by $\max\{M, 1 + |a|\}$.   ∎

Notice that by Example 2.3, the converse of Theorem 2.8 is false.

EXERCISES

2.1.0. Decide which of the following statements are true and which are false. Prove the true ones and provide a counterexample for the false ones.

   a) If x_n converges, then x_n/n also converges.
   b) If x_n does not converge, then x_n/n does not converge.
   c) If x_n converges and y_n is bounded, then $x_n y_n$ converges.
   d) If x_n converges to zero and $y_n > 0$ for all $n \in \mathbf{N}$, then $x_n y_n$ converges.

2.1.1. Using the method of Example 2.2i, prove that the following limits exist.

   a) $2 - 1/n \to 2$ as $n \to \infty$.
   b) $1 + \pi/\sqrt{n} \to 1$ as $n \to \infty$.
   c) $3(1 + 1/n) \to 3$ as $n \to \infty$.
   d) $(2n^2 + 1)/(3n^2) \to 2/3$ as $n \to \infty$.

2.1.2. Suppose that x_n is a sequence of real numbers that converges to 1 as $n \to \infty$. Using Definition 2.1, prove that each of the following limits exists.

   a) $1 + 2x_n \to 3$ as $n \to \infty$.
   b) $(\pi x_n - 2)/x_n \to \pi - 2$ as $n \to \infty$.
   c) $(x_n^2 - e)/x_n \to 1 - e$ as $n \to \infty$.

2.1.3. For each of the following sequences, find two convergent subsequences that have different limits.

   a) $3 - (-1)^n$
   b) $(-1)^{3n} + 2$
   c) $(n - (-1)^n n - 1)/n$

2.1.4. Suppose that $x_n \in \mathbf{R}$.

   a) Prove that $\{x_n\}$ is bounded if and only if there is a $C > 0$ such that $|x_n| \leq C$ for all $n \in \mathbf{N}$.
   b) Suppose that $\{x_n\}$ is bounded. Prove that $x_n/n^k \to 0$, as $n \to \infty$, for all $k \in \mathbf{N}$.

2.1.5. Let C be a fixed, positive constant. If $\{b_n\}$ is a sequence of nonnegative numbers that converges to 0, and $\{x_n\}$ is a real sequence that satisfies $|x_n - a| \leq Cb_n$ for large n, prove that x_n converges to a.

2.1.6. Let a be a fixed real number and define $x_n := a$ for $n \in \mathbf{N}$. Prove that the "constant" sequence x_n converges.

2.1.7. a) Suppose that $\{x_n\}$ and $\{y_n\}$ converge to the same real number. Prove that $x_n - y_n \to 0$ as $n \to \infty$.

b) Prove that the sequence $\{n\}$ does not converge.

c) Show that there exist unbounded sequences $x_n \neq y_n$ which satisfy the conclusion of part (a).

2.1.8. Suppose that $\{x_n\}$ is a sequence in **R**. Prove that x_n converges to a if and only if EVERY subsequence of x_n also converges to a.

2.2 LIMIT THEOREMS

One of the biggest challenges we face (both for theory and applications) is deciding whether or not a given sequence converges. Once we know that it converges, we can often use other techniques to approximate or evaluate its limit.

One way to identify convergent sequences is by comparing a sequence whose convergence is in doubt with another whose convergence property is already known (see Example 2.10). The following result is the first of many theorems that addresses this issue.

2.9 Theorem. [SQUEEZE THEOREM].
Suppose that $\{x_n\}$, $\{y_n\}$, and $\{w_n\}$ are real sequences.

i) *If $x_n \to a$ and $y_n \to a$ (the SAME a) as $n \to \infty$, and if there is an $N_0 \in \mathbf{N}$ such that*

$$x_n \leq w_n \leq y_n \quad for\ n \geq N_0,$$

then $w_n \to a$ as $n \to \infty$.

ii) *If $x_n \to 0$ as $n \to \infty$ and $\{y_n\}$ is bounded, then $x_n y_n \to 0$ as $n \to \infty$.*

Proof. i) Let $\varepsilon > 0$. Since x_n and y_n converge to a, use Definition 2.1 and Theorem 1.6 to choose $N_1, N_2 \in \mathbf{N}$ such that $n \geq N_1$ implies $-\varepsilon < x_n - a < \varepsilon$ and $n \geq N_2$ implies $-\varepsilon < y_n - a < \varepsilon$. Set $N = \max\{N_0, N_1, N_2\}$. If $n \geq N$, we have by hypothesis and the choice of N_1 and N_2 that

$$a - \varepsilon < x_n \leq w_n \leq y_n < a + \varepsilon;$$

that is, $|w_n - a| < \varepsilon$ for $n \geq N$. We conclude that $w_n \to a$ as $n \to \infty$.

ii) Suppose that $x_n \to 0$ and that there is an $M > 0$ such that $|y_n| \leq M$ for $n \in \mathbf{N}$. Let $\varepsilon > 0$ and choose an $N \in \mathbf{N}$ such that $n \geq N$ implies $|x_n| < \varepsilon/M$. Then $n \geq N$ implies

$$|x_n y_n| < M \frac{\varepsilon}{M} = \varepsilon.$$

We conclude that $x_n y_n \to 0$ as $n \to \infty$. ∎

The following example shows how to use the Squeeze Theorem to find the limit of a complicated sequence by ignoring its "less important" factors.

2.10 EXAMPLE.

Find $\lim_{n \to \infty} 2^{-n} \cos(n^3 - n^2 + n - 13)$.

Solution. The factor $\cos(n^3 - n^2 + n - 13)$ looks intimidating, but it is superfluous for finding the limit of this sequence. Indeed, since $|\cos x| \le 1$ for all $x \in \mathbf{R}$, the sequence $\{2^{-n} \cos(n^3 - n^2 + n - 13)\}$ is dominated by 2^{-n}. Since $2^n > n$, it is clear by Example 2.2i and the Squeeze Theorem that both $2^{-n} \to 0$ and $2^{-n} \cos(n^3 - n^2 + n - 13) \to 0$ as $n \to \infty$. ∎

The Squeeze Theorem can also be used to construct convergent sequences with certain properties. To illustrate how this works, we now establish a result that connects suprema and infima with convergent sequences.

2.11 Theorem. *Let $E \subset \mathbf{R}$. If E has a finite supremum (respectively, a finite infimum), then there is a sequence $x_n \in E$ such that $x_n \to \sup E$ (respectively, a sequence $y_n \in E$ such that $y_n \to \inf E$) as $n \to \infty$.*

Proof. Suppose that E has a finite supremum. For each $n \in \mathbf{N}$, choose (by the Approximation Property for Suprema) an $x_n \in E$ such that $\sup E - 1/n < x_n \le \sup E$. Then by the Squeeze Theorem and Example 2.2i, $x_n \to \sup E$ as $n \to \infty$. Similarly, there is a sequence $y_n \in E$ such that $y_n \to \inf E$. ∎

Here is another result that helps to evaluate limits of specific sequences. This one works by viewing complicated sequences in terms of simpler components.

2.12 Theorem. *Suppose that $\{x_n\}$ and $\{y_n\}$ are real sequences and that $\alpha \in \mathbf{R}$. If $\{x_n\}$ and $\{y_n\}$ are convergent, then*

i)
$$\lim_{n \to \infty} (x_n + y_n) = \lim_{n \to \infty} x_n + \lim_{n \to \infty} y_n,$$

ii)
$$\lim_{n \to \infty} (\alpha x_n) = \alpha \lim_{n \to \infty} x_n,$$

and

iii)
$$\lim_{n \to \infty} (x_n y_n) = (\lim_{n \to \infty} x_n)(\lim_{n \to \infty} y_n).$$

If, in addition, $y_n \ne 0$ and $\lim_{n \to \infty} y_n \ne 0$, then

iv)
$$\lim_{n \to \infty} \frac{x_n}{y_n} = \frac{\lim_{n \to \infty} x_n}{\lim_{n \to \infty} y_n}.$$

(In particular, all these limits exist.)

Proof. Suppose that $x_n \to x$ and $y_n \to y$ as $n \to \infty$.

i) Let $\varepsilon > 0$ and choose $N \in \mathbf{N}$ such that $n \geq N$ implies $|x_n - x| < \varepsilon/2$ and $|y_n - y| < \varepsilon/2$. Thus $n \geq N$ implies

$$|(x_n + y_n) - (x + y)| \leq |x_n - x| + |y_n - y| < \frac{\varepsilon}{2} + \frac{\varepsilon}{2} = \varepsilon.$$

ii) It suffices to show that $\alpha x_n - \alpha x \to 0$ as $n \to \infty$. But $x_n - x \to 0$ as $n \to \infty$, hence by the Squeeze Theorem, $\alpha(x_n - x) \to 0$ as $n \to \infty$.

iii) By Theorem 2.8, the sequence $\{x_n\}$ is bounded. Hence by the Squeeze Theorem the sequences $\{x_n(y_n - y)\}$ and $\{(x_n - x)y\}$ both converge to 0. Since

$$x_n y_n - xy = x_n(y_n - y) + (x_n - x)y,$$

it follows from part i) that $x_n y_n \to xy$ as $n \to \infty$. A similar argument establishes part iv) (see Exercise 2.2.4). ∎

Theorem 2.12 can be used to evaluate limits of sums, products, and quotients. Here is a typical example.

2.13 EXAMPLE.

Find $\lim_{n \to \infty}(n^3 + n^2 - 1)/(1 - 3n^3)$.

Solution. Multiplying the numerator and denominator by $1/n^3$, we find that

$$\frac{n^3 + n^2 - 1}{1 - 3n^3} = \frac{1 + (1/n) - (1/n^3)}{(1/n^3) - 3}.$$

By Example 2.2i and Theorem 2.12iii, $1/n^k = (1/n)^k \to 0$, as $n \to \infty$, for any $k \in \mathbf{N}$. Thus by Theorem 2.12i, ii, and iv,

$$\lim_{n \to \infty} \frac{n^3 + n^2 - 1}{1 - 3n^3} = \frac{1 + 0 - 0}{0 - 3} = -\frac{1}{3}.$$ ∎

The sequence $\{\log n\}_{n \in \mathbf{N}}$ fails to converge in a different way than $\{n(-1)^n\}_{n \in \mathbf{N}}$ does. Indeed, the terms $\log n$ get steadily larger as $n \to \infty$, but the terms $n(-1)^n$ bounce back and forth between large positive values and large negative values. It is sometimes convenient to emphasize this difference by generalizing limits to include extended real numbers.

2.14 Definition.

Let $\{x_n\}$ be a sequence of real numbers.

i) $\{x_n\}$ is said to *diverge* to $+\infty$ (notation: $x_n \to +\infty$ as $n \to \infty$ or $\lim_{n \to \infty} x_n = +\infty$) if and only if for each $M \in \mathbf{R}$ there is an $N \in \mathbf{N}$ such that

$$n \geq N \quad \text{implies} \quad x_n > M.$$

ii) $\{x_n\}$ is said to *diverge* to $-\infty$ (notation: $x_n \to -\infty$ as $n \to \infty$ or $\lim_{n\to\infty} x_n = -\infty$) if and only if for each $M \in \mathbf{R}$ there is an $N \in \mathbf{N}$ such that

$$n \geq N \quad \text{implies} \quad x_n < M.$$

Notice by Definition 2.14i that $x_n \to +\infty$ if and only if given $M \in \mathbf{R}$, x_n is greater than M for sufficiently large n; that is, eventually x_n exceeds every number M (no matter how large and positive M is). Similarly, $x_n \to -\infty$ if and only if x_n eventually is less than every number M (no matter how large and negative M is).

It is easy to see that the Squeeze Theorem can be extended to infinite limits (see Exercise 2.2.7). The following is an extension of Theorem 2.12.

2.15 Theorem. *Suppose that $\{x_n\}$ and $\{y_n\}$ are real sequences such that $x_n \to +\infty$ (respectively, $x_n \to -\infty$) as $n \to \infty$.*

i) *If y_n is bounded below (respectively, y_n is bounded above), then*

$$\lim_{n\to\infty} (x_n + y_n) = +\infty \quad (\text{respectively,} \ \lim_{n\to\infty} (x_n + y_n) = -\infty).$$

ii) *If $\alpha > 0$, then*

$$\lim_{n\to\infty} (\alpha x_n) = +\infty \quad (\text{respectively,} \ \lim_{n\to\infty} (\alpha x_n) = -\infty).$$

iii) *If $y_n > M_0$ for some $M_0 > 0$ and all $n \in \mathbf{N}$, then*

$$\lim_{n\to\infty} (x_n y_n) = +\infty \quad (\text{respectively,} \ \lim_{n\to\infty} (x_n y_n) = -\infty).$$

iv) *If $\{y_n\}$ is bounded and $x_n \neq 0$, then*

$$\lim_{n\to\infty} \frac{y_n}{x_n} = 0.$$

Proof. We suppose for simplicity that $x_n \to +\infty$ as $n \to \infty$.

i) By hypothesis, $y_n \geq M_0$ for some $M_0 \in \mathbf{R}$. Let $M \in \mathbf{R}$ and set $M_1 = M - M_0$. Since $x_n \to +\infty$, choose $N \in \mathbf{N}$ such that $n \geq N$ implies $x_n > M_1$. Then $n \geq N$ implies $x_n + y_n > M_1 + M_0 = M$.

ii) Let $M \in \mathbf{R}$ and set $M_1 = M/\alpha$. Choose $N \in \mathbf{N}$ such that $n \geq N$ implies $x_n > M_1$. Since $\alpha > 0$, we conclude that $\alpha x_n > \alpha M_1 = M$ for all $n \geq N$.

iii) Let $M \in \mathbf{R}$ and set $M_1 = M/M_0$. Choose $N \in \mathbf{N}$ such that $n \geq N$ implies $x_n > M_1$. Then $n \geq N$ implies $x_n y_n > M_1 M_0 = M$.

iv) Let $\varepsilon > 0$. Choose $M_0 > 0$ such that $|y_n| \leq M_0$ and $M_1 > 0$ so large that $M_0/M_1 < \varepsilon$. Choose $N \in \mathbf{N}$ such that $n \geq N$ implies $x_n > M_1$. Then $n \geq N$ implies

$$\left| \frac{y_n}{x_n} \right| = \frac{|y_n|}{x_n} < \frac{M_0}{M_1} < \varepsilon. \qquad \blacksquare$$

If we adopt the conventions

$$x + \infty = \infty, \quad x - \infty = -\infty, \qquad x \in \mathbf{R},$$
$$x \cdot \infty = \infty, \quad x \cdot (-\infty) = -\infty, \qquad x > 0,$$
$$x \cdot \infty = -\infty, \quad x \cdot (-\infty) = \infty, \qquad x < 0,$$
$$\infty + \infty = \infty, \quad -\infty - \infty = -\infty,$$
$$\infty \cdot \infty = (-\infty) \cdot (-\infty) = \infty, \quad \text{and} \quad \infty \cdot (-\infty) = (-\infty) \cdot \infty = -\infty,$$

then Theorem 2.15 contains the following corollary.

2.16 Corollary. *Let $\{x_n\}$, $\{y_n\}$ be real sequences and α, x, y be extended real numbers. If $x_n \to x$ and $y_n \to y$, as $n \to \infty$, then*

$$\lim_{n \to \infty} (x_n + y_n) = x + y$$

provided that the right side is not of the form $\infty - \infty$, and

$$\lim_{n \to \infty} (\alpha x_n) = \alpha x, \qquad \lim_{n \to \infty} (x_n y_n) = xy$$

provided that none of these products is of the form $0 \cdot \pm\infty$.

We have avoided the cases $\infty - \infty$ and $0 \cdot \pm\infty$ because they are "indeterminate." For a discussion of indeterminate forms, see l'Hôpital's Rule in Section 4.4.

Theorems 2.12 and 2.15 show how the limit sign interacts with the algebraic structure of **R**. (Namely, the limit of a sum (product, quotient) is the sum (product, quotient) of the limits.) The following theorem shows how the limit sign interacts with the order structure of **R**.

2.17 Theorem. [COMPARISON THEOREM].
Suppose that $\{x_n\}$ and $\{y_n\}$ are convergent sequences. If there is an $N_0 \in \mathbf{N}$ such that

$$x_n \leq y_n \quad \text{for } n \geq N_0, \tag{1}$$

then

$$\lim_{n \to \infty} x_n \leq \lim_{n \to \infty} y_n.$$

In particular, if $x_n \in [a, b]$ converges to some point c, then c must belong to $[a, b]$.

Proof. Suppose that the first statement is false; that is, that (1) holds but $x := \lim_{n \to \infty} x_n$ is greater than $y := \lim_{n \to \infty} y_n$. Set $\varepsilon = (x - y)/2$. Choose $N_1 > N_0$ such that $|x_n - x| < \varepsilon$ and $|y_n - y| < \varepsilon$ for $n \geq N_1$. Then for such an n,

$$x_n > x - \varepsilon = x - \left(\frac{x - y}{2}\right) = y + \left(\frac{x - y}{2}\right) = y + \varepsilon > y_n,$$

which contradicts (1). This proves the first statement.

We conclude by noting that the second statement follows from the first, since $a \leq x_n \leq b$ implies $a \leq c \leq b$. ∎

One way to remember this result is that it says the limit of an inequality is the inequality of the limits, provided these limits exist. We shall call this process "taking the limit of an inequality." Since $x_n < y_n$ implies $x_n \leq y_n$, the Comparison Theorem contains the following corollary: If $\{x_n\}$ and $\{y_n\}$ are convergent real sequences, then

$$x_n < y_n, \quad n \geq N_0, \quad \text{imply} \quad \lim_{n \to \infty} x_n \leq \lim_{n \to \infty} y_n.$$

It is important to notice that this result is false if $\leq$ is replaced by $<$; that is,

$$x_n < y_n, \quad n \geq N_0, \quad \text{does NOT imply that} \quad \lim_{n \to \infty} x_n < \lim_{n \to \infty} y_n.$$

For example, $1/n^2 < 1/n$, but the limits of these sequences are equal.

EXERCISES

2.2.0. Determine which of the following statements are true and which are false. Prove the true ones and provide counterexamples for the false ones.

a) If $x_n \to \infty$ and $y_n \to -\infty$, then $x_n + y_n \to 0$ as $n \to \infty$.
b) If $x_n \to -\infty$, then $1/x_n \to 0$ as $n \to \infty$.
c) If $x_n \to 0$, then $1/x_n \to \infty$ as $n \to \infty$.
d) If $x_n \to \infty$, then $(1/2)^{x_n} \to 0$ as $n \to \infty$.

2.2.1. Prove that each of the following sequences converges to zero.

a) $x_n = \sin(\log n + n^5 + e^{n^2})/n$
b) $x_n = 2n/(n^2 + \pi)$
c) $x_n = (\sqrt{2n} + 1)/(n + \sqrt{2})$
d) $x_n = n/2^n$

2.2.2. Use Definition 2.14 to prove that each of the following sequences diverges to $+\infty$ or to $-\infty$.

a) $x_n = n^2 - n$
b) $x_n = n - 3n^2$
c) $x_n = \dfrac{n^2 + 1}{n}$
d) $x_n = n^2(2 + \sin(n^3 + n + 1))$

2.2.3. Find the limit (if it exists) of each of the following sequences.

a) $x_n = (2 + 3n - 4n^2)/(1 - 2n + 3n^2)$
b) $x_n = (n^3 + n - 2)/(2n^3 + n - 2)$

c) $x_n = \sqrt{3n+2} - \sqrt{n}$

d) $x_n = (\sqrt{4n+1} - \sqrt{n-1})/(\sqrt{9n+1} - \sqrt{n+2})$

2.2.4. a) Prove Theorem 2.12iv.

b) Prove Corollary 2.16.

2.2.5. Suppose that $x \in \mathbf{R}$, $x_n \geq 0$, and $x_n \to x$ as $n \to \infty$. Prove that $\sqrt{x_n} \to \sqrt{x}$ as $n \to \infty$. [For the case $x = 0$, use inequality (8) in Section 1.2.]

2.2.6. Prove that given $x \in \mathbf{R}$ there is a sequence $r_n \in \mathbf{Q}$ such that $r_n \to x$ as $n \to \infty$.

2.2.7. Suppose that x and y are extended real numbers and that $\{x_n\}$, $\{y_n\}$, and $\{w_n\}$ are real sequences.

a) [SQUEEZE THEOREM FOR $\overline{\mathbf{R}}$]. If $x_n \to x$ and $y_n \to x$, as $n \to \infty$, and $x_n \leq w_n \leq y_n$ for $n \in \mathbf{N}$, prove that $w_n \to x$ as $n \to \infty$.

b) [COMPARISON THEOREM FOR $\overline{\mathbf{R}}$]. If $x_n \to x$ and $y_n \to y$, as $n \to \infty$, and $x_n \leq y_n$ for $n \in \mathbf{N}$, prove that $x \leq y$.

2.2.8. Using the result in Exercise 2.2.5, prove the following results.

a) Suppose that $0 \leq x_1 \leq 1$ and $x_{n+1} = 1 - \sqrt{1 - x_n}$ for $n \in \mathbf{N}$. If $x_n \to x$ as $n \to \infty$, then $x = 0$ or 1.

b) Suppose that $x_1 > 3$ and $x_{n+1} = 2 + \sqrt{x_n - 2}$ for $n \in \mathbf{N}$. If $x_n \to x$ as $n \to \infty$, then $x = 3$.

(c) Suppose that $x_1 \geq 0$ and $x_{n+1} = \sqrt{2 + x_n}$ for $n \in \mathbf{N}$. If $x_n \to x$ as $n \to \infty$, then $x = 2$. What happens if $x_1 > -2$?

$\boxed{\textbf{2.2.9}}$. **This exercise was used in Section 1.6.**

a) Suppose that $0 \leq y < 1/10^n$ for some integer $n \geq 0$. Prove that there is an integer $0 \leq w \leq 9$ such that

$$\frac{w}{10^{n+1}} \leq y < \frac{w}{10^{n+1}} + \frac{1}{10^{n+1}}.$$

b) Prove that given $x \in [0, 1)$ there exist integers $0 \leq x_k \leq 9$ such that for all $n \in \mathbf{N}$,

$$\sum_{k=1}^{n} \frac{x_k}{10^k} \leq x < \sum_{k=1}^{n} \frac{x_k}{10^k} + \frac{1}{10^n}.$$

c) Prove that given $x \in [0, 1)$ there exist integers $0 \leq x_k \leq 9$, $k \in \mathbf{N}$, such that

$$x = \lim_{n \to \infty} \sum_{k=1}^{n} \frac{x_k}{10^k}.$$

d) Using part c), prove that $0.5 = 0.4999\ldots$ and $1 = 0.999\ldots$.

NOTE: The numbers x_k are called *digits* of x, and $0.x_1x_2\ldots$ is called a *decimal expansion* of x. Unless x is a rational number whose denominator is of the form

$2^i 5^j$ for some integers $i \geq 0$, $j \geq 0$, this expansion is unique; that is, there is only one sequence of integers $\{x_k\}$ that satisfies part (c). On the other hand, if x is a rational number whose denominator is of the form $2^i 5^j$, then there are two sequences $\{x_k\}$ that satisfy part (c), one that satisfies $x_k = 0$ for large k and one that satisfies $x_k = 9$ for large k (see part d). We shall identify the second sequence by saying that it *terminates* in 9s.

2.3 BOLZANO–WEIERSTRASS THEOREM

Notice that although the sequence $\{(-1)^n\}$ does not converge, it has convergent subsequences. In this section we shall prove that this is a general principle. Namely, we shall establish the Bolzano–Weierstrass Theorem, which states that every bounded sequence has a convergent subsequence.

We begin with a special case (monotone sequences) for which the Bolzano–Weierstrass Theorem is especially transparent. Afterward, we shall use this special case to obtain the general result.

2.18 Definition.

Let $\{x_n\}_{n \in \mathbf{N}}$ be a sequence of real numbers.

i) $\{x_n\}$ is said to be *increasing* (respectively, *strictly increasing*) if and only if $x_1 \leq x_2 \leq \cdots$ (respectively, $x_1 < x_2 < \cdots$).
ii) $\{x_n\}$ is said to be *decreasing* (respectively, *strictly decreasing*) if and only if $x_1 \geq x_2 \geq \cdots$ (respectively, $x_1 > x_2 > \cdots$).
iii) $\{x_n\}$ is said to be *monotone* if and only if it is either increasing or decreasing.

(Some authors call decreasing sequences *nonincreasing* and increasing sequences *nondecreasing*.)

If $\{x_n\}$ is increasing (respectively, decreasing) and converges to a, we shall write $x_n \uparrow a$ (respectively, $x_n \downarrow a$), as $n \to \infty$. Clearly, every strictly increasing sequence is increasing, and every strictly decreasing sequence is decreasing. Also, $\{x_n\}$ is increasing if and only if the sequence $\{-x_n\}$ is decreasing.

By Theorem 2.8, any convergent sequence is bounded. We now establish the converse of this result for monotone sequences. (For an extension to extended real numbers, see Exercise 2.3.6.)

2.19 Theorem.   [MONOTONE CONVERGENCE THEOREM].
If $\{x_n\}$ is increasing and bounded above, or if $\{x_n\}$ is decreasing and bounded below, then $\{x_n\}$ converges to a finite limit.

Proof. Suppose first that $\{x_n\}$ is increasing and bounded above. By the Completeness Axiom, the supremum $a := \sup\{x_n : n \in \mathbf{N}\}$ exists and is finite. Let $\varepsilon > 0$. By the Approximation Property for Suprema, choose $N \in \mathbf{N}$ such that

$$a - \varepsilon < x_N \leq a.$$

Since $x_N \leq x_n$ for $n \geq N$ and $x_n \leq a$ for all $n \in \mathbf{N}$, it follows that $a - \varepsilon < x_n \leq a$ for all $n \geq N$. In particular, $x_n \uparrow a$ as $n \to \infty$.

If $\{x_n\}$ is decreasing with infimum $b := \inf\{x_n : n \in \mathbf{N}\}$, then $\{-x_n\}$ is increasing with supremum $-b$ (see Theorem 1.20). Hence, by the first case and Theorem 2.12ii,

$$b = -(-b) = - \lim_{n \to \infty} (-x_n) = \lim_{n \to \infty} x_n. \qquad \blacksquare$$

The Monotone Convergence Theorem is used most often to show that a limit exists. Once existence has been established, it is often easy to find the value of that limit by using Theorems 2.9 and 2.12. The following examples illustrate this fact.

2.20 EXAMPLE.

If $|a| < 1$, then $a^n \to 0$ as $n \to \infty$.

Proof. It suffices to prove that $|a|^n \to 0$ as $n \to \infty$. First, we notice that $|a|^n$ is monotone decreasing since by the Multiplicative Property, $|a| < 1$ implies $|a|^{n+1} < |a|^n$ for all $n \in \mathbf{N}$. Next, we observe that $|a|^n$ is bounded below (by 0). Hence by the Monotone Convergence Theorem, $L := \lim_{n \to \infty} |a|^n$ exists.

Take the limit of the algebraic identity $|a|^{n+1} = |a| \cdot |a|^n$, as $n \to \infty$. By Remark 2.6 and Theorem 2.12, we obtain $L = |a| \cdot L$. Thus either $L = 0$ or $|a| = 1$. Since $|a| < 1$ by hypothesis, we conclude that $L = 0$. $\blacksquare$

2.21 EXAMPLE.

If $a > 0$, then $a^{1/n} \to 1$ as $n \to \infty$.

Proof. We consider three cases.

Case 1. $a = 1$. Then $a^{1/n} = 1$ for all $n \in \mathbf{N}$, and it follows that $a^{1/n} \to 1$ as $n \to \infty$.

Case 2. $a > 1$. We shall apply the Monotone Convergence Theorem. To show that $\{a^{1/n}\}$ is decreasing, fix $n \in \mathbf{N}$ and notice that $a > 1$ implies $a^{n+1} > a^n$. Taking the $n(n+1)$st root of this inequality, we obtain $a^{1/n} > a^{1/(n+1)}$; that is, $a^{1/n}$ is decreasing. Since $a > 1$ implies $a^{1/n} > 1$, it follows that $a^{1/n}$ is decreasing and bounded below. Hence, by the Monotone Convergence Theorem, $L := \lim_{n \to \infty} a^{1/n}$ exists. To find its value, take the limit of the identity $(a^{1/(2n)})^2 = a^{1/n}$ as $n \to \infty$. We obtain $L^2 = L$; that is, $L = 0$ or 1. Since $a^{1/n} > 1$, the Comparison Theorem shows that $L \geq 1$. Hence $L = 1$.

Case 3. $0 < a < 1$. Then $1/a > 1$. It follows from Theorem 2.12 and Case 2 that

$$\lim_{n \to \infty} a^{1/n} = \lim_{n \to \infty} \frac{1}{1/a^{1/n}} = \frac{1}{\lim_{n \to \infty} (1/a)^{1/n}} = 1. \qquad \blacksquare$$

Next, we introduce a monotone property for sequences of sets.

2.22 Definition.

A sequence of sets $\{I_n\}_{n \in \mathbf{N}}$ is said to be *nested* if and only if

$$I_1 \supseteq I_2 \supseteq \cdots .$$

In Chapters 3, 8, and 9, we shall use this concept to study continuous functions. Here, we use it to establish the Bolzano–Weierstrass Theorem. All of these applications depend in a fundamental way on the following result.

2.23 Theorem.   [NESTED INTERVAL PROPERTY].
If $\{I_n\}_{n \in \mathbf{N}}$ *is a nested sequence of nonempty closed bounded intervals, then* $E := \bigcap_{n=1}^{\infty} I_n$ *is nonempty. Moreover, if the lengths of these intervals satisfy* $|I_n| \to 0$ *as* $n \to \infty$, *then* E *is a single point.*

Proof. Let $I_n = [a_n, b_n]$. Since $\{I_n\}$ is nested, the real sequence $\{a_n\}$ is increasing and bounded above by b_1, and $\{b_n\}$ is decreasing and bounded below by a_1 (see Figure 2.2). Thus by Theorem 2.19, there exist $a, b \in \mathbf{R}$ such that $a_n \uparrow a$ and $b_n \downarrow b$ as $n \to \infty$. Since $a_n \le b_n$ for all $n \in \mathbf{N}$, it also follows from the Comparison Theorem that $a_n \le a \le b \le b_n$. Hence, a number x belongs to I_n for all $n \in \mathbf{N}$ if and only if $a \le x \le b$; that is, if and only if $x \in [a, b]$. In particular, any $x \in [a, b]$ belongs to all the I_n's.

$$a_1 \qquad a_2 \qquad a_3 \ a_4 \ldots a \ b \ldots b_4 \ b_3 \qquad b_2 \qquad b_1$$

FIGURE 2.2

We have proved that there is exactly one number that belongs to all the I_n's if and only if $a = b$. But if $|I_n| \to 0$ as $n \to \infty$, then $b_n - a_n \to 0$ as $n \to \infty$. Hence, by Theorem 2.12, a does equal b when $|I_n| \to 0$ as $n \to \infty$.   ∎

The next two results show that neither of the hypotheses of Theorem 2.23 can be relaxed.

2.24 Remark. *The Nested Interval Property might not hold if "closed" is omitted.*

Proof. The intervals $I_n = (0, 1/n)$, $n \in \mathbf{N}$, are bounded and nested but not closed. If there were an $x \in I_n$ for all $n \in \mathbf{N}$, then $0 < x < 1/n$; that is, $n < 1/x$ for all $n \in \mathbf{N}$. Since this contradicts the Archimedean Principle, it follows that the intervals I_n have no point in common.   ∎

2.25 Remark. *The Nested Interval Property might not hold if "bounded" is omitted.*

Proof. The intervals $I_n = [n, \infty)$, $n \in \mathbf{N}$ are closed and nested but not bounded. Again, they have no point in common. ∎

We are now prepared to establish the main result of this section.

2.26 Theorem. [**BOLZANO–WEIERSTRASS THEOREM**].
Every bounded sequence of real numbers has a convergent subsequence.

Proof. We begin with a general observation. Let $\{x_n\}$ be any sequence. If $E = A \cup B$ are sets and E contains x_n for infinitely many values of n, then at least one of the sets A or B also contains x_n for infinitely many values of n. (If not, then E contains x_n for only finitely many n, a contradiction.)

Let $\{x_n\}$ be a bounded sequence. Choose $a, b \in \mathbf{R}$ such that $x_n \in [a, b]$ for all $n \in \mathbf{N}$, and set $I_0 = [a, b]$. Divide I_0 into two halves, say $I' = [a, (a + b)/2]$ and $I'' = [(a + b)/2, b]$. Since $I_0 = I' \cup I''$, at least one of these half-intervals contains x_n for infinitely many n. Call it I_1, and choose $n_1 > 1$ such that $x_{n_1} \in I_1$. Notice that $|I_1| = |I_0|/2 = (b - a)/2$.

Suppose that closed intervals $I_0 \supset I_1 \supset \ldots \supset I_m$ and natural numbers $n_1 < n_2 < \ldots < n_m$ have been chosen such that for each $0 \leq k \leq m$,

$$|I_k| = \frac{b - a}{2^k}, \quad x_{n_k} \in I_k, \quad \text{and} \quad x_n \in I_k \quad \text{for infinitely many } n. \tag{2}$$

To choose I_{m+1}, divide $I_m = [a_m, b_m]$ into two halves, say $I' = [a_m, (a_m + b_m)/2]$ and $I'' = [(a_m + b_m)/2, b_m]$. Since $I_m = I' \cup I''$, at least one of these half-intervals contains x_n for infinitely many n. Call it I_{m+1}, and choose $n_{m+1} > n_m$ such that $x_{n_{m+1}} \in I_{m+1}$. Since

$$|I_{m+1}| = \frac{|I_m|}{2} = \frac{b - a}{2^{m+1}},$$

it follows by induction that there is a nested sequence $\{I_k\}_{k \in \mathbf{N}}$ of nonempty closed bounded intervals that satisfy (2) for all $k \in \mathbf{N}$.

By the Nested Interval Property, there is an $x \in \mathbf{R}$ that belongs to I_k for all $k \in \mathbf{N}$. Since $x \in I_k$, we have by (2) that

$$0 \leq |x_{n_k} - x| \leq |I_k| \leq \frac{b - a}{2^k}$$

for all $k \in \mathbf{N}$. Hence by the Squeeze Theorem, $x_{n_k} \to x$ as $k \to \infty$. ∎

EXERCISES

2.3.0. Decide which of the following statements are true and which are false. Prove the true ones and provide counterexamples for the false ones.

a) If x_n is strictly decreasing and $0 \leq x_n < 1/2$, then $x_n \to 0$ as $n \to \infty$.

b) If

$$x_n = \frac{(n-1)\cos(n^2 + n + 1)}{2n - 1},$$

then x_n has a convergence subsequence.

c) If x_n is a strictly increasing sequence and

$$|x_n| < 1 + \frac{1}{n}$$

for $n = 1, 2, \ldots$, then $x_n \to 1$ as $n \to \infty$.

d) If x_n has a convergent subsequence, then x_n is bounded.

2.3.1. Suppose that $x_0 \in (-1, 0)$ and $x_n = \sqrt{x_{n-1} + 1} - 1$ for $n \in \mathbf{N}$. Prove that $x_n \uparrow 0$ as $n \to \infty$. What happens when $x_0 \in [-1, 0]$?

2.3.2. Suppose that $0 < x_1 < 1$ and $x_{n+1} = 1 - \sqrt{1 - x_n}$ for $n \in \mathbf{N}$. Prove that $x_n \downarrow 0$ as $n \to \infty$ and $x_{n+1}/x_n \to 1/2$, as $n \to \infty$. (Exercise 4.3 in Apostol [1].)

2.3.3. Suppose that $x_0 \geq 2$ and $x_n = 2 + \sqrt{x_{n-1} - 2}$ for $n \in \mathbf{N}$. Use the Monotone Convergence Theorem to prove that either $x_n \to 2$ or $x_n \to 3$ as $n \to \infty$.

2.3.4. Suppose that $x_0 \in \mathbf{R}$ and $x_n = (1 + x_{n-1})/2$ for $n \in \mathbf{N}$. Use the Monotone Convergence Theorem to prove that $x_n \to 1$ as $n \to \infty$.

2.3.5. Prove that

$$\lim_{n \to \infty} x^{1/(2n-1)} = \begin{cases} 1 & x > 0 \\ 0 & x = 0 \\ -1 & x < 0. \end{cases}$$

2.3.6. **This result is used in Section 6.3 and elsewhere**.

a) Suppose that $\{x_n\}$ is a monotone increasing sequence in $\mathbf{R}$ (not necessarily bounded above). Prove that there is an extended real number x such that $x_n \to x$ as $n \to \infty$.

b) State and prove an analogous result for decreasing sequences.

2.3.7. Suppose that $E \subset \mathbf{R}$ is a nonempty bounded set and that $\sup E \notin E$. Prove that there exists a strictly increasing sequence $\{x_n\}$ that converges to $\sup E$ such that $x_n \in E$ for all $n \in \mathbf{N}$.

2.3.8. Let $0 < y_1 < x_1$ and set

$$x_{n+1} = \frac{x_n + y_n}{2} \quad \text{and} \quad y_{n+1} = \sqrt{x_n y_n}, \qquad n \in \mathbf{N}.$$

a) Prove that $0 < y_n < x_n$ for all $n \in \mathbf{N}$.

b) Prove that y_n is increasing and bounded above, and that x_n is decreasing and bounded below.

c) Prove that $0 < x_{n+1} - y_{n+1} < (x_1 - y_1)/2^n$ for $n \in \mathbf{N}$.

d) Prove that $\lim_{n\to\infty} x_n = \lim_{n\to\infty} y_n$. (This common value is called the *arithmetic-geometric* mean of x_1 and y_1.)

2.3.9. Suppose that $x_0 = 1$, $y_0 = 0$,

$$x_n = x_{n-1} + 2y_{n-1}, \quad \text{and} \quad y_n = x_{n-1} + y_{n-1}$$

for $n \in \mathbf{N}$. Prove that $x_n^2 - 2y_n^2 = \pm 1$ for $n \in \mathbf{N}$ and

$$\frac{x_n}{y_n} \to \sqrt{2} \quad \text{as } n \to \infty.$$

2.3.10. [Archimedes] Suppose that $x_0 = 2\sqrt{3}$, $y_0 = 3$,

$$x_n = \frac{2x_{n-1}y_{n-1}}{x_{n-1} + y_{n-1}},$$

and

$$y_n = \sqrt{x_n y_{n-1}}$$

for $n \in \mathbf{N}$.
a) Prove that $x_n \downarrow x$ and $y_n \uparrow y$, as $n \to \infty$, for some $x, y \in \mathbf{R}$.
b) Prove that $x = y$ and

$$3.14155 < x < 3.14161.$$

(The actual value of x is π.)

2.4 CAUCHY SEQUENCES

In this section we introduce an extremely powerful and widely used concept.

By definition, if $\{x_n\}$ is a convergent sequence, then there is a point $a \in \mathbf{R}$ such that x_n is near a for large n. If the x_n's are near a, they are certainly near each other. This leads us to the following concept.

2.27 Definition.

A sequence of points $x_n \in \mathbf{R}$ is said to be *Cauchy* (in **R**) if and only if for every $\varepsilon > 0$ there is an $N \in \mathbf{N}$ such that

$$n, m \geq N \quad \text{imply} \quad |x_n - x_m| < \varepsilon. \tag{3}$$

The next two results show how this concept is related to convergence.

2.28 Remark. *If $\{x_n\}$ is convergent, then $\{x_n\}$ is Cauchy.*

Proof. Suppose that $x_n \to a$ as $n \to \infty$. Then by definition, given $\varepsilon > 0$ there is an $N \in \mathbf{N}$ such that $|x_n - a| < \varepsilon/2$ for all $n \geq N$. Hence if $n, m \geq N$, it follows from the triangle inequality that

$$|x_n - x_m| \leq |x_n - a| + |x_m - a| < \frac{\varepsilon}{2} + \frac{\varepsilon}{2} = \varepsilon. \qquad \blacksquare$$

The following result shows that the converse of Remark 2.28 is also true (for real sequences).

2.29 Theorem. [CAUCHY].
*Let $\{x_n\}$ be a sequence of real numbers. Then $\{x_n\}$ is Cauchy if and only if $\{x_n\}$ converges (to some point a in **R**).*

STRATEGY: By Remark 2.28 we need only show that every Cauchy sequence converges. Suppose that $\{x_n\}$ is Cauchy. Since the x_n's are near each other, the sequence $\{x_n\}$ should be bounded. Hence, by the Bolzano–Weierstrass Theorem, $\{x_n\}$ has a convergent subsequence, say x_{n_k}. This means that for large k, the x_{n_k}'s are near some point $a \in \mathbf{R}$. But since $\{x_n\}$ is Cauchy, the x_n's should be near the x_{n_k}'s for large n, hence also near a. Thus the full sequence should converge to that same point a. Here are the details.

Proof. Suppose that $\{x_n\}$ is Cauchy. Given $\varepsilon = 1$, choose $N \in \mathbf{N}$ such that $|x_N - x_m| < 1$ for all $m \geq N$. By the triangle inequality,

$$|x_m| < 1 + |x_N| \qquad \text{for } m \geq N.$$

Therefore, $\{x_n\}$ is bounded by $M = \max\{|x_1|, |x_2|, \ldots, |x_{N-1}|, 1 + |x_N|\}$.

By the Bolzano–Weierstrass Theorem, $\{x_n\}$ has a convergent subsequence, say $x_{n_k} \to a$ as $k \to \infty$. Let $\varepsilon > 0$. Since x_n is Cauchy, choose $N_1 \in \mathbf{N}$ such that

$$n, m \geq N_1 \quad \text{imply} \quad |x_n - x_m| < \frac{\varepsilon}{2}.$$

Since $x_{n_k} \to a$ as $k \to \infty$, choose $N_2 \in \mathbf{N}$ such that

$$k \geq N_2 \quad \text{implies} \quad |x_{n_k} - a| < \frac{\varepsilon}{2}.$$

Fix $k \geq N_2$ such that $n_k \geq N_1$. Then

$$|x_n - a| \leq |x_n - x_{n_k}| + |x_{n_k} - a| < \varepsilon$$

for all $n \geq N_1$. Thus $x_n \to a$ as $n \to \infty$. $\qquad \blacksquare$

This result is extremely useful because it is often easier to show that a sequence is Cauchy than to show that it converges. The reason for this, as the following example shows, is that we can prove that a sequence is Cauchy even when we have no idea what its limit is.

2.30 EXAMPLE.

Prove that any real sequence $\{x_n\}$ that satisfies

$$|x_n - x_{n+1}| \le \frac{1}{2^n}, \qquad n \in \mathbf{N},$$

is convergent.

Proof. If $m > n$, then

$$
\begin{aligned}
|x_n - x_m| &= |x_n - x_{n+1} + x_{n+1} - x_{n+2} + \cdots + x_{m-1} - x_m| \\
&\le |x_n - x_{n+1}| + |x_{n+1} - x_{n+2}| + \cdots + |x_{m-1} - x_m| \\
&\le \frac{1}{2^n} + \cdots + \frac{1}{2^{m-1}} \\
&= \frac{1}{2^{n-1}} \sum_{k=1}^{m-n} \frac{1}{2^k} = \frac{1}{2^{n-1}} \left(1 - \frac{1}{2^{m-n}} \right).
\end{aligned}
$$

(The last step uses Exercise 1.4.4c, for $a = 2$.) It follows that $|x_n - x_m| < 1/2^{n-1}$ for all integers $m > n \ge 1$. But given $\varepsilon > 0$, we can choose $N \in \mathbf{N}$ so large that $n \ge N$ implies $1/2^{n-1} < \varepsilon$. We have proved that $\{x_n\}$ is Cauchy. By Theorem 2.29, therefore, it converges to some real number. $\blacksquare$

The following result shows that a sequence is not necessarily Cauchy just because x_n is near x_{n+1} for large n.

2.31 Remark. *A sequence that satisfies $x_{n+1} - x_n \to 0$ is not necessarily Cauchy.*

Proof. Consider the sequence $x_n := \log n$. By basic properties of logarithms (see Exercise 5.3.7),

$$x_{n+1} - x_n = \log(n + 1) - \log n = \log((n + 1)/n) \to \log 1 = 0$$

as $n \to \infty$. $\{x_n\}$ cannot be Cauchy, however, because it does not converge; in fact, it diverges to $+\infty$ as $n \to \infty$. $\blacksquare$

EXERCISES

2.4.0. Decide which of the following statements are true and which are false. Prove the true ones and provide a counterexample for the false ones.

a) If $\{x_n\}$ is Cauchy and $\{y_n\}$ is bounded, then $\{x_n y_n\}$ is Cauchy.

b) If $\{x_n\}$ and $\{y_n\}$ are Cauchy and $y_n \ne 0$ for all $n \in \mathbf{N}$, then $\{x_n/y_n\}$ is Cauchy.

c) If $\{x_n\}$ and $\{y_n\}$ are Cauchy and $x_n + y_n > 0$ for all $n \in \mathbf{N}$, then $\{1/(x_n + y_n)\}$ cannot converge to zero.

d) If $\{x_n\}$ is a sequence of real numbers that satisfies $x_{2k} - x_{2k-1} \to 0$ as $k \to \infty$ and if $x_n = 0$ for all $n \neq 2^k$, $k \in \mathbf{N}$, then $\{x_n\}$ is Cauchy.

2.4.1. Prove that if $\{x_n\}$ is a sequence that satisfies

$$|x_n| \leq \frac{2n^2 + 3}{n^3 + 5n^2 + 3n + 1}$$

for all $n \in \mathbf{N}$, then $\{x_n\}$ is Cauchy.

2.4.2. Suppose that $x_n \in \mathbf{Z}$ for $n \in \mathbf{N}$. If $\{x_n\}$ is Cauchy, prove that x_n is eventually constant; that is, that there exist numbers $a \in \mathbf{Z}$ and $N \in \mathbf{N}$ such that $x_n = a$ for all $n \geq N$.

2.4.3. Suppose that x_n and y_n are Cauchy sequences in $\mathbf{R}$ and that $a \in \mathbf{R}$.

a) Without using Theorem 2.29, prove that ax_n is Cauchy.
b) Without using Theorem 2.29, prove that $x_n + y_n$ is Cauchy.
c) Without using Theorem 2.29, prove that $x_n y_n$ is Cauchy.

2.4.4. Let $\{x_n\}$ be a sequence of real numbers. Suppose that for each $\varepsilon > 0$ there is an $N \in \mathbf{N}$ such that $m \geq n \geq N$ implies $\left|\sum_{k=n}^{m} x_k\right| < \varepsilon$. Prove that

$$\lim_{n \to \infty} \sum_{k=1}^{n} x_k$$

exists and is finite.

2.4.5. Prove that $\lim_{n \to \infty} \sum_{k=1}^{n} (-1)^k / k$ exists and is finite.

2.4.6. Let $\{x_n\}$ be a sequence. Suppose that there is an $a \in (0, 1)$ such that

$$|x_{n+1} - x_n| \leq a^n$$

for all $n \in \mathbf{N}$. Prove that $x_n \to x$ for some $x \in \mathbf{R}$.

2.4.7. a) Let E be a subset of $\mathbf{R}$. A point $a \in \mathbf{R}$ is called a *cluster point* of E if $E \cap (a - r, a + r)$ contains infinitely many points for every $r > 0$. Prove that a is a cluster point of E if and only if for each $r > 0$, $E \cap (a - r, a + r) \setminus \{a\}$ is nonempty.

b) Prove that every bounded infinite subset of $\mathbf{R}$ has at least one cluster point.

2.4.8 a) A subset E of $\mathbf{R}$ is said to be *sequentially compact* if and only if every sequence $x_n \in E$ has a convergent subsequence whose limit belongs to E. Prove that every closed bounded interval is sequentially compact.

b) Prove that there exist bounded intervals in $\mathbf{R}$ that are not sequentially compact.

c) Prove that there exist closed intervals in $\mathbf{R}$ that are not sequentially compact.

*2.5   LIMITS SUPREMUM AND INFIMUM

This section uses no material from any other enrichment section.

In some situations (e.g., the Root Test in Section 6.3), we shall use the following generalization of limits.

2.32 Definition.

Let $\{x_n\}$ be a real sequence. Then the *limit supremum* of $\{x_n\}$ is the extended real number

$$\limsup_{n\to\infty} x_n := \lim_{n\to\infty} (\sup_{k\geq n} x_k), \qquad (4)$$

and the *limit infimum* of $\{x_n\}$ is the extended real number

$$\liminf_{n\to\infty} x_n := \lim_{n\to\infty} (\inf_{k\geq n} x_k).$$

Before we proceed, we must show that the limits in Definition 2.32 exist as extended real numbers. To this end, let $\{x_n\}$ be a sequence of real numbers and consider the sequences

$$s_n = \sup_{k\geq n} x_k := \sup\{x_k : k \geq n\} \quad \text{and} \quad t_n = \inf_{k\geq n} x_k := \inf\{x_k : k \geq n\}.$$

Each s_n and t_n is an extended real number, and by the Monotone Property, s_n is a decreasing sequence and t_n an increasing sequence of extended real numbers. In particular, there exist extended real numbers s and t such that $s_n \downarrow s$ and $t_n \uparrow t$ as $n \to \infty$ (see Exercise 2.3.6). These extended real numbers are, by Definition 2.32, the limit infimum and limit supremum of the sequence $\{x_n\}$.

Here are two examples of how to compute limits supremum and limits infimum.

2.33 EXAMPLE.

Find $\limsup_{n\to\infty} x_n$ and $\liminf_{n\to\infty} x_n$ if $x_n = (-1)^n$.

Solution. Since $\sup_{k\geq n} (-1)^k = 1$ for all $n \in \mathbf{N}$, it follows from Definition 2.32 that $\limsup_{n\to\infty} x_n = 1$. Similarly, $\liminf_{n\to\infty} x_n = -1$. ∎

2.34 EXAMPLE.

Find $\limsup_{n\to\infty} x_n$ and $\liminf_{n\to\infty} x_n$ if $x_n = 1 + 1/n$.

Solution. Since $\sup_{k\geq n}(1 + 1/k) = 1 + 1/n$ for all $n \in \mathbf{N}$, $\limsup_{n\to\infty} x_n = 1$. Since $\inf_{k\geq n}(1 + 1/k) = 1$ for all $n \in \mathbf{N}$, $\liminf_{n\to\infty} x_n = 1$. ∎

These examples suggest that there is a connection among limits supremum, limits infimum, and convergent subsequences. The next several results make this connection clear.

2.35 Theorem. *Let $\{x_n\}$ be a sequence of real numbers, $s = \limsup_{n\to\infty} x_n$, and $t = \liminf_{n\to\infty} x_n$. Then there are subsequences $\{x_{n_k}\}_{k\in\mathbf{N}}$ and $\{x_{\ell_j}\}_{j\in\mathbf{N}}$ such that $x_{n_k} \to s$ as $k \to \infty$ and $x_{\ell_j} \to t$ as $j \to \infty$.*

Proof. We will prove the result for the limit supremum. A similar argument establishes the result for the limit infimum. Let $s_n = \sup_{k\geq n} x_k$ and observe that $s_n \downarrow s$ as $n \to \infty$.

Case 1. $s = \infty$. Then by definition $s_n = \infty$ for all $n \in \mathbf{N}$. Since $s_1 = \infty$, there is an $n_1 \in \mathbf{N}$ such that $x_{n_1} > 1$. Since $s_{n_1+1} = \infty$, there is an $n_2 \geq n_1 + 1 > n_1$ such that $x_{n_2} > 2$. Continuing in this manner, we can choose a subsequence $\{x_{n_k}\}$ such that $x_{n_k} > k$ for all $k \in \mathbf{N}$. Hence, it follows from the Squeeze Theorem for $\overline{\mathbf{R}}$ (see Exercise 2.2.7) that $x_{n_k} \to \infty = s$ as $k \to \infty$.

Case 2. $s = -\infty$. Since $s_n \geq x_n$ for all $n \in \mathbf{N}$ and $s_n \to -\infty$ as $n \to \infty$, it follows from the Squeeze Theorem for $\overline{\mathbf{R}}$ that $x_n \to -\infty = s$ as $n \to \infty$.

Case 3. $-\infty < s < \infty$. Set $n_0 = 0$. By Theorem 1.14 (the Approximation Property for Suprema), there is an integer $n_1 \in \mathbf{N}$ such that $s_{n_0+1} - 1 < x_{n_1} \leq s_{n_0+1}$. Similarly, there is an integer $n_2 \geq n_1 + 1 > n_1$ such that $s_{n_1+1} - 1/2 < x_{n_2} \leq s_{n_1+1}$. Continuing in this manner, we can choose integers $n_1 < n_2 < \cdots$ such that

$$s_{n_{k-1}+1} - \frac{1}{k} < x_{n_k} \leq s_{n_{k-1}+1} \tag{5}$$

for $k \in \mathbf{N}$. Since $s_{n_{k-1}+1} \to s$ as $k \to \infty$, we conclude by the Squeeze Theorem that $x_{n_k} \to s$ as $k \to \infty$. ∎

This observation leads directly to a characterization of limits in terms of limits infimum and limits supremum.

2.36 Theorem. *Let $\{x_n\}$ be a real sequence and x be an extended real number. Then $x_n \to x$ as $n \to \infty$ if and only if*

$$\limsup_{n\to\infty} x_n = \liminf_{n\to\infty} x_n = x. \tag{6}$$

Proof. Suppose that $x_n \to x$ as $n \to \infty$. Then $x_{n_k} \to x$ as $k \to \infty$ for all subsequences $\{x_{n_k}\}$. Hence, by Theorem 2.35, $\limsup_{n\to\infty} x_n = x$ and $\liminf_{n\to\infty} x_n = x$; that is, (6) holds.

Conversely, suppose that (6) holds.

Case 1. $x = \pm\infty$. By considering $\pm x_n$ we may suppose that $x = \infty$. Thus given $M \in \mathbf{R}$ there is an $N \in \mathbf{N}$ such that $\inf_{k\geq N} x_k > M$. It follows that $x_n > M$ for all $n \geq N$; that is, $x_n \to \infty$ as $n \to \infty$.

Case 2. $-\infty < x < \infty$. Let $\varepsilon > 0$. Choose $N \in \mathbf{N}$ such that

$$\sup_{k\geq N} x_k - x < \frac{\varepsilon}{2} \quad \text{and} \quad x - \inf_{k\geq N} x_k < \frac{\varepsilon}{2}.$$

Let $n, m \geq N$ and suppose for simplicity that $x_n > x_m$. Then

$$|x_n - x_m| = x_n - x_m \leq \sup_{k \geq N} x_k - x + x - \inf_{k \geq N} x_k < \frac{\varepsilon}{2} + \frac{\varepsilon}{2} = \varepsilon.$$

Thus $\{x_n\}$ is Cauchy and converges to some finite real number. But by Theorem 2.35, some subsequence of $\{x_n\}$ converges to x. We conclude that $x_n \to x$ as $n \to \infty$. ∎

Theorem 2.35 also leads to the following geometric interpretation of limits supremum and limits infimum.

2.37 Theorem. *Let $\{x_n\}$ be a sequence of real numbers. Then $\limsup_{n \to \infty} x_n$ (respectively, $\liminf_{n \to \infty} x_n$) is the largest value (respectively, the smallest value) to which some subsequence of $\{x_n\}$ converges. Namely, if $x_{n_k} \to x$ as $k \to \infty$, then*

$$\liminf_{n \to \infty} x_n \leq x \leq \limsup_{n \to \infty} x_n. \tag{7}$$

Proof. Suppose that $x_{n_k} \to x$ as $k \to \infty$. Fix $N \in \mathbf{N}$ and choose K so large that $k \geq K$ implies $n_k \geq N$. Clearly,

$$\inf_{j \geq N} x_j \leq x_{n_k} \leq \sup_{j \geq N} x_j$$

for all $k \geq K$. Taking the limit of this inequality as $k \to \infty$, we obtain

$$\inf_{j \geq N} x_j \leq x \leq \sup_{j \geq N} x_j.$$

Taking the limit of this last inequality as $N \to \infty$ and applying Definition 2.32, we obtain (7). ∎

We close this section with several other properties of limits supremum and limits infimum.

2.38 Remark. *If $\{x_n\}$ is any sequence of real numbers, then*

$$\liminf_{n \to \infty} x_n \leq \limsup_{n \to \infty} x_n.$$

Proof. Since $\inf_{k \geq n} x_k \leq \sup_{k \geq n} x_k$ for all $n \in \mathbf{N}$, this inequality follows from Theorem 2.17 (the Comparison Theorem). ∎

The following result is an immediate consequence of Definition 2.32, the Comparison Theorem, and the Monotone Convergence Theorem.

2.39 Remark. *A real sequence $\{x_n\}$ is bounded above if and only if $\limsup_{n \to \infty} x_n < \infty$, and is bounded below if and only if $\liminf_{n \to \infty} x_n > -\infty$.*

The following result shows that we can take limits supremum and limits infimum of inequalities.

2.40 Theorem. *If $x_n \leq y_n$ for n large, then*

$$\limsup_{n\to\infty} x_n \leq \limsup_{n\to\infty} y_n \quad and \quad \liminf_{n\to\infty} x_n \leq \liminf_{n\to\infty} y_n. \tag{8}$$

Proof. If $x_k \leq y_k$ for $k \geq N$, then $\sup_{k\geq n} x_k \leq \sup_{k\geq n} y_k$ and $\inf_{k\geq n} x_k \leq \inf_{k\geq n} y_k$ for any $n \geq N$. Taking the limit of these inequalities as $n \to \infty$, we obtain (8). ∎

EXERCISES

2.5.1. Find the limit infimum and the limit supremum of each of the following sequences.

a) $x_n = 3 - (-1)^n$
b) $x_n = \cos(n\pi/2)$
c) $x_n = (-1)^{n+1} + (-1)^n/n$
d) $x_n = \sqrt{1 + n^2}/(2n - 5)$
e) $x_n = y_n/n$, where $\{y_n\}$ is any bounded sequence
f) $x_n = n(1 + (-1)^n) + n^{-1}((-1)^n - 1)$
g) $x_n = (n^3 + n^2 - n + 1)/(n^2 + 2n + 5)$

2.5.2. Suppose that $\{x_n\}$ is a real sequence. Prove that

$$- \limsup_{n\to\infty} x_n = \liminf_{n\to\infty}(-x_n)$$

and

$$- \liminf_{n\to\infty} x_n = \limsup_{n\to\infty}(-x_n).$$

2.5.3. Let $\{x_n\}$ be a real sequence and $r \in \mathbf{R}$.

a) Prove that

$$\limsup_{n\to\infty} x_n < r \quad \text{implies} \quad x_n < r$$

for n large.

b) Prove that

$$\limsup_{n\to\infty} x_n > r \quad \text{implies} \quad x_n > r$$

for infinitely many $n \in \mathbf{N}$.

2.5.4. Suppose that $\{x_n\}$ and $\{y_n\}$ are real sequences.

a) Prove that

$$\liminf_{n\to\infty} x_n + \liminf_{n\to\infty} y_n \leq \liminf_{n\to\infty}(x_n + y_n)$$

$$\leq \limsup_{n\to\infty} x_n + \liminf_{n\to\infty} y_n$$

$$\leq \limsup_{n\to\infty}(x_n + y_n) \leq \limsup_{n\to\infty} x_n + \limsup_{n\to\infty} y_n,$$

provided that none of these sums is of the form $\infty - \infty$.

b) Show that if $\lim_{n\to\infty} x_n$ exists, then

$$\liminf_{n\to\infty}(x_n + y_n) = \lim_{n\to\infty} x_n + \liminf_{n\to\infty} y_n$$

and

$$\limsup_{n\to\infty}(x_n + y_n) = \lim_{n\to\infty} x_n + \limsup_{n\to\infty} y_n.$$

c) Show by examples that each of the inequalities in part (a) can be strict.

2.5.5. Let $\{x_n\}$ and $\{y_n\}$ be real sequences.

a) Suppose that $x_n \geq 0$ and $y_n \geq 0$ for each $n \in \mathbf{N}$. Prove that

$$\limsup_{n\to\infty}(x_n y_n) \leq \left(\limsup_{n\to\infty} x_n\right)\left(\limsup_{n\to\infty} y_n\right),$$

provided that the product on the right is not of the form $0 \cdot \infty$. Show by example that this inequality can be strict.

b) Suppose that $x_n \leq 0 \leq y_n$ for $n \in \mathbf{N}$. Prove that

$$\left(\liminf_{n\to\infty} x_n\right)\left(\limsup_{n\to\infty} y_n\right) \leq \liminf_{n\to\infty}(x_n y_n),$$

provided that none of these products is of the form $0 \cdot \infty$.

2.5.6. Suppose that $x_n \geq 0$ and $y_n \geq 0$ for all $n \in \mathbf{N}$. Prove that if $x_n \to x$ as $n \to \infty$ (x may be an extended real number), then

$$\limsup_{n\to\infty}(x_n y_n) = x \limsup_{n\to\infty} y_n,$$

provided that none of these products is of the form $0 \cdot \infty$.

2.5.7. Prove that

$$\limsup_{n\to\infty} x_n = \inf_{n\in\mathbf{N}} \left(\sup_{k\geq n} x_k\right) \quad \text{and} \quad \liminf_{n\to\infty} x_n = \sup_{n\in\mathbf{N}} \left(\inf_{k\geq n} x_k\right)$$

for any real sequence $\{x_n\}$.

2.5.8. Suppose that $x_n \geq 0$ for $n \in \mathbf{N}$. Under the interpretation $1/0 = \infty$ and $1/\infty = 0$, prove that

$$\limsup_{n \to \infty} \frac{1}{x_n} = \frac{1}{\liminf_{n \to \infty} x_n} \quad \text{and} \quad \liminf_{n \to \infty} \frac{1}{x_n} = \frac{1}{\limsup_{n \to \infty} x_n}.$$

2.5.9. Let $x_n \in \mathbf{R}$. Prove that $x_n \to 0$ as $n \to \infty$ if and only if

$$\limsup_{n \to \infty} |x_n| = 0.$$

CHAPTER 3

Functions on R

3.1 TWO-SIDED LIMITS

In the preceding chapter we studied limits of real sequences. In this chapter we examine limits of real functions; that is, functions whose domains and ranges are subsets of **R**. To distinguish such functions from functions whose ranges include ∞ and/or $-\infty$, we shall sometimes refer to real functions as *finite valued*.

Recall from elementary calculus that a function $f(x)$ converges to a limit L, as x approaches a, if $f(x)$ is near L when x is near a. Here is a precise definition of this concept.

3.1 Definition.

Let $a \in \mathbf{R}$, let I be an open interval which contains a, and let f be a real function defined everywhere on I except possibly at a. Then $f(x)$ is said to *converge to L, as x approaches a*, if and only if for every $\varepsilon > 0$ there is a $\delta > 0$ (which in general depends on ε, f, I, and a) such that

$$0 < |x - a| < \delta \quad \text{implies} \quad |f(x) - L| < \varepsilon. \tag{1}$$

In this case we write

$$L = \lim_{x \to a} f(x) \quad \text{or} \quad f(x) \to L \quad \text{as} \quad x \to a,$$

and call L the *limit* of $f(x)$ as x approaches a.

As was the case for sequences, ε represents the maximal error allowed in the approximation $f(x)$ to L. In practice, the number δ represents the tolerance allowed in the measurement x of a which will produce an approximation $f(x)$ which is acceptably close to the value L.

According to Definition 3.1, to show that a function has a limit, we must begin with a general $\varepsilon > 0$ and describe how to choose a δ which satisfies (1).

3.2 EXAMPLE.

Suppose that $f(x) = mx + b$, where $m, b \in \mathbf{R}$. Prove that

$$f(a) = \lim_{x \to a} f(x)$$

for all $a \in \mathbf{R}$.

Proof. If $m = 0$, there is nothing to prove. Otherwise, given $\varepsilon > 0$, set $\delta = \varepsilon/|m|$. If $|x - a| < \delta$, then

$$|f(x) - f(a)| = |mx + b - (ma + b)| = |m|\,|x - a| < |m|\delta = \varepsilon.$$

Thus by definition, $f(x) \to f(a)$ as $x \to a$. ∎

Sometimes, in order to determine δ, we must break $f(x) - L$ into two factors, replacing the less important factor by an upper bound.

3.3 EXAMPLE.

If $f(x) = x^2 + x - 3$, prove that $f(x) \to -1$ as $x \to 1$.

Proof. Let $\varepsilon > 0$ and set $L = -1$. Notice that

$$f(x) - L = x^2 + x - 2 = (x - 1)(x + 2).$$

If $0 < \delta \leq 1$, then $|x - 1| < \delta$ implies $0 < x < 2$, so by the triangle inequality, $|x + 2| \leq |x| + 2 < 4$. Set $\delta = \min\{1, \varepsilon/4\}$. It follows that if $|x - 1| < \delta$, then

$$|f(x) - L| = |x - 1|\,|x + 2| < 4|x - 1| < 4\delta \leq \varepsilon.$$

Thus by definition, $f(x) \to L$ as $x \to 1$. ∎

Before continuing, we would like to draw your attention to two features of Definition 3.1: **Assumption 1**. The interval I is open; **Assumption 2**. $0 < |x - a|$. If $I = (c, d)$ is an open interval and $\delta_0 := \min\{a - c, d - a\}$, then $|x - a| < \delta_0$ implies $x \in I$. Hence, Assumption 1 guarantees that for $\delta > 0$ sufficiently small, $f(x)$ is defined for all $x \neq a$ satisfying $|x - a| < \delta$ (i.e., on BOTH sides of a). Since $|x - a| > 0$ is equivalent to $x \neq a$, Assumption 2 guarantees that f can have a limit at a without being defined at a. (This will be crucial for defining derivatives later.)

The next result shows that even when a function f is defined at a, the value of the limit of f at a is, in general, independent of the value $f(a)$.

3.4 Remark. *Let $a \in \mathbf{R}$, let I be an open interval which contains a, and let f, g be real functions defined everywhere on I except possibly at a. If $f(x) = g(x)$ for all $x \in I \setminus \{a\}$ and $f(x) \to L$ as $x \to a$, then $g(x)$ also has a limit as $x \to a$, and*

$$\lim_{x \to a} g(x) = \lim_{x \to a} f(x).$$

Proof. Let $\varepsilon > 0$ and choose $\delta > 0$ small enough so that (1) holds and $|x - a| < \delta$ implies $x \in I$. Suppose that $0 < |x - a| < \delta$. We have $f(x) = g(x)$ by hypothesis and $|f(x) - L| < \varepsilon$ by (1). It follows that $|g(x) - L| < \varepsilon$. ∎

Thus to prove that a function f has a limit, we may begin by simplifying f algebraically, even when that algebra is invalid at finitely many points.

3.5 *EXAMPLE.*

Prove that

$$g(x) = \frac{x^3 + x^2 - x - 1}{x^2 - 1}$$

has a limit as $x \to 1$.

Proof. Set $f(x) = x + 1$ and observe by Example 3.2 that $f(x) \to 2$ as $x \to 1$. Since

$$g(x) = \frac{x^3 + x^2 - x - 1}{x^2 - 1} = \frac{(x+1)(x^2 - 1)}{x^2 - 1} = f(x)$$

for $x \neq \pm 1$, it follows from Remark 3.4 that $g(x)$ has a limit as $x \to 1$ (and that limit is 2). ∎

There is a close connection between limits of functions and limits of sequences.

3.6 Theorem. [SEQUENTIAL CHARACTERIZATION OF LIMITS].

Let $a \in \mathbf{R}$, let I be an open interval which contains a, and let f be a real function defined everywhere on I except possibly at a. Then

$$L = \lim_{x \to a} f(x)$$

exists if and only if $f(x_n) \to L$ as $n \to \infty$ for every sequence $x_n \in I \setminus \{a\}$ which converges to a as $n \to \infty$.

Proof. Suppose that f converges to L as x approaches a. Then given $\varepsilon > 0$ there is a $\delta > 0$ such that (1) holds. If $x_n \in I \setminus \{a\}$ converges to a as $n \to \infty$, then choose an $N \in \mathbf{N}$ such that $n \geq N$ implies $|x_n - a| < \delta$. Since $x_n \neq a$, it follows from (1) that $|f(x_n) - L| < \varepsilon$ for all $n \geq N$. Therefore, $f(x_n) \to L$ as $n \to \infty$.

Conversely, suppose that $f(x_n) \to L$ as $n \to \infty$ for every sequence $x_n \in I \setminus \{a\}$ which converges to a. If f does not converge to L as x approaches a, then there is an $\varepsilon > 0$ (call it ε_0) such that the implication "$0 < |x - a| < \delta$ implies $|f(x) - L| < \varepsilon_0$" does not hold for any $\delta > 0$. Thus, for each $\delta = 1/n$, $n \in \mathbf{N}$, there is a point $x_n \in I$ which satisfies two conditions: $0 < |x_n - a| < 1/n$ and $|f(x_n) - L| \geq \varepsilon_0$. Now the first condition and the Squeeze Theorem (Theorem 2.9) imply that $x_n \neq a$ and $x_n \to a$ so by hypothesis, $f(x_n) \to L$, as $n \to \infty$. In particular, $|f(x_n) - L| < \varepsilon_0$ for n large, which contradicts the second condition. ∎

Thus to show that the limit of a function f does not exist as $x \to a$, we need only find two sequences converging to a whose images under f have different limits.

3.7 EXAMPLE.

Prove that

$$f(x) = \begin{cases} \sin \dfrac{1}{x} & x \neq 0 \\ 0 & x = 0 \end{cases}$$

has no limit as $x \to 0$.

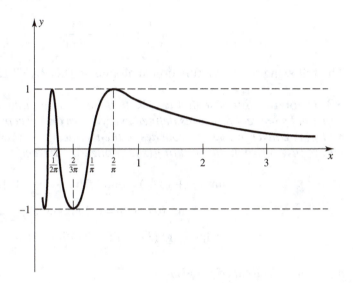

FIGURE 3.1

Proof. By examining the graph of $y = f(x)$ (see Figure 3.1), we are led to consider two extremes:

$$a_n := \frac{2}{(4n + 1)\pi} \quad \text{and} \quad b_n := \frac{2}{(4n + 3)\pi}, \qquad n \in \mathbf{N}.$$

Clearly, both a_n and b_n converge to 0 as $n \to \infty$. On the other hand, since $f(a_n) = 1$ and $f(b_n) = -1$ for all $n \in \mathbf{N}$, $f(a_n) \to 1$ and $f(b_n) \to -1$ as $n \to \infty$. Thus by Theorem 3.6, the limit of $f(x)$, as $x \to 0$, cannot exist. ∎

Theorem 3.6 also allows us to translate results about limits of sequences to results about limits of functions. The next three theorems illustrate this principle.

Before stating these results, we introduce an algebra of functions. Suppose that $f, g : E \to \mathbf{R}$. For each $x \in E$, the *pointwise sum*, $f + g$, of f and g is defined by

$$(f + g)(x) := f(x) + g(x),$$

the *scalar product*, αf, of a scalar $\alpha \in \mathbf{R}$ with f, by

$$(\alpha f)(x) := \alpha f(x),$$

the *pointwise product*, fg, of f and g, by

$$(fg)(x) := f(x)g(x),$$

and (when $g(x) \neq 0$) the *pointwise quotient*, f/g, of f and g, by

$$\left(\frac{f}{g}\right)(x) := \frac{f(x)}{g(x)}.$$

The following result is a function analogue of Theorem 2.12.

3.8 Theorem.   *Suppose that $a \in \mathbf{R}$, that I is an open interval which contains a, and that f, g are real functions defined everywhere on I except possibly at a. If $f(x)$ and $g(x)$ converge as x approaches a, then so do $(f + g)(x)$, $(fg)(x)$, $(\alpha f)(x)$, and $(f/g)(x)$ (when the limit of $g(x)$ is nonzero). In fact,*

$$\lim_{x \to a} (f + g)(x) = \lim_{x \to a} f(x) + \lim_{x \to a} g(x),$$

$$\lim_{x \to a} (\alpha f)(x) = \alpha \lim_{x \to a} f(x),$$

$$\lim_{x \to a} (fg)(x) = \lim_{x \to a} f(x) \lim_{x \to a} g(x),$$

and (when the limit of $g(x)$ is nonzero)

$$\lim_{x \to a} \left(\frac{f}{g}\right)(x) = \frac{\lim_{x \to a} f(x)}{\lim_{x \to a} g(x)}.$$

Proof. Let

$$L := \lim_{x \to a} f(x) \quad \text{and} \quad M := \lim_{x \to a} g(x).$$

If $x_n \in I \setminus \{a\}$ converges to a, then by Theorem 3.6, $f(x_n) \to L$ and $g(x_n) \to M$ as $n \to \infty$. By Theorem 2.12i, $f(x_n) + g(x_n) \to L + M$ as $n \to \infty$. Since this holds for any sequence $x_n \in I \setminus \{a\}$ which converges to a, we conclude by Theorem 3.6 that

$$\lim_{x \to a} (f + g)(x) = L + M = \lim_{x \to a} f(x) + \lim_{x \to a} g(x).$$

The other rules follow in an analogous way from Theorem 2.12ii through 2.12iv.   ∎

Similarly, the Sequential Characterization of Limits can be combined with the Squeeze and Comparison Theorems for sequences to establish the following results.

3.9 Theorem.   [SQUEEZE THEOREM FOR FUNCTIONS].
*Suppose that a $\in$ **R**, that I is an open interval which contains a, and that f,g,h are real functions defined everywhere on I except possibly at a.*

i) *If $g(x) \le h(x) \le f(x)$ for all $x \in I \setminus \{a\}$, and*

$$\lim_{x \to a} f(x) = \lim_{x \to a} g(x) = L,$$

then the limit of $h(x)$ exists, as $x \to a$, and

$$\lim_{x \to a} h(x) = L.$$

ii) *If $|g(x)| \le M$ for all $x \in I \setminus \{a\}$ and $f(x) \to 0$ as $x \to a$, then*

$$\lim_{x \to a} f(x)g(x) = 0.$$

The preceding result is illustrated in Figure 3.2.

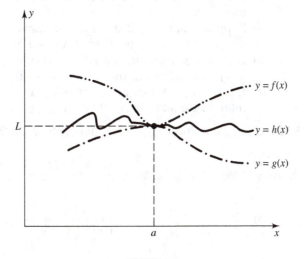

FIGURE 3.2

3.10 Theorem.   [COMPARISON THEOREM FOR FUNCTIONS].
*Suppose that $a \in$ **R**, that I is an open interval which contains a, and that f,g are real functions defined everywhere on I except possibly at a. If f and g have a limit as x approaches a and $f(x) \le g(x)$ for all $x \in I \setminus \{a\}$, then*

$$\lim_{x \to a} f(x) \le \lim_{x \to a} g(x).$$

We shall refer to this last result as taking the limit of an inequality.

The limit theorems (Theorems 3.8, 3.9, and 3.10) allow us to prove that limits exist without resorting to ε's and δ's.

3.11 *EXAMPLE.*

Prove that

$$\lim_{x \to 1} \frac{x - 1}{3x + 1} = 0.$$

Proof. By Example 3.2, $x - 1 \to 0$ and $3x + 1 \to 4$ as $x \to 1$. Hence, by Theorem 3.8, $(x - 1)/(3x + 1) \to 0/4 = 0$ as $x \to 1$. ∎

EXERCISES

3.1.0. Let $a \in \mathbf{R}$ and let f and g be real functions defined at all points x in some open interval containing a except possibly at $x = a$. Decide which of the following statements are true and which are false. Prove the true ones and give counterexamples for the false ones.

   a) For each $n \in \mathbf{N}$, the function $(x - a)^n \sin(f(x)(x - a)^{-n})$ has a limit as $x \to a$.
   b) Suppose that $\{x_n\}$ is a sequence converging to a with $x_n \neq a$. If $f(x_n) \to L$ as $n \to \infty$, then $f(x) \to L$ as $x \to a$.
   c) If f and g are finite valued on the open interval $(a - 1, a + 1)$ and $f(x) \to 0$ as $x \to a$, then $f(x)g(x) \to 0$ as $x \to a$.
   d) If $\lim_{x \to a} f(x)$ does not exist and $f(x) \leq g(x)$ for all x in some open interval I containing a, then $\lim_{x \to a} g(x)$ doesn't exist either.

3.1.1. Using Definition 3.1, prove that each of the following limits exist.

   a) $$\lim_{x \to 2} x^2 + 2x - 5 = 3$$

   b) $$\lim_{x \to 1} \frac{x^2 + x - 2}{x - 1} = 3$$

   c) $$\lim_{x \to 1} x^3 + 2x + 1 = 4$$

   d) $$\lim_{x \to 0} x^3 \sin(e^{x^2}) = 0$$

3.1.2. Decide which of the following limits exist and which do not. Prove that your answer is correct. (You can use well-known facts about the values of $\tan x$, $\cos x$, and $\log x$, e.g., that $\log x \to -\infty$ as $x \to 0+$.)

   a) $$\lim_{x \to 0} \tan\left(\frac{1}{x}\right)$$

b)
$$\lim_{x \to 0} x \cos \left(\frac{x^2 + 1}{x^3} \right)$$

c)
$$\lim_{x \to 1} \frac{1}{\log x}$$

3.1.3. Evaluate the following limits using results from this section. (You may assume that $\sin x$, $1 - \cos x$, $\tan x$, and $\sqrt[3]{x}$ converge to 0 as $x \to 0$.)

a)
$$\lim_{x \to 1} \frac{x^2 + 2x - 3}{x^3 - x}$$

b)
$$\lim_{x \to 1} \frac{x^n - 1}{x - 1}, \qquad n \in \mathbf{N}$$

c)
$$\lim_{x \to 1} \frac{\sqrt[3]{x^4 - 1}}{\cos(1 - x)}$$

d)
$$\lim_{x \to 0} \frac{2 \sin^2 x + 2x - 2x \cos^2 x}{1 - \cos^2(2x)}$$

e)
$$\lim_{x \to 0} \tan x \sin \left(\frac{1}{x^2} \right)$$

3.1.4. Prove Theorem 3.9.
3.1.5. Prove Theorem 3.10.
3.1.6. Suppose that f is a real function.

a) Prove that if
$$L = \lim_{x \to a} f(x)$$
exists, then $|f(x)| \to |L|$ as $x \to a$.
b) Show that there is a function such that, as $x \to a$, $|f(x)| \to |L|$ but the limit of $f(x)$ does not exist.

$\boxed{\textbf{3.1.7}}$. **This exercise is used in Sections 3.2 and 5.2.** For each real function f, define the *positive part* of f by

$$f^+(x) = \frac{|f(x)| + f(x)}{2}, \qquad x \in \text{Dom}\,(f)$$

and the *negative part* of f by

$$f^-(x) = \frac{|f(x)| - f(x)}{2}, \qquad x \in \text{Dom}\,(f).$$

a) Prove that $f^+(x) \geq 0$, $f^-(x) \geq 0$, $f(x) = f^+(x) - f^-(x)$, and $|f(x)| = f^+(x) + f^-(x)$ all hold for every $x \in \text{Dom}\,(f)$. (Compare with Exercise 1.2.3.)

b) Prove that if

$$L = \lim_{x \to a} f(x)$$

exists, then $f^+(x) \to L^+$ and $f^-(x) \to L^-$ as $x \to a$.

$\boxed{3.1.8}$. **This exercise is used in Sections 3.2 and 5.2.** Let f, g be real functions and for each $x \in \text{Dom}\,(f) \cap \text{Dom}\,(g)$ define

$$(f \vee g)(x) := \max\{f(x), g(x)\} \quad \text{and} \quad (f \wedge g)(x) := \min\{f(x), g(x)\}.$$

a) Prove that

$$(f \vee g)(x) = \frac{(f+g)(x) + |(f-g)(x)|}{2}$$

and

$$(f \wedge g)(x) = \frac{(f+g)(x) - |(f-g)(x)|}{2}$$

for all $x \in \text{Dom}\,(f) \cap \text{Dom}\,(g)$.

b) Prove that if

$$L = \lim_{x \to a} f(x) \quad \text{and} \quad M = \lim_{x \to a} g(x)$$

exist, then $(f \vee g)(x) \to L \vee M$ and $(f \wedge g)(x) \to L \wedge M$ as $x \to a$.

3.1.9. Suppose that $a \in \mathbf{R}$ and I is an open interval which contains a. If $f : I \to \mathbf{R}$ satisfies $f(x) \to f(a)$, as $x \to a$, and if there exist numbers M and m such that $m < f(a) < M$, prove that there exist positive numbers ε and δ such that

$$m + \varepsilon < f(x) < M - \varepsilon$$

for all x's which satisfy $|x - a| < \delta$.

3.2 ONE-SIDED LIMITS AND LIMITS AT INFINITY

In the preceding section we defined the limit of a real function. In this section we expand that definition to handle more general situations.

What is the limit of $f(x) := \sqrt{x - 1}$ as $x \to 1$? A reasonable answer is that the limit is zero. This function, however, does not satisfy Definition 3.1 because it is not defined on an OPEN interval containing $a = 1$. Indeed, f is only defined for $x \geq 1$. To handle such situations, we introduce "one-sided" limits.

3.12 Definition.

Let $a \in \mathbf{R}$ and f be a real function.

i) $f(x)$ is said to *converge to L as x approaches a from the right* if and only if f is defined on some open interval I with left endpoint a and for every $\varepsilon > 0$ there is a $\delta > 0$ (which in general depends on ε, f, I, and a) such that

$$a + \delta \in I \quad \text{and} \quad a < x < a + \delta \quad \text{imply} \quad |f(x) - L| < \varepsilon. \tag{2}$$

In this case we call L the *right-hand limit* of f at a, and denote it by

$$f(a+) := L =: \lim_{x \to a+} f(x).$$

ii) $f(x)$ is said to *converge to L as x approaches a from the left* if and only if f is defined on some open interval I with right endpoint a and for every $\varepsilon > 0$ there is a $\delta > 0$ (which in general depends on ε, f, I, and a) such that $a - \delta \in I$ and $a - \delta < x < a$ imply $|f(x) - L| < \varepsilon$. In this case we call L the *left-hand limit* of f at a and denote it by

$$f(a-) := L =: \lim_{x \to a-} f(x).$$

It is easy to check that when two-sided limits are replaced with one-sided limits, all the limit theorems from the preceding section hold. We shall use them as the need arises without further comment.

Existence of a one-sided limit can be established by these limit theorems or by appealing directly to the definition.

3.13 EXAMPLES.

i) Prove that

$$f(x) = \begin{cases} x + 1 & x \geq 0 \\ x - 1 & x < 0 \end{cases}$$

has one-sided limits at $a = 0$ but $\lim_{x \to 0} f(x)$ does not exist.

ii) Prove that

$$\lim_{x \to 0+} \sqrt{x} = 0.$$

Proof. i) Let $\varepsilon > 0$ and set $\delta = \varepsilon$. If $0 < x < \delta$, then $|f(x) - 1| = |x| < \delta = \varepsilon$. Hence $\lim_{x \to 0+} f(x)$ exists and equals 1. Similarly, $\lim_{x \to 0-} f(x)$ exists and equals -1. However, $x_n = (-1)^n/n \to 0$ but $f(x_n) = (-1)^n(1 + 1/n)$ does not converge as $n \to \infty$. Hence by the Sequential Characterization of Limits, $\lim_{x \to 0} f(x)$ does not exist.

ii) Let $\varepsilon > 0$ and set $\delta = \varepsilon^2$. If $0 < x < \delta$, then $|f(x)| = \sqrt{x} < \sqrt{\delta} = \varepsilon$. ∎

Not every function has one-sided limits (see Example 3.7). Examples 3.13 show that even when a function has one-sided limits, it may not have a two-sided limit. The following result, however, shows that if both one-sided limits, at a point a, exist and are EQUAL, then the two-sided limit at a exists.

3.14 Theorem. *Let f be a real function. Then the limit*

$$\lim_{x \to a} f(x)$$

exists and equals L if and only if

$$L = \lim_{x \to a+} f(x) = \lim_{x \to a-} f(x). \tag{3}$$

Proof. If the limit L of $f(x)$ exists as $x \to a$, then given $\varepsilon > 0$ choose $\delta > 0$ such that $0 < |x - a| < \delta$ implies $|f(x) - L| < \varepsilon$. Since any x which satisfies $a < x < a + \delta$ or $a - \delta < x < a$ also satisfies $0 < |x - a| < \delta$, it is clear that both the left and right limits of $f(x)$ exist as $x \to a$ and satisfy (3).

Conversely, suppose that (3) holds. Then given $\varepsilon > 0$ there exists a $\delta_1 > 0$ (respectively, a $\delta_2 > 0$) such that $a < x < a + \delta_1$ (respectively, $a - \delta_2 < x < a$) implies $|f(x) - L| < \varepsilon$. Set $\delta = \min\{\delta_1, \delta_2\}$. Then $0 < |x - a| < \delta$ implies $a < x < a + \delta_1$ or $a - \delta_2 < x < a$ (depending on whether x is to the right or to the left of a). Hence (1) holds; that is, $f(x) \to L$ as $x \to a$. ∎

The definition of limits of real functions can be expanded to include extended real numbers.

3.15 Definition.

Let $a, L \in \mathbf{R}$ and let f be a real function.

i) $f(x)$ is said to *converge* to L as $x \to \infty$ if and only if there exists a $c > 0$ such that $(c, \infty) \subset \mathrm{Dom}(f)$ and given $\varepsilon > 0$ there is an $M \in \mathbf{R}$ such that $x > M$ implies $|f(x) - L| < \varepsilon$, in which case we shall write

$$\lim_{x \to \infty} f(x) = L \quad \text{or} \quad f(x) \to L \text{ as } x \to \infty.$$

Similarly, $f(x)$ is said to *converge* to L as $x \to -\infty$ if and only if there exists a $c > 0$ such that $(-\infty, -c) \subset \mathrm{Dom}(f)$ and given $\varepsilon > 0$ there is an $M \in \mathbf{R}$ such that $x < M$ implies $|f(x) - L| < \varepsilon$, in which case we shall write

$$\lim_{x \to \infty} f(x) = L \quad \text{or} \quad f(x) \to L \text{ as } x \to \infty.$$

ii) The function $f(x)$ is said to *converge* to ∞ as $x \to a$ if and only if there is an open interval I containing a such that $I \setminus \{a\} \subset \mathrm{Dom}(f)$ and given $M \in \mathbf{R}$ there is a $\delta > 0$ such that $0 < |x - a| < \delta$ implies $f(x) > M$, in which case we shall write

$$\lim_{x \to a} f(x) = \infty \quad \text{or} \quad f(x) \to \infty \text{ as } x \to a.$$

Similarly, $f(x)$ is said to *converge* to $-\infty$ as $x \to a$ if and only if there is an open interval I containing a such that $I \setminus \{a\} \subset \mathrm{Dom}(f)$ and given $M \in \mathbf{R}$ there is a $\delta > 0$ such that $0 < |x - a| < \delta$ implies $f(x) < M$, in which case we shall write

$$\lim_{x \to a} f(x) = -\infty \qquad \text{or} \qquad f(x) \to -\infty \text{ as } x \to a.$$

Obvious modifications of Definition 3.15, which we leave to the reader, can be made to define $f(x) \to \pm\infty$ as $x \to a+$ and $x \to a-$, and $f(x) \to \pm\infty$ as $x \to \pm\infty$.

3.16 EXAMPLES.

i) Prove that $1/x \to 0$ as $x \to \infty$.
ii) Prove that

$$\lim_{x \to 1-} f(x) := \lim_{x \to 1-} \frac{x + 2}{2x^2 - 3x + 1} = -\infty.$$

Proof. i) Given $\varepsilon > 0$, set $M = 1/\varepsilon$. If $x > M$, then $|1/x| = 1/x < 1/M = \varepsilon$. Thus $1/x \to 0$ as $x \to \infty$.

ii) Let $M \in \mathbf{R}$. We must show that $f(x) < M$ for x near but to the left of 1 (no matter how large and negative M is). Without loss of generality, assume that $M < 0$. As x converges to 1 from the left, $2x^2 - 3x + 1$ is negative and converges to 0. (Observe that $2x^2 - 3x + 1$ is a parabola opening upward with roots 1/2 and 1.) Therefore, choose $\delta \in (0, 1)$ such that $1 - \delta < x < 1$ implies $2/M < 2x^2 - 3x + 1 < 0$; that is, $-1/(2x^2 - 3x + 1) > -M/2 > 0$. Since $0 < x < 1$ also implies $2 < x + 2 < 3$, it follows that $-(x+2)/(2x^2 - 3x + 1) > -M$; that is,

$$f(x) = \frac{x + 2}{2x^2 - 3x + 1} < M$$

for all $1 - \delta < x < 1$.   ∎

In order to unify the presentation of one-sided, two-sided, and infinite limits, we introduce the following notation. Let a be an extended real number, and let I be a nondegenerate open interval which either contains a or has a as one of its endpoints. Suppose further that f is a real function defined on I except possibly at a. If a is finite and I contains a, then

$$\lim_{\substack{x \to a \\ x \in I}} f(x) \qquad\qquad (4)$$

will denote $\lim_{x \to a} f(x)$ (when it exists); if a is a finite left endpoint of I, then (4) will denote $\lim_{x \to a+} f(x)$ (when it exists); if a is a finite right endpoint of I, then (4) will denote $\lim_{x \to a-} f(x)$ (when it exists); if $a = \pm\infty$ is an endpoint of I, then (4) will denote $\lim_{x \to \pm\infty} f(x)$ (when each exists).

Using this notation, we can state a Sequential Characterization of Limits valid for two-sided, one-sided, and infinite limits.

3.17 Theorem. *Let a be an extended real number, and let I be a nondegenerate open interval which either contains a or has a as one of its endpoints. Suppose further that f is a real function defined on I except possibly at a. Then*

$$\lim_{\substack{x \to a \\ x \in I}} f(x)$$

exists and equals L if and only if $f(x_n) \to L$ for all sequences $x_n \in I$ which satisfy $x_n \neq a$ and $x_n \to a$ as $n \to \infty$.

Proof. Since we have already proved this for two-sided limits, we must show it for the remaining eight cases which notation (4) represents. Since the proofs are similar, we shall give the details for only one of these cases, namely the case when a belongs to I and $L = \infty$. Thus we must prove that $f(x) \to \infty$ as $x \to a$ if and only if $f(x_n) \to \infty$ for any sequence $x_n \in I$ which converges to a and satisfies $x_n \neq a$ for $n \in \mathbf{N}$.

Suppose first that $f(x) \to \infty$ as $x \to a$. If $x_n \in I$, $x_n \to a$ as $n \to \infty$, and $x_n \neq a$, then given $M \in \mathbf{R}$ there is a $\delta > 0$ such that $0 < |x - a| < \delta$ implies $f(x) > M$, and there is an $N \in \mathbf{N}$ such that $n \geq N$ implies $|x_n - a| < \delta$. Consequently, $n \geq N$ implies $f(x_n) > M$; that is, $f(x_n) \to \infty$ as $n \to \infty$ as required.

Conversely, suppose to the contrary that $f(x_n) \to \infty$ for any sequence $x_n \in I$ which converges to a and satisfies $x_n \neq a$ but $f(x)$ does NOT converge to ∞ as $x \to a$. By the definition of "convergence" to ∞ there are numbers $M_0 \in \mathbf{R}$ and $x_n \in I$ such that $|x_n - a| < 1/n$ and $f(x_n) \leq M_0$ for all $n \in \mathbf{N}$. The first condition implies $x_n \to a$ but the second condition implies that $f(x_n)$ does not converge to ∞ as $n \to \infty$. This contradiction proves Theorem 3.17 in the case $a \in I$ and $L = \infty$. ∎

Using Theorem 3.17, we can prove limit theorems that are function analogues of Theorem 2.15 and Corollary 2.16. We leave this to the reader and will use these results as the need arises.

These limit theorems can be used to evaluate infinite limits and limits at $\pm\infty$.

3.18 EXAMPLE.

Prove that

$$\lim_{x \to \infty} \frac{2x^2 - 1}{1 - x^2} = -2.$$

Proof. Since the limit of a product is the product of the limits, we have by Example 3.16i that $1/x^m \to 0$ as $x \to \infty$ for any $m \in \mathbf{N}$. Multiplying numerator and denominator of the expression above by $1/x^2$, we obtain

$$\lim_{x \to \infty} \frac{2x^2 - 1}{1 - x^2} = \lim_{x \to \infty} \frac{2 - 1/x^2}{-1 + 1/x^2} = \frac{\lim_{x \to \infty}(2 - 1/x^2)}{\lim_{x \to \infty}(-1 + 1/x^2)} = \frac{2}{-1} = -2. \quad ∎$$

EXERCISES

3.2.0. Decide which of the following statements are true and which are false. Prove the true ones and provide counterexamples for the false ones.

a) If $f(x) \to \infty$ as $x \to \infty$ and $g(x) > 0$, then $g(x)/f(x) \to 0$ as $x \to \infty$.

b) If $f(x) \to 0$ as $x \to a+$ and $g(x) \geq 1$ for all $x \in \mathbf{R}$, then $g(x)/f(x) \to \infty$ as $x \to a+$.

c) If $f(x) \to \infty$ as $x \to \infty$, then $\sin(x^2 + x + 1)/f(x) \to 0$ as $x \to \infty$.

d) If P and Q are polynomials such that the degree of P is less than or equal to the degree of Q (see Exercise 3.2.3), then there is an $L \in \mathbf{R}$ such that

$$\lim_{x \to \infty} \frac{P(x)}{Q(x)} = \lim_{x \to -\infty} \frac{P(x)}{Q(x)} = L.$$

3.2.1. For each of the following, use definitions (rather than limit theorems) to prove that the limit exists. Identify the limit in each case.

a)
$$\lim_{x \to 0-} \frac{\sqrt{x^2}}{x}$$

b)
$$\lim_{x \to \infty} \frac{\sin x}{x^2}$$

c)
$$\lim_{x \to -1+} \frac{1}{x^2 - 1}$$

d)
$$\lim_{x \to 1+} \frac{x - 3}{3 - x - 2x^2}$$

e)
$$\lim_{x \to -\infty} \frac{\cos(\tan x)}{x + 1}$$

3.2.2. Assuming that $e^x \to e^a$, $\sin x \to \sin a$, and $\cos x \to \cos a$ as $x \to a$ for any $a \in \mathbf{R}$, evaluate the following limits when they exist.

a)
$$\lim_{x \to 2-} \frac{x^3 - x^2 - 4}{x^2 - 4}$$

b)
$$\lim_{x \to \infty} \frac{5x^2 + 3x - 2}{3x^2 - 2x + 1}$$

c)
$$\lim_{x \to -\infty} e^{-1/x^2}$$

d)
$$\lim_{x \to 0+} \frac{e^{x^2+2x-1}}{\sin x}$$

e)
$$\lim_{x \to 0-} \frac{\sin(x + \pi/2)}{\sqrt[3]{\cos x} - 1}$$

f)
$$\lim_{x \to 0+} \frac{\sqrt{1 - \cos x}}{\sin x}$$

3.2.3 . **This exercise is used many places**. Recall that a *polynomial of degree n* is a function of the form

$$P(x) = a_n x^n + a_{n-1}x^{n-1} + \cdots + a_1 x + a_0,$$

where $a_j \in \mathbf{R}$ for $j = 0, 1, \ldots, n$ and $a_n \neq 0$.

a) Prove that $\lim_{x \to a} x^n = a^n$ for $n = 0, 1, \cdots$ and $a \in \mathbf{R}$.
b) Prove that if P is a polynomial, then

$$\lim_{x \to a} P(x) = P(a)$$

for every $a \in \mathbf{R}$.

3.2.4. Prove the following comparison theorems for real functions f and g, and $a \in \mathbf{R}$.

a) If $f(x) \geq g(x)$ and $g(x) \to \infty$ as $x \to a$, then $f(x) \to \infty$ as $x \to a$.
b) If $f(x) \leq g(x) \leq h(x)$ and

$$L := \lim_{x \to \infty} f(x) = \lim_{x \to \infty} h(x),$$

then $g(x) \to L$ as $x \to \infty$.

3.2.5. Prove the following special case of Theorem 3.17: Suppose that $a \in \mathbf{R}$ and $f : [a, \infty) \to \mathbf{R}$ for some $a \in \mathbf{R}$. Then $f(x) \to L$ as $x \to \infty$ if and only if $f(x_n) \to L$ for any sequence $x_n \in (a, \infty)$ which converges to ∞ as $n \to \infty$.

3.2.6. Suppose that $f : [0, 1] \to \mathbf{R}$ and $f(a) = \lim_{x \to a} f(x)$ for all $a \in [0, 1]$. Prove that $f(q) = 0$ for all $q \in \mathbf{Q} \cap [0, 1]$ if and only if $f(x) = 0$ for all $x \in [0, 1]$.

3.2.7. Suppose that P is a polynomial and that $P(a) > 0$ for a fixed $a \in \mathbf{R}$. Prove that $P(x)/(x - a) \to \infty$ as $x \to a+$, $P(x)/(x - a) \to -\infty$ as $x \to a-$, but

$$\lim_{x \to a} \frac{P(x)}{x - a}$$

does not exist.

3.2.8. [CAUCHY] Suppose that $f : \mathbf{N} \to \mathbf{R}$. If

$$\lim_{n \to \infty} f(n+1) - f(n) = L,$$

prove that $\lim_{n \to \infty} f(n)/n$ exists and equals L.

3.3 CONTINUITY

In elementary calculus, a function is called *continuous* at a if $a \in \text{Dom } f$ and $f(x) \to f(a)$ as $x \to a$. In particular, it is tacitly assumed that f is defined on BOTH sides of a. Here, we introduce a more general concept of continuity which includes functions, such as $\sqrt{x}$ at $a = 0$, which are defined on only one side of some point in their domain.

3.19 Definition.

Let E be nonempty subset of $\mathbf{R}$ and $f : E \to \mathbf{R}$.

i) f is said to be *continuous at a point* $a \in E$ if and only if given $\varepsilon > 0$ there is a $\delta > 0$ (which in general depends on ε, f, and a) such that

$$|x - a| < \delta \quad \text{and} \quad x \in E \quad \text{imply} \quad |f(x) - f(a)| < \varepsilon. \tag{5}$$

ii) f is said to be *continuous on E* (notation: $f : E \to \mathbf{R}$ is continuous) if and only if f is continuous at every $x \in E$.

The following result shows that if E is an open interval which contains a, then "f is continuous at $a \in E$" means "$f(x) \to f(a)$ as $x \to a$." Therefore, we shall abbreviate "f is continuous at $a \in E$" by "f is continuous at a" when E is an open interval.

3.20 Remark. *Let I be an open interval which contains a point a and $f : I \to \mathbf{R}$. Then f is continuous at $a \in I$ if and only if*

$$f(a) = \lim_{x \to a} f(x).$$

Proof. Suppose that $I = (c, d)$ and set $\delta_0 := \min\{|c - a|, |d - a|\}$. If $\delta < \delta_0$, then $|x - a| < \delta$ implies $x \in I$. Therefore, condition (5) is identical to (1) when $f(a) = L$, $E = I$, and $\delta < \delta_0$. It follows that f is continuous at $a \in I$ if and only if $f(x) \to f(a)$ as $x \to a$. ∎

By repeating the proof of Theorem 3.6, we can establish a sequential characterization of continuity which is valid on any nonempty set.

3.21 Theorem. *Suppose that E is a nonempty subset of $\mathbf{R}$, that $a \in E$, and that $f : E \to \mathbf{R}$. Then the following statements are equivalent:*

i) *f is continuous at $a \in E$.*

ii) *If x_n converges to a and $x_n \in E$, then $f(x_n) \to f(a)$ as $n \to \infty$.*

In particular, $\sqrt{x}$ is continuous on $I = [0, \infty)$ by Exercise 2.2.5.
By combining Theorem 3.21 with Theorem 2.12, we obtain the following result.

3.22 Theorem. *Let E be a nonempty subset of **R** and $f, g : E \to$ **R**. If f,g are continuous at a point $a \in E$ (respectively, continuous on the set E), then so are $f + g$, fg, and αf (for any $\alpha \in$ **R**). Moreover, f/g is continuous at $a \in E$ when $g(a) \neq 0$ (respectively, on E when $g(x) \neq 0$ for all $x \in E$).*

It follows from Exercises 3.1.6, 3.1.7, and 3.1.8 that if f,g are continuous at a point $a \in E$ or on a set E, then so are $|f|$, f^+, f^-, $f \vee g$, and $f \wedge g$. We also notice by Exercise 3.2.3 that every polynomial is continuous on **R**.

Many complicated functions can be broken into simpler pieces, using sums, products, quotients, and the following operation.

3.23 Definition.

Suppose that A and B are subsets of **R**, that $f : A \to$ **R** and $g : B \to$ **R**. If $f(A) \subseteq B$ for every $x \in A$, then the *composition* of g with f is the function $g \circ f : A \to$ **R** defined by

$$(g \circ f)(x) := g(f(x)), \qquad x \in A.$$

The following result contains information about when a limit sign and something else (in this case, the evaluation of a function) can be interchanged. We shall return to this theme many times, identifying conditions under which we can interchange any two of the following objects: limits, integrals, derivatives, infinite summations, and computation of a function (see especially Sections 7.1, 7.2, and 11.1, and the entry "interchange the order of" in the Index).

3.24 Theorem. *Suppose that A and B are subsets of **R**, that $f : A \to$ **R** and $g : B \to$ **R**, and that $f(x) \in B$ for every $x \in A$.*

i) *If $A := I \setminus \{a\}$, where I is a nondegenerate interval which either contains a or has a as one of its endpoints, if*

$$L := \lim_{\substack{x \to a \\ x \in I}} f(x)$$

exists and belongs to B, and if g is continuous at $L \in B$, then

$$\lim_{\substack{x \to a \\ x \in I}} (g \circ f)(x) = g\left(\lim_{\substack{x \to a \\ x \in I}} f(x) \right).$$

ii) *If f is continuous at $a \in A$ and g is continuous at $f(a) \in B$, then $g \circ f$ is continuous at $a \in A$.*

Proof. Suppose that $x_n \in I \setminus \{a\}$ and that $x_n \to a$ as $n \to \infty$. Since $f(A) \subseteq B$, $f(x_n) \in B$. Also, by the Sequential Characterization of Limits (Theorem 3.17), $f(x_n) \to L$ as $n \to \infty$. Since g is continuous at $L \in B$, it follows from Theorem 3.21 that $g \circ f(x_n) := g(f(x_n)) \to g(L)$ as $n \to \infty$. Hence by Theorem 3.17, $g \circ f(x) \to g(L)$ as $x \to a$ in I. This proves i). A similar proof establishes part ii). ∎

For many applications, it is important to be able to find the maximum or minimum of a given function. As a first step in this direction, we introduce the following concept.

3.25 Definition.

Let E be a nonempty subset of $\mathbf{R}$. A function $f : E \to \mathbf{R}$ is said to be *bounded* on E if and only if there is an $M \in \mathbf{R}$ such that $|f(x)| \le M$ for all $x \in E$, in which case we shall say that f is *dominated* by M on E.

Notice that whether a function f is bounded or not on a set E depends on E as well as on f. For example, $f(x) = 1/x$ is dominated by 1 on $[1, \infty)$ but unbounded on $(0,2)$. Again, the function $f(x) = x^2$ is dominated by 4 on $(-2, 2)$ but unbounded on $[0, \infty)$.

The following result, which will be used often, shows that a continuous function on a closed, bounded interval is always bounded.

3.26 Theorem.   [EXTREME VALUE THEOREM].
If I is a closed, bounded interval and $f : I \to \mathbf{R}$ is continuous on I, then f is bounded on I. Moreover, if

$$M = \sup_{x \in I} f(x) \quad and \quad m = \inf_{x \in I} f(x),$$

then there exist points $x_m, x_M \in I$ such that

$$f(x_M) = M \quad and \quad f(x_m) = m. \tag{6}$$

Proof. Suppose first that f is not bounded on I. Then there exist $x_n \in I$ such that

$$|f(x_n)| > n, \qquad n \in \mathbf{N}. \tag{7}$$

Since I is bounded, we know (by the Bolzano–Weierstrass Theorem) that $\{x_n\}$ has a convergent subsequence, say $x_{n_k} \to a$ as $k \to \infty$. Since I is closed, we also know (by the Comparison Theorem) that $a \in I$. In particular, $f(a) \in \mathbf{R}$. On the other hand, substituting n_k for n in (7) and taking the limit of this inequality as $k \to \infty$, we have $|f(a)| = \infty$, a contradiction. Hence, the function f is bounded on I.

We have proved that both M and m are finite real numbers. To show that there is an $x_M \in I$ such that $f(x_M) = M$, suppose to the contrary that $f(x) < M$ for all $x \in I$. Then the function

$$g(x) = \frac{1}{M - f(x)}$$

is continuous, hence bounded on I. In particular, there is a $C > 0$ such that $|g(x)| = g(x) \le C$. It follows that

$$f(x) \le M - \frac{1}{C} \tag{8}$$

for all $x \in I$. Taking the supremum of (8) over all $x \in I$, we obtain $M \le M - 1/C < M$, a contradiction. Hence, there is an $x_M \in I$ such that $f(x_M) = M$. A similar argument proves that there is an $x_m \in I$ such that $f(x_m) = m$. ∎

We shall sometimes refer to (6) by saying that the supremum and infimum of f are *attained* on I. We shall also call the value M (respectively, m) the *maximum* (respectively, the *minimum*) of f on I.

Neither of the hypotheses on the interval I in Theorem 3.26 can be relaxed.

3.27 Remark. *The Extreme Value Theorem is false if either "closed" or "bounded" is dropped from the hypotheses.*

Proof. The interval $(0,1)$ is bounded but not closed, and the function $f(x) = 1/x$ is continuous and unbounded on $(0,1)$. The interval $[0, \infty)$ is closed but not bounded, and the function $f(x) = x$ is continuous and unbounded on $[0, \infty)$. ∎

What more can be said about continuous functions? One useful conceptualization of functions which are continuous on an interval is that their graphs have no holes or jumps (see Theorem 3.29 below). Our proof of this fact is based on the following elementary observation.

3.28 Lemma.
Suppose that $a < b$ and that $f : [a, b) \to \mathbf{R}$. If f is continuous at a point $x_0 \in [a, b)$ and $f(x_0) > 0$, then there exist a positive number ε and a point $x_1 \in [a, b)$ such that $x_1 > x_0$ and $f(x) > \varepsilon$ for all $x \in [x_0, x_1]$.

STRATEGY: The idea behind the proof is simple. If $f(x_0) > 0$, then $f(x) > f(x_0)/2$ for x near x_0. Here are the details.

Proof. Let $\varepsilon = f(x_0)/2$. Since $x_0 < b$, it is easy to see that $\delta_0 := (b - x_0)/2$ is positive and that $a \le x < x_0 + \delta_0$ implies $x \in [a, b)$. Use Definition 3.19 to choose $0 < \delta < \delta_0$ such that $x \in [a, b)$ and $|x - x_0| < \delta$ imply $|f(x) - f(x_0)| < \varepsilon$.

Fix $x_1 \in (x_0, x_0 + \delta)$ and suppose that $x \in [x_0, x_1]$. By the choice of ε and δ, it is clear that

$$-\frac{f(x_0)}{2} < f(x) - f(x_0) < \frac{f(x_0)}{2}.$$

Solving the left-hand inequality for $f(x)$, we conclude that $f(x) > f(x_0)/2 = \varepsilon$, as promised. ∎

A real number y_0 is said to *lie between* two numbers c and d if and only if $c < y_0 < d$ or $d < y_0 < c$.

3.29 Theorem.   [INTERMEDIATE VALUE THEOREM].
Suppose that $a < b$ and that $f : [a, b] \to \mathbf{R}$ is continuous. If y_0 lies between $f(a)$ and $f(b)$, then there is an $x_0 \in (a, b)$ such that $f(x_0) = y_0$.

Proof. We may suppose that $f(a) < y_0 < f(b)$. Consider the set $E = \{x \in [a, b] : f(x) < y_0\}$ (see Figure 3.3). Since $a \in E$ and $E \subseteq [a, b]$, E is a nonempty, bounded subset of $\mathbf{R}$. Hence, by the Completeness Axiom, $x_0 := \sup E$ is a finite real number. It remains to prove that $x_0 \in (a, b)$ and $f(x_0) = y_0$.

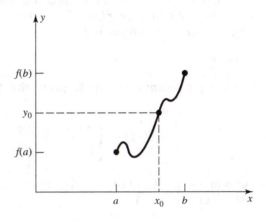

FIGURE 3.3

Choose by Theorem 2.11 a sequence $x_n \in E$ such that $x_n \to x_0$ as $n \to \infty$. Since $E \subseteq [a, b]$, it follows from Theorem 2.17 that $x_0 \in [a, b]$. Moreover, by the continuity of f and the definition of E, we have $f(x_0) = \lim_{n \to \infty} f(x_n) \le y_0$.

To show that $f(x_0) = y_0$, suppose to the contrary that $f(x_0) < y_0$. Then $y_0 - f(x)$ is a continuous function on the interval $[a, b)$ whose value at $x = x_0$ is positive. Hence, by Lemma 3.28, we can choose an ε and an $x_1 > x_0$ such that $y_0 - f(x_1) > \varepsilon > 0$. In particular, $x_1 \in E$ and $x_1 > \sup E$, a contradiction.

We have shown that $x_0 \in [a, b]$ and $y_0 = f(x_0)$. In view of our opening assumption, $f(a) < y_0 < f(b)$, it follows that x_0 cannot equal a or b. We conclude that $x_0 \in (a, b)$. ∎

Thus, if f is continuous on $[a, b]$ and $f(a) \le y_0 \le f(b)$, then there is an $x_0 \in [a, b]$ such that $f(x_0) = y_0$.

If f fails to be continuous at a point a, we say that f is *discontinuous* at a and call a a *point of discontinuity* of f. How badly can a function behave near a point of discontinuity? The following examples can be interpreted as answers to this question. (See also Exercise 9.6.9.)

3.30 *EXAMPLE.*

Prove that the function

$$f(x) = \begin{cases} \dfrac{|x|}{x} & x \neq 0 \\ 1 & x = 0 \end{cases}$$

is continuous on $(-\infty, 0)$ and $[0, \infty)$, discontinuous at 0, and that both $f(0+)$ and $f(0-)$ exist.

Proof. Since $f(x) = 1$ for $x \geq 0$, it is clear that $f(0+) = 1$ exists and $f(x) \to f(a)$ as $x \to a$ for any $a > 0$. In particular, f is continuous on $[0, \infty)$. Similarly, $f(0-) = -1$ and f is continuous on $(-\infty, 0)$. Finally, since $f(0+) \neq f(0-)$, the limit of $f(x)$ as $x \to 0$ does not exist by Theorem 3.14. Therefore, f is not continuous at 0. ∎

3.31 *EXAMPLE.*

Assuming that $\sin x$ is continuous on **R**, prove that the function

$$f(x) = \begin{cases} \sin \dfrac{1}{x} & x \neq 0 \\ 1 & x = 0 \end{cases}$$

is continuous on $(-\infty, 0)$ and $(0, \infty)$, discontinuous at 0, and neither $f(0+)$ nor $f(0-)$ exists. (See Figure 3.1.)

Proof. The function $1/x$ is continuous for $x \neq 0$ by Theorem 3.8. Hence, by Theorem 3.24, $f(x) = \sin(1/x)$ is continuous on $(-\infty, 0)$ and $(0, \infty)$. To prove that $f(0+)$ does not exist, let $x_n = 2/((2n + 1)\pi)$, and observe (see Appendix B) that $\sin(1/x_n) = (-1)^n$, $n \in \mathbf{N}$. Since $x_n \downarrow 0$ but $(-1)^n$ does not converge, it follows from Theorem 3.21 (the Sequential Characterization of Continuity) that $f(0+)$ does not exist. A similar argument proves that $f(0-)$ does not exist. ∎

3.32 *EXAMPLE.*

The *Dirichlet function* is defined on **R** by

$$f(x) := \begin{cases} 1 & x \in \mathbf{Q} \\ 0 & x \notin \mathbf{Q}. \end{cases}$$

Prove that every point $x \in \mathbf{R}$ is a point of discontinuity of f. (Such functions are called *nowhere continuous*.)

Proof. By Theorem 1.18 and Exercise 1.3.3 (Density of Rationals and Irrationals), given any $a \in \mathbf{R}$ and $\delta > 0$ we can choose $x_1 \in \mathbf{Q}$ and $x_2 \in \mathbf{R} \setminus \mathbf{Q}$ such that $|x_i - a| < \delta$ for $i = 1, 2$. Since $f(x_1) = 1$ and $f(x_2) = 0$, f cannot be continuous at a. ∎

3.33 EXAMPLE.

Prove that the function

$$f(x) = \begin{cases} \dfrac{1}{q} & x = \dfrac{p}{q} \in \mathbf{Q} \quad \text{(in reduced form)} \\ 0 & x \notin \mathbf{Q} \end{cases}$$

is continuous at every irrational in the interval $(0,1)$ but discontinuous at every rational in $(0,1)$.

Proof. Let a be a rational in $(0,1)$ and suppose that f is continuous at a. If x_n is a sequence of irrationals which converges to a, then $f(x_n) \to f(a)$; that is, $f(a) = 0$. But $f(a) \neq 0$ by definition. Hence, f is discontinuous at every rational in $(0,1)$.

Let a be an irrational in $(0,1)$. We must show that $f(x_n) \to f(a)$ for every sequence $x_n \in (0, 1)$ which satisfies $x_n \to a$ as $n \to \infty$. We may suppose that $x_n \in \mathbf{Q}$. For each $n \in \mathbf{N}$, write $x_n = p_n/q_n$ in reduced form. Since $f(a) = 0$, it suffices to show that $q_n \to \infty$ as $n \to \infty$. Suppose to the contrary that there exist integers $n_1 < n_2 < \ldots$ such that $|q_{n_k}| \leq M < \infty$ for $k \in \mathbf{N}$. Since $x_{n_k} \in (0, 1)$, it follows that the set

$$E := \left\{ x_{n_k} = \frac{p_{n_k}}{q_{n_k}} : k \in \mathbf{N} \right\}$$

contains only a finite number of points. Hence, the limit of any sequence in E must belong to E, a contradiction since a is such a limit and is irrational. ∎

To see how counterintuitive Example 3.33 is, try to draw a graph of $y = f(x)$. Stranger things can happen.

3.34 Remark. *The composition of two functions $g \circ f$ can be nowhere continuous, even though f is discontinuous only on $\mathbf{Q}$ and g is discontinuous at only one point.*

Proof. Let f be the function given in Example 3.33 and set

$$g(x) = \begin{cases} 1 & x \neq 0 \\ 0 & x = 0. \end{cases}$$

Clearly,

$$(g \circ f)(x) = \begin{cases} 1 & x \in \mathbf{Q} \\ 0 & x \notin \mathbf{Q}. \end{cases}$$

Hence, $g \circ f$ is the Dirichlet function, nowhere continuous by Example 3.32. ∎

In view of Example 3.33 and Remark 3.34, we must be skeptical of proofs which rely exclusively on geometric intuition. And although we shall use geometric intuition to suggest methods of proof for many results in subsequent chapters, these suggestions will always be followed by a careful rigorous proof which contains no fuzzy reasoning based on pictures or sketches no matter how plausible they seem.

EXERCISES

For these exercises, assume that $\sin x$, $\cos x$, *and* e^x *are continuous on* **R**.

3.3.0. Decide which of the following statements are true and which are false. Prove the true ones and provide counterexamples for the false ones.

a) If f is continuous on $[a,b]$ and $J := f([a, b])$, then J is a closed, bounded interval.

b) If f and g are continuous on $[a,b]$, if $f(a) < g(a)$ and $f(b) > g(b)$, then there is a $c \in [a, b]$ such that $f(c) = g(c)$.

c) Suppose that f and g are defined and finite valued on an open interval I which contains a, that f is continuous at a, and that $f(a) \neq 0$. Then g is continuous at a if and only if fg is continuous at a.

d) Suppose that f and g are defined and finite valued on **R**. If f and $g \circ f$ are continuous on **R**, then g is continuous on **R**.

3.3.1. Use limit theorems to show that the following functions are continuous on $[0,1]$.

a)
$$f(x) = \frac{e^{x^2}\sqrt{\sin x}}{\cos x}$$

b)
$$f(x) = \begin{cases} \dfrac{x^2 + x - 2}{x - 1} & x \neq 1 \\ 3 & x = 1 \end{cases}$$

c)
$$f(x) = \begin{cases} e^{-1/x} & x \neq 0 \\ 0 & x = 0 \end{cases}$$

d)
$$f(x) = \begin{cases} \sqrt{x}\sin\dfrac{1}{x} & x \neq 0 \\ 0 & x = 0 \end{cases}$$

3.3.2. For each of the following, prove that there is at least one $x \in \mathbf{R}$ which satisfies the given equation.

a) $e^x = x^3$
b) $e^x = 2\cos x + 1$
c) $2^x = 2 - 3x$

3.3.3. If $f : [a, b] \to \mathbf{R}$ is continuous, prove that $\sup_{x \in [a,b]} |f(x)|$ is finite.

3.3.4. If $f : [a, b] \to [a, b]$ is continuous, then f has a *fixed point*; that is, there is a $c \in [a, b]$ such that $f(c) = c$.

3.3.5. If f is a real function which is continuous at $a \in \mathbf{R}$ and if $f(a) < M$ for some $M \in \mathbf{R}$, prove that there is an open interval I containing a such that $f(x) < M$ for all $x \in I$.

3.3.6. Show that there exist nowhere continuous functions f and g whose sum $f + g$ is continuous on $\mathbf{R}$. Show that the same is true for the product of functions.

3.3.7. Suppose that $a \in \mathbf{R}$, that I is an open interval containing a, that $f, g : I \to \mathbf{R}$, and that f is continuous at a. Prove that g is continuous at a if and only if $f + g$ is continuous at a.

3.3.8. Suppose that $f : \mathbf{R} \to \mathbf{R}$ satisfies $f(x + y) = f(x) + f(y)$ for each $x, y \in \mathbf{R}$.

a) Show that $f(nx) = nf(x)$ for all $x \in \mathbf{R}$ and $n \in \mathbf{Z}$.
b) Prove that $f(qx) = qf(x)$ for all $x \in \mathbf{R}$ and $q \in \mathbf{Q}$.
c) Prove that f is continuous at 0 if and only if f is continuous on $\mathbf{R}$.
d) Prove that if f is continuous at 0, then there is an $m \in \mathbf{R}$ such that $f(x) = mx$ for all $x \in \mathbf{R}$.

3.3.9. **This exercise is used in Section 7.4.** Suppose that $f : \mathbf{R} \to (0, \infty)$ satisfies $f(x + y) = f(x)f(y)$. Modifying the outline in Exercise 3.3.8, show that if f is continuous at 0, then there is an $a \in (0, \infty)$ such that $f(x) = a^x$ for all $x \in \mathbf{R}$. (You may assume that the function a^x is continuous on $\mathbf{R}$.)

3.3.10. If $f : \mathbf{R} \to \mathbf{R}$ is continuous and

$$\lim_{x \to \infty} f(x) = \lim_{x \to -\infty} f(x) = \infty,$$

prove that f has a minimum on $\mathbf{R}$; that is, there is an $x_m \in \mathbf{R}$ such that

$$f(x_m) = \inf_{x \in \mathbf{R}} f(x) < \infty.$$

3.3.11. Let $a > 1$. Assume that $a^{p+q} = a^p a^q$ and $(a^p)^q = a^{pq}$ for all $p, q \in \mathbf{Q}$, and that $a^p < a^q$ for all $p, q \in \mathbf{Q}$ which satisfy $p < q$. (This is easy, but tedious, to prove using algebra, induction, and the definitions $a^0 = 1$, $a^{-n} = 1/a^n$, and $a^{m/n} = \sqrt[n]{a^m}$ for $n \in \mathbf{N}$ and $m \in \mathbf{Z}$. The hard part is proving that $\sqrt[n]{a^m}$ exists, and this requires the Completeness Axiom—see Appendix A.10.)

For each $x \in \mathbf{R}$, define

$$A(x) := \sup\{a^q : q \in \mathbf{Q} \text{ and } q \le x\}.$$

a) Prove that $A(x)$ exists and is finite for all $x \in \mathbf{R}$, and that $A(p) = a^p$ for all $p \in \mathbf{Q}$. Thus $a^x := A(x)$ extends the "power of a" function from $\mathbf{Q}$ to $\mathbf{R}$.

b) If $x, y \in \mathbf{R}$ with $x < y$, prove that $a^x < a^y$.

c) Use Example 2.21 to prove that the function a^x is continuous on **R**.

d) Prove that $a^{x+y} = a^x a^y$, $(a^x)^y = a^{xy}$, and $a^{-x} = 1/a^x$ for all x, $y \in \mathbf{R}$.

e) For $0 < b < 1$, define $b^x = (1/b)^{-x}$. Prove that c) and d) hold for b in place of a. State and prove an analogue of b) for b^x and b^y in place of a^x and a^y.

3.4   UNIFORM CONTINUITY

The following concept is very important and will be used many times in the rest of the book.

3.35 Definition.

Let E be a nonempty subset of **R** and $f : E \to \mathbf{R}$. Then f is said to be *uniformly continuous* on E (notation: $f : E \to \mathbf{R}$ is uniformly continuous) if and only if for every $\varepsilon > 0$ there is a $\delta > 0$ such that

$$|x - a| < \delta \quad \text{and} \quad x, a \in E \quad \text{imply} \quad |f(x) - f(a)| < \varepsilon. \tag{9}$$

Notice that the δ in Definition 3.35 depends on ε and f, but not on a and x. This issue needs to be addressed when we prove that a given function is uniformly continuous on a specific set (e.g., by determining δ before a is mentioned).

3.36 EXAMPLE.

Prove that $f(x) = x^2$ is uniformly continuous on the interval $(0,1)$.

Proof. Given $\varepsilon > 0$, set $\delta = \varepsilon/2$. If $x, a \in (0, 1)$, then $|x + a| \leq |x| + |a| \leq 2$. Therefore, if $x, a \in (0, 1)$ and $|x - a| < \delta$, then

$$|f(x) - f(a)| = |x^2 - a^2| = |x - a| \, |x + a| \leq 2|x - a| < 2\delta = \varepsilon. \qquad \blacksquare$$

The definitions of continuity and uniform continuity are very similar. In fact, the only difference is that for a continuous function, the parameter δ may depend on a, whereas for a uniformly continuous function, δ must be chosen independently of a. In particular, every function uniformly continuous on E is also continuous on E. The following example shows that the converse of this statement is false unless some restriction is made on E.

3.37 EXAMPLE.

Show that $f(x) = x^2$ is not uniformly continuous on **R**.

Proof. Suppose to the contrary that f is uniformly continuous on **R**. Then there is a $\delta > 0$ such that $|x - a| < \delta$ implies $|f(x) - f(a)| < 1$ for all $x, a \in \mathbf{R}$.

By the Archimedean Principle, choose $n \in \mathbf{N}$ so large that $n\delta > 1$. Set $a = n$ and $x = n + \delta/2$. Then $|x - a| < \delta$ and

$$1 > |f(x) - f(a)| = |x^2 - a^2| = n\delta + \frac{\delta^2}{4} > n\delta > 1.$$

This contradiction proves that f is not uniformly continuous on $\mathbf{R}$.   ∎

Here is a key which unlocks the difference between continuity and uniform continuity.

3.38 Lemma.
Suppose that $E \subseteq \mathbf{R}$ and that $f : E \to \mathbf{R}$ is uniformly continuous. If $x_n \in E$ is Cauchy, then $f(x_n)$ is Cauchy.

Proof. Let $\varepsilon > 0$ and choose $\delta > 0$ such that (9) holds. Since $\{x_n\}$ is Cauchy, choose $N \in \mathbf{N}$ such that $n, m \geq N$ implies $|x_n - x_m| < \delta$. Then $n, m \geq N$ implies $|f(x_n) - f(x_m)| < \varepsilon$.   ∎

Notice that $f(x) = 1/x$ is continuous on $(0,1)$ and $x_n = 1/n$ is Cauchy but $f(x_n)$ is not. In particular, $1/x$ is continuous but not uniformly continuous on the open interval $(0,1)$. Notice how the graph of $y = 1/x$ corroborates this fact. Indeed, as a gets closer to 0, the value of δ gets smaller (compare δ_1 to δ_0 in Figure 3.4) and hence cannot be chosen independently of a.

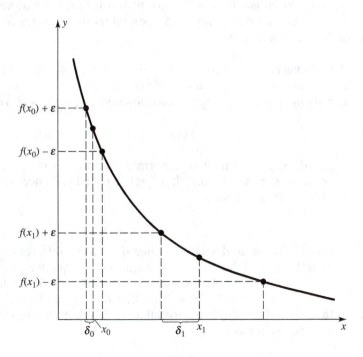

FIGURE 3.4

Thus on an open interval, continuity and uniform continuity are different, even if the interval is bounded. The following result shows that this is not the case for closed, bounded intervals. (This result is extremely important because uniform continuity is so strong. Indeed, we shall use it dozens of times before this book is finished.)

3.39 Theorem. *Suppose that I is a closed, bounded interval. If $f : I \to$ **R** is continuous on I, then f is uniformly continuous on I.*

Proof. Suppose to the contrary that f is continuous but not uniformly continuous on I. Then there is an $\varepsilon_0 > 0$ and points $x_n, y_n \in I$ such that $|x_n - y_n| < 1/n$ and

$$|f(x_n) - f(y_n)| \geq \varepsilon_0, \qquad n \in \mathbf{N}. \tag{10}$$

By the Bolzano–Weierstrass Theorem and the Comparison Theorem, the sequence $\{x_n\}$ has a subsequence, say x_{n_k}, which converges, as $k \to \infty$, to some $x \in I$. Similarly, the sequence $\{y_{n_k}\}_{k \in \mathbf{N}}$ has a convergent subsequence, say $y_{n_{k_j}}$, which converges, as $j \to \infty$, to some $y \in I$. Since $x_{n_{k_j}} \to x$ as $j \to \infty$ and f is continuous, it follows from (10) that $|f(x) - f(y)| \geq \varepsilon_0$; that is, $f(x) \neq f(y)$. But $|x_n - y_n| < 1/n$ for all $n \in \mathbf{N}$ so Theorem 2.9 (the Squeeze Theorem) implies $x = y$. Therefore, $f(x) = f(y)$, a contradiction. ∎

Our first application of this result is a useful but simple characterization of uniform continuity on bounded open intervals. (This result does NOT work for unbounded intervals.)

3.40 Theorem. *Suppose that $a < b$ and that $f : (a, b) \to$ **R**. Then f is uniformly continuous on (a, b) if and only if f can be continuously extended to $[a, b]$; that is, if and only if there is a continuous function $g : [a, b] \to$ **R** which satisfies*

$$f(x) = g(x), \qquad x \in (a, b). \tag{11}$$

Proof. Suppose that f is uniformly continuous on (a, b). Let $x_n \in (a, b)$ converge to b as $n \to \infty$. Then $\{x_n\}$ is Cauchy; hence, by Lemma 3.38, so is $\{f(x_n)\}$. In particular,

$$g(b) := \lim_{n \to \infty} f(x_n)$$

exists. This value does not change if we use a different sequence to approximate b. Indeed, let $y_n \in (a, b)$ be another sequence which converges to b as $n \to \infty$. Given $\varepsilon > 0$, choose $\delta > 0$ such that (9) holds for $E = (a, b)$. Since $x_n - y_n \to 0$, choose $N \in \mathbf{N}$ so that $n \geq N$ implies $|x_n - y_n| < \delta$. By (9), then, $|f(x_n) - f(y_n)| < \varepsilon$ for all $n \geq N$. Taking the limit of this inequality as $n \to \infty$, we obtain

$$\left| \lim_{n \to \infty} f(x_n) - \lim_{n \to \infty} f(y_n) \right| \leq \varepsilon$$

for all $\varepsilon > 0$. It follows from Theorem 1.9 that

$$\lim_{n \to \infty} f(x_n) = \lim_{n \to \infty} f(y_n).$$

Thus, $g(b)$ is well defined. A similar argument defines $g(a)$.

Set $g(x) = f(x)$ for $x \in (a, b)$. Then g is defined on $[a, b]$, satisfies (11), and is continuous on $[a, b]$ by the Sequential Characterization of Limits. Thus, f can be "continuously extended" to g as required.

Conversely, suppose that there is a function g continuous on $[a, b]$ which satisfies (11). By Theorem 3.39, g is uniformly continuous on $[a, b]$; hence, g is uniformly continuous on (a, b). We conclude that f is uniformly continuous on (a, b).   ∎

Let f be continuous on a bounded, open, nondegenerate interval (a, b). Notice that f is continuously extendable to $[a, b]$ if and only if the one-sided limits of f exist at a and b. Indeed, when they exist, we can always define g at a and b to be the values of these limits. In particular, we can prove that f is uniformly continuous without using ε's and δ's.

3.41 EXAMPLE.

Prove that $f(x) = (x - 1)/\log x$ is uniformly continuous on $(0,1)$.

Proof. It is clear that $f(x) \to 0$ as $x \to 0+$. Moreover, by l'Hôpital's Rule (see Theorem 4.27),

$$\lim_{x \to 1-} f(x) = \lim_{x \to 1-} \frac{1}{1/x} = 1.$$

Hence f is continuously extendable to $[0,1]$, so by Theorem 3.40, f is uniformly continuous on $(0,1)$.   ∎

EXERCISES

3.4.0. Decide which of the following statements are true and which are false. Prove the true ones and provide counterexamples for the false ones.

a) If f is uniformly continuous on $(0, \infty)$ and g is positive and bounded on $(0, \infty)$, then fg is uniformly continuous on $(0, \infty)$.

b) The function $x \log(1/x)$ is uniformly continuous on $(0,1)$.

c) The function

$$\frac{\cos x}{mx + b}$$

is uniformly continuous on $(0,1)$ for all nonzero $m, b \in \mathbf{R}$.

d) If f, g are uniformly continuous on an interval $[a, b]$ and $g(x) \neq 0$ for $x \in [a, b]$, then f/g is uniformly continuous on $[a, b]$.

3.4.1. Using Definition 3.35, prove that each of the following functions is uniformly continuous on $(0,1)$.

a) $f(x) = x^2 + x$
b) $f(x) = x^3 - x + 2$
c) $f(x) = x \sin 2x$

3.4.2. Prove that each of the following functions is uniformly continuous on $(0,1)$. (You may use l'Hôpital's Rule and assume that $\sin x$ and $\log x$ are continuous on their domains.)

a)
$$f(x) = \frac{\sin x}{x}$$

b)
$$f(x) = x \cos \frac{1}{x^2}$$

c)
$$f(x) = x \log x$$

d)
$$f(x) = (1 - x^2)^{1/x}$$

3.4.3. Assuming that $\sin x$ is continuous on **R**, find all real α such that $x^\alpha \sin(1/x)$ is uniformly continuous on the open interval $(0,1)$.

3.4.4. a) Suppose that $f : [0, \infty) \to$ **R** is continuous and that there is an $L \in$ **R** such that $f(x) \to L$ as $x \to \infty$. Prove that f is uniformly continuous on $[0, \infty)$.

b) Prove that $f(x) = 1/(x^2 + 1)$ is uniformly continuous on **R**.

3.4.5. Suppose that $\alpha \in$ **R**, that E is a nonempty subset of **R**, and that $f, g : E \to$ **R** are uniformly continuous on E.

a) Prove that $f + g$ and αf are uniformly continuous on E.

b) Suppose that f,g are bounded on E. Prove that fg is uniformly continuous on E.

c) Show that there exist functions f,g uniformly continuous on **R** such that fg is not uniformly continuous on **R**.

d) Suppose that f is bounded on E and that there is a positive constant ε_0 such that $g(x) \geq \varepsilon_0$ for all $x \in E$. Prove that f/g is uniformly continuous on E.

e) Show that there exist functions f,g, uniformly continuous on the interval $(0,1)$, with $g(x) > 0$ for all $x \in (0, 1)$, such that f/g is not uniformly continuous on $(0,1)$.

3.4.6. a) Let I be a bounded interval. Prove that if $f : I \to$ **R** is uniformly continuous on I, then f is bounded on I.

b) Prove that a) may be false if I is unbounded or if f is merely continuous.

3.4.7. Suppose that f is continuous on $[a, b]$. Prove that given $\varepsilon > 0$ there exist points $x_0 = a < x_1 < \cdots < x_n = b$ such that if $E_k := \{y : f(x) = y$ for some $x \in [x_{k-1}, x_k]\}$, then $\sup E_k - \inf E_k < \varepsilon$ for $k = 1, 2, \ldots, n$.

3.4.8. Let $E \subseteq \mathbf{R}$. A function $f : E \to \mathbf{R}$ is said to be increasing on E if and only if $x_1, x_2 \in E$ and $x_1 < x_2$ imply $f(x_1) \leq f(x_2)$. Suppose that f is increasing and bounded on an open, bounded, nonempty interval (a, b).

a) Prove that $f(a+)$ and $f(b-)$ both exist and are finite.

b) Prove that f is continuous on (a, b) if and only if f is uniformly continuous on (a, b).

c) Show that b) is false if f is unbounded. Indeed, find an increasing function $g : (0, 1) \to \mathbf{R}$ which is continuous on $(0,1)$ but not uniformly continuous on $(0,1)$.

3.4.9. Prove that a polynomial of degree n is uniformly continuous on $\mathbf{R}$ if and only if $n = 0$ or 1.

CHAPTER 4

Differentiability on R

4.1 THE DERIVATIVE

For many applications, we need to compute the slope of a tangent line to a curve. The following concept is useful in this regard.

4.1 Definition.

A real function f is said to be *differentiable* at a point $a \in \mathbf{R}$ if and only if f is defined on some open interval I containing a and

$$f'(a) := \lim_{h \to 0} \frac{f(a+h) - f(a)}{h} \tag{1}$$

exists. In this case $f'(a)$ is called the *derivative* of f at a.

The assumption that f be defined on an open interval containing a is made so that the quotients in (1) are defined for all $h \neq 0$ sufficiently small.

You may recall that the graph of $y = f(x)$ has a non-vertical tangent line at the point $(a, f(a))$ if and only if f has a derivative at a, in which case the slope of that tangent line is $f'(a)$. To see why this connection makes sense, let us consider a geometric interpretation of (1). Suppose that f is differentiable at a. A *secant line* of the graph $y = f(x)$ is a line passing through at least two points on the graph, and a *chord* is a line segment which runs from one point on the graph to another. Let $x = a + h$, and observe that the slope of the chord passing through the points $(x, f(x))$ and $(a, f(a))$ is given by $(f(x) - f(a))/(x - a)$. Now, since $x = a + h$, (1) becomes

$$f'(a) = \lim_{x \to a} \frac{f(x) - f(a)}{x - a}.$$

Hence, as $x \to a$ the slopes of the chords through $(x, f(x))$ and $(a, f(a))$ approximate the slope of the tangent line of $y = f(x)$ at $x = a$ (see Figure 4.1), and in the limit, the slope of the tangent line to $y = f(x)$ at $x = a$ is precisely $f'(a)$. Thus, we shall say that the graph of $y = f(x)$ *has a unique tangent line* at a point $(a, f(a))$ if and only if $f'(a)$ exists.

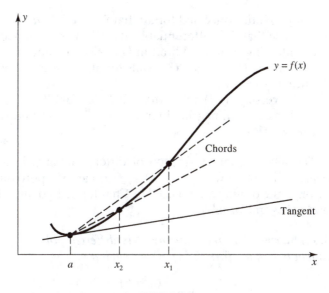

FIGURE 4.1

If f is differentiable at each point in a set E, then f' is a function on E. This function is denoted several ways:

$$D_x f = \frac{df}{dx} = f^{(1)} = f'.$$

When $y = f(x)$, we shall also use the notation dy/dx or y' for f'. Higher-order derivatives are defined recursively; that is, if $n \in \mathbf{N}$, then $f^{(n+1)}(a) := (f^{(n)})'(a)$, provided these derivatives exist. Higher-order derivatives are also denoted several ways, including $D_x^n f$, $d^n f/dx^n$, $f^{(n)}$, and by $d^n y/dx^n$ and $y^{(n)}$ when $y = f(x)$. The *second derivatives* $f^{(2)}$ (respectively, $y^{(2)}$) are usually written as f'' (respectively, y''), and when they exist at some point a, we shall say that f is *twice differentiable* at a.

Here are two characterizations of differentiability which we shall use to study derivatives. The first one, which characterizes the derivative in terms of the "chord function"

$$F(x) := \frac{f(x) - f(a)}{x - a} \qquad x \neq a, \tag{2}$$

will be used to establish the Chain Rule in Section 4.2.

4.2 Theorem. *A real function f is differentiable at some point $a \in \mathbf{R}$ if and only if there exist an open interval I and a function $F : I \to \mathbf{R}$ such that $a \in I$, f is defined on I, F is continuous at a, and*

$$f(x) = F(x)(x - a) + f(a) \tag{3}$$

holds for all $x \in I$, in which case $F(a) = f'(a)$.

Proof. Notice once and for all that for $x \in I \setminus \{a\}$, (2) and (3) are equivalent. Suppose that f is differentiable at a. Then f is defined on some open interval I containing a, and the limit in (1) exists. Define F on I by (2) if $x \neq a$, and by $F(a) := f'(a)$. Then (3) holds for all $x \in I$, and F is continuous at a by (2) since $f'(a)$ exists.

Conversely, if (3) holds, then (2) holds for all $x \in I$, $x \neq a$. Taking the limit of (2) as $x \to a$, bearing in mind that F is continuous at a, we conclude that $F(a) = f'(a)$. ∎

The second characterization of differentiability, in terms of *linear approximations* [i.e., how well $f(a + h) - f(a)$ can be approximated by a straight line through the origin] will be used in Chapter 11 to define the derivative of a function of several variables.

4.3 Theorem. *A real function f is differentiable at a if and only if there is a function T of the form $T(x) := mx$ such that*

$$\lim_{h \to 0} \frac{f(a + h) - f(a) - T(h)}{h} = 0. \tag{4}$$

Proof. Suppose that f is differentiable, and set $m := f'(a)$. Then by (1),

$$\frac{f(a + h) - f(a) - T(h)}{h} = \frac{f(a + h) - f(a)}{h} - f'(a) \to 0$$

as $h \to 0$.

Conversely, if (4) holds for $T(x) := mx$ and $h \neq 0$, then

$$\frac{f(a + h) - f(a)}{h} = m + \frac{f(a + h) - f(a) - mh}{h}$$

$$= m + \frac{f(a + h) - f(a) - T(h)}{h}.$$

By (4), the limit of this last expression is m. It follows that $(f(a + h) - f(a))/h \to m$, as $h \to 0$; that is, that $f'(a)$ exists and equals m. ∎

Our first application of Theorem 4.2 answers the question: Are differentiability and continuity related?

4.4 Theorem. *If f is differentiable at a, then f is continuous at a.*

Proof. Suppose that f is differentiable at a. By Theorem 4.2, there is an open interval I and a function F, continuous at a, such that $f(x) = f(a) + F(x)(x - a)$ for all $x \in I$. Taking the limit of this last expression as $x \to a$, we see that

$$\lim_{x \to a} f(x) = f(a) + F(a) \cdot 0 = f(a).$$

In particular, $f(x) \to f(a)$ as $x \to a$; that is, f is continuous at a. ∎

Thus any function which fails to be continuous at a cannot be differentiable at a. The following example shows that the converse of Theorem 4.4 is false.

4.5 EXAMPLE.

Show that $f(x) = |x|$ is continuous at 0 but not differentiable there.

Proof. Since $x \to 0$ implies $|x| \to 0$, f is continuous at 0. On the other hand, since $|h| = h$ when $h > 0$ and $|h| = -h$ when $h < 0$, we have

$$\lim_{h \to 0+} \frac{f(h) - f(0)}{h} = 1 \quad \text{and} \quad \lim_{h \to 0-} \frac{f(h) - f(0)}{h} = -1.$$

Since a limit exists if and only if its one-sided limits exist and are equal (Theorem 3.14), it follows that the limit in (1) does not exist when $a = 0$ and $f(x) = |x|$. Therefore, f is not differentiable at 0.   ∎

This example reflects the conventional wisdom about the difference between differentiable and continuous functions. Since a function differentiable at a always has a unique tangent line at $(a, f(a))$, the graph of a differentiable function on an interval is "smooth" with no corners, cusps, or kinks. On the contrary, although the graph of a continuous function on an interval is unbroken (has no holes or jumps), it may well have corners, cusps, or kinks. In particular, $f(x) = |x|$ is continuous but not differentiable at $x = 0$ and the graph of $y = |x|$ is unbroken but has a corner at the point $(0, 0)$ (see Figure 4.2).

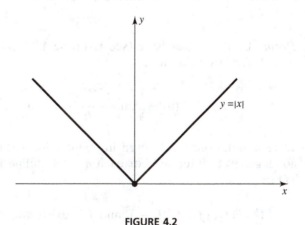

FIGURE 4.2

By Definition 4.1, if f is differentiable at a, then f must be defined on an open interval containing a (i.e., on both sides of a). As with the theory of limits, it is convenient to define "one-sided" derivatives to deal with functions whose domains are closed intervals (see Example 4.7 below). Here is a brief discussion of what it means for a real function to be differentiable *on* an interval (as opposed to being differentiable at every point *in* an interval). This concept will be used in Sections 5.3, 5.6, and 11.1.

4.6 Definition.

Let I be a nondegenerate interval.
i) A function $f : I \to \mathbf{R}$ is said to be *differentiable* on I if and only if

$$f'_I(a) := \lim_{\substack{x \to a \\ x \in I}} \frac{f(x) - f(a)}{x - a}$$

exists and is finite for every $a \in I$.
ii) f is said to be *continuously differentiable* on I if and only if f'_I exists and is continuous on I.

Notice that when a is not an endpoint of I, $f'_I(a)$ is the same as $f'(a)$. Because of this, we usually drop the subscript on f'_I. In particular, if f is differentiable on $[a, b]$, then

$$f'(a) := \lim_{h \to 0+} \frac{f(a + h) - f(a)}{h} \quad \text{and} \quad f'(b) := \lim_{h \to 0-} \frac{f(b + h) - f(b)}{h}.$$

The following example shows that Definition 4.6 enlarges the collection of differentiable functions.

4.7 EXAMPLE.

The function $f(x) = x^{3/2}$ is differentiable on $[0, \infty)$ and $f'(x) = 3\sqrt{x}/2$ for all $x \in [0, \infty)$.

Proof. By the Power Rule (see Exercise 4.2.7), $f'(x) = 3\sqrt{x}/2$ for all $x \in (0, \infty)$. And by definition,

$$f'(0) = \lim_{h \to 0+} \frac{h^{3/2} - 0}{h} = \lim_{h \to 0+} \sqrt{h} = 0. \qquad \blacksquare$$

Here is notation widely used in conjunction with Definition 4.6. Let I be a nondegenerate interval. For each $n \in \mathbf{N}$, define the collection of functions $\mathcal{C}^n(I)$ by

$$\mathcal{C}^n(I) := \{f : f : I \to \mathbf{R} \text{ and } f^{(n)} \text{ exists and is continuous on } I\}.$$

We shall denote the collection of f which belong to $\mathcal{C}^n(I)$ for all $n \in \mathbf{N}$ by $\mathcal{C}^\infty(I)$. Notice that $\mathcal{C}^1(I)$ is precisely the collection of real functions which are continuously differentiable on I. When dealing with specific intervals, we shall drop the outer set of parentheses; for example, we shall write $\mathcal{C}^n[a, b]$ for $\mathcal{C}^n([a, b])$.

By modifying the proof of Theorem 4.4, we can show that if f is differentiable on I, then f is continuous on I. Thus, $\mathcal{C}^\infty(I) \subset \mathcal{C}^m(I) \subset \mathcal{C}^n(I)$ for all integers $m > n > 0$.

The following example shows that not every function which is differentiable on **R** belongs to $C^1(\mathbf{R})$.

4.8 EXAMPLE.

The function

$$f(x) = \begin{cases} x^2 \sin(1/x) & x \neq 0 \\ 0 & x = 0 \end{cases}$$

is differentiable on **R** but not continuously differentiable on any interval which contains the origin.

Proof. By definition,

$$f'(0) = \lim_{h \to 0} h \sin\left(\frac{1}{h}\right) = 0 \quad \text{and} \quad f'(x) = 2x \sin\left(\frac{1}{x}\right) - \cos\left(\frac{1}{x}\right)$$

for $x \neq 0$. Thus f is differentiable on **R** but $\lim_{x \to 0} f'(x)$ does not exist. In particular, f' is not continuous on any interval which contains the origin. ∎

It is important to notice that a function which is differentiable on two sets is not necessarily differentiable on their union.

4.9 Remark. $f(x) = |x|$ *is differentiable on* $[0, 1]$ *and on* $[-1, 0]$ *but not on* $[-1, 1]$.

Proof. Since $f(x) = x$ when $x > 0$ and $= -x$ when $x < 0$, it is clear that f is differentiable on $[-1, 0) \cup (0, 1]$ [with $f'(x) = 1$ for $x > 0$ and $f'(x) = -1$ for $x < 0$]. By Example 4.5, f is not differentiable at $x = 0$. However,

$$f'_{[0,1]}(0) = \lim_{h \to 0+} \frac{|h|}{h} = 1 \quad \text{and} \quad f'_{[-1,0]}(0) = \lim_{h \to 0-} \frac{|h|}{h} = -1.$$

Therefore, f is differentiable on $[0, 1]$ and on $[-1, 0]$. ∎

EXERCISES

4.1.0. Suppose that $f, g : [a, b] \to \mathbf{R}$. Decide which of the following statements are true and which are false. Prove the true ones and provide counterexamples for the false ones.

a) If $f = g^2$ and f is differentiable on $[a, b]$, then g is differentiable on (a, b).

b) If f is differentiable on $[a, b]$, then f is uniformly continuous on $[a, b]$.

c) If f is differentiable on (a, b) and $f(a) = f(b) = 0$, then f is uniformly continuous on $[a, b]$.

d) If f is differentiable on $(a, b]$ and $f(x)/(x - a) \to 1$ as $x \to a+$, then f is uniformly continuous on $(a, b]$.

4.1.1. For each of the following real functions, use Definition 4.1 directly to prove that $f'(a)$ exists.

a) $f(x) = x^2 + x$, $a \in \mathbf{R}$
b) $f(x) = \sqrt{x}$, $a > 0$
c) $f(x) = 1/x$, $a \neq 0$

$\boxed{\textbf{4.1.2}}$. **This exercise is used in Section 4.2.**

a) Prove that $(x^n)' = nx^{n-1}$ for every $n \in \mathbf{N}$ and every $x \in \mathbf{R}$.
b) Prove that $(x^n)' = nx^{n-1}$ for every $n \in -\mathbf{N} \cup \{0\}$ and every $x \in (0, \infty)$.

4.1.3. Suppose that

$$f_\alpha(x) = \begin{cases} |x|^\alpha \sin \dfrac{1}{x} & x \neq 0 \\ 0 & x = 0. \end{cases}$$

Show that $f_\alpha(x)$ is continuous at $x = 0$ when $\alpha > 0$ and differentiable at $x = 0$ when $\alpha > 1$. Graph these functions for $\alpha = 1$ and $\alpha = 2$ and give a geometric interpretation of your results.

4.1.4. Let I be an open interval which contains 0 and $f : I \to \mathbf{R}$. If there exists an $\alpha > 1$ such that $|f(x)| \leq |x|^\alpha$ for all $x \in I$, prove that f is differentiable at 0. What happens when $\alpha = 1$?

4.1.5. a) Find all points (a, b) on the curve C, given by $y = x + \sin x$, so that the tangent lines to C at (a, b) are parallel to the line $y = x + 15$.

b) Find all points (a, b) on the curve C, given by $y = 3x^2 + 2$, so that the tangent lines to C at (a, b) pass through the point $(-1, -7)$.

4.1.6. Define f on $\mathbf{R}$ by

$$f(x) := \begin{cases} x^3 & x \geq 0 \\ 0 & x < 0. \end{cases}$$

Find all $n \in \mathbf{N}$ such that $f^{(n)}$ exists on all of $\mathbf{R}$.

4.1.7. Suppose that $f : (0, \infty) \to \mathbf{R}$ satisfies $f(x) - f(y) = f(x/y)$ for all $x, y \in (0, \infty)$ and $f(1) = 0$.

a) Prove that f is continuous on $(0, \infty)$ if and only if f is continuous at 1.

b) Prove that f is differentiable on $(0, \infty)$ if and only if f is differentiable at 1.

c) Prove that if f is differentiable at 1, then $f'(x) = f'(1)/x$ for all $x \in (0, \infty)$.

[**Note**: If $f'(1) = 1$, then $f(x) = \log x$.]

4.1.8. Let I be an open interval, $f : I \to \mathbf{R}$, and $c \in I$. The function f is said to have a *local maximum* at c if and only if there is a $\delta > 0$ such that $f(c) \geq f(x)$ holds for all $|x - c| < \delta$.

a) If f has a local maximum at c, prove that

$$\frac{f(c + u) - f(c)}{u} \leq 0 \quad \text{and} \quad \frac{f(c + t) - f(c)}{t} \geq 0$$

for $u > 0$ and $t < 0$ sufficiently small.

b) If f is differentiable at c and has a local maximum at c, prove that $f'(c) = 0$.

c) Make and prove analogous statements for local minima.

d) Show by example that the converses of the statements in parts b) and c) are false. Namely, find an f such that $f'(0) = 0$ but f has neither a local maximum nor a local minimum at 0.

4.1.9. Suppose that $I = (-a, a)$ for some $a > 0$. A function $f : I \to \mathbf{R}$ is said to be *even* if and only if $f(-x) = f(x)$ for all $x \in I$, and said to be *odd* if and only if $f(-x) = -f(x)$ for all $x \in I$.

a) Prove that if f is odd and differentiable on I, then f' is even on I.

b) Prove that if f is even and differentiable on I, then f' is odd on I.

4.2 DIFFERENTIABILITY THEOREMS

In this section we prove several familiar results about derivatives.

4.10 Theorem. *Let f and g be real functions and $\alpha \in \mathbf{R}$. If f and g are differentiable at a, then $f + g$, αf, $f \cdot g$, and [when $g(a) \neq 0$] f/g are all differentiable at a. In fact,*

$$(f + g)'(a) = f'(a) + g'(a), \tag{5}$$

$$(\alpha f)'(a) = \alpha f'(a), \tag{6}$$

$$(f \cdot g)'(a) = g(a)f'(a) + f(a)g'(a), \tag{7}$$

$$\left(\frac{f}{g}\right)'(a) = \frac{g(a)f'(a) - f(a)g'(a)}{g^2(a)}. \tag{8}$$

Proof. The proofs of these rules are similar. We provide the details only for (7). By adding and subtracting $f(a)g(x)$ in the numerator of the left side of the following expression, we can write

$$\frac{f(x)g(x) - f(a)g(a)}{x - a} = g(x)\frac{f(x) - f(a)}{x - a} + f(a)\frac{g(x) - g(a)}{x - a}.$$

This last expression is a product of functions. Since g is continuous (see Theorem 4.4), it follows from Definition 4.1 and Theorem 3.8 that

$$\lim_{x \to a} \frac{f(x)g(x) - f(a)g(a)}{x - a} = g(a)f'(a) + f(a)g'(a). \qquad \blacksquare$$

Formula (5) is called the *Sum Rule*, (6) is sometimes called the *Homogeneous Rule*, (7) is called the *Product Rule*, and (8) is called the *Quotient Rule*.

Next, we show what the derivative does to a composition of two functions.

4.11 Theorem.   [CHAIN RULE].
Let f and g be real functions. If f is differentiable at a and g is differentiable at $f(a)$, then $g \circ f$ is differentiable at a with

$$(g \circ f)'(a) = g'(f(a))f'(a). \qquad (9)$$

Proof. By Theorem 4.2, there exist open intervals I and J, and functions $F : I \to \mathbf{R}$, continuous at a, and $G : J \to \mathbf{R}$, continuous at $f(a)$, such that $F(a) = f'(a)$, $G(f(a)) = g'(f(a))$,

$$f(x) = F(x)(x - a) + f(a), \qquad x \in I, \qquad (10)$$

and

$$g(y) = G(y)(y - f(a)) + g(f(a)), \qquad y \in J. \qquad (11)$$

Since f is continuous at a, we may assume (by making I smaller if necessary) that $f(x) \in J$ for all $x \in I$.

Fix $x \in I$. Apply (11) to $y = f(x)$ and (10) to x to write

$$(g \circ f)(x) = g(f(x)) = G(f(x))(f(x) - f(a)) + g(f(a))$$
$$= G(f(x))F(x)(x - a) + (g \circ f)(a).$$

Set $H(x) = G(f(x))F(x)$ for $x \in I$. Since F is continuous at a and G is continuous at $f(a)$, it is clear that H is continuous at a. Moreover,

$$H(a) = G(f(a))F(a) = g'(f(a))f'(a).$$

It follows from Theorem 4.2, therefore, that $(g \circ f)'(a) = g'(f(a))f'(a)$. $\blacksquare$

EXERCISES

4.2.0. Suppose that I is an open interval containing a, and that $f, g, h : I \to \mathbf{R}$. Decide which of the following statements are true and which are false. Prove the true ones and provide counterexamples for the false ones.

a) If $f, g,$ and h are differentiable at a, then

$$(fgh)'(a) = f'(a)g(a)h(a) + f(a)g'(a)h(a) + f(a)g(a)h'(a).$$

b) If f is twice differentiable at a and g is twice differentiable at $f(a)$, then

$$(g \circ f)''(a) = g'(f(a))f''(a) + g''(f(a))(f'(a))^2.$$

c) If the nth-order derivatives $f^{(n)}(a)$ and $g^{(n)}(a)$ exist, then

$$(f + g)^{(n)}(a) = f^{(n)}(a) + g^{(n)}(a).$$

d) If the nth-order derivatives $f^{(n)}(a)$ and $g^{(n)}(a)$ exist and are nonzero, then

$$\left(\frac{f}{g}\right)^{(n)}(a) = \frac{g(a)f^{(n)}(a) + (-1)^n f(a)g^{(n)}(a)}{g^{n+1}(a)}.$$

4.2.1. Suppose that f and g are differentiable at 2 and 3 with $f'(2) = a$, $f'(3) = b$, $g'(2) = c$, and $g'(3) = d$. If $f(2) = 1$, $f(3) = 2$, $g(2) = 3$, and $g(3) = 4$, evaluate each of the following derivatives.

a) $(fg)'(2)$
b) $(f/g)'(3)$
c) $(g \circ f)'(3)$
d) $(f \circ g)'(2)$

4.2.2. Suppose that f is differentiable at 2 and 4 with $f(2) = 2$, $f(4) = 3$, $f'(2) = \pi$, and $f'(4) = e$.

a) If $g(x) = xf(x^2)$, find the value of $g'(2)$.
b) If $g(x) = f^2(\sqrt{x})$, find the value of $g'(4)$.
c) If $g(x) = x/f(x^3)$, find the value of $g'(\sqrt[3]{2})$.

4.2.3. [POWER RULE] Assume that $(e^x)' = e^x$ for $x \in \mathbf{R}$ and $(\log x)' = 1/x$ for $x > 0$. Use $x^\alpha := e^{\alpha \log x}$ to prove that $(x^\alpha)' = \alpha x^{\alpha-1}$ for all $x > 0$ and all $\alpha \in \mathbf{R}$.

4.2.4. Using Exercise 4.1.2, prove that every polynomial belongs to $C^\infty(\mathbf{R})$.

4.2.5. Suppose that f is differentiable at a and $f(a) \neq 0$.

a) Show that for h sufficiently small, $f(a + h) \neq 0$.
b) [RECIPROCAL RULE] Using Definition 4.1 directly, prove that $1/f(x)$ is differentiable at $x = a$ and

$$\left(\frac{1}{f}\right)'(a) = -\frac{f'(a)}{f^2(a)}.$$

c) Use the Product Rule and the Reciprocal Rule to prove the Quotient Rule directly.

4.2.6. Suppose that $n \in \mathbf{N}$ and f, g are real functions of a real variable whose nth derivatives $f^{(n)}$, $g^{(n)}$ exist at a point a. Prove Leibniz's generalization of the Product Rule:

$$(fg)^{(n)}(a) = \sum_{k=0}^{n} \binom{n}{k} f^{(k)}(a) g^{(n-k)}(a).$$

$\boxed{\textbf{4.2.7}}$. **This exercise is used in Section 5.3.**

a) Prove that if $q = n/m$ for $n \in \mathbf{Z}$ and $m \in \mathbf{N}$, then

$$x^n - a^n = (x^q - a^q)(x^{q(m-1)} + x^{q(m-2)}q + \cdots + qa^{q(m-2)} + a^{q(m-1)})$$

for every $x, a \in (0, \infty)$.

b) [POWER RULE] Use Exercise 4.1.2 and part a) to prove that x^q is differentiable on $(0, \infty)$ for every $q \in \mathbf{Q}$ and that $(x^q)' = qx^{q-1}$.

4.2.8. Assuming that e^x is differentiable on **R**, prove that

$$f(x) = \begin{cases} \dfrac{x}{1 + e^{1/x}} & x \neq 0 \\ 0 & x = 0 \end{cases}$$

is differentiable on $[0, \infty)$. Is f differentiable at 0?

4.2.9. Using elementary geometry and the definition of $\sin x, \cos x$, we can show that for every $x, y \in \mathbf{R}$ (see Appendix B)

i) $|\sin x| \leq 1$, $|\cos x| \leq 1$, $\sin(0) = 0$, $\cos(0) = 1$,

ii) $\sin(-x) = -\sin x$, $\cos(-x) = \cos x$,

iii) $\sin^2 x + \cos^2 x = 1$, $\cos x = 1 - 2\sin^2\left(\dfrac{x}{2}\right)$,

iv) $\sin(x \pm y) = \sin x \cos y \pm \cos x \sin y$.

Moreover, if x is measured in radians, then

v) $\cos x = \sin\left(\dfrac{\pi}{2} - x\right)$, $\sin x = \cos\left(\dfrac{\pi}{2} - x\right)$,

and

vi) $0 < x \cos x < \sin x < x$, $0 < x \leq \dfrac{\pi}{2}$.

Using these properties, prove each of the following statements.

a) The functions $\sin x$ and $\cos x$ are continuous at 0.

b) The functions $\sin x$ and $\cos x$ are continuous on **R**.

c) The limits

$$\lim_{x \to 0} \frac{\sin x}{x} = 1 \quad \text{and} \quad \lim_{x \to 0} \frac{1 - \cos x}{x} = 0$$

exist.
d) The function $\sin x$ is differentiable on **R** with $(\sin x)' = \cos x$.
e) The functions $\cos x$ and $\tan x := \sin x / \cos x$ are differentiable on **R** with $(\cos x)' = -\sin x$ and $(\tan x)' = \sec^2 x$.

4.3 THE MEAN VALUE THEOREM

The Mean Value Theorem makes a precise statement about the relationship between the derivative of a function and the slope of one of its chords. It was discovered by the following geometric reasoning. Suppose that f is differentiable on (a, b). Since the graph of f on (a, b) has a tangent at each of its points, it seems likely that the slope of the chord through the points $(a, f(a))$ and $(b, f(b))$ equals the slope $f'(c)$ for some value of $c \in (a, b)$ (see Figure 4.3).

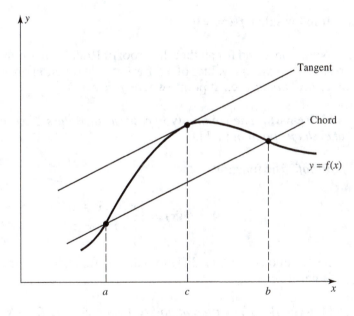

FIGURE 4.3

We begin with a special case.

4.12 Lemma. [ROLLE'S THEOREM].
Suppose that $a, b \in \mathbf{R}$ with $a < b$. If f is continuous on $[a, b]$, differentiable on (a, b), and if $f(a) = f(b)$, then $f'(c) = 0$ for some $c \in (a, b)$.

Proof. By the Extreme Value Theorem, f has a finite maximum M and a finite minimum m on $[a, b]$. If $M = m$, then f is constant on (a, b) and $f'(x) = 0$ for all $x \in (a, b)$.

Suppose that $M \neq m$. Since $f(a) = f(b)$, f must assume one of the values M or m at some point $c \in (a, b)$. By symmetry, we may suppose that $f(c) = M$. [That is, if we can prove the theorem when $f(c) = M$, then a similar proof establishes the theorem when $f(c) = m$.] Since M is the maximum of f on $[a, b]$, we have

$$f(c + h) - f(c) \leq 0$$

for all h which satisfy $c + h \in (a, b)$. In the case $h > 0$ this implies

$$f'(c) = \lim_{h \to 0+} \frac{f(c + h) - f(c)}{h} \leq 0,$$

and in the case $h < 0$ this implies

$$f'(c) = \lim_{h \to 0-} \frac{f(c + h) - f(c)}{h} \geq 0.$$

It follows that $f'(c) = 0$. ∎

Notice once and for all that the proof of Rolle's Theorem proves a well-known result: The extreme values of a differentiable function on an open interval occur at *critical points* (i.e., at points where f' is zero).

4.13 Remark. *The continuity hypothesis in Rolle's Theorem cannot be relaxed at even one point in $[a, b]$.*

Proof. The function

$$f(x) = \begin{cases} x & x \in [0, 1) \\ 0 & x = 1 \end{cases}$$

is continuous on $[0, 1)$, differentiable on $(0, 1)$, and $f(0) = f(1) = 0$, but $f'(x)$ is never zero. ∎

4.14 Remark. *The differentiability hypothesis in Rolle's Theorem cannot be relaxed at even one point in (a, b).*

Proof. The function $f(x) = |x|$ is continuous on $[-1, 1]$, differentiable on $(-1, 1) \setminus \{0\}$, and $f(-1) = f(1)$, but $f'(x)$ is never zero. ∎

We shall use Rolle's Theorem to obtain several useful results. The first is a pair of "Mean Value Theorems."

4.15 Theorem. *Suppose that $a, b \in \mathbf{R}$ with $a < b$.*

i) [GENERALIZED MEAN VALUE THEOREM] *If f, g are continuous on $[a, b]$ and differentiable on (a, b), then there is a $c \in (a, b)$ such that*

$$g'(c)(f(b) - f(a)) = f'(c)(g(b) - g(a)).$$

ii) [MEAN VALUE THEOREM] *If f is continuous on $[a, b]$ and differentiable on (a, b), then there is a $c \in (a, b)$ such that*

$$f(b) - f(a) = f'(c)(b - a).$$

Proof. i) Set $h(x) = f(x)(g(b) - g(a)) - g(x)(f(b) - f(a))$. Since $h'(x) = f'(x)(g(b) - g(a)) - g'(x)(f(b) - f(a))$, it is clear that h is continuous on $[a, b]$, differentiable on (a, b), and $h(a) = h(b)$. Thus, by Rolle's Theorem, $h'(c) = 0$ for some $c \in (a, b)$.

ii) Set $g(x) = x$ and apply part i). (For a geometric interpretation of this result, see the opening paragraph of this section and Figure 4.3.) ∎

The Generalized Mean Value Theorem is also called *Cauchy's Mean Value Theorem*. It is usually essential when comparing derivatives of two functions simultaneously, using higher-order derivatives to approximate functions, and studying certain kinds of generalized derivatives (e.g., see Taylor's Formula and l'Hôpital's Rule in the next section, and Remark 14.32).

The Mean Value Theorem is most often used to extract information about f from f' (see, e.g., Exercises 4.3.4, 4.3.5, and 4.3.9). Perhaps the best known result of this type is the criterion for deciding when a differentiable function increases. To prove this result, we begin with the following nomenclature.

4.16 Definition.

Let E be a nonempty subset of $\mathbf{R}$ and $f : E \to \mathbf{R}$.

i) f is said to be *increasing* (respectively, *strictly increasing*) on E if and only if $x_1, x_2 \in E$ and $x_1 < x_2$ imply $f(x_1) \leq f(x_2)$ [respectively, $f(x_1) < f(x_2)$].

ii) f is said to be *decreasing* (respectively, *strictly decreasing*) on E if and only if $x_1, x_2 \in E$ and $x_1 < x_2$ imply $f(x_1) \geq f(x_2)$ [respectively, $f(x_1) > f(x_2)$].

iii) f is said to be *monotone* (respectively, *strictly monotone*) on E if and only if f is either decreasing or increasing (respectively, either strictly decreasing or strictly increasing) on E.

Thus, although $f(x) = x^2$ is strictly monotone on $[0, 1]$, and on $[-1, 0]$, it is not monotone on $[-1, 1]$.

Monotone functions are important from both a theoretical and a practical point of view (e.g., see Theorem 5.34). Thus it will come as no surprise that the following result is very important and widely used.

4.17 Theorem. *Suppose that $a, b \in \mathbf{R}$, with $a < b$, that f is continuous on $[a, b]$, and that f is differentiable on (a, b).*

i) *If $f'(x) > 0$ [respectively, $f'(x) < 0$] for all $x \in (a, b)$, then f is strictly increasing (respectively, strictly decreasing) on $[a, b]$.*

ii) *If $f'(x) = 0$ for all $x \in (a, b)$, then f is constant on $[a, b]$.*

iii) *If g is continuous on $[a, b]$ and differentiable on (a, b), and if $f'(x) = g'(x)$ for all $x \in (a, b)$, then $f - g$ is constant on $[a, b]$.*

Proof. Let $a \leq x_1 < x_2 \leq b$. By the Mean Value Theorem, there is a $c \in (a, b)$ such that $f(x_2) - f(x_1) = f'(c)(x_2 - x_1)$. Thus, $f(x_2) > f(x_1)$ when $f'(c) > 0$ and $f(x_2) < f(x_1)$ when $f'(c) < 0$. This proves part i).

To prove part ii), notice that if $f' = 0$, then by the proof of part i), f is both increasing and decreasing, and hence constant on $[a, b]$. Finally, part iii) follows from part ii) applied to $f - g$. ∎

Theorem 4.17i is a great result. It makes checking a differentiable function for monotonicity a routine activity. However, there are many nondifferentiable functions which are monotone. For example, the *greatest integer function*,

$$f(x) = [x] := n, \qquad n \leq x < n + 1, \ n \in \mathbf{Z},$$

is increasing on **R** but not even continuous, much less differentiable.

How badly can these nondifferentiable, monotone functions behave? The following result shows that, just like the greatest integer function, any function which is monotone on an interval always has left and right limits (contrast with Examples 3.31 and 3.32). This is a function analogue of the Monotone Convergence Theorem.

4.18 Theorem. *Suppose that f is increasing on $[a, b]$.*

i) *If $c \in [a, b)$, then $f(c+)$ exists and $f(c) \leq f(c+)$.*

ii) *If $c \in (a, b]$, then $f(c-)$ exists and $f(c-) \leq f(c)$.*

Proof. By symmetry it suffices to show that $f(c-)$ exists and satisfies $f(c-) \leq f(c)$ for any fixed $c \in (a, b]$. Set $E = f((a, c))$ and $s = \sup E$. Since f is increasing, $f(c)$ is an upper bound of E. Hence, s is a finite real number which satisfies $s \leq f(c)$. Given $\varepsilon > 0$, choose by the Approximation Property an $x_0 \in (a, c)$ such that $s - \varepsilon < f(x_0) \leq s$. Since f is increasing,

$$s - \varepsilon < f(x_0) \leq f(x) \leq s$$

for all $x_0 < x < c$. Therefore, $f(c-)$ exists and satisfies $f(c-) = s \leq f(c)$. ∎

We have seen (Example 3.32) that a function can be nowhere continuous (i.e., can have uncountably many points of discontinuity). How many points of discontinuity can a monotone function have?

*4.19 Theorem. *If f is monotone on an interval I, then f has at most countably many points of discontinuity on I.*

Proof. Without loss of generality, we may suppose that f is increasing. Since the countable union of at most countable sets is at most countable (Theorem 1.42ii), it suffices to show that the set of points of discontinuity of f can be written as a countable union of at most countable sets. Since **R** is the union of closed intervals $[-n, n]$, $n \in \mathbf{N}$, we may suppose that I is a closed, bounded interval $[a, b]$.

Let E represent the set of points of discontinuity of f on (a, b). By Theorem 4.18, $f(x-) \leq f(x) \leq f(x+)$ for all $x \in (a, b)$. Thus, f is discontinuous at such an x if and only if $f(x+) - f(x-) > 0$. It follows that

$$E = \bigcup_{j=1}^{\infty} A_j,$$

where for each $j \in \mathbf{N}$, $A_j := \{x \in (a, b) : f(x+) - f(x-) \geq 1/j\}$. We will complete the proof by showing that each A_j is finite.

Suppose to the contrary that A_{j_0} is infinite for some j_0. Set $y_0 := j_0(f(b) - f(a))$ and observe that since f is finite valued on I, y_0 is a finite real number. On the other hand, since A_{j_0} is infinite, then by symmetry we may suppose that there exist $x_1 < x_2 < \ldots$ in $[a, b]$ such that $f(x_k+) - f(x_k-) \geq 1/j_0$ for $k \in \mathbf{N}$. Since f is monotone, it follows that

$$f(b) - f(a) \geq \sum_{k=1}^{n}(f(x_k+) - f(x_k-)) \geq \frac{n}{j_0};$$

that is, $y_0 = j_0(f(b) - f(a)) \geq n$ for all $n \in \mathbf{N}$. Taking the limit of this last inequality as $n \to \infty$, we see that $y_0 = +\infty$. With this contradiction, the proof of the theorem is complete. ∎

Theorem 4.17i can be used for less mundane tasks than finding intervals on which a given function is increasing. The following example shows how to use it to compare one function with another.

4.20 EXAMPLE.

Prove that $1 + x < e^x$ for all $x > 0$.

Proof. Let $f(x) = e^x - x$, and observe that $f'(x) = e^x - 1 > 0$ for all $x > 0$. It follows from Theorem 4.17i that $f(x)$ is strictly increasing on $(0, \infty)$. Thus $e^x - x = f(x) > f(0) = 1$ for $x > 0$. In particular, $e^x > x + 1$ for $x > 0$. ∎

We close this section with some optional results which further explore the mean value concept.

Our first result shows how $(1 + x)^\alpha$ is related to $1 + \alpha x$.

*4.21 Theorem.   [BERNOULLI'S INEQUALITY].
Let α be a positive real number. If $0 < \alpha \leq 1$, then $(1+x)^\alpha \leq 1 + \alpha x$ for all $x \in [-1, \infty)$, and if $\alpha \geq 1$, then $(1+x)^\alpha \geq 1 + \alpha x$ for all $x \in [-1, \infty)$.

Proof. The proofs of these inequalities are similar. We present the details only for the case $0 < \alpha \leq 1$. Fix $x \geq -1$ and let $f(t) = t^\alpha$, $t \in [0, \infty)$. Since $f'(t) = \alpha t^{\alpha-1}$, it follows from the Mean Value Theorem (applied to $a = 1$ and $b = 1 + x$) that

$$f(1+x) - f(1) = \alpha x c^{\alpha-1} \tag{12}$$

for some c between 1 and $1 + x$.

Case 1. $x > 0$. Then $c > 1$. Since $0 < \alpha \leq 1$ implies $\alpha - 1 \leq 0$, it follows that $c^{\alpha-1} \leq 1$, hence $xc^{\alpha-1} \leq x$. Therefore, we have by (12) that

$$(1+x)^\alpha = f(1+x) = f(1) + \alpha x c^{\alpha-1} \leq f(1) + \alpha x = 1 + \alpha x \tag{13}$$

as required.

Case 2. $-1 \leq x \leq 0$. Then $c \leq 1$ so $c^{\alpha-1} \geq 1$. But since $x \leq 0$, it follows that $xc^{\alpha-1} \leq x$ as before and we can repeat (13) to obtain the same conclusion. ∎

We will now use Bernoulli's Inequality to show once again that vague reasoning can produce wrong conclusions. To see why, assuming that

$$\lim_{n \to \infty} \left(1 + \frac{1}{n}\right)^n$$

exists, what do you think its limit is? Vague reasoning that well over half your class would agree with: Since $1 + 1/n$ gets near 1 and $1^\alpha = 1$ for all $\alpha \in \mathbf{R}$, the limit should be 1, right? Absolutely not.

*4.22 EXAMPLE.

Prove that the sequence $(1 + 1/n)^n$ is increasing, as $n \to \infty$, and its limit L satisfies $2 < L \leq 3$. (The limit L turns out to be an irrational number, the natural base $e = 2.718281828459 \cdots$.)

Proof. The sequence $(1 + 1/n)^n$ is increasing, since by Bernoulli's Inequality,

$$\left(1 + \frac{1}{n}\right)^{n/(n+1)} \leq \left(1 + \frac{1}{n+1}\right).$$

To prove that this sequence is bounded above, observe by the Binomial Formula that

$$\left(1 + \frac{1}{n}\right)^n = \sum_{k=0}^{n} \binom{n}{k} \left(\frac{1}{n}\right)^k.$$

Now,

$$\binom{n}{k}\left(\frac{1}{n}\right)^k = \frac{n(n-1)\ldots(n-k+1)}{n^k} \cdot \frac{1}{k!} \le \frac{1}{k!} \le \frac{1}{2^{k-1}}$$

for all $k \in \mathbf{N}$. It follows from Exercise 1.4.4c that

$$2 = \left(1 + \frac{1}{1}\right) < \left(1 + \frac{1}{n}\right)^n \le 1 + 1 + \sum_{k=1}^{n-1} \frac{1}{2^k} = 3 - \frac{1}{2^{n-1}} < 3$$

for $n > 1$. Hence, by the Monotone Convergence Theorem, the limit L exists and satisfies $2 < L \le 3$. ∎

The last result in this section shows that although a differentiable function might not be continuously differentiable, its derivative does satisfy an intermediate value theorem. (This result is sometimes called Darboux's Theorem.)

*4.23 Theorem. [INTERMEDIATE VALUE THEOREM FOR DERIVATIVES].
Suppose that f is differentiable on $[a, b]$ with $f'(a) \ne f'(b)$. If y_0 is a real number which lies between $f'(a)$ and $f'(b)$, then there is an $x_0 \in (a, b)$ such that $f'(x_0) = y_0$.

STRATEGY: Let $F(x) := f(x) - y_0 x$. We must find an $x_0 \in (a, b)$ such that $F'(x_0) := f'(x_0) - y_0 = 0$. Since local extrema of a differentiable function F occur only where the derivative of F is zero (e.g., see the proof of Rolle's Theorem), it suffices to show that F has a local extremum at some $x_0 \in (a, b)$.

Proof. Suppose that y_0 lies between $f'(a)$ and $f'(b)$. By symmetry, we may suppose that $f'(a) < y_0 < f'(b)$. Set $F(x) = f(x) - y_0 x$ for $x \in [a, b]$, and observe that F is differentiable on $[a, b]$. Hence, by the Extreme Value Theorem, F has an absolute minimum, say $F(x_0)$, on $[a, b]$. Now $F'(a) = f'(a) - y_0 < 0$, so $F(a + h) - F(a) < 0$ for $h > 0$ sufficiently small. Hence $F(a)$ is NOT the absolute minimum of F on $[a, b]$. Similarly, $F(b)$ is not the absolute minimum of F on $[a, b]$. Hence, the absolute minimum $F(x_0)$ must occur on (a, b); that is, $x_0 \in (a, b)$ and $F'(x_0) = 0$. ∎

EXERCISES

4.3.0. Suppose that $f, g : [a, b] \to \mathbf{R}$. Decide which of the following statements are true and which are false. Prove the true ones and provide counterexamples for the false ones.

a) If f and g are increasing on $[a, b]$, then $f + g$ is increasing on $[a, b]$.
b) If f and g are increasing on $[a, b]$, then fg is increasing on $[a, b]$.
c) If f is differentiable on (a, b) and $\lim_{x \to a+} f(x)$ exists and is finite, then for each $x \in (a, b)$ there is a c between a and x such that $f(x) - f(a+) = f'(c)(x - a)$.
d) If f and g are differentiable on $[a, b]$ and $|f'(x)| \le 1 \le |g'(x)|$ for all $x \in (a, b)$, then $|f(x) - f(a)| \le |g(x) - g(a)|$ for all $x \in [a, b]$.

4.3.1. Prove that each of the following inequalities holds.

    a) $2x + 0.7 < e^x$ for all $x \geq 1$.

    b) $\log x < \sqrt{x} - 0.6$ for all $x \geq 4$.

    c) $\sin^2 x \leq 2|x|$ for all $x \in \mathbf{R}$.

    d) $1 - \sin x \leq e^x$ for all $x \geq 0$.

4.3.2. Suppose that $I = (0, 2)$, that f is continuous at $x = 0$ and $x = 2$, and that f is differentiable on I. If $f(0) = 1$ and $f(2) = 3$, prove that $1 \in f'(I)$.

4.3.3. Let f be a real function and recall that an $r \in \mathbf{R}$ is called a *root* of a function f if and only if $f(r) = 0$. Show that if f is differentiable on $\mathbf{R}$, then its derivative f' has at least one root between any two roots of f.

4.3.4. Suppose that $a < b$ are extended real numbers and that f is differentiable on (a, b). If f' is bounded on (a, b), prove that f is uniformly continuous on (a, b).

4.3.5. Suppose that f is differentiable on $\mathbf{R}$. If $f(0) = 1$ and $|f'(x)| \leq 1$ for all $x \in \mathbf{R}$, prove that $|f(x)| \leq |x| + 1$ for all $x \in \mathbf{R}$.

4.3.6. Suppose that f is differentiable on (a, b), continuous on $[a, b]$, and that $f(a) = f(b) = 0$. Prove that if $f(c) \neq 0$ for some $c \in (a, b)$, then there exist $x_1, x_2 \in (a, b)$ such that $f'(x_1)$ is positive and $f'(x_2)$ is negative.

4.3.7. Suppose that f is continuous on $[a, b]$ and that

$$F(x) := \sup f([a, x]).$$

Prove that F is continuous on $[a, b]$.

4.3.8. Suppose that f is twice differentiable on (a, b) and that there are points $x_1 < x_2 < x_3$ in (a, b) such that $f(x_1) > f(x_2)$ and $f(x_3) > f(x_2)$. Prove that there is a point $c \in (a, b)$ such that $f''(c) > 0$.

4.3.9. Suppose that f is differentiable on $(0, \infty)$. If $L = \lim_{x \to \infty} f'(x)$ and $\lim_{n \to \infty} f(n)$ both exist and are finite, prove that $L = 0$.

4.3.10. Suppose that (a, b) is an open interval, that $f : (a, b) \to \mathbf{R}$ is differentiable on (a, b), and that $x_0 \in (a, b)$ is a proper local maximum of f (see Exercise 4.1.8).

    a) Prove that given $\delta > 0$, there exist $x_1 < x_0 < x_2$ such that $f'(x_1) > 0$, $f'(x_2) < 0$, and $|x_j - x_0| < \delta$ for $j = 1, 2$.

    b) Make and prove an analogous statement for a proper local minimum.

4.3.11. Suppose that $f : [a, b] \to \mathbf{R}$ is continuous and increasing. Prove that $\sup f(E) = f(\sup E)$ for every nonempty set $E \subseteq [a, b]$.

***4.3.12.** Suppose that f is differentiable at every point in a closed, bounded interval $[a, b]$. Prove that if f' is increasing on (a, b), then f' is continuous on (a, b).

4.4   TAYLOR'S THEOREM AND l'HÔPITAL'S RULE

In this section we use the Generalized Mean Value Theorem to obtain information about approximation.

To motivate the first result, notice by the Mean Value Theorem that if f is differentiable on (a, b), then for any points $x, x_0 \in (a, b)$, there is a c between x and x_0 such that

$$f(x) = f(x_0) + f'(c)(x - x_0), \qquad x \in (a, b).$$

Thus we have precise information about how closely $f(x)$ can be approximated by the constant function $y = f(x_0)$. Clearly, the values $f(x)$ of a function whose graph bends at the point $(x_0, f(x_0))$ cannot be closely approximated by a constant function unless x is near x_0. But a constant function is a polynomial of degree 0. If we used polynomials of higher degree (whose graphs do curve), might we be able to approximate $f(x)$ even when x is not so close to x_0? In fact, the next result contains precise information about how closely $f(x)$ can be approximated by a certain polynomial of degree n. (To understand how Taylor discovered this result, see the proof of Theorem 7.39.)

4.24 Theorem.   [TAYLOR'S FORMULA].
Let $n \in \mathbf{N}$ and let a, b be extended real numbers with $a < b$. If $f : (a, b) \to \mathbf{R}$, and if $f^{(n+1)}$ exists on (a, b), then for each pair of points $x, x_0 \in (a, b)$ there is a number c between x and x_0 such that

$$f(x) = f(x_0) + \sum_{k=1}^{n} \frac{f^{(k)}(x_0)}{k!}(x - x_0)^k + \frac{f^{(n+1)}(c)}{(n + 1)!}(x - x_0)^{n+1}.$$

Proof. Without loss of generality, suppose that $x_0 < x$. Define

$$F(t) := \frac{(x - t)^{n+1}}{(n + 1)!} \quad \text{and} \quad G(t) := f(x) - f(t) - \sum_{k=1}^{n} \frac{f^{(k)}(t)}{k!}(x - t)^k$$

for each $t \in (a, b)$, and observe that the theorem will be proved if we can show that there is a c between x and x_0 such that

$$G(x_0) = F(x_0) \cdot f^{(n+1)}(c). \tag{14}$$

This looks like a job for the Generalized Mean Value Theorem.

To verify that F and G satisfy the hypotheses of the Generalized Mean Value Theorem, notice that

$$\frac{d}{dt}\left(\frac{f^{(k)}(t)}{k!}(x - t)^k\right) = \frac{f^{(k+1)}(t)}{k!}(x - t)^k - \frac{f^{(k)}(t)}{(k - 1)!}(x - t)^{k-1}$$

for $t \in (a, b)$ and $k \in \mathbf{N}$. Telescoping, we obtain

$$G'(t) = -\frac{f^{(n+1)}(t)}{n!}(x-t)^n$$

for $t \in (a, b)$. On the other hand, by the Chain Rule

$$F'(t) = -\frac{(x-t)^n}{n!}$$

for $t \in \mathbf{R}$. Thus F and G are differentiable on (x_0, x), continuous on $[x_0, x]$, and satisfy

$$\frac{G'(t)}{F'(t)} = f^{(n+1)}(t), \qquad t \neq x. \tag{15}$$

By the Generalized Mean Value Theorem, there is a number $c \in (x_0, x)$ such that

$$(F(x) - F(x_0))G'(c) = (G(x) - G(x_0))F'(c). \tag{16}$$

Since $F(x) = G(x) = 0$ and $x \neq c$, it follows that $-F(x_0)G'(c) = -G(x_0)F'(c)$; that is, $G(x_0) = F(x_0) \cdot G'(c)/F'(c)$. We conclude by (15) that (14) holds, as promised. ∎

We shall use this result in Chapter 7 to show that most of the functions you've used in calculus classes before are very nearly polynomials themselves. To lay some ground work for these results, we introduce some additional notation.

Define $0! = 1$ and $f^{(0)}(x) = f(x)$, and notice that $f(x_0) = f^{(0)}(x_0)/0!$. We shall call

$$P_n^{f, x_0}(x) := \sum_{k=0}^{n} \frac{f^{(k)}(x_0)}{k!}(x - x_0)^k$$

the *Taylor polynomial of order n generated by f centered at* x_0. Clearly, for each $f \in C^\infty(a, b)$, Taylor's Formula gives us an estimate of how well Taylor polynomials approximate f. In fact, since Taylor's Formula implies

$$|f(x) - P_n^{f, x_0}(x)| \leq \left| \frac{f^{(n+1)}(c)}{(n+1)!}(x - x_0)^{n+1} \right|, \tag{17}$$

for some c between x and x_0 and the fraction $1/(n+1)!$ gets smaller as n gets larger, we see that when the derivatives of f are bounded, the higher-order Taylor polynomials approximate f better than the lower-order ones do.

Let's look at two specific examples to see how this works out in practice.

4.25 EXAMPLE.

Let $f(x) = e^x$ and $n \in \mathbf{N}$.

a) Find the Taylor polynomial $P_n := P_n^{f,0}$.
b) Prove that if $x \in [-1, 1]$, then

$$|e^x - P_n(x)| \le \frac{3}{(n+1)!}.$$

c) Find an n so large that P_n approximates e^x on $[-1, 1]$ to four decimal places.

Proof. a) Since $f^{(k)}(x) = e^x$ for all $x \in \mathbf{R}$ and $k = 0, 1, \ldots$, it is clear that $f^{(k)}(0) = 1$ for all $k \ge 0$; that is, that

$$P_n^{e^x,0}(x) = \sum_{k=0}^{n} \frac{x^k}{k!}. \tag{18}$$

b) Let $c, x \in [-1, 1]$. Clearly, $|e^c| \le e^1 < 3$ and $|x^n| \le 1$ for all $n \in \mathbf{N}$. But if c lies between x and 0, then $c \in [-1, 1]$. Thus it follows from (17) that $|e^x - P_n(x)| \le |e^c x^{n+1}|/(n+1)! < 3/(n+1)!$.

c) To get four-place accuracy, we want $|e^x - P_n(x)| \le .00005$. By part b), this will hold when $3/(n+1)! < 0.00005$; that is, when $(n+1)! \ge 60,000$. According to my calculator, this occurs when $n + 1 \ge 9$, so set $n = 8$. ∎

4.26 EXAMPLE.

Let $f(x) = \sin x$ and $n \in \mathbf{N}$.

a) Find the Taylor polynomial $P_{2n+1} := P_{2n+1}^{f,0}$.
b) Prove that if $x \in [-1, 1]$, then

$$|\sin x - P_{2n+1}(x)| \le \frac{1}{(2n+2)!}.$$

c) Find an n so large that P_{2n+1} approximates $\sin x$ on $[-1, 1]$ to three decimal places.

Proof. a) Observe that $f(x) = \sin x$, $f'(x) = \cos x$, $f''(x) = -\sin x$, $f^{(3)}(x) = -\cos x$, and $f^{(4)}(x) = \sin x$, right back where we started from. Thus it is clear that $f^{(2k)}(x) = (-1)^k \sin x$ and $f^{(2k+1)}(x) = (-1)^k \cos x$ for $k = 0, 1, \ldots$. It follows that $f^{(2k)}(0) = 0$ and $f^{(2k+1)}(0) = (-1)^k$ for $k \ge 0$; that is, that

$$P_{2n+1}^{\sin x,0}(x) = \sum_{k=0}^{n} \frac{(-1)^k x^{2k+1}}{(2k+1)!}. \tag{19}$$

b) Let $c, x \in [-1, 1]$. Clearly, $|f^{2n+2}(c)| \le 1$ and $|x^{2n+2}| \le 1^{2n+2} = 1$ for all $n \in \mathbf{N}$. Thus it follows from (17) that $|\sin x - P_{2n+1}(x)| \le 1/(2n+2)!$.

c) To get three-place accuracy, we want $|\sin x - P_{2n+2}(x)| \leq .0005$. By part b), this will hold when $1/(2n+2)! < 0.0005$; that is, when $(2n+2)! \geq 2000$. According to my calculator, this occurs when $2n + 2 \geq 7$, so set $n = 3$. ∎

The next result is a widely known technique for evaluating limits of the form $0/0$ or ∞/∞. Since it involves using information about derivatives to draw conclusions about the functions themselves, it should come as no surprise that the proof uses the Mean Value Theorem. (Notice that our statement is general enough to include one-sided limits and limits at infinity.)

4.27 Theorem.   [L'HÔPITAL'S RULE].
Let a be an extended real number and I be an open interval which either contains a or has a as an endpoint. Suppose that f and g are differentiable on $I \backslash \{a\}$ and that $g(x) \neq 0 \neq g'(x)$ for all $x \in I \backslash \{a\}$. Suppose further that

$$A := \lim_{\substack{x \to a \\ x \in I}} f(x) = \lim_{\substack{x \to a \\ x \in I}} g(x)$$

is either 0 or ∞. If

$$B := \lim_{\substack{x \to a \\ x \in I}} \frac{f'(x)}{g'(x)}$$

exists as an extended real number, then

$$\lim_{\substack{x \to a \\ x \in I}} \frac{f(x)}{g(x)} = \lim_{\substack{x \to a \\ x \in I}} \frac{f'(x)}{g'(x)}.$$

Proof. Let $x_k \in I$ be distinct points with $x_k \to a$ as $k \to \infty$ such that either $x_k < a$ or $x_k > a$ for all $k \in \mathbf{N}$. By the Sequential Characterization of Limits and by the characterization of two-sided limits in terms of one-sided limits, it suffices to show that $f(x_k)/g(x_k) \to B$ as $k \to \infty$.

We suppose for simplicity that $B \in \mathbf{R}$. (For the cases $B = \pm\infty$, see Exercise 4.4.10.) Notice once and for all, since g' is never zero on I, that by Mean Value Theorem the differences $g(x) - g(y)$ are never zero for $x, y \in I$, $x \neq y$, provided either $x, y > a$ or $x, y < a$. Hence, we can divide by these differences at will.

Case 1. $A = 0$ and $a \in \mathbf{R}$. Extend f and g to $I \cup \{a\}$ by $f(a) := 0 =: g(a)$. By hypothesis, f and g are continuous on $I \cup \{a\}$ and differentiable on $I \backslash \{a\}$. Hence by the Generalized Mean Value Theorem, there is a c_k between x_k and $y := a$ such that

$$\frac{f(x_k) - f(y)}{g(x_k) - g(y)} = \frac{f'(c_k)}{g'(c_k)}. \tag{20}$$

Since $f(y) = g(y) = 0$, it follows that

$$\frac{f(x_k)}{g(x_k)} = \frac{f(x_k) - f(y)}{g(x_k) - g(y)} = \frac{f'(c_k)}{g'(c_k)}. \tag{21}$$

Let $k \to \infty$. Since c_k lies between x_k and a, c_k also converges to a as $k \to \infty$. Hence hypothesis and (21) imply $f(x_k)/g(x_k) \to B$ as $k \to \infty$.

Case 2. $A = \pm\infty$ and $a \in \mathbf{R}$. We suppose by symmetry that $A = +\infty$. For each $k, n \in \mathbf{N}$, apply the Generalized Mean Value Theorem to choose a $c_{k,n}$ between x_k and x_n such that (20) holds for x_n in place of y and $c_{k,n}$ in place of c_k. Thus

$$\frac{f(x_n)}{g(x_n)} - \frac{f(x_k)}{g(x_n)} = \frac{f(x_n) - f(x_k)}{g(x_n)} = \frac{1}{g(x_n)} \cdot (g(x_n) - g(x_k)) \cdot \frac{f'(c_{k,n})}{g'(c_{k,n})}$$

$$= \left(1 - \frac{g(x_k)}{g(x_n)}\right) \frac{f'(c_{k,n})}{g'(c_{k,n})};$$

that is,

$$\frac{f(x_n)}{g(x_n)} = \frac{f(x_k)}{g(x_n)} - \frac{g(x_k)}{g(x_n)} \cdot \frac{f'(c_{k,n})}{g'(c_{k,n})} + \frac{f'(c_{k,n})}{g'(c_{k,n})}. \tag{22}$$

Since $A = \infty$, it is clear that $1/g(x_n) \to 0$ as $n \to \infty$, and since $c_{k,n}$ lies between x_k and x_n, it is also clear that $c_{k,n} \to a$, as $k, n \to \infty$. Thus (22) and hypothesis should imply that $f(x_n)/g(x_n) \approx 0 - 0 + B = B$ for large n and k. Specifically, let $0 < \varepsilon < 1$. Since $c_{k,n} \to a$ as $k, n \to \infty$, choose an N_0 so large that $n \geq N_0$ implies $|f'(c_{N_0,n})/g'(c_{N_0,n}) - B| < \varepsilon/3$. Since $g(x_n) \to \infty$, choose an $N > N_0$ such that $|f(x_{N_0})/g(x_n)|$ and the product $|g(x_{N_0})/g(x_n)| \cdot |f'(c_{N_0,n})/g'(c_{N_0,n})|$ are both less than $\varepsilon/3$ for all $n \geq N$. It follows from (22) that for any $n \geq N$,

$$\left|\frac{f(x_n)}{g(x_n)} - B\right| \leq \left|\frac{f(x_{N_0})}{g(x_n)}\right| + \left|\frac{g(x_{N_0})}{g(x_n)} \frac{f'(c_{N_0,n})}{g'(c_{N_0,n})}\right| + \left|\frac{f'(c_{N_0,n})}{g'(c_{N_0,n})} - B\right| < \varepsilon.$$

Hence, $f(x_n)/g(x_n) \to B$ as $n \to \infty$.

Case 3. $a = \pm\infty$. We suppose by symmetry that $a = +\infty$. Choose $c > 0$ such that $I \supset (c, \infty)$. For each $y \in (0, 1/c)$, set $\phi(y) = f(1/y)$ and $\psi(y) = g(1/y)$. Notice that by the Chain Rule,

$$\frac{\phi'(y)}{\psi'(y)} = \frac{f'(1/y)(-1/y^2)}{g'(1/y)(-1/y^2)} = \frac{f'(1/y)}{g'(1/y)}.$$

Thus, for $x = 1/y \in (c, \infty)$, $f'(x)/g'(x) = \phi'(y)/\psi'(y)$. Since $x \to \infty$ if and only if $y = 1/x \to 0+$, it follows that ϕ and ψ satisfy the hypotheses of Cases 1 or 2 for $a = 0$ and $I = (0, 1/c)$. In particular,

$$\lim_{x \to \infty} \frac{f'(x)}{g'(x)} = \lim_{y \to 0+} \frac{\phi'(y)}{\psi'(y)} = \lim_{y \to 0+} \frac{\phi(y)}{\psi(y)} = \lim_{x \to \infty} \frac{f(x)}{g(x)}. \qquad \blacksquare$$

L'Hôpital's Rule can be used to compare the relative rates of growth of two functions. For example, the next result shows that as $x \to \infty$, e^x converges to ∞ much faster than x^2 does.

4.28 EXAMPLE.

Prove that $\lim_{x \to \infty} x^2/e^x = 0$.

Proof. Since the limits of x^2/e^x and x/e^x are of the form ∞/∞, we apply l'Hôpital's Rule twice to verify

$$\lim_{x \to \infty} \frac{x^2}{e^x} = \lim_{x \to \infty} \frac{2x}{e^x} = \lim_{x \to \infty} \frac{2}{e^x} = 0. \qquad \blacksquare$$

For each subsequent application of l'Hôpital's Rule, it is important to check that the hypotheses still hold. For example,

$$\lim_{x \to 0} \frac{x^2}{x^2 + \sin x} = \lim_{x \to 0} \frac{2x}{2x + \cos x} = 0 \neq 1 = \lim_{x \to 0} \frac{2}{2 - \sin x}.$$

Notice that the middle limit is not of the form $0/0$.

l'Hôpital's Rule can be used to evaluate limits of the form $0 \cdot \infty = -0(-\infty)$.

4.29 EXAMPLE.

Find $\lim_{x \to 0+} x \log x$.

Solution. By writing x as $1/(1/x)$, we see that the limit in question is of the form ∞/∞. Hence, by l'Hôpital's Rule,

$$\lim_{x \to 0+} x \log x = \lim_{x \to 0+} \frac{\log x}{1/x} = \lim_{x \to 0+} \frac{1/x}{-1/x^2} = 0. \qquad \blacksquare$$

The next two examples show that l'Hôpital's Rule can also be used to evaluate limits of the form 1^∞ and 0^0.

4.30 EXAMPLE.

Find $L = \lim_{x \to 0+} (1 + 3x)^{1/x}$.

Solution. If the limit exists, then by a law of logarithms and the fact that $\log x$ is continuous, we have $\log L = \lim_{x \to 0+} \log(1 + 3x)/x$. Thus it follows from l'Hôpital's Rule and the Chain Rule that

$$\log L = \lim_{x \to 0+} \frac{\log(1 + 3x)}{x} = \lim_{x \to 0+} \frac{3/(1 + 3x)}{1} = 3.$$

In particular, the limit exists by l'Hôpital's Rule and $L = e^{\log L} = e^3$.   ∎

4.31 EXAMPLE.

Find $L = \lim_{x \to 1+} (\log x)^{1-x}$.

Solution. If the limit $L > 0$ exists, then $\log L = \lim_{x \to 1}(1 - x)\log \log x$ is of the form $0 \cdot \infty$. Hence, by l'Hôpital's Rule,

$$\log L = \lim_{x \to 1} \frac{\log \log x}{1/(1 - x)} = \lim_{x \to 1} \frac{1/(x \log x)}{1/(1 - x)^2} = \lim_{x \to 1} \frac{-2(1 - x)}{1 + \log x} = 0.$$

Therefore, the limit exists by l'Hôpital's Rule and $L = e^0 = 1$.   ∎

EXERCISES

4.4.0. Decide which of the following statements are true and which are false. Prove the true ones and provide counterexamples for the false ones.

a) $x/\log x \to 0$ as $x \to 0$.

b) If $n \in \mathbf{N}$, then $\sin(1/x)/x^n \to 0$ as $x \to \infty$.

c) $x^{\log x} \to 0$ as $x \to 0+$.

d) If there is a $\beta > 0$ such that $f'(x) \geq \beta$ for all $x \in (0, \infty)$, then $x^2/f(x) \to \infty$ as $x \to \infty$.

4.4.1. Let $f(x) = \cos x$ and $n \in \mathbf{N}$.

a) Find the Taylor polynomial $P_{2n} := P_{2n}^{f,0}$.

b) Prove that if $x \in [-1, 1]$, then

$$|\cos x - P_{2n}(x)| \leq \frac{1}{(2n + 1)!}.$$

c) Find an n so large that P_{2n} approximates $\cos x$ on $[-1, 1]$ to seven decimal places.

4.4.2. Let $f(x) = \log x$ and $n \in \mathbf{N}$.

a) Find the Taylor polynomial $P_n := P_n^{f,1}$.

b) Prove that if $x \in [1, 2]$, then

$$|\log x - P_n(x)| \leq \frac{1}{n + 1}.$$

c) Find an n so large that P_n approximates $\log x$ on $[1, 2]$ to three decimal places.

4.4.3. Prove that

$$1 + x + \frac{x^2}{2!} + \cdots + \frac{x^n}{n!} < e^x$$

for every $x > 0$ and every $n \in \mathbb{N}$.

4.4.4. Prove that

$$x - \frac{x^3}{3!} + \frac{x^5}{5!} - \cdots - \frac{x^{4n-1}}{(4n-1)!} < \sin x < x - \frac{x^3}{3!} + \frac{x^5}{5!} - \cdots + \frac{x^{4n+1}}{(4n+1)!}$$

for every $n \in \mathbb{N}$ and $x \in (0, \pi)$.

4.4.5. Evaluate the following limits.

a) $\lim_{x \to 0} \dfrac{\sin^2(5x)}{x^2}$

b) $\lim_{x \to 0+} \dfrac{\cos x - e^x}{\log(1 + x^2)}$

c) $\lim_{x \to 0} \left(\dfrac{x}{\sin x} \right)^{1/x^2}$

d) $\lim_{x \to 0+} (1 - x^2)^{1/x}$

e) $\lim_{x \to 1} \dfrac{\log x}{\sin(\pi x)}$

f) $\lim_{x \to 0+} |\log x|^x$

g) $\lim_{x \to \infty} \dfrac{\sqrt{x^2 + 2} - \sqrt{x^2}}{\sqrt{2x^2 - 1} - \sqrt{2x^2}}$

h) $\lim_{x \to \infty} \dfrac{\sqrt{x + 4} - \sqrt{x + 1}}{\sqrt{x + 3} - \sqrt{x + 1}}$

4.4.6 . **This exercise is used in Sections 5.4, 6.3, and elsewhere.** Let $\alpha > 0$ and recall that $(x^\alpha)' = \alpha x^{\alpha-1}$ and $(\log x)' = 1/x$ for all $x > 0$.

a) Prove that $\log x \le x^\alpha$ for x large. Prove that there exists a constant C_α such that $\log x \le C_\alpha x^\alpha$ for all $x \in [1, \infty)$, $C_\alpha \to \infty$ as $\alpha \to 0+$, and $C_\alpha \to 0$ as $\alpha \to \infty$.

b) Obtain an analogue of part a) valid for e^x and x^α in place of $\log x$ and x^α.

4.4.7 . **This exercise is used in Sections 7.4 and 12.5.** Assume that e^x is differentiable on $\mathbf{R}$ with $(e^x)' = e^x$.

a) Show that the following function is differentiable on $\mathbf{R}$ with $f'(0) = 0$:

$$f(x) = \begin{cases} e^{-1/x^2} & x \neq 0 \\ 0 & x = 0. \end{cases}$$

b) Do analogous statements hold for $f^{(n)}(x)$ when $n = 2, 3, \ldots$?

4.4.8. Suppose that $n \in \mathbf{N}$ is odd and $f^{(n)}$ exists on $[a, b]$. If $f^{(k)}(a) = f^{(k)}(b) = 0$ for all $k = 0, 1, \ldots, n - 1$ and $f(c) \neq 0$ for some $c \in (a, b)$, prove that there exist $x_1, x_2 \in (a, b)$ such that $f^{(n)}(x_1)$ is positive and $f^{(n)}(x_2)$ is negative.

4.4.9. a) Prove that $|\delta + \sin(\delta + \pi)| \leq \delta^3/3!$ for all $0 < \delta \leq 1$.
b) Prove that if $|x - \pi| \leq \delta \leq 1$, then $|x + \sin x - \pi| \leq \delta^3/3!$.

4.4.10. Prove l'Hôpital's Rule for the case $|B| = \infty$ by first proving that $g(x)/f(x) \to 0$ when $f(x)/g(x) \to \pm\infty$, as $x \to a$.

4.4.11. Suppose that f and g are differentiable on an open interval I and that $a \in \mathbf{R}$ either belongs to I or is an endpoint of I. Suppose further that g and g' are never zero on $I \setminus \{a\}$ and that

$$\lim_{x \to a} \frac{f(x)}{g(x)}$$

is of the form $0/0$. If there is an $M \in \mathbf{R}$ such that $|f'(x)/g'(x)| \leq M$ for all $x \in I \setminus \{a\}$, prove that $|f(x)/g(x)| \leq M$ for all $x \in I \setminus \{a\}$.
Is this result true if the limit of $f(x)/g(x)$ is of the form ∞/∞?

4.5 INVERSE FUNCTION THEOREMS

In this section, we explore the continuity and differentiability of inverse functions.

Recall that $f : X \to Y$ has an inverse function f^{-1} if and only if f is 1–1 and onto (Theorem 1.30), in which case $f^{-1}(f(x)) = x$ for all $x \in X$ and $f(f^{-1}(y)) = y$ for all $y \in Y$. Since $(x, f(x)) = (f^{-1}(y), y)$, this means that the graph of $y = f^{-1}(x)$ is a reflection of the graph of $y = f(x)$ about the line $y = x$ (see Figure 4.4). In particular, it is not difficult to imagine that f^{-1} is as smooth as f. This is the subject of the next two results.

4.32 Theorem. *Let I be a nondegenerate interval and suppose that $f : I \to \mathbf{R}$ is 1–1. If f is continuous on I, then $J := f(I)$ is an interval, f is strictly monotone on I, and f^{-1} is continuous and strictly monotone on J.*

Proof. Since f is 1–1 from I onto J, Theorem 1.30 implies that f^{-1} exists and takes J onto I.

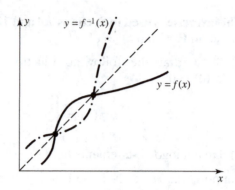

FIGURE 4.4

To show that J is an interval, since I contains at least two points, so does J. Let $c, d \in J$ with $c < d$. By the definition of an interval, it suffices to prove that every $y_0 \in (c, d)$ belongs to J. Since f takes I onto J, there exist points $a, b \in I$ such that $f(a) = c$ and $f(b) = d$. Since y_0 lies between $f(a)$ and $f(b)$, we can use the Intermediate Value Theorem to choose an x_0 between a and b such that $y_0 = f(x_0)$. Since $x_0 \in I$ and f takes I onto J, $y_0 = f(x_0)$ must belong to J, as required.

Suppose that f is not strictly monotone on I. Then there exist points $a, b, c \in I$ such that $a < c < b$ but $f(c)$ does not lie between $f(a)$ and $f(b)$. Since f is 1–1, $f(a) \neq f(b)$, so by symmetry we may suppose that $f(a) < f(b)$. Since $f(c)$ does not lie between $f(a)$ and $f(b)$, it follows that either $f(c) < f(a) < f(b)$ or $f(a) < f(b) < f(c)$. Hence by the Intermediate Value Theorem, there is an $x_1 \in (a, b)$ such that either $f(x_1) = f(a)$ or $f(x_1) = f(b)$. Since f is 1–1, we conclude that either $x_1 = a$ or $x_1 = b$, both contradictions. Therefore, f is strictly monotone on I.

By symmetry, suppose that f is strictly increasing on I. To prove that f^{-1} is strictly increasing on J, suppose to the contrary that there exist $y_1, y_2 \in J$ such that $y_1 < y_2$ but $f^{-1}(y_1) \geq f^{-1}(y_2)$. Then $x_1 := f^{-1}(y_1)$ and $x_2 := f^{-1}(y_2)$ satisfy $x_1 \geq x_2$ and $x_1, x_2 \in I$. Since f is strictly increasing on I, it follows that $y_1 = f(x_1) \geq f(x_2) = y_2$, a contradiction. Thus, f^{-1} is strictly increasing on J.

It remains to prove that f^{-1} is continuous from the left and from the right at each $y_0 \in J$. We will provide the details for continuity from the right. To this end, suppose that f^{-1} is not continuous from the right at some $y_0 \in J$; that is, that there exist $y_n \in J$ such that $y_n > y_0$, $y_n \to y_0$ as $n \to \infty$, but that

$$f^{-1}(y_n) \geq a_0 > f^{-1}(y_0) \tag{23}$$

for some number a_0. Since I is an interval and f^{-1} takes J onto I, it follows that a_0 belongs to I and there is a $b_0 \in J$ such that $a_0 = f^{-1}(b_0)$. Substituting this into (23), we see that $f^{-1}(y_n) \geq f^{-1}(b_0) > f^{-1}(y_0)$. Since f is strictly increasing, we conclude that $y_n \geq b_0 > y_0$; that is, y_n cannot converge to y_0, a contradiction. A similar argument verifies that f^{-1} is continuous from the left at each $y_0 \in J$. Thus f^{-1} is continuous on J. ∎

Our final result addresses the differentiability of an inverse function.

4.33 Theorem. [**INVERSE FUNCTION THEOREM**].
Let I be an open interval and $f : I \to \mathbf{R}$ be 1–1 and continuous. If $b = f(a)$ for some $a \in I$ and if $f'(a)$ exists and is nonzero, then f^{-1} is differentiable at b and $(f^{-1})'(b) = 1/f'(a)$.

Proof. By Theorem 4.32, f is strictly monotone, say strictly increasing on I, and f^{-1} exists and is both continuous and strictly increasing on the range $f(I)$. Moreover, since $a := f^{-1}(b) \in I$ and I is open, we can choose $c, d \in \mathbf{R}$ such that $a \in (c, d) \subset I$.

Let E_0 be the range of f on (c, d); that is, $E_0 = f((c, d))$. By Theorem 4.32, E_0 must be an interval. Since f is strictly increasing, it follows that $E_0 = (f(c), f(d))$. Hence, we can choose $\delta > 0$ so small that $0 < |h| < \delta$ implies $b + h \in E_0$. In particular, $f^{-1}(b + h)$ is defined for all $0 < |h| < \delta$.

Fix such an h and set $x = f^{-1}(b+h)$. Observe that $f(x) - f(a) = b + h - b = h$. Since f^{-1} is continuous, $x \to a$ if and only if $h \to 0$. Therefore, by direct substitution, we conclude that

$$\lim_{h \to 0} \frac{f^{-1}(b + h) - f^{-1}(b)}{h} = \lim_{x \to a} \frac{x - a}{f(x) - f(a)} = \frac{1}{f'(a)}. \qquad \blacksquare$$

This theorem is usually presented in elementary calculus texts in a form more easily remembered: If $y = f(x)$ and $x = f^{-1}(y)$, then

$$\frac{dx}{dy} = \frac{1}{dy/dx}.$$

Notice that, by using this formula, we do not need to solve explicitly for f^{-1} to be able to compute $(f^{-1})'$.

4.34 EXAMPLE.

If $f(x) = x^5 + x^4 + x^3 + x^2 + x + 1$, prove that $f^{-1}(x)$ exists at $x = 6$ and find a value for $(f^{-1})'(6)$.

Solution. Observe that $f(1) = 6$ and $f'(x) > 0$ for all $x > 0$. Thus f is strictly increasing on $(0, \infty)$, and hence 1–1 there.

Let $I = (0, 2)$, $a = 1$, and $b = 6$. Then $f(a) = b$ and $f'(a) = 15 \neq 0$. Hence, it follows from the Inverse Function Theorem that $(f^{-1})'(6) = 1/f'(1) = 1/15$. ∎

EXERCISES

4.5.0. Decide which of the following statements are true and which are false. Prove the true ones and provide counterexamples for the false ones.

a) Suppose that $I \subseteq \mathbf{R}$ is nonempty. If $f : I \to \mathbf{R}$ is 1–1 and continuous, then f is strictly monotone on I.

b) Suppose that I is an open interval which contains 0 and that $f : I \to \mathbf{R}$ is 1–1 and differentiable. If f and f' are never zero on I, then the derivative of f^{-1} has at least one root in $f(I)$; that is, there is an $a \in I$ such that $(f^{-1})'(a) = 0$.

c) Suppose that f and g are 1–1 on **R**. If f and $g \circ f$ are continuous on **R**, then g is continuous on **R**.

d) Suppose that I is an open interval and that $a \in I$. Suppose further that $f : I \to \mathbf{R}$ and $g : f(I) \to \mathbf{R}$ are both 1–1 and continuous and that $b := f(a)$. If $f'(a)$ and $g'(b)$ both exist and are nonzero, then $(g \circ f)^{-1}(x)$ is differentiable at $x = g(b)$, and $((g \circ f)^{-1})'(g(b)) = (f'(a) \cdot g'(b))^{-1}$.

4.5.1. Suppose that f and g are 1–1 and continuous on **R**. If $f(0) = 2$, $g(1) = 2$, $f'(0) = \pi$, and $g'(1) = e$, compute the following derivatives.

a) $(f^{-1})'(2)$
b) $(g^{-1})'(2)$
c) $(f^{-1} \cdot g^{-1})'(2)$

4.5.2. Let $f(x) = x^2 e^{x^2}$, and assume that $(e^x)' = e^x$ for all $x \in \mathbf{R}$.

a) Show that f^{-1} exists and is differentiable on $(0, \infty)$.
b) Compute $(f^{-1})'(e)$.

4.5.3. Using the Inverse Function Theorem, prove that $(\arcsin x)' = 1/\sqrt{1 - x^2}$ for $x \in (-1, 1)$ and $(\arctan x)' = 1/(1 + x^2)$ for $x \in (-\infty, \infty)$.

4.5.4. Suppose that f' exists and is continuous on a nonempty, open interval (a, b) with $f'(x) \neq 0$ for all $x \in (a, b)$.

a) Prove that f is 1–1 on (a, b) and takes (a, b) onto some open interval (c, d).
b) Show that $f^{-1} \in \mathcal{C}^1(c, d)$.
c) Using the function $f(x) = x^3$, show that b) is false if the assumption $f'(x) \neq 0$ fails to hold for some $x \in (a, b)$.
d) Sketch the graphs of $y = \tan x$ and $y = \arctan x$ to see that c and d in part b) might be infinite.

4.5.5. Suppose that $a := \lim_{x \to \infty}(1 + 1/x)^x$ exists and is greater than 1 (see Example 4.22). Assume that $a^x : \mathbf{R} \to (0, \infty)$ is onto, continuous, strictly increasing, and satisfies $a^x a^y = a^{x+y}$ and $(a^x)^y = a^{xy}$ for all $x, y \in \mathbf{R}$ (see Exercise 3.3.11). Let $L(x)$ denote the inverse function of a^x.

a) Prove that $tL(1 + 1/t) \to 1$ as $t \to \infty$.

b) Prove that $(a^h - 1)/h \to 1$ as $h \to 0$.

c) Prove that a^x is differentiable on **R** and $(a^x)' = a^x$ for all $x \in$ **R**.

d) Prove that $L'(x) = 1/x$ for all $x > 0$.

[**Note:** a is the natural base e and $L(x)$ is the natural logarithm $\log x$.]

4.5.6. Suppose that I is a nondegenerate interval, that $f : I \to$ **R** is differentiable, and that $f'(x) \neq 0$ for all $x \in I$.

a) Prove that f^{-1} exists and is differentiable on $f(I)$.

b) Suppose further that I is a closed, bounded interval and that f' is continuous. Prove that $(f^{-1})'$ is bounded on $f(I)$.

4.5.7. Suppose that $f : [a, b] \to [c, d]$ is differentiable and onto. If f' is never zero on $[a, b]$ and $d - c \geq 2$, prove that for every $x \in [c, d]$ there exist $x_1 \in [a, b]$ and $x_2 \in [c, d]$ such that $|f'(x_1)(f^{-1}(x) - f^{-1}(x_2))| = 1$.

4.5.8. Suppose that f is differentiable on a closed, bounded interval $[a, b]$. If $f[a, b] = [a, b]$ and f' is never zero on $[a, b]$, prove that for every $x \in [a, b]$ there exist $x_1, x_2 \in (a, b)$ such that

$$f(x) = f'(x_1)f'(x_2)(f^{-1}(x) - f^{-1}(a)) + f(a).$$

4.5.9. Let $[a, b]$ be a closed, bounded, nondegenerate interval. Find all functions f which satisfy the following conditions for some fixed $\alpha > 0$: f is continuous and 1–1 on $[a, b]$, $f'(x) \neq 0$ and $f'(x) = \alpha(f^{-1})'(f(x))$ for all $x \in (a, b)$.

4.5.10. Suppose that f is C^1 on an interval (a, b). If $f'(x_0) \neq 0$ for some $x_0 \in (a, b)$, prove that there exist intervals I and J such that f is 1–1 from I onto J and f^{-1} is continuously differentiable on J.

*4.5.11. Suppose that f is differentiable at every point in a closed, bounded interval $[a, b]$. Prove that if f' is 1–1 on $[a, b]$, then f' is strictly monotone on $[a, b]$.

CHAPTER 5

Integrability on R

5.1 THE RIEMANN INTEGRAL

In this chapter we shall study integration of real functions. We begin our discussion by introducing the following terminology.

5.1 Definition.

Let $a, b \in \mathbf{R}$ with $a < b$.

i) A *partition* of the interval $[a, b]$ is a set of points $P = \{x_0, x_1, \ldots x_n\}$ such that

$$a = x_0 < x_1 < \cdots < x_n = b.$$

ii) The *norm* of a partition $P = \{x_0, x_1, \ldots x_n\}$ is the number

$$\|P\| = \max_{1 \le j \le n} |x_j - x_{j-1}|.$$

iii) A *refinement* of a partition $P = \{x_0, x_1, \ldots x_n\}$ is a partition Q of $[a, b]$ which satisfies $Q \supseteq P$. In this case we say that Q is *finer* than P.

5.2 EXAMPLE. [THE DYADIC PARTITION].

Prove that for each $n \in \mathbf{N}$, $P_n = \{j/2^n : j = 0, 1, \ldots, 2^n\}$ is a partition of the interval $[0, 1]$, and P_m is finer than P_n when $m > n$.

Proof. Fix $n \in \mathbf{N}$. If $x_j = j/2^n$, then $0 = x_0 < x_1 < \cdots < x_{2^n} = 1$. Thus, P_n is a partition of $[0, 1]$. Let $m > n$ and set $p = m - n$. If $0 \le j \le 2^n$, then $j/2^n = j2^p/2^m$ and $0 \le j2^p \le 2^m$. Thus P_m is finer than P_n. ∎

It is clear that by definition, if P and Q are partitions of $[a, b]$, then $P \cup Q$ is finer than both P and Q. (Note that *finer* does not rule out the possibility that $P \cup Q = Q$, which would be the case if Q were a refinement of P.) And if Q is a refinement of P, then $\|Q\| \le \|P\|$. We shall use these observations often.

Let f be nonnegative on an interval $[a, b]$. You may recall that the integral of f over $[a, b]$ (when this integral exists) is the area of the region bounded by the curves $y = f(x)$, $y = 0$, $x = a$, and $x = b$. This area, A, can be approximated by rectangles whose bases lie in $[a, b]$ and whose heights approximate f (see Figure 5.1). If the tops of these rectangles lie above the curve $y = f(x)$, the resulting approximation is larger than A. If the tops of these rectangles lie below

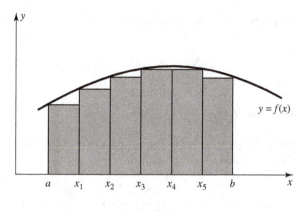

FIGURE 5.1

the curve $y = f(x)$, the resulting approximation is smaller than A. Hence, we make the following definition.

5.3 Definition.

Let $a, b \in \mathbf{R}$ with $a < b$, let $P = \{x_0, x_1, \ldots x_n\}$ be a partition of the interval $[a, b]$, set $\Delta x_j := x_j - x_{j-1}$ for $j = 1, 2, \ldots, n$, and suppose that $f : [a, b] \to \mathbf{R}$ is bounded.

i) The *upper Riemann sum* of f over P is the number

$$U(f, P) := \sum_{j=1}^{n} M_j(f)\, \Delta x_j,$$

where

$$M_j(f) := \sup f([x_{j-1}, x_j]) := \sup_{t \in [x_{j-1}, x_j]} f(t).$$

ii) The *lower Riemann sum* of f over P is the number

$$L(f, P) := \sum_{j=1}^{n} m_j(f)\, \Delta x_j,$$

where

$$m_j(f) := \inf f([x_{j-1}, x_j]) := \inf_{t \in [x_{j-1}, x_j]} f(t).$$

(**Note:** Since we assumed that f is bounded, the numbers $M_j(f)$ and $m_j(f)$ exist and are finite.)

Some specific upper and lower Riemann sums can be evaluated with the help of the following elementary observation.

5.4 Remark. *If* $g : \mathbf{N} \to \mathbf{R}$, *then*

$$\sum_{k=m}^{n} (g(k+1) - g(k)) = g(n+1) - g(m)$$

for all $n \geq m$ *in* **N**.

Proof. The proof is by induction on n. The formula holds for $n = m$. If it holds for some $n - 1 \geq m$, then

$$\sum_{k=m}^{n} (g(k+1) - g(k)) = (g(n) - g(m)) + (g(n+1) - g(n)) = g(n+1) - g(m).$$

∎

We shall refer to this algebraic identity by saying the sum *telescopes* to $g(n+1) - g(m)$. In particular, if $P = \{x_0, x_1, \ldots, x_n\}$ is a partition of $[a, b]$, the sum $\sum_{j=1}^{n} \Delta x_j$ telescopes to $x_n - x_0 = b - a$.

Before we define what it means for a function to be integrable, we make several elementary observations concerning upper and lower sums.

5.5 Remark. *If* $f(x) = \alpha$ *is constant on* $[a, b]$, *then*

$$U(f, P) = L(f, P) = \alpha(b - a)$$

for all partitions P *of* $[a, b]$.

Proof. Since $M_j(f) = m_j(f) = \alpha$ for all j, the sums $U(f, P)$ and $L(f, P)$ telescope to $\alpha(b - a)$. ∎

5.6 Remark. $L(f, P) \leq U(f, P)$ *for all partitions* P *and all bounded functions* f.

Proof. By definition, $m_j(f) \leq M_j(f)$ for all j. ∎

The next result shows that as the partitions get finer, the upper and lower Riemann sums get nearer each other.

5.7 Remark. *If* P *is any partition of* $[a, b]$ *and* Q *is a refinement of* P, *then*

$$L(f, P) \leq L(f, Q) \leq U(f, Q) \leq U(f, P).$$

Proof. Let $P = \{x_0, x_1, \ldots x_n\}$ be a partition of $[a, b]$. Since Q is finer than P, Q can be obtained from P in a finite number of steps by adding one point at a time. Hence it suffices to prove the inequalities above for the special case $Q = \{c\} \bigcup P$ for some $c \in (a, b)$. Moreover, by symmetry and Remark 5.6, we need only show $U(f, Q) \leq U(f, P)$.

We may suppose that $c \notin P$. Hence, there is a unique index j_0 such that $x_{j_0-1} < c < x_{j_0}$. By definition, it is clear that

$$U(f, Q) - U(f, P) = M^{(\ell)}(c - x_{j_0-1}) + M^{(r)}(x_{j_0} - c) - M \, \Delta x_{j_0},$$

where

$$M^{(\ell)} = \sup f([x_{j_0-1}, c]), \quad M^{(r)} = \sup f([c, x_{j_0}]), \quad \text{and}$$
$$M = \sup f([x_{j_0-1}, x_{j_0}]).$$

By the Monotone Property of Suprema, $M^{(\ell)}$ and $M^{(r)}$ are both less than or equal to M. Therefore,

$$U(f, Q) - U(f, P) \leq M(c - x_{j_0-1}) + M(x_{j_0} - c) - M \, \Delta x_{j_0} = 0. \quad \blacksquare$$

5.8 Remark. *If P and Q are any partitions of $[a, b]$, then*

$$L(f, P) \leq U(f, Q).$$

Proof. Since $P \cup Q$ is a refinement of P and Q, it follows from Remark 5.7 that

$$L(f, P) \leq L(f, P \cup Q) \leq U(f, P \cup Q) \leq U(f, Q)$$

for any pair of partitions P, Q, whether Q is a refinement of P or not. $\quad \blacksquare$

We now use the connection between area and integration to motivate the definition of *integrable*. Suppose that $f(x)$ is nonnegative on $[a, b]$ and that the region bounded by the curves $y = f(x)$, $y = 0$, $x = a$, and $x = b$ has a well-defined area A. By Definition 5.3, every upper Riemann sum is an over-estimate of A, and every lower Riemann sum is an underestimate of A (see Figure 5.1). Since the estimates $U(f, P)$ and $L(f, P)$ should get nearer to A as P gets finer, the differences $U(f, P) - L(f, P)$ should get smaller. [The shaded area in Figure 5.2 represents the difference $U(f, P) - L(f, P)$ for a particular P.] This leads us to the following definition (see also Exercise 5.1.9).

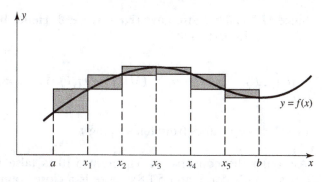

FIGURE 5.2

5.9 Definition.

Let $a, b \in \mathbf{R}$ with $a < b$. A function $f : [a, b] \to \mathbf{R}$ is said to be (*Riemann*) *integrable* on $[a, b]$ if and only if f is bounded on $[a, b]$, and for every $\varepsilon > 0$ there is a partition P of $[a, b]$ such that $U(f, P) - L(f, P) < \varepsilon$.

Notice that this definition makes sense whether or not f is nonnegative. The connection between nonnegative functions and area was only a convenient vehicle to motivate Definition 5.9. Also notice that, by Remark 5.6, $U(f, P) - L(f, P) = |U(f, P) - L(f, P)|$ for all partitions P. Hence, $U(f, P) - L(f, P) < \varepsilon$ is equivalent to $|U(f, P) - L(f, P)| < \varepsilon$.

This section provides a good illustration of how mathematics works. The connection between area and integration leads directly to Definition 5.9. This definition, however, is not easy to apply in concrete situations. Thus, we search for conditions which imply integrability *and* are easy to apply. In view of Figure 5.2, it seems reasonable that a function is integrable if its graph does not jump around too much (so that it can be covered by thinner and thinner rectangles). Since the graph of a continuous function does not jump at all, we are led to the following simple criterion that is sufficient (but not necessary) for integrability.

5.10 Theorem. *Suppose that $a, b \in \mathbf{R}$ with $a < b$. If f is continuous on the interval $[a, b]$, then f is integrable on $[a, b]$.*

Proof. Let $\varepsilon > 0$. Since f is uniformly continuous on $[a, b]$, choose $\delta > 0$ such that

$$|x - y| < \delta \quad \text{implies} \quad |f(x) - f(y)| < \frac{\varepsilon}{b - a}. \tag{1}$$

Let $P = \{x_0, x_1, \ldots, x_n\}$ be any partition of $[a, b]$ which satisfies $\|P\| < \delta$. Fix an index j and notice, by the Extreme Value Theorem, that there are points x_m and x_M in $[x_{j-1}, x_j]$ such that

$$f(x_m) = m_j(f) \quad \text{and} \quad f(x_M) = M_j(f).$$

Since $\|P\| < \delta$, we also have $|x_M - x_m| < \delta$. Hence by (1), $M_j(f) - m_j(f) < \varepsilon/(b - a)$. In particular,

$$U(f, P) - L(f, P) = \sum_{j=1}^{n} (M_j(f) - m_j(f)) \, \Delta x_j < \frac{\varepsilon}{b - a} \sum_{j=1}^{n} \Delta x_j = \varepsilon.$$

(The last step comes from telescoping.) ∎

Although the converse of Theorem 5.10 is false (see Example 5.12 and Exercises 5.1.3, 5.1.6, and 5.1.8), there is a close connection between integrability and continuity. Indeed, we shall see (Theorem 9.49) that a function is integrable if and only if it has relatively few discontinuities. This principle is illustrated by the following examples. The nonintegrable function in Example 5.11 is nowhere continuous (hence has many discontinuities) but the integrable function in Example 5.12 has only one discontinuity (hence has few discontinuities).

5.11 EXAMPLE.

The Dirichlet function

$$f(x) = \begin{cases} 1 & x \in \mathbf{Q} \\ 0 & x \notin \mathbf{Q} \end{cases}$$

is not Riemann integrable on $[0, 1]$.

Proof. Clearly, f is bounded on $[0, 1]$. By Theorem 1.18 and Exercise 1.3.3 (Density of Rationals and Irrationals), the supremum of f over any nondegenerate interval is 1, and the infimum of f over any nondegenerate interval is 0. Therefore, $U(f, P) - L(f, P) = 1 - 0 = 1$ for any partition P of the interval $[0, 1]$; that is, f is not integrable on $[0, 1]$. ∎

5.12 EXAMPLE.

The function

$$f(x) = \begin{cases} 0 & 0 \le x < 1/2 \\ 1 & 1/2 \le x \le 1 \end{cases}$$

is integrable on $[0, 1]$.

Proof. Let $\varepsilon > 0$. Choose $0 < x_1 < 0.5 < x_2 < 1$ such that $x_2 - x_1 < \varepsilon$. The set

$$P := \{0, x_1, x_2, 1\}$$

is a partition of $[0, 1]$. Since $m_1(f) = 0 = M_1(f)$, $m_2(f) = 0 < 1 = M_2(f)$, and $m_3(f) = 1 = M_3(f)$, it is easy to see that $U(f, P) - L(f, P) = x_2 - x_1 < \varepsilon$. Therefore, f is integrable on $[0, 1]$. ∎

We have defined integrability, but not the value of the integral. We remedy this situation by using the Riemann sums $U(f, P)$ and $L(f, P)$ to define upper and lower integrals.

5.13 Definition.

Let $a, b \in \mathbf{R}$ with $a < b$, and $f : [a, b] \to \mathbf{R}$ be bounded.

i) The *upper integral* of f on $[a, b]$ is the number

$$(U) \int_a^b f(x)\, dx := \inf\{U(f, P) : P \text{ is a partition of } [a, b]\}.$$

5.13 Definition. (Continued)

---

ii) The *lower integral* of f on $[a, b]$ is the number

$$(L) \int_a^b f(x)\, dx := \sup\{L(f, P) : P \text{ is a partition of } [a, b]\}.$$

iii) If the upper and lower integrals of f on $[a, b]$ are equal, we define the *integral* of f on $[a, b]$ to be the common value

$$\int_a^b f(x)\, dx := (U) \int_a^b f(x)\, dx = (L) \int_a^b f(x)\, dx.$$

This defines integration over nondegenerate intervals. Motivated by the interpretation of integration as area, we define the integral of any bounded function f on $[a, a]$ to be zero; that is,

$$\int_a^a f(x)\, dx := 0.$$

Although a bounded function might not be integrable (see Example 5.11 above), the following result shows that the upper and lower integrals of a bounded function always exist.

5.14 Remark. *If $f : [a, b] \to \mathbf{R}$ is bounded, then its upper and lower integrals exist and are finite, and satisfy*

$$(L) \int_a^b f(x)\, dx \leq (U) \int_a^b f(x)\, dx.$$

Proof. By Remark 5.8, $L(f, P) \leq U(f, Q)$ for all partitions P and Q of $[a, b]$. Taking the supremum of this inequality over all partitions P of $[a, b]$, we have

$$(L) \int_a^b f(x)\, dx \leq U(f, Q);$$

that is, the lower integral exists and is finite. Taking the infimum of this last inequality over all partitions Q of $[a, b]$, we conclude that the upper integral is also finite and greater than or equal to the lower integral. ∎

Suppose that f is bounded and nonnegative on $[a, b]$. Since the upper and lower sums of f approximate the "area" of the region bounded by the curves $y = f(x)$, $y = 0$, $x = a$, and $x = b$, we guess that f is integrable if and only if the upper and lower integrals of f are equal. The following result shows that this guess is true whether or not f is nonnegative.

5.15 Theorem. *Let $a, b \in \mathbf{R}$ with $a < b$, and $f : [a, b] \to \mathbf{R}$ be bounded. Then f is integrable on $[a, b]$ if and only if*

$$(L) \int_a^b f(x)\, dx = (U) \int_a^b f(x)\, dx. \tag{2}$$

Proof. Suppose that f is integrable. Let $\varepsilon > 0$ and choose a partition P of $[a, b]$ such that

$$U(f, P) - L(f, P) < \varepsilon. \tag{3}$$

By definition, $(U) \int_a^b f(x)dx \le U(f, P)$ and the opposite inequality holds for the lower integral and the lower sum $L(f, P)$. Therefore, it follows from Remark 5.14 and (3) that

$$\left| (U) \int_a^b f(x)\, dx - (L) \int_a^b f(x)\, dx \right| = (U) \int_a^b f(x)\, dx - (L) \int_a^b f(x)\, dx$$
$$\le U(f, P) - L(f, P) < \varepsilon.$$

Since this is valid for all $\varepsilon > 0$, (2) holds as promised.

Conversely, suppose that (2) holds. Let $\varepsilon > 0$ and choose, by the Approximation Property, partitions P_1 and P_2 of $[a, b]$ such that

$$U(f, P_1) < (U) \int_a^b f(x)\, dx + \frac{\varepsilon}{2}$$

and

$$L(f, P_2) > (L) \int_a^b f(x)\, dx - \frac{\varepsilon}{2}.$$

Set $P = P_1 \cup P_2$. Since P is a refinement of both P_1 and P_2, it follows from Remark 5.7, the choices of P_1 and P_2, and (2) that

$$U(f, P) - L(f, P) \le U(f, P_1) - L(f, P_2)$$
$$\le (U) \int_a^b f(x)\, dx + \frac{\varepsilon}{2} - (L) \int_a^b f(x)\, dx + \frac{\varepsilon}{2} = \varepsilon. \quad \blacksquare$$

Since the integral has been defined only on intervals $[a, b]$, we have tacitly assumed that $a \le b$. We shall use the convention

$$\int_b^a f(x)\, dx = - \int_a^b f(x)\, dx$$

to extend the integral to the case $a > b$. In particular, if $f(x)$ is integrable and nonpositive on $[a, b]$, then the area of the region bounded by the curves $y = f(x)$, $y = 0$, $x = a$, and $x = b$ is given by $\int_b^a f(x)dx$.

In the next section we shall use the machinery of upper and lower sums to prove several familiar theorems about the Riemann integral. We close this section with one more result which reinforces the connection between integration and area.

5.16 Theorem. *If $f(x) = \alpha$ is constant on $[a, b]$, then*

$$\int_a^b f(x)\, dx = \alpha(b - a).$$

Proof. By Theorem 5.10, f is integrable on $[a, b]$. Hence, it follows from Theorem 5.15 and Remark 5.5 that

$$\int_a^b f(x)\, dx = (U) \int_a^b f(x)\, dx = \inf_P U(f, P) = \alpha(b - a). \qquad \blacksquare$$

EXERCISES

5.1.0. Suppose that $a < b < c$. Decide which of the following statements are true and which are false. Prove the true ones and give counterexamples for the false ones.

a) If f is Riemann integrable on $[a, b]$, then f is continuous on $[a, b]$.
b) If $|f|$ is Riemann integrable on $[a, b]$, then f is Riemann integrable on $[a, b]$.
c) For all bounded functions $f : [a, b] \to \mathbf{R}$,

$$(L) \int_a^b f(x)\, dx \le \int_a^b f(x)\, dx \le (U) \int_a^b f(x)\, dx.$$

d) If f is continuous on $[a, b)$ and on $[b, c]$, then f is Riemann integrable on $[a, c]$.

5.1.1. For each of the following, compute $U(f, P)$, $L(f, P)$, and $\int_0^2 f(x)dx$, where

$$P = \left\{ 0, \frac{1}{2}, 1, 2 \right\}.$$

Find out whether the lower sum or the upper sum is a better approximation to the integral. Graph f and explain why this is so.

a) $f(x) = x^3$
b) $f(x) = 3 - x^2$
c) $f(x) = \sin(x/5)$

5.1.2. a) Prove that for each $n \in \mathbf{N}$,

$$P_n := \left\{ \frac{j}{n} : j = 0, 1, \ldots, n \right\}$$

is a partition of $[0, 1]$.

b) Prove that a bounded function f is integrable on $[0, 1]$ if

(*) $\qquad\qquad I_0 := \lim_{n \to \infty} L(f, P_n) = \lim_{n \to \infty} U(f, P_n),$

in which case $\int_0^1 f(x)dx$ equals I_0.

c) For each of the following functions, use Exercise 1.4.4 to find formulas for the upper and lower sums of f on P_n, and use them to compute the value of $\int_0^1 f(x)dx$.

α) $\qquad\qquad\qquad\qquad f(x) = x$

β) $\qquad\qquad\qquad\qquad f(x) = x^2$

γ) $\qquad\qquad f(x) = \begin{cases} 0 & 0 \le x < 1/2 \\ 1 & 1/2 \le x \le 1 \end{cases}$

5.1.3. Let $E = \{1/n : n \in \mathbf{N}\}$. Prove that the function

$$f(x) = \begin{cases} 1 & x \in E \\ 0 & \text{otherwise} \end{cases}$$

is integrable on $[0, 1]$. What is the value of $\int_0^1 f(x)dx$?

| **5.1.4** |. **This exercise is used in Section *14.2.** Suppose that $a < b$ and that $f : [a, b] \to \mathbf{R}$ is bounded.

a) Prove that if f is continuous at $x_0 \in [a, b]$ and $f(x_0) \neq 0$, then

$$(L) \quad \int_a^b |f(x)| \, dx > 0.$$

b) Show that if f is continuous on $[a, b]$, then $\int_a^b |f(x)|dx = 0$ if and only if $f(x) = 0$ for all $x \in [a, b]$.

c) Does part b) hold if the absolute values are removed? If it does, prove it. If it does not, provide a counterexample.

5.1.5. Suppose that $a < b$ and that $f : [a, b] \to \mathbf{R}$ is continuous. Show that

$$\int_a^c f(x) \, dx = 0$$

for all $c \in [a, b]$ if and only if $f(x) = 0$ for all $x \in [a, b]$. (Compare with Exercise 5.1.4, and notice that f need not be nonnegative here.)

5.1.6. Let f be integrable on $[a, b]$ and E be a finite subset of $[a, b]$. Show that if g is a bounded function which satisfies $g(x) = f(x)$ for all $x \in [a, b] \setminus E$, then g is integrable on $[a, b]$ and

$$\int_a^b g(x)\, dx = \int_a^b f(x)\, dx.$$

$\boxed{\textbf{5.1.7}}$. **This exercise is used in Section 12.3.** Let f, g be bounded on $[a, b]$.

a) Prove that

$$(U) \int_a^b (f(x) + g(x))\, dx \le (U) \int_a^b f(x)\, dx + (U) \int_a^b g(x)\, dx$$

and

$$(L) \int_a^b (f(x) + g(x))\, dx \ge (L) \int_a^b f(x)\, dx + (L) \int_a^b g(x)\, dx.$$

b) Prove that

$$(U) \int_a^b f(x)\, dx = (U) \int_a^c f(x)\, dx + (U) \int_c^b f(x)\, dx$$

and

$$(L) \int_a^b f(x)\, dx = (L) \int_a^c f(x)\, dx + (L) \int_c^b f(x)\, dx$$

for $a < c < b$.

$\boxed{\textbf{5.1.8}}$. **This exercise is used in Sections *5.5, 6.2, and *7.5.**

a) If f is increasing on $[a, b]$ and $P = \{x_0, \dots, x_n\}$ is any partition of $[a, b]$, prove that

$$\sum_{j=1}^n (M_j(f) - m_j(f))\, \Delta x_j \le (f(b) - f(a))\, \|P\|.$$

b) Prove that if f is monotone on $[a, b]$, then f is integrable on $[a, b]$. [**Note:** By Theorem 4.19, f has at most countably many (i.e., relatively few) discontinuities on $[a, b]$. This has nothing to do with the proof of part b), but points out a general principle which will be discussed in Section 9.6.]

5.1.9. Let $a < b$ and $0 < c < d$ be real numbers and $f : [a, b] \to [c, d]$. If f is Riemann integrable on $[a, b]$, prove that $\sqrt{f}$ is Riemann integrable on $[a, b]$.

5.1.10. Let f be bounded on a nondegenerate interval $[a, b]$. Prove that f is integrable on $[a, b]$ if and only if given $\varepsilon > 0$ there is a partition P_ε of $[a, b]$ such that

$$P \supseteq P_\varepsilon \quad \text{implies} \quad |U(f, P) - L(f, P)| < \varepsilon.$$

5.2 RIEMANN SUMS

There is another definition of the Riemann integral frequently found in elementary calculus texts.

5.17 Definition.

Let $f : [a, b] \to \mathbf{R}$.

i) A *Riemann sum* of f with respect to a partition $P = \{x_0, \dots, x_n\}$ of $[a, b]$ generated by samples $t_j \in [x_{j-1}, x_j]$ is a sum

$$\mathcal{S}(f, P, t_j) := \sum_{j=1}^{n} f(t_j) \, \Delta x_j.$$

ii) The Riemann sums of f are said to *converge* to $I(f)$ as $\|P\| \to 0$ if and only if given $\varepsilon > 0$ there is a partition P_ε of $[a, b]$ such that

$$P = \{x_0, \dots, x_n\} \supseteq P_\varepsilon \quad \text{implies} \quad \left| \mathcal{S}(f, P, t_j) - I(f) \right| < \varepsilon$$

for all choices of $t_j \in [x_{j-1}, x_j]$, $j = 1, 2, \dots, n$. In this case we shall use the notation

$$I(f) = \lim_{\|P\| \to 0} \mathcal{S}(f, P, t_j) := \lim_{\|P\| \to 0} \sum_{j=1}^{n} f(t_j) \, \Delta x_j.$$

The following result shows that this definition of the Riemann integral is the same as the one using upper and lower integrals.

5.18 Theorem. *Let $a, b \in \mathbf{R}$ with $a < b$, and suppose that $f : [a, b] \to \mathbf{R}$. Then f is Riemann integrable on $[a, b]$ if and only if*

$$I(f) = \lim_{\|P\| \to 0} \sum_{j=1}^{n} f(t_j) \, \Delta x_j$$

exists, in which case $I(f) = \int_a^b f(x)dx$.

Proof. Suppose that f is integrable on $[a, b]$ and that $\varepsilon > 0$. By the Approximation Property, there is a partition P_ε of $[a, b]$ such that

$$L(f, P_\varepsilon) > \int_a^b f(x)\, dx - \varepsilon \quad \text{and} \quad U(f, P_\varepsilon) < \int_a^b f(x)\, dx + \varepsilon. \qquad (4)$$

Let $P = \{x_0, x_1, \ldots, x_n\} \supseteq P_\varepsilon$. Then (4) holds with P in place of P_ε. But $m_j(f) \le f(t_j) \le M_j(f)$ for any choice of $t_j \in [x_{j-1}, x_j]$. Hence,

$$\int_a^b f(x)\, dx - \varepsilon < L(f, P) \le \sum_{j=1}^n f(t_j)\, \Delta x_j \le U(f, P) < \int_a^b f(x)\, dx + \varepsilon;$$

that is, $-\varepsilon < \sum_{j=1}^n f(t_j)\, \Delta x_j - \int_a^b f(x)\, dx < \varepsilon$. We conclude that

$$\left| \sum_{j=1}^n f(t_j)\, \Delta x_j - \int_a^b f(x)\, dx \right| < \varepsilon$$

for all partitions $P \supseteq P_\varepsilon$ and all choices of $t_j \in [x_{j-1}, x_j]$, $j = 1, 2, \ldots, n$.

Conversely, suppose that the Riemann sums of f converge to $I(f)$. Let $\varepsilon > 0$ and choose a partition $P = \{x_0, x_1, \ldots, x_n\}$ of $[a, b]$ such that

$$\left| \sum_{j=1}^n f(t_j)\, \Delta x_j - I(f) \right| < \frac{\varepsilon}{3} \qquad (5)$$

for all choices of $t_j \in [x_{j-1}, x_j]$. Since f is bounded on $[a, b]$ (see Exercise 5.2.11), use the Approximation Property to choose $t_j, u_j \in [x_{j-1}, x_j]$ such that $f(t_j) - f(u_j) > M_j(f) - m_j(f) - \varepsilon/(3(b-a))$. By (5) and telescoping, we have

$$U(f, P) - L(f, P) = \sum_{j=1}^n (M_j(f) - m_j(f))\, \Delta x_j$$

$$< \sum_{j=1}^n (f(t_j) - f(u_j))\, \Delta x_j + \frac{\varepsilon}{3(b-a)} \sum_{j=1}^n \Delta x_j$$

$$\le \left| \sum_{j=1}^n f(t_j)\, \Delta x_j - I(f) \right|$$

$$+ \left| I(f) - \sum_{j=1}^n f(u_j)\, \Delta x_j \right| + \frac{\varepsilon}{3(b-a)} \sum_{j=1}^n \Delta x_j$$

$$< \frac{2\varepsilon}{3} + \frac{\varepsilon}{3} = \varepsilon.$$

Therefore, f is integrable on $[a, b]$.   ∎

The next two results show that Riemann integrals of complicated functions can be broken into simpler pieces.

5.19 Theorem.   [LINEAR PROPERTY].
If f, g are integrable on $[a, b]$ and $\alpha \in \mathbf{R}$, then $f + g$ and αf are integrable on $[a, b]$. In fact,

$$\int_a^b (f(x) + g(x))\, dx = \int_a^b f(x)\, dx + \int_a^b g(x)\, dx \tag{6}$$

and

$$\int_a^b (\alpha f(x))\, dx = \alpha \int_a^b f(x)\, dx. \tag{7}$$

Proof. Let $\varepsilon > 0$ and choose P_ε such that for any partition $P = \{x_0, x_1, \dots, x_n\} \supseteq P_\varepsilon$ of $[a, b]$ and any choice of $t_j \in [x_{j-1}, x_j]$, we have

$$\left| \sum_{j=1}^n f(t_j)\, \Delta x_j - \int_a^b f(x)\, dx \right| < \frac{\varepsilon}{2}$$

and

$$\left| \sum_{j=1}^n g(t_j)\, \Delta x_j - \int_a^b g(x)\, dx \right| < \frac{\varepsilon}{2}.$$

By the Triangle Inequality,

$$\left| \sum_{j=1}^n f(t_j)\Delta x_j + \sum_{j=1}^n g(t_j)\Delta x_j - \int_a^b f(x)\, dx - \int_a^b g(x)\, dx \right| < \varepsilon$$

for any choice of $t_j \in [x_{j-1}, x_j]$. Hence, (6) follows directly from Theorem 5.18.

To prove (7), we may suppose that $\alpha \neq 0$. Choose P_ε such that if $P = \{x_0, \dots, x_n\}$ is finer than P_ε, then

$$\left| \sum_{j=1}^n f(t_j)\, \Delta x_j - \int_a^b f(x)\, dx \right| < \frac{\varepsilon}{|\alpha|}$$

for any choice of $t_j \in [x_{j-1}, x_j]$. Multiplying this inequality by $|\alpha|$, we obtain

$$\left| \sum_{j=1}^n \alpha f(t_j)\, \Delta x_j - \alpha \int_a^b f(x)\, dx \right| < |\alpha| \frac{\varepsilon}{|\alpha|} = \varepsilon$$

for any choice of $t_j \in [x_{j-1}, x_j]$. We conclude by Theorem 5.18 that (7) holds. ∎

5.20 Theorem. *If f is integrable on $[a, b]$, then f is integrable on each subinterval $[c, d]$ of $[a, b]$. Moreover,*

$$\int_a^b f(x)\,dx = \int_a^c f(x)\,dx + \int_c^b f(x)\,dx \tag{8}$$

for all $c \in (a, b)$.

Proof. We may suppose that $a < b$. Let $\varepsilon > 0$ and choose a partition P of $[a, b]$ such that

$$U(f, P) - L(f, P) < \varepsilon. \tag{9}$$

Let $P' = P \cup \{c\}$ and $P_1 = P' \cap [a, c]$. Since P_1 is a partition of $[a, c]$ and P' is a refinement of P, we have by (9) that

$$U(f, P_1) - L(f, P_1) \le U(f, P') - L(f, P') \le U(f, P) - L(f, P) < \varepsilon.$$

Therefore, f is integrable on $[a, c]$. A similar argument proves that f is integrable on any subinterval $[c, d]$ of $[a, b]$.

To verify (8), suppose that P is any partition of $[a, b]$. Let $P_0 = P \cup \{c\}$, $P_1 = P_0 \cap [a, c]$, and $P_2 = P_0 \cap [c, b]$. Then $P_0 = P_1 \cup P_2$ and by definition

$$U(f, P) \ge U(f, P_0) = U(f, P_1) + U(f, P_2)$$

$$\ge (U) \int_a^c f(x)\,dx + (U) \int_c^b f(x)\,dx = \int_a^c f(x)\,dx + \int_c^b f(x)\,dx.$$

(This last equality follows from the fact that f is integrable on both $[a, c]$ and $[c, b]$.) Taking the infimum of

$$U(f, P) \ge \int_a^c f(x)\,dx + \int_c^b f(x)\,dx$$

over all partitions P of $[a, b]$, we obtain

$$\int_a^b f(x)\,dx = (U) \int_a^b f(x)\,dx \ge \int_a^c f(x)\,dx + \int_c^b f(x)\,dx.$$

A similar argument using lower integrals shows that

$$\int_a^b f(x)\,dx \le \int_a^c f(x)\,dx + \int_c^b f(x)\,dx. \qquad \blacksquare$$

Using the conventions

$$\int_a^b f(x)\,dx = -\int_b^a f(x)\,dx \quad \text{and} \quad \int_a^a f(x)\,dx = 0,$$

it is easy to see that (8) holds whether or not c lies between a and b, provided f is integrable on the union of these intervals (see Exercise 5.2.4).

5.21 Theorem.   [**COMPARISON THEOREM FOR INTEGRALS**].
If f, g are integrable on $[a, b]$ and $f(x) \le g(x)$ for all $x \in [a, b]$, then

$$\int_a^b f(x)\,dx \le \int_a^b g(x)\,dx.$$

In particular, if $m \le f(x) \le M$ for $x \in [a, b]$, then

$$m(b-a) \le \int_a^b f(x)\,dx \le M(b-a).$$

Proof. Let P be a partition of $[a, b]$. By hypothesis, $M_j(f) \le M_j(g)$ whence $U(f, P) \le U(g, P)$. It follows that

$$\int_a^b f(x)\,dx = (U)\int_a^b f(x)\,dx \le U(g, P)$$

for all partitions P of $[a, b]$. Taking the infimum of this inequality over all partitions P of $[a, b]$, we obtain

$$\int_a^b f(x)\,dx \le \int_a^b g(x)\,dx.$$

If $m \le f(x) \le M$, then (by what we just proved and by Theorem 5.16)

$$m(b-a) = \int_a^b m\,dx \le \int_a^b f(x)\,dx \le \int_a^b M\,dx = M(b-a). \qquad \blacksquare$$

We shall use the following result nearly every time we need to estimate an integral.

5.22 Theorem.   *If f is (Riemann) integrable on $[a, b]$, then $|f|$ is integrable on $[a, b]$ and*

$$\left| \int_a^b f(x)\,dx \right| \le \int_a^b |f(x)|\,dx.$$

Proof. Let $P = \{x_0, x_1, \ldots, x_n\}$ be a partition of $[a, b]$. We claim that

$$M_j(|f|) - m_j(|f|) \le M_j(f) - m_j(f) \tag{10}$$

holds for $j = 1, 2, \ldots, n$. Indeed, let $x, y \in [x_{j-1}, x_j]$. If $f(x)$, $f(y)$ have the same sign, say both are nonnegative, then

$$|f(x)| - |f(y)| = f(x) - f(y) \leq M_j(f) - m_j(f).$$

If $f(x)$, $f(y)$ have opposite signs, say $f(x) \geq 0 \geq f(y)$, then $m_j(f) \leq 0$ and, hence,

$$|f(x)| - |f(y)| = f(x) + f(y) \leq M_j(f) + 0 \leq M_j(f) - m_j(f).$$

Thus in either case, $|f(x)| \leq M_j(f) - m_j(f) + |f(y)|$. Taking the supremum of this last inequality for $x \in [x_{j-1}, x_j]$ and then the infimum as $y \in [x_{j-1}, x_j]$, we see that (10) holds, as promised.

Let $\varepsilon > 0$ and choose a partition P of $[a, b]$ such that $U(f, P) - L(f, P) < \varepsilon$. Since (10) implies $U(|f|, P) - L(|f|, P) \leq U(f, P) - L(f, P)$, it follows that

$$U(|f|, P) - L(|f|, P) < \varepsilon.$$

Thus $|f|$ is integrable on $[a, b]$. Since $-|f(x)| \leq f(x) \leq |f(x)|$ holds for any $x \in [a, b]$, we conclude by Theorem 5.21 that

$$-\int_a^b |f(x)| \, dx \leq \int_a^b f(x) \, dx \leq \int_a^b |f(x)| \, dx. \qquad \blacksquare$$

By Theorem 5.19, the sum of integrable functions is integrable. What about the product?

5.23 Corollary. *If f and g are (Riemann) integrable on $[a, b]$, then so is fg.*

Proof. Suppose for a moment that the square of any integrable function is integrable. Then, by hypothesis, f^2, g^2, and $(f + g)^2$ are integrable on $[a, b]$. Since

$$fg = \frac{(f+g)^2 - f^2 - g^2}{2},$$

it follows from Theorem 5.19 that fg is integrable on $[a, b]$.

It remains to prove that f^2 is integrable on $[a, b]$. Let P be a partition of $[a, b]$. Since $M_j(f^2) = (M_j(|f|))^2$ and $m_j(f^2) = (m_j(|f|))^2$, it is clear that

$$\begin{aligned}
M_j(f^2) - m_j(f^2) &= (M_j(|f|))^2 - (m_j(|f|))^2 \\
&= (M_j(|f|) + m_j(|f|))(M_j(|f|) - m_j(|f|)) \\
&\leq 2M(M_j(|f|) - m_j(|f|)),
\end{aligned}$$

where $M = \sup |f|([a, b])$; that is, $|f(x)| \leq M$ for all $x \in [a, b]$. Multiplying the displayed inequality by Δx_j and summing over $j = 1, 2, \ldots, n$, we have

$$U(f^2, P) - L(f^2, P) \leq 2M(U(|f|, P) - L(|f|, P)).$$

Hence, it follows from Theorem 5.22 that f^2 is integrable on $[a, b]$.   ∎

We close this section with two integral analogues of the Mean Value Theorem.

5.24 Theorem.   [FIRST MEAN VALUE THEOREM FOR INTEGRALS].
Suppose that f and g are integrable on $[a, b]$ with $g(x) \geq 0$ for all $x \in [a, b]$. If

$$m = \inf f[a, b] \quad and \quad M = \sup f[a, b],$$

then there is a number $c \in [m, M]$ such that

$$\int_a^b f(x)g(x)\, dx = c \int_a^b g(x)\, dx.$$

In particular, if f is continuous on $[a, b]$, then there is an $x_0 \in [a, b]$ which satisfies

$$\int_a^b f(x)g(x)\, dx = f(x_0) \int_a^b g(x)\, dx.$$

Proof. Since $g \geq 0$ on $[a, b]$, Theorem 5.21 implies

$$m \int_a^b g(x)\, dx \leq \int_a^b f(x)g(x)\, dx \leq M \int_a^b g(x)\, dx.$$

If $\int_a^b g(x)dx = 0$, then $\int_a^b f(x)g(x)dx = 0$ and there is nothing to prove. Otherwise, set

$$c = \frac{\int_a^b f(x)g(x)\, dx}{\int_a^b g(x)\, dx}$$

and note that $c \in [m, M]$. If f is continuous, then (by the Intermediate Value Theorem) we can choose $x_0 \in [a, b]$ such that $f(x_0) = c$.   ∎

Before we state the Second Mean Value Theorem, we introduce an idea that will be used in the next section to prove the Fundamental Theorem of Calculus. If f is integrable on $[a, b]$, then f can be used to define a new function

$$F(x) := \int_a^x f(t)\, dt, \qquad x \in [a, b].$$

5.25 *EXAMPLE.*

Find $F(x) = \int_0^x f(t)\, dt$ if

$$f(x) = \begin{cases} 1 & x \geq 0 \\ -1 & x < 0. \end{cases}$$

Solution. By Theorem 5.16,

$$F(x) = \int_0^x f(t)\, dt = \begin{cases} x & x \geq 0 \\ -x & x < 0. \end{cases}$$

Hence, $F(x) = |x|$.   ∎

Notice in Example 5.25 that the integral F of f is continuous even though f itself is not. The following result shows that this is a general principle.

5.26 Theorem. *If f is (Riemann) integrable on $[a, b]$, then $F(x) = \int_a^x f(t)\, dt$ exists and is continuous on $[a, b]$.*

Proof. By Theorem 5.20, $F(x)$ exists for all $x \in [a, b]$. To prove that F is continuous on $[a, b]$, it suffices to show that $F(x+) = F(x)$ for all $x \in [a, b)$ and $F(x-) = F(x)$ for all $x \in (a, b]$. Fix $x_0 \in [a, b)$. By definition, f is bounded on $[a, b]$. Thus, choose $M \in \mathbf{R}$ such that $|f(t)| \leq M$ for all $t \in [a, b]$. Let $\varepsilon > 0$ and set $\delta = \varepsilon / M$. If $0 \leq x - x_0 < \delta$, then by Theorem 5.22,

$$|F(x) - F(x_0)| = \left| \int_{x_0}^x f(t)\, dt \right| \leq \int_{x_0}^x |f(t)|\, dt \leq M|x - x_0| < \varepsilon.$$

Hence, $F(x_0+) = F(x_0)$. A similar argument shows that $F(x_0-) = F(x_0)$ for all $x_0 \in (a, b]$.   ∎

5.27 Theorem. [SECOND MEAN VALUE THEOREM FOR INTEGRALS].
Suppose that f, g are integrable on $[a, b]$, that g is nonnegative on $[a, b]$, and that m, M are real numbers which satisfy $m \leq \inf f([a, b])$ and $M \geq \sup f([a, b])$. Then there is an $c \in [a, b]$ such that

$$\int_a^b f(x)g(x)\, dx = m \int_a^c g(x)\, dx + M \int_c^b g(x)\, dx.$$

In particular, if f is also nonnegative on $[a, b]$, then there is an $c \in [a, b]$ which satisfies

$$\int_a^b f(x)g(x)\, dx = M \int_c^b g(x)\, dx.$$

Proof. The second statement follows from the first since we may use $m = 0$ when $f \geq 0$. To prove the first statement, set

$$F(x) = m \int_a^x g(t)\, dt + M \int_x^b g(t)\, dt$$

for $x \in [a, b]$, and observe by Theorem 5.26 that F is continuous on $[a, b]$. Since g is nonnegative, we also have $mg(t) \leq f(t)g(t) \leq Mg(t)$ for all $t \in [a, b]$. Hence, it follows from the Comparison Theorem (Theorem 5.21) that

$$F(b) = m \int_a^b g(t)\, dt \leq \int_a^b f(t)g(t)\, dt \leq M \int_a^b g(t)\, dt = F(a).$$

Since F is continuous, we conclude by the Intermediate Value Theorem that there is an $c \in [a, b]$ such that

$$F(c) = \int_a^b f(t)g(t)\, dt.$$

When $g(x) = 1$ and $f(x) \geq 0$, these mean value theorems have simple geometric interpretations. Indeed, let A represent the area bounded by the curves $y = f(x)$, $y = 0$, $x = a$, and $x = b$. By the First Mean Value Theorem, there is a $c \in [m, M]$ such that the area of the rectangle of height c and base $b - a$ equals A (see Figure 5.3). And by the Second Mean Value Theorem, if M is the maximum value of f on $[a, b]$, then there is an $c \in [a, b]$ such that the area of the rectangle of height M and base $b - c$ equals A (see Figure 5.4).

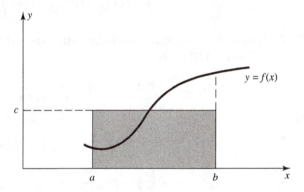

FIGURE 5.3

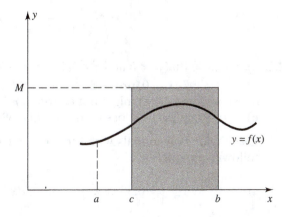

FIGURE 5.4

EXERCISES

5.2.0. Suppose that $a < b$. Decide which of the following statements are true and which are false. Prove the true ones and give counterexamples for the false ones.

a) If f and g are Riemann integrable on $[a, b]$, then $f - g$ is Riemann integrable on $[a, b]$.

b) If f is Riemann integrable on $[a, b]$ and P is any polynomial on $\mathbf{R}$, then $P \circ f$ is Riemann integrable on $[a, b]$.

c) If f and g are nonnegative real functions on $[a, b]$, with f continuous and g Riemann integrable on $[a, b]$, then there exist $x_0, x_1 \in [a, b]$ such that

$$\int_a^b f(x)g(x)\, dx = f(x_0) \int_{x_1}^b g(x)\, dx.$$

d) If f and g are Riemann integrable on $[a, b]$ and f is continuous, then there is an $x_0 \in [a, b]$ such that

$$\int_a^b f(x)g(x)\, dx = f(x_0) \int_a^b g(x)\, dx.$$

5.2.1. Using the connection between integrals and area, evaluate each of the following integrals.

a)
$$\int_{-2}^2 |x + 1|\, dx$$

b)
$$\int_{-2}^2 (|x + 1| + |x|)\, dx$$

c)
$$\int_{-a}^a \sqrt{a^2 - x^2}\, dx, \qquad a > 0$$

d)
$$\int_0^2 (5 + \sqrt{2x - x^2})\, dx$$

5.2.2. a) Suppose that $a < b$ and $n \in \mathbf{N}$ is even. If f is continuous on $[a, b]$ and $\int_a^b f(x)x^n dx = 0$, prove that $f(x) = 0$ for at least one $x \in [a, b]$.

b) Show that part a) might not be true if n is odd.

c) Prove that part a) does hold for odd n when $a \geq 0$.

5.2.3. Use Taylor polynomials with three or four nonzero terms to verify the following inequalities.

a)
$$0.3095 < \int_0^1 \sin(x^2)\, dx < 0.3103$$

(The value of this integral is approximately 0.3102683.)

b) $$1.4571 < \int_0^1 e^{x^2} \, dx < 1.5704$$

(The value of this integral is approximately 1.4626517.)

5.2.4. Suppose that $f : [0, \infty) \to [0, \infty)$ is integrable on every closed interval $[a, b] \subset [0, \infty)$. If

$$F(x) := \int_0^x e^{-y^2} f(y) \, dy, \quad x \in [0, \infty),$$

then there is a function $g : [0, \infty) \to [0, \infty)$ such that $F(x) = \int_{g(x)}^x f(y) \, dy$ for all $x \in [0, \infty)$.

5.2.5. Prove that if f is integrable on $[0, 1]$ and $\beta > 0$, then

$$\lim_{n \to \infty} n^\alpha \int_0^{1/n^\beta} f(x) \, dx = 0$$

for all $\alpha < \beta$.

5.2.6. a) Suppose that $g_n \geq 0$ is a sequence of integrable functions which satisfies

$$\lim_{n \to \infty} \int_a^b g_n(x) \, dx = 0.$$

Show that if $f : [a, b] \to \mathbf{R}$ is integrable on $[a, b]$, then

$$\lim_{n \to \infty} \int_a^b f(x) g_n(x) \, dx = 0.$$

b) Prove that if f is integrable on $[0, 1]$, then

$$\lim_{n \to \infty} \int_0^1 x^n f(x) \, dx = 0.$$

5.2.7. Suppose that f is integrable on $[a, b]$, that $x_0 = a$, and that x_n is a sequence of numbers in $[a, b]$ such that $x_n \uparrow b$ as $n \to \infty$. Prove that

$$\int_a^b f(x) \, dx = \lim_{n \to \infty} \sum_{k=0}^n \int_{x_k}^{x_{k+1}} f(x) \, dx.$$

5.2.8. Let f be continuous on a closed, nondegenerate interval $[a, b]$ and set

$$M = \sup_{x \in [a,b]} |f(x)|.$$

a) Prove that if $M > 0$ and $p > 0$, then for every $\varepsilon > 0$ there is a nondegenerate interval $I \subset [a, b]$ such that

$$(M - \varepsilon)^p |I| \leq \int_a^b |f(x)|^p \, dx \leq M^p(b - a).$$

b) Prove that

$$\lim_{p \to \infty} \left(\int_a^b |f(x)|^p \, dx \right)^{1/p} = M.$$

5.2.9. Let $f : [a, b] \to \mathbf{R}$, $a = x_0 < x_1 < \cdots < x_n = b$, and suppose that $f(x_k+)$ exists and is finite for $k = 0, 1, \ldots, n-1$ and $f(x_k-)$ exists and is finite for $k = 1, \ldots, n$. Show that if f is continuous on each subinterval (x_{k-1}, x_k), then f is integrable on $[a, b]$ and

$$\int_a^b f(x) \, dx = \sum_{k=1}^n \int_{x_{k-1}}^{x_k} f(x) \, dx.$$

5.2.10. Prove that if f and g are integrable on $[a, b]$, then so are $f \vee g$ and $f \wedge g$ (see Exercise 3.1.8).

5.2.11. Suppose that $f : [a, b] \to \mathbf{R}$.

a) If f is not bounded above on $[a, b]$, then given any partition P of $[a, b]$ and $M > 0$, there exist $t_j \in [x_{j-1}, x_j]$ such that $S(f, P, t_j) > M$.

b) If the Riemann sums of f converge to a finite number $I(f)$, as $\|P\| \to 0$, then f is bounded on $[a, b]$.

5.3   THE FUNDAMENTAL THEOREM OF CALCULUS

Let f be integrable on $[a, b]$ and $F(x) = \int_a^x f(t) \, dt$. By Theorem 5.26, F is continuous on $[a, b]$. The next result shows that if f is continuous, then F is continuously differentiable. Thus "indefinite integration" improves the behavior of the function.

5.28 Theorem.   [FUNDAMENTAL THEOREM OF CALCULUS].
Let $[a, b]$ be nondegenerate and suppose that $f : [a, b] \to \mathbf{R}$.

i) *If f is continuous on $[a, b]$ and $F(x) = \int_a^x f(t) \, dt$, then $F \in C^1[a, b]$ and*

$$\frac{d}{dx} \int_a^x f(t) \, dt := F'(x) = f(x)$$

for each $x \in [a, b]$.

ii) *If f is differentiable on $[a, b]$ and f' is integrable on $[a, b]$, then*

$$\int_a^x f'(t) \, dt = f(x) - f(a)$$

for each $x \in [a, b]$.

Proof. i) For $x \in [a, b]$, set $F(x) = \int_a^x f(t) \, dt$. By symmetry, it suffices to show that if $f(x_0+) = f(x_0)$ for some $x_0 \in [a, b)$, then

$$\lim_{h \to 0+} \frac{F(x_0 + h) - F(x_0)}{h} = f(x_0) \tag{11}$$

(see Definition 4.6). Let $\varepsilon > 0$ and choose a $\delta > 0$ such that $x_0 \leq t < x_0 + \delta$ implies $|f(t) - f(x_0)| < \varepsilon$. Fix $0 < h < \delta$. Notice that by Theorem 5.20,

$$F(x_0 + h) - F(x_0) = \int_{x_0}^{x_0+h} f(t) \, dt$$

and that by Theorem 5.16,

$$f(x_0) = \frac{1}{h} \int_{x_0}^{x_0+h} f(x_0) \, dt.$$

Therefore,

$$\frac{F(x_0 + h) - F(x_0)}{h} - f(x_0) = \frac{1}{h} \int_{x_0}^{x_0+h} (f(t) - f(x_0)) \, dt.$$

Since $0 < h < \delta$, it follows from Theorem 5.22 and the choice of δ that

$$\left| \frac{F(x_0 + h) - F(x_0)}{h} - f(x_0) \right| \leq \frac{1}{h} \int_{x_0}^{x_0+h} |f(t) - f(x_0)| \, dt \leq \varepsilon.$$

This verifies (11) and the proof of part i) is complete.

ii) We may suppose that $x = b$. Let $\varepsilon > 0$. Since f' is integrable, choose a partition $P = \{x_0, x_1, \ldots, x_n\}$ of $[a, b]$ such that

$$\left| \sum_{j=1}^{n} f'(t_j) \, \Delta x_j - \int_a^b f'(t) \, dt \right| < \varepsilon$$

for any choice of points $t_j \in [x_{j-1}, x_j]$. Use the Mean Value Theorem to choose points $t_j \in [x_{j-1}, x_j]$ such that $f(x_j) - f(x_{j-1}) = f'(t_j) \, \Delta x_j$. It follows by telescoping that

$$\left| f(b) - f(a) - \int_a^b f'(t) \, dt \right| = \left| \sum_{j=1}^{n} (f(x_j) - f(x_{j-1})) - \int_a^b f'(t) \, dt \right| < \varepsilon. \quad \blacksquare$$

5.29 Remark. The hypotheses of the Fundamental Theorem of Calculus cannot be relaxed.

Proof. i) Define f on $[-1, 1]$ by

$$f(x) = \begin{cases} -1 & x < 0 \\ 1 & x \geq 0; \end{cases}$$

then f is integrable on $[-1, 1]$, but $F(x) := \int_{-1}^{x} f(x) \, dx = |x| - 1$ is not differentiable at $x = 0$.

ii) Define f on $[0, 1]$ by $f(x) := x^2 \sin(1/x^2)$ when $x \neq 0$ and $f(0) = 0$. Then f is differentiable on $[0, 1]$, but

$$f'(x) = 2x \sin\left(\frac{1}{x^2}\right) - \frac{2}{x} \cdot \cos\left(\frac{1}{x^2}\right), \qquad x \neq 0,$$

is not even bounded on $(0, 1]$, much less integrable on $[0, 1]$. ∎

By the Fundamental Theorem of Calculus, integration is the inverse of differentiation in the following sense. If f' is integrable, then

$$\int_{a}^{b} f'(x) \, dx = f(x) \Big|_{a}^{b} := f(b) - f(a).$$

In particular,

$$\int_{a}^{b} x^\alpha \, dx = \frac{x^{\alpha+1}}{\alpha + 1} \Big|_{a}^{b}$$

for each $\alpha \geq 0$, and for each $\alpha < 0$, provided $\alpha \neq -1$ and $[a, b]$ is a subset of $(0, \infty)$ (see Exercises 4.2.3 and 5.3.7). This result is sometimes called the *Power Rule*.

These observations can be used to evaluate many integrals.

5.30 EXAMPLES.

i) Find $\int_{0}^{1} (3x - 2)^2 \, dx$.
ii) Find $\int_{0}^{\pi/2} (1 + \sin x) \, dx$.

Solution. i) Since $(3x - 2)^2 = 9x^2 - 12x + 4$, we have by the Power Rule that

$$\int_{0}^{1} (3x - 2)^2 \, dx = 3x^3 - 6x^2 + 4x \Big|_{0}^{1} = 1.$$

ii) Since $(\cos x)' = -\sin x$, we have by the Fundamental Theorem of Calculus that

$$\int_{0}^{\pi/2} (1 + \sin x) \, dx = x - \cos x \Big|_{0}^{\pi/2} = \frac{\pi}{2} + 1.$$ ∎

Combining the Product Rule and the Fundamental Theorem of Calculus, we have another tool for evaluating integrals.

5.31 Theorem.   [INTEGRATION BY PARTS].
Suppose that f, g are differentiable on [a, b] *with* f', g' *integrable on* [a, b]. *Then*

$$\int_a^b f'(x)g(x)\,dx = f(b)g(b) - f(a)g(a) - \int_a^b f(x)g'(x)\,dx.$$

Proof. By the Product Rule, $(f(x)g(x))' = f'(x)g(x) + f(x)g'(x)$ for $x \in [a, b]$. Since f, g are continuous on $[a, b]$ and f', g' are integrable on $[a, b]$, it follows that $(fg)'$ is a sum of products of integrable functions and, hence, integrable on $[a, b]$. Thus, by the Fundamental Theorem of Calculus,

$$f(b)g(b) - f(a)g(a) = \int_a^b f'(x)g(x)\,dx + \int_a^b f(x)g'(x)\,dx. \qquad \blacksquare$$

This rule is sometimes abbreviated as

$$\int u\,dv = uv - \int v\,du,$$

where it is understood that if $w = h(x)$ for some differentiable function h, then the *Leibnizian differential dw* is defined by $dw = h'(x)\,dx$.

Integration by parts can be used to reduce the exponent n on an expression of the form $(ax + b)^n f(x)$ when f is integrable.

5.32 EXAMPLE.

Find $\int_0^{\pi/2} x \sin x\,dx$.

Solution. Let $u = x$ and $dv = \sin x\,dx$. Then $du = dx$ and $v = -\cos x$. Hence, by parts,

$$\int_0^{\pi/2} x \sin x\,dx = -x \cos x \Big|_0^{\pi/2} - \int_0^{\pi/2} (-\cos x)\,dx = \sin x \Big|_0^{\pi/2} = 1. \qquad \blacksquare$$

Integration by parts is also very effective on integrals involving products of polynomials and logarithms.

5.33 EXAMPLE.

Find $\int_1^3 \log x \, dx$.

Solution. Let $u = \log x$ and $dv = dx$. Then $du = dx/x$ and $v = x$. Hence, by parts,

$$\int_1^3 \log x \, dx = x \log x \Big|_1^3 - \int_1^3 dx = 3 \log 3 - 2. \qquad \blacksquare$$

Complicated problems can frequently be reduced to simpler ones by changing variables. The following result shows how to change variables in a Riemann integral on **R**.

5.34 Theorem.   [CHANGE OF VARIABLES].
Let ϕ be continuously differentiable on a closed, nondegenerate interval $[a, b]$. If

$$\phi' \text{ is nonzero on } [a, b], \tag{12}$$

and if

$$f \text{ is integrable on } [c, d] := \phi[a, b], \tag{13}$$

then $f \circ \phi \cdot |\phi'|$ is integrable on $[a, b]$, and

$$\int_c^d f(t) \, dt = \int_a^b f(\phi(x)) \cdot |\phi'(x)| \, dx. \tag{14}$$

STRATEGY: By the Mean Value Theorem, hypothesis (12) implies that ϕ is 1–1 on $[a, b]$. Hence by Theorem 4.32, ϕ is strictly monotone on $[a, b]$ and $[c, d] := \phi[a, b]$ is a closed interval.

Suppose that ϕ is strictly increasing on $[a, b]$; that is, $|\phi'| = \phi'$ and $[c, d] = [\phi(a), \phi(b)]$. By Theorem 4.32, ϕ^{-1} is increasing on $[c, d]$. Thus if $P = \{t_0, t_1, \ldots, t_n\}$ is a partition of $[c, d]$ and $x_j = \phi^{-1}(t_j)$, then $\tilde{P} := \{x_0, x_1, \ldots, x_n\}$ is a partition of $[a, b]$. A Riemann sum of the right side of (14) looks like

$$S(f \circ \phi \cdot |\phi'|, \tilde{P}, s_j) = \sum_{j=1}^n f(\phi(s_j))|\phi'(s_j)| \, \Delta t_j. \tag{15}$$

On the other hand, a typical term of a Riemann sum of the left side of (14) looks like

$$f(u_j) \, \Delta x_j = f(u_j)(\phi(t_j) - \phi(t_{j-1})).$$

Since ϕ', hence ϕ, is continuous, we can use the Intermediate Value Theorem to choose $s_j \in [x_{j-1}, x_j]$ such that $u_j = \phi(s_j)$, and the Mean Value Theorem to choose $c_j \in [x_{j-1}, x_j]$ such that $\phi(x_j) - \phi(x_{j-1}) = \phi'(c_j)\Delta x_j$. It follows that a Riemann sum the left side of (14) looks like

$$S(f, \widetilde{P}, u_j) = \sum_{j=1}^{n} f(\phi(s_j))\phi'(c_j)\,\Delta x_j.$$

The only difference between this last sum and (15) is that s_j has been replaced by c_j. Since c_j and s_j both belong to the interval $[x_{j-1}, x_j]$ and ϕ' is continuous, making this replacement should not change S much if the norm of P is small enough. Hence, a Riemann sum of the left side of (14) is approximately equal to a Riemann sum of the right side of (14). This means the integrals in (14) should be equal. Here are the details.

Case 1. Suppose that ϕ is strictly increasing on $[a, b]$. Let $\varepsilon > 0$. Since f is bounded, choose $M \in (0, \infty)$ such that $|f(x)| \le M$ for all $x \in [c, d]$. Since ϕ' is uniformly continuous on $[a, b]$, choose $\delta > 0$ such that

$$|\phi'(s_j) - \phi'(c_j)| < \frac{\varepsilon}{2M(b-a)};$$

that is,

$$|f(\phi(s_j))(\phi'(s_j) - \phi'(c_j))| < \frac{\varepsilon}{2(b-a)} \tag{16}$$

for all $s_j, c_j \in [a, b]$ with $|s_j - c_j| < \delta$.

Next, use the Inverse Function Theorem to verify that ϕ^{-1} is continuously differentiable on $[c, d]$. Thus there is an $\eta > 0$ such that if $s, c \in [c, d]$ and $|s - c| < \eta$, then $|\phi^{-1}(s) - \phi^{-1}(c)| < \delta$.

Finally, since f is integrable on $[c, d] = [\phi(a), \phi(b)]$, choose a partition $P = \{t_0, t_1, \dots t_n\}$ of $[c, d]$ such that $\|P\| < \eta$ and

$$\left| S(f, P, u_j) - \int_{\phi(a)}^{\phi(b)} f(t)\,dt \right| < \frac{\varepsilon}{2} \tag{17}$$

holds for any choice of $u_j \in [t_{j-1}, t_j]$.

Set $x_j = \phi^{-1}(t_j)$ and observe (by the choice of η) that $\widetilde{P} := \{x_0, \dots, x_n\}$ is a partition of $[a, b]$ which satisfies $\|\widetilde{P}\| < \delta$.

Let $s_j \in [x_{j-1}, x_j]$, set $u_j = \phi(s_j)$, and apply the Mean Value Theorem to choose $c_j \in [x_{j-1}, x_j]$ such that $\phi(x_j) - \phi(x_{j-1}) = \phi'(c_j)\,\Delta x_j$. Then, by the choices of c_j, u_j, and t_j, we have $u_j \in [t_{j-1}, t_j]$ and

$$f(\phi(s_j))\phi'(c_j)\,\Delta x_j = f(u_j)(\phi(x_j) - \phi(x_{j-1})) = f(u_j)(t_j - t_{j-1}).$$

Hence, it follows from (16) and (17) that

$$
\left| \sum_{j=1}^{n} f(\phi(s_j))\phi'(s_j)\,\Delta x_j - \int_{\phi(a)}^{\phi(b)} f(t)\,dt \right|
$$

$$
\leq \left| \sum_{j=1}^{n} f(\phi(s_j))(\phi'(s_j) - \phi'(c_j))\,\Delta x_j \right|
$$

$$
+ \left| \sum_{j=1}^{n} f(u_j)(t_j - t_{j-1}) - \int_{\phi(a)}^{\phi(b)} f(t)\,dt \right|
$$

$$
< \frac{\varepsilon}{2(b-a)} \sum_{j=1}^{n} \Delta x_j + \frac{\varepsilon}{2} < \varepsilon.
$$

We obtained this estimate for the fixed partition $\widetilde{P}$ of $[a, b]$, but the same steps also verify this estimate for any partition finer than $\widetilde{P}$. We conclude by Theorem 5.18 that $(f \circ \phi) \cdot |\phi'|$ is integrable on $[a, b]$ and (14) holds.

Case 2. ϕ is strictly decreasing on $[a, b]$. Repeat the proof in case 1. The only changes are $\widetilde{P} = \{\phi^{-1}(t_n), \ldots, \phi^{-1}(t_0)\}$ and $|\phi'| = -\phi'$. Thus the Mean Value Theorem implies that

$$
\phi(x_{j-1}) - \phi(x_j) = \phi'(c_j)(x_{j-1} - x_j) = |\phi'(c_j)|\,\Delta x_j.
$$

Estimating the Riemann sums as above, we again conclude that

$$
\int_{c}^{d} f(t)\,dt = \int_{a}^{b} f(\phi(x)) \cdot |\phi'(x)|\,dx. \qquad \blacksquare
$$

The proof of Theorem 5.34 also establishes the following more familiar form of the Change of Variables Formula: If ϕ is C^1 on $[a, b]$, if ϕ' is never zero on $[a, b]$, and if f is integrable on $\phi[a, b]$, then

$$
\int_{\phi(a)}^{\phi(b)} f(t)\,dt = \int_{a}^{b} f(\phi(x))\phi'(x)\,dx.
$$

The difficult part of Theorem 5.34 was verifying that $f \circ \phi \cdot |\phi'|$ is integrable on $[a, b]$ when f is integrable on $[c, d]$. If we assume that f is continuous, the proof is a lot easier.

5.35 Theorem. [CHANGE OF VARIABLES FOR CONTINUOUS INTEGRANDS].
If ϕ is continuously differentiable on a closed, nondegenerate interval $[a, b]$ and f is continuous on $\phi([a, b])$, then

$$
\int_{\phi(a)}^{\phi(b)} f(t)\,dt = \int_{a}^{b} f(\phi(x))\phi'(x)\,dx.
$$

Proof. Set

$$G(x) := \int_a^x f(\phi(t))\phi'(t)\,dt, \quad x \in [a, b], \quad \text{and}$$

$$F(u) := \int_{\phi(a)}^u f(t)\,dt, \quad u \in \phi([a, b]),$$

and observe that if m is the infimum of $\phi([a, b])$, then $F(u) = \int_m^u f(t)dt - \int_m^{\phi(a)} f(t)dt$. It follows from the Fundamental Theorem of Calculus that $G'(x) = f(\phi(x))\phi'(x)$ and $F'(u) = f(u)$. Hence, by the Chain Rule,

$$\frac{d}{dx}(G(x) - F(\phi(x))) = 0$$

for all $x \in [a, b]$. It follows from Theorem 4.17ii that $G(x) - F(\phi(x))$ is constant on $[a, b]$. Evaluation at $x = a$ shows that this constant is zero. Thus $G(x) = F(\phi(x))$ for all $x \in [a, b]$, in particular, when $x = b$. ∎

These Change of Variables Formulas can be remembered as a substitution if we use the Leibnizian differentials introduced above: $u = \phi(x)$ implies $du = \phi'(x)dx$.

Besides the usual applications to finding exact values for integrals of compositions of functions, the Change of Variables Formula can also be used to estimate the value of an integral. Since energy, power, force, and many other physical quantities can be computed by integration, this technique has practical implications. For example, it sometimes allows one to use data from a particular prototype to estimate what would happen if the prototype were redesigned, without going to the expense of building another prototype.

5.36 *EXAMPLE.*

Suppose that f is an unknown function which is nonnegative and continuous on $[2, 5]$. If data are collected that can be interpreted as $\int_2^5 f(x)dx = 3$, find an upper bound for the integral

$$I := \int_1^2 f(x^2 + 1)\,dx.$$

Solution. Let $u = x^2 + 1$. Then $du = 2x\,dx$. Unlike textbook-style problems, we do not have a du term already in I. However, since $x \in [1, 2]$ implies $x \geq 1$, and since f is nonnegative, it is clear that $f(x^2 + 1) \leq 2xf(x^2 + 1)/2$. Therefore,

$$I = \int_1^2 f(x^2 + 1)\,dx \leq \frac{1}{2}\int_1^2 2xf(x^2 + 1)\,dx = \frac{1}{2}\int_2^5 f(u)\,du = \frac{3}{2}. \quad \blacksquare$$

EXERCISES

5.3.0. Suppose that $a < b$. Decide which of the following statements are true and which are false. Prove the true ones and give counterexamples for the false ones.

a) If f is continuous and nonnegative on $[a, b]$ and $g : [a, b] \to [a, b]$ is differentiable and increasing on $[a, b]$, then

$$F(x) := \int_a^{g(x)} f(t)\, dt$$

is increasing on $[a, b]$.

b) If f and g are differentiable on $[a, b]$, if f' and g' are Riemann integrable on $[a, b]$, and if $f(a) = 0$ but g is never zero on $[a, b]$, then

$$f(x) = \int_a^x g(t) \left(\frac{f(t)}{g(t)} \right)' dt + \int_a^x \frac{f(t)g'(t)}{g(t)}\, dt$$

for all $x \in [a, b]$.

c) If f and g are differentiable on $[a, b]$, and if f' and g' are Riemann integrable on $[a, b]$, then

$$\int_a^b f'(x)g(x)\, dx + \int_a^b f(x)g'(x)\, dx = 0$$

if and only if $f(a)g(a) = f(b)g(b)$.

d) If f and g are continuously differentiable on $[a, b]$, and if h is continuous on $[a, b]$, then

$$\int_{g(f(a))}^{g(f(b))} h(x)\, dx = \int_a^b h(g(f(x)))g'(f(x))f'(x)\, dx.$$

5.3.1. If $f : \mathbf{R} \to \mathbf{R}$ is continuous, find $F'(x)$ for each of the following functions.

a) $$F(x) = \int_{x^2}^1 f(t)\, dt$$

b) $$F(x) = \int_{x^2}^{x^3} f(t)\, dt$$

c) $$F(x) = \int_0^{x \cos x} t f(t)\, dt$$

d) $$F(x) = \int_0^x f(t - x)\, dt$$

5.3.2. Suppose that f is nonnegative and continuous on $[1, 2]$ and that $\int_1^2 x^k f(x)\,dx = 5 + k^2$ for $k = 0, 1, 2$. Prove that each of the following statements is correct.

a)
$$\int_1^4 f(\sqrt{x})\,dx \leq 20$$

b)
$$\int_{\sqrt{2}/2}^1 f\left(\frac{1}{x^2}\right) dx \leq \frac{5}{2}$$

c)
$$\int_0^1 x^2 f(x + 1)\,dx = 2$$

5.3.3. Suppose that f is integrable on $[0.5, 2]$ and that

$$\int_{0.5}^1 x^k f(x)\,dx = \int_1^2 x^k f(x)\,dx + 2k^2 = 3 + k^2$$

for $k = 0, 1, 2$. Compute the exact values of each of the following integrals.

a)
$$\int_0^1 x^3 f(x^2 + 1)\,dx$$

b)
$$\int_0^{\sqrt{3}/2} \frac{x^3}{\sqrt{1 - x^2}} f\left(\sqrt{1 - x^2}\right) dx$$

5.3.4. Suppose that f and g are differentiable on $[0, e]$ and that f' and g' are integrable on $[0, e]$.

a) If $\int_1^e f(x)/x\,dx < f(e)$, prove that

$$\int_1^e f'(x) \log x\,dx > 0.$$

b) If $f(0) = f(1) = 0$, prove that

$$\int_0^1 e^x(f(x) + f'(x))\,dx = 0.$$

c) If $0 \in \{f(0), g(0)\} \cap \{f(e), g(e)\}$, prove that

$$\int_0^e f(x)g'(x)\,dx = -\int_0^e g(x)f'(x)\,dx.$$

5.3.5. Use the First Mean Value Theorem for Integrals to prove the following version of the Mean Value Theorem for Derivatives. If $f \in C^1[a, b]$, then there is an $x_0 \in [a, b]$ such that

$$f(b) - f(a) = (b - a)f'(x_0).$$

5.3.6. If f is continuous on $[a, b]$ and there exist numbers $\alpha \neq \beta$ such that

$$\alpha \int_a^c f(x)\, dx + \beta \int_c^b f(x)\, dx = 0$$

holds for all $c \in (a, b)$, prove that $f(x) = 0$ for all $x \in [a, b]$.

5.3.7 . **This exercise is used in Sections 5.4 and 6.1.** Define $L : (0, \infty) \to \mathbf{R}$ by

$$L(x) = \int_1^x \frac{dt}{t}.$$

a) Prove that L is differentiable and strictly increasing on $(0, \infty)$, with $L'(x) = 1/x$ and $L(1) = 0$.

b) Prove that $L(x) \to \infty$ as $x \to \infty$ and $L(x) \to -\infty$ as $x \to 0+$. (You may wish to prove

$$L(2^n) = \sum_{k=1}^n \int_{2^{k-1}}^{2^k} \frac{dt}{t} > \sum_{k=1}^n 2^{-k} \left(2^k - 2^{k-1}\right) = \frac{n}{2}$$

for all $n \in \mathbf{N}$.)

c) Using the fact that $(x^q)' = qx^{q-1}$ for $x > 0$ and $q \in \mathbf{Q}$ (see Exercise 4.2.7), prove that $L(x^q) = qL(x)$ for all $q \in \mathbf{Q}$ and $x > 0$.

d) Prove that $L(xy) = L(x) + L(y)$ for all $x, y \in (0, \infty)$.

e) Suppose that $e = \lim_{n \to \infty}(1 + 1/n)^n$ exists. (It does—see Example 4.22.) Use l'Hôpital's Rule to show that $L(e) = 1$. [$L(x)$ is the *natural logarithm* function $\log x$.]

5.3.8 . **This exercise was used in Section 4.3.** Let $E = L^{-1}$, where L is defined in Exercise 5.3.7.

a) Use the Inverse Function Theorem to show that E is differentiable and strictly increasing on **R** with $E'(x) = E(x)$, $E(0) = 1$, and $E(1) = e$.

b) Prove that $E(x) \to \infty$ as $x \to \infty$ and $E(x) \to 0$ as $x \to -\infty$.

c) Prove that $E(xq) = (E(x))^q$ and $E(q) = e^q$ for all $q \in \mathbf{Q}$ and $x \in \mathbf{R}$.

d) Prove that $E(x + y) = E(x)E(y)$ for all $x, y \in \mathbf{R}$.

e) For each $\alpha \in \mathbf{R}$ define $e^\alpha = E(\alpha)$. Let $x > 0$ and define $x^\alpha = e^{\alpha \log x} := E(\alpha L(x))$. Prove that $0 < x < y$ implies $x^\alpha < y^\alpha$ for $\alpha > 0$ and $x^\alpha > y^\alpha$ for $\alpha < 0$. Also prove that

$$x^{\alpha+\beta} = x^\alpha x^\beta, \qquad x^{-\alpha} = \frac{1}{x^\alpha}, \qquad \text{and} \quad (x^\alpha)' = \alpha x^{\alpha-1}$$

for all $\alpha, \beta \in \mathbf{R}$ and $x > 0$.

5.3.9. Suppose that $f : [a, b] \to \mathbf{R}$ is continuously differentiable and 1–1 on $[a, b]$. Prove that

$$\int_a^b f(x)\, dx + \int_{f(a)}^{f(b)} f^{-1}(x)\, dx = bf(b) - af(a).$$

5.3.10. Suppose that ϕ is C^1 on $[a, b]$ and f is integrable on $[c, d] := \phi[a, b]$. If ϕ' is never zero on $[a, b]$, prove that $f \circ \phi$ is integrable on $[a, b]$.

5.3.11. Let $q \in \mathbf{Q}$. Suppose that $a < b, 0 < c < d$, and that $f : [a, b] \to [c, d]$. If f is integrable on $[a, b]$, then prove that

$$\left(\int_a^x f^q(t)\, dt \right)' = f^q(x)$$

for all $x \in [a, b]$.

5.3.12. For each $n \in \mathbf{N}$, define

$$a_n := \left(\frac{(2n)!}{n! n^n} \right)^{1/n}.$$

Prove that $a_n \to 4/e$.

5.4 IMPROPER RIEMANN INTEGRATION

To extend the Riemann integral to unbounded intervals or unbounded functions, we begin with an elementary observation.

5.37 Remark. *If f is integrable on $[a, b]$, then*

$$\int_a^b f(x)\, dx = \lim_{c \to a+} \left(\lim_{d \to b-} \int_c^d f(x)\, dx \right).$$

Proof. By Theorem 5.26,

$$F(x) = \int_a^x f(t)\, dt$$

is continuous on $[a, b]$. Thus

$$\int_a^b f(x)\, dx = F(b) - F(a) = \lim_{c \to a+} (\lim_{d \to b-} (F(d) - F(c)))$$

$$= \lim_{c \to a+} \left(\lim_{d \to b-} \int_c^d f(x)\, dx \right). \qquad \blacksquare$$

This leads to the following generalization of the Riemann integral.

> **5.38 Definition.**
> ___
>
> Let (a, b) be a nonempty, open (possibly unbounded) interval and $f : (a, b) \to \mathbf{R}$.
>
> i) f is said to be *locally integrable* on (a, b) if and only if f is integrable on each closed subinterval $[c, d]$ of (a, b).
> ii) f is said to be *improperly integrable* on (a, b) if and only if f is locally integrable on (a, b) and
>
> $$\int_a^b f(x)\, dx := \lim_{c \to a+} \left(\lim_{d \to b-} \int_c^d f(x)\, dx \right) \qquad (18)$$
>
> exists and is finite. This limit is called the *improper (Riemann) integral* of f over (a, b).

5.39 Remark. *The order of the limits in* (18) *does not matter. In particular, if the limit in* (18) *exists, then*

$$\int_a^b f(x)\, dx = \lim_{d \to b-} \left(\lim_{c \to a+} \int_c^d f(x)\, dx \right).$$

Proof. Let $x_0 \in (a, b)$ be fixed. By Theorems 5.20 and 3.8,

$$\lim_{c \to a+} \left(\lim_{d \to b-} \int_c^d f(x)\, dx \right) = \lim_{c \to a+} \left(\int_c^{x_0} f(x)\, dx + \lim_{d \to b-} \int_{x_0}^d f(x)\, dx \right)$$

$$= \lim_{c \to a+} \int_c^{x_0} f(x)\, dx + \lim_{d \to b-} \int_{x_0}^d f(x)\, dx$$

$$= \lim_{d \to b-} \left(\lim_{c \to a+} \int_c^d f(x)\, dx \right). \qquad \blacksquare$$

Thus we shall use the notation

$$\lim_{\substack{c \to a+ \\ d \to b-}} \int_c^d f(x)\, dx$$

to represent the limit in (18). If the integral is not improper at one of the endpoints—for example, if f is Riemann integrable on closed subintervals of $(a, b]$—we shall say that f is improperly integrable on $(a, b]$ and simplify the notation even further by writing

$$\int_a^b f(x)\, dx = \lim_{c \to a+} \int_c^b f(x)\, dx.$$

The following example shows that an improperly integrable function need not be bounded.

5.40 EXAMPLE.

Show that $f(x) = 1/\sqrt{x}$ is improperly integrable on $(0, 1]$.

Solution. By definition,

$$\int_0^1 \frac{1}{\sqrt{x}} \, dx = \lim_{a \to 0+} \int_a^1 \frac{1}{\sqrt{x}} \, dx = \lim_{a \to 0+} (2 - 2\sqrt{a}) = 2. \qquad \blacksquare$$

The following example shows that a function can be improperly integrable on an unbounded interval.

5.41 EXAMPLE.

Show that $f(x) = 1/x^2$ is improperly integrable on $[1, \infty)$.

Solution. By definition,

$$\int_1^\infty \frac{1}{x^2} \, dx = \lim_{d \to \infty} \int_1^d \frac{1}{x^2} \, dx = \lim_{d \to \infty} \left(1 - \frac{1}{d} \right) = 1. \qquad \blacksquare$$

Because an improper integral is a limit of Riemann integrals, many of the results we proved earlier in this chapter have analogues for the improper integral. The next two results illustrate this principle.

5.42 Theorem. *If f, g are improperly integrable on (a, b) and $\alpha, \beta \in \mathbf{R}$, then $\alpha f + \beta g$ is improperly integrable on (a, b) and*

$$\int_a^b (\alpha f(x) + \beta g(x)) \, dx = \alpha \int_a^b f(x) \, dx + \beta \int_a^b g(x) \, dx.$$

Proof. By Theorem 5.19 (the Linear Property for Riemann Integrals),

$$\int_c^d (\alpha f(x) + \beta g(x)) \, dx = \alpha \int_c^d f(x) \, dx + \beta \int_c^d g(x) \, dx$$

for all $a < c < d < b$. Taking the limit as $c \to a+$ and $d \to b-$ finishes the proof. $\qquad \blacksquare$

5.43 Theorem. **[COMPARISON THEOREM FOR IMPROPER INTEGRALS].**
Suppose that f, g are locally integrable on (a, b). If $0 \le f(x) \le g(x)$ for $x \in (a, b)$, and g is improperly integrable on (a, b), then f is improperly integrable on (a, b) and

$$\int_a^b f(x) \, dx \le \int_a^b g(x) \, dx.$$

Proof. Fix $c \in (a, b)$. Let $F(d) = \int_c^d f(x)dx$ and $G(d) = \int_c^d g(x)dx$ for $d \in [c, b)$. By the Comparison Theorem for Integrals, $F(d) \le G(d)$. Since $f \ge 0$, the function F is increasing on $[c, b]$; hence $F(b-)$ exists (see Theorem 4.18). Thus, by definition, f is improperly integrable on (c, b) and

$$\int_c^b f(x) \, dx = F(b-) \le G(b-) = \int_c^b g(x) \, dx.$$

A similar argument works for the case $c \to a+$. ∎

This test is frequently used in conjunction with the following inequalities: $|\sin x| \le |x|$ for all $x \in \mathbf{R}$ (see Appendix B); for every $\alpha > 0$ there exists a constant $B_\alpha > 1$ such that $|\log x| \le x^\alpha$ for all $x \ge B_\alpha$ (see Exercise 4.4.6). Here are two typical examples.

5.44 EXAMPLE.

Prove that $f(x) = \sin x / \sqrt{x^3}$ is improperly integrable on $(0, 1]$.

Proof. Since f is continuous on $(0, 1]$, f is locally integrable there as well. Since f is nonnegative on $(0, 1]$, it is clear that $0 \le f(x) = |\sin x / \sqrt{x^3}| \le |x|/x^{3/2} = 1/\sqrt{x}$ on $(0, 1]$. Since this last function is improperly integrable on $(0, 1]$ by Example 5.40, it follows from the Comparison Test that $f(x)$ is improperly integrable on $(0, 1]$. ∎

5.45 EXAMPLE.

Prove that $f(x) = \log x / \sqrt{x^5}$ is improperly integrable on $[1, \infty)$.

Proof. Since f is continuous on $(0, \infty)$, f is integrable on $[1, C]$ for any $C \in [1, \infty)$. By Exercise 4.4.6, there is a constant $C > 1$ such that $0 \le f(x) = \log x / \sqrt{x^5} \le x^{1/2}/x^{5/2} = 1/x^2$ for $x \ge C$. Since this last function is improperly integrable on $[1, \infty)$ by Example 5.41, it follows from the Comparison Theorem that $f(x)$ is improperly integrable on $[1, \infty)$. ∎

Although improperly integrable functions are not closed under multiplication (see Exercise 5.4.5), the Comparison Theorem can be used to show that some kinds of products are improperly integrable.

5.46 Remark. *If f is bounded and locally integrable on (a, b) and $|g|$ is improperly integrable on (a, b), then $|fg|$ is improperly integrable on (a, b).*

Proof. Let $M = \sup_{x \in (a,b)} |f(x)|$. Then $0 \le |f(x)g(x)| \le M|g(x)|$ for all $x \in (a, b)$. Hence, by Theorem 5.43, $|fg|$ is improperly integrable on (a, b). ∎

For the Riemann integral, we proved that $|f|$ is integrable when f is (see Theorem 5.22). This is not the case for the improper integral (see Example 5.49 below). For this reason we introduce the following concepts.

5.47 Definition.

Let (a, b) be a nonempty, open interval and $f : (a, b) \to \mathbf{R}$.

i) f is said to be *absolutely integrable* on (a, b) if and only if f is locally integrable and $|f|$ is improperly integrable on (a, b).

ii) f is said to be *conditionally integrable* on (a, b) if and only if f is improperly integrable but not absolutely integrable on (a, b).

The following result, an analogue of Theorem 5.22 for absolute integrable functions, shows that absolute integrability implies improper integrability.

5.48 Theorem. *If f is absolutely integrable on (a, b), then f is improperly integrable on (a, b) and*

$$\left| \int_a^b f(x)\, dx \right| \le \int_a^b |f(x)|\, dx.$$

Proof. Since $0 \le |f(x)| + f(x) \le 2|f(x)|$, we have by Theorem 5.43 that $|f| + f$ is improperly integrable on $[a, b]$. Hence, by Theorem 5.42, so is $f = (|f| + f) - |f|$. Moreover,

$$\left| \int_c^d f(x)\, dx \right| \le \int_c^d |f(x)|\, dx$$

for every $a < c < d < b$. We finish the proof by taking the limit of this last inequality as $c \to a+$ and $d \to b-$. ∎

The converse of Theorem 5.48, however, is false.

5.49 EXAMPLE.

Prove that the function $\sin x / x$ is conditionally integrable on $[1, \infty)$.

Proof. Integrating by parts, we have

$$\int_1^d \frac{\sin x}{x}\, dx = -\frac{\cos x}{x}\Big|_1^d - \int_1^d \frac{\cos x}{x^2}\, dx$$

$$= \cos(1) - \frac{\cos d}{d} - \int_1^d \frac{\cos x}{x^2}\, dx.$$

Since $1/x^2$ is absolutely integrable on $[1, \infty)$, it follows from Remark 5.46 that $\cos x / x^2$ is absolutely integrable on $[1, \infty)$. Therefore, $\sin x / x$ is improperly integrable on $[1, \infty)$ and

$$\int_1^\infty \frac{\sin x}{x}\, dx = \cos(1) - \int_1^\infty \frac{\cos x}{x^2}\, dx.$$

To show that $\sin x / x$ is not absolutely integrable on $[1, \infty)$, notice that

$$\int_1^{n\pi} \frac{|\sin x|}{x}\, dx \geq \sum_{k=2}^{n} \int_{(k-1)\pi}^{k\pi} \frac{|\sin x|}{x}\, dx$$

$$\geq \sum_{k=2}^{n} \frac{1}{k\pi} \int_{(k-1)\pi}^{k\pi} |\sin x|\, dx$$

$$= \sum_{k=2}^{n} \frac{2}{k\pi} = \frac{2}{\pi} \sum_{k=2}^{n} \frac{1}{k}$$

for each $n \in \mathbf{N}$. Since

$$\sum_{k=2}^{n} \frac{1}{k} \geq \sum_{k=2}^{n} \int_k^{k+1} \frac{1}{x}\, dx = \int_2^{n+1} \frac{1}{x}\, dx = \log(n+1) - \log 2 \to \infty$$

as $n \to \infty$, it follows from the Squeeze Theorem that

$$\lim_{n \to \infty} \int_1^{n\pi} \frac{|\sin x|}{x}\, dx = \infty.$$

Thus, $\sin x / x$ is not absolutely integrable on $[1, \infty)$. ∎

EXERCISES

5.4.0. Suppose that $a < b$. Decide which of the following statements are true and which are false. Prove the true ones and give counterexamples for the false ones.

  a) If f is bounded on $[a, b]$, if g is absolutely integrable on (a, b), and if $|f(x)| \leq g(x)$ for all $x \in (a, b)$, then f is absolutely integrable on (a, b).

  b) Suppose that h is absolutely integrable on (a, b). If f is continuous on (a, b), if g is continuous and never zero on $[a, b]$, and if $|f(x)| \leq h(x)$ for all $x \in [a, b]$, then f/g is absolutely integrable on (a, b).

  c) If $f : (a, b) \to [0, \infty)$ is continuous and absolutely integrable on (a, b) for some $a, b \in \mathbf{R}$, then $\sqrt{f}$ is absolutely integrable on (a, b).

  d) If f and g are absolutely integrable on (a, b), then $\max\{f, g\}$ and $\min\{f, g\}$ are both absolutely integrable on (a, b).

5.4.1. Evaluate the following improper integrals.

  a)
$$\int_1^\infty \frac{1+x}{x^3}\, dx$$

  b)
$$\int_{-\infty}^0 x^2 e^{x^3}\, dx$$

c)
$$\int_0^{\pi/2} \frac{\cos x}{\sqrt[3]{\sin x}}\, dx$$

d)
$$\int_0^1 \log x \, dx$$

5.4.2. For each of the following, find all values of $p \in \mathbf{R}$ for which f is improperly integrable on I.

a) $f(x) = 1/x^p$, $I = (1, \infty)$
b) $f(x) = 1/x^p$, $I = (0, 1)$
c) $f(x) = 1/(x \log^p x)$, $I = (e, \infty)$
d) $f(x) = 1/(1 + x^p)$, $I = (0, \infty)$
e) $f(x) = \log^a x / x^p$, where $a > 0$ is fixed, and $I = (1, \infty)$

5.4.3. Let $p > 0$. Show that $\sin x / x^p$ is improperly integrable on $[1, \infty)$ and that $\cos x / \log^p x$ is improperly integrable on $[e, \infty)$.

5.4.4. Decide which of the following functions are improperly integrable on I.

a) $f(x) = \sin x$, $I = (0, \infty)$
b) $f(x) = 1/x^2$, $I = [-1, 1]$
c) $f(x) = x^{-1} \sin(x^{-1})$, $I = (1, \infty)$
d) $f(x) = \log(\sin x)$, $I = (0, 1)$
e) $f(x) = (1 - \cos x)/x^2$, $I = (0, \infty)$

5.4.5. Use the examples provided by Exercise 5.4.2b to show that the product of two improperly integrable functions might not be improperly integrable.

5.4.6. Suppose that f, g are nonnegative and locally integrable on $[a, b)$ and that

$$L := \lim_{x \to b-} \frac{f(x)}{g(x)}$$

exists as an extended real number.

a) Show that if $0 \le L < \infty$ and g is improperly integrable on $[a, b)$, then so is f.
b) Show that if $0 < L \le \infty$ and g is not improperly integrable on $[a, b)$, then neither is f.

5.4.7. a) Suppose that f is improperly integrable on $[0, \infty)$. Prove that if $L = \lim_{x \to \infty} f(x)$ exists, then $L = 0$.

b) Let

$$f(x) = \begin{cases} 1 & n \le x < n + 2^{-n}, \ n \in \mathbf{N} \\ 0 & \text{otherwise.} \end{cases}$$

Prove that f is improperly integrable on $[0, \infty)$ but $\lim_{x \to \infty} f(x)$ does not exist.

5.4.8. Prove that if f is absolutely integrable on $[1, \infty)$, then

$$\lim_{n \to \infty} \int_1^\infty f(x^n)\, dx = 0.$$

5.4.9. Assuming $e = \lim_{n \to \infty} \sum_{k=0}^n 1/k!$ (see Example 7.45), prove that

$$\lim_{n \to \infty} \left(\frac{1}{n!} \int_1^\infty x^n e^{-x}\, dx \right) = 1.$$

5.4.10. a) Prove that

$$\int_0^{\pi/2} e^{-a \sin x}\, dx \le \frac{2}{a}$$

for all $a > 0$.

b) What happens if $\cos x$ replaces $\sin x$?

*5.5 FUNCTIONS OF BOUNDED VARIATION

This section uses no material from any other enrichment section.

In this section we study functions which do not wiggle too much. These functions, which play a prominent role in the theory of Fourier series (see Sections *14.3 and *14.4) and probability theory, are important tools for theoretical as well as applied mathematics.

Let $\phi : [a, b] \to \mathbf{R}$. To measure how much ϕ wiggles on an interval $[a, b]$, set

$$V(\phi, P) = \sum_{j=1}^n |\phi(x_j) - \phi(x_{j-1})|$$

for each partition $P = \{x_0, x_1, \ldots, x_n\}$ of $[a, b]$. The *variation* of ϕ is defined by

$$\mathrm{Var}(\phi) := \sup\{V(\phi, P) : P \text{ is a partition of } [a, b]\}. \tag{19}$$

5.50 Definition.

Let $[a, b]$ be a closed, nondegenerate interval and $\phi : [a, b] \to \mathbf{R}$. Then ϕ is said to be of *bounded variation* on $[a, b]$ if and only if $\mathrm{Var}(\phi) < \infty$.

The following three remarks show how the collection of functions of bounded variation is related to other collections of functions we have studied.

5.51 Remark. *If $\phi \in C^1[a, b]$, then ϕ is of bounded variation on $[a, b]$. However, there exist functions of bounded variation which are not continuously differentiable.*

Proof. Let $P = \{x_0, x_1, \ldots, x_n\}$ be a partition of $[a, b]$. By the Extreme Value Theorem, there is an $M > 0$ such that $|\phi'(x)| \le M$ for all $x \in [a, b]$. Therefore, it follows from the Mean Value Theorem that for each k between 1 and n there is a point c_k between x_{k-1} and x_k such that

$$|\phi(x_k) - \phi(x_{k-1})| = |\phi'(c_k)|(x_k - x_{k-1}) \le M(x_k - x_{k-1}).$$

By telescoping, we obtain $V(\phi, P) \le M(b - a)$ for any partition P of $[a, b]$. Therefore,

$$\text{Var}(\phi) \le M(b - a).$$

On the other hand, $x^2 \sin(1/x)$ is of bounded variation on $[0, 1]$ (see Exercise 5.5.2) but does not belong to $\mathcal{C}^1[0, 1]$ (see Example 4.8).   ∎

5.52 Remark. *If ϕ is monotone on $[a, b]$, then ϕ is of bounded variation on $[a, b]$. However, there exist functions of bounded variation which are not monotone.*

Proof. Let ϕ be increasing on $[a, b]$ and $P = \{x_0, x_1, \dots, x_n\}$ be a partition of $[a, b]$. Then, by telescoping,

$$\sum_{j=1}^{n} |\phi(x_j) - \phi(x_{j-1})| = \sum_{j=1}^{n} (\phi(x_j) - \phi(x_{j-1}))$$

$$= \phi(x_n) - \phi(x_0) = \phi(b) - \phi(a) =: M < \infty.$$

Thus, $\text{Var}(f) = M$. On the other hand, by Remark 5.51, $\phi(x) = x^2$ is of bounded variation on $[-1, 1]$.   ∎

5.53 Remark. *If ϕ is of bounded variation on $[a, b]$, then ϕ is bounded on $[a, b]$. However, there exist bounded functions which are not of bounded variation.*

Proof. Let $x \in [a, b]$ and note by definition that

$$|\phi(x) - \phi(a)| \le |\phi(x) - \phi(a)| + |\phi(b) - \phi(x)| \le \text{Var}(\phi).$$

Hence, by the Triangle Inequality,

$$|\phi(x)| \le |\phi(a)| + \text{Var}(\phi).$$

To find a bounded function which is not of bounded variation, consider

$$\phi(x) := \begin{cases} \sin(1/x) & x \ne 0 \\ 0 & x = 0. \end{cases}$$

Clearly, ϕ is bounded by 1. On the other hand, if

$$x_j = \begin{cases} 0 & j = 0 \\ \dfrac{2}{(n - j + 1)\pi} & 1 \le j \le n, \end{cases}$$

then

$$\sum_{j=1}^{n} |\phi(x_j) - \phi(x_{j-1})| = n \to \infty$$

as $n \to \infty$. Thus ϕ is not of bounded variation on $[0, 2/\pi]$.   ∎

The following result and Exercise 5.5.3 are partial answers to the question, Is the class of functions of bounded variation preserved by algebraic operations?

5.54 Theorem. *If ϕ and ψ are of bounded variation on a closed interval $[a, b]$, then so are $\phi + \psi$ and $\phi - \psi$.*

Proof. Let $a = x_0 < x_1 < \cdots < x_n = b$. Then

$$\sum_{j=1}^{n} |\phi(x_j) \pm \psi(x_j) - (\phi(x_{j-1}) \pm \psi(x_{j-1}))|$$

$$\leq \sum_{j=1}^{n} |\phi(x_j) - \phi(x_{j-1})| + \sum_{j=1}^{n} |\psi(x_j) - \psi(x_{j-1})|$$

$$\leq \text{Var}(\phi) + \text{Var}(\psi).$$

Therefore, $\text{Var}(\phi \pm \psi) \leq \text{Var}(\phi) + \text{Var}(\psi)$.   ∎

It turns out that there is a close connection between functions of bounded variation and monotone functions (see Corollary 5.57 below). To make this connection clear, we introduce the following concept.

5.55 Definition.

Let ϕ be of bounded variation on a closed interval $[a, b]$. The *total variation* of ϕ is the function Φ defined on $[a, b]$ by

$$\Phi(x) := \sup \left\{ \sum_{j=1}^{k} |\phi(x_j) - \phi(x_{j-1})| : \{x_0, x_1, \ldots, x_k\} \text{ is a partition of } [a, x] \right\}.$$

5.56 Theorem. *Let ϕ be of bounded variation on $[a, b]$ and Φ be its total variation. Then*

i) $|\phi(y) - \phi(x)| \leq \Phi(y) - \Phi(x)$ *for all $a \leq x < y \leq b$,*
ii) Φ *and* $\Phi - \phi$ *are increasing on $[a, b]$, and*
iii) $\text{Var}(\phi) \leq \text{Var}(\Phi)$.

Proof. i) Let $x < y$ belong to $[a, b]$ and $\{x_0, x_1, \ldots, x_k\}$ be a partition of $[a, x]$. Then $\{x_0, x_1, \ldots, x_k, y\}$ is a partition of $[a, y]$, and we have by Definition 5.55 that

$$\sum_{j=1}^{k} |\phi(x_j) - \phi(x_{j-1})| \leq \sum_{j=1}^{k} |\phi(x_j) - \phi(x_{j-1})| + |\phi(y) - \phi(x)| \leq \Phi(y).$$

Taking the supremum of this inequality over all partitions $\{x_0, x_1, \ldots, x_k\}$ of $[a, x]$, we obtain

$$\Phi(x) \leq \Phi(x) + |\phi(y) - \phi(x)| \leq \Phi(y).$$

ii) By the Monotone Property of Suprema, Φ is increasing on $[a, b]$. To show that $\Phi - \phi$ also increases, suppose that $a \leq x < y \leq b$. By part i),

$$\phi(y) - \phi(x) \leq |\phi(y) - \phi(x)| \leq \Phi(y) - \Phi(x).$$

Therefore, $\Phi(x) - \phi(x) \leq \Phi(y) - \phi(y)$.

iii) Let $P = \{x_0, x_1, \ldots, x_n\}$ be a partition of $[a, b]$. By part i) and Definition 5.50,

$$\sum_{j=1}^{n} |\phi(x_j) - \phi(x_{j-1})| \leq \sum_{j=1}^{n} |\Phi(x_j) - \Phi(x_{j-1})| \leq \text{Var}(\Phi).$$

Taking the supremum of this inequality over all partitions P of $[a, b]$, we obtain $\text{Var}(\phi) \leq \text{Var}(\Phi)$.   ∎

5.57 Corollary. *Let $[a, b]$ be a closed interval. Then ϕ is of bounded variation on $[a, b]$ if and only if there exist increasing functions f, g on $[a, b]$ such that*

$$\phi(x) = f(x) - g(x), \qquad x \in [a, b].$$

Proof. Suppose that ϕ is of bounded variation and let Φ represent the total variation of ϕ. Set $f = \Phi$ and $g = \Phi - \phi$. By Theorem 5.56, f and g are increasing, and by construction, $\phi = f - g$.

Conversely, suppose that $\phi = f - g$ for some increasing f, g on $[a, b]$. Then by Remark 5.52 and Theorem 5.54, ϕ is of bounded variation on $[a, b]$.   ∎

In particular, if f is of bounded variation on $[a, b]$, then

i) $f(x+)$ exists for each $x \in [a, b)$ and $f(x-)$ exists for each $x \in (a, b]$ (see Theorem 4.18),
ii) f has no more than countably many points of discontinuity in $[a, b]$ (see Theorem 4.19), and
iii) f is integrable on $[a, b]$ (see Exercise 5.1.8).

EXERCISES

5.5.1. a) Show that $4k/(4k^2 - 1) > 1/k$ for $k \in \mathbf{N}$.

b) Prove that

$$\sum_{k=1}^{2^n-1} \frac{1}{k} > \int_1^{2^n} \frac{1}{x}\, dx = \log(2^n)$$

for all $n \in \mathbf{N}$.

c) Prove that

$$\phi(x) = \begin{cases} x^2 \sin \dfrac{1}{x^2} & x \neq 0 \\ 0 & x = 0 \end{cases}$$

is not of bounded variation on $[0, 1]$.

5.5.2. a) Show that $(8k^2 + 2)/(4k^2 - 1)^2 < 1/k^2$ for $k = 2, 3, \dots$.

b) Prove that

$$\sum_{k=1}^{n} \frac{1}{k^2} \leq 1 + \int_1^{n} \frac{1}{x^2}\, dx = 2 - \frac{1}{n}$$

for $n \in \mathbf{N}$.

c) Prove that

$$\phi(x) = \begin{cases} x^2 \sin \dfrac{1}{x} & x \neq 0 \\ 0 & x = 0 \end{cases}$$

is of bounded variation on $[0, 1]$.

5.5.3. **This exercise is used in Section *14.3.** Suppose that ϕ and ψ are of bounded variation on a closed interval $[a, b]$.

a) Prove that $\alpha\phi$ is of bounded variation on $[a, b]$ for every $\alpha \in \mathbf{R}$.

b) Prove that $\phi\psi$ is of bounded variation on $[a, b]$.

c) If there is an $\varepsilon_0 > 0$ such that

$$\phi(x) \geq \varepsilon_0, \qquad x \in [a, b],$$

prove that $1/\phi$ is of bounded variation on $[a, b]$.

5.5.4. Suppose that ϕ is of bounded variation on a closed, bounded interval $[a, b]$. Prove that ϕ is continuous on (a, b) if and only if ϕ is uniformly continuous on (a, b).

5.5.5. a) If ϕ is continuous on a closed nondegenerate interval $[a, b]$, differentiable on (a, b), and if ϕ' is bounded on (a, b), prove that ϕ is of bounded variation on $[a, b]$.

b) Show that $\phi(x) = \sqrt[3]{x}$ is of bounded variation on $[-1, 1]$ but ϕ' is unbounded on $(-1, 1)$.

5.5.6. Let P be a polynomial of degree N.

    a) Show that P is of bounded variation on any closed interval $[a, b]$.

    b) Obtain an estimate for $\text{Var}(P)$ on $[a, b]$, using values of the derivative $P'(x)$ at no more than N points.

5.5.7. Let ϕ be a function of bounded variation on $[a, b]$ and Φ be its total variation function. Prove that if Φ is continuous at some point $x_0 \in (a, b)$, then ϕ is continuous at x_0.

$\boxed{\textbf{5.5.8}}$. **This exercise is used in Section *14.4**. If f is integrable on $[a, b]$, prove that

$$F(x) = \int_a^x f(t)\, dt$$

is of bounded variation on $[a, b]$.

5.5.9. Suppose that f' exists and is integrable on $[a, b]$. Prove that f is of bounded variation and

$$\text{Var}(f) = \int_a^b |f'(x)|\, dx.$$

If f' is bounded rather than integrable, how do the upper and lower integrals of f' compare to the variation of f?

*5.6 CONVEX FUNCTIONS

The last two results of this section use enrichment Theorems 4.19 and 4.23.

    In this section we examine another collection of functions which is important for certain applications, especially for Fourier analysis, functional analysis, numerical analysis, and probability theory.

5.58 Definition.

Let I be an interval and $f : I \to \mathbf{R}$.

i) f is said to be *convex* on I if and only if

$$f(\alpha x + (1 - \alpha)y) \le \alpha f(x) + (1 - \alpha)f(y)$$

    for all $0 \le \alpha \le 1$ and all $x, y \in I$.

ii) f is said to be *concave* on I if and only if $-f$ is convex on I.

    Notice that, by definition, a function f is convex on an interval I if and only if f is convex on every closed subinterval of I.

    It is easy to check that $f(x) = mx + b$ is both convex and concave on any interval (see also Exercise 5.6.3) but in general it is difficult to apply Definition 5.58 directly. For this reason, we include the following simple geometric characterizations of convexity.

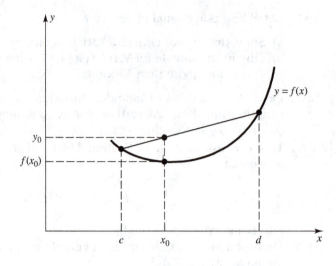

FIGURE 5.5

5.59 Remark. *Let I be an interval and $f : I \to$ **R**. Then f is convex on I if and only if given any $[c, d] \subseteq I$, the chord through the points $(c, f(c))$, $(d, f(d))$ lies on or above the graph of $y = f(x)$ for all $x \in [c, d]$. (See Figure 5.5.)*

Proof. Suppose that f is convex on I and that $x_0 \in [c, d]$. Choose $0 \le \alpha \le 1$ such that $x_0 = \alpha c + (1 - \alpha)d$. The chord from $(c, f(c))$ to $(d, f(d))$ has slope $(f(d) - f(c))/(d - c)$. Hence, the point on this chord which has the form (x_0, y_0) must satisfy $y_0 = \alpha f(c) + (1 - \alpha)f(d)$. Since f is convex, it follows that $f(x_0) \le y_0$; that is, the point (x_0, y_0) lies on or above the point $(x_0, f(x_0))$. A similar argument establishes the reverse implication. ∎

Thus both $f(x) = |x|$ and $f(x) = x^2$ are convex on any interval.

5.60 Remark. *A function f is convex on a nonempty, open interval (a, b) if and only if the slope of the chord always increases on (a, b); that is, if and only if*

$$a < c < x < d < b \quad \text{implies} \quad \frac{f(x) - f(c)}{x - c} \le \frac{f(d) - f(x)}{d - x}.$$

Proof. Fix $a < c < x < d < b$ and let $\lambda(x)$ be the equation of the chord to f through the points $(c, f(c))$ and $(d, f(d))$. If f is convex, then $f(x) \le \lambda(x)$ (see Figure 5.6). Therefore,

$$\frac{f(x) - f(c)}{x - c} \le \frac{\lambda(x) - \lambda(c)}{x - c} = \frac{\lambda(d) - \lambda(x)}{d - x} \le \frac{f(d) - f(x)}{d - x}.$$

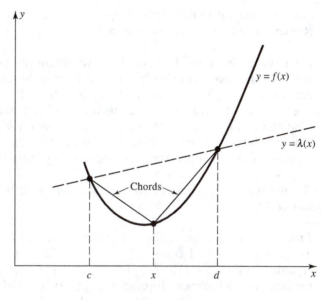

FIGURE 5.6

Conversely, if f is not convex, then $\lambda(x) < f(x)$ for some $x \in (c, d)$. It follows that

$$\frac{f(x) - f(c)}{x - c} > \frac{\lambda(x) - \lambda(c)}{x - c} = \frac{\lambda(d) - \lambda(x)}{d - x} > \frac{f(d) - f(x)}{d - x}.$$

Therefore, the slope of the chord does not increase on (a, b).   ∎

This leads us to a characterization of differentiable convex functions.

5.61 Theorem. *Suppose that f is differentiable on a nonempty, open interval I. Then f is convex on I if and only if f' is increasing on I.*

Proof. Suppose that f is convex on $I =: (a, b)$ and that $c, d \in (a, b)$ satisfy $c < d$. Choose $h > 0$ so small that $c + h < d$ and $d + h < b$. Then by Remark 5.60,

$$\frac{f(c + h) - f(c)}{h} \leq \frac{f(d + h) - f(d)}{h}.$$

In particular, $f'(c) \leq f'(d)$.

Conversely, suppose that f' is increasing on (a, b). Let $a < c < x < d < b$ and use the Mean Value Theorem to choose x_0 (between c and x) and x_1 (between x and d) such that

$$\frac{f(x) - f(c)}{x - c} = f'(x_0) \quad \text{and} \quad \frac{f(d) - f(x)}{d - x} = f'(x_1).$$

Since $x_0 < x_1$, it follows that $f'(x_0) \leq f'(x_1)$. In particular, we conclude by Remark 5.60 that f is convex on (a, b). ∎

Combining Theorems 4.17 and 5.61, we obtain the usual convexity criterion in terms of the second derivative: If f is twice differentiable on (a, b), then f is convex on (a, b) if and only if $f''(x) \geq 0$ for all $x \in (a, b)$. In particular, convexity is what elementary calculus texts call *concave upward* and concavity is what elementary calculus texts call *concave downward*.

On open intervals, convex functions are always continuous. (The statements and proofs of the next two results come from Zygmund [15].)

5.62 Theorem. *If f is convex on some nonempty, open interval I, then f is continuous on I.*

Proof. Let $x_0 \in I =: (a, b)$. By symmetry, it suffices to show that $f(x) \to f(x_0)$ as $x \to x_0+$. Let $a < c < x_0 < x < d < b$, $y = g(x)$ represent the equation of the chord through $(c, f(c))$, $(x_0, f(x_0))$, and $y = h(x)$ represent the equation of the chord through $(x_0, f(x_0))$, $(d, f(d))$. Since f is convex, we have by Remark 5.59 that $f(x) \leq h(x)$. Since $f(x_0)$ lies on or below the chord from $(c, f(c))$ to $(x, f(x))$, we also have that $g(x) \leq f(x)$. Consequently,

$$g(x) \leq f(x) \leq h(x), \qquad x \in (x_0, b).$$

Both chords $y = g(x)$ and $y = h(x)$ pass through the point $(x_0, f(x_0))$, so $g(x) \to f(x_0)$ and $h(x) \to f(x_0)$ as $x \to x_0+$. Hence, it follows from the Squeeze Theorem that $f(x) \to f(x_0)$ as $x \to x_0+$. ∎

Theorem 5.62 does not hold for closed intervals $[a, b]$. Indeed, the function

$$f(x) := \begin{cases} 0 & 0 \leq x < 1 \\ 1 & x = 1 \end{cases}$$

is convex on $[0, 1]$ but not continuous there.

A function f is said to have a *proper maximum* (respectively, *proper minimum*) at x_0 if and only if there exists a $\delta > 0$ such that $f(x) < f(x_0)$ [respectively, $f(x) > f(x_0)$] for all $0 < |x - x_0| < \delta$. As far as proper extrema are concerned, convex functions behave like strictly increasing functions.

5.63 Theorem.

i) *If f is convex on a nonempty, open interval (a, b), then f has no proper maximum on (a, b).*

ii) *If f is convex on $[0, \infty)$ and has a proper minimum, then $f(x) \to \infty$ as $x \to \infty$.*

Proof. i) Suppose that $x_0 \in (a, b)$ and that $f(x_0)$ is a proper maximum of f. Then there exist $c < x_0 < d$ such that $f(x) < f(x_0)$ for $c < x < d$. In particular, the chord through $(c, f(c))$, $(d, f(d))$ must lie below $f(x_0)$ for c, d near x_0, a contradiction.

ii) Suppose that $x_0 \in (a, b)$ and that $f(x_0)$ is a proper minimum of f. Fix $x_1 > x_0$. Let $y = g(x)$ represent the equation of the chord through $(x_0, f(x_0))$ and $(x_1, f(x_1))$. Since $f(x_0)$ is a proper minimum, $f(x_1) > f(x_0)$; hence, g has positive slope. Moreover, by the proof of Theorem 5.62, $g(x) \le f(x)$ for all $x \in (x_1, \infty)$. Since $g(x) \to \infty$ as $x \to \infty$, we conclude that $f(x) \to \infty$ as $x \to \infty$. ∎

Another important result about convex functions addresses the question, What happens when we interchange the order of a convex function and an integral sign?

5.64 Theorem.   [**JENSEN'S INEQUALITY**].
Let ϕ be convex on a closed interval $[a, b]$ and $f : [0, 1] \to [a, b]$. If f and $\phi \circ f$ are integrable on $[0, 1]$, then

$$\phi \left(\int_0^1 f(x)\, dx \right) \le \int_0^1 (\phi \circ f)(x)\, dx. \tag{20}$$

Proof. Set

$$c = \int_0^1 f(x)\, dx$$

and observe that

$$\phi \left(\int_0^1 f(x)\, dx \right) = \phi(c) + s \left(\int_0^1 f(x)\, dx - c \right) \tag{21}$$

for all $s \in \mathbf{R}$. (*Note:* Since $a \le f(x) \le b$ for each $x \in [0, 1]$, c must belong to the interval $[a, b]$ by the Comparison Theorem for Integrals. Thus $\phi(c)$ is defined.)

Let

$$s = \sup_{x \in [a, c)} \frac{\phi(c) - \phi(x)}{c - x}.$$

By Remark 5.60, $s \le (\phi(u) - \phi(c))/(u - c)$ for all $u \in (c, b]$; that is,

$$\phi(c) + s(u - c) \le \phi(u) \tag{22}$$

for all $u \in [c, b]$. On the other hand, if $u \in [a, c)$, we have by the definition of s that

$$s \ge \frac{\phi(c) - \phi(u)}{c - u}.$$

Thus (22) holds for all $u \in [a, b]$. Applying (22) to $u = f(x)$, we obtain

$$\phi(c) + s(f(x) - c) \leq (\phi \circ f)(x).$$

Integrating this inequality as x runs from 0 to 1, we obtain

$$\phi(c) + s \left(\int_0^1 f(x) \, dx - c \right) \leq \int_0^1 (\phi \circ f)(x) \, dx.$$

Combining this inequality with (21), we conclude that (20) holds. ∎

What about differentiability of convex functions? To answer this question we introduce the following concepts (compare with Definition 4.6).

5.65 Definition.

Let $f : (a, b) \to \mathbf{R}$ and $x \in (a, b)$.

i) f is said to have a *right-hand derivative* at x if and only if

$$D_R f(x) := \lim_{h \to 0+} \frac{f(x + h) - f(x)}{h},$$

exists as an extended real number.

ii) f is said to have a *left-hand derivative* at x if and only if

$$D_L f(x) := \lim_{h \to 0-} \frac{f(x + h) - f(x)}{h},$$

exists as an extended real number.

The following result is a simple consequence of the definition of differentiability and the characterization of two-sided limits by one-sided limits (see Theorem 3.14).

5.66 Remark. *A real function f is differentiable at x if and only if both $D_R f(x)$ and $D_L f(x)$ exist, are finite, and are equal, in which case $f'(x) = D_R f(x) = D_L f(x)$.*

The next result shows that the left-hand and right-hand derivatives of a convex function are remarkably well-behaved.

5.67 Theorem. *Let f be convex on an open interval (a, b). Then the left-hand and right-hand derivatives of f exist, are increasing on (a, b), and satisfy*

$$-\infty < D_L f(x) \leq D_R f(x) < \infty$$

for all $x \in (a, b)$.

Proof. Let $h < 0$ and notice that the slope of the chord through the points $(x, f(x))$ and $(x+h, f(x+h))$ is $(f(x+h) - f(x))/h$. By Remark 5.60, these slopes increase as $h \to 0-$. Since increasing functions have a limit (which may be $+\infty$), it follows that $D_L f(x)$ exists and satisfies $-\infty < D_L f(x) \le \infty$. Similarly, $D_R f(x)$ exists and satisfies $-\infty \le D_R f(x) < \infty$. Remark 5.60 also implies

$$D_L f(x) \le D_R f(x). \tag{23}$$

Hence, both numbers are finite, and by symmetry it remains to show that $D_R f(x)$ is increasing on (a, b).

Let $x_1 < u < t < x_2$ be points which belong to (a, b). Then

$$\frac{f(u) - f(x_1)}{u - x_1} \le \frac{f(x_2) - f(t)}{x_2 - t}.$$

Taking the limit of this inequality as $u \to x_1+$ and $t \to x_2-$, we conclude by (23) that

$$D_R f(x_1) \le D_L f(x_2) \le D_R f(x_2). \tag{24}$$

■

The next proof uses enrichment Theorem 4.19.

*5.68 Corollary. *If f is convex on an open interval (a, b), then f is differentiable at all but countably many points of (a, b); that is, there is an at most countable set $E \subset (a, b)$ such that $f'(x)$ exists for all $x \in (a, b) \setminus E$.*

Proof. Let E be the set where either $D_L f(x)$ or $D_R f(x)$ is discontinuous. By Theorems 5.67 and 4.19, the set E is at most countable. Suppose that $x_0 \in (a, b) \setminus E$ and that $x < x_0$. By (24),

$$D_R f(x) \le D_L f(x_0) \le D_R f(x_0).$$

Let $x \to x_0$. Since both $D_L f(x)$ and $D_R f(x)$ are continuous at x_0, we obtain $D_R f(x_0) \le D_L f(x_0) \le D_R f(x_0)$. In particular, $f'(x_0)$ exists for all $x_0 \in (a, b) \setminus E$. ■

How useful is a statement about $f'(x)$ which holds for all but countably many points x? We address this question by proving a generalization of Theorem 4.17. (The proof here uses enrichment Theorem 4.23.)

*5.69 Theorem. *Suppose that f is continuous on a closed interval $[a, b]$ and differentiable on (a, b). If $f'(x) \ge 0$ for all but countably many $x \in (a, b)$, then f is increasing on $[a, b]$.*

Proof. Suppose that $f'(x_1) < 0$ for some $x_1 \in (a, b)$ and let $y \in (f'(x_1), 0)$. By Theorem 4.23 (the Intermediate Value Theorem for derivatives), there is an

$x = x(y) \in (a, b)$ such that $f'(x) = y < 0$. It follows that if $f'(x) < 0$ for one $x \in (a, b)$, then $f'(x) < 0$ for uncountably many $x \in (a, b)$, a contradiction. Therefore, $f'(x) \geq 0$ for all $x \in (a, b)$; hence, by Theorem 4.17, f is increasing on (a, b).   ∎

*5.70 Corollary. *If f is continuous on a closed interval $[a, b]$ and differentiable on (a, b) with $f'(x) = 0$ for all but countably many $x \in (a, b)$, then f is constant on $[a, b]$.*

EXERCISES

5.6.1. Suppose that f, g are convex on an interval I. Prove that $f + g$ and cf are convex on I for any $c \geq 0$.

5.6.2. Suppose that f_n is a sequence of functions convex on an interval I and that

$$f(x) = \lim_{n \to \infty} f_n(x)$$

exists for each $x \in I$. Prove that f is convex on I.

5.6.3. Prove that a function f is both convex and concave on I if and only if there exist $m, b \in \mathbf{R}$ such that $f(x) = mx + b$ for $x \in I$.

5.6.4. Prove that $f(x) = x^p$ is convex on $[0, \infty)$ for $p \geq 1$, and concave on $[0, \infty)$ for $0 < p \leq 1$.

5.6.5. Show that if f is increasing on $[a, b]$, then

$$F(x) = \int_a^x f(t)\, dt$$

is convex on $[a, b]$. (Recall that by Exercise 5.1.8, f is integrable on $[a, b]$.)

5.6.6. If $f : [a, b] \to \mathbf{R}$ is integrable on $[a, b]$, prove that

$$\int_a^b |f(x)|\, dx \leq (b - a)^{1/2} \left(\int_a^b f^2(x)\, dx \right)^{1/2}.$$

5.6.7. Suppose that $f : [0, 1] \to [a, b]$ is integrable on $[0, 1]$. Assume that $e^{f(x)}$ and $|f(x)|^p$ are integrable for all $0 < p < \infty$ (see Exercise 12.2.11).

a) Prove that

$$e^{\int_0^1 f(x)\, dx} \leq \int_0^1 e^{f(x)}\, dx \quad \text{and} \quad \left(\int_0^1 |f(x)|^r\, dx \right)^{1/r} \leq \int_0^1 |f(x)|\, dx$$

for all $0 < r \leq 1$.

b) If $0 < p < q$, prove that

$$\left(\int_0^1 |f(x)|^p\, dx \right)^{1/p} \leq \left(\int_0^1 |f(x)|^q\, dx \right)^{1/q}.$$

c) State and prove analogues of these results for improper integrals.

*5.6.8. Let f be continuous on a closed, bounded interval $[a, b]$ and suppose that $D_R f(x)$ exists for all $x \in (a, b)$.

a) Show that if $f(b) < y_0 < f(a)$, then

$$x_0 := \sup\{x \in [a, b] : f(x) > y_0\}$$

satisfies $f(x_0) = y_0$ and $D_R f(x_0) \leq 0$.

b) Prove that if $f(b) < f(a)$, then there are uncountably many points x which satisfy $D_R f(x) \leq 0$.

c) Prove that if $D_R f(x) > 0$ for all but countably many points $x \in (a, b)$, then f is increasing on $[a, b]$.

d) Prove that if $D_R f(x) \geq 0$ and $g(x) = f(x) + x/n$ for some $n \in \mathbf{N}$, then $D_R g(x) > 0$.

e) Prove that if $D_R f(x) \geq 0$ for all but countably many points $x \in (a, b)$, then f is increasing on $[a, b]$.

CHAPTER 6

Infinite Series of Real Numbers

Infinite series are one of the most widely used tools of analysis. They are used to approximate numbers and functions. (Series of Ramanujan type have been used to compute billions of digits of the decimal expansion of π.) They are used to approximate solutions of differential equations. (You may have used power series to solve ordinary differential equations with nonconstant coefficients.) They even form the basis for some very practical applications, including pattern recognition (e.g., reading zip codes), image enhancement (e.g., removing raindrop clutter from a radar scan), and data compression (e.g., transmission of hundreds of TV programs through a single, photonic, fiber optic cable). Other applications of infinite series can be found in Section 7.5. In view of the variety of these applications, it should come as no surprise that the subject matter of this chapter (and the next) is of fundamental importance.

6.1 INTRODUCTION

Let $\{a_k\}_{k\in\mathbf{N}}$ be a sequence of numbers. We shall call an expression of the form

$$\sum_{k=1}^{\infty} a_k \tag{1}$$

an *infinite series* with *terms* a_k. (No convergence is assumed at this point. This is merely a formal expression.)

6.1 Definition.

Let $S = \sum_{k=1}^{\infty} a_k$ be an infinite series with terms a_k.

i) For each $n \in \mathbf{N}$, the *partial sum of S of order n* is defined by

$$s_n := \sum_{k=1}^{n} a_k.$$

ii) *S* is said to *converge* if and only if its sequence of partial sums $\{s_n\}$ converges to some $s \in \mathbf{R}$ as $n \to \infty$; that is, if and only if for every $\varepsilon > 0$ there is an $N \in \mathbf{N}$ such that $n \geq N$ implies $|s_n - s| < \varepsilon$. In this case we shall write

184

$$\sum_{k=1}^{\infty} a_k = s \tag{2}$$

and call s the *sum*, or *value*, of the series $\sum_{k=1}^{\infty} a_k$.

iii) S is said to *diverge* if and only if its sequence of partial sums $\{s_n\}$ does not converge as $n \to \infty$. When s_n diverges to $+\infty$ as $n \to \infty$, we shall also write

$$\sum_{k=1}^{\infty} a_k = \infty.$$

(We shall deal with series of functions in Chapter 7.)

You are already familiar with one type of infinite series, decimal expansions. Every decimal expansion of a number $x \in (0, 1)$ is a series of the form $\sum_{k=1}^{\infty} x_k/10^k$, where the x_k's are integers in $[0, 9]$. For example, when we write $1/3 = 0.333\ldots$ we mean

$$\frac{1}{3} = \sum_{k=1}^{\infty} \frac{3}{10^k}.$$

In particular, the partial sums 0.3, 0.33, $0.333,\ldots$ are approximations to $1/3$ which get closer and closer to $1/3$ as more terms of the decimal expansion are taken.

One way to determine if a given series converges is to find a formula for its partial sums simple enough so that we can decide whether or not they converge. Here are two examples.

6.2 EXAMPLE.

Prove that $\sum_{k=1}^{\infty} 2^{-k} = 1$.

Proof. By induction, we can show that the partial sums $s_n = \sum_{k=1}^{n} 1/2^k$ satisfy $s_n = 1 - 2^{-n}$ for $n \in \mathbf{N}$. Thus $s_n \to 1$ as $n \to \infty$. ∎

6.3 EXAMPLE.

Prove that $\sum_{k=1}^{\infty}(-1)^k$ diverges.

Proof. The partial sums $s_n = \sum_{k=1}^{n}(-1)^k$ satisfy

$$s_n = \begin{cases} -1 & \text{if } n \text{ is odd} \\ 0 & \text{if } n \text{ is even.} \end{cases}$$

Thus s_n does not converge as $n \to \infty$. ∎

Another way to show that a series diverges is to estimate its partial sums.

6.4 *EXAMPLE.* [THE HARMONIC SERIES].

Prove that the sequence $1/k$ converges but the series $\sum_{k=1}^{\infty} 1/k$ diverges to $+\infty$.

Proof. The sequence $1/k$ converges to zero (by Example 2.2i). On the other hand, by the Comparison Theorem for Integrals,

$$s_n = \sum_{k=1}^{n} \frac{1}{k} \geq \sum_{k=1}^{n} \int_{k}^{k+1} \frac{1}{x} \, dx = \int_{1}^{n+1} \frac{1}{x} \, dx = \log(n+1).$$

We conclude that $s_n \to \infty$ as $n \to \infty$. ∎

This example shows that the terms of a divergent series may converge. In particular, a series does not converge just because its terms converge. On the other hand, the following result shows that a series cannot converge if its terms do not converge to zero.

6.5 Theorem. [DIVERGENCE TEST].

Let $\{a_k\}_{k \in \mathbb{N}}$ be a sequence of real numbers. If a_k does not converge to zero, then the series $\sum_{k=1}^{\infty} a_k$ diverges.

Proof. Suppose to the contrary that $\sum_{k=1}^{\infty} a_k$ converges to some $s \in \mathbf{R}$. By definition, the sequence of partial sums $s_n := \sum_{k=1}^{n} a_k$ converges to s as $n \to \infty$. Therefore, $a_k = s_k - s_{k-1} \to s - s = 0$ as $k \to \infty$, a contradiction. ∎

The proof of this result establishes a property interesting in its own right: If $\sum_{k=1}^{\infty} a_k$ converges, then $a_k \to 0$ as $k \to \infty$. It is important to realize from the beginning that the converse of this statement is false; that is, Theorem 6.5 is a test for divergence, not a test for convergence. Indeed, the harmonic series is a divergent series whose terms converge to zero.

Finding the sum of a convergent series is usually difficult. The following two results show that this is not the case for two special kinds of series.

6.6 Theorem. [TELESCOPIC SERIES].

If $\{a_k\}$ is a convergent real sequence, then

$$\sum_{k=1}^{\infty} (a_k - a_{k+1}) = a_1 - \lim_{k \to \infty} a_k.$$

Proof. By telescoping, we have

$$s_n := \sum_{k=1}^{n}(a_k - a_{k+1}) = a_1 - a_{n+1}.$$

Hence, $s_n \to a_1 - \lim_{k\to\infty} a_k$ as $n \to \infty$.   ∎

It's clear how to modify Definition 6.1 to accommodate series that start at some index other than $k = 1$. We use this concept in the following very important result.

6.7 Theorem.   [GEOMETRIC SERIES].
Suppose that $x \in \mathbf{R}$, that $N \in \{0, 1, \ldots\}$, and that 0^0 is interpreted to be 1. Then the series $\sum_{k=N}^{\infty} x^k$ converges if and only if $|x| < 1$, in which case

$$\sum_{k=N}^{\infty} x^k = \frac{x^N}{1 - x}.$$

In particular,

$$\sum_{k=0}^{\infty} x^k = \frac{1}{1 - x}, \qquad |x| < 1.$$

Proof. If $|x| \geq 1$, then $\sum_{k=N}^{\infty} x^k$ diverges by the Divergence Test. If $|x| < 1$, then set $s_n = \sum_{k=1}^{n} x^k$ and observe by telescoping that

$$(1 - x)s_n = (1 - x)(x + x^2 + \cdots + x^n)$$
$$= x + x^2 + \cdots + x^n - x^2 - x^3 - \cdots - x^{n+1} = x - x^{n+1}.$$

Hence,

$$s_n = \frac{x}{1 - x} - \frac{x^{n+1}}{1 - x}$$

for all $n \in \mathbf{N}$. Since $x^{n+1} \to 0$ as $n \to \infty$ for all $|x| < 1$ (see Example 2.20), we conclude that $s_n \to x/(1 - x)$ as $n \to \infty$.

For general N, we may suppose that $|x| < 1$ and $x \neq 0$. Hence,

$$\sum_{k=N}^{n} x^k = x^N + \cdots + x^n = x^{N-1} \sum_{k=1}^{n-N+1} x^k.$$

Hence, it follows from Definition 6.1 and what we've already proved that

$$\sum_{k=N}^{\infty} x^k = \lim_{n\to\infty} \sum_{k=N}^{n} x^k = \lim_{n\to\infty} x^{N-1} \sum_{k=1}^{n-N+1} x^k = \frac{x^N}{1 - x}.$$ ∎

In everyday speech, the words *sequence* and *series* are considered synonyms. Example 6.4 shows that in mathematics, this is not the case. In particular, you must not apply a result valid for sequences to series and vice versa. Nevertheless, because convergence of an infinite series is defined in terms of convergence of its sequence of partial sums, any result about sequences contains a result about infinite series. The following three theorems illustrate this principle.

6.8 Theorem.   [THE CAUCHY CRITERION].
Let $\{a_k\}$ *be a real sequence. Then the infinite series* $\sum_{k=1}^{\infty} a_k$ *converges if and only if for every* $\varepsilon > 0$ *there is an* $N \in \mathbf{N}$ *such that*

$$m \geq n \geq N \quad imply \quad \left| \sum_{k=n}^{m} a_k \right| < \varepsilon.$$

Proof. Let s_n represent the sequence of partial sums of $\sum_{k=1}^{\infty} a_k$ and set $s_0 = 0$. By Cauchy's Theorem (Theorem 2.29), s_n converges if and only if given $\varepsilon > 0$ there is an $N \in \mathbf{N}$ such that $m, n \geq N$ imply $|s_m - s_{n-1}| < \varepsilon$. Since

$$s_m - s_{n-1} = \sum_{k=n}^{m} a_k$$

for all integers $m \geq n \geq 1$, the proof is complete.   ∎

6.9 Corollary. *Let* $\{a_k\}$ *be a real sequence. Then the infinite series* $\sum_{k=1}^{\infty} a_k$ *converges if and only if given* $\varepsilon > 0$ *there is an* $N \in \mathbf{N}$ *such that*

$$n \geq N \quad implies \quad \left| \sum_{k=n}^{\infty} a_k \right| < \varepsilon.$$

6.10 Theorem.   *Let* $\{a_k\}$ *and* $\{b_k\}$ *be real sequences. If* $\sum_{k=1}^{\infty} a_k$ *and* $\sum_{k=1}^{\infty} b_k$ *are convergent series, then*

$$\sum_{k=1}^{\infty} (a_k + b_k) = \sum_{k=1}^{\infty} a_k + \sum_{k=1}^{\infty} b_k$$

and

$$\sum_{k=1}^{\infty} (\alpha a_k) = \alpha \sum_{k=1}^{\infty} a_k$$

for any $\alpha \in \mathbf{R}$.

Proof. Both identities are corollaries of Theorem 2.12; we provide the details only for the first identity.

Let s_n represent the partial sums of $\sum_{k=1}^{\infty} a_k$ and t_n represent the partial sums of $\sum_{k=1}^{\infty} b_k$. Since real addition is commutative, we have

$$\sum_{k=1}^{n}(a_k + b_k) = s_n + t_n, \qquad n \in \mathbf{N}.$$

Taking the limit of this identity as $n \to \infty$, we conclude by Theorem 2.12 that

$$\sum_{k=1}^{\infty}(a_k + b_k) = \lim_{n\to\infty} s_n + \lim_{n\to\infty} t_n = \sum_{k=1}^{\infty} a_k + \sum_{k=1}^{\infty} b_k. \qquad \blacksquare$$

EXERCISES

6.1.0. Let $\{a_k\}$ and $\{b_k\}$ be real sequences. Decide which of the following statements are true and which are false. Prove the true ones and give counterexamples to the false ones.

a) If a_k is strictly decreasing and $a_k \to 0$ as $k \to \infty$, then $\sum_{k=1}^{\infty} a_k$ converges.

b) If $a_k \neq b_k$ for all $k \in \mathbf{N}$ and if $\sum_{k=1}^{\infty}(a_k + b_k)$ converges, then either $\sum_{k=1}^{\infty} a_k$ converges or $\sum_{k=1}^{\infty} b_k$ converges.

c) Suppose that $\sum_{k=1}^{\infty}(a_k + b_k)$ converges. Then $\sum_{k=1}^{\infty} a_k$ converges if and only if $\sum_{k=1}^{\infty} b_k$ converges.

d) If $a_k \to a$ as $k \to \infty$, then

$$\sum_{k=1}^{\infty}(a_k - a_{k+2}) = a_1 + a_2 - 2a.$$

6.1.1. Prove that each of the following series converges and find its value.

a) $\displaystyle\sum_{k=1}^{\infty} \frac{(-1)^{k+1}}{e^{k-1}}$

b) $\displaystyle\sum_{k=0}^{\infty} \frac{(-1)^{k-1}}{\pi^{2k}}$

c) $\displaystyle\sum_{k=2}^{\infty} \frac{4^{k+1}}{9^{k-1}}$

d) $\displaystyle\sum_{k=0}^{\infty} \frac{5^{k+1} + (-3)^k}{7^{k+2}}$

6.1.2. Represent each of the following series as a telescopic series and find its value.

a) $\displaystyle\sum_{k=1}^{\infty} \frac{1}{k(k+1)}$

b) $\displaystyle\sum_{k=1}^{\infty} \frac{12}{(k+2)(k+3)}$

c) $\displaystyle\sum_{k=2}^{\infty} \log\left(\frac{k(k+2)}{(k+1)^2}\right)$

d) $\displaystyle\sum_{k=1}^{\infty} 2\sin\left(\frac{1}{k} - \frac{1}{k+1}\right)\cos\left(\frac{1}{k} + \frac{1}{k+1}\right)$

6.1.3. Prove that each of the following series diverges.

a) $\displaystyle\sum_{k=1}^{\infty} \cos\left(\frac{1}{k^2}\right)$

b) $\displaystyle\sum_{k=1}^{\infty}\left(1 - \frac{1}{k}\right)^k$

c) $\displaystyle\sum_{k=1}^{\infty} \frac{k+1}{k^2}$

6.1.4. Let $a_0, a_1, \ldots$ be a sequence of real numbers. If $a_k \to L$ as $k \to \infty$, does

$$\sum_{k=1}^{\infty}(a_{k+1} - 2a_k + a_{k-1})$$

converge? If so, what is its value?

6.1.5. Find all $x \in \mathbf{R}$ for which

$$\sum_{k=1}^{\infty}(x^k - x^{k-1})(x^k + x^{k-1})$$

converges. For each such x, find the value of this series.

6.1.6. a) Prove that if $\sum_{k=1}^{\infty} a_k$ converges, then its partial sums s_n are bounded.

b) Show that the converse of part a) is false. Namely, show that a series $\sum_{k=1}^{\infty} a_k$ may have bounded partial sums and still diverge.

6.1.7. Suppose that I is a closed interval and $x_0 \in I$. Suppose further that f is differentiable on $\mathbf{R}$, that $f'(a) \neq 0$ for some $a \in \mathbf{R}$, that the function

$$F(x) := x - \frac{f(x)}{f'(a)} \qquad x \in \mathbf{R}$$

satisfies $F(I) \subseteq I$, and that there is a number $0 < r < 1$ such that $f'(x)/f'(a) \in [1 - r, 1]$ for all $x \in I$.

a) Prove that $|F(x) - F(y)| \leq r|x - y|$ for all $x, y \in I$.

b) If $x_n := F(x_{n-1})$ for $n \in \mathbf{N}$, prove that $|x_{n+1} - x_n| \leq r^n|x_1 - x_0|$ for all $n \in \mathbf{N}$.

c) If $x_n = x_{n-1} - f(x_{n-1})/f'(a)$ for $n \in \mathbf{N}$, prove that

$$b := \lim_{n\to\infty} x_n$$

exists, belongs to I, and is a root of f; that is, that $f(b) = 0$.

6.1.8. a) Suppose that $\{a_k\}$ is a decreasing sequence of real numbers. Prove that if $\sum_{k=1}^{\infty} a_k$ converges, then $ka_k \to 0$ as $k \to \infty$.

b) Let $s_n = \sum_{k=1}^{n}(-1)^{k+1}/k$ for $n \in \mathbf{N}$. Prove that s_{2n} is strictly increasing, s_{2n+1} is strictly decreasing, and $s_{2n+1} - s_{2n} \to 0$ as $n \to \infty$.

c) Prove that part a) is false if *decreasing* is removed.

6.1.9. Let $\{b_k\}$ be a real sequence and $b \in \mathbf{R}$.

a) Suppose that there are $M, N \in \mathbf{N}$ such that $|b - b_k| \le M$ for all $k \ge N$. Prove that

$$\left| nb - \sum_{k=1}^{n} b_k \right| \le \sum_{k=1}^{N} |b_k - b| + M(n - N)$$

for all $n > N$.

b) Prove that if $b_k \to b$ as $k \to \infty$, then

$$\frac{b_1 + b_2 + \cdots + b_n}{n} \to b$$

as $n \to \infty$.

c) Show that the converse of b) is false.

6.1.10. A series $\sum_{k=0}^{\infty} a_k$ is said to be *Cesàro summable* to an $L \in \mathbf{R}$ if and only if

$$\sigma_n := \sum_{k=0}^{n-1} \left(1 - \frac{k}{n} \right) a_k$$

converges to L as $n \to \infty$.

a) Let $s_n = \sum_{k=0}^{n-1} a_k$. Prove that

$$\sigma_n = \frac{s_1 + \cdots + s_n}{n}$$

for each $n \in \mathbf{N}$.

b) Prove that if $a_k \in \mathbf{R}$ and $\sum_{k=0}^{\infty} a_k = L$ converges, then $\sum_{k=0}^{\infty} a_k$ is Cesàro summable to L.

c) Prove that $\sum_{k=0}^{\infty}(-1)^k$ is Cesàro summable to $1/2$; hence the converse of b) is false.

d) [TAUBER'S THEOREM]. Prove that if $a_k \ge 0$ for $k \in \mathbf{N}$ and $\sum_{k=0}^{\infty} a_k$ is Cesàro summable to L, then $\sum_{k=0}^{\infty} a_k = L$.

6.1.11. Suppose that $a_k \ge 0$ for k large and that $\sum_{k=1}^{\infty} a_k/k$ converges. Prove that

$$\lim_{j \to \infty} \sum_{k=1}^{\infty} \frac{a_k}{j + k} = 0.$$

6.1.12. If $\sum_{k=1}^{n} ka_k = (n+1)/(n+2)$ for $n \in \mathbf{N}$, prove that

$$\sum_{k=1}^{\infty} a_k = \frac{3}{4}.$$

6.2 SERIES WITH NONNEGATIVE TERMS

Although we obtained exact values in the preceding section for telescopic series and geometric series, finding exact values of a given series is frequently difficult, if not impossible. Fortunately, for many applications it is not as important to be able to find the value of a series as it is to know that the series converges. When it does converge, we can use its partial sums to approximate its value as accurately as we wish (up to the limitations of whatever computing device we are using). Therefore, much of this chapter is devoted to establishing tests which can be used to decide whether a given series converges or whether it diverges.

The partial sums of a divergent series may be bounded [like $\sum_{k=1}^{\infty}(-1)^k$] or unbounded [like $\sum_{k=1}^{\infty} 1/k$]. When the terms of a divergent series are nonnegative, the former cannot happen.

6.11 Theorem. *Suppose that $a_k \geq 0$ for large k. Then $\sum_{k=1}^{\infty} a_k$ converges if and only if its sequence of partial sums $\{s_n\}$ is bounded; that is, if and only if there exists a finite number $M > 0$ such that*

$$\left| \sum_{k=1}^{n} a_k \right| \leq M \text{ for all } n \in \mathbf{N}.$$

Proof. Set $s_n = \sum_{k=1}^{n} a_k$ for $n \in \mathbf{N}$. If $\sum_{k=1}^{\infty} a_k$ converges, then s_n converges as $n \to \infty$. Since every convergent sequence is bounded (Theorem 2.8), $\sum_{k=1}^{\infty} a_k$ has bounded partial sums.

Conversely, suppose that $|s_n| \leq M$ for $n \in \mathbf{N}$. Recall from Section 2.1 that $a_k \geq 0$ for large k means that there is an $N \in \mathbf{N}$ such that $a_k \geq 0$ for $k \geq N$. It follows that s_n is an increasing sequence when $n \geq N$. Hence by the Monotone Convergence Theorem (Theorem 2.19), s_n converges. ∎

If $a_k \geq 0$ for large k, we shall write $\sum_{k=1}^{\infty} a_k < \infty$ when the series is convergent and $\sum_{k=1}^{\infty} a_k = \infty$ when the series is divergent.

In some cases, integration can be used to test convergence of a series. The idea behind this test is that

$$\int_{1}^{\infty} f(x)\, dx = \sum_{k=1}^{\infty} \int_{k}^{k+1} f(x)\, dx \approx \sum_{k=1}^{\infty} f(k)$$

when f is almost constant on each interval $[k, k+1]$. This will surely be the case for large k if $f(k) \downarrow 0$ as $k \to \infty$ (see Figure 6.1). This observation leads us to the following result.

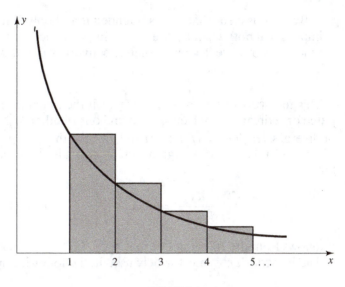

FIGURE 6.1

6.12 Theorem.   [INTEGRAL TEST].

Suppose that $f : [1, \infty) \to \mathbf{R}$ is positive and decreasing on $[1, \infty)$. Then $\sum_{k=1}^{\infty} f(k)$ converges if and only if f is improperly integrable on $[1, \infty)$; that is, if and only if

$$\int_{1}^{\infty} f(x)\, dx < \infty.$$

Proof. Let $s_n = \sum_{k=1}^{n} f(k)$ and $t_n = \int_{1}^{n} f(x)dx$ for $n \in \mathbf{N}$. Since f is decreasing, f is locally integrable on $[1, \infty)$ (see Exercise 5.1.8) and $f(k+1) \le f(x) \le f(k)$ for all $x \in [k, k + 1]$. Hence, by the Comparison Theorem for Integrals,

$$f(k + 1) \le \int_{k}^{k+1} f(x)dx \le f(k)$$

for $k \in \mathbf{N}$. Summing over $k = 1, \dots, n - 1$, we obtain

$$s_n - f(1) = \sum_{k=2}^{n} f(k) \le \int_{1}^{n} f(x)\, dx = t_n \le \sum_{k=1}^{n-1} f(k) = s_n - f(n)$$

for all $n \ge N$. In particular,

$$f(n) \le \sum_{k=1}^{n} f(k) - \int_{1}^{n} f(x)\, dx \le f(1) \text{ for } n \in \mathbf{N}. \tag{3}$$

By (3) it is clear that $\{s_n\}$ is bounded if and only if $\{t_n\}$ is. Since $f(x) \geq 0$ implies that both s_n and t_n are increasing sequences, it follows from the Monotone Convergence Theorem that s_n converges if and only if t_n converges, as $n \to \infty$. ∎

This test works best on series for which the integral of f can be easily computed or estimated. For example, to find out whether $\sum_{k=1}^{\infty} 1/(1+k^2)$ converges or diverges, let $f(x) = 1/(1+x^2)$ and observe that f is positive on $[1, \infty)$. Since $f'(x) = -2x/(1+x^2)^2$ is negative on $[1, \infty)$, it is also clear that f is decreasing. Since

$$\int_1^\infty \frac{dx}{1+x^2} = \arctan x \Big|_1^\infty = \frac{\pi}{2} - \arctan(1) < \infty,$$

it follows from the Integral Test that $\sum_{k=1}^{\infty} 1/(1+k^2)$ converges.

The Integral Test is most widely used in the following special case.

6.13 Corollary. [p- SERIES TEST]. *The series*

$$\sum_{k=1}^{\infty} \frac{1}{k^p} \tag{4}$$

converges if and only if $p > 1$.

Proof. If $p = 1$ or $p \leq 0$, the series diverges. If $p > 0$ and $p \neq 1$, set $f(x) = x^{-p}$ and observe that $f'(x) = -px^{-p-1} < 0$ for all $x \in [1, \infty)$. Hence, f is nonnegative and decreasing on $[1, \infty)$. Since

$$\int_1^\infty \frac{dx}{x^p} = \lim_{n \to \infty} \frac{x^{1-p}}{1-p} \Big|_1^n = \lim_{n \to \infty} \frac{n^{1-p} - 1}{1 - p}$$

has a finite limit if and only if $1 - p < 0$, it follows from the Integral Test that (4) converges if and only if $p > 1$. ∎

The Integral Test, which requires f to satisfy some very restrictive hypotheses, has limited applications. The following test can be used in a much broader context.

6.14 Theorem. [COMPARISON TEST].
Suppose that $0 \leq a_k \leq b_k$ for large k.

i) *If $\sum_{k=1}^{\infty} b_k < \infty$, then $\sum_{k=1}^{\infty} a_k < \infty$.*

ii) *If $\sum_{k=1}^{\infty} a_k = \infty$, then $\sum_{k=1}^{\infty} b_k = \infty$.*

Proof. By hypothesis, choose $N \in \mathbf{N}$ so large that $0 \leq a_k \leq b_k$ for $k > N$. Set $s_n = \sum_{k=1}^{n} a_k$ and $t_n = \sum_{k=1}^{n} b_k$, $n \in \mathbf{N}$. Then $0 \leq s_n - s_N \leq t_n - t_N$ for all $n \geq N$. Since N is fixed, it follows that s_n is bounded when t_n is, and t_n is unbounded when s_n is. Apply Theorem 6.11 and the proof of the theorem is complete. ∎

The Comparison Test is used to compare one series with another whose convergence property is already known (e.g., a p-series or a geometric series). Frequently, the inequalities $|\sin x| \leq |x|$ for all $x \in \mathbf{R}$ (see Appendix B) and $|\log x| \leq x^{\alpha}$ for each $\alpha > 0$ provided x is sufficiently large (see Exercise 4.4.6) are helpful in this regard. Although there is no simple algorithm for this process, the idea is to examine the terms of the given series, ignoring the superfluous factors, and dominating the more complicated factors by simpler ones. Here is a typical example.

6.15 EXAMPLE.

Determine whether the series

$$\sum_{k=1}^{\infty} \frac{3k}{k^2 + k} \sqrt{\frac{\log k}{k}} \tag{5}$$

converges or diverges.

Solution. The kth term of this series can be written by using three factors:

$$\frac{1}{k} \frac{3k}{k+1} \sqrt{\frac{\log k}{k}}.$$

The factor $3k/(k + 1)$ is dominated by 3. Since $\log k \leq \sqrt{k}$ for large k, the factor $\sqrt{\log k / k}$ satisfies

$$\sqrt{\frac{\log k}{k}} \leq \sqrt{\frac{\sqrt{k}}{k}} = \frac{1}{\sqrt[4]{k}}$$

for large k. Therefore, the terms of (5) are dominated by $3/k^{5/4}$. Since $\sum_{k=1}^{\infty} 3/k^{5/4}$ converges by the p-Series Test, it follows from the Comparison Test that (5) converges. ∎

The Comparison Test may not be easy to apply to a given series, even when we know which series it should be compared with, because the process of comparison often involves use of delicate inequalities. For situations like this, the following test is usually more efficient.

6.16 Theorem. **[LIMIT COMPARISON TEST].**

Suppose that $a_k \geq 0$, that $b_k > 0$ for large k, and that $L := \lim_{n \to \infty} a_n / b_n$ exists as an extended real number.

i) *If $0 < L < \infty$, then $\sum_{k=1}^{\infty} a_k$ converges if and only if $\sum_{k=1}^{\infty} b_k$ converges.*

ii) *If $L = 0$ and $\sum_{k=1}^{\infty} b_k$ converges, then $\sum_{k=1}^{\infty} a_k$ converges.*

iii) *If $L = \infty$ and $\sum_{k=1}^{\infty} b_k$ diverges, then $\sum_{k=1}^{\infty} a_k$ diverges.*

Proof. i) If L is finite and nonzero, then there is an $N \in \mathbf{N}$ such that

$$\frac{L}{2} b_k < a_k < \frac{3L}{2} b_k$$

for $k \geq N$. Hence, part i) follows immediately from the Comparison Test and Theorem 6.10. Similar arguments establish parts ii) and iii)—see Exercise 6.2.6. ∎

In general, the Limit Comparison Test is used to replace a series $\sum_{k=1}^{\infty} a_k$ by $\sum_{k=1}^{\infty} b_k$ when $a_k \approx C b_k$ for k large and some absolute fixed constant C. For example, to determine whether or not the series

$$S := \sum_{k=1}^{\infty} \frac{k}{\sqrt{4k^4 + k^2 + 5k}}$$

converges, notice that its terms are approximately $1/(2k)$ for k large. This leads us to compare S with the harmonic series $\sum_{k=1}^{\infty} 1/k$. Since the harmonic series diverges and since

$$\frac{k / (\sqrt{4k^4 + k^2 + 5k})}{1/k} = \frac{k^2}{\sqrt{4k^4 + k^2 + 5k}} \to \frac{1}{2} > 0$$

as $k \to \infty$, it follows from the Limit Comparison Test that S diverges.

Here is another application of the Limit Comparison Test.

6.17 EXAMPLE.

Let $a_k \to 0$ as $k \to \infty$. Prove that $\sum_{k=1}^{\infty} \sin |a_k|$ converges if and only if $\sum_{k=1}^{\infty} |a_k|$ converges.

Proof. By l'Hôpital's Rule,

$$\lim_{k \to \infty} \frac{\sin |a_k|}{|a_k|} = \lim_{x \to 0+} \frac{\sin x}{x} = 1.$$

Hence, by the Limit Comparison Test, $\sum_{k=1}^{\infty} \sin |a_k|$ converges if and only if $\sum_{k=1}^{\infty} |a_k|$ converges. ∎

EXERCISES

6.2.0. Let $\{a_k\}$ and $\{b_k\}$ be real sequences. Decide which of the following statements are true and which are false. Prove the true ones and give counterexamples to the false ones.

a) If $\sum_{k=1}^{\infty} a_k$ converges and $a_k/b_k \to 0$ as $k \to \infty$, then $\sum_{k=1}^{\infty} b_k$ converges.

b) Suppose that $0 < a < 1$. If $a_k \geq 0$ and $\sqrt[k]{a_k} \leq a$ for all $k \in \mathbf{N}$, then $\sum_{k=1}^{\infty} a_k$ converges.

c) Suppose that $a_k \to 0$ as $k \to \infty$. If $a_k \geq 0$ and $\sqrt{a_{k+1}} \leq a_k$ for all $k \in \mathbf{N}$, then $\sum_{k=1}^{\infty} a_k$ converges.

d) Suppose that $a_k = f(k)$ for some continuous function $f : [1, \infty) \to [0, \infty)$ which satisfies $f(x) \to 0$ as $x \to \infty$. If $\sum_{k=1}^{\infty} a_k$ converges, then $\int_1^{\infty} f(x)dx$ converges.

6.2.1. Prove that each of the following series converges.

a) $\displaystyle\sum_{k=1}^{\infty} \frac{2k+5}{3k^3 + 2k - 1}$

b) $\displaystyle\sum_{k=1}^{\infty} \frac{k-1}{k2^k}$

c) $\displaystyle\sum_{k=1}^{\infty} \frac{\log k}{k^p}, \quad p > 1$

d) $\displaystyle\sum_{k=1}^{\infty} \frac{k^3 \log^2 k}{e^k}$

e) $\displaystyle\sum_{k=1}^{\infty} \frac{\sqrt{k} + \pi}{2 + \sqrt[5]{k^8}}$

f) $\displaystyle\sum_{k=1}^{\infty} \frac{1}{k^{\log k}}$

6.2.2. Prove that each of the following series diverges.

a) $\displaystyle\sum_{k=1}^{\infty} \frac{3k^3 + k - 4}{5k^4 - k^2 + 1}$

b) $\displaystyle\sum_{k=1}^{\infty} \frac{\sqrt[k]{k}}{k}$

c) $\displaystyle\sum_{k=1}^{\infty} \left(\frac{k+1}{k}\right)^k$

d) $\displaystyle\sum_{k=2}^{\infty} \frac{1}{k \log^p k}, \quad p \leq 1$

6.2.3. If $a_k \geq 0$ is a bounded sequence, prove that

$$\sum_{k=1}^{\infty} \frac{a_k}{(k+1)^p}$$

converges for all $p > 1$.

6.2.4. Find all $p \geq 0$ such that the following series converges:

$$\sum_{k=1}^{\infty} \frac{1}{k \log^p (k+1)}.$$

6.2.5. If $\sum_{k=1}^{\infty} |a_k|$ converges, prove that

$$\sum_{k=1}^{\infty} \frac{|a_k|}{k^p}$$

converges for all $p \geq 0$. What happens if $p < 0$?

6.2.6. Prove Theorem 6.16ii and iii.

6.2.7. Suppose that a_k and b_k are nonnegative for all $k \in \mathbf{N}$. Prove that if $\sum_{k=1}^{\infty} a_k$ and $\sum_{k=1}^{\infty} b_k$ converge, then $\sum_{k=1}^{\infty} a_k b_k$ also converges.

6.2.8. Suppose that $a, b \in \mathbf{R}$ satisfy $b/a \in \mathbf{R} \backslash \mathbf{Z}$. Find all $q > 0$ such that

$$\sum_{k=1}^{\infty} \frac{1}{(ak+b)q^k}$$

converges.

6.2.9. Suppose that $a_k \to 0$. Prove that $\sum_{k=1}^{\infty} a_k$ converges if and only if the series $\sum_{k=1}^{\infty} (a_{2k} + a_{2k+1})$ converges.

6.2.10. Find all $p \in \mathbf{R}$ such that

$$\sum_{k=2}^{\infty} \frac{1}{(\log(\log k))^{p \log k}}$$

converges.

6.3 ABSOLUTE CONVERGENCE

In this section we investigate what happens to a convergent series when its terms are replaced by their absolute values. We begin with some terminology.

6.18 Definition.

Let $S = \sum_{k=1}^{\infty} a_k$ be an infinite series.

i) S is said to *converge absolutely* if and only if $\sum_{k=1}^{\infty} |a_k| < \infty$.
ii) S is said to *converge conditionally* if and only if S converges but not absolutely.

The Cauchy Criterion gives us the following test for absolute convergence.

6.19 Remark. *A series $\sum_{k=1}^{\infty} a_k$ converges absolutely if and only if for every $\varepsilon > 0$ there is an $N \in \mathbf{N}$ such that*

$$m > n \geq N \quad \text{implies} \quad \sum_{k=n}^{m} |a_k| < \varepsilon. \tag{6}$$

As was the case for improper integrals, absolute convergence is stronger than convergence.

6.20 Remark. *If $\sum_{k=1}^{\infty} a_k$ converges absolutely, then $\sum_{k=1}^{\infty} a_k$ converges, but not conversely. In particular, there exist conditionally convergent series.*

Proof. Suppose that $\sum_{k=1}^{\infty} a_k$ converges absolutely. Given $\varepsilon > 0$, choose $N \in \mathbf{N}$ so that (6) holds. Then

$$\left| \sum_{k=n}^{m} a_k \right| \leq \sum_{k=n}^{m} |a_k| < \varepsilon$$

for $m > n \geq N$. Hence, by the Cauchy Criterion, $\sum_{k=1}^{\infty} a_k$ converges.

We shall finish the proof by showing that $S := \sum_{k=1}^{\infty} (-1)^k / k$ converges conditionally. Since the harmonic series diverges, S does not converge absolutely. On the other hand, the tails of S look like

$$\sum_{j=k}^{\infty} \frac{(-1)^j}{j} = (-1)^k \left(\frac{1}{k} - \frac{1}{k+1} + \frac{1}{k+2} - \frac{1}{k+3} + \cdots \right).$$

By grouping pairs of terms together, it is easy to see that the sum inside the parentheses is greater than 0 but less than $1/k$; that is,

$$\left| \sum_{j=k}^{\infty} \frac{(-1)^j}{j} \right| < \frac{1}{k}.$$

Hence $\sum_{k=1}^{\infty} (-1)^k / k$ converges by Corollary 6.9. ∎

We shall see below that it is important to be able to identify absolutely convergent series. Since every result about series with nonnegative terms can be applied to the series $\sum_{k=1}^{\infty} |a_k|$, we already have three tests for absolute convergence (the Integral Test, the Comparison Test, and the Limit Comparison Test). We now develop two additional tests for absolute convergence which are arguably the most practical tests presented in this chapter.

Before we state these tests, we need to introduce another concept. (If you covered Section 2.5, you may proceed directly to Theorem 6.23.)

6.21 Definition.

The *limit supremum* of a sequence of real numbers $\{x_k\}$ is defined to be

$$\limsup_{k \to \infty} x_k := \lim_{n \to \infty} \left(\sup_{k > n} x_k \right).$$

NOTE: Unlike the limit, the limit supremum of a sequence always exists as an extended real number. Indeed, let $s_n := \sup_{k > n} x_k$. If $s_n = \infty$ for all n, then $s_n \to \infty$ as $n \to \infty$, so, by definition, the limit supremum of x_k is ∞. On the other hand, if s_n is finite for some n, the Monotone Property for Suprema implies that the sequence s_n is decreasing. Hence, by the Monotone Convergence Theorem, $\lim_{n \to \infty} s_n$ exists. (It might be $-\infty$, e.g., when $x_k = -k$.)

In practice, the limit supremum of a sequence is usually easy to find by inspection. For example, since $(-1)^k$ is 1 when k is even and -1 when k is odd, it is clear that $\sup_{k > n}(-1)^k = 1$ for all $n \in \mathbf{N}$. Hence the limit supremum of $(-1)^k$ is 1. Similarly,

$$\limsup_{k \to \infty}(3 + (-1)^k) = 4 \quad \text{and} \quad \limsup_{k \to \infty} \frac{2^k + (-2)^k}{k} = \infty.$$

The only thing we need to know about limits supremum (for now) is the following result.

6.22 Remark. *Let $x \in \mathbf{R}$ and $\{x_k\}$ be a real sequence.*

i) *If $\limsup_{k \to \infty} x_k < x$, then $x_k < x$ for large k.*
ii) *If $\limsup_{k \to \infty} x_k > x$, then $x_k > x$ for infinitely many k's.*
iii) *If $x_k \to x$ as $k \to \infty$, then $\limsup_{k \to \infty} x_k = x$.*

Proof. Let $s := \limsup_{k \to \infty} x_k$ and $s_n := \sup_{k > n} x_k$ and recall by Definition 6.21 that $s_n \to s$ as $n \to \infty$.

i) If $s < x$, then there is an $N \in \mathbf{N}$ such that $s_N < x$. In particular, $x_k < x$ for all $k > N$.

ii) If $s > x$, then $s_n > x$ for all n (because s_n is decreasing). Since $s_1 > x$, there is a $k_1 > 1$ such that $x_{k_1} > x$. Suppose that k_j has been chosen so that $x_{k_j} > x$. Since $s_{k_j} > x$, there is a $k_{j+1} > k_j$ such that $x_{k_{j+1}} > x$. In particular, there is an increasing sequence of positive integers k_j such that $x_{k_j} > x$ for all $j \in \mathbf{N}$. It follows that $x_k > x$ for infinitely many k's.

iii) If x_k converges to x, given $\varepsilon > 0$ there is an $N \in \mathbf{N}$ such that $k \geq N$ implies $|x_k - x| < \varepsilon$. In particular, for any $n \geq N$, $x_k > x - \varepsilon$ for $k > n$. Taking the supremum of this last inequality over $k > n$, we see that $s_n > x - \varepsilon$ for $n \geq N$. Hence, the limit of the s_n's satisfies $s \geq x - \varepsilon$. Thus $s \geq x$. A similar argument proves that $s \leq x$. ∎

The limit supremum gives a very useful and efficient test for absolute convergence.

6.23 Theorem.   [ROOT TEST].
Let $a_k \in \mathbf{R}$ and $r := \limsup_{k \to \infty} |a_k|^{1/k}$.

i) *If $r < 1$, then $\sum_{k=1}^{\infty} a_k$ converges absolutely.*

ii) *If $r > 1$, then $\sum_{k=1}^{\infty} a_k$ diverges.*

Proof. i) Suppose that $r < 1$. Let $r < x < 1$ and notice that the geometric series $\sum_{k=1}^{\infty} x^k$ converges. By Remark 6.22i (or by Exercise 2.5.3),

$$|a_k|^{1/k} < x$$

for large k. Hence, $|a_k| < x^k$ for large k and it follows from the Comparison Test that $\sum_{k=1}^{\infty} |a_k|$ converges.

ii) Suppose that $r > 1$. By Remark 6.22ii (or by Exercise 2.5.3),

$$|a_k|^{1/k} > 1$$

for infinitely many $k \in \mathbf{N}$. Hence, $|a_k| > 1$ for infinitely many k and it follows from the Divergence Test that $\sum_{k=1}^{\infty} a_k$ diverges. ∎

Note by Remark 6.22iii or Theorem 2.36 that if $r := \lim_{k \to \infty} |a_k|^{1/k}$ exists, then (by the Root Test) $\sum_{k=1}^{\infty} a_k$ converges absolutely when $r < 1$ and diverges when $r > 1$.

The following test is weaker than the Root Test (see Exercise 6.3.8) but is easier to use when the terms of $\sum_{k=1}^{\infty} a_k$ are made up of products (e.g., of factorials).

6.24 Theorem.   [RATIO TEST].
Let $a_k \in \mathbf{R}$ with $a_k \neq 0$ for large k and suppose that

$$r = \lim_{k \to \infty} \frac{|a_{k+1}|}{|a_k|}$$

exists as an extended real number.

i) *If $r < 1$, then $\sum_{k=1}^{\infty} a_k$ converges absolutely.*

ii) *If $r > 1$, then $\sum_{k=1}^{\infty} a_k$ diverges.*

Proof. If $r > 1$, then $|a_{k+1}| \geq |a_k|$ for k large and thus a_k cannot converge to zero. Hence, by the Divergence Test, $\sum_{k=1}^{\infty} a_k$ diverges.

If $r < 1$, then observe for any $x \in (r, 1)$ that

$$\frac{|a_{k+1}|}{|a_k|} < x = \frac{x^{k+1}}{x^k}$$

for k large. Hence, the sequence $|a_k|/x^k$ is decreasing for large k and thus bounded. In particular, there is an $M > 0$ such that $|a_k| \leq Mx^k$ for all $k \in \mathbf{N}$. Since $x < 1$, it follows from the Comparison Test that $\sum_{k=1}^{\infty} |a_k|$ converges. ∎

6.25 Remark. *The Root and Ratio Tests are inconclusive when $r = 1$.*

For example, apply the Root and Ratio Tests to $\sum_{k=1}^{\infty} 1/k$ and $\sum_{k=1}^{\infty} 1/k^2$. In all four cases, $r = 1$. Nevertheless, the first series diverges whereas the second converges absolutely.

How should we proceed when the Root and Ratio Tests are inconclusive (e.g., when $r = 1$)? We can always try to use one of the Comparison Tests. Since this can be technically daunting, there are other ways to cope with the case $r = 1$. If the ratios of terms of a series converge to 1 rapidly enough, then the series converges. (For three tests of this type, see the results and exercises of Section 6.6.) If the terms of a series have $k!$ as a factor, then there is a very useful asymptotic estimate of $k!$ (called Stirling's Formula—see Theorem 12.73) that can be used in conjunction with the Comparison Test (e.g., see Exercises 6.3.3f and 6.6.2c).

It is natural to assume that the usual laws of algebra hold for infinite series (e.g., associativity and commutativity). Is this assumption warranted? We have "inserted parentheses" (i.e., grouped terms together) to aid evaluation of some series [e.g., to evaluate some telescopic series and to prove that $\sum_{k=1}^{\infty}(-1)^k/k$ converges conditionally]. This is valid for convergent series (absolutely or conditionally) because if the sequence of partial sums s_n converges to s, then any subsequence s_{n_k} also converges to s. The situation is more complicated when we start changing the order of the terms (compare Theorem 6.27 with Theorem 6.29). To describe what happens, we introduce the following terminology.

6.26 Definition.

A series $\sum_{j=1}^{\infty} b_j$ is called a *rearrangement* of a series $\sum_{k=1}^{\infty} a_k$ if and only if there is a 1–1 function f from $\mathbf{N}$ onto $\mathbf{N}$ such that

$$b_{f(k)} = a_k, \qquad k \in \mathbf{N}.$$

The following result demonstrates why absolutely convergent series are so important.

6.27 Theorem. *If $\sum_{k=1}^{\infty} a_k$ converges absolutely and $\sum_{j=1}^{\infty} b_j$ is any rearrangement of $\sum_{k=1}^{\infty} a_k$, then $\sum_{j=1}^{\infty} b_j$ converges and*

$$\sum_{k=1}^{\infty} a_k = \sum_{j=1}^{\infty} b_j.$$

Proof. Let $\varepsilon > 0$. Set $s_n = \sum_{k=1}^{n} a_k$, $s = \sum_{k=1}^{\infty} a_k$, and $t_m = \sum_{j=1}^{m} b_j$, $n, m \in \mathbf{N}$. Since $\sum_{k=1}^{\infty} a_k$ converges absolutely, we can choose $N \in \mathbf{N}$ (see Corollary 6.9) such that

$$\sum_{k=N+1}^{\infty} |a_k| < \frac{\varepsilon}{2}. \tag{9}$$

Thus

$$|s_N - s| = \left| \sum_{k=N+1}^{\infty} a_k \right| \leq \sum_{k=N+1}^{\infty} |a_k| < \frac{\varepsilon}{2}. \tag{10}$$

Let f be a 1–1 function from $\mathbf{N}$ onto $\mathbf{N}$ which satisfies

$$b_{f(k)} = a_k, \qquad k \in \mathbf{N}$$

and set $M = \max\{f(1), \ldots, f(N)\}$. Notice that

$$\{a_1, \ldots, a_N\} \subseteq \{b_1, \ldots, b_M\}.$$

Let $m \geq M$. Then $t_m - s_N$ contains only a_k's whose indices satisfy $k > N$. Thus, it follows from (9) that

$$|t_m - s_N| \leq \sum_{k=N+1}^{\infty} |a_k| < \frac{\varepsilon}{2}.$$

Hence, by (10),

$$|t_m - s| \leq |t_m - s_N| + |s_N - s| < \frac{\varepsilon}{2} + \frac{\varepsilon}{2} = \varepsilon$$

for $m \geq M$. Therefore,

$$s = \sum_{j=1}^{\infty} b_j. \qquad \blacksquare$$

The rest of this section, which is used nowhere else in this book, is optional.

We now show that Theorem 6.27 fails in a catastrophic way for conditionally convergent series (see Theorem 6.29 below). To facilitate our discussion, recall (see Exercise 1.2.3) that the *positive and negative parts* of an $a \in \mathbf{R}$ are defined by

$$a^+ := \frac{|a| + a}{2} = \begin{cases} a & a \geq 0 \\ 0 & a < 0 \end{cases}$$

and

$$a^- := \frac{|a| - a}{2} = \begin{cases} 0 & a \geq 0 \\ -a & a < 0. \end{cases}$$

Notice that

$$a^+ \geq 0, \quad a^- \geq 0, \tag{7}$$

and

$$a = a^+ - a^-, \quad |a| = a^+ + a^- \tag{8}$$

for all $a \in \mathbf{R}$.

*6.28 Lemma.
Suppose that $a_k \in \mathbf{R}$ for $k \in \mathbf{N}$.

i) *If $\sum_{k=1}^{\infty} a_k$ converges absolutely, then so do $\sum_{k=1}^{\infty} a_k^+$ and $\sum_{k=1}^{\infty} a_k^-$. In fact,*

$$\sum_{k=1}^{\infty} |a_k| = \sum_{k=1}^{\infty} a_k^+ + \sum_{k=1}^{\infty} a_k^- \quad and \quad \sum_{k=1}^{\infty} a_k = \sum_{k=1}^{\infty} a_k^+ - \sum_{k=1}^{\infty} a_k^-.$$

ii) *If $\sum_{k=1}^{\infty} a_k$ converges conditionally, then*

$$\sum_{k=1}^{\infty} a_k^+ = \sum_{k=1}^{\infty} a_k^- = \infty.$$

Proof. By definition, $a_k^+ = (|a_k| + a_k)/2$. Since both $\sum_{k=1}^{\infty} |a_k|$ and $\sum_{k=1}^{\infty} a_k$ converge, it follows from Theorem 6.10 that

$$\sum_{k=1}^{\infty} a_k^+ = \frac{1}{2} \sum_{k=1}^{\infty} |a_k| + \frac{1}{2} \sum_{k=1}^{\infty} a_k$$

converges. Similarly,

$$\sum_{k=1}^{\infty} a_k^- = \frac{1}{2} \sum_{k=1}^{\infty} |a_k| - \frac{1}{2} \sum_{k=1}^{\infty} a_k$$

converges. This proves part i).

Suppose that part ii) is false. By symmetry we may suppose that $\sum_{k=1}^{\infty} a_k^+$ converges. Since $\sum_{k=1}^{\infty} a_k$ converges, it follows from (8) that

$$\sum_{k=1}^{\infty} a_k^- = \sum_{k=1}^{\infty} a_k^+ - \sum_{k=1}^{\infty} a_k$$

converges. Thus,

$$\sum_{k=1}^{\infty} |a_k| = \sum_{k=1}^{\infty} a_k^+ + \sum_{k=1}^{\infty} a_k^-$$

converges, a contradiction. ∎

We are prepared to show that Theorem 6.27 is false if the hypothesis "absolutely convergent" is dropped. In fact, as the following result shows, rearrangements of conditionally convergent series can converge to anything one wishes (see also Exercise 6.3.10).

*6.29 Theorem.   [RIEMANN].
Let $x \in \mathbf{R}$. If $\sum_{k=1}^{\infty} a_k$ is conditionally convergent, then there is a rearrangement of $\sum_{k=1}^{\infty} a_k$ which converges to x.

STRATEGY: The idea behind the proof is simple. Since $\sum_{k=1}^{\infty} a_k^+ = \sum_{k=1}^{\infty} a_k^- = \infty$ by Lemma 6.28, begin by adding enough a_k^+'s until the resulting partial sum is $> x$. Then subtract enough a_k^-'s until the resulting partial sum is $< x$, and continue adding and subtracting. Since $a_k \to 0$ as $k \to \infty$, the resulting partial sums should be getting closer to x. We now make this precise.

Proof. Since $\sum_{k=1}^{\infty} a_k^+ = \infty$, let k_1 be the smallest integer which satisfies $a_1^+ + a_2^+ + \cdots + a_{k_1}^+ > x$. Since k_1 is least, $a_1^+ + \cdots + a_{k_1-1}^+ \le x$, so $a_1^+ + a_2^+ + \cdots + a_{k_1}^+ \le x + a_{k_1}^+$. Set $r_0 = 0$ and observe that $s_{k_1+r_0} := a_1^+ + \cdots + a_{k_1+r_0}^+ \le x + a_{k_1}^+$.

Suppose for some $j \ge 1$ that integers $r_0 < r_1 < \cdots < r_{j-1}$ and $k_1 < k_2 < \cdots < k_j$ have been chosen such that a partial sum s_{k_j}, of a_k^+'s and a_k^-'s, satisfies

$$s_{k_j+r_{j-1}} \le x + a_{k_j}^+. \tag{11}$$

Since $\sum_{k=1}^{\infty} a_k^- = \infty$, let $r_j > r_{j-1}$ be the smallest integer which satisfies $s_{k_j} - a_{r_{j-1}+1}^- - \cdots - a_{r_j}^- < x$. For $k_j + r_{j-1} < n \le k_j + r_j$, set $s_n := s_{k_j} - a_{r_{j-1}+1}^- - \cdots - a_{n-k_j}^-$. It is easy to see that $s_{k_j+r_{j-1}} \ge s_{k_j+r_{j-1}+1} \ge \cdots \ge s_{k_j+r_j}$. Since r_j is least, we also have $s_{k_j+r_j} = s_{k_j+r_j-1} - a_{r_j}^- \ge x - a_{r_j}^-$. It follows from (11) that

$$|s_n - x| \le \max\{a_{r_j}^-, a_{k_j}^+\} \quad \text{for} \quad k_j + r_{j-1} < n \le k_j + r_j. \tag{12}$$

Similarly, if we let k_{j+1} be the smallest integer which satisfies

$$s_{k_j+r_j} + a_{k_j+1}^+ + \cdots + a_{k_{j+1}}^+ > x$$

and set $s_n := s_{k_j+r_j} + a_{k_j+1}^+ + \cdots + a_{n-r_j}^+$, for $k_j + r_j < n \le k_{j+1} + r_j$, then

$$|s_n - x| \le \max\{a_{r_j}^-, a_{k_{j+1}}^+\} \quad \text{for} \quad k_j + r_j < n \le k_{j+1} + r_j. \tag{13}$$

Let $\varepsilon > 0$. Since each a_k^+ and $-a_k^-$ is either a_k or 0, it is clear (after deleting the zero terms) that the s_n's are partial sums of a rearrangement of $\sum_{k=1}^{\infty} a_k$. Moreover, since $a_k \to 0$ as $k \to \infty$, we can choose an N so large that $j \ge N$ implies that $a_{r_j}^-$ and $a_{k_j}^+$ are both less than ε. We conclude by (12) and (13) that if $n > k_N + r_N$, then $|s_n - x| < \varepsilon$; that is, that $s_n \to x$ as $n \to \infty$. ∎

EXERCISES

6.3.0. Let $\{a_k\}$ and $\{b_k\}$ be real sequences. Decide which of the following statements are true and which are false. Prove the true ones and give counterexamples to the false ones.

a) Suppose that $0 < \alpha < \infty$. If $|a_k|^{\alpha/k} \to a_0$, where $a_0 < 1$, then $\sum_{k=1}^{\infty} a_k^{\alpha}$ is absolutely convergent.

b) If $\sum_{k=1}^{\infty} a_k$ is absolutely convergent and $a_k \downarrow 0$ as $k \to \infty$, then

$$\limsup_{k \to \infty} |a_k|^{1/k} < 1.$$

c) If $a_k \le b_k$ for all $k \in \mathbf{N}$ and $\sum_{k=1}^{\infty} b_k$ is absolutely convergent, then $\sum_{k=1}^{\infty} a_k$ converges.

d) If $\sum_{k=1}^{\infty} a_k$ is absolutely convergent, then $\sum_{k=1}^{\infty} a_k^2$ is absolutely convergent.

6.3.1. Prove that each of the following series converges.

a) $\displaystyle\sum_{k=1}^{\infty} \frac{1}{k!}$

b) $\displaystyle\sum_{k=1}^{\infty} \frac{1}{k^k}$

c) $\displaystyle\sum_{k=1}^{\infty} \frac{\pi^k}{k!}$

d) $\displaystyle\sum_{k=1}^{\infty} \left(\frac{k}{k+1}\right)^{k^2}$

6.3.2. Decide, using results covered so far in this chapter, which of the following series converge and which diverge.

a) $\displaystyle\sum_{k=1}^{\infty} \frac{k^3}{(k+1)^{\log k}}$

b) $\displaystyle\sum_{k=1}^{\infty} \frac{k^{100}}{e^k}$

c) $\displaystyle\sum_{k=1}^{\infty} \left(\frac{k+1}{2k+3}\right)^k$

d) $\displaystyle\sum_{k=1}^{\infty} \frac{1 \cdot 3 \cdots (2k-1)}{(2k)!}$

e) $\displaystyle\sum_{k=1}^{\infty} \left(\frac{(k-1)!}{k!+1}\right)^k$

f) $\displaystyle\sum_{k=1}^{\infty} \left(\frac{3+(-1)^k}{5}\right)^k$

g) $\displaystyle\sum_{k=1}^{\infty} \frac{(3-(-1)^k)^k}{\pi^k}$

6.3.3. For each of the following, find all values of $p \in \mathbf{R}$ for which the given series converges absolutely.

a) $\displaystyle\sum_{k=2}^{\infty} \frac{1}{k \log^p k}$

b) $\displaystyle\sum_{k=2}^{\infty} \frac{1}{\log^p k}$

c) $\displaystyle\sum_{k=1}^{\infty} \frac{k^p}{p^k}$

d) $\displaystyle\sum_{k=2}^{\infty} \frac{1}{\sqrt{k}(k^p - 1)}$

e) $\displaystyle\sum_{k=0}^{\infty} (\sqrt{k^{2p} + 1} - k^p)$

*f) $\displaystyle\sum_{k=1}^{\infty} \frac{2^{kp} k!}{k^k}$

6.3.4. Suppose that $a_k \geq 0$ and that $a_k^{1/k} \to a$ as $k \to \infty$. Prove that $\sum_{k=1}^{\infty} a_k x^k$ converges absolutely for all $|x| < 1/a$ if $a \neq 0$ and for all $x \in \mathbf{R}$ if $a = 0$.

6.3.5. Define a_k recursively by $a_1 = 1$ and

$$a_k = (-1)^k \left(1 + k \sin\left(\frac{1}{k}\right)\right)^{-1} a_{k-1}, \qquad k > 1.$$

Prove that $\sum_{k=1}^{\infty} a_k$ converges absolutely.

6.3.6. Suppose that $a_{kj} \geq 0$ for $k, j \in \mathbf{N}$. Set

$$A_k = \sum_{j=1}^{\infty} a_{kj}$$

for each $k \in \mathbf{N}$, and suppose that $\sum_{k=1}^{\infty} A_k$ converges.

a) Prove that

$$\sum_{j=1}^{\infty} \left(\sum_{k=1}^{\infty} a_{kj}\right) \leq \sum_{k=1}^{\infty} \left(\sum_{j=1}^{\infty} a_{kj}\right).$$

b) Show that

$$\sum_{j=1}^{\infty} \left(\sum_{k=1}^{\infty} a_{kj}\right) = \sum_{k=1}^{\infty} \left(\sum_{j=1}^{\infty} a_{kj}\right).$$

c) Prove that b) may not hold if a_{kj} has both positive and negative values.

Hint: Consider

$$a_{kj} = \begin{cases} 1 & j = k \\ -1 & j = k+1 \\ 0 & \text{otherwise.} \end{cases}$$

6.3.7. a) Suppose that $\sum_{k=1}^{\infty} a_k$ converges absolutely. Prove that $\sum_{k=1}^{\infty} |a_k|^p$ converges for all $p \geq 1$.

b) Suppose that $\sum_{k=1}^{\infty} a_k$ converges conditionally. Prove that $\sum_{k=1}^{\infty} k^p a_k$ diverges for all $p > 1$.

6.3.8. For any real sequence $\{x_k\}$, define

$$\liminf_{k \to \infty} x_k := \lim_{n \to \infty} \left(\inf_{k > n} x_k \right).$$

a) Prove that if $\liminf_{k \to \infty} x_k > x$ for some $x \in \mathbf{R}$, then $x_k > x$ for k large.

b) Prove that if $x_k \to x$ as $k \to \infty$, for some $x \in \mathbf{R}$, then $\liminf_{k \to \infty} x_k = x$.

c) If $a_k > 0$ for all $k \in \mathbf{N}$, prove that

$$\liminf_{n \to \infty} \frac{a_{k+1}}{a_k} \leq \liminf_{n \to \infty} \sqrt[k]{a_k} \leq \limsup_{k \to \infty} \sqrt[k]{a_k} \leq \limsup_{k \to \infty} \frac{a_{k+1}}{a_k}.$$

d) Prove that if $b_n \in \mathbf{R} \setminus \{0\}$ and $|b_{n+1}/b_n| \to r$ as $n \to \infty$, for some $r > 0$, then $|b_n|^{1/n} \to r$ as $n \to \infty$.

6.3.9. Given that $\sum_{k=1}^{\infty} 1/k^2 = \pi^2/6$ (see Exercise 14.3.7), find the exact value of

$$\sum_{k=1}^{\infty} \frac{1}{(2k-1)^2}.$$

*6.3.10. Let $x \leq y$ be any pair of extended real numbers. Prove that if $\sum_{k=1}^{\infty} a_k$ is conditionally convergent, then there is a rearrangement $\sum_{j=1}^{\infty} b_j$ of $\sum_{k=1}^{\infty} a_k$ whose partial sums s_n satisfy

$$\liminf_{n \to \infty} s_n = x \quad \text{and} \quad \limsup_{n \to \infty} s_n = y.$$

6.3.11. a) Using Exercise 4.4.4, prove that

$$\sin x = \sum_{k=0}^{\infty} \frac{(-1)^k x^{2k+1}}{(2k+1)!}$$

for all $x \in [0, \pi/2]$.

b) Prove that

$$\cos x = \sum_{k=0}^{\infty} \frac{(-1)^k x^{2k}}{(2k)!}$$

for $x \in [0, \pi/2]$.

6.4 ALTERNATING SERIES

We have identified many tests for absolute convergence but have said little about conditionally convergent series. In this section we derive two tests to use on series whose terms have mixed signs.

Both tests rely on the following algebraic observation. (This result will also be used in Chapter 7 to prove that limits of power series are continuous.)

6.30 Theorem. [ABEL'S FORMULA].
Let $\{a_k\}_{k\in\mathbf{N}}$ and $\{b_k\}_{k\in\mathbf{N}}$ be real sequences, and for each pair of integers $n \geq m \geq 1$ set

$$A_{n,m} := \sum_{k=m}^{n} a_k.$$

Then

$$\sum_{k=m}^{n} a_k b_k = A_{n,m} b_n - \sum_{k=m}^{n-1} A_{k,m}(b_{k+1} - b_k)$$

for all integers $n > m \geq 1$.

Proof. Since $A_{k,m} - A_{(k-1),m} = a_k$ for $k > m$ and $A_{m,m} = a_m$, we have

$$\sum_{k=m}^{n} a_k b_k = a_m b_m + \sum_{k=m+1}^{n} (A_{k,m} - A_{(k-1),m}) b_k$$

$$= a_m b_m + \sum_{k=m+1}^{n} A_{k,m} b_k - \sum_{k=m}^{n-1} A_{k,m} b_{k+1}$$

$$= a_m b_m + \sum_{k=m+1}^{n-1} A_{k,m} b_k + A_{n,m} b_n - \sum_{k=m+1}^{n-1} A_{k,m} b_{k+1} - A_{m,m} b_{m+1}$$

$$= A_{n,m} b_n - A_{m,m}(b_{m+1} - b_m) - \sum_{k=m+1}^{n-1} A_{k,m}(b_{k+1} - b_k)$$

$$= A_{n,m} b_n - \sum_{k=m}^{n-1} A_{k,m}(b_{k+1} - b_k). \qquad \blacksquare$$

This result is somewhat easier to remember using the following analogy. If $f : [1, N] \to \mathbf{R}$ for some $N \in \mathbf{N}$, then the *summation* $\sum_{k=1}^{N-1} f(k)$ is an approximation to $\int_1^N f(x)dx$ and the *finite difference* $f(k+1) - f(k)$ is an approximation to $f'(k)$ for $k = 1, 2, \ldots, N - 1$. In particular, summation is an analogue of integration and finite difference is an analogue of differentiation. In this context, Abel's Formula can be interpreted as a discrete analogue of integration by parts.

Our first application of Abel's Formula is the following test. (Notice that it does not require that the a_k's be nonnegative.)

6.31 Theorem.   [DIRICHLET'S TEST].
Let $a_k, b_k \in \mathbf{R}$ for $k \in \mathbf{N}$. If the sequence of partial sums $s_n = \sum_{k=1}^n a_k$ is bounded and $b_k \downarrow 0$ as $k \to \infty$, then $\sum_{k=1}^\infty a_k b_k$ converges.

Proof. Choose $M > 0$ such that

$$|s_n| = \left| \sum_{k=1}^n a_k \right| \le \frac{M}{2}, \qquad n \in \mathbf{N}.$$

By the triangle inequality,

$$|A_{n,m}| = \left| \sum_{k=m}^n a_k \right| = |s_n - s_{m-1}| \le \frac{M}{2} + \frac{M}{2} = M$$

for $n > m > 1$.

Let $\varepsilon > 0$ and choose $N \in \mathbf{N}$ so that $|b_k| < \varepsilon/M$ for $k \ge N$. Since $\{b_k\}$ is decreasing and nonnegative, we find, by Abel's Formula, the choice of M, and by telescoping that

$$\left| \sum_{k=m}^n a_k b_k \right| \le |A_{n,m}| \, |b_n| + \sum_{k=m}^{n-1} |A_{k,m}| \, (b_k - b_{k+1})$$

$$\le M b_n + M(b_m - b_n) = M b_m < \varepsilon$$

for all $n > m \ge N$. ∎

The following special case of Dirichlet's Test is widely used.

6.32 Corollary. [ALTERNATING SERIES TEST]. *If $a_k \downarrow 0$ as $k \to \infty$, then*

$$\sum_{k=1}^\infty (-1)^k a_k$$

converges.

Proof. Since the partial sums of $\sum_{k=1}^{\infty}(-1)^k$ are bounded, $\sum_{k=1}^{\infty}(-1)^k a_k$ converges by Dirichlet's Test. ∎

We note that the series $\sum_{k=1}^{\infty}(-1)^k/k$, used in Remark 6.20, is an alternating series. Here is another example.

6.33 EXAMPLE.

Prove that $\sum_{k=1}^{\infty}(-1)^k/\log k$ converges.

Proof. Since $1/\log k \downarrow 0$ as $k \to \infty$, this follows immediately from the Alternating Series Test. ∎

The Dirichlet Test can be used for more than just alternating series.

*6.34 EXAMPLE.

Prove that $S(x) = \sum_{k=1}^{\infty} \sin(kx)/k$ converges for each $x \in \mathbf{R}$.

Proof. Since $\phi(x) = \sin(kx)$ is *periodic* of period 2π [i.e., $\phi(x + 2\pi) = \phi(x)$ for all $x \in \mathbf{R}$] and has value identically zero when $x = 0$ or 2π, we need only show that $S(x)$ converges for each $x \in (0, 2\pi)$. By Dirichlet's Test, it suffices to show that

$$\tilde{D}_n(x) := \sum_{k=1}^{n} \sin(kx), \qquad n \in \mathbf{N} \tag{14}$$

is a bounded sequence for each fixed $x \in (0, 2\pi)$.

This proof, originally discovered by Dirichlet, involves a clever trick which leads to a formula for $\tilde{D}_n$. Indeed, applying a sum angle formula (see Appendix B) and telescoping, we have

$$2 \sin\left(\frac{x}{2}\right) \tilde{D}_n(x) = \sum_{k=1}^{n} 2 \sin\left(\frac{x}{2}\right) \sin(kx)$$

$$= \sum_{k=1}^{n} \left(\cos\left(\left(k - \frac{1}{2}\right)x\right) - \cos\left(\left(k + \frac{1}{2}\right)x\right)\right)$$

$$= \cos\left(\frac{x}{2}\right) - \cos\left(\left(n + \frac{1}{2}\right)x\right).$$

Therefore,

$$|\tilde{D}_n(x)| = \left|\frac{\cos\left(\frac{x}{2}\right) - \cos\left(\left(n + \frac{1}{2}\right)x\right)}{2 \sin\left(\frac{x}{2}\right)}\right| \leq \frac{1}{\left|\sin\left(\frac{x}{2}\right)\right|}$$

for all $n \in \mathbf{N}$. ∎

REVIEW. We have introduced more than a dozen tests for convergence/divergence. With such a wealth of options, students can sometimes be overwhelmed. Here is one suggestion for an order in which to apply these tests to a series $S := \sum_{k=1}^{\infty} a_k$.

i) Try (but not too hard) to find $L = \lim_{k \to \infty} a_k$. If $L \neq 0$ or L doesn't exist, S diverges by the Divergence Test. If $L = 0$ or L is too hard to find, continue.

ii) If a_k is geometric or a p-series, use Theorem 6.7 or Corollary 6.13 to determine convergence properties of $\sum_{k=1}^{\infty} a_k$. If a_k looks a lot like some geometric or a p-series b_k, use the Limit Comparison Test, replace a_k by b_k, and apply 6.7 and 6.13 to b_k.

iii) Try to find

$$r = \lim_{k \to \infty} \frac{|a_{k+1}|}{|a_k|} \text{ or } r = \limsup_{k \to \infty} |a_k|^{1/k}.$$

If $r < 1$, then S converges absolutely. If $r > 1$, then S diverges. If $r = 1$ or these limits are too hard to evaluate, continue.

iv) If the series "alternates" [has factors that oscillate between positive and negative values, i.e. $(-1)^k$, $\sin k$ or $\cos(2k+1)$], try to use the Alternating Series, the Dirichlet, or Abel's Test (see Exercise 6.4.4).

v) If $|a_k| \approx b_k$, where b_k is some nonnegative sequence such that the convergence property of $\sum_{k=1}^{\infty} b_k$ is known, try the Limit Comparison Test or the Comparison Test. If $|a_k|$ is "integrable with respect to k," try the Integral Test.

As long as you don't spend too much time on any one step, you should converge (no pun intended) to an answer fairly quickly. If you get to the end of the process and still haven't arrived at a conclusion, repeat the steps again, trying a little harder this time. Most series, especially the ones that come up in practice, will succumb to this process sooner rather than later.

EXERCISES

6.4.0. Let $\{a_k\}$ and $\{b_k\}$ be real sequences. Decide which of the following statements are true and which are false. Prove the true ones and give counterexamples to the false ones.

a) If $a_k \downarrow 0$, as $k \to \infty$, and $\sum_{k=1}^{\infty} b_k$ converges conditionally, then $\sum_{k=1}^{\infty} a_k b_k$ converges.

b) If $a_k \to 0$, as $k \to \infty$, then $\sum_{k=1}^{\infty} (-1)^k a_k$ converges.

c) If $a_k \to 0$, as $k \to \infty$, and $a_k \geq 0$ for all $k \in \mathbf{N}$, then $\sum_{k=1}^{\infty} (-1)^k a_k$ converges.

d) If $a_k \to 0$, as $k \to \infty$, and $\sum_{k=1}^{\infty} (-1)^k a_k$ converges, then $a_k \downarrow 0$ as $k \to \infty$.

6.4.1. Prove that each of the following series converges.

a) $\displaystyle\sum_{k=1}^{\infty} \frac{(-1)^k}{k^p}$, $p > 0$

b) $\displaystyle\sum_{k=1}^{\infty} \frac{\sin(kx)}{k^p}$ $x \in \mathbf{R}$, $p > 0$

c) $\displaystyle\sum_{k=1}^{\infty} \frac{1 - \cos(1/k)}{(-1)^k}$

d) $\displaystyle\sum_{k=0}^{\infty} \frac{(-1)^{k+1}k}{3^k}$

e) $\displaystyle\sum_{k=1}^{\infty} (-1)^k \left(\frac{\pi}{2} - \arctan k\right)$

6.4.2. For each of the following, find all values $x \in \mathbf{R}$ for which the given series converges.

a) $\displaystyle\sum_{k=1}^{\infty} \frac{x^k}{k}$

b) $\displaystyle\sum_{k=1}^{\infty} \frac{x^{3k}}{2^k}$

c) $\displaystyle\sum_{k=1}^{\infty} \frac{(-1)^k x^k}{\sqrt{k^2 + 1}}$

d) $\displaystyle\sum_{k=1}^{\infty} \frac{(x + 2)^k}{k\sqrt{k + 1}}$

6.4.3. Using any test covered in this chapter so far, find out which of the following series converge absolutely, which converge conditionally, and which diverge.

a) $\displaystyle\sum_{k=1}^{\infty} \frac{(-1)^k k^3}{(k + 1)!}$

b) $\displaystyle\sum_{k=1}^{\infty} \frac{(-1)(-3)\ldots(1 - 2k)}{1 \cdot 4 \ldots (3k - 2)}$

c) $\displaystyle\sum_{k=1}^{\infty} \frac{(k + 1)^k}{p^k k!}$, $p > e$

d) $\displaystyle\sum_{k=1}^{\infty} \frac{(-1)^{k+1}\sqrt{k}}{k + 1}$

e) $\displaystyle\sum_{k=1}^{\infty} \frac{(-1)^k \sqrt{k + 1}}{\sqrt{k}\, k^k}$

6.4.4. [ABEL'S TEST] Suppose that $\sum_{k=1}^{\infty} a_k$ converges and that $b_k \downarrow b$ as $k \to \infty$. Prove that $\sum_{k=1}^{\infty} a_k b_k$ converges.

6.4.5. Show that under the hypotheses of Dirichlet's Test,

$$\sum_{k=1}^{\infty} a_k b_k = \sum_{k=1}^{\infty} s_k(b_k - b_{k+1}).$$

6.4.6. Suppose that $\{a_k\}$ and $\{b_k\}$ are real sequences such that $a_k \to 0$ as $k \to \infty$,

$$\sum_{k=1}^{\infty} |a_{k+1} - a_k| < \infty, \quad \text{and} \quad \left| \sum_{k=1}^{n} b_k \right| \le M \qquad n \in \mathbf{N}.$$

Prove that $\sum_{k=1}^{\infty} a_k b_k$ converges.

6.4.7. Suppose that $\sum_{k=1}^{\infty} a_k$ converges. Prove that if $b_k \uparrow \infty$ and $\sum_{k=1}^{\infty} a_k b_k$ converges, then

$$b_m \sum_{k=m}^{\infty} a_k \to 0$$

as $m \to \infty$.

*6.4.8. Prove that

$$\sum_{k=1}^{\infty} a_k \cos(kx)$$

converges for every $x \in (0, 2\pi)$ and every $a_k \downarrow 0$. What happens when $x = 0$?

*6.4.9. Suppose that $a_k \downarrow 0$ as $k \to \infty$. Prove that

$$\sum_{k=1}^{\infty} a_k \sin((2k+1)x)$$

converges for all $x \in \mathbf{R}$.

***6.5 ESTIMATION OF SERIES**

In practice, one estimates a convergent series by truncation (i.e., by adding finitely many terms of the given series). In this section we show how to estimate the error associated with such a truncation.

The proofs of several of our earlier tests actually contain estimates of the truncation error. Here is what we can get from the Integral Test.

6.35 Theorem. *Suppose that $f : [1, \infty) \to \mathbf{R}$ is positive and decreasing on $[1, \infty)$. Then*

$$f(n) \le \sum_{k=1}^{n} f(k) - \int_{1}^{n} f(x)\,dx \le f(1) \ \text{for } n \in \mathbf{N}.$$

Moreover, if $\sum_{k=1}^{\infty} f(k)$ converges, then

$$0 \le \sum_{k=1}^{n} f(k) + \int_{n}^{\infty} f(x)\,dx - \sum_{k=1}^{\infty} f(k) \le f(n)$$

for all $n \in \mathbf{N}$.

Proof. The first set of inequalities has already been verified [see (3) in the proof of Theorem 6.12]. To establish the second set, let $u_k = s_k - t_k$ for $k \in \mathbf{N}$, and observe, since f is decreasing, that

$$0 \le u_k - u_{k+1} = \int_k^{k+1} f(x)\,dx - f(k+1) \le f(k) - f(k+1).$$

Summing these inequalities over $k \ge n$ and telescoping, we have

$$0 \le u_n - \lim_{j \to \infty} u_j = \sum_{k=n}^\infty (u_k - u_{k+1}) \le \sum_{k=n}^\infty (f(k) - f(k+1)) = f(n).$$

Since $u_j \to \sum_{k=1}^\infty f(k) - \int_1^\infty f(x)dx$ as $j \to \infty$, we conclude that

$$0 \le \sum_{k=1}^n f(k) + \int_n^\infty f(x)\,dx - \sum_{k=1}^\infty f(k) \le f(n). \qquad \blacksquare$$

The following example shows how to use this result to estimate the accuracy of a truncation of a series to which the Integral Test applies.

6.36 EXAMPLE.

Prove that $\sum_{k=1}^\infty ke^{-k^2}$ converges and estimate its value to three decimal places.

Proof. Let $f(x) = xe^{-x^2}$. Since $f'(x) = e^{-x^2}(1 - 2x^2) \le 0$ for $x \ge 1$, f is decreasing on $[1, \infty)$. Since

$$\int_1^\infty xe^{-x^2}\,dx = \frac{1}{2}\int_1^\infty e^{-u}\,du = \frac{1}{2e} < \infty,$$

it follows from the Integral Test that $\sum_{k=1}^\infty ke^{-k^2}$ converges.

By Theorem 6.35, the error of replacing s by $\sum_{k=1}^n f(k) + \int_n^\infty f(x)dx$ is dominated by $f(n)$. By the rounding process, this estimate to s will be accurate to three decimal places if the error $f(n)$ is ≤ 0.0005. Since $f(2) = 0.036631$ and $f(3) = 0.000370$, it follows that we should use $n = 3$. Since

$$\sum_{k=1}^3 ke^{-k^2} + \int_3^\infty xe^{-x^2}\,dx = \frac{1}{e} + \frac{2}{e^4} + \frac{3}{e^9} + \frac{1}{2e^9} \approx 0.4049427,$$

we conclude that a three-place estimate to s is given by 0.405. $\blacksquare$

The next example shows that Theorem 6.35 can be used to estimate divergent series as well.

6.37 EXAMPLE.

Prove that there exist numbers $C_n \in (0, 1]$ such that

$$\sum_{k=1}^{n} \frac{1}{k} = \log n + C_n$$

for all $n \in \mathbf{N}$.

Proof. Clearly, $f(x) = 1/x$ is positive, decreasing, and locally integrable on $[1, \infty)$. Hence, by Theorem 6.35,

$$\frac{1}{n} \leq \sum_{k=1}^{n} \frac{1}{k} - \int_{1}^{n} \frac{1}{x} \, dx = \sum_{k=1}^{n} \frac{1}{k} - \log n \leq 1. \qquad \blacksquare$$

Next, we see what the Alternating Series Test has to say about truncation error.

6.38 Theorem. *Suppose that $a_k \downarrow 0$ as $k \to \infty$. If $s = \sum_{k=1}^{\infty}(-1)^k a_k$ and $s_n = \sum_{k=1}^{n}(-1)^k a_k$, then $0 \leq |s - s_n| \leq a_{n+1}$ for all $n \in \mathbf{N}$.*

Proof. Suppose first that n is even, say $n = 2m$. Then

$$0 \geq (-a_{2m+1} + a_{2m+2}) + (-a_{2m+3} + a_{2m+4}) + \cdots$$
$$= \sum_{k=2m+1}^{\infty} (-1)^k a_k = s - s_n$$
$$= -a_{2m+1} + (a_{2m+2} - a_{2m+3}) + (a_{2m+4} - a_{2m+5}) + \cdots$$
$$\geq -a_{2m+1};$$

that is, $0 \geq s - s_n \geq -a_{n+1}$. A similar argument proves that $0 \leq s - s_n \leq a_{n+1}$ when n is odd. $\qquad \blacksquare$

This result can be used to estimate the error of a truncation of any alternating series.

6.39 EXAMPLE.

For each $\alpha > 0$, prove that the series $\sum_{k=1}^{\infty}(-1)^k k/(k^2 + \alpha)$ converges. If s_n represents its nth partial sum and s its value, find an n so large that s_n approximates s to two decimal places.

Proof. Let $f(x) = x/(x^2 + \alpha)$ and note that $f(x) \to 0$ as $x \to \infty$. Since $f'(x) = (\alpha - x^2)/(x^2 + \alpha)^2$ is negative for $x > \sqrt{|\alpha|}$, it follows that $k/(k^2 + \alpha) \downarrow 0$ as $k \to \infty$. Hence, the given series converges by the Alternating Series Test.

By Theorem 6.38, s_n will estimate s to two decimal places if $f(n) < 0.005$ (i.e., if $n^2 - 200n + \alpha > 0$). When $\alpha > 10^4$, this last quadratic has no real roots; hence, the inequality is always satisfied and we may choose $n = 1$. When $\alpha \le 10^4$, the quadratic has roots $100 \pm \sqrt{10^4 - \alpha}$. Hence, choose any n which satisfies $n > 100 + \sqrt{10^4 - \alpha}$.   ∎

Finally, we examine what information the proofs of the Root and Ratio Tests contain about accuracy of truncations.

6.40 Theorem.   *Suppose that $\sum_{k=1}^{\infty} a_k$ converges absolutely and that s is the value of $\sum_{k=1}^{\infty} |a_k|$.*

i) *If there exist numbers $x \in (0, 1)$ and $N \in \mathbf{N}$ such that*

$$|a_k|^{1/k} \le x$$

for all $k > N$, then

$$0 \le s - \sum_{k=1}^{n} |a_k| \le \frac{x^{n+1}}{1 - x}$$

for all $n \ge N$.

ii) *If there exist numbers $x \in (0, 1)$ and $N \in \mathbf{N}$ such that*

$$\frac{|a_{k+1}|}{|a_k|} \le x$$

for $k > N$, then

$$0 \le s - \sum_{k=1}^{n} |a_k| \le \frac{|a_N| x^{n-N+1}}{1 - x}$$

for all $n \ge N$.

Proof. Let $n \ge N$. Since $|a_k| \le x^k$ for $k > N$, we have, by summing a geometric series, that

$$0 \le s - \sum_{k=1}^{n} |a_k| = \sum_{k=n+1}^{\infty} |a_k| \le \sum_{k=n+1}^{\infty} x^k = \frac{x^{n+1}}{1 - x}$$

for all $n \ge N$. This proves part i). The proof of part ii) is left as an exercise.   ∎

6.41 *EXAMPLE.*

Prove that $\sum_{k=1}^{\infty} k^{2k}/(3k^2 + k)^k$ converges absolutely. If s_n represents its nth partial sum and s its value, find an n so large that s_n approximates s to an accuracy of 10^{-2}.

Solution. Since

$$
\left(\frac{k^{2k}}{(3k^2 + k)^k} \right)^{1/k} = \frac{k^2}{3k^2 + k} \le \frac{1}{3}
$$

for all $k \ge N := 1$, the series converges absolutely by the Root Test. Since $(1/3)^{n+1}/(1 - 1/3) \le 10^{-2}$ for $n \ge 4$, we conclude by Theorem 6.40i that it takes at most four terms to approximate the value of this series to an accuracy of 10^{-2}. ∎

EXERCISES

6.5.1. For each of the following series, let s_n represent its partial sums and s its value. Prove that s is finite and find an n so large that s_n approximates s to an accuracy of 10^{-2}.

a) $\displaystyle\sum_{k=1}^{\infty} (-1)^k \left(\frac{\pi}{2} - \arctan k \right)$

b) $\displaystyle\sum_{k=1}^{\infty} \frac{(-1)^k k^2}{2^k}$

c) $\displaystyle\sum_{k=1}^{\infty} \frac{(-1)^k}{k^2} \frac{2 \cdot 4 \cdots (2k)}{1 \cdot 3 \cdots (2k - 1)}$

6.5.2. a) Find all $p \ge 0$ such that the following series converges:

$$
\sum_{k=1}^{\infty} \frac{1}{k \log^p (k + 1)}.
$$

b) For each such p, prove that the partial sums of this series s_n and its value s satisfy

$$
|s - s_n| \le \frac{n + p - 1}{n(p - 1)} \left(\frac{1}{\log^{p-1}(n)} \right)
$$

for all $n \ge 2$.

6.5.3. For each of the following series, let s_n represent its partial sums, and let s represent its value. Prove that s is finite and find an n so large that s_n approximates s to three decimal places.

a) $\displaystyle\sum_{k=1}^{\infty} \frac{1}{k!}$

b) $\displaystyle\sum_{k=1}^{\infty} \frac{1}{k^k}$

c) $\displaystyle\sum_{k=1}^{\infty} \frac{2^k}{k!}$

d) $\displaystyle\sum_{k=1}^{\infty} \left(\frac{k}{k + 1} \right)^{k^2}$

6.5.4. Prove Theorem 6.40ii.

*6.6 ADDITIONAL TESTS

If the Ratio or Root Test yields a value $r = 1$, then no conclusion can be made. There are some tests designed to handle just that situation (see Exercise 6.6.3). We cover two of them in this section (see also Exercises 6.6.4 and 6.6.5).

The first test compares the growth of the terms of a series with the growth of the logarithm function.

6.42 Theorem. [THE LOGARITHMIC TEST].
Suppose that $a_k \neq 0$ for large k and that

$$p = \lim_{k \to \infty} \frac{\log(1/|a_k|)}{\log k}$$

exists as an extended real number. If $p > 1$, then $\sum_{k=1}^{\infty} a_k$ converges absolutely. If $p < 1$, then $\sum_{k=1}^{\infty} |a_k|$ diverges.

Proof. Suppose that $p > 1$. Fix $q \in (1, p)$ and choose $N \in \mathbf{N}$ so that $k \geq N$ implies $\log(1/|a_k|) > q \log k = \log(k^q)$. Since the logarithm function is monotone increasing, it follows that $1/|a_k| > k^q$; that is, that $|a_k| < k^{-q}$ for $k \geq N$. Hence, by the Comparison Test, $\sum_{k=1}^{\infty} |a_k|$ converges.

Similarly, if $p < 1$, then $|a_k| > 1/k$ for large k. Hence, by the Comparison Test, $\sum_{k=1}^{\infty} |a_k|$ diverges. ∎

Our final test works by examining how rapidly the ratios of a_{k+1}/a_k converge to $r = 1$ (see also Exercise 6.6.5 below). Its proof uses Bernoulli's Inequality.

*6.43 Theorem. [RAABE'S TEST].
Suppose that there is a constant C and a parameter p such that

$$\left| \frac{a_{k+1}}{a_k} \right| \leq 1 - \frac{p}{k + C} \tag{15}$$

for large k. If $p > 1$, then $\sum_{k=1}^{\infty} a_k$ converges absolutely.

Proof. Set $x_k = k + C - 1$ for $k \in \mathbf{N}$ and choose $N \in \mathbf{N}$ such that $x_k > 1$ and (15) hold for $k \geq N$. By the p-Series Test and the Limit Comparison Test,

$$\sum_{k=N}^{\infty} x_k^{-p} < \infty. \tag{16}$$

By (15) and Bernoulli's Inequality,

$$\left| \frac{a_{k+1}}{a_k} \right| \leq 1 - \frac{p}{x_{k+1}} \leq \left(1 - \frac{1}{x_{k+1}} \right)^p = \frac{x_k^p}{x_{k+1}^p}.$$

Hence, the sequence $\{|a_k| |x_k^p|\}_{k=N}^{\infty}$ is decreasing and bounded above. In particular, there is an $M > 0$ such that $|a_k| \leq M x_k^{-p}$ for $k \geq N$. We conclude by (16) that $\sum_{k=1}^{\infty} a_k$ converges. ∎

EXERCISES

6.6.1. Using any test covered in this chapter, find out which of the following series converge absolutely, which converge conditionally, and which diverge.

a) $\displaystyle\sum_{k=1}^{\infty} \frac{3 \cdot 5 \cdots (2k+1)}{2 \cdot 4 \cdots 2k}$

b) $\displaystyle\sum_{k=1}^{\infty} \frac{1 \cdot 3 \cdots (2k-1)}{5 \cdot 7 \cdots (2k+3)}$

c) $\displaystyle\sum_{k=2}^{\infty} \frac{1}{(\log k)^{\log \log k}}$

d) $\displaystyle\sum_{k=1}^{\infty} \left(\frac{\sqrt{k}-1}{\sqrt{k}} \right)^k$

6.6.2. For each of the following, find all values of $p \in \mathbf{R}$ for which the given series converges absolutely, for which it converges conditionally, and for which it diverges.

a) $\displaystyle\sum_{k=1}^{\infty} ke^{-kp}$

b) $\displaystyle\sum_{k=2}^{\infty} \frac{1}{(\log k)^{p \log k}}$

c) $\displaystyle\sum_{k=1}^{\infty} \frac{(pk)^k}{k!}$

*6.6.3. a) Prove that the Root Test applied to the series

$$\sum_{k=2}^{\infty} \frac{1}{(\log k)^{\log k}}$$

yields $r = 1$. Use the Logarithmic Test to prove that this series converges.

b) Prove that the Ratio Test applied to the series

$$\sum_{k=1}^{\infty} \frac{1 \cdot 3 \cdots (2k-1)}{4 \cdot 6 \cdots (2k+2)}$$

yields $r = 1$. Use Raabe's Test to prove that this series converges.

6.6.4. Suppose that $f : \mathbf{R} \to (0, \infty)$ is differentiable, that $f(x) \to 0$ as $x \to \infty$, and that

$$\alpha := \lim_{x \to \infty} \frac{xf'(x)}{f(x)}$$

exists. If $\alpha < -1$, prove that $\sum_{k=1}^{\infty} f(k)$ converges.

6.6.5. Suppose that $\{a_k\}$ is a sequence of nonzero real numbers and that

$$p = \lim_{k \to \infty} k \left(1 - \left| \frac{a_{k+1}}{a_k} \right| \right)$$

exists as an extended real number. Prove that $\sum_{k=1}^{\infty} a_k$ converges absolutely when $p > 1$.

CHAPTER 7

Infinite Series of Functions

7.1 UNIFORM CONVERGENCE OF SEQUENCES

You are familiar with what it means for a sequence of numbers to converge. In this section we examine what it means for a sequence of functions to converge. It turns out there are several different ways to define *convergence* of a sequence of functions. We begin with the simplest way.

7.1 Definition.

Let E be a nonempty subset of $\mathbf{R}$. A sequence of functions $f_n : E \to \mathbf{R}$ is said to *converge pointwise* on E (notation: $f_n \to f$ pointwise on E as $n \to \infty$) if and only if $f(x) = \lim_{n \to \infty} f_n(x)$ exists for each $x \in E$.

Because $\{f_n\}$ converges pointwise on a set E if and only if the sequence of real numbers $\{f_n(x)\}$ converges for each $x \in E$, every result about convergence of real numbers contains a result about pointwise convergence of functions. Here is a typical example.

7.2 Remark. *Let E be a nonempty subset of $\mathbf{R}$. Then a sequence of functions f_n converges pointwise on E, as $n \to \infty$, if and only if for every $\varepsilon > 0$ and $x \in E$ there is an $N \in \mathbf{N}$ (which may depend on x as well as ε) such that*

$$n \geq N \quad implies \quad |f_n(x) - f(x)| < \varepsilon.$$

Proof. By Definition 7.1, $f_n \to f$ pointwise on E if and only if $f_n(x) \to f(x)$ for all $x \in E$. This occurs, by Definition 2.1, if and only if for every $\varepsilon > 0$ and $x \in E$ there is an $N \in \mathbf{N}$ such that $n \geq N$ implies $|f_n(x) - f(x)| < \varepsilon$. ∎

If $f_n \to f$ pointwise on $[a, b]$, it is natural to ask, What does f inherit from f_n? The next four remarks show that, in general, the answer to this question is "not much."

7.3 Remark. *The pointwise limit of continuous (respectively, differentiable) functions is not necessarily continuous (respectively, differentiable).*

Proof. Let $f_n(x) = x^n$ and set

$$f(x) = \begin{cases} 0 & 0 \leq x < 1 \\ 1 & x = 1. \end{cases}$$

Then $f_n \to f$ pointwise on $[0, 1]$ (see Example 2.20), each f_n is continuous and differentiable on $[0, 1]$, but f is neither differentiable nor continuous at $x = 1$. $\blacksquare$

7.4 Remark. *The pointwise limit of integrable functions is not necessarily integrable.*

Proof. Set

$$f_n(x) = \begin{cases} 1 & x = p/m \in \mathbf{Q}, \text{ written in reduced form, where } m \leq n \\ 0 & \text{otherwise,} \end{cases}$$

for $n \in \mathbf{N}$ and

$$f(x) = \begin{cases} 1 & x \in \mathbf{Q} \\ 0 & \text{otherwise.} \end{cases}$$

Then $f_n \to f$ pointwise on $[0, 1]$, each f_n is integrable on $[0, 1]$ (with integral zero), but f is not integrable on $[0, 1]$ (see Example 5.11). $\blacksquare$

7.5 Remark. *There exist differentiable functions f_n and f such that $f_n \to f$ pointwise on $[0, 1]$ but*

$$\lim_{n \to \infty} f_n'(x) \neq \left(\lim_{n \to \infty} f_n(x) \right)' \tag{1}$$

for $x = 1$.

Proof. Let $f_n(x) = x^n/n$ and set $f(x) = 0$. Then $f_n \to f$ pointwise on $[0, 1]$, each f_n is differentiable with $f_n'(x) = x^{n-1}$. Thus the left side of (1) is 1 at $x = 1$ but the right side of (1) is zero. $\blacksquare$

7.6 Remark. *There exist continuous functions f_n and f such that $f_n \to f$ pointwise on $[0, 1]$ but*

$$\lim_{n \to \infty} \int_0^1 f_n(x)\, dx \neq \int_0^1 \left(\lim_{n \to \infty} f_n(x) \right) dx. \tag{2}$$

Proof. Let $f_1(x) = 1$ and, for $n > 1$, let f_n be a sequence of functions whose graphs are triangles with bases $2/n$ and altitudes n (see Figure 7.1). By the point-slope form, formulas for these f_n's can be given by

$$f_n(x) = \begin{cases} n^2 x & 0 \leq x < 1/n \\ 2n - n^2 x & 1/n \leq x < 2/n \\ 0 & 2/n \leq x \leq 1. \end{cases}$$

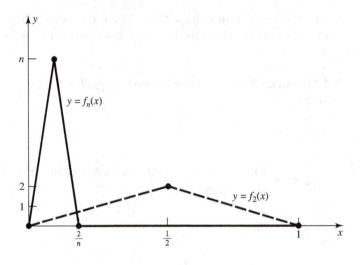

FIGURE 7.1

Then $f_n \to 0$ pointwise on $[0, 1]$ and, since the area of a triangle is one-half base times altitude, $\int_0^1 f_n(x)\,dx = 1$ for all $n \in \mathbf{N}$. Thus, the left side of (2) is 1 but the right side is zero. ∎

In view of the preceding examples, it is clear that pointwise convergence is of limited value for the calculus of limits of sequences. It turns out that the following concept, discovered independently by Stokes, Cauchy, and Weierstrass around 1850, is much more useful in this context.

7.7 Definition.

Let E be a nonempty subset of $\mathbf{R}$. A sequence of functions $f_n : E \to \mathbf{R}$ is said to *converge uniformly* on E to a function f (notation: $f_n \to f$ uniformly on E as $n \to \infty$) if and only if for every $\varepsilon > 0$ there is an $N \in \mathbf{N}$ such that

$$n \geq N \quad \text{implies} \quad |f_n(x) - f(x)| < \varepsilon$$

for all $x \in E$.

Comparing Definition 7.7 with Remark 7.2 above, we see that the only difference between uniform convergence and pointwise convergence is that, for uniform convergence, the integer N must be chosen independently of x (see Figure 7.2). Notice that this is similar to the difference between uniform continuity and continuity (see the discussion following Example 3.36).

By definition, if f_n converges uniformly on E, then f_n converges pointwise on E. The following example shows that the converse of this statement is false. [This example also shows how to prove that $f_n \to f$ uniformly on a set E:

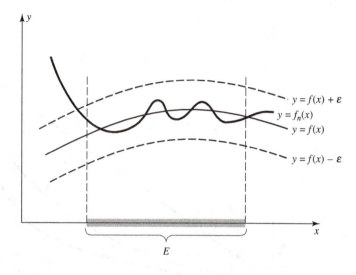

$y = f(x) + \varepsilon$
$y = f_n(x)$
$y = f(x)$
$y = f(x) - \varepsilon$

E

FIGURE 7.2

dominate $|f_n(x) - f(x)|$ by constants b_n, independent of $x \in E$, which converge to zero as $n \to \infty$.]

7.8 EXAMPLE.

Prove that $x^n \to 0$ uniformly on $[0, b]$ for any $b < 1$, and pointwise, but not uniformly, on $[0, 1)$.

Proof. By Example 2.20, $x^n \to 0$ pointwise on $[0, 1)$. Let $b < 1$. Given $\varepsilon > 0$, choose $N \in \mathbf{N}$ such that $n \geq N$ implies $b^n < \varepsilon$. Then $x \in [0, b]$ and $n \geq N$ imply $|x^n| \leq b^n < \varepsilon$; that is, $x^n \to 0$ uniformly for $x \in [0, b]$.

Does x^n converge to 0 uniformly on $[0, 1)$? If it does, then given $0 < \varepsilon < 1/2$, there is an $N \in \mathbf{N}$ such that $|x^N| < \varepsilon$ for all $x \in [0, 1)$. But $x^N \to 1$ as $x \to 1-$ so we can choose an $x_0 \in (0, 1)$ such that $x_0^N > \varepsilon$ (see Figure 7.3). Thus $\varepsilon < x_0^N < \varepsilon$, a contradiction. ∎

The next several results show that if $f_n \to f$ or $f_n' \to f'$ uniformly, then f inherits much from f_n.

7.9 Theorem. *Let E be a nonempty subset of $\mathbf{R}$ and suppose that $f_n \to f$ uniformly on E, as $n \to \infty$. If each f_n is continuous at some $x_0 \in E$, then f is continuous at $x_0 \in E$.*

Proof. Let $\varepsilon > 0$ and choose $N \in \mathbf{N}$ such that

$$n \geq N \quad \text{and} \quad x \in E \quad \text{imply} \quad |f_n(x) - f(x)| < \frac{\varepsilon}{3}.$$

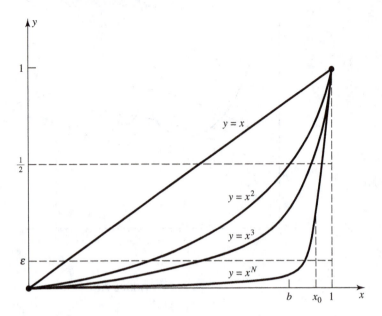

FIGURE 7.3

Since f_N is continuous at $x_0 \in E$, choose $\delta > 0$ such that

$$|x - x_0| < \delta \quad \text{and} \quad x \in E \quad \text{imply} \quad |f_N(x) - f_N(x_0)| < \frac{\varepsilon}{3}.$$

Suppose that $|x - x_0| < \delta$ and that $x \in E$. Then

$$|f(x) - f(x_0)| \le |f(x) - f_N(x)| + |f_N(x) - f_N(x_0)| + |f_N(x_0) - f(x_0)| < \varepsilon.$$

Thus f is continuous at $x_0 \in E$. ■

(For a generalization of this result, see Exercise 7.1.6. For a converse of this result when the sequence f_n is pointwise monotone, see Theorem 9.40.)

Here is an important theorem about interchanging a limit sign and an integral sign (compare with Remark 7.6).

7.10 Theorem. *Suppose that $f_n \to f$ uniformly on a closed interval $[a, b]$. If each f_n is integrable on $[a, b]$, then so is f and*

$$\lim_{n \to \infty} \int_a^b f_n(x)\, dx = \int_a^b \left(\lim_{n \to \infty} f_n(x) \right) dx.$$

In fact, $\lim_{n \to \infty} \int_a^x f_n(t)\, dt = \int_a^x f(t)\, dt$ uniformly for $x \in [a, b]$.

Proof. By Exercise 7.1.3, f is bounded on $[a, b]$. To prove that f is integrable, let $\varepsilon > 0$ and choose $N \in \mathbf{N}$ such that

$$n \geq N \quad \text{implies} \quad |f(x) - f_n(x)| < \frac{\varepsilon}{3(b - a)} \tag{3}$$

for all $x \in [a, b]$. Using this inequality for $n = N$, we see that by the definition of upper and lower sums,

$$U(f - f_N, P) \leq \frac{\varepsilon}{3} \quad \text{and} \quad L(f - f_N, P) \geq -\frac{\varepsilon}{3}$$

for any partition P of $[a, b]$. Since f_N is integrable, choose a partition P such that

$$U(f_N, P) - L(f_N, P) < \frac{\varepsilon}{3}.$$

It follows that

$$U(f, P) - L(f, P) \leq U(f - f_N, P) + U(f_N, P) - L(f_N, P) - L(f - f_N, P)$$
$$< \frac{\varepsilon}{3} + \frac{\varepsilon}{3} + \frac{\varepsilon}{3} = \varepsilon;$$

that is, f is integrable on $[a, b]$. We conclude by Theorem 5.22 and (3) that

$$\left| \int_a^x f_n(t)\, dt - \int_a^x f(t)\, dt \right| \leq \int_a^x |f_n(t) - f(t)|\, dt \leq \frac{\varepsilon(x - a)}{3(b - a)} < \varepsilon$$

for all $x \in [a, b]$ and $n \geq N$.   ∎

Here is a Cauchy Criterion for uniform convergence.

7.11 Lemma. [UNIFORM CAUCHY CRITERION].
Let E be a nonempty subset of $\mathbf{R}$ and let $f_n : E \to \mathbf{R}$ be a sequence of functions. Then f_n converges uniformly on E if and only if for every $\varepsilon > 0$ there is an $N \in \mathbf{N}$ such that

$$n, m \geq N \quad \text{imply} \quad |f_n(x) - f_m(x)| < \varepsilon \tag{4}$$

for all $x \in E$.

Proof. Suppose first that $f_n \to f$ uniformly on E as $n \to \infty$. Let $\varepsilon > 0$ and choose $N \in \mathbf{N}$ such that

$$n \geq N \quad \text{implies} \quad |f_n(x) - f(x)| < \frac{\varepsilon}{2}$$

for $x \in E$. Since $|f_n(x) - f_m(x)| \leq |f_n(x) - f(x)| + |f(x) - f_m(x)|$, it is clear that (4) holds for all $x \in E$.

Conversely, if (4) holds for $x \in E$, then $\{f_n(x)\}_{n \in \mathbb{N}}$ is Cauchy for each $x \in E$. Hence, by Cauchy's Theorem for sequences (Theorem 2.29),

$$f(x) := \lim_{n \to \infty} f_n(x)$$

exists for each $x \in E$. Take the limit of the second inequality in (4) as $m \to \infty$. We obtain $|f_n(x) - f(x)| \le \varepsilon/2 < \varepsilon$ for all $n \ge N$ and $x \in E$. Hence, by definition, $f_n \to f$ uniformly on E. ∎

Here is a result about interchanging a limit sign and the derivative sign (compare with Remark 7.5). The proof presented here comes from Apostol [1].

7.12 Theorem. *Let (a, b) be a bounded interval and suppose that f_n is a sequence of functions which converges at some $x_0 \in (a, b)$. If each f_n is differentiable on (a, b), and f_n' converges uniformly on (a, b) as $n \to \infty$, then f_n converges uniformly on (a, b) and*

$$\lim_{n \to \infty} f_n'(x) = \left(\lim_{n \to \infty} f_n(x) \right)'$$

for each $x \in (a, b)$.

Proof. Fix $c \in (a, b)$ and define

$$g_n(x) = \begin{cases} \dfrac{f_n(x) - f_n(c)}{x - c} & x \ne c \\ f_n'(c) & x = c \end{cases}$$

for $n \in \mathbb{N}$. Clearly,

$$f_n(x) = f_n(c) + (x - c)g_n(x) \tag{5}$$

for $n \in \mathbb{N}$ and $x \in (a, b)$.

We claim that for any $c \in (a, b)$, the sequence g_n converges uniformly on (a, b). Let $\varepsilon > 0$, $n, m \in \mathbb{N}$, and $x \in (a, b)$ with $x \ne c$. By the Mean Value Theorem, there is a ξ between x and c such that

$$g_n(x) - g_m(x) = \frac{f_n(x) - f_m(x) - (f_n(c) - f_m(c))}{x - c} = f_n'(\xi) - f_m'(\xi).$$

Since f_n' converges uniformly on (a, b), it follows that there is an $N \in \mathbb{N}$ such that

$$n, m \ge N \quad \text{implies} \quad |g_n(x) - g_m(x)| < \varepsilon$$

for $x \in (a, b)$ with $x \ne c$. This implication also holds for $x = c$ because $g_n(c) = f_n'(c)$ for all $n \in \mathbb{N}$. This proves the claim.

To show that f_n converges uniformly on (a, b), notice that by the claim, g_n converges uniformly as $n \to \infty$ and (5) holds for $c = x_0$. Since $f_n(x_0)$

converges as $n \to \infty$ by hypothesis, it follows from (5) and $b - a < \infty$ that f_n converges uniformly on (a, b) as $n \to \infty$.

Fix $c \in (a, b)$. Define f, g on (a, b) by $f(x) := \lim_{n \to \infty} f_n(x)$ and $g(x) := \lim_{n \to \infty} g_n(x)$. We need to show that

$$f'(c) = \lim_{n \to \infty} f_n'(c). \tag{6}$$

Since each g_n is continuous at c, the claim implies g is continuous at c. Since $g_n(c) = f_n'(c)$, it follows that the right side of (6) can be written as

$$\lim_{n \to \infty} f_n'(c) = \lim_{n \to \infty} g_n(c) = g(c) = \lim_{x \to c} g(x).$$

On the other hand, if $x \neq c$ we have by definition that

$$\frac{f(x) - f(c)}{x - c} = \lim_{n \to \infty} \frac{f_n(x) - f_n(c)}{x - c} = \lim_{n \to \infty} g_n(x) = g(x).$$

Therefore, the left side of (6) also reduces to

$$f'(c) = \lim_{x \to c} \frac{f(x) - f(c)}{x - c} = \lim_{x \to c} g(x).$$

This verifies (6), and the proof of the theorem is complete. ∎

EXERCISES

7.1.1. a) Prove that $x/n \to 0$ uniformly, as $n \to \infty$, on any closed interval $[a, b]$.

   b) Prove that $1/(nx) \to 0$ pointwise but not uniformly on $(0, 1)$ as $n \to \infty$.

7.1.2. Prove that the following limits exist and evaluate them.

   a) $\lim_{n \to \infty} \int_1^3 \dfrac{nx^{99} + 5}{x^3 + nx^{66}} \, dx$

   b) $\lim_{n \to \infty} \int_0^2 e^{x^2/n} \, dx$

   c) $\lim_{n \to \infty} \int_0^3 \sqrt{\sin \dfrac{x}{n} + x + 1} \, dx$

7.1.3. A sequence of functions f_n is said to be *uniformly bounded* on a set E if and only if there is an $M > 0$ such that $|f_n(x)| \leq M$ for all $x \in E$ and all $n \in \mathbf{N}$.

   Suppose that for each $n \in \mathbf{N}$, $f_n : E \to \mathbf{R}$ is bounded. If $f_n \to f$ uniformly on E, as $n \to \mathbf{N}$, prove that $\{f_n\}$ is uniformly bounded on E and f is a bounded function on E.

7.1.4. Let $[a, b]$ be a closed bounded interval, $f : [a, b] \to \mathbf{R}$ be bounded, and $g : [a, b] \to \mathbf{R}$ be continuous with $g(a) = g(b) = 0$. Let f_n be a uniformly bounded sequence of functions on $[a, b]$ (see Exercise 7.1.3). Prove that if $f_n \to f$ uniformly on all closed intervals $[c, d] \subset (a, b)$, then $f_n g \to f g$ uniformly on $[a, b]$.

7.1.5. Suppose that $f_n \to f$ and $g_n \to g$, as $n \to \infty$, uniformly on some set $E \subseteq \mathbf{R}$.

a) Prove that $f_n + g_n \to f + g$ and $\alpha f_n \to \alpha f$, as $n \to \infty$, uniformly on E for all $\alpha \in \mathbf{R}$.

b) Prove that $f_n g_n \to f g$ pointwise on E.

c) Prove that if f and g are bounded on E, then $f_n g_n \to f g$ uniformly on E.

d) Show that c) may be false when g is unbounded.

7.1.6. Suppose that E is a nonempty subset of $\mathbf{R}$ and that $f_n \to f$ uniformly on E. Prove that if each f_n is uniformly continuous on E, then f is uniformly continuous on E.

7.1.7. Suppose that f is uniformly continuous on $\mathbf{R}$. If $y_n \to 0$ as $n \to \infty$ and $f_n(x) := f(x + y_n)$ for $x \in \mathbf{R}$, prove that f_n converges uniformly on $\mathbf{R}$.

7.1.8. Suppose that $b > a > 0$. Prove that

$$\lim_{n \to \infty} \int_a^b \left(1 + \frac{x}{n}\right)^n e^{-x} \, dx = b - a.$$

7.1.9. Let f, g be continuous on a closed bounded interval $[a, b]$ with $|g(x)| > 0$ for $x \in [a, b]$. Suppose that $f_n \to f$ and $g_n \to g$ as $n \to \infty$, uniformly on $[a, b]$.

a) Prove that $1/g_n$ is defined for large n and $f_n/g_n \to f/g$ uniformly on $[a, b]$ as $n \to \infty$.

b) Show that a) is false if $[a, b]$ is replaced by (a, b).

7.1.10. Let E be a nonempty subset of $\mathbf{R}$ and f be a real-valued function defined on E. Suppose that f_n is a sequence of bounded functions on E which converges to f uniformly on E. Prove that

$$\frac{f_1(x) + \cdots + f_n(x)}{n} \to f(x)$$

uniformly on E as $n \to \infty$ (compare with Exercise 6.1.9).

7.1.11. Let f_n be integrable on $[0, 1]$ and $f_n \to f$ uniformly on $[0, 1]$. Show that if $b_n \uparrow 1$ as $n \to \infty$, then

$$\lim_{n \to \infty} \int_0^{b_n} f_n(x) \, dx = \int_0^1 f(x) \, dx.$$

7.2 UNIFORM CONVERGENCE OF SERIES

In this section we extend the concepts introduced in Section 7.1 from sequences to series.

7.13 Definition.

Let f_k be a sequence of real functions defined on some set E and set

$$s_n(x) := \sum_{k=1}^{n} f_k(x), \qquad x \in E, \ n \in \mathbf{N}.$$

i) The series $\sum_{k=1}^{\infty} f_k$ is said to *converge pointwise* on E if and only if the sequence $s_n(x)$ converges pointwise on E as $n \to \infty$.

ii) The series $\sum_{k=1}^{\infty} f_k$ is said to *converge uniformly* on E if and only if the sequence $s_n(x)$ converges uniformly on E as $n \to \infty$.

iii) The series $\sum_{k=1}^{\infty} f_k$ is said to *converge absolutely (pointwise)* on E if and only if $\sum_{k=1}^{\infty} |f_k(x)|$ converges for each $x \in E$.

Since convergence of series is defined in terms of convergence of sequences of partial sums, every result about convergence of sequences of functions contains a result about convergence of series of functions. For example, the following result is an immediate consequence of Theorems 7.9, 7.10, and 7.12.

7.14 Theorem. *Let E be a nonempty subset of $\mathbf{R}$ and let $\{f_k\}$ be a sequence of real functions defined on E.*

i) *Suppose that $x_0 \in E$ and that each f_k is continuous at $x_0 \in E$. If $f = \sum_{k=1}^{\infty} f_k$ converges uniformly on E, then f is continuous at $x_0 \in E$.*

ii) [TERM-BY-TERM INTEGRATION]. *Suppose that $E = [a, b]$ and that each f_k is integrable on $[a, b]$. If $f = \sum_{k=1}^{\infty} f_k$ converges uniformly on $[a, b]$, then f is integrable on $[a, b]$ and*

$$\int_a^b \sum_{k=1}^{\infty} f_k(x) \, dx = \sum_{k=1}^{\infty} \int_a^b f_k(x) \, dx.$$

iii) [TERM-BY-TERM DIFFERENTIATION]. *Suppose that E is a bounded, open interval and that each f_k is differentiable on E. If $\sum_{k=1}^{\infty} f_k$ converges at some $x_0 \in E$, and $\sum_{k=1}^{\infty} f_k'$ converges uniformly on E, then $f := \sum_{k=1}^{\infty} f_k$ converges uniformly on E, f is differentiable on E, and*

$$\left(\sum_{k=1}^{\infty} f_k(x) \right)' = \sum_{k=1}^{\infty} f_k'(x)$$

for $x \in E$.

Here are two much-used tests for uniform convergence of series. (The second test, and its example, is optional because we do not use it elsewhere in this text.)

7.15 Theorem. [WEIERSTRASS M-TEST].
Let E be a nonempty subset of $\mathbf{R}$, let $f_k : E \to \mathbf{R}$, $k \in \mathbf{N}$, and suppose that $M_k \geq 0$ satisfies $\sum_{k=1}^{\infty} M_k < \infty$. If $|f_k(x)| \leq M_k$ for $k \in \mathbf{N}$ and $x \in E$, then $\sum_{k=1}^{\infty} f_k$ converges absolutely and uniformly on E.

Proof. Let $\varepsilon > 0$ and use the Cauchy Criterion to choose $N \in \mathbf{N}$ such that $m \geq n \geq N$ implies $\sum_{k=n}^{m} M_k < \varepsilon$. Thus, by hypothesis,

$$\left| \sum_{k=n}^{m} f_k(x) \right| \leq \sum_{k=n}^{m} |f_k(x)| \leq \sum_{k=n}^{m} M_k < \varepsilon$$

for $m \geq n \geq N$ and $x \in E$. Hence, the partial sums of $\sum_{k=1}^{\infty} f_k$ are uniformly Cauchy and the partial sums of $\sum_{k=1}^{\infty} |f_k(x)|$ are Cauchy for each $x \in E$. ∎

*7.16 Theorem. [DIRICHLET'S TEST FOR UNIFORM CONVERGENCE].
Let E be a nonempty subset of $\mathbf{R}$ and suppose that $f_k, g_k : E \to \mathbf{R}$, $k \in \mathbf{N}$. If

$$\left| \sum_{k=1}^{n} f_k(x) \right| \leq M < \infty$$

for $n \in \mathbf{N}$ and $x \in E$, and if $g_k \downarrow 0$ uniformly on E as $k \to \infty$, then $\sum_{k=1}^{\infty} f_k g_k$ converges uniformly on E.

Proof. Let

$$F_{n,m}(x) = \sum_{k=m}^{n} f_k(x), \qquad m, n \in \mathbf{N}, \; n \geq m, \; x \in E$$

and fix integers $n > m > 0$. By Abel's Formula and hypothesis,

$$\left| \sum_{k=m}^{n} f_k(x)g_k(x) \right| = \left| F_{n,m}(x)g_n(x) + \sum_{k=m}^{n-1} F_{k,m}(x)(g_k(x) - g_{k+1}(x)) \right|$$

$$\leq 2Mg_n(x) + 2M \sum_{k=m}^{n-1} (g_k(x) - g_{k+1}(x))$$

$$= 2Mg_m(x)$$

for all $x \in E$. Since $g_m(x) \to 0$ uniformly on E, as $m \to \infty$, it follows from the uniform Cauchy Criterion that $\sum_{k=1}^{\infty} f_k(x)g_k(x)$ converges uniformly on E. ∎

Here is a typical application of Dirichlet's Test.

*7.17 EXAMPLE.

Prove that if $a_k \downarrow 0$ as $k \to \infty$, then $\sum_{k=0}^{\infty} a_k \cos kx$ converges uniformly on any closed subinterval $[a, b]$ of $(0, 2\pi)$.

Proof. Let $f_k(x) = \cos kx$ and $g_k(x) = a_k$ for $k \in \mathbf{N}$. By the technique used in Example 6.34, we can show that

$$D_n(x) := \sum_{k=0}^{n} \cos kx = \frac{\sin\left(\frac{x}{2}\right) + \sin\left(\left(n + \frac{1}{2}\right)x\right)}{2\sin\left(\frac{x}{2}\right)}$$

for $n \in \mathbf{N}$ and $x \in (0, 2\pi)$. Hence the partial sums of $\sum_{k=0}^{\infty} f_k(x)$ satisfy

$$|D_n(x)| = \left| \frac{\sin\left(\frac{x}{2}\right) + \sin\left(\left(n + \frac{1}{2}\right)x\right)}{2\sin\left(\frac{x}{2}\right)} \right| \leq \frac{1}{\left|\sin\left(\frac{x}{2}\right)\right|}$$

for $x \in (0, 2\pi)$. If $\delta = \min\{2\pi - b, a\}$ and $x \in [a, b]$, then $\sin(x/2) \geq \sin(\delta/2)$ (see Figure 7.4). Therefore, $\sum_{k=1}^{\infty} a_k \cos kx$ converges uniformly on $[a, b]$ by Dirichlet's Test. ∎

This example can be used to show that uniform convergence of a series alone is not sufficient for term-by-term differentiation. Indeed, although $\sum_{k=1}^{\infty} \cos kx/k$ converges uniformly on $[\pi/2, 3\pi/2]$, its term-by-term derivative $\sum_{k=1}^{\infty} (-\sin kx)$ converges at only one point in $[\pi/2, 3\pi/2]$.

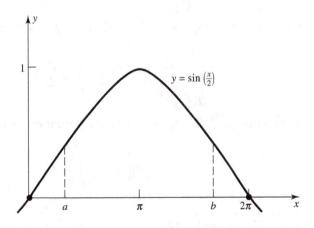

FIGURE 7.4

A *double series* is a series of numbers or functions of the form

$$\sum_{k=1}^{\infty}\left(\sum_{j=1}^{\infty}a_{kj}\right).$$

Such a double series is said to converge if and only if $\sum_{j=1}^{\infty}a_{kj}$ converges for each $k \in \mathbf{N}$ and

$$\sum_{k=1}^{\infty}\sum_{j=1}^{\infty}a_{kj} := \lim_{N\to\infty}\sum_{k=1}^{N}\left(\sum_{j=1}^{\infty}a_{kj}\right)$$

exists and is finite.

When working with double series, one frequently wants to be able to change the order of summation. We already know that the order of summation can be changed when $a_{kj} \geq 0$ (see Exercise 6.3.6). We now prove a more general result. (The elegant proof given here, which comes from Rudin [11],[1] uses uniform convergence.)

7.18 Theorem. *Let $a_{kj} \in \mathbf{R}$ for $k, j \in \mathbf{N}$ and suppose that*

$$A_j = \sum_{k=1}^{\infty}|a_{kj}| < \infty$$

for each $j \in \mathbf{N}$. If $\sum_{j=1}^{\infty}A_j$ converges (i.e., the double sum converges absolutely), then

$$\sum_{k=1}^{\infty}\sum_{j=1}^{\infty}a_{kj} = \sum_{j=1}^{\infty}\sum_{k=1}^{\infty}a_{kj}.$$

Proof. Let $E = \{0, 1, \frac{1}{2}, \frac{1}{3}, \ldots\}$. For each $j \in \mathbf{N}$, define a function f_j on E by

$$f_j(0) = \sum_{k=1}^{\infty}a_{kj}, \quad f_j\left(\frac{1}{n}\right) = \sum_{k=1}^{n}a_{kj}, \quad n \in \mathbf{N}.$$

By hypothesis, $f_j(0)$ exists and by the definition of series convergence,

$$\lim_{n\to\infty}f_j\left(\frac{1}{n}\right) = f_j(0);$$

---

[1]Walter Rudin, *Principles of Mathematical Analysis*, 3rd ed. (New York: McGraw-Hill Book Co., 1976). Reprinted with permission of McGraw-Hill Book Co.

that is, f_j is continuous at $0 \in E$ for each $j \in \mathbf{N}$. Moreover, since $|f_j(x)| \le A_j$ for all $x \in E$ and $j \in \mathbf{N}$, the Weierstrass M-Test implies that

$$f(x) := \sum_{j=1}^{\infty} f_j(x)$$

converges uniformly on E. Thus f is continuous at $0 \in E$ by Theorem 7.9. It follows from the sequential characterization of continuity (Theorem 3.21) that $f(1/n) \to f(0)$ as $n \to \infty$. Therefore,

$$\sum_{k=1}^{\infty}\sum_{j=1}^{\infty} a_{kj} = \lim_{n \to \infty} \sum_{k=1}^{n}\sum_{j=1}^{\infty} a_{kj} = \lim_{n \to \infty} \sum_{j=1}^{\infty}\sum_{k=1}^{n} a_{kj}$$

$$= \lim_{n \to \infty} \sum_{j=1}^{\infty} f_j\left(\frac{1}{n}\right) = \lim_{n \to \infty} f\left(\frac{1}{n}\right) = f(0) = \sum_{j=1}^{\infty}\sum_{k=1}^{\infty} a_{kj}. \quad \blacksquare$$

EXERCISES

7.2.1. a) Prove that $\sum_{k=1}^{\infty} \sin(x/k^2)$ converges uniformly on any bounded interval in $\mathbf{R}$.

   b) Prove that $\sum_{k=0}^{\infty} e^{-kx}$ converges uniformly on any closed subinterval of $(0, \infty)$.

7.2.2. Prove that the geometric series

$$\sum_{k=0}^{\infty} x^k = \frac{1}{1-x}$$

converges uniformly on any closed interval $[a, b] \subset (-1, 1)$.

7.2.3. Let $E(x) = \sum_{k=0}^{\infty} x^k/k!$.

   a) Prove that the series defining $E(x)$ converges uniformly on any closed interval $[a, b]$.

   b) Prove that

$$\int_a^b E(x)\, dx = E(b) - E(a)$$

   for all $a, b \in \mathbf{R}$.

   c) Prove that the function $y = E(x)$ satisfies the initial value problem

$$y' - y = 0, \qquad y(0) = 1.$$

[We shall see in Section 7.4 that $E(x) = e^x$.]

7.2.4. Suppose that

$$f(x) = \sum_{k=1}^{\infty} \frac{\cos(kx)}{k^2}.$$

Prove that

$$\int_0^{\pi/2} f(x)\, dx = \sum_{k=0}^{\infty} \frac{(-1)^k}{(2k+1)^3}.$$

7.2.5. Show that

$$f(x) = \sum_{k=1}^{\infty} \frac{1}{k} \sin\left(\frac{x}{k+1}\right)$$

converges, pointwise on **R** and uniformly on each bounded interval in **R**, to a differentiable function f which satisfies

$$|f(x)| \le |x| \quad \text{and} \quad |f'(x)| \le 1$$

for all $x \in \mathbf{R}$.

7.2.6. Prove that

$$\left| \sum_{k=1}^{\infty} (1 - \cos(1/k)) \right| \le 2.$$

7.2.7. Suppose that $f = \sum_{k=1}^{\infty} f_k$ converges uniformly on a set $E \subseteq \mathbf{R}$. If g_1 is bounded on E and $g_k(x) \ge g_{k+1}(x) \ge 0$ for all $x \in E$ and $k \in \mathbf{N}$, prove that $\sum_{k=1}^{\infty} f_k g_k$ converges uniformly on E.

7.2.8. Let $n \ge 0$ be a fixed nonnegative integer and recall that $0! := 1$. The *Bessel function* of order n is the function defined by

$$B_n(x) := \sum_{k=0}^{\infty} \frac{(-1)^k}{(k!)(n+k)!} \left(\frac{x}{2}\right)^{n+2k}.$$

a) Show that $B_n(x)$ converges pointwise on **R** and uniformly on any closed interval $[a, b]$.

b) Prove that $y = B_n(x)$ satisfies the differential equation

$$x^2 y'' + xy' + (x^2 - n^2)y = 0$$

for $x \in \mathbf{R}$.

c) Prove that

$$(x^n B_n(x))' = x^n B_{n-1}(x)$$

for $n \in \mathbf{N}$ and $x \in \mathbf{R}$.

*7.2.9. Suppose that $a_k \downarrow 0$ as $k \to \infty$. Prove that $\sum_{k=1}^{\infty} a_k \sin kx$ converges uniformly on any closed interval $[a, b] \subset (0, 2\pi)$.

7.2.10. Suppose that $f_1, f_2, \ldots$ are continuous real functions defined on a closed, bounded interval $[a, b]$. If $0 \leq f_k(x) \leq f_{k+1}(x)$ for all $k \in \mathbf{N}$ and $x \in [a, b]$, and if $f_k \to f$ uniformly on $[a, b]$, prove that

$$\lim_{n \to \infty} \int_a^b \left(\sum_{k=1}^n f_k^n(x) \right)^{1/n} dx = \int_a^b f(x) \, dx.$$

7.3 POWER SERIES

Polynomials are functions of the form $P(x) = \sum_{k=0}^n a_k x^k$, where $a_k \in \mathbf{R}$ and $n \geq 0$. In this section we investigate a natural generalization of polynomials, namely, series of the form $\sum_{k=0}^\infty a_k x^k$.

Actually, we shall consider a slightly more general class of series. A *power series* (centered at x_0) is a series of the form

$$S(x) := \sum_{k=0}^\infty a_k (x - x_0)^k,$$

where we use the convention that $(x - x_0)^0 = 1$. In fact, although 0^0 is in general indeterminate, when dealing with power series we always interpret $0^0 = 1$.

Since $S(x)$ is identically a_0 when $x = x_0$, it is clear that every power series converges at at least one point. The following result shows that this may be the only point.

7.19 Remark. *There exist power series which converge only at one point.*

Proof. For each $x \neq 0$, $(k^k |x|^k)^{1/k} = k|x| \to \infty$ as $k \to \infty$. Therefore, by the Root Test, the series $\sum_{k=1}^\infty k^k x^k$ diverges when $x \neq 0$. $\blacksquare$

In general, a series of functions can converge at several isolated points. [For example, the series $\sum_{k=1}^\infty \sin(kx)$ converges only when $x = n\pi$ for some $n \in \mathbf{Z}$.] We shall see (Theorem 7.21 below) that this cannot happen for power series. Hence, we introduce the following concept.

7.20 Definition.

An extended real number R is said to be the *radius of convergence* of a power series $S(x) := \sum_{k=0}^\infty a_k (x - x_0)^k$ if and only if $S(x)$ converges absolutely for $|x - x_0| < R$ and $S(x)$ diverges for $|x - x_0| > R$.

The extreme cases are $R = 0$ and $R = \infty$. When $R = 0$, the power series $S(x)$ converges only when $x = x_0$. When $R = \infty$, the power series $S(x)$ converges absolutely for every $x \in \mathbf{R}$.

The next result shows that every power series S has a radius of convergence which can be computed using roots of the coefficients of S.

7.21 Theorem. *Let $S(x) = \sum_{k=0}^{\infty} a_k(x - x_0)^k$ be a power series centered at x_0. If $R = 1/\limsup_{k\to\infty}|a_k|^{1/k}$, with the convention that $1/\infty = 0$ and $1/0 = \infty$, then R is the radius of convergence of S. In fact,*

i) *$S(x)$ converges absolutely for each $x \in (x_0 - R, x_0 + R)$,*
ii) *$S(x)$ converges uniformly on any closed interval $[a, b] \subset (x_0 - R, x_0 + R)$,*
iii) *and (when R is finite), $S(x)$ diverges for each $x \notin [x_0 - R, x_0 + R]$.*

Proof. Fix $x \in \mathbf{R}$, $x \neq x_0$, and set $\rho := 1/\limsup_{k\to\infty}|a_k|^{1/k}$, with the convention that $1/\infty = 0$ and $1/0 = \infty$. To apply the Root Test to $S(x)$, consider

$$r(x) := \limsup_{k\to\infty} |a_k(x - x_0)^k|^{1/k} = |x - x_0| \cdot \limsup_{k\to\infty} |a_k|^{1/k}.$$

Case 1. $\rho = 0$. By our convention, $\rho = 0$ implies $r(x) = \infty > 1$, so by the Root Test, $S(x)$ does not converge for any $x \neq x_0$. Hence, the radius of convergence of S is $R = 0 = \rho$.

Case 2. $\rho = \infty$. Then $r(x) = 0 < 1$, so by the Root Test, $S(x)$ converges absolutely for all $x \in \mathbf{R}$. Hence, the radius of convergence of S is $R = \infty = \rho$.

Case 3. $\rho \in (0, \infty)$. Then $r(x) = |x - x_0|/\rho$. Since $r(x) < 1$ if and only if $|x - x_0| < \rho$, it follows from the Root Test that $S(x)$ converges absolutely when $x \in (x_0 - \rho, x_0 + \rho)$. Similarly, since $r(x) > 1$ if and only if $|x - x_0| > \rho$, we also have that $S(x)$ diverges when $x \notin [x_0 - \rho, x_0 + \rho]$. This proves that ρ is the radius of convergence of S, and that parts i) and iii) hold.

To prove part ii), let $[a, b] \subset (x_0 - R, x_0 + R)$. Choose an $x_1 \in (x_0 - R, x_0 + R)$ such that $x \in [a, b]$ implies $|x - x_0| \leq |x_1 - x_0|$ (see Figure 7.5). Set $M_k = |a_k| |x_1 - x_0|^k$ and observe by part i) that $\sum_{k=0}^{\infty} M_k$ converges. Since $|a_k(x - x_0)^k| \leq M_k$ for $x \in [a, b]$ and $k \in \mathbf{N}$, it follows from the Weierstrass M-Test that $S(x)$ converges uniformly on $[a, b]$. $\blacksquare$

FIGURE 7.5

The following result, which is weaker than Theorem 7.21 (see Exercise 6.3.8), provides another way to compute the radius of convergence of some power series (see also Exercise 7.3.8). This way is easier when a_k contains products (e.g., factorials).

7.22 Theorem. *If the limit*

$$R = \lim_{k\to\infty} \frac{|a_k|}{|a_{k+1}|}$$

exists as an extended real number, then R is the radius of convergence of the power series $S(x) = \sum_{k=0}^{\infty} a_k(x - x_0)^k$.

Proof. Repeat the proof of Theorem 7.21, using the Ratio Test instead of the Root Test, to find that $S(x)$ converges absolutely on $(x_0 - R, x_0 + R)$ and diverges for each $x \notin [x_0 - R, x_0 + R]$. By Definition 7.20, R must be the radius of convergence of $S(x)$. ∎

7.23 Definition.

The *interval of convergence* of a power series $S(x)$ is the largest interval on which $S(x)$ converges.

By Theorem 7.21, for a given power series $S = \sum_{k=0}^{\infty} a_k(x - x_0)^k$, there are only three possibilities:

i) $R = \infty$, in which case the interval of convergence of S is $(-\infty, \infty)$,
ii) $R = 0$, in which case the interval of convergence of S is $\{x_0\}$, and
iii) $0 < R < \infty$, in which case the interval of convergence of S is $(x_0 - R, x_0 + R)$, $[x_0 - R, x_0 + R)$, $(x_0 - R, x_0 + R]$, or $[x_0 - R, x_0 + R]$.

To find the interval of convergence of a power series, therefore, one needs to compute the radius of convergence R first. If $0 < R < \infty$, one must also check both endpoints, $x_0 - R$ and $x_0 + R$, to see whether the interval of convergence is closed, open, or half open/closed. Notice once and for all that the Ratio and Root Tests cannot be used to test the endpoints, since it was the Ratio and Root Tests which gave us R to begin with.

7.24 EXAMPLE.

Find the interval of convergence of $S(x) = \sum_{k=1}^{\infty} x^k / \sqrt{k}$.

Solution. By Theorem 7.22,

$$R = \lim_{k \to \infty} \frac{\sqrt{k + 1}}{\sqrt{k}} = \sqrt{\lim_{k \to \infty} \frac{k + 1}{k}} = 1.$$

Thus, the interval of convergence has endpoints 1 and -1. $S(x)$ diverges at $x = 1$ by the p-Series Test and converges at $x = -1$ by the Alternating Series Test. Thus, the interval of convergence of $S(x)$ is $[-1, 1)$. ∎

7.25 Remark. *The interval of convergence may contain none, one, or both its endpoints.*

Proof. By Theorem 7.22, the radius of convergence of each of the series

$$\sum_{k=1}^{\infty} x^k, \qquad \sum_{k=1}^{\infty} \frac{x^k}{k}, \qquad \sum_{k=1}^{\infty} \frac{x^k}{k^2}$$

is 1, but by the Divergence Test, the Alternating Series Test, and the p-Series Test, the intervals of convergence of these series are $(-1, 1)$, $[-1, 1)$, and $[-1, 1]$, respectively. ∎

We now pass from convergence properties of power series to the calculus of power series. The next several results answer the question, What properties (e.g., continuity, differentiability, integrability) does the limit of a power series satisfy?

7.26 Theorem. *If $f(x) = \sum_{k=0}^{\infty} a_k(x - x_0)^k$ is a power series with positive radius of convergence R, then f is continuous on $(x_0 - R, x_0 + R)$.*

Proof. Let $x \in (x_0 - R, x_0 + R)$ and choose $a, b \in \mathbf{R}$ such that $x \in (a, b)$ and $[a, b] \subset (x_0 - R, x_0 + R)$. By Theorems 7.21ii and 7.14i, f is continuous on (a, b) and hence at x. ∎

The following result shows that continuity of the limit extends to the endpoints when they belong to the interval of convergence.

7.27 Theorem. [ABEL'S THEOREM].
Suppose that $[a, b]$ is nondegenerate. If $f(x) := \sum_{k=0}^{\infty} a_k(x - x_0)^k$ converges on $[a, b]$, then $f(x)$ is continuous and converges uniformly on $[a, b]$.

Proof. By Theorems 7.21ii and 7.26, we may suppose that f has a positive, finite radius of convergence R, and, by symmetry, that $a = x_0$ and $b = x_0 + R$. Thus, suppose that $f(x)$ converges at $x = x_0 + R$ and fix $x_1 \in (x_0, x_0 + R]$. Set $b_k = a_k R^k$ and $c_k = (x_1 - x_0)^k / R^k$ for $k \in \mathbf{N}$. By hypothesis, $\sum_{k=1}^{\infty} b_k$ converges. Hence, given $\varepsilon > 0$, there is an integer $N > 1$ such that

$$k > m \geq N \quad \text{imply} \quad \left| \sum_{j=m}^{k} b_j \right| < \varepsilon.$$

Since $0 < x_1 - x_0 \leq R$, the sequence $\{c_k\}$ is decreasing. Applying Abel's Formula and telescoping, we have

$$\left| \sum_{k=m}^{n} a_k(x_1 - x_0)^k \right| = \left| \sum_{k=m}^{n} b_k c_k \right|$$

$$= \left| c_n \sum_{k=m}^{n} b_k + \sum_{k=m}^{n-1} (c_k - c_{k+1}) \sum_{j=m}^{k} b_j \right|$$

$$< c_n \varepsilon + (c_m - c_n)\varepsilon = c_m \varepsilon.$$

Since $c_m \le c_1 \le R/R = 1$, it follows that

$$\left| \sum_{k=m}^{n} a_k (x_1 - x_0)^k \right| < \varepsilon$$

for all $x_1 \in (x_0, x_0 + R]$. Since this inequality also holds for $x_1 = x_0$, we conclude that $\sum_{k=0}^{\infty} a_k (x - x_0)^k$ converges uniformly on $[x_0, x_0 + R]$. ■

7.28 Remark. *If a power series $S(x) = \sum_{k=0}^{\infty} a_k (x - x_0)^k$ converges at some $x_1 > x_0$, then $S(x)$ converges uniformly on $[x_0, x_1]$ and absolutely on $[x_0, x_1)$. It might not converge absolutely at $x = x_1$.*

Proof. By Theorems 7.21 and 7.27, $S(x)$ converges uniformly on $[x_0, x_1]$ and absolutely on $[x_0, x_1)$. The power series $\sum_{k=1}^{\infty} (-x)^k / k$ converges uniformly on $[0, 1]$ but not absolutely at $x = 1$. ■

To discuss differentiability of the limit of a power series, we first show that the radius of convergence of a power series is not changed by term-by-term differentiation (compare with Exercise 2.5.6).

7.29 Lemma.
If $a_n \in \mathbf{R}$ for $n \in \mathbf{N}$, then

$$\limsup_{n \to \infty} (n|a_n|)^{1/n} = \limsup_{n \to \infty} |a_n|^{1/n}.$$

Proof. Let $\varepsilon > 0$. Since $n^{1/n} \to 1$ as $n \to \infty$, choose $N \in \mathbf{N}$ so that $n \ge N$ implies $1 - \varepsilon < n^{1/n} < 1 + \varepsilon$; that is,

$$(1 - \varepsilon)|a_n|^{1/n} < (n|a_n|)^{1/n} < (1 + \varepsilon)|a_n|^{1/n}.$$

It follows that if $n \ge N$, then $\sup_{k>n} (k|a_k|)^{1/k} \le (1 + \varepsilon) \sup_{k>n} |a_k|^{1/k}$. Taking the limit of this last inequality, as $n \to \infty$, we have by definition that

$$x := \limsup_{n \to \infty} (n|a_n|)^{1/n} \le (1 + \varepsilon) \limsup_{n \to \infty} |a_n|^{1/n} =: (1 + \varepsilon)y.$$

Taking the limit of this inequality as $\varepsilon \to 0$, we obtain $x \le y$. A similar argument, using $(1 - \varepsilon)$ in place of $(1 + \varepsilon)$, proves that $x \ge y$. We conclude that $x = y$ as promised. ■

We use this result to prove that each power series with a positive radius of convergence is term-by-term differentiable.

7.30 Theorem. *If $f(x) = \sum_{k=0}^{\infty} a_k (x - x_0)^k$ is a power series with positive radius of convergence R, then $f'(x) = \sum_{k=1}^{\infty} k a_k (x - x_0)^{k-1}$ for $x \in (x_0 - R, x_0 + R)$.*

Proof. Let $I := (x_0 - R, x_0 + R)$ and suppose that $a, b \in \mathbf{R}$ satisfy $[a, b] \subset I$. By Lemma 7.29 and hypothesis, the radius of convergence of the series $g(x) := \sum_{k=0}^{\infty} ka_k(x - x_0)^k$ is R. Thus g converges absolutely on I and uniformly on $[a, b]$.

Consider the derived series $S^*(x) := \sum_{k=1}^{\infty} ka_k(x - x_0)^{k-1}$. Since $S^*(x_0)$ has only one nonzero term, the series $S^*(x_0)$ converges absolutely. If $x \in I \setminus \{x_0\}$, then $S^*(x) = g(x)/(x - x_0)$, so again, $S^*(x)$ converges absolutely. It follows that the radius of convergence of S^* is at least R. Hence, by Theorems 7.21 and 7.14iii (term-by-term differentiation), f is differentiable on $[a, b]$ and $S^*(x) = f'(x)$ for all $x \in [a, b]$. Since any $x \in I$ belongs to some $[a, b] \subset I$, we conclude that $f'(x) = S^*(x)$ for all $x \in I$. ∎

NOTE: A similar proof shows that S^* diverges for all $x \notin [x_0 - R, x_0 + R]$, so the radius of convergence of S^* is exactly R.

Recall that for each nonempty, open interval (a, b), $\mathcal{C}^{\infty}(a, b)$ represents the set of functions f such that $f^{(k)}$ exists and is continuous on (a, b) for all $k \in \mathbf{N}$. The following result generalizes Theorem 7.30.

7.31 Corollary. *If $f(x) = \sum_{k=0}^{\infty} a_k(x - x_0)^k$ has a positive radius of convergence R, then $f \in \mathcal{C}^{\infty}(x_0 - R, x_0 + R)$ and*

$$f^{(k)}(x) = \sum_{n=k}^{\infty} \frac{n!}{(n-k)!} a_n (x - x_0)^{n-k} \tag{7}$$

for $x \in (x_0 - R, x_0 + R)$ and $k \in \mathbf{N}$.

Proof. The proof is by induction on k. By Theorem 7.30 and the fact that $0! := 1$, (7) holds for $k = 1$ and $x \in (x_0 - R, x_0 + R)$. If (7) holds for some $k \in \mathbf{N}$ and all $x \in (x_0 - R, x_0 + R)$, then $f^{(k)}$ is a power series with radius of convergence R. It follows from Theorem 7.30 that

$$f^{(k+1)}(x) = (f^{(k)}(x))' = \left(\sum_{n=k}^{\infty} \frac{n!}{(n-k)!} a_n (x - x_0)^{n-k} \right)'$$

$$= \sum_{n=k+1}^{\infty} \frac{n!}{(n-k-1)!} a_n (x - x_0)^{n-k-1}$$

for all $x \in (x_0 - R, x_0 + R)$. Hence, (7) holds for $k + 1$ in place of k. ∎

The following result shows that each power series with a positive radius of convergence can also be integrated term by term.

7.32 Theorem. *Let* $f(x) = \sum_{k=0}^{\infty} a_k (x - x_0)^k$ *be a power series and let* $a, b \in \mathbf{R}$ *with* $a < b$.

i) *If* $f(x)$ *converges on* $[a, b]$, *then* f *is integrable on* $[a, b]$ *and*

$$\int_a^b f(x)\, dx = \sum_{k=0}^{\infty} a_k \int_a^b (x - x_0)^k\, dx.$$

ii) If $f(x)$ *converges on* $[a, b)$ *and if* $\sum_{k=0}^{\infty} a_k (b - x_0)^{k+1}/(k+1)$ *converges, then* f *is improperly integrable on* $[a, b)$ *and*

$$\int_a^b f(x)\, dx = \sum_{k=0}^{\infty} a_k \int_a^b (x - x_0)^k\, dx.$$

Proof. i) By Abel's Theorem, $f(x)$ converges uniformly on $[a, b]$. Hence, by Theorem 7.14ii, $f(x)$ is term-by-term integrable on $[a, b]$.

ii) Let $a < t < b$ and set $A = \sum_{k=0}^{\infty} a_k (a - x_0)^{k+1}/(k+1)$. By part i),

$$\int_a^t f(x)\, dx = \sum_{k=0}^{\infty} a_k \int_a^t (x - x_0)^k\, dx = \sum_{k=0}^{\infty} \frac{a_k}{k+1} (t - x_0)^{k+1} - A.$$

The leftmost term of this last difference is a power series which by hypothesis converges at $t = b$. Thus, by the definition of improper integration and Abel's Theorem,

$$\int_a^b f(x)\, dx = \lim_{t \to b-} \int_a^t f(x)\, dx$$

$$= \lim_{t \to b-} \sum_{k=0}^{\infty} \frac{a_k}{k+1} (t - x_0)^{k+1} - A$$

$$= \sum_{k=0}^{\infty} \frac{a_k}{k+1} (b - x_0)^{k+1} - A = \sum_{k=0}^{\infty} a_k \int_a^b (x - x_0)^k\, dx. \quad \blacksquare$$

The following result shows that the product of two power series is a power series. (For a result on the division of power series, see Taylor [13], p. 619.)

7.33 Theorem. *If* $f(x) = \sum_{k=0}^{\infty} a_k x^k$ *and* $g(x) = \sum_{k=0}^{\infty} b_k x^k$ *converge on* $(-r, r)$ *and*

$$c_k = \sum_{j=0}^{k} a_j b_{k-j}, \qquad k = 0, 1, \ldots,$$

then $\sum_{k=0}^{\infty} c_k x^k$ *converges on* $(-r, r)$ *and converges to* $f(x)g(x)$.

Proof. Fix $x \in (-r, r)$ and for each $n \in \mathbf{N}$, set

$$f_n(x) = \sum_{k=0}^{n} a_k x^k, \qquad g_n(x) = \sum_{k=0}^{n} b_k x^k, \quad \text{and} \quad h_n(x) = \sum_{k=0}^{n} c_k x^k.$$

By changing the order of summation, we see that

$$h_n(x) = \sum_{k=0}^{n} \sum_{j=0}^{k} a_j b_{k-j} x^j x^{k-j} = \sum_{j=0}^{n} a_j x^j \sum_{k=j}^{n} b_{k-j} x^{k-j}$$

$$= \sum_{j=0}^{n} a_j x^j g_{n-j}(x) = g(x) f_n(x) + \sum_{j=0}^{n} a_j x^j (g_{n-j}(x) - g(x)).$$

Thus, it suffices to show that

$$\lim_{n \to \infty} \sum_{j=0}^{n} a_j x^j (g_{n-j}(x) - g(x)) = 0.$$

Let $\varepsilon > 0$. Since $f(x)$ converges absolutely and $g_n(x)$ converges as $n \to \infty$, choose $M > 0$ such that $\sum_{k=0}^{\infty} |a_k x^k| < M$ and

$$|g_{n-j}(x) - g(x)| \le M$$

for all integers $n > j > 0$. Similarly, choose $N \in \mathbf{N}$ such that

$$\ell \ge N \quad \text{implies} \quad |g_\ell(x) - g(x)| < \frac{\varepsilon}{2M} \quad \text{and} \quad \sum_{j=N+1}^{\infty} |a_j x^j| < \frac{\varepsilon}{2M}.$$

Let $n > 2N$. Then

$$\left| \sum_{j=0}^{n} a_j x^j (g_{n-j}(x) - g(x)) \right|$$

$$= \left| \sum_{j=0}^{N} a_j x^j (g_{n-j}(x) - g(x)) + \sum_{j=N+1}^{n} a_j x^j (g_{n-j}(x) - g(x)) \right|$$

$$< \frac{\varepsilon}{2M} \sum_{j=0}^{N} |a_j x^j| + M \sum_{j=N+1}^{n} |a_j x^j| < \frac{\varepsilon}{2} + \frac{\varepsilon}{2} = \varepsilon. \qquad \blacksquare$$

7.34 Corollary. *Suppose that $a_k, b_k \in \mathbf{R}$ and that $c_k := \sum_{j=0}^{k} a_j b_{k-j}$ for $k = 0, 1, \ldots$. If either*

i) *$\sum_{k=0}^{\infty} a_k$ and $\sum_{k=0}^{\infty} b_k$ both converge, and at least one of them converges absolutely,*

ii) *or, if $\sum_{k=0}^{\infty} a_k$, $\sum_{k=0}^{\infty} b_k$, and $\sum_{k=0}^{\infty} c_k$ all converge,*

then

$$\sum_{k=0}^{\infty} c_k = \left(\sum_{k=0}^{\infty} a_k\right)\left(\sum_{k=0}^{\infty} b_k\right). \tag{8}$$

Proof. i) Repeat the proof of Theorem 7.33 with $x = 1$.

ii) By hypothesis, the radii of convergence of $\sum_{k=0}^{\infty} a_k x^k$, $\sum_{k=0}^{\infty} b_k x^k$, and $\sum_{k=0}^{\infty} c_k x^k$ are all at least 1; hence, by Theorem 7.33,

$$\sum_{k=0}^{\infty} c_k x^k = \left(\sum_{k=0}^{\infty} a_k x^k\right)\left(\sum_{k=0}^{\infty} b_k x^k\right) \tag{9}$$

for $x \in (-1, 1)$. But by Abel's Theorem (Theorem 7.27), the limit of (9) as $x \uparrow 1$ is (8). ∎

The hypotheses of Corollary 7.34 cannot be relaxed.

*7.35 EXAMPLE.

If $a_k = b_k = (-1)^k / \sqrt{k}$ for $k \in \mathbf{N}$ and $a_0 = b_0 = 0$, then $\sum_{k=0}^{\infty} c_k$ diverges.

Proof. If $\sum_{k=0}^{\infty} c_k$ converges, then $c_k \to 0$ as $k \to \infty$. But for $k > 1$ odd,

$$|c_k| = \sum_{j=1}^{k-1} \frac{1}{\sqrt{j}\sqrt{k-j}} = 2 \sum_{j=1}^{(k-1)/2} \frac{1}{\sqrt{j}\sqrt{k-j}}$$

$$\geq 2\left(\frac{k-1}{2}\right)\left(\frac{1}{\sqrt{(k-1)/2}}\right)\left(\frac{1}{\sqrt{(k-1)}}\right) = \sqrt{2}.$$

Thus c_k cannot converge to zero, a contradiction. ∎

We close this section with some optional material on finding exact values of convergent power series. Namely, we show how term-by-term differentiation and integration can be used in conjunction with the geometric series to obtain simple formulas for certain kinds of power series. Such formulas are called *closed forms.*

***7.36 EXAMPLE.**

Find a closed form of the power series

$$f(x) = \sum_{k=1}^{\infty} kx^k.$$

Solution. Since the interval of convergence of this power series is $(-1, 1)$, we have by Theorems 7.32 and 6.7 (the Geometric Series) that

$$\int_0^x \frac{f(t)}{t}\, dt = \sum_{k=1}^{\infty} k \int_0^x t^{k-1}\, dt = \sum_{k=1}^{\infty} x^k = \frac{x}{1-x}$$

for each $x \in (-1, 1)$. [Note that $f(x)/x$ is defined at $x = 0$ and has value 1.] Hence, by the Fundamental Theorem of Calculus,

$$\frac{f(x)}{x} = \left(\frac{x}{1-x}\right)' = \frac{1}{(1-x)^2}$$

and it follows that

$$f(x) = \frac{x}{(1-x)^2}, \qquad x \in (-1, 1). \qquad\blacksquare$$

***7.37 EXAMPLE.**

Find a closed form of the power series

$$g(x) = \sum_{k=0}^{\infty} \frac{x^k}{k+1}.$$

Solution. Since the interval of convergence of this power series is $[-1, 1)$, we have by Theorem 7.30 that

$$(xg(x))' = \sum_{k=0}^{\infty} \left(\frac{x^{k+1}}{k+1}\right)' = \sum_{k=0}^{\infty} x^k = \frac{1}{1-x}$$

for $x \in (-1, 1)$. Hence, by the Fundamental Theorem of Calculus,

$$xg(x) = \int_0^x \frac{dt}{1-t} = -\log(1-x)$$

for $x \in (-1, 1)$. Since $g(-1)$ exists and $\log(1-x)$ is continuous at $x = -1$, we conclude by Abel's Theorem that

$$g(x) = -\frac{\log(1-x)}{x}, \qquad x \in [-1, 1) \setminus \{0\}, \quad \text{and} \quad g(0) = 1. \qquad\blacksquare$$

EXERCISES

7.3.1. Find the interval of convergence of each of the following power series.

a) $\displaystyle\sum_{k=0}^{\infty} \frac{kx^k}{(2k+1)^2}$

b) $\displaystyle\sum_{k=0}^{\infty} (2+(-1)^k)^k x^k$

c) $\displaystyle\sum_{k=0}^{\infty} 3^{k^2} x^{k^2}$

d) $\displaystyle\sum_{k=0}^{\infty} k^{k^2} x^{k^3}$

7.3.2. Find the interval of convergence of each of the following power series.

a) $\displaystyle\sum_{k=0}^{\infty} \frac{x^k}{2^k}$

b) $\displaystyle\sum_{k=0}^{\infty} ((-1)^k + 3)^k (x-1)^k$

c) $\displaystyle\sum_{k=1}^{\infty} \log\left(\frac{k+1}{k}\right) x^k$

*d) $\displaystyle\sum_{k=1}^{\infty} \frac{1 \cdot 3 \ldots (2k-1)}{(k+1)!} x^{2k}$

7.3.3. Suppose that $\sum_{k=0}^{\infty} a_k x^k$ has radius of convergence $R \in (0, \infty)$.

a) Find the radius of convergence of $\sum_{k=0}^{\infty} a_k x^{2k}$.
b) Find the radius of convergence of $\sum_{k=0}^{\infty} a_k^2 x^k$.

7.3.4. Suppose that $|a_k| \le |b_k|$ for large k. Prove that if $\sum_{k=0}^{\infty} b_k x^k$ converges on an open interval I, then $\sum_{k=0}^{\infty} a_k x^k$ also converges on I. Is this result true if *open* is omitted?

7.3.5. Suppose that $\{a_k\}_{k=0}^{\infty}$ is a bounded sequence of real numbers. Prove that

$$f(x) := \sum_{k=0}^{\infty} a_k x^k$$

has a positive radius of convergence.

7.3.6. A series $\sum_{k=0}^{\infty} a_k$ is said to be *Abel summable* to L if and only if

$$\lim_{r \to 1-} \sum_{k=0}^{\infty} a_k r^k = L.$$

a) Prove that if $\sum_{k=0}^{\infty} a_k$ converges to L, then $\sum_{k=0}^{\infty} a_k$ is Abel summable to L.
b) Find the Abel sum of $\sum_{k=0}^{\infty} (-1)^k$.

*7.3.7. Find a closed form for each of the following series and the largest set on which this formula is valid.

a) $\displaystyle\sum_{k=1}^{\infty} 3x^{3k-1}$

b) $\displaystyle\sum_{k=2}^{\infty} kx^{k-2}$

c) $\displaystyle\sum_{k=1}^{\infty} \frac{2k}{k+1}(1-x)^k$

d) $\displaystyle\sum_{k=0}^{\infty} \frac{x^{3k}}{k+1}$

*7.3.8. If $\sum_{k=1}^{\infty} a_k x^k$ has radius of convergence R and $a_k \neq 0$ for large k, prove that

$$\liminf_{k\to\infty} \left|\frac{a_k}{a_{k+1}}\right| \leq R \leq \limsup_{k\to\infty} \left|\frac{a_k}{a_{k+1}}\right|.$$

*7.3.9. Prove that

$$f(x) = \sum_{k=0}^{\infty} \left(\frac{x}{(-1)^k + 4}\right)^k$$

is differentiable on $(-3, 3)$ and

$$|f'(x)| \leq \frac{3}{(3-x)^2}$$

for $0 \leq x < 3$.

7.3.10. Suppose that $a_k \downarrow 0$ as $k \to \infty$. Prove that given $\varepsilon > 0$ there is a $\delta > 0$ such that

$$\left|\sum_{k=0}^{\infty} (-1)^k a_k (x^k - y^k)\right| < \varepsilon$$

for all $x, y \in [0, 1]$ which satisfy $|x - y| < \delta$.

*7.3.11. a) Prove the following weak form of Stirling's Formula (compare with Theorem 12.73):

$$\frac{n^n}{e^{n-1}} < n! < \frac{n^{n+1}}{e^{n-1}}.$$

b) Find all $x \in \mathbf{R}$ for which the power series

$$\sum_{k=0}^{\infty} \frac{k^k}{k!} x^k$$

converges absolutely.

7.4 ANALYTIC FUNCTIONS

In this section we study functions which can be represented by power series. (For a discussion of how to represent functions by trigonometric series instead of power series, see Chapter 14.) We begin with the following definition.

7.38 Definition.

A real-valued function f is said to be (real) *analytic* on a nonempty, open interval (a, b) if and only if given $x_0 \in (a, b)$ there is a power series centered at x_0 which converges to f near x_0; that is, if and only if there exist coefficients $\{a_k\}_{k=0}^{\infty}$ and points $c, d \in (a, b)$ such that $c < x_0 < d$ and

$$f(x) = \sum_{k=0}^{\infty} a_k (x - x_0)^k$$

for all $x \in (c, d)$.

We shall develop several techniques for showing that a given function is analytic. To simplify statements of results, we continue to use the conventions $f^{(0)} := f$ and $0! := 1$.

First, it is important to realize that if f is analytic on an open interval I, then for each center x_0 there is one and only one power series that represents f near x_0, and that power series has the same coefficients that the Taylor polynomials have.

7.39 Theorem. [UNIQUENESS].
Let c, d be extended real numbers with $c < d$, let $x_0 \in (c, d)$, and suppose that $f : (c, d) \to \mathbf{R}$. If $f(x) = \sum_{k=0}^{\infty} a_k (x - x_0)^k$ for $x \in (c, d)$, then $f \in C^{\infty}(c, d)$ and

$$a_k = \frac{f^{(k)}(x_0)}{k!}, \qquad k = 0, 1, \cdots .$$

Proof. Clearly, $f(x_0) = a_0$. Fix $k \in \mathbf{N}$. By hypothesis, the radius of convergence R of the power series $\sum_{k=0}^{\infty} a_k (x - x_0)^k$ is positive and $(c, d) \subseteq (x_0 - R, x_0 + R)$. Hence, by Corollary 7.31, $f \in C^{\infty}(c, d)$ and

$$f^{(k)}(x) = \sum_{n=k}^{\infty} \frac{n!}{(n - k)!} a_n (x - x_0)^{n-k} \tag{10}$$

for $x \in (c, d)$. Apply this to $x = x_0$. The terms on the right side of (10) are zero when $n > k$ and $k! a_k$ when $n = k$. Hence, $f^{(k)}(x_0) = k! a_k$ for each $k \in \mathbf{N}$. ∎

This "locally unique" power series has a name.

7.40 Definition.

Let $f \in C^\infty(a, b)$ and let $x_0 \in (a, b)$. The *Taylor expansion* (or *Taylor series*) of f *centered at* x_0 is the series

$$\sum_{k=0}^{\infty} \frac{f^{(k)}(x_0)}{k!}(x - x_0)^k.$$

(No convergence is implied or assumed.) The Taylor expansion of f centered at $x_0 = 0$ is usually called the *Maclaurin expansion* (or *Maclaurin series*) of f.

Theorem 7.39 not only says that the power series representation of an analytic function is locally unique. It also says that every analytic function is a C^∞ function. The next remark shows that the converse of this statement is false.

7.41 Remark. [CAUCHY]. *The function*

$$f(x) = \begin{cases} e^{-1/x^2} & x \neq 0 \\ 0 & x = 0 \end{cases}$$

belongs to $C^\infty(-\infty, \infty)$ but is not analytic on any interval which contains $x = 0$.

Proof. It is easy to see (Exercise 4.4.7) that $f \in C^\infty(-\infty, \infty)$ and $f^{(k)}(0) = 0$ for all $k \in \mathbf{N}$. Thus the Taylor expansion of f about the point $x_0 = 0$ is identically zero but $f(x) = 0$ only when $x = 0$. ∎

One of our aims in this section is to prove that many of the classical C^∞ functions used in elementary calculus are analytic on their domain. Since, by Theorem 7.39, a C^∞ function f is analytic on an open interval I if and only if its Taylor expansion at each $x_0 \in I$ converges to f near x_0, the following concept is useful in this regard.

7.42 Definition.

Let $f \in C^\infty(a, b)$ and $x_0 \in (a, b)$. The *remainder term of order n* of the Taylor expansion of f centered at x_0 is the function

$$R_n(x) = R_n^{f, x_0}(x) := f(x) - \sum_{k=0}^{n-1} \frac{f^{(k)}(x_0)}{k!}(x - x_0)^k.$$

In fact, by Theorem 7.39 and Definition 7.42, a function $f \in C^\infty(a, b)$ is analytic on (a, b) if and only if for each $x_0 \in (a, b)$ there is an interval (c, d) containing

x_0 such that $R_n^{f,x_0} \to 0$, as $n \to \infty$, for every $x \in (c, d)$. We shall use this observation frequently below.

By Taylor's Formula (Theorem 4.24) the remainder term of an $f \in C^\infty(a, b)$ satisfies

$$R_n^{f,x_0}(x) = \frac{f^{(n)}(c)}{n!}(x - x_0)^n$$

for some c between x_0 and x (note the index shift from $n + 1$ to n.) Therefore, it should come as no surprise that there are several results that state the following: If the nth derivative of f satisfies a certain condition, then f is analytic on (a, b). Here is a particularly simple but useful result of this type.

7.43 Theorem. *Let $f \in C^\infty(a, b)$. If there is an $M > 0$ such that*

$$\left| f^{(n)}(x) \right| \le M^n$$

for all $x \in (a, b)$ and $n \in \mathbf{N}$, then f is analytic on (a, b). In fact, for each $x_0 \in (a, b)$,

$$f(x) = \sum_{k=0}^{\infty} \frac{f^{(k)}(x_0)}{k!}(x - x_0)^k$$

holds for all $x \in (a, b)$.

Proof. Fix $x_0 \in (a, b)$ and set $C = \max\{M|a - x_0|, M|b - x_0|\}$. By Taylor's Formula,

$$|R_n^{f,x_0}(x)| = \frac{|f^{(n)}(c)|}{n!}|x - x_0|^n \le \frac{M^n}{n!}|x - x_0|^n \le \frac{C^n}{n!}$$

for all $n \in \mathbf{N}$. But $C^n/n! \to 0$ as $n \to \infty$ for any $C \in \mathbf{R}$ (being terms of a convergent series by the Ratio Test). Thus, by the Squeeze Theorem, the remainder term $R_n^{f,x_0}(x)$ converges to zero for every $x \in (a, b)$. ∎

Here are three examples of Theorem 7.43 in practice.

7.44 EXAMPLE.

Prove that $\sin x$ and $\cos x$ are analytic on $\mathbf{R}$ and have Maclaurin expansions

$$\sin x = \sum_{k=0}^{\infty} \frac{(-1)^k x^{2k+1}}{(2k + 1)!}, \qquad \cos x = \sum_{k=0}^{\infty} \frac{(-1)^k x^{2k}}{(2k)!}. \tag{11}$$

Proof. In Example 4.26 [see (19) there], we proved that the Taylor series of $f(x) := \sin x$ centered at $x_0 = 0$ is $S(x) := \sum_{k=0}^{\infty}(-1)^k x^{2k+1}/(2k + 1)!$. Since $f^{(n)}(x)$ is $\pm \sin x$ or $\pm \cos x$, it is clear that $|f^{(n)}(x)| \le 1$ for all $x \in \mathbf{R}$.

Therefore, it follows from Theorem 7.43 that $\sin x$ is analytic on $\mathbf{R}$ and that the left side of (11) holds everywhere on $\mathbf{R}$. A similar argument proves the right side of (11). ∎

7.45 EXAMPLE.

Prove that e^x is analytic on $\mathbf{R}$ and has Maclaurin expansion

$$e^x = \sum_{k=0}^{\infty} \frac{x^k}{k!}. \tag{12}$$

Proof. In Example 4.25 [see (18) there], we proved that the Taylor series of $f(x) := e^x$ centered at $x_0 = 0$ is $S(x) := \sum_{k=0}^{\infty} x^k/k!$.

Fix $C > 0$ and notice that $|f^{(n)}(x)| = |e^x| \le e^C =: M \le M^n$ for all $n \in \mathbf{N}$ and $x \in [-C, C]$. It follows from Theorem 7.43 that e^x is analytic on $[-C, C]$ and that $S(x)$ converges to e^x everywhere on $[-C, C]$. Since $C > 0$ was arbitrary, we conclude that (12) holds for all $x \in \mathbf{R}$. ∎

Sometimes, it is impractical to get the kind of global estimates on the derivatives of f necessary to apply Theorem 7.43. The following result, which shows that the center of a power series can be changed within its interval of convergence, is sometimes used to circumvent this problem.

7.46 Theorem. *Suppose that I is an open interval centered at c and that*

$$f(x) = \sum_{k=0}^{\infty} a_k (x - c)^k, \qquad x \in I.$$

If $x_0 \in I$ and $r > 0$ satisfy $(x_0 - r, x_0 + r) \subseteq I$, then

$$f(x) = \sum_{k=0}^{\infty} \frac{f^{(k)}(x_0)}{k!} (x - x_0)^k$$

for all $x \in (x_0 - r, x_0 + r)$. In particular, if f is a C^∞ function whose Taylor series expansion converges to f on some open interval J, then f is analytic on J.

Proof. It suffices to prove the first statement. By making the change of variables $w = x - c$, we may suppose that $c = 0$ and $I = (-R, R)$; that is, that $f(x) = \sum_{k=0}^{\infty} a_k x^k$, for all $x \in (-R, R)$. Suppose that $(x_0 - r, x_0 + r) \subseteq (-R, R)$ and fix $x \in (x_0 - r, x_0 + r)$. By hypothesis and the Binomial Formula,

$$f(x) = \sum_{k=0}^{\infty} a_k x^k = \sum_{k=0}^{\infty} a_k ((x - x_0) + x_0)^k = \sum_{k=0}^{\infty} a_k \sum_{j=0}^{k} \binom{k}{j} x_0^{k-j} (x - x_0)^j. \tag{13}$$

Since $\sum_{k=0}^{\infty} a_k y^k$ converges absolutely at $y := |x - x_0| + |x_0| < R$, we have

$$\sum_{k=0}^{\infty} \left| a_k \sum_{j=0}^{k} \binom{k}{j} x_0^{k-j} (x - x_0)^j \right| \leq \sum_{k=0}^{\infty} |a_k| \sum_{j=0}^{k} \binom{k}{j} |x_0|^{k-j} |x - x_0|^j$$

$$= \sum_{k=0}^{\infty} |a_k| (|x - x_0| + |x_0|)^k < \infty.$$

Hence, by (13), Theorem 7.18, and Corollary 7.31,

$$f(x) = \sum_{k=0}^{\infty} a_k \sum_{j=0}^{k} \binom{k}{j} x_0^{k-j} (x - x_0)^j$$

$$= \sum_{j=0}^{\infty} \left(\sum_{k=j}^{\infty} \binom{k}{j} a_k x_0^{k-j} \right) (x - x_0)^j$$

$$= \sum_{j=0}^{\infty} \left(\sum_{k=j}^{\infty} \frac{k!}{(k-j)!} a_k (x_0 - 0)^{k-j} \right) \frac{(x - x_0)^j}{j!} = \sum_{j=0}^{\infty} \frac{f^{(j)}(x_0)}{j!} (x - x_0)^j. \quad \blacksquare$$

7.47 EXAMPLE.

Prove that $\arctan x$ is analytic on $(-1, 1)$ and has Maclaurin expansion

$$\arctan x = \sum_{k=0}^{\infty} \frac{(-1)^k x^{2k+1}}{2k + 1} \qquad x \in (-1, 1).$$

Proof. For each $0 < x < 1$, the geometric series $\sum_{k=0}^{\infty} (-1)^k t^{2k}$ converges uniformly on $[-x, x]$ to $1/(1 + t^2)$. Thus, by Theorem 7.32,

$$\arctan x = \int_0^x \frac{dt}{1 + t^2} = \int_0^x \sum_{k=0}^{\infty} (-1)^k t^{2k} \, dt = \sum_{k=0}^{\infty} \frac{(-1)^k x^{2k+1}}{2k + 1}.$$

By uniqueness, this is the Maclaurin expansion of $\arctan x$. Since this expansion converges on $(-1, 1)$, it follows from Theorem 7.46 that $\arctan x$ is analytic on $(-1, 1)$. $\quad \blacksquare$

In Examples 7.44 and 7.45, we found the Taylor expansion of a given f by computing the derivatives of f and estimating the remainder term. In the preceding example, we found the Taylor expansion of $\arctan x$ without computing its derivatives. This can be done in general, using term-by-term differentiation or integration or products of power series, when the function in question can be written as an integral or derivative or product of functions whose Taylor series are known. Here are two more examples of this type.

7.48 EXAMPLE.

Find the Maclaurin expansion of $\arctan x/(1-x)$.

Proof. By Theorem 7.33 and Example 7.47, for each $|x| < 1$,

$$
\left(\frac{\arctan x}{1-x} \right) = \left(\sum_{k=0}^{\infty} x^k \right) \left(\sum_{k=0}^{\infty} \frac{(-1)^k x^{2k+1}}{2k+1} \right)
$$

$$
= \sum_{k=0}^{\infty} \left(\sum_{j \in A_k} \frac{(-1)^j}{2j+1} \right) x^k,
$$

where $A_k := \{ j \in \mathbf{N} : 0 \le j \le (k-1)/2 \}$. ∎

7.49 EXAMPLE.

Show that the Taylor expansion of $\log x$ centered at $x_0 = 1$ is

$$
\log x = \sum_{k=1}^{\infty} \frac{(-1)^{k+1}}{k} (x-1)^k \qquad x \in (0,2).
$$

Proof. By Theorem 7.32, for each $x \in (0,2)$,

$$
\log x = \int_1^x \frac{dt}{t} = \int_1^x \frac{dt}{1-(1-t)}
$$

$$
= \int_1^x \sum_{k=0}^{\infty} (1-t)^k \, dt = \sum_{k=1}^{\infty} \frac{(-1)^{k+1}}{k} (x-1)^k. \quad ∎
$$

In some situations it is useful to have an integral form of the remainder term. This requires a slightly stronger hypothesis than Taylor's Formula but can yield a sharper estimate.

7.50 Theorem. [LAGRANGE].
Let $n \in \mathbf{N}$. If $f \in C^n(a,b)$, then

$$
R_n(x) := R_n^{f,x_0}(x) = \frac{1}{(n-1)!} \int_{x_0}^x (x-t)^{n-1} f^{(n)}(t) \, dt
$$

for all $x, x_0 \in (a,b)$.

Proof. The proof is by induction on n. If $n = 1$, the formula holds by the Fundamental Theorem of Calculus.

Suppose that the formula holds for some $n \in \mathbf{N}$. Since

$$R_{n+1}(x) = R_n(x) - \frac{f^{(n)}(x_0)}{n!}(x - x_0)^n \quad \text{and}$$

$$\frac{(x - x_0)^n}{n!} = \frac{1}{(n-1)!} \int_{x_0}^x (x - t)^{n-1} \, dt,$$

it follows that

$$R_{n+1}(x) = \frac{1}{(n-1)!} \int_{x_0}^x (x - t)^{n-1} \left(f^{(n)}(t) - f^{(n)}(x_0) \right) \, dt.$$

Let $u = f^{(n)}(t) - f^{(n)}(x_0)$, $dv = (x - t)^{n-1}$ and integrate the right side of the identity above by parts. Since $u(x_0) = 0$ and $v(x) = 0$, we have

$$R_{n+1}(x) = -\frac{1}{(n-1)!} \int_{x_0}^x u'(t)v(t) \, dt = \frac{1}{n!} \int_{x_0}^x (x - t)^n f^{(n+1)}(t) \, dt.$$

Hence, the formula holds for $n + 1$. ∎

The rest of this section contains some additional (but optional) material on analytic functions.

In order to generalize the Binomial Formula from integer exponents to real exponents (compare Theorem 1.26 with Theorem 7.52 below), we introduce the following notation. Let $\alpha \in \mathbf{R}$ and k be a nonnegative integer. The *generalized binomial coefficient α over k* is defined by

$$\binom{\alpha}{k} := \begin{cases} \dfrac{\alpha(\alpha - 1) \ldots (\alpha - k + 1)}{k!} & k \neq 0 \\ 1 & k = 0. \end{cases}$$

Notice that when $\alpha \in \mathbf{N}$, these generalized binomial coefficients coincide with the usual binomial coefficients, because in this case $\binom{\alpha}{k} = 0$ for $k > \alpha$.

*7.51 Lemma.
Suppose that $\alpha, \beta \in \mathbf{R}$. Then

$$\sum_{j=0}^k \binom{\alpha}{k-j} \binom{\beta}{j} = \binom{\alpha + \beta}{k} \qquad k = 0, 1, \ldots.$$

Proof. The formula holds for $k = 0$ and $k = 1$. If it holds for some $k \geq 1$, then by the inductive hypothesis and the definition of the generalized binomial coefficients,

$$
\binom{\alpha + \beta}{k + 1} = \binom{\alpha + \beta}{k} \frac{\alpha + \beta - k}{k + 1}
$$

$$
= \sum_{j=0}^{k} \binom{\alpha}{k - j} \binom{\beta}{j} \left(\frac{\alpha - k + j}{k + 1} + \frac{\beta - j}{k + 1} \right)
$$

$$
= \sum_{j=0}^{k} \left(\frac{k - j + 1}{k + 1} \right) \binom{\alpha}{k - j + 1} \binom{\beta}{j} + \left(\frac{j + 1}{k + 1} \right) \binom{\alpha}{k - j} \binom{\beta}{j + 1}
$$

$$
= \binom{\alpha}{k + 1} + \sum_{j=1}^{k} \left(\frac{k - j + 1}{k + 1} + \frac{j}{k + 1} \right) \binom{\alpha}{k - j + 1} \binom{\beta}{j} + \binom{\beta}{k + 1}
$$

$$
= \sum_{j=0}^{k+1} \binom{\alpha}{k + 1 - j} \binom{\beta}{j}. \qquad \blacksquare
$$

With this ugly calculation out of the way, we are prepared to generalize the Binomial Formula.

*7.52 Theorem.   [THE BINOMIAL SERIES].
If $\alpha \in \mathbf{R}$ and $|x| < 1$, then

$$
(1 + x)^\alpha = \sum_{k=0}^{\infty} \binom{\alpha}{k} x^k.
$$

In particular, $(1 + x)^\alpha$ is analytic on $(-1, 1)$ for all $\alpha \in \mathbf{R}$.

Proof. Fix $|x| < 1$ and consider the series $F(\alpha) := \sum_{k=0}^{\infty} \binom{\alpha}{k} x^k$. Since

$$
\lim_{k \to \infty} \left| \frac{\binom{\alpha}{k + 1} x^{k+1}}{\binom{\alpha}{k} x^k} \right| = \lim_{k \to \infty} \left| \frac{\alpha - k}{k + 1} \right| |x| = |x| < 1
$$

is independent of α, it follows from the proof of the Ratio Test that F converges absolutely and uniformly on $\mathbf{R}$. Hence, F is continuous. Moreover, by Theorem 7.33 and Lemma 7.51,

$$F(\alpha)F(\beta) = \sum_{k=0}^{\infty} \binom{\alpha}{k} x^k \sum_{k=0}^{\infty} \binom{\beta}{k} x^k$$

$$= \sum_{k=0}^{\infty} \sum_{j=0}^{k} \binom{\alpha}{k-j} \binom{\beta}{j} x^k$$

$$= \sum_{k=0}^{\infty} \binom{\alpha+\beta}{k} x^k = F(\alpha+\beta).$$

Hence, it follows from Exercise 3.3.9 that $F(\alpha) = F(1)^{\alpha}$. Since

$$F(1) = \sum_{k=0}^{\infty} \binom{1}{k} x^k = 1 + x,$$

we conclude that $F(\alpha) = (1+x)^{\alpha}$ for all $|x| < 1$.   ∎

Lagrange's Theorem gives us another condition on the derivatives of f suffi-cient to conclude that f is analytic.

*7.53 Theorem.   [BERNSTEIN].
If $f \in C^{\infty}(a, b)$ and $f^{(n)}(x) \geq 0$ for all $x \in (a, b)$ and $n \in \mathbf{N}$, then f is analytic on (a, b). In fact, if $x_0 \in (a, b)$ and $f^{(n)}(x) \geq 0$ for $x \in [x_0, b)$ and $n \in \mathbf{N}$, then

$$f(x) = \sum_{k=0}^{\infty} \frac{f^{(k)}(x_0)}{k!} (x - x_0)^k \tag{14}$$

for all $x \in [x_0, b)$.

Proof. Fix $x_0 < x < b$ and $n \in \mathbf{N}$. Use Lagrange's Theorem and a change of variables $t = (x - x_0)u + x_0$ to write

$$R_n(x) = R_n^{f, x_0}(x) = \frac{(x - x_0)^n}{(n-1)!} \int_0^1 (1 - u)^{n-1} f^{(n)}((x - x_0)u + x_0) \, du. \tag{15}$$

Since $f^{(n)} \geq 0$, (15) implies $R_n(x) \geq 0$. On the other hand, by definition and hypothesis,

$$R_n(x) = f(x) - \sum_{k=0}^{n-1} \frac{f^{(k)}(x_0)}{k!} (x - x_0)^k \leq f(x).$$

Therefore,

$$0 \leq R_n(x) \leq f(x) \tag{16}$$

for all $x \in (x_0, b)$.

Let $b_0 \in (x_0, b)$ and notice that it suffices to verify (14) for $x_0 \leq x < b_0$. (We introduce the parameter b_0 in order to handle the cases $b \in \mathbf{R}$ and $b = \infty$ simultaneously.) Since $R_n(x_0) = 0$ for all $n \in \mathbf{N}$, we need only show that $R_n(x) \to 0$ as $n \to \infty$ for each $x \in (x_0, b_0)$.

By hypothesis, $f^{(n+1)}(t) \geq 0$ for $t \in [x_0, b)$, so $f^{(n)}$ is increasing on $[x_0, b)$. Since $x < b_0 < b$, we have by (15) and (16) that

$$
\begin{aligned}
0 \leq R_n(x) &= \frac{(x - x_0)^n}{(n-1)!} \int_0^1 (1 - u)^{n-1} f^{(n)}((x - x_0)u + x_0) \, du \\
&\leq \frac{(x - x_0)^n}{(n-1)!} \int_0^1 (1 - u)^{n-1} f^{(n)}((b_0 - x_0)u + x_0) \, du \\
&= \left(\frac{x - x_0}{b_0 - x_0} \right)^n R_n(b_0).
\end{aligned}
$$

Since $(x - x_0)/(b_0 - x_0) < 1$ and, by (16), $R_n(b_0) \leq f(b_0)$, we conclude by the Squeeze Theorem that $R_n(x) \to 0$ as $n \to \infty$. ∎

*7.54 EXAMPLE.

Prove that a^x is analytic on $\mathbf{R}$ for each $a > 0$.

Proof. First suppose that $a \geq 1$. Since $f^{(n)}(x) = (\log a)^n \cdot a^x \geq 0$ for all $x \in \mathbf{R}$ and $n \in \mathbf{N}$, a^x is analytic on $\mathbf{R}$ by Bernstein's Theorem. If $0 < a < 1$, then by what we just proved and a change of variables,

$$
a^x = (a^{-1})^{-x} = \sum_{k=0}^{\infty} \frac{\log^k (a^{-1})(-x)^k}{k!} = \sum_{k=0}^{\infty} \frac{\log^k a \cdot x^k}{k!}.
$$

Hence by Theorem 7.46, a^x is analytic on $\mathbf{R}$. ∎

Our final theorem shows that an analytic function cannot be extended in an arbitrary way to produce another analytic function. We first prove the following special case.

*7.55 Lemma.

Suppose that f, g are analytic on an open interval (c, d) and that $x_0 \in (c, d)$. If $f(x) = g(x)$ for $x \in (c, x_0)$, then there is a $\delta > 0$ such that $f(x) = g(x)$ for all $x \in (x_0 - \delta, x_0 + \delta)$.

Proof. By Theorem 7.39 and Definition 7.38, there is a $\delta > 0$ such that

$$
f(x) = \sum_{k=0}^{\infty} \frac{f^{(k)}(x_0)}{k!} (x - x_0)^k \quad \text{and} \quad g(x) = \sum_{k=0}^{\infty} \frac{g^{(k)}(x_0)}{k!} (x - x_0)^k \tag{17}
$$

for all $x \in (x_0 - \delta, x_0 + \delta)$. By hypothesis, f, g are continuous at x_0 and

$$f(x_0) = \lim_{x \to x_0-} f(x) = \lim_{x \to x_0-} g(x) = g(x_0). \tag{18}$$

Similarly, $f^{(k)}(x_0) = g^{(k)}(x_0)$ for $k \in \mathbf{N}$. We conclude from (17) that $f(x) = g(x)$ for all $x \in (x_0 - \delta, x_0 + \delta)$. ∎

*7.56 Theorem. [**ANALYTIC CONTINUATION**].
Suppose that I and J are open intervals, that f is analytic on I, that g is analytic on J, and that a < b are points in I ∩ J. If f(x) = g(x) for x ∈ (a, b), then f(x) = g(x) for all x ∈ I ∩ J.

Proof. We assume for simplicity that I and J are bounded intervals. Since $I \cap J \neq \emptyset$, choose $c, d \in \mathbf{R}$ such that $I \cap J = (c, d)$ (see Figure 7.6).

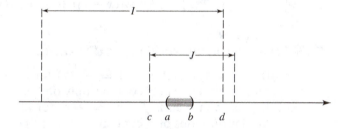

FIGURE 7.6

Consider the set $E = \{t \in (a, d) : f(x) = g(x) \text{ for all } x \in (a, t)\}$. By our assumption, $d < \infty$ and by hypothesis $b \in E$. Thus E is bounded and nonempty. Let $x_0 = \sup E$. If $x_0 < d$, then by Lemma 7.55 there is a $\delta > 0$ such that $f(x) = g(x)$ for all $x \in (x_0 - \delta, x_0 + \delta)$. This contradicts the choice of x_0. Therefore, $x_0 = d$; that is, $f(x) = g(x)$ for all $x \in (a, d)$. A similar argument proves that $f(x) = g(x)$ for all $x \in (c, b)$. ∎

EXERCISES

7.4.1. Prove that each of the following functions is analytic on $\mathbf{R}$ and find its Maclaurin expansion.

a) $x^2 + \cos(2x)$
b) $x^2 3^x$
c) $\cos^2 x - \sin^2 x$
d) $\dfrac{e^x - 1}{x}$

7.4.2. Prove that each of the following functions is analytic on $(-1, 1)$ and find its Maclaurin expansion.

a) $\dfrac{x}{x^5 + 1}$

b) $\dfrac{e^x}{1+x}$

c) $\log\left(\dfrac{1}{|x^2-1|}\right)$

*d) $\arcsin x$

7.4.3. For each of the following functions, find its Taylor expansion centered at $x_0 = 1$ and determine the largest interval on which it converges.

a) e^x
b) $\log_2(x^5)$
c) $x^3 - x + 5$
*d) $\sqrt{x}$

7.4.4. Prove that if P is a polynomial of degree n and $x_0 \in \mathbf{R}$, then there are numbers $\beta_k \in \mathbf{R}$ such that

$$P(x) = \beta_0 + \beta_1(x - x_0) + \cdots + \beta_n(x - x_0)^n$$

for all $x \in \mathbf{R}$.

7.4.5 Let $a > 0$ and suppose that $f \in C^\infty(-a, a)$.

a) If f is odd [i.e., if $f(-x) = -f(x)$ for all $x \in (-a, a)$], then the Maclaurin series of f contains only odd powers of x.
b) If f is even [i.e., if $f(-x) = f(x)$ for all $x \in (-a, a)$], then the Maclaurin series of f contains only even powers of x.

7.4.6. Suppose that $f \in C^\infty(-\infty, \infty)$ and that

$$\lim_{n \to \infty} \frac{1}{n!} \int_0^a x^n f^{(n+1)}(a - x)\, dx = 0$$

for all $a \in \mathbf{R}$. Prove that f is analytic on $(-\infty, \infty)$ and

$$f(x) = \sum_{k=0}^{\infty} \frac{f^{(k)}(0)}{k!} x^k, \qquad x \in \mathbf{R}.$$

7.4.7. a) Prove that

$$\left| \int_0^1 e^{x^2}\, dx - \sum_{k=0}^{n-1} \frac{1}{(2k+1)k!} \right| \le \frac{3}{n!}$$

for $n \in \mathbf{N}$.

b) Show that

$$2.9253 < \int_{-1}^{1} e^{x^2}\, dx < 2.9254.$$

7.4.8. Let $f \in C^\infty(a, b)$. Prove that f is analytic on (a, b) if and only if f' is analytic on (a, b).

7.4.9. Suppose that I is a nonempty open interval and that f is bounded and C^∞ on I. If there is an $M > 0$ such that $|f^{(k)}(x)| \le Mk$ for all $x \in I$ and all positive integers k sufficiently large, and if there exist $a, b \in I$ such that

$$\int_a^b f(x)\, x^n\, dx = 0$$

for $n = 0, 1, 2, \ldots$, then prove that f is zero on $[a, b]$.

*7.4.10. Suppose that f is analytic on $(-\infty, \infty)$ and that

$$\int_a^b |f(x)|\, dx = 0$$

for some $a \ne b$ in $\mathbf{R}$. Prove that $f(x) = 0$ for all $x \in \mathbf{R}$.

*7.4.11. Prove that

$$\left(\sum_{k=1}^\infty |a_k|^\beta \right)^{1/\beta} \le \sum_{k=1}^\infty |a_k|$$

for all $a_k \in \mathbf{R}$ and all $\beta > 1$.

***7.5 APPLICATIONS**

This section uses no material from any other enrichment section.

The theory of infinite series is a potent tool for both pure and applied mathematics. In this section we give several examples to back up this claim.

We begin with a nontrivial theorem from number theory. Recall that an integer $n \ge 2$ is called *prime* if the only factors of n in $\mathbf{N}$ are 1 and n. Also recall that given $n \in \mathbf{N}$ there are primes $p_1, p_2, \ldots, p_k$ and exponents $\alpha_1, \alpha_2, \ldots, \alpha_k$ such that

$$n = p_1^{\alpha_1} p_2^{\alpha_2} \ldots p_k^{\alpha_k}.$$

7.57 Theorem. [EUCLID'S THEOREM; EULER'S PROOF].
There are infinitely many primes in $\mathbf{N}$.

Proof. Suppose to the contrary that $p_1, p_2, \ldots, p_k$ represent all the primes in $\mathbf{N}$. Fix $N \in \mathbf{N}$ and set $\alpha = \sup\{\alpha_1, \ldots, \alpha_k\}$, where this supremum is taken over all α_j's which satisfy $n = p_1^{\alpha_1} p_2^{\alpha_2} \ldots p_k^{\alpha_k}$ for some $n \le N$. Since every integer $j \in [1, N]$ must have the form $j = p_1^{e_1} \ldots p_k^{e_k}$ for some choice of integers $0 \le e_i \le \alpha$, we have

$$\left(1 + \frac{1}{p_1} + \cdots + \frac{1}{p_1^\alpha} \right) \left(1 + \frac{1}{p_2} + \cdots + \frac{1}{p_2^\alpha} \right) \cdots \left(1 + \frac{1}{p_k} + \cdots + \frac{1}{p_k^\alpha} \right)$$

$$= \sum_{0 \le e_i \le \alpha} 1 \cdot \frac{1}{p_1^{e_1}} \cdots \frac{1}{p_k^{e_k}} \ge \sum_{j=1}^N \frac{1}{j}.$$

On the other hand, for each integer $i \in [1, k]$, we have by Theorem 6.7 that

$$1 + \frac{1}{p_i} + \cdots + \frac{1}{p_i^\alpha} \le \sum_{\ell=1}^\infty \left(\frac{1}{p_i}\right)^\ell = \frac{p_i}{p_i - 1}.$$

Consequently,

$$\sum_{j=1}^N \frac{1}{j} \le \left(\frac{p_1}{p_1 - 1}\right) \cdots \left(\frac{p_k}{p_k - 1}\right) = M < \infty.$$

Taking the limit of this inequality as $N \to \infty$, we conclude that $\sum_{j=1}^\infty 1/j \le M < \infty$, a contradiction. ∎

Our next application, a result used to approximate roots of twice differentiable functions, shows that if an initial guess x_0 is close enough to a root of a suitably well-behaved function f, then the sequence x_n generated by (19) converges to a root of f.

7.58 Theorem. [**NEWTON-RAPHSON**].
Suppose that $f : [a, b] \to \mathbf{R}$ is continuous on $[a, b]$ and that $f(c) = 0$ for some $c \in (a, b)$. If f'' exists and is bounded on (a, b) and there is an $\varepsilon_0 > 0$ such that $|f'(x)| \ge \varepsilon_0$ for all $x \in (a, b)$, then there is a closed interval $I \subseteq (a, b)$ containing c such that given $x_0 \in I$, the sequence $\{x_n\}_{n \in \mathbf{N}}$ defined by

$$x_n = x_{n-1} - \frac{f(x_{n-1})}{f'(x_{n-1})}, \qquad n \in \mathbf{N}, \tag{19}$$

satisfies $x_n \in I$ and $x_n \to c$ as $n \to \infty$.

Proof. Choose $M > 0$ such that $|f''(x)| \le M$ for $x \in (a, b)$. Choose $r_0 \in (0, 1)$ so small that $I = [c - r_0, c + r_0]$ is a subinterval of (a, b) and $r_0 < \varepsilon_0/M$. Suppose that $x_0 \in I$ and define the sequence $\{x_n\}$ by (19). Set $r := r_0 M/\varepsilon_0$ and observe by the choice of r_0 that $r < 1$. Thus it suffices to show that

$$|x_n - c| \le r^n |x_0 - c| \tag{20}$$

and

$$|x_n - c| \le r_0 \tag{21}$$

hold for all $n \in \mathbf{N}$.

The proof is by induction on n. Clearly, (20) and (21) hold for $n = 0$. Fix $n \in \mathbf{N}$ and suppose that

$$|x_{n-1} - c| \le r^{n-1} |x_0 - c| \tag{22}$$

and that

$$|x_{n-1} - c| \leq r_0. \tag{23}$$

Use Taylor's Formula to choose a point ξ between c and x_{n-1} such that

$$-f(x_{n-1}) = f(c) - f(x_{n-1}) = f'(x_{n-1})(c - x_{n-1}) + \frac{1}{2}f''(\xi)(c - x_{n-1})^2.$$

Since (19) implies $-f(x_{n-1}) = f'(x_{n-1})(x_n - x_{n-1})$, it follows that

$$f'(x_{n-1})(x_n - c) = \frac{1}{2}f''(\xi)(c - x_{n-1})^2.$$

Solving this equation for $x_n - c$, we have by the choice of M and ε_0 that

$$|x_n - c| = \left|\frac{f''(\xi)}{2f'(x_{n-1})}\right| |x_{n-1} - c|^2 \leq \frac{M}{2\varepsilon_0}|x_{n-1} - c|^2. \tag{24}$$

Since $M/\varepsilon_0 < 1/r_0$, it follows from (24) and (23) that

$$|x_n - c| \leq \frac{M}{\varepsilon_0}|x_{n-1} - c|^2 \leq \frac{1}{r_0}|r_0|^2 = r_0.$$

This proves (21). Again, by (24), (22), and the choice of r, we have

$$|x_n - c| \leq \frac{M}{\varepsilon_0}(r^{n-1}|x_0 - c|)^2 = \frac{r}{r_0}(r^{2n-2}|x_0 - c|^2) \leq r^{2n-1}|x_0 - c|.$$

Since $r < 1$ and $2n - 1 \geq n$ imply $r^{2n-1} \leq r^n$, we conclude that $|x_n - c| \leq r^{2n-1}|x_0 - c| \leq r^n|x_0 - c|$. ∎

Notice if x_{n-1} and x_n satisfy (19), then x_n is the x-intercept of the tangent line to $y = f(x)$ at the point $(x_{n-1}, f(x_{n-1}))$ (see Exercise 7.5.4). Thus, Newton's method is based on a simple geometric principle (see Figure 7.7). Also notice that, by (24), this method converges very rapidly. Indeed, the number of decimal places of accuracy nearly doubles with each successive approximation.

As a general rule, it is extremely difficult to show that a given nonalgebraic number is irrational. The next result shows how to use infinite series to give an easy proof that certain kinds of numbers are irrational.

7.59 Theorem. [EULER].
The number e is irrational.

Proof. Suppose to the contrary that $e = p/q$ for some $p, q \in \mathbf{N}$. By Example 7.45,

$$\frac{q}{p} = e^{-1} = \sum_{k=0}^{\infty} \frac{(-1)^k}{k!}.$$

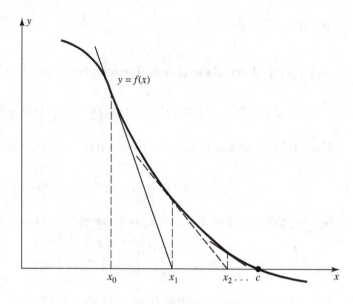

FIGURE 7.7

Breaking this sum into two pieces and multiplying by $(-1)^{p+1} p!$, we have

$$x := (-1)^{p+1} \left(q(p-1)! - \sum_{k=0}^{p} \frac{(-1)^k p!}{k!} \right) = y := \sum_{k=p+1}^{\infty} (-1)^{k+p+1} \frac{p!}{k!}.$$

Since $p!/k! \in \mathbf{N}$ for all integers $k \le p$, the number x must be an integer. On the other hand,

$$y = \frac{1}{p+1} - \frac{1}{(p+1)(p+2)} + \frac{1}{(p+1)(p+2)(p+3)} - \cdots$$

lies between $1/(p+1)$ and $1/(p+1) - 1/(p+1)(p+2)$. Therefore, y is a number which satisfies $0 < y < 1$. In particular, $x \ne y$, a contradiction. ∎

We know that a continuous function can fail to be differentiable at one point [e.g., $f(x) = |x|$]. Hence, it is not difficult to see that, given any finite set of points E, there is a continuous function which fails to be differentiable at every point in E. We shall now show that there is a continuous function which fails to be differentiable at all points in $\mathbf{R}$. Once again, here is a clear indication that, although we use sketches to motivate proofs and to explain results, we cannot rely on sketches to give a complete picture of the general situation.

7.60 Theorem. [WEIERSTRASS].
There is a function f continuous on $\mathbf{R}$ *which is not differentiable at any point in* $\mathbf{R}$.
(**Note:** Such functions are called *nowhere differentiable*.)

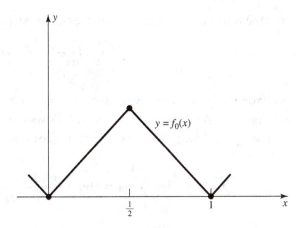

FIGURE 7.8

Proof. Let

$$f_0(x) = \begin{cases} x & 0 \le x < 1/2 \\ 1 - x & 1/2 \le x < 1 \end{cases}$$

and extend f_0 to $\mathbf{R}$ by periodicity of period 1, that is, so that $f_0(x) = f_0(x+1)$ for all $x \in \mathbf{R}$ (see Figure 7.8). Set $f_k(x) = f_0(2^k x)/2^k$ for $x \in \mathbf{R}$ and $k \in \mathbf{N}$ and consider the function

$$f(x) = \sum_{k=0}^{\infty} f_k(x), \qquad x \in \mathbf{R}.$$

Normalizing f_k by 2^k has two consequences. First, since $f_0'(y) = \pm 1$ for each y which satisfies $2y \notin \mathbf{Z}$, it is easy to see that

$$f_k'(y) = \pm 1 \quad \text{for each } y \text{ which satisfies } 2^{k+1}y \notin \mathbf{Z}. \tag{25}$$

Second, by the Weierstrass M-Test, f converges uniformly and, hence, is continuous on $\mathbf{R}$.

Since f is periodic of period 1, it suffices to show that f is not differentiable at any $x \in [0, 1)$. Suppose to the contrary that f is differentiable at some $x \in [0, 1)$. For each $n \in \mathbf{N}$, choose $p \in \mathbf{Z}$ such that $x \in [\alpha_n, \beta_n)$ for $\alpha_n = p/2^n$ and $\beta_n = (p+1)/2^n$. Since each f_k is linear on $[\alpha_{k+1}, \beta_{k+1}]$ and $[\alpha_n, \beta_n] \subseteq [\alpha_{k+1}, \beta_{k+1}]$ for $n > k$, it is clear that

$$c_k := \frac{f_k(\beta_n) - f_k(\alpha_n)}{\beta_n - \alpha_n}$$

depends only on k and not on n when $n > k$. Moreover, by (25), it is also clear that each $c_k = \pm 1$. Therefore, $\sum_{k=0}^{\infty} c_k$ cannot be convergent.

On the other hand, since f is differentiable at x,

$$f'(x) = \lim_{n\to\infty} \frac{f(\beta_n) - f(\alpha_n)}{\beta_n - \alpha_n} \tag{26}$$

(see Exercise 7.5.7). However, since $f_0(y) = 0$ if and only if $y \in \mathbf{Z}$, we also have $f_k(\beta_n) = f_k(\alpha_n) = 0$ for $k \geq n$. It follows that $f(\beta_n) = \sum_{k=0}^{n-1} f_k(\beta_n)$ and $f(\alpha_n) = \sum_{k=0}^{n-1} f_k(\alpha_n)$. We conclude from (26) that

$$\sum_{k=0}^{\infty} c_k = \lim_{n\to\infty} \sum_{k=0}^{n-1} c_k = \lim_{n\to\infty} \frac{f(\beta_n) - f(\alpha_n)}{\beta_n - \alpha_n} = f'(x)$$

is convergent, a contradiction.   ∎

EXERCISES

7.5.1. Using a calculator and Theorem 7.58, approximate all real roots of $f(x) = x^3 + 3x^2 + 4x + 1$ to five decimal places.

7.5.2. a) Using the proof of Theorem 7.58, prove that (20) holds if $r/2$ replaces r.

  Use part a) to estimate the difference $|x_4 - \pi|$, where $x_0 = 3$, $f(x) = \sin x$, and x_n is defined by (19). Evaluate x_4 directly, and verify that x_4 is actually closer than the theory predicts.

7.5.3. Prove that given any $n \in \mathbf{N}$, there is a function $f \in C^n(\mathbf{R})$ such that $f^{(n+1)}(x)$ does not exist for any $x \in \mathbf{R}$.

7.5.4. Prove that if x_{n-1}, x_n satisfy (19), then x_n is the x-intercept of the tangent line to $y = f(x)$ at the point $(x_{n-1}, f(x_{n-1}))$.

7.5.5. Prove that $\cos(1)$ is irrational.

7.5.6. Suppose that $f : \mathbf{R} \to \mathbf{R}$. If f'' exists and is bounded on $\mathbf{R}$, and there is an $\varepsilon_0 > 0$ such that $|f'(x)| \geq \varepsilon_0$ for all $x \in \mathbf{R}$, prove that there exists a $\delta > 0$ such that if $|f(x_0)| \leq \delta$ for some $x_0 \in \mathbf{R}$, then f has a root; that is, that $f(c) = 0$ for some $c \in \mathbf{R}$.

7.5.7. Let $x \in [0, 1)$ and α_n, β_n be defined as in Theorem 7.60.

a) If $f : [0, 1) \to \mathbf{R}$ and $\gamma \in \mathbf{R}$, prove that

$$\frac{f(\beta_n) - f(\alpha_n)}{\beta_n - \alpha_n} - \gamma = \left(\frac{f(\beta_n) - f(x)}{\beta_n - x} - \gamma \right) \left(\frac{\beta_n - x}{\beta_n - \alpha_n} \right)$$
$$+ \left(\frac{f(x) - f(\alpha_n)}{x - \alpha_n} - \gamma \right) \left(\frac{x - \alpha_n}{\beta_n - \alpha_n} \right).$$

b) If f is differentiable at x, prove that (26) holds.

CHAPTER 8

Euclidean Spaces

The world we live in is at least four dimensional: three spatial dimensions together with the time dimension. Moreover, certain problems from engineering, physics, chemistry, and economics force us to consider even higher dimensions. For example, guidance systems for missiles frequently require as many as 100 variables (longitude, latitude, altitude, velocity, time after launch, pitch, yaw, fuel on board, etc.). Another example, the state of a gas in a closed container, can best be described by a function of $6m$ variables, where m is the number of molecules in the system. (Six enters the picture because each molecule of gas is described by three space variables and three momentum variables.) Thus, there are practical reasons for studying functions of more than one variable.

8.1 ALGEBRAIC STRUCTURE

For each $n \in \mathbf{N}$, let $\mathbf{R}^n$ denote the n-fold cartesian product of $\mathbf{R}$ with itself; that is,

$$\mathbf{R}^n := \{(x_1, x_2, \ldots, x_n) : x_j \in \mathbf{R} \text{ for } j = 1, 2, \ldots, n\}.$$

By a *Euclidean space* we shall mean $\mathbf{R}^n$ together with the "Euclidean inner product" defined in Definition 8.1 below. The integer n is called the *dimension* of $\mathbf{R}^n$, elements $\mathbf{x} = (x_1, x_2, \ldots, x_n)$ of $\mathbf{R}^n$ are called *points* or *vectors* or *ordered n-tuples*, and the numbers x_j are called *coordinates*, or *components*, of $\mathbf{x}$. Two vectors $\mathbf{x}, \mathbf{y}$ are said to be *equal* if and only if their components are equal; that is, if and only if $x_j = y_j$ for $j = 1, 2, \ldots, n$. The *zero vector* is the vector whose components are all zero; that is, $\mathbf{0} := (0, 0, \ldots, 0)$. When $n = 2$ (respectively, $n = 3$), we usually denote the components of $\mathbf{x}$ by x, y (respectively, by x, y, z).

You have already encountered the sets $\mathbf{R}^n$ for small n. $\mathbf{R}^1 = \mathbf{R}$ is the real line; we shall call its elements *scalars*. $\mathbf{R}^2$ is the xy-plane used to graph functions of the form $y = f(x)$. And $\mathbf{R}^3$ is the xyz-space used to graph functions of the form $z = f(x, y)$.

We have called elements of $\mathbf{R}^n$ points and vectors. In general, we make no distinction between points and vectors, but in each situation we adopt the interpretation which proves most useful.

In earlier courses, vectors were (most likely) directed line segments, but our vectors look like points in $\mathbf{R}^n$. What is going on? When we call an $\mathbf{a} \in \mathbf{R}^n$ a vector, we are thinking of the directed line segment which starts at the origin and ends at the point $\mathbf{a}$.

What about directed line segments which begin at arbitrary points? Two arbitrary directed line segments are said to be equivalent if and only if they have the same length and same direction. Thus every directed line segment V is equivalent to a directed line segment *in standard position*; that is, one which points in the same direction as V, has the same length as V, but whose "tail" sits at the origin and whose "head," $\mathbf{a}$, is a point in $\mathbf{R}^n$. If we identify V with $\mathbf{a}$, then we can represent any arbitrary directed line segment in $\mathbf{R}^n$ by a point in $\mathbf{R}^n$.

Identifying arbitrary vectors in $\mathbf{R}^n$ with vectors in standard position and, in turn, with points in $\mathbf{R}^n$ may sound confusing and sloppy, but it is no different from letting 1/2 represent 2/4, 3/6, 4/8, and so on. (In both cases, there is an underlying equivalence relation, and we are using one member of an equivalence class to represent all of its members. For vectors, we are using the representative which lies in standard position; for rational numbers, we are using the representative which is in reduced form.)

We began our study of functions of one variable by examining the algebraic structure of $\mathbf{R}$. In this section we begin our study of functions of several variables by examining the algebraic structure of $\mathbf{R}^n$. That structure is described in the following definition.

8.1 Definition.

Let $\mathbf{x} = (x_1, \ldots, x_n)$, $\mathbf{y} = (y_1, \ldots, y_n) \in \mathbf{R}^n$, and $\alpha \in \mathbf{R}$.
  i) The *sum* of the vectors $\mathbf{x}$ and $\mathbf{y}$ is the vector

$$\mathbf{x} + \mathbf{y} := (x_1 + y_1, x_2 + y_2, \ldots, x_n + y_n).$$

  ii) The *difference* of the vectors $\mathbf{x}$ and $\mathbf{y}$ is the vector

$$\mathbf{x} - \mathbf{y} := (x_1 - y_1, x_2 - y_2, \ldots, x_n - y_n).$$

  iii) The *product* of the scalar α and the vector $\mathbf{x}$ is the vector

$$\alpha\mathbf{x} := (\alpha x_1, \alpha x_2, \ldots, \alpha x_n).$$

  iv) The (*Euclidean*) *dot product* (or *scalar product* or *inner product*) of the vectors $\mathbf{x}$ and $\mathbf{y}$ is the scalar

$$\mathbf{x} \cdot \mathbf{y} := x_1 y_1 + x_2 y_2 + \cdots + x_n y_n.$$

These algebraic operations are analogues of addition, subtraction, and multiplication on $\mathbf{R}$. It is natural to ask, Do the usual laws of algebra hold in $\mathbf{R}^n$? An answer to this question is contained in the following result.

8.2 Theorem. *Let* $\mathbf{x}, \mathbf{y}, \mathbf{z} \in \mathbf{R}^n$ *and* $\alpha, \beta \in \mathbf{R}$. *Then*

$$\alpha \mathbf{0} = \mathbf{0}, \quad 0\mathbf{x} = \mathbf{0}, \quad \mathbf{0} \cdot \mathbf{x} = 0, \quad 1\mathbf{x} = \mathbf{x}, \quad \mathbf{0} + \mathbf{x} = \mathbf{x}, \quad \mathbf{x} - \mathbf{x} = \mathbf{0},$$

$$\alpha(\beta\mathbf{x}) = \beta(\alpha\mathbf{x}) = (\alpha\beta)\mathbf{x}, \qquad \alpha(\mathbf{x} \cdot \mathbf{y}) = (\alpha\mathbf{x}) \cdot \mathbf{y} = \mathbf{x} \cdot (\alpha\mathbf{y}),$$

$$\mathbf{x} + \mathbf{y} = \mathbf{y} + \mathbf{x}, \quad \mathbf{x} + (\mathbf{y} + \mathbf{z}) = (\mathbf{x} + \mathbf{y}) + \mathbf{z}, \quad \mathbf{x} \cdot \mathbf{y} = \mathbf{y} \cdot \mathbf{x},$$

$$\alpha(\mathbf{x} + \mathbf{y}) = \alpha\mathbf{x} + \alpha\mathbf{y}, \quad and \quad \mathbf{x} \cdot (\mathbf{y} + \mathbf{z}) = \mathbf{x} \cdot \mathbf{y} + \mathbf{x} \cdot \mathbf{z}.$$

Proof. These properties are direct consequences of Definition 8.1 and corresponding properties of real numbers. We will prove that vector addition is associative, and leave the proof of the rest of these properties as an exercise.

By definition and associativity of addition on $\mathbf{R}$ (see Postulate 1 in Section 1.2),

$$\begin{aligned}\mathbf{x} + (\mathbf{y} + \mathbf{z}) &= (x_1, \ldots, x_n) + (y_1 + z_1, \ldots, y_n + z_n) \\ &= (x_1 + (y_1 + z_1), \ldots, x_n + (y_n + z_n)) \\ &= ((x_1 + y_1) + z_1, \ldots, (x_n + y_n) + z_n) = (\mathbf{x} + \mathbf{y}) + \mathbf{z}. \quad \blacksquare\end{aligned}$$

Thus (with the exception of the closure of the dot product and the existence of the multiplicative identity and multiplicative inverses), $\mathbf{R}^n$ satisfies the same algebraic laws, listed in Postulate 1, that $\mathbf{R}$ does. This means one can use instincts developed in high school algebra to compute with these vector operations. For example, just as $(x - y)^2 = x^2 - 2xy + y^2$ holds for real numbers x and y, even so,

$$(\mathbf{x} - \mathbf{y}) \cdot (\mathbf{x} - \mathbf{y}) = \mathbf{x} \cdot \mathbf{x} - 2\mathbf{x} \cdot \mathbf{y} + \mathbf{y} \cdot \mathbf{y} \tag{1}$$

holds for any vectors $\mathbf{x}, \mathbf{y} \in \mathbf{R}^n$.

In the first four chapters, we used algebra together with the absolute value to define convergence of sequences and functions in $\mathbf{R}$. Is there an analogue of the absolute value for $\mathbf{R}^n$? The following definition illustrates the fact that there are many such analogues.

8.3 Definition.

Let $\mathbf{x} \in \mathbf{R}^n$.

i) The (*Euclidean*) *norm* (or *magnitude*) of $\mathbf{x}$ is the scalar

$$\|\mathbf{x}\| := \sqrt{\sum_{k=1}^{n} |x_k|^2}.$$

ii) The ℓ^1-*norm* (read L-one-norm) of $\mathbf{x}$ is the scalar

$$\|\mathbf{x}\|_1 := \sum_{k=1}^{n} |x_k|.$$

8.3 Definition. (Continued)

---

iii) The *sup-norm* of **x** is the scalar

$$\|\mathbf{x}\|_\infty := \max\{|x_1|, \dots, |x_n|\}.$$

iv) The (*Euclidean*) *distance* between two points $\mathbf{a}, \mathbf{b} \in \mathbf{R}^n$ is the scalar

$$\mathrm{dist}(\mathbf{a}, \mathbf{b}) := \|\mathbf{a} - \mathbf{b}\|.$$

(**Note**: For relationships between these three norms, see Remark 8.7 below. The subscript ∞ is frequently used for supremum norms because the supremum of a continuous function on an interval $[a, b]$ can be computed by taking the limit of $(\int_a^b |f(x)|^p dx)^{1/p}$ as $p \to \infty$—see Exercise 5.2.8.)

Since $\|x\| = \|x\|_1 = \|x\|_\infty = |x|$, when $n = 1$, each norm defined above is an extension of the absolute value from $\mathbf{R}$ to $\mathbf{R}^n$. The most important, and in some senses the most natural, of these norms is the Euclidean norm. This is true for at least two reasons. First, by definition,

$$\|\mathbf{x}\|^2 = \mathbf{x} \cdot \mathbf{x} \quad \text{for all} \quad \mathbf{x} \in \mathbf{R}^n.$$

(This aids in many calculations; see, for example, the proofs of Theorems 8.5 and 8.6 below.) Second, if Δ is the triangle in $\mathbf{R}^2$ with vertices $(0, 0)$, $\mathbf{x} := (a, b)$, and $(a, 0)$, then by the Pythagorean Theorem, the hypotenuse of Δ, $\sqrt{a^2 + b^2}$, is exactly the norm of **x**. In particular, the Euclidean norm of a vector has a simple geometric interpretation in $\mathbf{R}^2$.

The algebraic structure of $\mathbf{R}^n$ also has a simple geometric interpretation in $\mathbf{R}^2$ which gives us another very useful way to think about vectors. To describe it, fix vectors $\mathbf{a} = (a_1, a_2)$ and $\mathbf{b} = (b_1, b_2)$ and let $\mathcal{P}(\mathbf{a}, \mathbf{b})$ denote *parallelogram* associated with **a** and **b** (i.e., the parallelogram whose sides are given by **a** and **b**). (We are assuming that this parallelogram is not degenerate—see Figure 8.1.) Then the vector sum of **a** and **b**, $(a_1 + b_1, a_2 + b_2)$, is evidently the diagonal of $\mathcal{P}(\mathbf{a}, \mathbf{b})$; that is, $\mathbf{a} + \mathbf{b}$ is the vector which begins at the origin and ends at the opposite vertex of $\mathcal{P}(\mathbf{a}, \mathbf{b})$. Similarly, the difference $\mathbf{a} - \mathbf{b}$ can be identified with the other diagonal of $\mathcal{P}(\mathbf{a}, \mathbf{b})$ (see Figure 8.1). The scalar product of t and **a**, (ta_1, ta_2), evidently stretches or compresses the vector **a**, but leaves it in the same straight line which passes through **0** and **a**. Indeed, if $t > 0$, then $t\mathbf{a}$ has the same direction as **a**, but its magnitude, $|t| \|\mathbf{a}\|$, is $\geq$ or $<$ the magnitude of **a**, depending on whether $t \geq 1$ or $t < 1$. When t is negative, $t\mathbf{a}$ points in the opposite direction from **a** but is again stretched or compressed depending on the size of $|t|$.

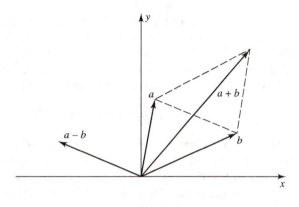

FIGURE 8.1

Using $\mathbf{R}^2$ as a guide, we can extend concepts from $\mathbf{R}^2$ to $\mathbf{R}^n$. Here are five examples.

1) Every $(a, b) \in \mathbf{R}^2$ can be written as $(a, b) = a(1, 0) + b(0, 1)$. Using this as a guide, we define the *usual basis* of $\mathbf{R}^n$ to be the collection $\{\mathbf{e}_1, \ldots, \mathbf{e}_n\}$, where $\mathbf{e}_j$ is the point in $\mathbf{R}^n$ whose jth coordinate is 1, and all other coordinates are 0. Notice by definition that each $\mathbf{x} = (x_1, \ldots, x_n) \in \mathbf{R}^n$ can be written as a linear combination of the $\mathbf{e}_j$'s:

$$\mathbf{x} = \sum_{j=1}^{n} x_j \mathbf{e}_j.$$

We shall not discuss other bases of $\mathbf{R}^n$ or the more general concept of "vector spaces," which can be introduced using postulates similar in spirit to Postulate 1 in Chapter 1. Instead, we have introduced just enough algebraic machinery in $\mathbf{R}^n$ to develop the calculus of multivariable functions. For more information about $\mathbf{R}^n$ and abstract vector spaces, see Noble and Daniel [9].

Note: In $\mathbf{R}^2$ or $\mathbf{R}^3$, $\mathbf{e}_1$ is denoted by $\mathbf{i}$, $\mathbf{e}_2$ is denoted by $\mathbf{j}$, and, in $\mathbf{R}^3$, $\mathbf{e}_3$ is denoted by $\mathbf{k}$. Thus, in $\mathbf{R}^3$, $\mathbf{i} := (1, 0, 0)$, $\mathbf{j} := (0, 1, 0)$, and $\mathbf{k} := (0, 0, 1)$.

2) Let $t \in \mathbf{R}$ and $\mathbf{a}, \mathbf{b} \in \mathbf{R}^2$ with $\mathbf{b}$ nonzero. By the geometric interpretation of vector addition, $\phi(t) := \mathbf{a} + t\mathbf{b}$ is a point on the line passing through $\mathbf{a}$ in the direction of $\mathbf{b}$. Using this as a guide, we define the *straight line* in $\mathbf{R}^n$ which passes through a point $\mathbf{a} \in \mathbf{R}^n$ in the *direction* $\mathbf{b} \in \mathbf{R}^n \setminus \{\mathbf{0}\}$ to be the set of points

$$\ell_{\mathbf{a}}(\mathbf{b}) := \{\mathbf{a} + t\mathbf{b} : t \in \mathbf{R}\}.$$

In particular, it is easy to see that the parallelogram $\mathcal{P}(\mathbf{a}, \mathbf{b})$ determined by nonzero vectors $\mathbf{a}$ and $\mathbf{b}$ in $\mathbf{R}^n$ can be described as

$$\mathcal{P}(\mathbf{a}; \mathbf{b}) := \{u\mathbf{a} + v\mathbf{b} : u, v \in [0, 1]\}.$$

3) Fix $\mathbf{a} \neq \mathbf{b}$ in $\mathbf{R}^2$, and set $\psi(t) := (1 - t)\mathbf{a} + t\mathbf{b}$, for $t \in \mathbf{R}$. Since $\psi(t) = \mathbf{a} + t(\mathbf{b} - \mathbf{a})$, it is evident that ψ describes the line $\ell_{\mathbf{a}}(\mathbf{b} - \mathbf{a})$. This line passes through the

points $\psi(0) = \mathbf{a}$ and $\psi(1) = \mathbf{b}$. In fact, by the geometric interpretation of vector subtraction, as t ranges from 0 to 1, the points $\psi(t)$ trace out the diagonal of $\mathcal{P}(\mathbf{a}, \mathbf{b})$ that does not contain the origin (see Figure 8.1). It begins at $\mathbf{a}$ and ends at $\mathbf{b}$. Using this as a guide, we define the *line segment* from $\mathbf{a} \in \mathbf{R}^n$ to $\mathbf{b} \in \mathbf{R}^n$ to be the set of points

$$L(\mathbf{a}; \mathbf{b}) := \{(1 - t)\mathbf{a} + t\mathbf{b} : t \in [0, 1]\}.$$

4) The angle between two nonzero vectors $\mathbf{a}, \mathbf{b} \in \mathbf{R}^2$ can be computed by the following process. If $\triangle$ is the triangle determined by the points $\mathbf{0}, \mathbf{a}$, and $\mathbf{b}$, then the sides of $\triangle$ have length $\|\mathbf{a}\|$, $\|\mathbf{b}\|$, and $\|\mathbf{a} - \mathbf{b}\|$. If we let θ be the angle between $\mathbf{a}$ and $\mathbf{b}$ [i.e., the angle in $\triangle$ at the vertex $(0, 0)$], then by the Law of Cosines (see Appendix B),

$$\|\mathbf{a} - \mathbf{b}\|^2 = \|\mathbf{a}\|^2 + \|\mathbf{b}\|^2 - 2\|\mathbf{a}\|\,\|\mathbf{b}\|\cos\theta.$$

Since Theorem 8.2 implies $\|\mathbf{a} - \mathbf{b}\|^2 = (\mathbf{a} - \mathbf{b}) \cdot (\mathbf{a} - \mathbf{b}) = \|\mathbf{a}\|^2 - 2\mathbf{a} \cdot \mathbf{b} + \|\mathbf{b}\|^2$, it follows that $-2\mathbf{a} \cdot \mathbf{b} = -2\|\mathbf{a}\|\,\|\mathbf{b}\|\cos\theta$. Since neither $\mathbf{a}$ nor $\mathbf{b}$ is zero, we conclude that

$$\cos\theta = \frac{\mathbf{a} \cdot \mathbf{b}}{\|\mathbf{a}\|\,\|\mathbf{b}\|}. \tag{2}$$

Using this as a guide, we define the *angle* between two nonzero vectors $\mathbf{a}, \mathbf{b} \in \mathbf{R}^n$ (for any $n \in \mathbf{N}$) to be the number $\theta \in [0, \pi]$ determined by (2). (Our next result, the Cauchy–Schwarz Inequality, shows that the right side of (2) always belongs to the interval $[-1, 1]$. Hence, for each pair of nonzero vectors $\mathbf{a}, \mathbf{b} \in \mathbf{R}^n$, there is a unique angle $\theta \in [0, \pi]$ which satisfies (2).)

5) Two vectors in $\mathbf{R}^2$ are parallel when one is a multiple of the other, and orthogonal when the angle, θ, between them is $\pi/2$; that is, when $\mathbf{a} \cdot \mathbf{b} = \cos\theta\|\mathbf{a}\|\,\|\mathbf{b}\| = 0$. Using this as a guide, we make the following definition in $\mathbf{R}^n$.

8.4 Definition.

Let $\mathbf{a}$ and $\mathbf{b}$ be nonzero vectors in $\mathbf{R}^n$.

i) $\mathbf{a}$ and $\mathbf{b}$ are said to be *parallel* if and only if there is a scalar $t \in \mathbf{R}$ such that $\mathbf{a} = t\mathbf{b}$.

ii) $\mathbf{a}$ and $\mathbf{b}$ are said to be *orthogonal* if and only if $\mathbf{a} \cdot \mathbf{b} = 0$.

Notice that the usual basis $\{\mathbf{e}_j\}$ consists of pairwise orthogonal vectors; that is, $\mathbf{e}_j \cdot \mathbf{e}_k = 0$ when $j \neq k$. In particular, the usual basis is an *orthogonal basis*.

We note in passing that Definition 8.4 is consistent with formula (2)—see Exercise 8.1.4b. Indeed, if θ is the angle between two nonzero vectors $\mathbf{a}$ and $\mathbf{b}$ in $\mathbf{R}^n$, then $\mathbf{a}$ and $\mathbf{b}$ are parallel if and only if $\theta = 0$ or $\theta = \pi$, and $\mathbf{a}$ and $\mathbf{b}$ are orthogonal if and only if $\theta = \pi/2$.

We shall see below that in addition to suggesting definitions for $\mathbf{R}^n$, the geometry of $\mathbf{R}^2$ can also be used to help suggest proof strategies in $\mathbf{R}^n$.

Let's return to the analogy between **R** and **R**^{n}. Surely, if we are going to develop a calculus of several variables, we need to know more about the Euclidean norm on **R**^{n}. The next two results answer the question, How many properties do the absolute value and the Euclidean norm share?

Although the norm is not multiplicative, the following fundamental inequality can be used as a replacement for the multiplicative property in most proofs. (Some authors call this the Cauchy–Schwarz–Bunyakovsky Inequality.)

8.5 Theorem.   [CAUCHY–SCHWARZ INEQUALITY].
If **x, y** $\in$ **R**^{n}, *then*

$$|\mathbf{x} \cdot \mathbf{y}| \leq \|\mathbf{x}\| \, \|\mathbf{y}\|.$$

STRATEGY: Using the fact that the dot product of a vector with itself is the square of the norm of the vector and the square of any real number is nonnegative, identity (1) becomes $0 \leq \|\mathbf{x} - \mathbf{y}\|^2 = \|\mathbf{x}\|^2 - 2\mathbf{x} \cdot \mathbf{y} + \|\mathbf{y}\|^2$. We could solve this inequality to get an estimate of the dot product of $\mathbf{x} \cdot \mathbf{y}$, but this estimate might be very crude if $\|\mathbf{x} - \mathbf{y}\|$ were much larger than zero. But $\mathbf{x} - \mathbf{y}$ is only one point on the line $\ell_\mathbf{x}(\mathbf{y})$. We might get a better estimate of the dot product $\mathbf{x} \cdot \mathbf{y}$ by using the inequality

$$0 \leq \|\mathbf{x} - t\mathbf{y}\|^2 = (\mathbf{x} - t\mathbf{y}) \cdot (\mathbf{x} - t\mathbf{y}) = \|\mathbf{x}\|^2 - 2t(\mathbf{x} \cdot \mathbf{y}) + t^2\|\mathbf{y}\|^2 \qquad (3)$$

for other values of t. In fact, if we draw a picture in **R**^{2} (see Figure 8.2), we see that the norm of $\|\mathbf{x} - t\mathbf{y}\|$ is smallest for the value of t which makes $\mathbf{x} - t\mathbf{y}$ orthogonal to **y**; that is, when

$$0 = (\mathbf{x} - t\mathbf{y}) \cdot \mathbf{y} = \mathbf{x} \cdot \mathbf{y} - t\mathbf{y} \cdot \mathbf{y} = \mathbf{x} \cdot \mathbf{y} - t\|\mathbf{y}\|^2.$$

This suggests using $t = \mathbf{x} \cdot \mathbf{y}/\|\mathbf{y}\|^2$ when $\mathbf{y} \neq \mathbf{0}$. It turns out that this value of t is exactly the one which reproduces the Cauchy–Schwarz Inequality. Here are the details.

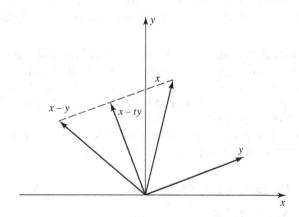

FIGURE 8.2

Proof. The Cauchy–Schwarz Inequality is trivial when $\mathbf{y} = \mathbf{0}$. If $\mathbf{y} \neq \mathbf{0}$, substitute $t = (\mathbf{x} \cdot \mathbf{y})/\|\mathbf{y}\|^2$ into (3) to obtain

$$0 \leq \|\mathbf{x}\|^2 - t(\mathbf{x} \cdot \mathbf{y}) = \|\mathbf{x}\|^2 - \frac{(\mathbf{x} \cdot \mathbf{y})^2}{\|\mathbf{y}\|^2}.$$

It follows that $0 \leq \|\mathbf{x}\|^2 - (\mathbf{x} \cdot \mathbf{y})^2/\|\mathbf{y}\|^2$. Solving this inequality for $(\mathbf{x} \cdot \mathbf{y})^2$, we conclude that

$$(\mathbf{x} \cdot \mathbf{y})^2 \leq \|\mathbf{x}\|^2 \|\mathbf{y}\|^2. \qquad \blacksquare$$

The analogy between the absolute value and the Euclidean norm is further reinforced by the following result (compare with Theorem 1.7). (See also Exercise 8.1.10.)

8.6 Theorem. *Let* $\mathbf{x}, \mathbf{y} \in \mathbf{R}^n$. *Then*

i) $\|\mathbf{x}\| \geq 0$ *with equality only when* $\mathbf{x} = \mathbf{0}$,
ii) $\|\alpha\mathbf{x}\| = |\alpha| \|\mathbf{x}\|$ *for all scalars* α,
iii) [TRIANGLE INEQUALITIES]. $\|\mathbf{x} + \mathbf{y}\| \leq \|\mathbf{x}\| + \|\mathbf{y}\|$ *and* $\|\mathbf{x} - \mathbf{y}\| \geq \|\mathbf{x}\| - \|\mathbf{y}\|$.

Proof. Statements i) and ii) are easy to verify.
To prove iii), observe that by Definition 8.3, Theorem 8.2, and the Cauchy–Schwarz Inequality,

$$\|\mathbf{x} + \mathbf{y}\|^2 = (\mathbf{x} + \mathbf{y}) \cdot (\mathbf{x} + \mathbf{y}) = \mathbf{x} \cdot \mathbf{x} + 2\mathbf{x} \cdot \mathbf{y} + \mathbf{y} \cdot \mathbf{y}$$
$$= \|\mathbf{x}\|^2 + 2\mathbf{x} \cdot \mathbf{y} + \|\mathbf{y}\|^2 \leq \|\mathbf{x}\|^2 + 2\|\mathbf{x}\| \|\mathbf{y}\| + \|\mathbf{y}\|^2 = (\|\mathbf{x}\| + \|\mathbf{y}\|)^2.$$

This establishes the first inequality in iii). By modifying the proof of Theorem 1.7, we can also establish the second inequality in iii). $\qquad \blacksquare$

Notice that the Triangle Inequality has a simple geometric interpretation. Indeed, since $\|\mathbf{x}\|$ is the magnitude of the vector $\mathbf{x}$, the inequality $\|\mathbf{x} + \mathbf{y}\| \leq \|\mathbf{x}\| + \|\mathbf{y}\|$ states that the length of one side of a triangle (namely, the triangle whose vertices are $\mathbf{0}, \mathbf{x}$, and $\mathbf{x} + \mathbf{y}$) is less than or equal to the sum of the lengths of its other two sides.

For some estimates, it is convenient to relate the Euclidean norm to the ℓ^1-norm and the sup-norm.

8.7 Remark. *Let* $\mathbf{x} \in \mathbf{R}^n$. *Then*

i) $\|\mathbf{x}\|_\infty \leq \|\mathbf{x}\| \leq \sqrt{n}\|\mathbf{x}\|_\infty$, *and*
ii) $\|\mathbf{x}\| \leq \|\mathbf{x}\|_1 \leq \sqrt{n}\|\mathbf{x}\|$.

Proof. i) Let $1 \leq j \leq n$. By definition,

$$|x_j|^2 \leq \|\mathbf{x}\|^2 = x_1^2 + \cdots + x_n^2 \leq n \left(\max_{1 \leq \ell \leq n} |x_\ell| \right)^2 = n \|\mathbf{x}\|_\infty^2;$$

that is, $|x_j| \leq \|\mathbf{x}\|$ and $\|\mathbf{x}\| \leq \sqrt{n}\|\mathbf{x}\|_\infty$. Taking the supremum of the first of these inequalities, over all $1 \leq j \leq n$, we also have $\|\mathbf{x}\|_\infty \leq \|\mathbf{x}\|$.

ii) Let $A = \{(i, j) : 1 \leq i, j \leq n$ and $i < j\}$. To verify the first inequality, observe by algebra that

$$\|\mathbf{x}\|_1^2 = \left(\sum_{i=1}^{n} |x_i| \right)^2 = \sum_{i=1}^{n} |x_i|^2 + 2 \sum_{(i,j)\in A} |x_i|\,|x_j| = \|\mathbf{x}\|^2 + 2 \sum_{(i,j)\in A} |x_i|\,|x_j|.$$

Since $\sum_{(i,j)\in A} |x_i|\,|x_j| \geq 0$, it follows that $\|\mathbf{x}\|^2 \leq \|\mathbf{x}\|_1^2$.

On the other hand,

$$0 \leq \sum_{(i,j)\in A} (|x_i| - |x_j|)^2 = \sum_{i=1}^{n} (n-1)|x_i|^2 - 2 \sum_{(i,j)\in A} |x_i|\,|x_j|$$

$$= n\|\mathbf{x}\|^2 - \left(\sum_{i=1}^{n} |x_i|^2 + 2 \sum_{(i,j)\in A} |x_i|\,|x_j| \right) = n\|\mathbf{x}\|^2 - \|\mathbf{x}\|_1^2.$$

This proves the second inequality. ∎

Since $\mathbf{x}\cdot\mathbf{y}$ is a scalar, the dot product in $\mathbf{R}^n$ does not satisfy the closure property for any $n > 1$. Here is another product, defined only on $\mathbf{R}^3$, which does satisfy the closure property. (As we shall see below, this product allows us to exploit the geometry of $\mathbf{R}^3$ in several unique ways.)

8.8 Definition.

The *cross product* of two vectors $\mathbf{x} = (x_1, x_2, x_3)$ and $\mathbf{y} = (y_1, y_2, y_3)$ in $\mathbf{R}^3$ is the vector defined by

$$\mathbf{x} \times \mathbf{y} := (x_2 y_3 - x_3 y_2, \, x_3 y_1 - x_1 y_3, \, x_1 y_2 - x_2 y_1).$$

Using the usual basis $\mathbf{i} = \mathbf{e}_1$, $\mathbf{j} = \mathbf{e}_2$, $\mathbf{k} = \mathbf{e}_3$, and the determinant operator (see Appendix C), we can give the cross product a more easily remembered form:

$$\mathbf{x} \times \mathbf{y} = \det \begin{bmatrix} \mathbf{i} & \mathbf{j} & \mathbf{k} \\ x_1 & x_2 & x_3 \\ y_1 & y_2 & y_3 \end{bmatrix}.$$

The following result shows that the cross product satisfies some, but not all, of the usual laws of algebra. (Specifically, notice that although the cross product satisfies the distributive property, it satisfies neither the commutative property nor the associative property.)

8.9 Theorem. *Let $\mathbf{x}, \mathbf{y}, \mathbf{z} \in \mathbf{R}^3$ be vectors and α be a scalar. Then*

i)
$$\mathbf{x} \times \mathbf{x} = \mathbf{0}, \qquad \mathbf{x} \times \mathbf{y} = -\mathbf{y} \times \mathbf{x},$$

ii)
$$(\alpha\mathbf{x}) \times \mathbf{y} = \alpha(\mathbf{x} \times \mathbf{y}) = \mathbf{x} \times (\alpha\mathbf{y}),$$

iii)
$$\mathbf{x} \times (\mathbf{y} + \mathbf{z}) = (\mathbf{x} \times \mathbf{y}) + (\mathbf{x} \times \mathbf{z}),$$

iv)
$$(\mathbf{x} \times \mathbf{y}) \cdot \mathbf{z} = \mathbf{x} \cdot (\mathbf{y} \times \mathbf{z}) = \det \begin{bmatrix} x_1 & x_2 & x_3 \\ y_1 & y_2 & y_3 \\ z_1 & z_2 & z_3 \end{bmatrix},$$

v)
$$\mathbf{x} \times (\mathbf{y} \times \mathbf{z}) = (\mathbf{x} \cdot \mathbf{z})\mathbf{y} - (\mathbf{x} \cdot \mathbf{y})\mathbf{z},$$

and

vi)
$$\|\mathbf{x} \times \mathbf{y}\|^2 = (\mathbf{x} \cdot \mathbf{x})(\mathbf{y} \cdot \mathbf{y}) - (\mathbf{x} \cdot \mathbf{y})^2.$$

vii)   *Moreover, if $\mathbf{x} \times \mathbf{y} \neq \mathbf{0}$, then the vector $\mathbf{x} \times \mathbf{y}$ is orthogonal to $\mathbf{x}$ and $\mathbf{y}$.*

Proof. These properties follow immediately from the definitions. We will prove properties iv), v), and vii) and leave the rest as an exercise.
    iv) Notice that by definition,

$$(\mathbf{x} \times \mathbf{y}) \cdot \mathbf{z} = (x_2 y_3 - x_3 y_2)z_1 + (x_3 y_1 - x_1 y_3)z_2 + (x_1 y_2 - x_2 y_1)z_3$$
$$= x_1(y_2 z_3 - y_3 z_2) + x_2(y_3 z_1 - y_1 z_3) + x_3(y_1 z_2 - y_2 z_1).$$

Since this last expression is both the scalar $\mathbf{x} \cdot (\mathbf{y} \times \mathbf{z})$ and the value of the determinant on the right side of iv) (expanded along the first row), this verifies iv).
    v) Since $\mathbf{x} \times (\mathbf{y} \times \mathbf{z}) = (x_1, x_2, x_3) \times (y_2 z_3 - y_3 z_2, y_3 z_1 - y_1 z_3, y_1 z_2 - y_2 z_1)$, the first component of $\mathbf{x} \times (\mathbf{y} \times \mathbf{z})$ is

$$x_2 y_1 z_2 - x_2 y_2 z_1 - x_3 y_3 z_1 + x_3 y_1 z_3 = (x_1 z_1 + x_2 z_2 + x_3 z_3)y_1 - (x_1 y_1 + x_2 y_2 + x_3 y_3)z_1.$$

This proves that the first components of $\mathbf{x} \times (\mathbf{y} \times \mathbf{z})$ and $(\mathbf{x} \cdot \mathbf{z})\mathbf{y} - (\mathbf{x} \cdot \mathbf{y})\mathbf{z}$ are equal. A similar argument shows that the second and third components are also equal.
    vii) By parts i) and iv), $(\mathbf{x} \times \mathbf{y}) \cdot \mathbf{x} = -(\mathbf{y} \times \mathbf{x}) \cdot \mathbf{x} = -\mathbf{y} \cdot (\mathbf{x} \times \mathbf{x}) = -\mathbf{y} \cdot \mathbf{0} = 0$. Thus $\mathbf{x} \times \mathbf{y}$ is orthogonal to $\mathbf{x}$. A similar calculation shows that $\mathbf{x} \times \mathbf{y}$ is orthogonal to $\mathbf{y}$. ∎

Part vii) is illustrated in Figure 8.3. Notice that $\mathbf{x} \times \mathbf{y}$ satisfies the "right-hand" rule. Indeed, if one puts the fingers of the right hand along $\mathbf{x}$ and the palm of the right hand along $\mathbf{y}$, then the thumb points in the direction of $\mathbf{x} \times \mathbf{y}$.

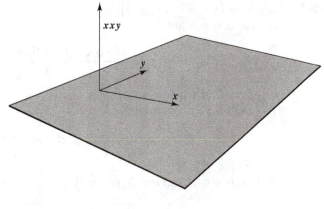

FIGURE 8.3

By (2), there is a close connection between dot products and cosines. The following result shows that there is a similar connection between cross products and sines.

8.10 Remark. *Let* $\mathbf{x}, \mathbf{y}$ *be nonzero vectors in* $\mathbf{R}^3$ *and* θ *be the angle between* $\mathbf{x}$ *and* $\mathbf{y}$. *Then*

$$\|\mathbf{x} \times \mathbf{y}\| = \|\mathbf{x}\| \, \|\mathbf{y}\| \, \sin\theta.$$

Proof. By Theorem 8.9vi and (2),

$$\|\mathbf{x} \times \mathbf{y}\|^2 = (\|\mathbf{x}\| \, \|\mathbf{y}\|)^2 - (\|\mathbf{x}\| \, \|\mathbf{y}\| \cos\theta)^2$$
$$= (\|\mathbf{x}\| \, \|\mathbf{y}\|)^2 (1 - \cos^2\theta) = (\|\mathbf{x}\| \, \|\mathbf{y}\|)^2 \sin^2\theta. \qquad \blacksquare$$

This observation can be used to establish a connection between cross products and area or volume (see Exercise 8.2.7).

EXERCISES

8.1.1. Let $\mathbf{x}, \mathbf{y}, \mathbf{z} \in \mathbf{R}^n$.

    a) If $\|\mathbf{x} - \mathbf{z}\| < 2$ and $\|\mathbf{y} - \mathbf{z}\| < 3$, prove that $\|\mathbf{x} - \mathbf{y}\| < 5$.

    b) If $\|\mathbf{x}\| < 2$, $\|\mathbf{y}\| < 3$, and $\|\mathbf{z}\| < 4$, prove that $|\mathbf{x} \cdot \mathbf{y} - \mathbf{x} \cdot \mathbf{z}| < 14$.

    c) If $\|\mathbf{x} - \mathbf{y}\| < 2$ and $\|\mathbf{z}\| < 3$, prove that $|\mathbf{x} \cdot (\mathbf{y} - \mathbf{z}) - \mathbf{y} \cdot (\mathbf{x} - \mathbf{z})| < 6$.

    d) If $\|2\mathbf{x} - \mathbf{y}\| < 2$ and $\|\mathbf{y}\| < 1$, prove that $\big| \, \|\mathbf{x} - \mathbf{y}\|^2 - \mathbf{x} \cdot \mathbf{x} \, \big| < 2$.

    e) If $n = 3$, $\|\mathbf{x} - \mathbf{y}\| < 2$, and $\|\mathbf{z}\| < 3$, prove that $\|\mathbf{x} \times \mathbf{z} - \mathbf{y} \times \mathbf{z}\| < 6$.

    f) If $n = 3$, $\|\mathbf{x}\| < 1$, $\|\mathbf{y}\| < 2$, and $\|\mathbf{z}\| < 3$, prove that $\|\mathbf{x} \cdot (\mathbf{y} \times \mathbf{z})\| < 6$.

8.1.2. Let $B := \{\mathbf{x} \in \mathbf{R}^n : \|\mathbf{x}\| \leq 1\}$.

a) If $\mathbf{a}, \mathbf{b}, \mathbf{c} \in B$ and

$$\mathbf{v} := \frac{(\mathbf{a} \cdot \mathbf{b})\mathbf{c} + (\mathbf{a} \cdot \mathbf{c})\mathbf{b} + (\mathbf{c} \cdot \mathbf{b})\mathbf{a}}{3},$$

prove that $\mathbf{v}$ belongs to B.

b) If $\mathbf{a}, \mathbf{b} \in B$, prove that

$$|\mathbf{a} \cdot \mathbf{c} - \mathbf{b} \cdot \mathbf{d}| \leq \|\mathbf{b} - \mathbf{c}\| + \|\mathbf{a} - \mathbf{d}\|$$

for all $\mathbf{c}, \mathbf{d} \in \mathbf{R}^n$.

c) If $\mathbf{a}, \mathbf{b}, \mathbf{c} \in B$ and $n = 3$, prove that

$$\sqrt{|\mathbf{a} \cdot (\mathbf{b} \times \mathbf{c})|^2 + |\mathbf{a} \cdot \mathbf{b}|^2} \leq 1.$$

8.1.3. Use the proof of Theorem 8.5 to show that equality in the Cauchy–Schwarz Inequality holds if and only if $\mathbf{x} = \mathbf{0}$, $\mathbf{y} = \mathbf{0}$, or $\mathbf{x}$ is parallel to $\mathbf{y}$.

8.1.4. Let $\mathbf{a}$ and $\mathbf{b}$ be nonzero vectors in $\mathbf{R}^n$.

a) If $\phi(t) = \mathbf{a} + t\mathbf{b}$ for $t \in \mathbf{R}$, show that for each $t_0, t_1, t_2 \in \mathbf{R}$ with $t_1, t_2 \neq t_0$, the angle between $\phi(t_1) - \phi(t_0)$ and $\phi(t_2) - \phi(t_0)$ is 0 or π.

b) If θ is the angle between $\mathbf{a}$ and $\mathbf{b}$, show that $\mathbf{a}$ and $\mathbf{b}$ are parallel according to Definition 8.4 if and only if $\theta = 0$ or π, and that $\mathbf{a}$ and $\mathbf{b}$ are orthogonal according to Definition 8.4 if and only if $\theta = \pi/2$.

8.1.5. The midpoint of a side of a triangle in $\mathbf{R}^3$ is the point that bisects that side (i.e., that divides it into two equal pieces). Let $\triangle$ be a triangle in $\mathbf{R}^3$ with sides A, B, and C and let L denote the line segment between the midpoints of A and B. Prove that L is parallel to C and that the length of L is one-half the length of C.

8.1.6. a) Prove that $(1, 2, 3)$, $(4, 5, 6)$, and $(0, 4, 2)$ are vertices of a right triangle in $\mathbf{R}^3$.

b) Find all nonzero vectors orthogonal to $(1, -1, 0)$ which lie in the plane $z = x$.

c) Find all nonzero vectors orthogonal to the vector $(3, 2, -5)$ whose components sum to 4.

8.1.7. Let $a < b$ be real numbers. The Cartesian product $[a, b] \times [a, b]$ is obviously a square in $\mathbf{R}^2$. Define a cube Q in $\mathbf{R}^n$ to be the n-fold Cartesian product of $[a, b]$ with itself; that is, $Q := [a, b] \times \cdots \times [a, b]$. Find a formula of the angle between the longest diagonal of Q and any of its edges. Show that when $n = 3$, this angle is approximately 54.74 degrees.

8.1.8. a) Using Postulate 1 in Section 1.2 and Definition 8.1, prove Theorem 8.2.

b) Prove Theorem 8.9, parts i) through iii) and vi).

c) Prove that if $\mathbf{x}, \mathbf{y} \in \mathbf{R}^3$, then $\|\mathbf{x} \times \mathbf{y}\| \leq \|\mathbf{x}\| \|\mathbf{y}\|$.

8.1.9. Suppose that $\{a_k\}$ and $\{b_k\}$ are sequences of real numbers which satisfy

$$\sum_{k=1}^{\infty} a_k^2 < \infty \quad \text{and} \quad \sum_{k=1}^{\infty} b_k^2 < \infty.$$

Prove that the infinite series $\sum_{k=1}^{\infty} a_k b_k$ converges absolutely.

8.1.10. Prove that the ℓ^1-norm and the sup-norm also satisfy Theorem 8.6.

8.2   PLANES AND LINEAR TRANSFORMATIONS

A plane Π in $\mathbf{R}^3$ is a set of points that is "flat" in some sense. What do we mean by flat? Any vector that lies in Π is orthogonal to a common direction, called the *normal*, which we will denote by $\mathbf{b}$. Fix a point $\mathbf{a} \in \Pi$. Since the vector $\mathbf{x} - \mathbf{a}$ lies in Π for all $\mathbf{x} \in \Pi$ and since two vectors are orthogonal when their dot product is zero, we see that $(\mathbf{x} - \mathbf{a}) \cdot \mathbf{b} = 0$ for all $\mathbf{x} \in \Pi$ (see Figure 8.4).

Using this three-dimensional case as a guide, for any $\mathbf{a}, \mathbf{b} \in \mathbf{R}^n$ with $\mathbf{b} \neq \mathbf{0}$, we call the set

$$\Pi_{\mathbf{b}}(\mathbf{a}) := \{\mathbf{x} \in \mathbf{R}^n : (\mathbf{x} - \mathbf{a}) \cdot \mathbf{b} = 0\}$$

the *hyperplane* in $\mathbf{R}^n$ passing through a point $\mathbf{a} \in \mathbf{R}^n$ with normal $\mathbf{b}$. (We call it a *plane* when $n = 3$.) In particular, $\Pi_{\mathbf{b}}(\mathbf{a})$ is the set of all points $\mathbf{x}$ such that $\mathbf{x} - \mathbf{a}$ is orthogonal to $\mathbf{b}$.

There is nothing unique about "the normal" of a hyperplane: Any nonzero vector $\mathbf{c}$ parallel to $\mathbf{b}$ will define the same hyperplane. Indeed, if $\mathbf{b}$ and $\mathbf{c}$ are parallel, then, by definition, $\mathbf{b} = t\mathbf{c}$ for some nonzero $t \in \mathbf{R}$; hence $(\mathbf{x} - \mathbf{a}) \cdot \mathbf{b} = 0$ if and only if $(\mathbf{x} - \mathbf{a}) \cdot \mathbf{c} = 0$. Nevertheless, many properties of hyperplane can be

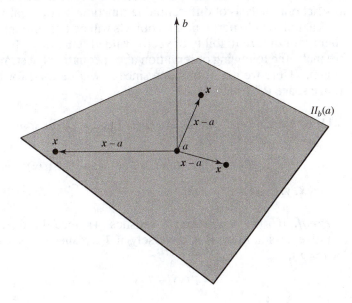

FIGURE 8.4

determined by their normals. For example, the *angle between two hyperplanes* with respective normals **b** and **c** is defined to be the angle between the normals **b** and **c**.

By an *equation* of a hyperplane Π we mean an expression of the form $F(\mathbf{x}) = 0$, where $F : \mathbf{R}^n \to \mathbf{R}$ is a function determined by the following property: A point **x** belongs to Π if and only if $F(\mathbf{x}) = 0$. By definition, then, an equation of the hyperplane $\Pi_{\mathbf{b}}(\mathbf{a})$ [i.e., the hyperplane passing through the point $\mathbf{a} = (a_1, \dots, a_n)$ with normal $\mathbf{b} = (b_1, \dots, b_n)$] is given by

$$\mathbf{b} \cdot \mathbf{x} = \mathbf{b} \cdot \mathbf{a}.$$

This form is sometimes referred to as the *point-normal form*. It can also be written in the form

$$b_1 x_1 + b_2 x_2 + \cdots + b_n x_n = d,$$

where $d = b_1 a_1 + b_2 a_2 + \cdots + b_n a_n$ is a constant determined by **a** and **b** (and related to the distance from $\Pi_{\mathbf{b}}(\mathbf{a})$ to the origin—see Exercise 8.2.8). In particular, planes in $\mathbf{R}^3$ have equations of the form

$$ax + by + cz = d.$$

Notice that a "hyperplane" in $\mathbf{R}^2$ is by definition a straight line. Just as straight lines through the origin played a prominent role in characterizing differentiability of functions of one variable (see Theorem 4.3), even so hyperplane-like objects will play a crucial role in defining differentiability of functions of several variables. Why hyperplane-like objects and not just hyperplanes themselves? Equations of hyperplanes are by definition real valued and we do not want to restrict our analysis of differentiable functions to the real-valued case.

What kind of hyperplane-like objects will be rich enough to develop a general theory for differentiability of vector-valued functions? To answer this question, we make the following observation about equations of straight lines through the origin. (Here we use s for slope since m will be used for the dimension of the range space $\mathbf{R}^m$.)

8.11 Remark. *Let $T : \mathbf{R} \to \mathbf{R}$. Then $T(x) = sx$ for some $s \in \mathbf{R}$ if and only if T satisfies*

$$T(x + y) = T(x) + T(y) \quad \text{and} \quad T(\alpha x) = \alpha T(x) \qquad (4)$$

for all $x, y, \alpha \in \mathbf{R}$.

Proof. If $T(x) = sx$, then T satisfies (4) since the distributive and commutative laws hold on $\mathbf{R}$. Conversely, if T satisfies (4), set $s := T(1)$. Then (let $\alpha = x$),

$$T(x) = T(x \cdot 1) = xT(1) = sx$$

for all $x \in \mathbf{R}$. ∎

Accordingly, we introduce the following concept.

8.12 Definition.

A function $\mathbf{T} : \mathbf{R}^n \to \mathbf{R}^m$ is said to be *linear* [notation: $\mathbf{T} \in \mathcal{L}(\mathbf{R}^n; \mathbf{R}^m)$] if and only if it satisfies

$$\mathbf{T}(\mathbf{x} + \mathbf{y}) = \mathbf{T}(\mathbf{x}) + \mathbf{T}(\mathbf{y}) \quad \text{and} \quad \mathbf{T}(\alpha\mathbf{x}) = \alpha\mathbf{T}(\mathbf{x})$$

for all $\mathbf{x}, \mathbf{y} \in \mathbf{R}^n$ and all scalars α.

When $m = 1$ (i.e., when the range of $\mathbf{T}$ is $\mathbf{R}$), we shall often drop the boldface notation (i.e., write T for $\mathbf{T}$).

Notice once and for all that if $\mathbf{T}$ is a linear function, then

$$\mathbf{T}(\mathbf{0}) = \mathbf{0}. \tag{5}$$

Indeed, by definition, $\mathbf{T}(\mathbf{0}) = \mathbf{T}(\mathbf{0}+\mathbf{0}) = \mathbf{T}(\mathbf{0})+\mathbf{T}(\mathbf{0})$. Hence (5) can be obtained by subtracting $\mathbf{T}(\mathbf{0})$ from both sides of this last equation. Also notice that if $F(\mathbf{x}) = 0$ is the equation of a hyperplane passing through the origin, then $F(\mathbf{x}) = a_1 x_1 + \cdots + a_n x_n$. In particular, $F \in \mathcal{L}(\mathbf{R}^n; \mathbf{R})$.

Functions in $\mathcal{L}(\mathbf{R}^n; \mathbf{R}^m)$ are sometimes called *linear transformations* or *linear operators* because of the fundamental role they play in the theory of change of variables in $\mathbf{R}^n$. We shall take up this connection in Chapter 12.

According to Remark 8.11, linear transformations of one variable [i.e., objects $T \in \mathcal{L}(\mathbf{R}; \mathbf{R})$] can be identified with $\mathbf{R}$ by representing T by its slope s. Is there an analogue of slope which can be used to represent linear transformations of several variables? To answer this question, we use the following half page to review some elementary linear algebra.

Recall that an $m \times n$ *matrix* B is a rectangular array which has m rows and n columns:

$$B = [b_{ij}]_{m \times n} := \begin{bmatrix} b_{11} & b_{12} & \cdots & b_{1n} \\ b_{21} & b_{22} & \cdots & b_{2n} \\ \vdots & \vdots & \ddots & \vdots \\ b_{m1} & b_{m2} & \cdots & b_{mn} \end{bmatrix}.$$

For us, the *entries* b_{ij} of a matrix B will usually be numbers or real-valued functions. Let $B = [b_{ij}]_{m \times n}$ and $C = [c_{vk}]_{p \times q}$ be such matrices. Recall that the *product* of B and a scalar α is defined by

$$\alpha B = [\alpha b_{ij}]_{m \times n},$$

the *sum* of B and C is defined (when $m = p$ and $n = q$) by

$$B + C = [b_{ij} + c_{ij}]_{m \times n},$$

and the *product* of B and C is defined (when $n = p$) by

$$BC = \left[\sum_{\nu=1}^{n} b_{i\nu} c_{\nu j} \right]_{m \times q}.$$

Also recall that most of the usual laws of algebra hold for addition and multiplication of matrices (see Theorem C.1 in Appendix C). One glaring exception is that matrix multiplication is not commutative.

We shall identify points $\mathbf{x} = (x_1, x_2, \ldots, x_n) \in \mathbf{R}^n$ with $1 \times n$ row matrices or $n \times 1$ column matrices by setting

$$[\mathbf{x}] = [x_1 \quad x_2 \quad \ldots \quad x_n] \quad \text{or} \quad [\mathbf{x}] = [x_1 \quad x_2 \quad \ldots \quad x_n]^T := \begin{bmatrix} x_1 \\ x_2 \\ \vdots \\ x_n \end{bmatrix},$$

where B^T represents the transpose of a matrix B (see Appendix C). Abusing the notation slightly, we shall usually represent the product of an $m \times n$ matrix B and an $n \times 1$ column matrix $[\mathbf{x}]$ by $B\mathbf{x}$. This notation is justified, as the following result shows, since the function $\mathbf{x} \longmapsto [\mathbf{x}]$ takes vector addition to matrix addition, the dot product to matrix multiplication, and scalar multiplication to scalar multiplication.

8.13 Remark. *If* $\mathbf{x}, \mathbf{y} \in \mathbf{R}^n$ *and* α *is a scalar, then*

$$[\mathbf{x} + \mathbf{y}] = [\mathbf{x}] + [\mathbf{y}], \qquad [\mathbf{x} \cdot \mathbf{y}] = [\mathbf{x}][\mathbf{y}]^T, \quad \text{and} \quad [\alpha \mathbf{x}] = \alpha[\mathbf{x}].$$

Proof. These laws follow immediately from the definitions of addition and multiplication of matrices and vectors. For example,

$$[\mathbf{x} + \mathbf{y}] = [x_1 + y_1 \quad x_2 + y_2 \quad \ldots \quad x_n + y_n]$$
$$= [x_1 \quad x_2 \quad \ldots \quad x_n] + [y_1 \quad y_2 \quad \ldots \quad y_n] = [\mathbf{x}] + [\mathbf{y}]. \qquad \blacksquare$$

The following result shows that each $m \times n$ matrix gives rise to a linear function from $\mathbf{R}^n$ to $\mathbf{R}^m$.

8.14 Remark. *Let* $B = [b_{ij}]$ *be an* $m \times n$ *matrix whose entries are real numbers and let* $\mathbf{e}_1, \ldots, \mathbf{e}_n$ *represent the usual basis of* $\mathbf{R}^n$. *If*

$$\mathbf{T}(\mathbf{x}) = B\mathbf{x}, \qquad \mathbf{x} \in \mathbf{R}^n, \tag{6}$$

then **T** *is a linear function from* $\mathbf{R}^n$ *to* $\mathbf{R}^m$ *and the jth column of B can be obtained by evaluating* **T** *at* $\mathbf{e}_j$:

$$(b_{1j}, b_{2j}, \ldots, b_{mj}) = T(\mathbf{e}_j), \qquad j = 1, 2, \ldots, n. \tag{7}$$

Proof. Notice, first, that (7) holds by (6) and the definition of matrix multiplication. Next, observe by Remark 8.13 and the distributive law of matrix multiplication (see Theorem C.1) that

$$\mathbf{T}(\mathbf{x} + \mathbf{y}) = B[\mathbf{x} + \mathbf{y}] = B([\mathbf{x}] + [\mathbf{y}]) = B[\mathbf{x}] + B[\mathbf{y}] = \mathbf{T}(\mathbf{x}) + \mathbf{T}(\mathbf{y})$$

for all $\mathbf{x}, \mathbf{y} \in \mathbf{R}^n$. Similarly, $\mathbf{T}(\alpha \mathbf{x}) = B[\alpha \mathbf{x}] = B(\alpha [\mathbf{x}]) = \alpha B[\mathbf{x}] = \alpha \mathbf{T}(\mathbf{x})$ for all $\mathbf{x} \in \mathbf{R}^n$ and $\alpha \in \mathbf{R}$. Thus $T \in \mathcal{L}(\mathbf{R}^n; \mathbf{R}^m)$. ∎

Remark 8.14 would barely be worth mentioning were it not the case that ALL linear functions from $\mathbf{R}^n$ to $\mathbf{R}^m$ have this form. Here, then, is the multidimensional analogue of Remark 8.11.

8.15 Theorem. *For each* $\mathbf{T} \in \mathcal{L}(\mathbf{R}^n; \mathbf{R}^m)$ *there is a matrix* $B = [b_{ij}]_{m \times n}$ *such that (6) holds. Moreover, the matrix B is unique. Specifically, for each fixed* **T** *there is only one B which satisfies (6), and the columns of B are defined by (7).*

Proof. Uniqueness has been established in Remark 8.14. To prove existence, suppose that $\mathbf{T} \in \mathcal{L}(\mathbf{R}^n; \mathbf{R}^m)$. Define B by (7). Then

$$\mathbf{T}(\mathbf{x}) = \mathbf{T}\left(\sum_{j=1}^{n} x_j \mathbf{e}_j\right)$$

$$= \sum_{j=1}^{n} x_j \mathbf{T}(\mathbf{e}_j) = \sum_{j=1}^{n} x_j (b_{1j}, b_{2j}, \ldots, b_{mj})$$

$$= \left(\sum_{j=1}^{n} x_j b_{1j}, \sum_{j=1}^{n} x_j b_{2j}, \ldots, \sum_{j=1}^{n} x_j b_{mj}\right) = B\mathbf{x}. \qquad ∎$$

The unique matrix B which satisfies (6) is called the *matrix which represents* **T**.

In Chapter 11 we shall use this point of view to define what it means for a function from $\mathbf{R}^n$ into $\mathbf{R}^m$ to be differentiable. At that point, we shall show that many of the one-dimensional results about differentiation remain valid in the multidimensional setting. Since the one-dimensional theory relied on estimates using the absolute value of various functions, we expect the theory in $\mathbf{R}^n$ to rely on estimates using the norms of various functions. Since some of those functions will be linear, the following concept will be useful in this regard.

8.16 Definition.

Let $\mathbf{T} \in \mathcal{L}(\mathbf{R}^n; \mathbf{R}^m)$. The *operator norm* of $\mathbf{T}$ is the extended real number

$$\|\mathbf{T}\| := \sup_{\|\mathbf{x}\| \neq 0} \frac{\|\mathbf{T}(\mathbf{x})\|}{\|\mathbf{x}\|}.$$

One interesting corollary of Theorem 8.15 is that the operator norm of a linear function is always finite.

8.17 Theorem. *Let $\mathbf{T} \in \mathcal{L}(\mathbf{R}^n; \mathbf{R}^m)$. Then the operator norm of $\mathbf{T}$ is finite and satisfies*

$$\|\mathbf{T}(\mathbf{x})\| \leq \|\mathbf{T}\| \, \|\mathbf{x}\| \tag{8}$$

for all $\mathbf{x} \in \mathbf{R}^n$.

Proof. Since $\mathbf{T}(\mathbf{0}) = \mathbf{0}$, (8) holds for $\mathbf{x} = \mathbf{0}$. On the other hand, by Definition 8.16, (8) holds for $\mathbf{x} \neq \mathbf{0}$. It remains to prove that the extended real number $\|\mathbf{T}\|$ is finite.

Let B be the $m \times n$ matrix which represents $\mathbf{T}$, and suppose that the rows of $\mathbf{T}$ are given by $\mathbf{b}_1, \ldots, \mathbf{b}_m$. By the definition of matrix multiplication and our identification of $\mathbf{R}^m$ with $m \times 1$ matrices,

$$\mathbf{T}(\mathbf{x}) = (\mathbf{b}_1 \cdot \mathbf{x}, \ldots, \mathbf{b}_m \cdot \mathbf{x}).$$

If $B = O$, then $\|\mathbf{T}\| = 0$ and (8) is an equality. If $B \neq O$, then, by the Cauchy–Schwarz Inequality, the square of the Euclidean norm of $\mathbf{T}(\mathbf{x})$ satisfies

$$\|\mathbf{T}(\mathbf{x})\|^2 = (\mathbf{b}_1 \cdot \mathbf{x})^2 + \cdots + (\mathbf{b}_m \cdot \mathbf{x})^2$$
$$\leq (\|\mathbf{b}_1\| \, \|\mathbf{x}\|)^2 + \cdots + (\|\mathbf{b}_m\| \, \|\mathbf{x}\|)^2$$
$$\leq m \cdot \max\{\|\mathbf{b}_j\|^2 : 1 \leq j \leq m\} \, \|\mathbf{x}\|^2 =: C \, \|\mathbf{x}\|^2.$$

Therefore, the quotients $\|\mathbf{T}(\mathbf{x})\|/\|\mathbf{x}\|$ are bounded (by $\sqrt{C}$). It follows from the Completeness Axiom that $\|\mathbf{T}\|$ exists and is finite. ∎

Theorem 8.17, an analogue of the Cauchy–Schwarz Inequality, will be used to estimate differentiable functions of several variables. If B is the matrix which represents a linear transformation $\mathbf{T}$, we will refer to the number $\|\mathbf{T}\|$ as the *operator norm* of B, and denote it by $\|B\|$. (For two other ways to calculate this norm, see Exercise 8.2.11.)

We close this section with an optional result which shows that under the identification of linear functions with matrices, function composition is taken to matrix multiplication. This, in fact, is why matrix multiplication is defined the way it is.

*8.18 Remark. *If* $T : R^n \to R^m$ *and* $U : R^m \to R^p$ *are linear, then so is* $U \circ T$. *In fact, if* B *is the* $m \times n$ *matrix which represents* T, *and* C *is the* $p \times m$ *matrix which represents* U, *then* CB *is the matrix which represents* $U \circ T$.

Proof. Let $e_1, \ldots, e_n$ be the usual basis of R^n, $u_1, \ldots, u_m$ be the usual basis of R^m, and $w_1, \ldots, w_p$ be the usual basis of R^p. If $B = [b_{ij}]_{m \times n}$ represents T and $C = [c_{vk}]_{p \times m}$ represents U, then, by Theorem 8.15,

$$\sum_{k=1}^{m} b_{kj} u_k = (b_{1j}, \ldots, b_{mj}) = T(e_j), \qquad j = 1, 2, \ldots, n,$$

and

$$\sum_{v=1}^{p} c_{vk} w_v = (c_{1k}, \ldots, c_{pk}) = U(u_k), \qquad k = 1, 2, \ldots, m.$$

Hence

$$(U \circ T)(e_j) = U(T(e_j)) = U \left(\sum_{k=1}^{m} b_{kj} u_k \right) = \sum_{k=1}^{m} b_{kj} U(u_k)$$

$$= \sum_{k=1}^{m} \sum_{v=1}^{p} b_{kj} c_{vk} w_v = \left(\sum_{k=1}^{m} b_{kj} c_{1k}, \ldots, \sum_{k=1}^{m} b_{kj} c_{pk} \right)$$

for each $1 \le j \le n$. Since this last vector is the jth column of the matrix CB, it follows that CB is the matrix which represents $U \circ T$.   ■

EXERCISES

8.2.1. Let $a, b, c \in R^3$.

a) Prove that if a, b, and c do not all lie on the same line, then an equation of the plane through these points is given by $(x, y, z) \cdot d = a \cdot d$, where
$$d := (a - b) \times (a - c).$$

b) Prove that if c does not lie on the line $\phi(t) = ta + b$, $t \in R$, then an equation of the plane that contains this line and the point c is given by $(x, y, z) \cdot d = b \cdot d$, where $d := a \times (b - c)$.

8.2.2. a) Find an equation of the hyperplane through the points $(1, 0, 0, 0)$, $(2, 1, 0, 0)$, $(0, 1, 1, 0)$, and $(0, 4, 0, 1)$.

b) Find an equation of the hyperplane that contains the lines $\phi(t) = (t, t, t, 1)$ and $\psi(t) = (1, t, 1 + t, t)$, $t \in R$.

c) Find an equation of the plane parallel to the hyperplane $x_1 + \cdots + x_n = \pi$ passing through the point $(1, 2, \ldots, n)$.

8.2.3. Find two lines in $\mathbf{R}^3$ which are not parallel but do not intersect.

8.2.4. Suppose that $\mathbf{T} \in \mathcal{L}(\mathbf{R}^n; \mathbf{R}^m)$ for some $n, m \in \mathbf{N}$.

    a) Find the matrix representative of $\mathbf{T}$ if $\mathbf{T}(x, y, z, w) = (0, x + y, x - z, x + y + w)$.

    b) Find the matrix representative of T if $T(x, y, z) = x - y + z$.

    c) Find the matrix representative of $\mathbf{T}$ if $\mathbf{T}(x_1, x_2, \ldots, x_n) = (x_1 - x_n, x_n - x_1)$.

8.2.5. Suppose that $\mathbf{T} \in \mathcal{L}(\mathbf{R}^n; \mathbf{R}^m)$ for some $n, m \in \mathbf{N}$.

    a) If $\mathbf{T}(1, 1) = (3, \pi, 0)$ and $\mathbf{T}(0, 1) = (4, 0, 1)$, find the matrix representative of $\mathbf{T}$.

    b) If $\mathbf{T}(1, 1, 0) = (e, \pi)$, $\mathbf{T}(0, -1, 1) = (1, 0)$, and $\mathbf{T}(1, 1, -1) = (1, 2)$, find the matrix representative of $\mathbf{T}$.

    c) If $\mathbf{T}(0, 1, 1, 0) = (3, 5)$, $\mathbf{T}(0, 1, -1, 0) = (5, 3)$, and $\mathbf{T}(0, 0, 0, -1) = (\pi, 3)$, find all possible matrix representatives of $\mathbf{T}$.

    d) If $\mathbf{T}(1, 1, 0, 0) = (5, 4, 1)$, $\mathbf{T}(0, 0, 1, 0) = (1, 2, 0)$, and $\mathbf{T}(0, 0, 0, -1) = (\pi, 3, -1)$, find all possible matrix representatives of $\mathbf{T}$.

8.2.6. Suppose that $\mathbf{a}, \mathbf{b}, \mathbf{c} \in \mathbf{R}^3$ are three points which do not lie on the same straight line and that Π is the plane which contains the points $\mathbf{a}, \mathbf{b}, \mathbf{c}$. Prove that an equation of Π is given by

$$\det \begin{bmatrix} x - a_1 & y - a_2 & z - a_3 \\ b_1 - a_1 & b_2 - a_2 & b_3 - a_3 \\ c_1 - a_1 & c_2 - a_2 & c_3 - a_3 \end{bmatrix} = 0.$$

8.2.7. **This exercise is used in Appendix E.** Recall that the area of a parallelogram with base b and altitude h is given by bh, and the volume of a parallelepiped is given by the area of its base times its altitude.

    a) Let $\mathbf{a}, \mathbf{b} \in \mathbf{R}^3$ be nonzero vectors and $\mathcal{P}$ represent the parallelogram

$$\{(x, y, z) = u\mathbf{a} + v\mathbf{b} : u, v \in [0, 1]\}.$$

    Prove that the area of $\mathcal{P}$ is $\|\mathbf{a} \times \mathbf{b}\|$.

    b) Let $\mathbf{a}, \mathbf{b}, \mathbf{c} \in \mathbf{R}^3$ be nonzero vectors and $\mathcal{P}$ represent the parallelepiped

$$\{(x, y, z) = t\mathbf{a} + u\mathbf{b} + v\mathbf{c} : t, u, v \in [0, 1]\}.$$

    Prove that the volume of $\mathcal{P}$ is $|(\mathbf{a} \times \mathbf{b}) \cdot \mathbf{c}|$.

8.2.8. The distance from a point $\mathbf{x}_0 = (x_0, y_0, z_0)$ to a plane Π in $\mathbf{R}^3$ is defined to be

$$\text{dist}(\mathbf{x}_0, \Pi) := \begin{cases} 0 & \mathbf{x}_0 \in \Pi \\ \|\mathbf{v}\| & \mathbf{x}_0 \notin \Pi, \end{cases}$$

where $\mathbf{v} := (x_0 - x_1, y_0 - y_1, z_0 - z_1)$ for some $(x_1, y_1, z_1) \in \Pi$, and $\mathbf{v}$ is orthogonal to Π (i.e., parallel to its normal). Sketch Π and $\mathbf{x}_0$ for a typical plane Π, and convince yourself that this is the correct definition. Prove that this definition does not depend on the choice of $\mathbf{v}$, by showing that the distance from $\mathbf{x}_0 = (x_0, y_0, z_0)$ to the plane Π described by $ax + by + cz = d$ is

$$\text{dist}(\mathbf{x}_0, \Pi) = \frac{|ax_0 + by_0 + cz_0 - d|}{\sqrt{a^2 + b^2 + c^2}}.$$

8.2.9. [ROTATIONS IN $\mathbf{R}^2$]. **This exercise is used in Section *15.1.** Let

$$B = \begin{bmatrix} \cos\theta & -\sin\theta \\ \sin\theta & \cos\theta \end{bmatrix}$$

for some $\theta \in \mathbf{R}$.

a) Prove that $\|B(x, y)\| = \|(x, y)\|$ for all $(x, y) \in \mathbf{R}^2$.
b) Let $(x, y) \in \mathbf{R}^2$ be a nonzero vector and φ represent the angle between $B(x, y)$ and (x, y). Prove that $\cos\varphi = \cos\theta$. Thus, show that B rotates $\mathbf{R}^2$ through an angle θ. (When $\theta > 0$, we shall call B *counterclockwise rotation* about the origin through the angle θ.)

8.2.10. For each of the following functions $\mathbf{f}$, find the matrix representative of a linear transformation $\mathbf{T} \in \mathcal{L}(\mathbf{R}; \mathbf{R}^m)$ which satisfies

$$\lim_{h \to 0} \frac{\|\mathbf{f}(x + h) - \mathbf{f}(x) - \mathbf{T}(h)\|}{h} = 0.$$

a) $\mathbf{f}(x) = (x^2, \sin x)$
b) $\mathbf{f}(x) = (e^x, \sqrt[3]{x}, 1 - x^2)$
c) $\mathbf{f}(x) = (1, 2, 3, x^2 + x, x^2 - x)$

8.2.11. Fix $\mathbf{T} \in \mathcal{L}(\mathbf{R}^n; \mathbf{R}^m)$. Set

$$M_1 := \sup_{\|\mathbf{x}\|=1} \|\mathbf{T}(\mathbf{x})\| \quad \text{and}$$

$$M_2 := \inf\{C > 0 : \|\mathbf{T}(\mathbf{x})\| \le C\|\mathbf{x}\| \text{ for all } \mathbf{x} \in \mathbf{R}^n\}.$$

a) Prove that $M_1 \le \|\mathbf{T}\|$.
b) Using the linear property of $\mathbf{T}$, prove that if $\mathbf{x} \ne \mathbf{0}$, then

$$\frac{\|\mathbf{T}(\mathbf{x})\|}{\|\mathbf{x}\|} \le M_1.$$

c) Prove that $M_1 = M_2 = \|\mathbf{T}\|$.

8.3 TOPOLOGY OF R^n

If you want a more abstract introduction to the topology of Euclidean spaces, skip the rest of this chapter and the next, and begin Chapter 10 now.

Topology, a study of geometric objects which emphasizes how they are put together over their exact shape and proportion, is based on the fundamental concepts of open and closed sets, a generalization of open and closed intervals. In this section we introduce these concepts in $\mathbf{R}^n$ and identify their most basic properties. In the next chapter, we shall explore how they can be used to characterize limits and continuity without using distance explicitly. This additional step in abstraction will yield powerful benefits, as we shall see in Section 9.4 and in Chapter 11 when we begin to study the calculus of functions of several variables.

We begin with a natural generalization of intervals to $\mathbf{R}^n$.

8.19 Definition.

Let $\mathbf{a} \in \mathbf{R}^n$.

i) For each $r > 0$, the *open ball* centered at $\mathbf{a}$ of radius r is the set of points

$$B_r(\mathbf{a}) := \{\mathbf{x} \in \mathbf{R}^n : \|\mathbf{x} - \mathbf{a}\| < r\}.$$

ii) For each $r \geq 0$, the *closed ball* centered at $\mathbf{a}$ of radius r is the set of points

$$\{\mathbf{x} \in \mathbf{R}^n : \|\mathbf{x} - \mathbf{a}\| \leq r\}.$$

Notice that when $n = 1$, the open ball centered at a of radius r is the open interval $(a - r, a + r)$, and the corresponding closed ball is the closed interval $[a - r, a + r]$. Also notice that the open ball (respectively, the closed ball) centered at $\mathbf{a}$ of radius r contains none of its (respectively, all of its) "boundary" $\{\mathbf{x} : \|\mathbf{x} - \mathbf{a}\| = r\}$. Accordingly, we will draw pictures of balls in $\mathbf{R}^2$ with the following conventions: Open balls will be drawn with dashed "boundaries" and closed balls will be drawn with solid "boundaries" (see Figure 8.5).

To generalize the concept of open and closed intervals even further, observe that each element of an open interval I lies "inside" I (i.e., is surrounded by other points in I). On the other hand, although closed intervals do NOT satisfy this property, their complements do. Accordingly, we make the following definition.

8.20 Definition.

Let $n \in \mathbf{N}$.

i) A subset V of $\mathbf{R}^n$ is said to be *open* (in $\mathbf{R}^n$) if and only if for every $\mathbf{a} \in V$ there is an $\varepsilon > 0$ such that $B_\varepsilon(\mathbf{a}) \subseteq V$.

ii) A subset E of $\mathbf{R}^n$ is said to be *closed* (in $\mathbf{R}^n$) if and only if $E^c := \mathbf{R}^n \setminus E$ is open.

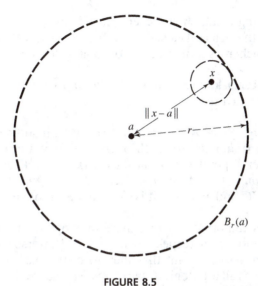

FIGURE 8.5

The following result shows that every "open" ball is open. (Closed balls are also closed—see Exercise 8.3.2.)

8.21 Remark. *For every $\mathbf{x} \in B_r(\mathbf{a})$ there is an $\varepsilon > 0$ such that $B_\varepsilon(\mathbf{x}) \subseteq B_r(\mathbf{a})$.*

Proof. Let $\mathbf{x} \in B_r(\mathbf{a})$. Using Figure 8.5 for guidance, we set $\varepsilon = r - \|\mathbf{x} - \mathbf{a}\|$. If $\mathbf{y} \in B_\varepsilon(\mathbf{x})$, then by the Triangle Inequality, assumption, and the choice of ε,

$$\|\mathbf{y} - \mathbf{a}\| \le \|\mathbf{y} - \mathbf{x}\| + \|\mathbf{x} - \mathbf{a}\| < \varepsilon + \|\mathbf{x} - \mathbf{a}\| = r.$$

Thus, by definition, $\mathbf{y} \in B_r(\mathbf{a})$. In particular, $B_\varepsilon(\mathbf{x}) \subseteq B_r(\mathbf{a})$. ∎

(Once again, drawing diagrams in $\mathbf{R}^2$ led us to a proof valid for all Euclidean spaces.)

Here are more examples of open sets and closed sets.

8.22 Remark. *If $\mathbf{a} \in \mathbf{R}^n$, then $\mathbf{R}^n \setminus \{\mathbf{a}\}$ is open and $\{\mathbf{a}\}$ is closed.*

Proof. By Definition 8.20, it suffices to prove that the complement of every *singleton* $E := \{\mathbf{a}\}$ is open. Let $\mathbf{x} \in E^c$ and set $\varepsilon = \|\mathbf{x} - \mathbf{a}\|$. Then, by definition, $\mathbf{a} \notin B_\varepsilon(\mathbf{x})$, so $B_\varepsilon(\mathbf{x}) \subseteq E^c$. Therefore, E^c is open by Definition 8.20. ∎

Students sometimes mistakenly believe that every set is either open or closed. Some sets are neither open nor closed (like the interval $[0, 1)$). And, as the following result shows, every Euclidean space contains two special sets which are both open and closed. (We shall see below that these are the only subsets of $\mathbf{R}^n$ which are simultaneously open and closed in $\mathbf{R}^n$.)

8.23 Remark. *For each $n \in \mathbf{N}$, the empty set $\emptyset$ and the whole space $\mathbf{R}^n$ are both open and closed.*

Proof. Since $\mathbf{R}^n = \emptyset^c$ and $\emptyset = (\mathbf{R}^n)^c$, it suffices by Definition 8.20 to prove that $\emptyset$ and $\mathbf{R}^n$ are both open. Because the empty set contains no points, "every" point $\mathbf{x} \in \emptyset$ satisfies $B_\varepsilon(\mathbf{x}) \subseteq \emptyset$. (This is called the *vacuous implication*.) Therefore, $\emptyset$ is open. On the other hand, since $B_\varepsilon(\mathbf{x}) \subseteq \mathbf{R}^n$ for all $\mathbf{x} \in \mathbf{R}^n$ and all $\varepsilon > 0$, it is clear that $\mathbf{R}^n$ is open. ∎

It is important to recognize that open sets and closed sets behave very differently with respect to unions and intersections. (In fact, these properties are so important that they form the basis of an axiomatic system which describes all topological spaces, even those for which measurement of distance is impossible.)

8.24 Theorem. *Let $n \in \mathbf{N}$.*

i) *If $\{V_\alpha\}_{\alpha \in A}$ is any collection of open subsets of $\mathbf{R}^n$, then*

$$\bigcup_{\alpha \in A} V_\alpha$$

is open.

ii) *If $\{V_k : k = 1, 2, \ldots, p\}$ is a finite collection of open subsets of $\mathbf{R}^n$, then*

$$\bigcap_{k=1}^{p} V_k := \bigcap_{k \in \{1,2,\ldots,p\}} V_k$$

is open.

iii) *If $\{E_\alpha\}_{\alpha \in A}$ is any collection of closed subsets of $\mathbf{R}^n$, then*

$$\bigcap_{\alpha \in A} E_\alpha$$

is closed.

iv) *If $\{E_k : k = 1, 2, \ldots, p\}$ is a finite collection of closed subsets of $\mathbf{R}^n$, then*

$$\bigcup_{k=1}^{p} E_k := \bigcup_{k \in \{1,2,\ldots,p\}} E_k$$

is closed.

v) *If V is open and E is closed, then $V \setminus E$ is open and $E \setminus V$ is closed.*

Proof. i) Let $\mathbf{x} \in \bigcup_{\alpha \in A} V_\alpha$. Then $\mathbf{x} \in V_\alpha$ for some $\alpha \in A$. Since V_α is open, it follows that there is an $r > 0$ such that $B_r(\mathbf{x}) \subseteq V_\alpha$. Thus $B_r(\mathbf{x}) \subseteq \bigcup_{\alpha \in A} V_\alpha$; that is, this union is open.

ii) Let $\mathbf{x} \in \bigcap_{k=1}^{p} V_k$. Then $\mathbf{x} \in V_k$ for $k = 1, 2, \ldots, p$. Since each V_k is open, it follows that there are numbers $r_k > 0$ such that $B_{r_k}(\mathbf{x}) \subseteq V_k$. Let $r = \min\{r_1, \ldots, r_p\}$. Then $r > 0$ and $B_r(\mathbf{x}) \subseteq V_k$ for all $k = 1, 2, \ldots, p$; that is, $B_r(\mathbf{x}) \subseteq \bigcap_{k=1}^{p} V_k$. Hence, this intersection is open.

iii) By DeMorgan's Law (Theorem 1.36) and part i),

$$\left(\bigcap_{\alpha \in A} E_\alpha \right)^c = \bigcup_{\alpha \in A} E_\alpha^c$$

is open, so $\bigcap_{\alpha \in A} E_\alpha$ is closed.

iv) By DeMorgan's Law and part ii),

$$\left(\bigcup_{k=1}^{p} E_k \right)^c = \bigcap_{k=1}^{p} E_k^c$$

is open, so $\bigcup_{k=1}^{p} E_k$ is closed.

v) Since $V \setminus E = V \cap E^c$ and $E \setminus V = E \cap V^c$, the former is open by part ii), and the latter is closed by part iii). ∎

The finiteness hypotheses in Theorem 8.24 are crucial, even for the case $n = 1$.

8.25 Remark. *Statements ii) and iv) of Theorem 8.24 are false if arbitrary collections are used in place of finite collections.*

Proof. In the Euclidean space $\mathbf{R}$,

$$\bigcap_{k \in \mathbf{N}} \left(-\frac{1}{k}, \frac{1}{k} \right) = \{0\}$$

is closed and

$$\bigcup_{k \in \mathbf{N}} \left[\frac{1}{k+1}, \frac{k}{k+1} \right] = (0, 1)$$

is open. ∎

To see why open sets are so important to analysis, we reexamine the definition of continuity using open sets. By Definition 3.19, a function $f : E \to \mathbf{R}$ is continuous at $a \in E$ if and only if given $\varepsilon > 0$ there is a $\delta > 0$ such that $|x - a| < \delta$ and $x \in E$ imply $|f(x) - f(a)| < \varepsilon$. Put in "ball language," this says that f is continuous at $a \in E$ if and only if $f(E \cap B_\delta(a)) \subseteq B_\varepsilon(f(a))$; that is, $E \cap B_\delta(a) \subseteq f^{-1}(B_\varepsilon(f(a)))$. In particular, f is continuous at $a \in E$ if and only if for all

$a \in \mathbf{R}$, the inverse image under f of every open ball centered at $f(a)$ contains the intersection of E and an open ball centered at a.

Intersecting the ball centered at a with E adds a complication. It would be simpler if the inverse image of an open ball under a continuous function just contained another open ball. Can we discard the set E like that? To answer this question, we consider two functions, $f(x) = 1/x$ and $g(x) = 1 + \sqrt{x - 1}$, and one open ball, $(-1, 3)$, centered at 1. Notice that $f^{-1}(-1, 3) = (-\infty, -1) \cup (1/3, \infty)$ does contain an open ball centered at $a = 1$ but $g^{-1}(-1, 3) = [1, 5)$ does not. It is merely the intersection of an open ball and the domain of g: $g^{-1}(-1, 3) = [1, 5) = [1, \infty) \cap (-5, 5)$. Evidently, we cannot discard the domain E of a continuous function when restating the definition of continuity using open balls, unless E is open (see Exercise 9.4.3).

Accordingly, we modify the definition of *open* and *closed* along the following lines.

8.26 Definition.

Let $E \subseteq \mathbf{R}^n$.

i) A set $U \subseteq E$ is said to be *relatively open* in E if and only if there is an open set A such that $U = E \cap A$.

ii) A set $C \subseteq E$ is said to be *relatively closed* in E if and only if there is a closed set B such that $C = E \cap B$.

We postpone using these concepts to study continuous functions on $\mathbf{R}^n$ until Chapter 9. Meanwhile, we shall use relatively open sets to introduce *connectivity*, a concept which generalizes to $\mathbf{R}^n$ an important property of intervals which played a role in the proof of the Intermediate Value Theorem, and which will be used several times in our development of the calculus of functions of several variables. First, we explore the analogy between relatively open sets and open sets.

8.27 Remark. *Let $U \subseteq E \subseteq \mathbf{R}^n$.*

i) *Then U is relatively open in E if and only if for each $\mathbf{a} \in U$ there is an $r > 0$ such that $B_r(\mathbf{a}) \cap E \subset U$.*

ii) *If E is open, then U is relatively open in E if and only if U is (plain old vanilla) open (in the usual sense).*

Proof. i) If U is relatively open in E, then $U = E \cap A$ for some open set A. Since A is open, there is an $r > 0$ such that $B_r(\mathbf{a}) \subset A$. Hence, $B_r(\mathbf{a}) \cap E \subset A \cap E = U$.

Conversely, for each $\mathbf{a} \in U$ choose an $r(\mathbf{a}) > 0$ such that $B_{r(\mathbf{a})}(\mathbf{a}) \cap E \subset U$. Then $\bigcup_{\mathbf{a} \in U} B_{r(\mathbf{a})}(\mathbf{a}) \cap E \subseteq U$. Since the union is taken over all $\mathbf{a} \in U$, the reverse set inequality is also true. Thus $\bigcup_{\mathbf{a} \in U} B_{r(\mathbf{a})}(\mathbf{a}) \cap E = U$. Since the union of these open balls is open by Theorem 8.24, it follows that U is relatively open in E.

ii) Suppose that U is relatively open in E. If E and A are open, then $U = E \cap A$ is open. Thus U is open in the usual sense. Conversely, if U is open, then $E \cap U = U$ is open. Thus every open subset of E is relatively open in E. ∎

Next, we introduce connectivity.

8.28 Definition.

Let E be a subset of $\mathbf{R}^n$.

i) A pair of sets U, V is said to *separate E* if and only if U and V are nonempty, relatively open in E, $E = U \cup V$, and $U \cap V = \emptyset$.

ii) E is said to be *connected* if and only if E cannot be separated by any pair of relatively open sets U, V.

Loosely speaking, a connected set is all in one piece (i.e., cannot be broken into smaller, nonempty, relatively open pieces which do not share any common points).

The empty set is connected, since it can never be written as the union of nonempty sets. Every singleton $E = \{\mathbf{a}\}$ is also connected, since if $E = U \cup V$, where $U \cap V = \emptyset$ and both U and V are nonempty, then E has at least two points. More complicated connected sets can be found in the exercises.

Notice that by Definitions 8.26 and 8.28, a set E is not connected if there are open sets A, B such that $E \cap A$, $E \cap B$ are nonempty, $E = (E \cap A) \cup (E \cap B)$, and $A \cap B = \emptyset$. Is this statement valid if we replace $E = (E \cap A) \cup (E \cap B)$ by $E \subseteq A \cup B$?

8.29 Remark. *Let $E \subseteq \mathbf{R}^n$. If there exists a pair of open sets A,B such that $E \cap A \neq \emptyset$, $E \cap B \neq \emptyset$, $E \subseteq A \cup B$, and $A \cap B = \emptyset$, then E is not connected.*

Proof. Set $U = E \cap A$ and $V = E \cap B$. By hypothesis and Definition 8.26, U and V are relatively open in E and nonempty. Since $U \cap V \subseteq A \cap B = \emptyset$, it suffices by Definition 8.28 to prove that $E = U \cup V$. But E is a subset of $A \cup B$, so $E \subseteq U \cup V$. On the other hand, both U and V are subsets of E, so $E \supseteq U \cup V$. We conclude that $E = U \cup V$. ∎

(The converse of this result is also true, but harder to prove—see Theorem 8.38 below.)

In practice, Remark 8.29 is often easier to apply than Definition 8.28. Here are several examples. The set $\mathbf{Q}$ is not connected: set $A = (-\infty, \sqrt{2})$ and $B = (\sqrt{2}, \infty)$. The "bow-tie-shaped set" $\{(x, y) : -1 \leq x \leq 1 \text{ and } -|x| < y < |x|\}$ is not connected (see Figure 8.6): set $A = \{(x, y) : x < 0\}$ and $B = \{(x, y) : x > 0\}$.

Is there a simple description of all connected subsets of $\mathbf{R}$?

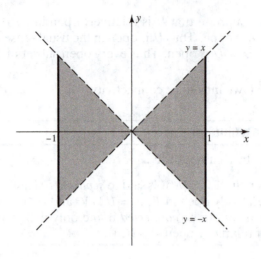

FIGURE 8.6

8.30 Theorem. *A subset E of* **R** *is connected if and only if E is an interval.*

Proof. Suppose that E is a connected subset of **R**. If E is empty or if E contains only one point c, then E is one of the intervals (c, c) or $[c, c]$.

Suppose that E contains at least two points. Set $a = \inf E$ and $b = \sup E$, and observe that $-\infty \le a < b \le \infty$. If $a \in E$ set $a_k = a$, and if $b \in E$ set $b_k = b$, $k \in$ **N**. Otherwise, use the Approximation Property to choose $a_k, b_k \in E$ such that $a_k \downarrow a$ and $b_k \uparrow b$ as $k \to \infty$. Notice that in all cases, E contains each $[a_k, b_k]$. Indeed, if not, say there is an $x \in [a_k, b_k] \backslash E$, then $a_k \in E \cap (-\infty, x)$, $b_k \in E \cap (x, \infty)$, and $E \subseteq (-\infty, x) \cup (x, \infty)$. Hence, by Remark 8.29, E is not connected, a contradiction. Therefore, $E \supseteq [a_k, b_k]$ for all $k \in$ **N**. It follows from construction that

$$E = \bigcup_{k=1}^{\infty} [a_k, b_k].$$

Since this last union is either (a, b), $[a, b)$, $(a, b]$, or $[a, b]$, we conclude that E is an interval.

Conversely, suppose that E is an interval which is not connected. Then there are sets U, V, relatively open in E, which separate E (i.e., $E = U \cup V$, $U \cap V = \emptyset$), and there exist points $x_1 \in U$ and $x_2 \in V$. We may suppose that $x_1 < x_2$. Since $x_1, x_2 \in E$ and E is an interval, $I_0 := [x_1, x_2] \subseteq E$. Define f on I_0 by

$$f(x) = \begin{cases} 0 & x \in U \\ 1 & x \in V. \end{cases}$$

Since $U \cap V = \emptyset$, f is well defined. We claim that f is continuous on I_0. Indeed, fix $x_0 \in [x_1, x_2]$. Since $U \cup V = E \supseteq I_0$, it is evident that $x_0 \in U$

or $x_0 \in V$. We may suppose the former. Let $y_k \in I_0$ and suppose that $y_k \to x_0$ as $k \to \infty$. Since U is relatively open, there is an $\varepsilon > 0$ such that $(x_0 - \varepsilon, x_0 + \varepsilon) \cap E \subset U$. Since $y_k \in E$ and $y_k \to x_0$, it follows that $y_k \in U$ for large k. Hence $f(y_k) = 0 = f(x_0)$ for large k. Therefore, f is continuous at x_0 by the Sequential Characterization of Continuity.

We have proved that f is continuous on I_0. Hence by the Intermediate Value Theorem (Theorem 3.29), f must take on the value $1/2$ somewhere on I_0. This is a contradiction, since by construction, f takes on only the values 0 or 1. ∎

We shall use this result later to prove that a real function is continuous on a closed, bounded interval if and only if its graph is closed and connected (see Theorem 9.51).

EXERCISES

8.3.1. Sketch each of the following sets. Identify which of the following sets are open, which are closed, and which are neither. Also discuss the connectivity of each set.

a) $E = \{(x, y) : y \neq 0\}$
b) $E = \{(x, y) : x^2 + 4y^2 \leq 1\}$
c) $E = \{(x, y) : y \geq x^2, \ 0 \leq y < 1\}$
d) $E = \{(x, y) : x^2 - y^2 > 1, \ -1 < y < 1\}$
e) $E = \{(x, y) : x^2 - 2x + y^2 = 0\} \cup \{(x, 0) : x \in [2, 3]\}$

8.3.2. Let $n \in \mathbf{N}$, let $\mathbf{a} \in \mathbf{R}^n$, let $s, r \in \mathbf{R}$ with $s < r$, and set

$$V = \{\mathbf{x} \in \mathbf{R}^n : s < \|\mathbf{x} - \mathbf{a}\| < r\} \quad \text{and} \quad E = \{\mathbf{x} \in \mathbf{R}^n : s \leq \|\mathbf{x} - \mathbf{a}\| \leq r\}.$$

Prove that V is open and E is closed.

8.3.3. a) Let $a \leq b$ and $c \leq d$ be real numbers. Sketch a graph of the rectangle

$$[a, b] \times [c, d] := \{(x, y) : x \in [a, b], y \in [c, d]\},$$

and decide whether this set is connected. Explain your answers.

b) Sketch a graph of set

$$B_1(-2, 0) \cup B_1(2, 0) \cup \{(x, 0) : -1 < x < 1\},$$

and decide whether this set is connected. Explain your answers.

8.3.4. a) Set $E_1 := \{(x, y) : y \geq 0\}$ and $E_2 := \{(x, y) : x^2 + 2y^2 < 6\}$, and sketch a graph of the set

$$U := \{(x, y) : x^2 + 2y^2 < 6 \quad \text{and} \quad y \geq 0\}.$$

    b) Decide whether U is relatively open or relatively closed in E_1. Explain your answer.

    c) Decide whether U is relatively open or relatively closed in E_2. Explain your answer.

8.3.5. a) Let E_1 denote the closed ball centered at $(0, 0)$ of radius 1 and $E_2 := B_{\sqrt{2}}(2, 0)$, and sketch a graph of the set

$$U := \{(x, y) : x^2 + y^2 \le 1 \quad \text{and} \quad x^2 - 4x + y^2 + 2 < 0\}.$$

    b) Decide whether U is relatively open or relatively closed in E_1. Explain your answer.

    c) Decide whether U is relatively open or relatively closed in E_2. Explain your answer.

8.3.6. Suppose that $E \subseteq \mathbf{R}^n$ and that C is a subset of E.

    a) Prove that if E is closed, then C is relatively closed in E if and only if C is (plain old vanilla) closed (in the usual sense).

    b) Prove that C is relatively closed in E if and only if $E \backslash C$ is relatively open in E.

8.3.7. a) If A and B are connected in $\mathbf{R}^n$ and $A \cap B \ne \emptyset$, prove that $A \cup B$ is connected.

    b) If $\{E_\alpha\}_{\alpha \in A}$ is a collection of connected sets in $\mathbf{R}^n$ and $\bigcap_{\alpha \in A} E_\alpha \ne \emptyset$, prove that

$$E = \bigcup_{\alpha \in A} E_\alpha$$

is connected.

    c) If A and B are connected in $\mathbf{R}$ and $A \cap B \ne \emptyset$, prove that $A \cap B$ is connected.

    d) Show that part c) is no longer true if $\mathbf{R}^2$ replaces $\mathbf{R}$.

8.3.8. Let V be a subset of $\mathbf{R}^n$.

    a) Prove that V is open if and only if there is a collection of open balls $\{B_\alpha : \alpha \in A\}$ such that

$$V = \bigcup_{\alpha \in A} B_\alpha.$$

    b) What happens to this result when *open* is replaced by *closed*?

8.3.9. Show that if E is closed in $\mathbf{R}^n$ and $\mathbf{a} \notin E$, then

$$\inf_{\mathbf{x} \in E} \|\mathbf{x} - \mathbf{a}\| > 0.$$

8.3.10. Graph generic open balls in $\mathbf{R}^2$ with respect to each of the "non-Euclidean" norms $\| \cdot \|_1$ and $\| \cdot \|_\infty$. What shape are they?

8.4   INTERIOR, CLOSURE, AND BOUNDARY

To prove that every set contains a largest open set and is contained in a smallest closed set, we introduce the following topological operations.

8.31 Definition.

Let E be a subset of a Euclidean space $\mathbf{R}^n$.

i) The *interior* of E is the set

$$E^o := \bigcup \{V : V \subseteq E \text{ and } V \text{ is open in } \mathbf{R}^n\}.$$

ii) The *closure* of E is the set

$$\overline{E} := \bigcap \{B : B \supseteq E \text{ and } B \text{ is closed in } \mathbf{R}^n\}.$$

Notice that every set E contains the open set $\emptyset$ and is contained in the closed set $\mathbf{R}^n$. Hence, the sets E^o and $\overline{E}$ are well defined. Also notice by Theorem 8.24 that the interior of a set is always open and the closure of a set is always closed.

The following result shows that E^o is the largest open set contained in E, and $\overline{E}$ is the smallest closed set which contains E.

8.32 Theorem. *Let $E \subseteq \mathbf{R}^n$. Then*

i) $E^o \subseteq E \subseteq \overline{E}$,
ii) *if V is open and $V \subseteq E$, then $V \subseteq E^o$, and*
iii) *if C is closed and $C \supseteq E$, then $C \supseteq \overline{E}$.*

Proof. Since every open set V in the union defining E^o is a subset of E, it is clear that the union of these V's is a subset of E. Thus $E^o \subseteq E$. A similar argument establishes $E \subseteq \overline{E}$. This proves i).

By Definition 8.31, if V is an open subset of E, then $V \subseteq E^o$, and if C is a closed set containing E, then $\overline{E} \subseteq C$. This proves ii) and iii). ∎

In particular, the interior of a bounded interval with endpoints a and b is (a, b), and its closure is $[a, b]$. In fact, it is evident by parts ii) and iii) that $E = E^o$ if and only if E is open, and $E = \overline{E}$ if and only if E is closed. We shall use this observation many times below.

Let us examine these concepts in the concrete setting $\mathbf{R}^2$.

8.33 EXAMPLES.

a) Find the interior and closure of the set $E = \{(x, y) : -1 \le x \le 1 \text{ and } -|x| < y < |x|\}$.

b) Find the interior and closure of the set $E = B_1(-2, 0) \cup B_1(2, 0) \cup \{(x, 0) : -1 \le x \le 1\}$.

Solution.

a) Graph $y = |x|$ and $x = \pm 1$, and observe that E is a bow tie-shaped region with "solid" vertical edges (see Figure 8.6). Now, by Definition 8.20, any open set in $\mathbf{R}^2$ must contain a disk around each of its points. Since E^o is the largest open set inside E, it is clear that

$$E^o = \{(x, y) : -1 < x < 1 \text{ and } -|x| < y < |x|\}.$$

Similarly,

$$\overline{E} = \{(x, y) : -1 \leq x \leq 1 \text{ and } -|x| \leq y \leq |x|\}.$$

b) Draw a graph of this region. It turns out to be "dumbbell shaped": two open disks joined by a straight line. Thus $E^o = B_1(-2, 0) \cup B_1(2, 0)$, and

$$\overline{E} = \overline{B_1(-2, 0)} \cup \overline{B_1(2, 0)} \cup \{(x, 0) : -1 \leq x \leq 1\}. \qquad \blacksquare$$

These examples illustrate the fact that the interior of a nice enough set E in $\mathbf{R}^2$ can be obtained by removing all its "edges," and the closure of E by adding all its "edges."

One of the most important results from Chapter 5 is the Fundamental Theorem of Calculus. It states that the behavior of a derivative f' on an interval $[a, b]$, as measured by its integral, is determined by the values of f at the endpoints of $[a, b]$. What shall we use for "endpoints" of an arbitrary set in $\mathbf{R}^n$? Notice that the endpoints a, b are the only points which lie near both $[a, b]$ and the complement of $[a, b]$. Using this as a cue, we introduce the following concept.

8.34 Definition.

Let $E \subseteq \mathbf{R}^n$. The *boundary* of E is the set

$$\partial E := \{\mathbf{x} \in \mathbf{R}^n : \text{for all } r > 0, \quad B_r(\mathbf{x}) \cap E \neq \emptyset \text{ and } B_r(\mathbf{x}) \cap E^c \neq \emptyset\}.$$

[We will refer to the last two conditions in the definition of ∂E by saying that $B_r(\mathbf{x})$ *intersects E and E^c.*]

8.35 EXAMPLE.

Describe the boundary of the set

$$E = \{(x, y) : x^2 + y^2 \leq 9 \text{ and } (x - 1)(y + 2) > 0\}.$$

Solution. Graph the relations $x^2 + y^2 = 9$ and $(x - 1)(y + 2) = 0$ to see that E is a region with solid curved edges and dashed straight edges (see Figure 8.7). By definition, then, the boundary of E is the union of these curved and straight edges (all made solid). Rather than describing ∂E analytically (which would

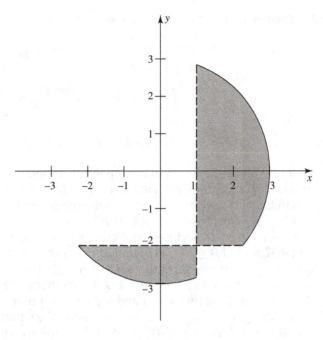

FIGURE 8.7

involve solving for the intersection points of the straight lines $x = 1$, $y = -2$, and the circle $x^2 + y^2 = 9$), it is easier to describe ∂E by using set algebra.

$$\partial E = \{(x, y) : x^2 + y^2 \leq 9 \text{ and } (x - 1)(y + 2) \geq 0\}$$
$$\setminus \{(x, y) : x^2 + y^2 < 9 \text{ and } (x - 1)(y + 2) > 0\} \qquad \blacksquare$$

It turns out that set algebra can be used to describe the boundary of any set.

8.36 Theorem. *Let $E \subseteq \mathbf{R}^n$. Then $\partial E = \overline{E} \setminus E^o$.*

Proof. By Definition 8.34, it suffices to show that

$$\mathbf{x} \in \overline{E} \text{ if and only if } B_r(\mathbf{x}) \cap E \neq \emptyset \text{ for all } r > 0, \text{ and} \qquad (9)$$

$$\mathbf{x} \notin E^o \text{ if and only if } B_r(\mathbf{x}) \cap E^c \neq \emptyset \text{ for all } r > 0. \qquad (10)$$

We will provide the details for (9) and leave the proof of (10) as an exercise. Suppose that $\mathbf{x} \in \overline{E}$ but that $B_{r_0}(\mathbf{x}) \cap E = \emptyset$ for some $r_0 > 0$. Then $(B_{r_0}(\mathbf{x}))^c$ is a closed set which contains E; hence, by Theorem 8.32iii, $\overline{E} \subseteq (B_{r_0}(\mathbf{x}))^c$. It follows that $\overline{E} \cap B_{r_0}(\mathbf{x}) = \emptyset$ (e.g., $\mathbf{x} \notin \overline{E}$), a contradiction.

Conversely, suppose that $\mathbf{x} \notin \overline{E}$. Since $(\overline{E})^c$ is open, there is an $r_0 > 0$ such that $B_{r_0}(\mathbf{x}) \subseteq (\overline{E})^c$. In particular, $\emptyset = B_{r_0}(\mathbf{x}) \cap \overline{E} \supseteq B_{r_0}(\mathbf{x}) \cap E$ for some $r_0 > 0$. $\qquad \blacksquare$

We have introduced topological operations (interior, closure, and boundary). The following result answers the question, How do these operations interact with the set operations (union and intersection)?

8.37 Theorem. *Let $A, B \subseteq \mathbf{R}^n$. Then*

i)
$$(A \cup B)^o \supseteq A^o \cup B^o, \quad (A \cap B)^o = A^o \cap B^o,$$

ii)
$$\overline{A \cup B} = \overline{A} \cup \overline{B}, \qquad \overline{A \cap B} \subseteq \overline{A} \cap \overline{B},$$

iii)
$$\partial(A \cup B) \subseteq \partial A \cup \partial B, \quad and \quad \partial(A \cap B) \subseteq \partial A \cup \partial B.$$

Proof. i) Since the union of two open sets is open, $A^o \cup B^o$ is an open subset of $A \cup B$. Hence, by Theorem 8.32ii, $A^o \cup B^o \subseteq (A \cup B)^o$.

Similarly, $(A \cap B)^o \supseteq A^o \cap B^o$. On the other hand, if $V \subset A \cap B$, then $V \subset A$ and $V \subset B$. Thus $(A \cap B)^o \subseteq A^o \cap B^o$.

ii) Since $\overline{A} \cup \overline{B}$ is closed and contains $A \cup B$, it is clear that, by Theorem 8.32iii, $\overline{A \cup B} \subseteq \overline{A} \cup \overline{B}$. Similarly, $\overline{A \cap B} \subseteq \overline{A} \cap \overline{B}$. To prove the reverse inequality for union, suppose that $\mathbf{x} \notin \overline{A \cup B}$. Then, by Definition 8.31, there is a closed set E which contains $A \cup B$ such that $\mathbf{x} \notin E$. Since E contains both A and B, it follows that $\mathbf{x} \notin \overline{A}$ and $\mathbf{x} \notin \overline{B}$. This proves part ii).

iii) Let $\mathbf{x} \in \partial(A \cup B)$; that is, suppose that $B_r(\mathbf{x})$ intersects $A \cup B$ and $(A \cup B)^c$ for all $r > 0$. Since $(A \cup B)^c = A^c \cap B^c$, it follows that $B_r(\mathbf{x})$ intersects both A^c and B^c for all $r > 0$. Thus $B_r(\mathbf{x})$ intersects A and A^c for all $r > 0$, or $B_r(\mathbf{x})$ intersects B and B^c for all $r > 0$ (i.e., $\mathbf{x} \in \partial A \cup \partial B$). This proves the first set inequality in part iii). A similar argument establishes the second inequality in part iii). ∎

The second inequality in part iii) can be improved (see Exercise 8.4.10d).

Finally, we note (Exercise 8.4.11) that relatively open sets in E can be divided into two kinds: those inside E^o, which contain none of their boundary, and those which intersect ∂E, which contain only that part of their boundary which intersects ∂E. (See Figures 15.3 and 15.4 for an illustration of both types.)

We close this section by showing that the converse of Remark 8.30 is also true. This result is optional because we do not use it anywhere else.

*8.38 Theorem. *Let $E \subseteq \mathbf{R}^n$. If there exist nonempty, relatively open sets U, V which separate E, then there is a pair of open sets A, B such that $A \cap E \neq \emptyset$, $B \cap E \neq \emptyset$, $A \cap B = \emptyset$, and $E \subseteq A \cup B$.*

Proof. We first show that

$$\overline{U} \cap V = \emptyset. \tag{11}$$

Indeed, since V is relatively open in E, there is a set Ω, open in $\mathbf{R}^n$, such that $V = E \cap \Omega$. Since $U \cap V = \emptyset$, it follows that $U \subset \Omega^c$. This last set is closed in $\mathbf{R}^n$. Therefore,

$$\overline{U} \subseteq \overline{\Omega^c} = \Omega^c;$$

that is, (11) holds.

Next, we use (11) to construct the open set B. Set

$$\delta_{\mathbf{x}} := \inf\{\|\mathbf{x} - \mathbf{u}\| : \mathbf{u} \in \overline{U}\}, \quad \mathbf{x} \in V, \quad \text{and} \quad B = \bigcup_{\mathbf{x} \in V} B_{\delta_{\mathbf{x}}/2}(\mathbf{x}).$$

Clearly, B is open in $\mathbf{R}^n$. Since $\delta_{\mathbf{x}} > 0$ for each $\mathbf{x} \notin \overline{U}$ (see Exercise 8.3.9), B contains V; hence $B \cap E \supseteq V$. The reverse inequality also holds, since by construction $B \cap U = \emptyset$ and by hypothesis $E = U \cup V$. Therefore, $B \cap E = V$. Similarly, we can construct an open set A such that $A \cap E = U$ by setting

$$\varepsilon_{\mathbf{y}} := \inf\{\|\mathbf{v} - \mathbf{y}\| : \mathbf{v} \in \overline{V}\}, \quad \mathbf{y} \in U \quad \text{and} \quad A = \bigcup_{\mathbf{y} \in U} B_{\varepsilon_{\mathbf{y}}/2}(\mathbf{y}).$$

In particular, A and B are nonempty open sets which satisfy $E \subseteq A \cup B$.

It remains to prove that $A \cap B = \emptyset$. Suppose, to the contrary, that there is a point $\mathbf{a} \in A \cap B$. Then $\mathbf{a} \in B_{\delta_{\mathbf{x}}/2}(\mathbf{x})$ for some $\mathbf{x} \in V$ and $\mathbf{a} \in B_{\varepsilon_{\mathbf{y}}/2}(\mathbf{y})$ for some $\mathbf{y} \in U$. We may suppose that $\delta_{\mathbf{x}} \leq \varepsilon_{\mathbf{y}}$. Then

$$\|\mathbf{x} - \mathbf{y}\| \leq \|\mathbf{x} - \mathbf{a}\| + \|\mathbf{a} - \mathbf{y}\| < \frac{\delta_{\mathbf{x}}}{2} + \frac{\varepsilon_{\mathbf{y}}}{2} \leq \varepsilon_{\mathbf{y}}.$$

Therefore, $\|\mathbf{x} - \mathbf{y}\| < \inf\{\|\mathbf{v} - \mathbf{y}\| : \mathbf{v} \in \overline{V}\}$. Since $\mathbf{x} \in V$, this is impossible. We conclude that $A \cap B = \emptyset$. ∎

EXERCISES

8.4.1. Find the interior, closure, and boundary of each of the following subsets of $\mathbf{R}$.

a) $E = \{1/n : n \in \mathbf{N}\}$

b) $E = \bigcup_{n=1}^{\infty} \left(\dfrac{1}{n+1}, \dfrac{1}{n} \right)$

c) $E = \bigcup_{n=1}^{\infty} (-n, n)$

d) $E = \mathbf{Q}$

8.4.2. For each of the following sets, sketch E^o, $\overline{E}$, and ∂E.

a) $E = \{(x, y) : x^2 + 4y^2 \leq 1\}$

b) $E = \{(x, y) : x^2 - 2x + y^2 = 0\} \cup \{(x, 0) : x \in [2, 3]\}$

c) $E = \{(x, y) : y \geq x^2, \ 0 \leq y < 1\}$

d) $E = \{(x, y) : x^2 - y^2 < 1, \ -1 < y < 1\}$

$\boxed{\textbf{8.4.3}}$. **This exercise is used in Section 12.1.** Suppose that $A \subseteq B \subseteq \mathbf{R}^n$. Prove that

$$\overline{A} \subseteq \overline{B} \quad \text{and} \quad A^o \subseteq B^o.$$

8.4.4. Let E be a subset of $\mathbf{R}^n$.

a) Prove that every subset $A \subseteq E$ contains a set B which is the largest subset of A that is relatively open in E.

    b) Prove that every subset $A \subseteq E$ is contained in a set B which is the smallest closed set containing A that is relatively closed in E.

8.4.5. Complete the proof of Theorem 8.36 by verifying (10).

8.4.6. Prove that if $E \subseteq \mathbf{R}$ is connected, then E^o is also connected. Show that this is false if "$\mathbf{R}$" is replaced by "$\mathbf{R}^2$."

8.4.7. Suppose that $E \subset \mathbf{R}^n$ is connected and that $E \subseteq A \subseteq \overline{E}$. Prove that A is connected.

8.4.8. A set A is called *clopen* if and only if it is both open and closed.

    a) Prove that every Euclidean space has at least two clopen sets.

    b) Prove that a proper subset E of $\mathbf{R}^n$ is connected if and only if it contains exactly two relatively clopen sets.

    c) Prove that every nonempty proper subset of $\mathbf{R}^n$ has a nonempty boundary.

8.4.9. Show that Theorem 8.37 is best possible in the following sense.

    a) There exist sets A, B in $\mathbf{R}$ such that $(A \cup B)^o \neq A^o \cup B^o$.

    b) There exist sets A, B in $\mathbf{R}$ such that $\overline{A \cap B} \neq \overline{A} \cap \overline{B}$.

    c) There exist sets A, B in $\mathbf{R}$ such that $\partial(A \cup B) \neq \partial A \cup \partial B$ and $\partial(A \cap B) \neq \partial A \cup \partial B$.

8.4.10. Let A and B be subsets of $\mathbf{R}^n$.

    a) Show that $\partial(A \cap B) \cap (A^c \cup (\partial B)^c) \subseteq \partial A$.

    b) Show that if $\mathbf{x} \in \partial(A \cap B)$ and $\mathbf{x} \notin (A \cap \partial B) \cup (B \cap \partial A)$, then $\mathbf{x} \in \partial A \cap \partial B$.

    c) Prove that $\partial(A \cap B) \subseteq (A \cap \partial B) \cup (B \cap \partial A) \cup (\partial A \cap \partial B)$.

    d) Show that even in $\mathbf{R}$, there exist sets A and B such that $\partial(A \cap B) \neq (A \cap \partial B) \cup (B \cap \partial A) \cup (\partial A \cap \partial B)$.

8.4.11. Let $E \subseteq \mathbf{R}^n$ and U be relatively open in E.

    a) If $U \subseteq E^o$, then $U \cap \partial U = \emptyset$.

    b) If $U \cap \partial E \neq \emptyset$, then $U \cap \partial U = U \cap \partial E$.

CHAPTER 9

Convergence in $\mathbf{R}^n$

In this chapter we generalize the concepts of limits and continuity from $\mathbf{R}$ to $\mathbf{R}^n$. We begin, as we did in Chapter 2, with sequences.

9.1 LIMITS OF SEQUENCES

Using the analogy between the Euclidean norm and the absolute value, we can define what it means for a sequence in $\mathbf{R}^n$ to be convergent, bounded, or Cauchy in the following way.

9.1 Definition.

Let $\{\mathbf{x}_k\}$ be a sequence of points in $\mathbf{R}^n$.

i) $\{\mathbf{x}_k\}$ is said to *converge* to some point $\mathbf{a} \in \mathbf{R}^n$ (called the *limit* of $\mathbf{x}_k$) if and only if for every $\varepsilon > 0$ there is an $N \in \mathbf{N}$ such that

$$k \geq N \quad \text{implies} \quad \|\mathbf{x}_k - \mathbf{a}\| < \varepsilon.$$

Notation: $\mathbf{x}_k \to \mathbf{a}$ as $k \to \infty$ or $\mathbf{a} = \lim_{k\to\infty} \mathbf{x}_k$.
ii) $\{\mathbf{x}_k\}$ is said to be *bounded* if and only if there is an $M > 0$ such that $\|\mathbf{x}_k\| \leq M$ for all $k \in \mathbf{N}$.
iii) $\{\mathbf{x}_k\}$ is said to be *Cauchy* if and only if for every $\varepsilon > 0$ there is an $N \in \mathbf{N}$ such that

$$k, m \geq N \quad \text{imply} \quad \|\mathbf{x}_k - \mathbf{x}_m\| < \varepsilon.$$

The following result shows that to evaluate the limit of a specific sequence in $\mathbf{R}^n$ we need only take the limits of the component sequences.

9.2 Theorem. *Let* $\mathbf{a} = (a_1, a_2, \ldots, a_n) \in \mathbf{R}^n$ *and* $\{\mathbf{x}_k = (x_k^{(1)}, x_k^{(2)}, \ldots, x_k^{(n)})\}_{k\in\mathbf{N}}$ *be a sequence in* $\mathbf{R}^n$. *Then* $\mathbf{x}_k \to \mathbf{a}$, *as* $k \to \infty$, *if and only if for each* $j \in \{1, 2, \ldots, n\}$, *the component sequence* $x_k^{(j)} \to a_j$ *as* $k \to \infty$.

Proof. Fix $j \in \{1, \ldots, n\}$. By Remark 8.7,

$$|x_k^{(j)} - a_j| \leq \|\mathbf{x}_k - \mathbf{a}\| \leq \sqrt{n} \max_{1\leq\ell\leq n} |x_k^{(\ell)} - a_\ell|.$$

Hence, by the Squeeze Theorem, $x_k^{(j)} \to a_j$ as $k \to \infty$ if and only if $\|\mathbf{x}_k - \mathbf{a}\| \to 0$ as $k \to \infty$. Since $\|\mathbf{x}_k - \mathbf{a}\| \to 0$ if and only if $\mathbf{x}_k \to \mathbf{a}$, as $k \to \infty$, the proof of the theorem is complete. ∎

This result can be used to obtain the following analogue of the Density of Rationals (Theorem 1.18).

9.3 Theorem. *Let* $\mathbf{Q}^n := \{\mathbf{x} \in \mathbf{R}^n : x_j \in \mathbf{Q} \text{ for } j = 1, 2, \dots, n\}$. *For each* $\mathbf{a} \in \mathbf{R}^n$ *there is a sequence* $\mathbf{x}_k \in \mathbf{Q}^n$ *such that* $\mathbf{x}_k \to \mathbf{a}$ *as* $k \to \infty$.

Proof. Let $\mathbf{a} := (a_1, \dots, a_n) \in \mathbf{R}^n$. For each $1 \le j \le n$, choose by Theorem 1.18 sequences $r_k^{(j)} \in \mathbf{Q}$ such that $r_k^{(j)} \to a_j$ (in $\mathbf{R}$) as $k \to \infty$. By Theorem 9.2, $\mathbf{x}_k := (r_k^{(1)}, \dots, r_k^{(n)})$ converges to $\mathbf{a}$ (in $\mathbf{R}^n$) as $k \to \infty$. Moreover, $\mathbf{x}_k \in \mathbf{Q}^n$ for each $k \in \mathbf{N}$. ∎

A set E is said to be *separable* if and only if there exists an at most countable subset Z of E such that to each $\mathbf{a} \in E$ there corresponds a sequence $\mathbf{x}_k \in Z$ such that $\mathbf{x}_k \to \mathbf{a}$ as $k \to \infty$. Since $\mathbf{Q}^n$ is countable (just iterate Theorem 1.42i), it follows from Theorem 9.3 that $\mathbf{R}^n$ is separable.

Theorem 9.3 illustrates a general principle. As long as we stay away from results about monotone sequences (which have no analogue in $\mathbf{R}^n$ when $n > 1$), we can extend most of the results found in Chapter 2 from $\mathbf{R}$ to $\mathbf{R}^n$. Since the proofs of these results require little more than replacing $|x - y|$ in the real case by $\|\mathbf{x} - \mathbf{y}\|$ in the vector case, we will summarize what is true and leave most of the details to the reader.

9.4 Theorem. *Let* $n \in \mathbf{N}$.

i) *A sequence in* $\mathbf{R}^n$ *can have at most one limit.*

ii) *If* $\{\mathbf{x}_k\}_{k \in \mathbf{N}}$ *is a sequence in* $\mathbf{R}^n$ *which converges to* $\mathbf{a}$ *and* $\{\mathbf{x}_{k_j}\}_{j \in \mathbf{N}}$ *is any subsequence of* $\{\mathbf{x}_k\}_{k \in \mathbf{N}}$, *then* $\mathbf{x}_{k_j}$ *converges to* $\mathbf{a}$ *as* $j \to \infty$.

iii) *Every convergent sequence in* $\mathbf{R}^n$ *is bounded, but not conversely.*

iv) *Every convergent sequence in* $\mathbf{R}^n$ *is Cauchy.*

v) *If* $\{\mathbf{x}_k\}$ *and* $\{\mathbf{y}_k\}$ *are convergent sequences in* $\mathbf{R}^n$ *and* $\alpha \in \mathbf{R}$, *then*

$$\lim_{k \to \infty} (\mathbf{x}_k + \mathbf{y}_k) = \lim_{k \to \infty} \mathbf{x}_k + \lim_{k \to \infty} \mathbf{y}_k,$$

$$\lim_{k \to \infty} (\alpha \mathbf{x}_k) = \alpha \lim_{k \to \infty} \mathbf{x}_k,$$

and

$$\lim_{k \to \infty} (\mathbf{x}_k \cdot \mathbf{y}_k) = (\lim_{k \to \infty} \mathbf{x}_k) \cdot (\lim_{k \to \infty} \mathbf{y}_k).$$

Moreover, when $n = 3$,

$$\lim_{k \to \infty} (\mathbf{x}_k \times \mathbf{y}_k) = (\lim_{k \to \infty} \mathbf{x}_k) \times (\lim_{k \to \infty} \mathbf{y}_k).$$

Notice once and for all that (since $\|\mathbf{x}_k\|^2 = \mathbf{x}_k \cdot \mathbf{x}_k$), the penultimate equation above contains the following corollary. If $\mathbf{x}_k$ converges, then

$$\lim_{k\to\infty} \|\mathbf{x}_k\| = \| \lim_{k\to\infty} \mathbf{x}_k \|.$$

As in the real case, the converse of part iv) is also true. In order to prove that, we need an n-dimensional version of the Bolzano–Weierstrass Theorem.

9.5 Theorem.   [BOLZANO–WEIERSTRASS THEOREM FOR $\mathbf{R}^n$].
Every bounded sequence in $\mathbf{R}^n$ has a convergent subsequence.

Proof. Let $\mathbf{x}_k := (x_k^{(1)}, x_k^{(2)}, \ldots, x_k^{(n)})$ be bounded in $\mathbf{R}^n$. Then, by Remark 8.7i, the real sequence $\{x_k^{(j)}\}_{k\in\mathbf{N}}$ is bounded in $\mathbf{R}$ for each $j = 1, 2, \ldots, n$.

Let $j = 1$. By the one-dimensional Bolzano–Weierstrass Theorem, there is a sequence of integers $1 \le k(1, 1) < k(1, 2) < \cdots$ and a number a_1 such that $x_{k(1,\nu)}^{(1)} \to a_1$ as $\nu \to \infty$.

Let $j = 2$. Again, since the sequence $\{x_{k(1,\nu)}^{(2)}\}_{\nu\in\mathbf{N}}$ is bounded in $\mathbf{R}$, there is a subsequence $\{k(2, \nu)\}_{\nu\in\mathbf{N}}$ of $\{k(1, \nu)\}_{\nu\in\mathbf{N}}$ and a number a_2 such that $x_{k(2,\nu)}^{(2)} \to a_2$ as $\nu \to \infty$. Since $\{k(2, \nu)\}_{\nu\in\mathbf{N}}$ is a subsequence of $\{k(1, \nu)\}_{\nu\in\mathbf{N}}$, we also have $x_{k(2,\nu)}^{(1)} \to a_1$ as $\nu \to \infty$. Thus, $x_{k(2,\nu)}^{(\ell)} \to a_\ell$ as $\nu \to \infty$ for all $1 \le \ell \le j = 2$.

Continuing this process until $j = n$, we choose a subsequence $k_\nu = k(n, \nu)$ and points a_ℓ such that

$$\lim_{\nu\to\infty} x_{k_\nu}^{(\ell)} = a_\ell$$

for $1 \le \ell \le j = n$. Set $\mathbf{a} = (a_1, a_2, \ldots, a_n)$. Then, by Theorem 9.2, $\mathbf{x}_{k_\nu}$ converges to $\mathbf{a}$ as $\nu \to \infty$. ∎

Since the Bolzano–Weierstrass Theorem holds for $\mathbf{R}^n$, we can modify proof of Theorem 2.29 to establish the following result.

9.6 Theorem. *A sequence $\{\mathbf{x}_k\}$ in $\mathbf{R}^n$ is Cauchy if and only if it converges.*

Thus sequences in $\mathbf{R}^n$ behave pretty much the same as sequences in $\mathbf{R}$. We now turn our attention to something new. How does the limit of sequences interact with the topological structure of $\mathbf{R}^n$? Answers to this question contain a surprising bonus. The ε's begin to disappear from the theory.

9.7 Theorem. *Let $\mathbf{x}_k \in \mathbf{R}^n$. Then $\mathbf{x}_k \to \mathbf{a}$ as $k \to \infty$ if and only if for every open set V which contains $\mathbf{a}$ there is an $N \in \mathbf{N}$ such that $k \ge N$ implies $\mathbf{x}_k \in V$.*

Proof. Suppose that $\mathbf{x}_k \to \mathbf{a}$ and let V be an open set which contains $\mathbf{a}$. By Definition 8.20, there is an $\varepsilon > 0$ such that $B_\varepsilon(\mathbf{a}) \subseteq V$. Given this ε, use Definition 9.1 to choose an $N \in \mathbf{N}$ such that $k \ge N$ implies $\mathbf{x}_k \in B_\varepsilon(\mathbf{a})$. By the choice of ε, $\mathbf{x}_k \in V$ for all $k \ge N$.

Conversely, let $\varepsilon > 0$ and set $V = B_\varepsilon(\mathbf{a})$. Then V is an open set which contains $\mathbf{a}$; hence, by hypothesis, there is an $N \in \mathbf{N}$ such that $k \geq N$ implies $\mathbf{x}_k \in V$. In particular, $\|\mathbf{x}_k - \mathbf{a}\| < \varepsilon$ for all $k \geq N$. ∎

This is a first step toward developing a "distance-less" theory of convergence. The next result, which we shall use many times, shows that convergent sequences characterize closed sets.

9.8 Theorem. *Let $E \subseteq \mathbf{R}^n$. Then E is closed if and only if E contains all its limit points; that is, if and only if $\mathbf{x}_k \in E$ and $\mathbf{x}_k \to \mathbf{x}$ imply that $\mathbf{x} \in E$.*

Proof. The theorem is vacuously satisfied if E is the empty set.

Suppose that $E \neq \emptyset$ is closed but some sequence $\mathbf{x}_k \in E$ converges to a point $\mathbf{x} \in E^c$. Since E is closed, E^c is open. Thus, by Theorem 9.7, there is an $N \in \mathbf{N}$ such that $k \geq N$ implies $\mathbf{x}_k \in E^c$, a contradiction.

Conversely, suppose that E is a nonempty set which contains all its limit points. If E is not closed, then, by Remark 8.23, $E \neq \mathbf{R}^n$ and by definition E^c is nonempty and not open. Thus, there is at least one point $\mathbf{x} \in E^c$ such that no ball $B_r(\mathbf{x})$ is contained in E^c. Let $\mathbf{x}_k \in B_{1/k}(\mathbf{x}) \cap E$ for $k = 1, 2, \ldots$. Then $\mathbf{x}_k \in E$ and $\|\mathbf{x}_k - \mathbf{x}\| < 1/k$ for all $k \in \mathbf{N}$. Now by the Squeeze Theorem, $\|\mathbf{x}_k - \mathbf{x}\| \to 0$ (i.e., $\mathbf{x}_k \to \mathbf{x}$ as $k \to \infty$). Thus, by hypothesis, $\mathbf{x} \in E$, a contradiction. ∎

EXERCISES

9.1.1. Using Definition 9.1i, prove that the following limits exist.

a)
$$\mathbf{x}_k = \left(\frac{1}{k}, 1 - \frac{1}{k^2} \right)$$

b)
$$\mathbf{x}_k = \left(\frac{k}{k+1}, \frac{\sin k^3}{k} \right)$$

c)
$$\mathbf{x}_k = \left(\log(k+1) - \log k, 2^{-k} \right)$$

9.1.2. Using limit theorems, find the limit of each of the following vector sequences.

a)
$$\mathbf{x}_k = \left(\frac{1}{k}, \frac{2k^2 - k + 1}{k^2 + 2k - 1} \right)$$

b)
$$\mathbf{x}_k = \left(1, \sin \pi k, \cos \frac{1}{k} \right)$$

c)
$$\mathbf{x}_k = \left(k - \sqrt{k^2 + k}, k^{1/k}, \frac{1}{k} \right)$$

9.1.3. Suppose that $\mathbf{x}_k \to \mathbf{0}$ in $\mathbf{R}^n$ as $k \to \infty$ and that $\mathbf{y}_k$ is bounded in $\mathbf{R}^n$.

    a) Prove that $\mathbf{x}_k \cdot \mathbf{y}_k \to 0$ as $k \to \infty$.
    b) If $n = 3$, prove that $\mathbf{x}_k \times \mathbf{y}_k \to \mathbf{0}$ as $k \to \infty$.

9.1.4. Suppose that $\mathbf{a} \in \mathbf{R}^n$, that $\mathbf{x}_k \to \mathbf{a}$, and that $\mathbf{x}_k - \mathbf{y}_k \to \mathbf{0}$, as $k \to \infty$. Prove that $\mathbf{y}_k \to \mathbf{a}$ as $k \to \infty$.

9.1.5. a) Prove Theorem 9.4i and ii.
    b) Prove Theorem 9.4iii and iv.
    c) Prove Theorem 9.4v.
    d) Prove Theorem 9.6.

9.1.6. Let E be a nonempty subset of $\mathbf{R}^n$.

    a) Show that a sequence $\mathbf{x}_k \in E$ converges to some point $\mathbf{a} \in E$ if and only if for every set U, which is relatively open in E and contains $\mathbf{a}$, there is an $N \in \mathbf{N}$ such that $\mathbf{x}_k \in U$ for $k \geq N$.
    b) Prove that a set $C \subseteq E$ is relatively closed in E if and only if the limit of every sequence $\mathbf{x}_k \in C$ which converges to a point in E satisfies $\lim_{k \to \infty} \mathbf{x}_k \in C$.

9.1.7. a) A subset E of $\mathbf{R}^n$ is said to be *sequentially compact* if and only if every sequence $\mathbf{x}_k \in E$ has a convergent subsequence whose limit belongs to E. Prove that every closed ball in $\mathbf{R}^n$ is sequentially compact.
    b) Prove that $\mathbf{R}^n$ is not sequentially compact.

9.1.8. a) Let E be a subset of $\mathbf{R}^n$. A point $\mathbf{a} \in \mathbf{R}^n$ is called a *cluster point* of E if $E \cap B_r(\mathbf{a})$ contains infinitely many points for every $r > 0$. Prove that $\mathbf{a}$ is a cluster point of E if and only if for each $r > 0$, $E \cap B_r(\mathbf{a}) \backslash \{\mathbf{a}\}$ is nonempty.
    b) Prove that every bounded infinite subset of $\mathbf{R}^n$ has at least one cluster point.

9.2 HEINE–BOREL THEOREM

In this section, we use the theory of sequences developed above to prove the Heine–Borel Theorem. It is difficult to overestimate the usefulness of this powerful result, which allows us to extend local results to global ones in an almost effortless manner (e.g., see Example 9.12).

We begin with the following "covering" lemma.

9.9 *Lemma.* [BOREL COVERING LEMMA].
Let E be a closed, bounded subset of $\mathbf{R}^n$. If r is any function from E into $(0, \infty)$, then there exist finitely many points $\mathbf{y}_1, \ldots, \mathbf{y}_N \in E$ such that

$$E \subseteq \bigcup_{j=1}^{N} B_{r(\mathbf{y}_j)}(\mathbf{y}_j).$$

STRATEGY: Since $r(\mathbf{y}) > 0$ and $\mathbf{y} \in B_{r(\mathbf{y})}(\mathbf{y})$ for each $\mathbf{y} \in E$, it is clear that $E \subseteq \bigcup_{\mathbf{y} \in E} B_{r(\mathbf{y})}(\mathbf{y})$. By moving the centers a little bit, we might be able to assume that the centers are rational; that is, that $E \subseteq \bigcup_{\mathbf{y} \in E \cap \mathbf{Q}^n} B_{r(\mathbf{y})}(\mathbf{y})$. Since $\mathbf{Q}^n$ is countable (see Theorem 1.42i and Remark 1.43), it would follow that there exist $\mathbf{y}_j \in E \cap \mathbf{Q}^n$ and $r_j := r(\mathbf{y}_j)$ such that $E \subseteq \bigcup_{j=1}^{\infty} B_{r_j}(\mathbf{y}_j)$. Hence, if the covering lemma is false, then there exist $\mathbf{x}_k \in E$ such that $\mathbf{x}_k \notin \bigcup_{j=1}^{k} B_{r_j}(\mathbf{y}_j)$ for $k = 1, 2, \dots$. Since E is closed and bounded, it follows from the Bolzano–Weierstrass Theorem and Theorem 9.8 that some subsequence $\mathbf{x}_{k_\nu}$ converges to a point $\mathbf{x} \in E$ as $\nu \to \infty$. Since E is a subset of the union of balls $B_{r_j}(\mathbf{y}_j)$, this $\mathbf{x}$ must belong to some $B_{r_{j_0}}(\mathbf{y}_{j_0})$. Hence by Theorem 9.7, $\mathbf{x}_{k_\nu} \in B_{r_j}(\mathbf{y}_j)$ for large ν. But this contradicts the fact that if $k \geq j$, then $\mathbf{x}_k \notin B_{r_j}(\mathbf{y}_j)$. Here are the details.

Proof. *Step 1: Change the centers.* Fix $\mathbf{y}_0 \in E$. By Theorem 9.3, choose $\mathbf{a} \in \mathbf{Q}^n$ and $\rho := \rho(\mathbf{y}_0, \mathbf{a}) \in \mathbf{Q}$ such that $\|\mathbf{y}_0 - \mathbf{a}\| < r(\mathbf{y}_0)/4$ and $r(\mathbf{y}_0)/4 < \rho < r(\mathbf{y}_0)/2$. Since $\|\mathbf{y}_0 - \mathbf{a}\| < r(\mathbf{y}_0)/4 < \rho$, we have $\mathbf{y}_0 \in B_\rho(\mathbf{a})$. On the other hand, $\mathbf{y} \in B_\rho(\mathbf{a})$ implies $\|\mathbf{y}_0 - \mathbf{y}\| \leq \|\mathbf{y}_0 - \mathbf{a}\| + \|\mathbf{a} - \mathbf{y}\| < \rho + \rho < r(\mathbf{y}_0)$; that is, $B_\rho(\mathbf{a}) \subset B_{r(\mathbf{y}_0)}(\mathbf{y}_0)$.

Step 2: Construct the sequence. We just proved that to each $\mathbf{y}_0 \in E$ there correspond $\mathbf{a} \in \mathbf{Q}^n$ and $\rho(\mathbf{y}_0, \mathbf{a}) \in \mathbf{Q}$ such that $\mathbf{y}_0 \in B_{\rho(\mathbf{y}_0, \mathbf{a})}(\mathbf{a}) \subset B_{r(\mathbf{y}_0)}(\mathbf{y}_0)$. Since $\mathbf{Q}$ and $\mathbf{Q}^n$ are countable, it follows that there exist $\mathbf{a}_j \in \mathbf{Q}^n$ and $\rho_j \in \mathbf{Q}$ such that

$$E \subseteq \bigcup_{j=1}^{\infty} B_{\rho_j}(\mathbf{a}_j).$$

Suppose for a moment that E is not a subset of any of the finite unions $\bigcup_{j=1}^{k} B_{\rho_j}(\mathbf{a}_j), k \in \mathbf{N}$. For each k, choose $\mathbf{x}_k \in E \backslash \bigcup_{j=1}^{k} B_{\rho_j}(\mathbf{a}_j)$. By Theorems 9.5, 9.8, and 9.7 there is a subsequence $\mathbf{x}_{k_\nu}$ and an index j_0 such that $\mathbf{x}_{k_\nu} \in B_{\rho_{j_0}}(\mathbf{a}_{j_0})$ for ν large. But by construction, if $k_\nu > j_0$, then $\mathbf{x}_{k_\nu} \notin \bigcup_{j=1}^{j_0} B_{\rho_j}(\mathbf{a}_j)$; in particular, $\mathbf{x}_{k_\nu}$ cannot belong to $B_{\rho_{j_0}}(\mathbf{a}_{j_0})$ for large ν. This contradiction proves that there is an $N \in \mathbf{N}$ such that

$$E \subseteq \bigcup_{j=1}^{N} B_{\rho_j}(\mathbf{a}_j).$$

Step 3: Finish the proof. By Step 1, given $j \in \mathbf{N}$ there is a point in E, say $\mathbf{y}_j$, such that $B_{\rho_j}(\mathbf{a}_j) \subset B_{r(\mathbf{y}_j)}(\mathbf{y}_j)$. We conclude by Step 2 that

$$E \subseteq \bigcup_{j=1}^{N} B_{\rho_j}(\mathbf{a}_j) \subset \bigcup_{j=1}^{N} B_{r(\mathbf{y}_j)}(\mathbf{y}_j). \qquad \blacksquare$$

In conjunction with this important result, we introduce the following concepts. (For a more complete treatment, see Section 9.4.)

9.10 Definition.

Let E be a subset of $\mathbf{R}^n$.

i) An *open covering* of E is a collection of sets $\{V_\alpha\}_{\alpha \in A}$ such that each V_α is open and

$$E \subseteq \bigcup_{\alpha \in A} V_\alpha.$$

ii) The set E is said to be *compact* if and only if every open covering of E has a *finite subcovering*; that is, if and only if given any open covering $\{V_\alpha\}_{\alpha \in A}$ of E, there is a finite subset $A_0 = \{\alpha_1, \dots, \alpha_N\}$ of A such that

$$E \subseteq \bigcup_{j=1}^{N} V_{\alpha_j}.$$

This definition is sufficiently abstract to make students uneasy when first introduced to it. And with good reason! It's not obvious whether a particular open covering has a finite subcovering. Is there an easy way to recognize when a set is compact?

For Euclidean spaces, the answer to this question is yes. In fact, we shall use the Borel Covering Lemma to establish the following simple but important characterization of compact sets.

9.11 Theorem. [**Heine–Borel Theorem**].
Let E be a subset of $\mathbf{R}^n$. Then E is compact if and only if E is closed and bounded.

Proof. Suppose that E is compact. Since $\{B_k(\mathbf{0})\}_{k \in \mathbf{N}}$ is an open covering of $\mathbf{R}^n$, hence of E, there is an $N \in \mathbf{N}$ such that

$$E \subseteq \bigcup_{k=1}^{N} B_k(\mathbf{0}).$$

In particular, E is bounded by N.

To verify that E is closed, suppose not. Then E is nonempty and (by Theorem 9.8) there is a convergent sequence $\mathbf{x}_k \in E$ whose limit $\mathbf{x}$ does not belong to E. For each $\mathbf{y} \in E$, set $r(\mathbf{y}) := \|\mathbf{x} - \mathbf{y}\|/2$. Since $\mathbf{x}$ does not belong to E, $r(\mathbf{y}) > 0$. Thus each $B_{r(\mathbf{y})}(\mathbf{y})$ is open and contains $\mathbf{y}$; that is, $\{B_{r(\mathbf{y})}(\mathbf{y}) : \mathbf{y} \in E\}$ is an open covering of E. Since E is compact, we can choose points $\mathbf{y}_j$ and radii $r_j := r(\mathbf{y}_j)$, for $j = 1, 2, \dots, M$ such that

$$E \subseteq \bigcup_{j=1}^{M} B_{r_j}(\mathbf{y}_j).$$

Set $r := \min\{r_1, \ldots, r_M\}$. (This is a finite set of positive numbers, so r is also positive.) Since $\mathbf{x}_k \to \mathbf{x}$ as $k \to \infty$, $\mathbf{x}_k \in B_r(\mathbf{x})$ for large k. But $\mathbf{x}_k \in B_r(\mathbf{x}) \cap E$ implies $\mathbf{x}_k \in B_{r_j}(\mathbf{y}_j)$ for some $j \in \mathbf{N}$. Therefore, it follows from the choices of r_j and r, and from the Triangle Inequality, that

$$
\begin{aligned}
r_j \geq \|\mathbf{x}_k - \mathbf{y}_j\| &\geq \|\mathbf{x} - \mathbf{y}_j\| - \|\mathbf{x}_k - \mathbf{x}\| \\
&= 2r_j - \|\mathbf{x}_k - \mathbf{x}\| > 2r_j - r \geq 2r_j - r_j = r_j,
\end{aligned}
$$

a contradiction.

Conversely, suppose that E is closed and bounded. Let $\{V_\alpha\}_{\alpha \in A}$ be an open covering of E. Let $\mathbf{x} \in E$. Since $\{V_\alpha\}_{\alpha \in A}$ is an open covering of E, there exists an $r(\mathbf{x}) > 0$ such that $B_{r(\mathbf{x})}(\mathbf{x}) \subset V_\alpha$. Thus by the Borel Covering Lemma, there exist finitely many points $\mathbf{x}_1, \ldots, \mathbf{x}_N$ and radii $r_j := r(\mathbf{x}_j)$ such that

$$
E \subseteq \bigcup_{j=1}^N B_{r_j}(\mathbf{x}_j).
$$

But by construction, for each j there is an index $\alpha_j \in A$ such that $B_{r_j}(\mathbf{x}_j) \subset V_{\alpha_j}$. We conclude that $\{V_{\alpha_j}\}_{j=1}^N$ is a finite subcovering of E. ∎

It is important to recognize that the Heine–Borel Theorem no longer holds if either *closed* or *bounded* is dropped from the hypothesis, even when $n = 1$ and E is an interval. Indeed, $(0, 1)$ is bounded but not closed and

$$
(0, 1) = \bigcup_{n \in \mathbf{N}} \left(\frac{1}{n}, 1 - \frac{1}{n}\right)
$$

has no finite subcovering. And $[1, \infty)$ is closed but not bounded and

$$
[1, \infty) \subset \bigcup_{n \in \mathbf{N}} \left(1 - \frac{1}{n}, n\right)
$$

has no finite subcovering.

As promised above, we can use the Heine–Borel Theorem to extend local results to global ones.

9.12 EXAMPLE.

Suppose that E is a closed, bounded subset of $\mathbf{R}$. If for every $x \in E$ there exist a nonnegative function $f = f_x$ and a number $r = r(x) > 0$ such that f is differentiable on $\mathbf{R}$, $f(t) > 0$ for $t \in (x - r, x + r)$, and $f(t) = 0$ for $t \notin (x - 2r, x + 2r)$, prove that there exist a differentiable function f and an open set V which contains E such that f is nonzero and bounded on E and $f(x) = 0$ for $x \notin V$.

Proof. For each $x \in E$, choose $r = r_x > 0$ and $f_x \geq 0$ such that f_x is differentiable on $\mathbf{R}$, $f_x(t) > 0$ for $t \in I_r(x) := (x - r, x + r)$, and $f_x(t) = 0$ for

$t \notin J_r(x) := (x - 2r, x + 2r)$. Since $\{I_r(x)\}_{x \in E}$ covers E, which is compact by the Heine–Borel Theorem, there exist finitely many x_j's in E such that

$$E \subset \bigcup_{j=1}^{N} I_{r_j}(x_j)$$

for $r_j = r(x_j)$. Set $f = \sum_{k=1}^{N} f_{x_j}$ and $V = \bigcup_{j=1}^{N} J_{r_j}(x_j)$. Then f is differentiable since it is a finite sum of differentiable functions. Clearly, V contains E. V is open since it is a union of open intervals. If $x \in E$, then $x \in I_{r_j}(x_j)$ for some j, so $f_{x_j}(x) > 0$. Thus $f(x) \geq 0 + \cdots + f_{x_j}(x) + \cdots + 0 > 0$ for all $x \in E$. Moreover, since f_{x_k} is continuous on $H := \bigcup_{k=1}^{N} [x_k - r_k, x_k + r_k]$, the Extreme Value Theorem implies that there are constants M_k that $|f_{x_k}| \leq M_k$ on H for all k. Thus $|f(x)| \leq M_1 + \cdots + M_N =: M$ for all $x \in H \supset E$. Finally, if $x \notin V$, then $x \notin J_{r_j}(x_j)$ for all j. Thus $f(x) = 0 + 0 + \cdots + 0 = 0$. ∎

EXERCISES

9.2.1. Suppose that K is compact in $\mathbf{R}^n$ and $E \subseteq K$. Prove that E is compact if and only if E is closed.

9.2.2. Suppose that E is a bounded noncompact subset of $\mathbf{R}^n$ and that $f : E \to (0, \infty)$. If there is a $g : E \to \mathbf{R}$ such that $g(\mathbf{x}) > f(\mathbf{x})$ for all $\mathbf{x} \in E$, then does there exist $\mathbf{x}_1, \ldots, \mathbf{x}_N \in E$ such that

$$E \subset \bigcup_{j=1}^{N} B_{g(\mathbf{x}_j)}(\mathbf{x}_j) \, ?$$

9.2.3. Suppose that E is a compact subset of $\mathbf{R}$. If for every $x \in E$ there exist a nonnegative function $f = f_x$ and an $r = r(x) > 0$ such that f is C^∞ on $\mathbf{R}$, $f(t) = 1$ for $t \in (x - r, x + r)$, and $f(t) = 0$ for $t \notin (x - 2r, x + 2r)$, prove that there exist a differentiable function f, a nonzero constant M, and a bounded, open set V which contains E such that $1 \leq f(x) \leq M$ for all $x \in E$ and $f(x) = 0$ for $x \notin V$.

9.2.4. Suppose that K is compact in $\mathbf{R}^n$ and that for every $\mathbf{x} \in K$ there is an $r = r(\mathbf{x}) > 0$ such that $B_r(\mathbf{x}) \cap K = \{\mathbf{x}\}$. Prove that K is a finite set.

9.2.5. Let E be closed and bounded in $\mathbf{R}$, and suppose that for each $x \in E$ there is a function f_x, nonnegative, nonconstant, increasing, and C^∞ on $\mathbf{R}$, such that $f_x(x) > 0$ and $f_x'(y) = 0$ for $y \notin E$. Prove that there exists a nonnegative, nonconstant, increasing C^∞ function f on $\mathbf{R}$ such that $f(y) > 0$ for all $y \in E$ and $f'(y) = 0$ for all $y \notin E$.

9.2.6. Suppose that $\mathbf{f} : \mathbf{R}^n \to \mathbf{R}^m$ and that $\mathbf{a} \in K$, where K is a compact, connected subset of $\mathbf{R}^n$. Suppose further that for each $\mathbf{x} \in K$ there is a $\delta_{\mathbf{x}} > 0$ such that $\mathbf{f}(\mathbf{x}) = \mathbf{f}(\mathbf{y})$ for all $\mathbf{y} \in B_{\delta_{\mathbf{x}}}(\mathbf{x})$. Prove that $\mathbf{f}$ is constant on K; that is, if $\mathbf{a} \in K$, then $\mathbf{f}(\mathbf{x}) = \mathbf{f}(\mathbf{a})$ for all $\mathbf{x} \in K$.

9.2.7. Define the distance between two nonempty subsets A and B of $\mathbf{R}^n$ by

$$\text{dist}(A, B) := \inf\{\|\mathbf{x} - \mathbf{y}\| : \mathbf{x} \in A \quad \text{and} \quad \mathbf{y} \in B\}.$$

a) Prove that if A and B are compact sets which satisfy $A \cap B = \emptyset$, then $\text{dist}(A, B) > 0$.

b) Show that there exist nonempty, closed sets A, B in $\mathbf{R}^2$ such that $A \cap B = \emptyset$ but $\text{dist}(A, B) = 0$.

9.2.8. Suppose that E and V are subsets of $\mathbf{R}$ with E bounded, V open, and $\overline{E} \subset V$. Prove that there is a C^∞ function $f : E \to \mathbf{R}$ such that $f(x) > 0$ for $x \in E$ and $f(x) = 0$ for $x \notin V$.

9.3 LIMITS OF FUNCTIONS

We now turn our attention to limits of functions. By a *vector function (from n variables to m variables)* we shall mean a function $\mathbf{f}$ of the form $\mathbf{f} : A \to \mathbf{R}^m$, where $A \subseteq \mathbf{R}^n$ and m, n are fixed positive integers. Since $\mathbf{f}(\mathbf{x}) \in \mathbf{R}^m$ for each $\mathbf{x} \in A$, there are functions $f_j : A \to \mathbf{R}$ (called the *coordinate* or *component functions* of $\mathbf{f}$) such that $\mathbf{f}(\mathbf{x}) = (f_1(\mathbf{x}), \dots, f_m(\mathbf{x}))$ for each $\mathbf{x} \in A$. When $m = 1$, $\mathbf{f}$ has only one component and we shall call $\mathbf{f}$ *real valued*. Sometimes, to emphasize the fact that a function is real valued (as opposed to vector valued), we will denote real functions without boldface (i.e., $f : \mathbf{R}^n \to \mathbf{R}$).

If $\mathbf{f} = (f_1, \dots, f_m)$ is a vector function where the f_j's have intrinsic domains (e.g., the f_j's might be defined by formulas), then the *maximal domain* of $\mathbf{f}$ is defined to be the intersection of the domains of the f_j's. The following examples illustrate this idea.

9.13 EXAMPLES.

i) Find the maximal domain of

$$\mathbf{f}(x, y) = (\log(xy - y + 2x - 2), \sqrt{9 - x^2 - y^2}).$$

ii) Find the maximal domain of

$$\mathbf{g}(x, y) = (\sqrt{1 - x^2}, \log(x^2 - y^2), \sin x \cos y).$$

Solution.

i) This function has two components: $f_1(x, y) = \log(xy - y + 2x - 2)$ and $f_2(x, y) = \sqrt{9 - x^2 - y^2}$. Since the logarithm is real valued only when its argument is positive, the domain of f_1 is the set of points (x, y) which satisfy

$$0 < xy - y + 2x - 2 = (x - 1)(y + 2).$$

Since the square root function is real valued if and only if its argument is nonnegative, the domain of f_2 is the set of points (x, y) which satisfy $x^2 + y^2 \le 9$. Thus the maximal domain of **f** is

$$\{(x, y) : x^2 + y^2 \le 9 \quad \text{and} \quad (x - 1)(y + 2) > 0\}.$$

(This set was shown in Figure 8.7.)

ii) This function has three component functions: $g_1(x, y) = \sqrt{1 - x^2}$, $g_2(x, y) = \log(x^2 - y^2)$, and $g_3(x, y) = \sin x \cos y$. g_1 is real valued when $1 - x^2 \ge 0$; that is, $-1 \le x \le 1$. g_2 is real valued when $x^2 - y^2 > 0$; that is, when $-|x| < y < |x|$. The domain of g_3 is all of $\mathbf{R}^2$. Thus the maximal domain of **g** is

$$\{(x, y) : -1 \le x \le 1 \text{ and } - |x| < y < |x|\}.$$

(This set was shown in Figure 8.6.) ∎

To set up notation for the algebra of vector functions, let $E \subseteq \mathbf{R}^n$ and suppose that $\mathbf{f}, \mathbf{g} : E \to \mathbf{R}^m$. For each $\mathbf{x} \in E$, the *scalar product* of an $\alpha \in \mathbf{R}$ with **f** is defined by

$$(\alpha \mathbf{f})(\mathbf{x}) := \alpha \mathbf{f}(\mathbf{x}),$$

the *sum* of **f** and **g** is defined by

$$(\mathbf{f} + \mathbf{g})(\mathbf{x}) := \mathbf{f}(\mathbf{x}) + \mathbf{g}(\mathbf{x}),$$

the (*Euclidean*) *dot product* of **f** and **g** is defined by

$$(\mathbf{f} \cdot \mathbf{g})(\mathbf{x}) := \mathbf{f}(\mathbf{x}) \cdot \mathbf{g}(\mathbf{x}),$$

and (when $m = 3$) the *cross product* of **f** and **g** is defined by

$$(\mathbf{f} \times \mathbf{g})(\mathbf{x}) := \mathbf{f}(\mathbf{x}) \times \mathbf{g}(\mathbf{x}).$$

(Notice that when $m = 1$, the dot product of two functions is the pointwise product defined in Section 3.1.)

Here is the multivariable analogue of two-sided limits (compare with Definition 3.1).

9.14 Definition.

Let $n, m \in \mathbf{N}$ and $\mathbf{a} \in \mathbf{R}^n$, let V be an open set which contains $\mathbf{a}$, and suppose that $\mathbf{f} : V \setminus \{\mathbf{a}\} \to \mathbf{R}^m$. Then $\mathbf{f}(\mathbf{x})$ is said to *converge to* **L**, *as* $\mathbf{x}$ *approaches* **a**, if and only if for every $\varepsilon > 0$ there is a $\delta > 0$ (which in general depends on ε, **f**, V, and **a**) such that

$$0 < \|\mathbf{x} - \mathbf{a}\| < \delta \quad \text{implies} \quad \|\mathbf{f}(\mathbf{x}) - \mathbf{L}\| < \varepsilon.$$

9.14 Definition. (Continued)

In this case we write $\mathbf{f}(\mathbf{x}) \to \mathbf{L}$ as $\mathbf{x} \to \mathbf{a}$ or

$$\mathbf{L} = \lim_{\mathbf{x} \to \mathbf{a}} \mathbf{f}(\mathbf{x})$$

and call $\mathbf{L}$ the *limit* of $\mathbf{f}(\mathbf{x})$ as $\mathbf{x}$ approaches $\mathbf{a}$.

Using the analogy between the norm on $\mathbf{R}^n$ and the absolute value on $\mathbf{R}$, we can extend much of the theory of limits of functions developed in Chapter 3 to the Euclidean space setting. Here is a brief summary of what is true.

9.15 Theorem. *Let $\mathbf{a} \in \mathbf{R}^n$, let V be an open set which contains $\mathbf{a}$, and suppose that $\mathbf{f}, \mathbf{g} : V \setminus \{\mathbf{a}\} \to \mathbf{R}^m$.*

i) *If $\mathbf{f}(\mathbf{x}) = \mathbf{g}(\mathbf{x})$ for all $\mathbf{x} \in V \setminus \{\mathbf{a}\}$ and if $\mathbf{f}(\mathbf{x})$ has a limit as $\mathbf{x} \to \mathbf{a}$, then $\mathbf{g}(\mathbf{x})$ has a limit as $\mathbf{x} \to \mathbf{a}$, and*

$$\lim_{\mathbf{x} \to \mathbf{a}} \mathbf{g}(\mathbf{x}) = \lim_{\mathbf{x} \to \mathbf{a}} \mathbf{f}(\mathbf{x}).$$

ii) *[SEQUENTIAL CHARACTERIZATION OF LIMITS]. $\mathbf{L} = \lim_{\mathbf{x} \to \mathbf{a}} \mathbf{f}(\mathbf{x})$ exists if and only if $\mathbf{f}(\mathbf{x}_k) \to \mathbf{L}$ as $k \to \infty$ for every sequence $\mathbf{x}_k \in V \setminus \{\mathbf{a}\}$ which converges to $\mathbf{a}$ as $k \to \infty$.*

iii) *Suppose that $\alpha \in \mathbf{R}$. If $\mathbf{f}(\mathbf{x})$ and $\mathbf{g}(\mathbf{x})$ have limits, as $\mathbf{x}$ approaches $\mathbf{a}$, then so do $(\mathbf{f} + \mathbf{g})(\mathbf{x})$, $(\alpha\mathbf{f})(\mathbf{x})$, $(\mathbf{f} \cdot \mathbf{g})(\mathbf{x})$, and $\|f(\mathbf{x})\|$. In fact,*

$$\lim_{\mathbf{x} \to \mathbf{a}} (\mathbf{f} + \mathbf{g})(\mathbf{x}) = \lim_{\mathbf{x} \to \mathbf{a}} \mathbf{f}(\mathbf{x}) + \lim_{\mathbf{x} \to \mathbf{a}} \mathbf{g}(\mathbf{x}),$$

$$\lim_{\mathbf{x} \to \mathbf{a}} (\alpha\mathbf{f})(\mathbf{x}) = \alpha \lim_{\mathbf{x} \to \mathbf{a}} \mathbf{f}(\mathbf{x}),$$

$$\lim_{\mathbf{x} \to \mathbf{a}} (\mathbf{f} \cdot \mathbf{g})(\mathbf{x}) = \left(\lim_{\mathbf{x} \to \mathbf{a}} \mathbf{f}(x) \right) \cdot \left(\lim_{\mathbf{x} \to \mathbf{a}} \mathbf{g}(\mathbf{x}) \right),$$

and

$$\left\| \lim_{\mathbf{x} \to \mathbf{a}} \mathbf{f}(\mathbf{x}) \right\| = \lim_{\mathbf{x} \to \mathbf{a}} \|\mathbf{f}(\mathbf{x})\|.$$

Moreover, when $m = 3$,

$$\lim_{\mathbf{x} \to \mathbf{a}} (\mathbf{f} \times \mathbf{g})(\mathbf{x}) = \left(\lim_{\mathbf{x} \to \mathbf{a}} \mathbf{f}(\mathbf{x}) \right) \times \left(\lim_{\mathbf{x} \to \mathbf{a}} \mathbf{g}(\mathbf{x}) \right),$$

and when $m = 1$ and the limit of $\mathbf{g}$ is nonzero,

$$\lim_{\mathbf{x} \to \mathbf{a}} \mathbf{f}(\mathbf{x})/\mathbf{g}(\mathbf{x}) = \left(\lim_{\mathbf{x} \to \mathbf{a}} \mathbf{f}(\mathbf{x}) \right) / \left(\lim_{\mathbf{x} \to \mathbf{a}} \mathbf{g}(\mathbf{x}) \right).$$

iv) *[SQUEEZE THEOREM FOR FUNCTIONS]. Suppose that $f, g, h : V \setminus \{\mathbf{a}\} \to \mathbf{R}$ and that $g(\mathbf{x}) \leq h(\mathbf{x}) \leq f(\mathbf{x})$ for all $\mathbf{x} \in V \setminus \{\mathbf{a}\}$. If*

$$\lim_{\mathbf{x} \to \mathbf{a}} f(\mathbf{x}) = \lim_{\mathbf{x} \to \mathbf{a}} g(\mathbf{x}) = L,$$

then the limit of h also exists, as $\mathbf{x} \to \mathbf{a}$, *and*

$$\lim_{\mathbf{x} \to \mathbf{a}} h(\mathbf{x}) = L.$$

v) *Suppose that* U *is open in* $\mathbf{R}^m$, *that* $\mathbf{L} \in U$, *and that* $\mathbf{h} : U \to \mathbf{R}^p$ *for some* $p \in \mathbf{N}$. *If* $\mathbf{L} = \lim_{\mathbf{x} \to \mathbf{a}} \mathbf{g}(\mathbf{x})$ *and* $\mathbf{h}$ *is continuous at* $\mathbf{L}$. *Then*

$$\lim_{\mathbf{x} \to \mathbf{a}} (\mathbf{h} \circ \mathbf{g})(\mathbf{x}) = \mathbf{h}(\mathbf{L}).$$

How do we actually compute the limit of a given vector-valued function? The following result shows that evaluation of such limits reduces to the real-valued case (i.e., the case where the range is one dimensional). Consequently, our examples will be almost exclusively real-valued.

9.16 Theorem. *Let* $\mathbf{a} \in \mathbf{R}^n$, *let* V *be an open set which contains* $\mathbf{a}$, *and suppose that* $\mathbf{f} = (f_1, \ldots, f_m) : V \setminus \{\mathbf{a}\} \to \mathbf{R}^m$. *Then*

$$\lim_{\mathbf{x} \to \mathbf{a}} \mathbf{f}(\mathbf{x}) = \mathbf{L} := (L_1, L_2, \ldots, L_m) \tag{1}$$

exists in $\mathbf{R}^m$ *if and only if*

$$\lim_{\mathbf{x} \to \mathbf{a}} f_j(\mathbf{x}) = L_j \tag{2}$$

exists in $\mathbf{R}$ *for each* $j = 1, 2, \ldots, m$.

Proof. By the Sequential Characterization of Limits, we must show that for all sequences $\mathbf{x}_k \in V \setminus \{\mathbf{a}\}$ which converge to $\mathbf{a}$, $\mathbf{f}(\mathbf{x}_k) \to \mathbf{L}$ as $k \to \infty$ if and only if $f_j(\mathbf{x}_k) \to L_j$, as $k \to \infty$, for each $1 \le j \le n$. But this last statement is obviously true by Theorem 9.2. Therefore, (1) holds if and only if (2) holds. ∎

Using Theorem 9.15, it is easy to see that if f_j are real functions continuous at points a_j, for $j = 1, 2, \ldots, n$, then $F(x_1, x_2, \ldots, x_n) := f_1(x_1) + f_2(x_2) + \cdots + f_n(x_n)$ and $G(x_1, x_2, \ldots, x_n) := f_1(x_1) f_2(x_2) \cdots f_n(x_n)$ both have limits at the point $\mathbf{a} := (a_1, a_2, \ldots, a_n)$. In fact (see Exercise 9.3.6),

$$\lim_{\mathbf{x} \to \mathbf{a}} F(\mathbf{x}) = F(\mathbf{a}) \quad \text{and} \quad \lim_{\mathbf{x} \to \mathbf{a}} G(\mathbf{x}) = G(\mathbf{a}).$$

This observation is often used in conjunction with Theorem 9.16 to evaluate simple limits.

9.17 EXAMPLES.

i) Find

$$\lim_{(x,y) \to (0,0)} (3xy + 1, e^y + 2).$$

ii) Prove that the function

$$f(x, y) = \frac{2 + x - y}{1 + 2x^2 + 3y^2}$$

has a limit as $(x, y) \to (0, 0)$.

Solution.

i) By Theorem 9.16, this limit is $(0 + 1, e^0 + 2) = (1, 3)$.
ii) The polynomial $2 + x - y$ (respectively, $1 + 2x^2 + 3y^2$) converges to 2 (respectively, to 1) as $(x, y) \to (0, 0)$. Hence, by Theorem 9.15,

$$\lim_{(x,y) \to (0,0)} \frac{2 + x - y}{1 + 2x^2 + 3y^2} = \frac{2}{1} = 2. \qquad \blacksquare$$

The application of Theorem 9.15 in Example 9.17ii is legitimate because the limit quotient was not of the form 0/0. For the multidimensional case, l'Hôpital's Rule does not work (see the paragraph following Example 9.19). Hence, proving that a limit of the form 0/0 exists in several variables often involves showing that the absolute value of the function minus its supposed limit, $|f(\mathbf{x}) - L|$, is dominated by (i.e., less than or equal to) some nonnegative function g which satisfies $g(\mathbf{x}) \to 0$ as $\mathbf{x} \to \mathbf{a}$. Here is a typical example.

9.18 EXAMPLE.

Prove that

$$f(x, y) = \frac{3x^2 y}{x^2 + y^2}$$

converges as $(x, y) \to (0, 0)$.

Proof. Since the numerator is a polynomial of degree 3 (see Exercise 9.3.4) and the denominator is a polynomial of degree 2, we expect the numerator to overpower the denominator; that is, the limit to be 0 as $(x, y) \to (0, 0)$. To prove this, we must estimate $f(x, y)$ near $(0,0)$. Since $2|xy| \le x^2 + y^2$ for all $(x, y) \in \mathbf{R}^2$, it is easy to check that

$$|f(x, y)| \le \frac{3}{2}|x| < 2|x|$$

for all $(x, y) \ne (0, 0)$. Let $\varepsilon > 0$ and set $\delta = \varepsilon/2$. If $0 < \|(x, y)\| < \delta$, then $|f(x, y)| < 2|x| \le 2\|(x, y)\| < 2\delta = \varepsilon$. Thus, by definition,

$$\lim_{(x,y) \to (0,0)} f(x, y) = 0. \qquad \blacksquare$$

It is important to realize that by Definition 9.14, if $\mathbf{f}$ converges to $\mathbf{L}$ as $\mathbf{x} \to \mathbf{a}$, then $\|f(\mathbf{x}) - \mathbf{L}\|$ is small for all $\mathbf{x}$ near $\mathbf{a}$. In particular, $\mathbf{f}(\mathbf{x}) \to \mathbf{L}$ as $\mathbf{x} \to \mathbf{a}$, no matter what path **x** takes. The next two examples show how to use this observation to prove that a limit does not exist.

9.19 EXAMPLE.

Prove that the function

$$f(x, y) = \frac{2xy}{x^2 + y^2}$$

has no limit as $(x, y) \to (0, 0)$.

> **Proof.** Let $g(x, y) = 2xy$ and $h(x, y) = x^2 + y^2$ and suppose that f has a limit L, as $(x, y) \to (0, 0)$. If (x, y) approaches $(0,0)$ along a vertical path (e.g., if $x = 0$ and $y \to 0$, $y \neq 0$), then $g(0, y) = 0$ but $h(0, y) \neq 0$ so L would have to be 0. On the other hand, if (x, y) approaches $(0,0)$ along a "diagonal" path (e.g., if $y = x$ and $x \to 0$, $x \neq 0$), then $g(x, x) = h(x, x) = 2x^2$ so L would have to be 1. Since $0 \neq 1$, f has no limit at $(0,0)$. ∎

Recall that if f is a function of two variables then f_x denotes the partial derivative of f with respect to x and f_y denotes the partial derivative of f with respect to y. Sometimes students guess that an analogue of l'Hôpital's Rule holds for $\mathbf{R}^2$; for example, that

$$\lim_{(x,y) \to (a,b)} \frac{g(x, y)}{h(x, y)} \stackrel{?}{=} \lim_{(x,y) \to (a,b)} \frac{g_x(x, y) + g_y(x, y)}{h_x(x, y) + h_y(x, y)}.$$

Example 9.19 shows that this guess is wrong. Indeed, $f = g/h$ satisfies $g_x + g_y = 2x + 2y = h_x + h_y$, but the limit of f is NOT $(2x + 2y)/(2x + 2y) = 1$. It has no limit. In particular, be careful about applying one-dimensional results to functions of several variables unless the analogue has been proved.

In the solution to Example 9.19, the diagonal path was chosen so that the denominator of $f(x, y)$ would collapse to a single term. This same strategy is used in the next example.

9.20 EXAMPLE.

Determine whether

$$f(x, y) = \frac{xy^2}{x^2 + y^4}$$

has a limit as $(x, y) \to (0, 0)$.

> **Solution.** The vertical path $x = 0$ gives $f(0, y) = 0$ even before we take the limit as $y \to 0$. On the other hand, the parabolic path $x = y^2$ gives

$$f(y^2, y) = \frac{y^4}{2y^4} = \frac{1}{2} \neq 0.$$

Therefore, f cannot have a limit as $(x, y) \to (0, 0)$. ∎

[Notice that if $y = mx$, then

$$f(x, y) = \frac{m^2 x^3}{x^2 + m^4 x^4} \to 0$$

as $x \to 0$. Thus, Example 9.20 shows that the two-dimensional limit of a function might not exist even when its limit along every linear path exists and gives the same value.]

When asked whether the limit of a function $f(\mathbf{x})$ exists, it is natural to begin by taking the limit as each variable moves independently. Comparing Examples 9.17 and 9.19, we see that this strategy works for some functions but not all. To look at this problem more closely, we introduce the following terminology. Let V be an open set in $\mathbf{R}^2$, let $(a, b) \in V$, and suppose that $\mathbf{f} : V \setminus \{(a, b)\} \to \mathbf{R}^m$. The *iterated limits* of $\mathbf{f}$ at (a, b) are defined to be

$$\lim_{x \to a} \lim_{y \to b} \mathbf{f}(x, y) := \lim_{x \to a} \left(\lim_{y \to b} \mathbf{f}(x, y) \right) \text{ and } \lim_{y \to b} \lim_{x \to a} \mathbf{f}(x, y) := \lim_{y \to b} \left(\lim_{x \to a} \mathbf{f}(x, y) \right),$$

when they exist.

The iterated limits of a given function might not exist. Even when they do, we cannot be sure that the corresponding two-dimensional limit exists. Indeed, although the iterated limits of the function f in Example 9.19 above exist and are both zero at $(0,0)$, f has no limit as $(x, y) \to (0, 0)$.

It is even possible for both iterated limits to exist but give different values.

9.21 EXAMPLE.

Evaluate the iterated limits of

$$f(x, y) = \frac{x^2}{x^2 + y^2}$$

at $(0,0)$.

Solution. For each $x \neq 0$, $x^2/(x^2 + y^2) \to 1$ as $y \to 0$. Therefore,

$$\lim_{x \to 0} \lim_{y \to 0} \frac{x^2}{x^2 + y^2} = \lim_{x \to 0} \frac{x^2}{x^2} = 1.$$

On the other hand,

$$\lim_{y \to 0} \lim_{x \to 0} \frac{x^2}{x^2 + y^2} = \lim_{y \to 0} \frac{0}{y^2} = 0. \qquad \blacksquare$$

This leads us to ask, When are the iterated limits equal? The following result shows that if f has a limit as $(x, y) \to (a, b)$ and both iterated limits exist, then these limits must be equal.

9.22 Remark. *Suppose that I and J are open intervals, that* $a \in I$ *and* $b \in J$, *and that* $f : (I \times J) \setminus \{(a, b)\} \to \mathbf{R}$. *If*

$$g(x) := \lim_{y \to b} f(x, y)$$

exists for each $x \in I \setminus \{a\}$, *if* $\lim_{x \to a} f(x, y)$ *exists for each* $y \in J \setminus \{b\}$, *and if* $f(x, y) \to L$ *as* $(x, y) \to (a, b)$ *(in* $\mathbf{R}^2$*), then*

$$L = \lim_{x \to a} \lim_{y \to b} f(x, y) = \lim_{y \to b} \lim_{x \to a} f(x, y).$$

Proof. Let $\varepsilon > 0$. By hypothesis, choose $\delta > 0$ such that

$$0 < \|(x, y) - (a, b)\| < \delta \quad \text{implies} \quad |f(x, y) - L| < \varepsilon.$$

Suppose that $x \in I$ and that $0 < |x - a| < \delta/\sqrt{2}$. Then for any y which satisfies $0 < |y - b| < \delta/\sqrt{2}$, we have $0 < \|(x, y) - (a, b)\| < \delta$; hence

$$|g(x) - L| \le |g(x) - f(x, y)| + |f(x, y) - L| < |g(x) - f(x, y)| + \varepsilon.$$

Taking the limit of this inequality as $y \to b$, we find that $|g(x) - L| \le \varepsilon$ for all $x \in I$ which satisfy $0 < |x - a| < \delta/\sqrt{2}$. It follows that $g(x) \to L$ as $x \to a$; that is,

$$L = \lim_{x \to a} \lim_{y \to b} f(x, y).$$

A similar argument proves that the other iterated limit also exists and equals L. ∎

Notice by Example 9.21 that the conclusion of Remark 9.22 might not hold if the hypothesis "$f(x, y) \to L$ as $(x, y) \to (a, b)$" is omitted. In particular, if the limit of a function does not exist, we must be careful about changing the order of an iterated limit.

EXERCISES

9.3.1. For each of the following functions, find the maximal domain of **f**, prove that the limit of f exists as $(x, y) \to (a, b)$, and find the value of that limit. (*Note:* You can prove that the limit exists without using ε's and δ's—see Example 9.17.)

a) $\qquad \mathbf{f}(x, y) = \left(\dfrac{x - 1}{y - 1}, x + 2 \right), \qquad (a, b) = (1, -1)$

b) $\qquad \mathbf{f}(x, y) = \left(\dfrac{y \sin x}{x}, \tan \dfrac{x}{y}, x^2 + y^2 - xy \right), \qquad (a, b) = (0, 1)$

c) $$\mathbf{f}(x, y) = \left(\frac{x^4 + y^4}{x^2 + y^2}, \frac{\sqrt{|xy|}}{\sqrt[3]{x^2 + y^2}} \right),   \qquad (a, b) = (0, 0)$$

d) $$\mathbf{f}(x, y) = \left(\frac{x^2 - 1}{y^2 + 1}, \frac{x^2 y - 2xy + y - (x - 1)^2}{x^2 + y^2 - 2x - 2y + 2} \right),   \qquad (a, b) = (1, 1)$$

9.3.2. Compute the iterated limits at $(0,0)$ of each of the following functions. Determine which of these functions has a limit as $(x, y) \to (0, 0)$ in $\mathbf{R}^2$, and prove that the limit exists.

a) $$f(x, y) = \frac{\sin x \sin y}{x^2 + y^2}$$

b) $$f(x, y) = \frac{x^2 + y^4}{x^2 + 2y^4}$$

c) $$f(x, y) = \frac{x - y}{(x^2 + y^2)^\alpha},   \qquad \alpha < \frac{1}{2}$$

9.3.3. Prove that each of the following functions has a limit as $(x, y) \to (0, 0)$.

a) $$f(x, y) = \frac{x^3 - y^3}{x^2 + y^2},   \qquad (x, y) \neq (0, 0)$$

b) $$f(x, y) = \frac{|x|^\alpha y^4}{x^2 + y^4},   \qquad (x, y) \neq (0, 0),$$

where α is ANY positive number.

9.3.4. A *polynomial* on $\mathbf{R}^n$ of *degree* N is a function of the form

$$P(x_1, x_2, \ldots, x_n) = \sum_{j_1 = 0}^{N_1} \cdots \sum_{j_n = 0}^{N_n} a_{j_1, \ldots, j_n} x_1^{j_1} \ldots x_n^{j_n},$$

where $a_{j_1, \ldots, j_n}$ are scalars, $N_1, \ldots, N_n$ are nonnegative integers, and $N = N_1 + N_2 + \cdots + N_n$. Prove that if P is a polynomial on $\mathbf{R}^n$ and $\mathbf{a} \in \mathbf{R}^n$, then $\lim_{\mathbf{x} \to \mathbf{a}} P(\mathbf{x}) = P(\mathbf{a})$.

9.3.5. Suppose that $\mathbf{a} \in \mathbf{R}^n$, that $\mathbf{L} \in \mathbf{R}^m$, and that $\mathbf{f} : \mathbf{R}^n \to \mathbf{R}^m$. Prove that if $\mathbf{f}(\mathbf{x}) \to \mathbf{L}$ as $\mathbf{x} \to \mathbf{a}$, then there is an open set V containing $\mathbf{a}$ and a constant $M > 0$ such that $\|\mathbf{f}(\mathbf{x})\| \leq M$ for all $\mathbf{x} \in V$.

9.3.6. Suppose that $\mathbf{a} = (a_1, \ldots, a_n) \in \mathbf{R}^n$, that $f_j : \mathbf{R} \to \mathbf{R}$ for $j = 1, 2, \ldots, n$, and that $g(x_1, x_2, \ldots, x_n) := f_1(x_1) \cdots f_n(x_n)$.

a) Prove that if $f_j(t) \to f_j(a_j)$ as $t \to a_j$, for each $j = 1, \ldots, n$, then $g(\mathbf{x}) \to f_1(a_1) \cdots f_n(a_n)$ as $\mathbf{x} \to \mathbf{a}$.

b) Show that the limit of g might not exist if, even for one j, the hypothesis "$f_j(t) \to f_j(a_j)$" is replaced by "$f_j(t) \to L_j$" for some $L_j \in \mathbf{R}$.

9.3.7. Suppose that $g : \mathbf{R} \to \mathbf{R}$ is differentiable and that $g'(x) > 1$ for all $x \in \mathbf{R}$. Prove that if $g(1) = 0$ and $f(x, y) = (x - 1)^2(y + 1)/(yg(x))$, then there is an $L \in \mathbf{R}$ such that $f(x, y) \to L$ as $(x, y) \to (1, b)$ for all $b \in \mathbf{R} \setminus \{0\}$.

9.3.8. a) Prove Theorem 9.15i.
   b) Prove Theorem 9.15ii.
   c) Prove Theorem 9.15iii.
   d) Prove Theorem 9.15iv.

9.4 CONTINUOUS FUNCTIONS

In this section we define what it means for a vector function to be continuous, obtain analogues of many results in Sections 3.3 and 3.4, and examine how open sets, closed sets, and connected sets behave under images and inverse images by continuous functions. We shall use these results many times in the subsequent chapters.

9.23 Definition.

Let E be a nonempty subset of $\mathbf{R}^n$ and $\mathbf{f} : E \to \mathbf{R}^m$.

i) $\mathbf{f}$ is said to be *continuous at* $\mathbf{a} \in E$ if and only if for every $\varepsilon > 0$ there is a $\delta > 0$ (which in general depends on $\varepsilon, \mathbf{f}, E$, and $\mathbf{a}$) such that

$$\|x - \mathbf{a}\| < \delta \quad \text{and} \quad \mathbf{x} \in E \quad \text{imply} \quad \|\mathbf{f(x)} - \mathbf{f(a)}\| < \varepsilon. \qquad (3)$$

ii) $\mathbf{f}$ is said to be *continuous on* E (notation: $\mathbf{f} : E \to \mathbf{R}^m$ is continuous) if and only if $\mathbf{f}$ is continuous at every $\mathbf{x} \in E$.

Suppose that E is a nonempty subset of $\mathbf{R}^n$. It is easy to verify that $\mathbf{f}$ is continuous at $\mathbf{a} \in E$ if and only if $\mathbf{f(x}_k) \to \mathbf{f(a)}$ for all $\mathbf{x}_k \in E$ which converge to $\mathbf{a}$. Hence, by Theorem 9.4, if $\mathbf{f}$ and $\mathbf{g}$ are continuous at a point $\mathbf{a} \in E$ (respectively, continuous on E), then so are $\mathbf{f} + \mathbf{g}$, $\alpha\mathbf{f}$ (for $\alpha \in \mathbf{R}$), $\mathbf{f} \cdot \mathbf{g}$, $\|\mathbf{f}\|$, and (when $m = 3$) $\mathbf{f} \times \mathbf{g}$. Moreover, if $\mathbf{f} : E \to \mathbf{R}^m$ is continuous at $\mathbf{a} \in E$ and $\mathbf{g} : f(E) \to \mathbf{R}^p$ is continuous at $\mathbf{f(a)} \in f(E)$, then $\mathbf{g} \circ \mathbf{f}$ is continuous at $\mathbf{a} \in E$.

We shall frequently need a stronger version of continuity.

9.24 Definition.

Let E be a nonempty subset of $\mathbf{R}^n$ and $\mathbf{f} : E \to \mathbf{R}^m$. Then $\mathbf{f}$ is said to be *uniformly continuous* on E (notation: $\mathbf{f} : E \to \mathbf{R}^m$ is uniformly continuous) if and only if for every $\varepsilon > 0$ there is a $\delta > 0$ such that

$$\|\mathbf{x} - \mathbf{a}\| < \delta \quad \text{and} \quad \mathbf{x}, \mathbf{a} \in E \quad \text{imply} \quad \|\mathbf{f(x)} - \mathbf{f(a)}\| < \varepsilon.$$

As in the real case, continuity and uniform continuity of a vector function are equivalent on closed, bounded sets. By the powerful Heine–Borel Theorem, we need only verify this result for compact sets. The definition of compact sets allows us to construct a direct proof (compare with the proof of Theorem 3.39).

9.25 Theorem. *Let E be a nonempty compact subset of $\mathbf{R}^n$. If $\mathbf{f}$ is continuous on E, then $\mathbf{f}$ is uniformly continuous on E.*

Proof. Suppose that $\mathbf{f}$ is continuous on E. Given $\varepsilon > 0$ and $\mathbf{a} \in E$, choose $\delta(\mathbf{a}) > 0$ such that

$$\mathbf{x} \in B_{\delta(\mathbf{a})}(\mathbf{a}) \quad \text{and} \quad \mathbf{x} \in E \quad \text{imply} \quad \|\mathbf{f}(\mathbf{x}) - \mathbf{f}(\mathbf{a})\| < \frac{\varepsilon}{2}.$$

Since $\delta(\mathbf{a})/2$ is positive for all $\mathbf{a} \in E$, the collection $\{B_{\delta(\mathbf{a})/2}\}_{\mathbf{a} \in E}$ is an open covering of E. By the definition of compact sets, there exist finitely many points $\mathbf{a}_j \in E$ and numbers $\delta_j := \delta(\mathbf{a}_j)/2$ such that

$$E \subset \bigcup_{j=1}^{N} B_{\delta_j}(\mathbf{a}_j). \tag{4}$$

Set $\delta := \min\{\delta_1, \ldots, \delta_N\}$. Clearly, $\delta > 0$.

Suppose that $\mathbf{x}, \mathbf{a} \in E$ with $\|\mathbf{x} - \mathbf{a}\| < \delta$. By (4), $\mathbf{x}$ belongs to $B_{\delta_j}(\mathbf{a}_j)$ for some $1 \le j \le N$. Hence, $\|\mathbf{a} - \mathbf{a}_j\| \le \|\mathbf{a} - \mathbf{x}\| + \|\mathbf{x} - \mathbf{a}_j\| < \delta_j + \delta_j = 2\delta_j = \delta(\mathbf{a}_j)$; that is, $\mathbf{a}$ also belongs to $B_{\delta(\mathbf{a}_j)}(\mathbf{a}_j)$. It follows, therefore, from the choice of $\delta(\mathbf{a}_j)$ that

$$\|\mathbf{f}(\mathbf{x}) - \mathbf{f}(\mathbf{a})\| \le \|\mathbf{f}(\mathbf{x}) - \mathbf{f}(\mathbf{a}_j)\| + \|\mathbf{f}(\mathbf{a}_j) - \mathbf{f}(\mathbf{a})\| < \frac{\varepsilon}{2} + \frac{\varepsilon}{2} = \varepsilon.$$

This proves that $\mathbf{f}$ is uniformly continuous on E. ∎

Thus continuous vector functions behave much the same as continuous real functions.

When we turn our attention to how continuous functions interact with the topological structure of $\mathbf{R}^n$, we again find a surprising bonus. The ε's and δ's disappear.

9.26 Theorem. *Suppose that $E \subseteq \mathbf{R}^n$ and that $\mathbf{f} : E \to \mathbf{R}^m$. Then $\mathbf{f}$ is continuous on E if and only if $\mathbf{f}^{-1}(V)$ is relatively open in E for every V open in $\mathbf{R}^m$.*

Proof. Suppose that $\mathbf{f}$ is continuous on E and that V is open in $\mathbf{R}^m$. Since $\varnothing$ is open, we may suppose that some $\mathbf{a} \in \mathbf{f}^{-1}(V)$. By Remark 8.27, to show that $\mathbf{f}^{-1}(V)$ is relatively open in E we need to find a $\delta > 0$ such that $B_\delta(\mathbf{a}) \cap E \subset \mathbf{f}^{-1}(V)$. But $\mathbf{f}(\mathbf{a}) \in V$ and V is open, so there is a $\varepsilon > 0$ such that $B_\varepsilon(\mathbf{f}(\mathbf{a})) \subset V$. Since $\mathbf{f}$ is continuous at $\mathbf{a} \in E$, choose $\delta > 0$ such that $\|\mathbf{x} - \mathbf{a}\| < \delta$ and $\mathbf{x} \in E$

implies $\|\mathbf{f}(\mathbf{x}) - \mathbf{f}(\mathbf{a})\| < \varepsilon$; that is, $\mathbf{x} \in B_\delta(\mathbf{a}) \cap E$ implies $f(\mathbf{x}) \in B_\varepsilon(\mathbf{f}(\mathbf{a}))$. It follows that $\mathbf{f}(B_\delta(\mathbf{a}) \cap E) \subseteq B_\varepsilon(\mathbf{f}(\mathbf{a})) \subset V$; that is, $B_\delta(\mathbf{a}) \cap E \subset f^{-1}(V)$.

Conversely, if $\mathbf{a} \in E$ and $\varepsilon > 0$, then $B_\varepsilon(\mathbf{f}(\mathbf{a}))$ is open in $\mathbf{R}^m$. By hypothesis $\mathbf{f}^{-1}(B_\varepsilon(\mathbf{f}(\mathbf{a})))$ is relatively open in E; that is, by Remark 8.27, there is a $\delta > 0$ such that $B_\delta(\mathbf{a}) \cap E \subset \mathbf{f}^{-1}(B_\varepsilon(\mathbf{f}(\mathbf{a})))$. We conclude that if $\|\mathbf{x} - \mathbf{a}\| < \delta$ and $\mathbf{x} \in E$, then $\|\mathbf{f}(\mathbf{x}) - \mathbf{f}(\mathbf{a})\| < \varepsilon$ (i.e., that $\mathbf{f}$ is continuous at $\mathbf{a} \in E$). ∎

By Theorem 9.26, when $\mathbf{f}$ is continuous on E, $\mathbf{f}^{-1}$ takes open sets to relatively open sets in E. If the domain E is open (e.g., if the domain is $\mathbf{R}^n$), then the word *relatively* can be dropped (see Exercise 9.4.3). We shall refer to this property by saying that open sets are *invariant* under inverse images by continuous functions.

Analogues of these results also hold for closed sets (see Exercises 9.4.5a and 9.4.4). In particular, if $\mathbf{f}$ is continuous and its domain is a closed set, then $\mathbf{f}^{-1}$ takes closed sets to closed sets. Thus closed sets are invariant under inverse images by continuous functions.

It is natural to ask whether bounded sets or connected sets are invariant under inverse images by continuous functions. The following examples show that the answers to these questions are no, even when the range and domain are one dimensional.

9.27 EXAMPLES.

i) If $f(x) = 1/(x^2 + 1)$ and $E = (0, 1]$, then f is continuous on $\mathbf{R}$ and E is bounded, but $f^{-1}(E) = (-\infty, \infty)$ is not bounded.

ii) If $f(x) = x^2$ and $E = (1, 4)$, then f is continuous on $\mathbf{R}$ and E is connected, but $f^{-1}(E) = (-2, -1) \cup (1, 2)$ is not connected.

We now turn our attention from inverse images of sets to images of sets. Are open sets and closed sets invariant under images by continuous functions? The following examples show that the answers to these questions are also no.

9.28 EXAMPLES.

i) If $f(x) = x^2$ and $V = (-1, 1)$, then f is continuous on V and V is open, but $f(V) = [0, 1)$ is neither open nor closed.

ii) If $f(x) = 1/x$ and $E = [1, \infty)$, then f is continuous on E and E is closed, but $f(E) = (0, 1]$ is neither open nor closed.

As the next result shows, however, if a set is both closed and bounded (i.e., compact), then so is its image under any continuous function. This innocent-looking result has far-reaching consequences which we shall exploit on many occasions.

9.29 Theorem. *If H is compact in $\mathbf{R}^n$ and $\mathbf{f} : H \to \mathbf{R}^m$ is continuous on H, then $\mathbf{f}(H)$ is compact in $\mathbf{R}^m$.*

Proof. Suppose that $\{V_\alpha\}_{\alpha \in A}$ is an open covering of $f(H)$. Then, by Theorem 1.37, parts iii and v, $\{\mathbf{f}^{-1}(V_\alpha)\}_{\alpha \in A}$ covers H. But by Theorem 9.26, $\mathbf{f}^{-1}(V_\alpha)$ are relatively open in H; that is, there exist open sets O_α such that $\mathbf{f}^{-1}(V_\alpha) = O_\alpha \cap H$. Since $\{O_\alpha\}_{\alpha \in A}$ is an open covering of H and H is compact, there exist $\alpha_j \in A$ such that $H \subset \bigcup_{j=1}^N O_{\alpha_j}$. We conclude by Theorem 1.37, parts i and v, that

$$
\mathbf{f}(H) = \mathbf{f}\left(\bigcup_{j=1}^N O_{\alpha_j} \cap H \right) = \bigcup_{j=1}^N \mathbf{f}\left(\mathbf{f}^{-1}(V_{\alpha_j}) \right) = \bigcup_{j=1}^N V_{\alpha_j};
$$

that is, $\mathbf{f}(H)$ is compact by definition. ∎

Connected sets are also invariant under images by continuous functions.

9.30 Theorem. *If E is connected in $\mathbf{R}^n$ and $\mathbf{f} : E \to \mathbf{R}^m$ is continuous on E, then $\mathbf{f}(E)$ is connected in $\mathbf{R}^m$.*

Proof. Suppose that $\mathbf{f}(E)$ is not connected. By Definition 8.28, there exist a pair of relatively open sets U, V in $\mathbf{f}(E)$ which separates $\mathbf{f}(E)$; that is, $U \cap \mathbf{f}(E) \neq \emptyset$, $V \cap \mathbf{f}(E) \neq \emptyset$, $\mathbf{f}(E) = U \cup V$, and $U \cap V = \emptyset$. Set $A := \mathbf{f}^{-1}(U)$ and $B := \mathbf{f}^{-1}(V)$. By Exercise 9.4.5b, A and B are relatively open in E. Since $\mathbf{f}(E) = U \cup V$ and both $\mathbf{f}^{-1}(U)$ and $\mathbf{f}^{-1}(V)$ are subsets of E, we also have (see Theorem 1.37iii)

$$
E = \mathbf{f}^{-1}(U) \cup \mathbf{f}^{-1}(V) = A \cup B.
$$

Finally, $U \cap V = \emptyset$ implies $\mathbf{f}^{-1}(U) \cap \mathbf{f}^{-1}(V) = \emptyset$ (i.e., $A \cap B = \emptyset$). Thus A, B is a pair of relatively open sets which separates E (i.e., E is not connected, a contradiction). ∎

Keeping track of which kind of sets are invariant under images and inverse images by continuous functions is a powerful tool. To illustrate this fact, we offer the following four results.

9.31 Remark. *The graph $y = f(x)$ of a continuous real function f on an interval $[a, b]$ is compact and connected.*

Proof. The function $F(x) = (x, f(x))$ is continuous from $[a, b]$ into $\mathbf{R}^2$, and the graph of $y = f(x)$ for $x \in [a, b]$ is the image of $[a, b]$ under F. Hence the graph of f is compact and connected by Theorems 9.29 and 9.30. ∎

It is interesting to note that this property actually characterizes continuity of real functions (see Theorem 9.51).

To appreciate the perspective that the topological point of view gives, compare the following simple proof with that of its one-dimensional analogue (Theorem 3.26).

9.32 Theorem.   [EXTREME VALUE THEOREM].
Suppose that H is a nonempty subset of $\mathbf{R}^n$ and that $f : H \to \mathbf{R}$. If H is compact, and f is continuous on H, then

$$M := \sup\{f(\mathbf{x}) : \mathbf{x} \in H\} \quad and \quad m := \inf\{f(\mathbf{x}) : \mathbf{x} \in H\}$$

are finite real numbers. Moreover, there exist points $\mathbf{x}_M$, $\mathbf{x}_m \in H$ such that $M = f(\mathbf{x}_M)$ and $m = f(\mathbf{x}_m)$.

Proof. By symmetry, it suffices to prove the result for M. Since H is compact, $f(H)$ is compact by Theorem 9.29. Thus $f(H)$ is closed and bounded by the Heine–Borel Theorem. Since $f(H)$ is bounded, M is finite. By the Approximation Property, choose $\mathbf{x}_k \in H$ such that $f(\mathbf{x}_k) \to M$ as $k \to \infty$. Since $f(H)$ is closed, $M \in f(H)$. Therefore, there is an $\mathbf{x}_M \in H$ such that $M = f(\mathbf{x}_M)$. ∎

(For a multidimensional analogue of Theorem 3.29, see Exercise 9.4.9.)
The following analogue of Theorem 4.32 will be used in Chapter 13 to examine change of parametrizations of curves and surfaces.

9.33 Theorem. *If H is a compact subset of $\mathbf{R}^n$ and $\mathbf{f} : H \to \mathbf{R}^m$ is 1–1 and continuous, then $\mathbf{f}^{-1}$ is continuous on $\mathbf{f}(H)$.*

Proof. By Theorem 9.29 and the Heine–Borel Theorem, $\mathbf{f}(H)$ is closed. Thus, by Exercise 9.4.4, it suffices to show that $(\mathbf{f}^{-1})^{-1}$ takes closed sets to closed sets. To this end, let E be closed in $\mathbf{R}^n$. Since the domain of $\mathbf{f}^{-1}$ is $\mathbf{f}(H)$, we have by definition that

$$(\mathbf{f}^{-1})^{-1}(E) = \{\mathbf{x} \in \mathbf{f}(H) : \mathbf{f}^{-1}(\mathbf{x}) = y \quad \text{for some } \mathbf{y} \in E\}.$$

Since $\mathbf{f}$ is 1–1, $\mathbf{f}^{-1}(\mathbf{x}) = \mathbf{y} \in E$ implies that $\mathbf{x} \in \mathbf{f}(E)$. Thus $(\mathbf{f}^{-1})^{-1}(E) = \mathbf{f}(E \cap H)$. But $E \cap H$ is closed (see Theorem 8.24) and bounded (by "the bound" of H), so by Theorem 9.29 and the Heine–Borel Theorem, $\mathbf{f}(E \cap H)$ is closed and bounded. In particular, $(\mathbf{f}^{-1})^{-1}(E) = \mathbf{f}(E \cap H)$ is closed. ∎

The final result of this section shows that "rectangles" are connected in $\mathbf{R}^n$.

9.34 Remark. *If $a_j \le b_j$ for $j = 1, 2, \ldots, n$, then*

$$R := \{(x_1, \ldots, x_n) : a_j \le x_j \le b_j\}$$

is connected.

Proof. Suppose not. Choose nonempty sets U and V, relatively open in R, such that $R = U \cup V$ and $U \cap V = \emptyset$. Let $\mathbf{a} \in U$ and $\mathbf{b} \in V$, and consider the line segment $E := \{t\mathbf{a} + (1 - t)\mathbf{b} : t \in [0, 1]\}$. Since E is a continuous image of the interval $[0, 1]$, we have by Theorems 8.30 and 9.30 that E is connected. On the other hand, since $E \subset R$ by the definition of R, it is easy to check that $U_0 := U \cap E$ and $V_0 := V \cap E$ are nonempty sets which are relatively open in E and satisfy $E = U_0 \cup V_0$ and $U_0 \cap V_0 = \emptyset$. It follows that E is not connected, a contradiction. ∎

EXERCISES

9.4.1. Define f and g on $\mathbf{R}$ by $f(x) = \sin x$ and $g(x) = x/|x|$ if $x \neq 0$ and $g(0) = 0$.

a) Find $f(E)$ and $g(E)$ for $E = (0, \pi)$, $E = [0, \pi]$, $E = (-1, 1)$, and $E = [-1, 1]$. Compare your answers with what Theorems 9.26, 9.29, and 9.30 predict. Explain any differences you notice.

b) Find $f^{-1}(E)$ and $g^{-1}(E)$ for $E = (0, 1)$, $E = [0, 1]$, $E = (-1, 1)$, and $E = [-1, 1]$. Compare your answers with what Theorems 9.26, 9.29, and 9.30 predict. Explain any differences you notice.

9.4.2. Define f on $[0, \infty)$ and g on $\mathbf{R}$ by $f(x) = \sqrt{x}$ and $g(x) = 1/x$ if $x \neq 0$ and $g(0) = 0$.

a) Find $f(E)$ and $g(E)$ for $E = (0, 1)$, $E = [0, 1)$, and $E = [0, 1]$. Compare your answers with what Theorems 9.26, 9.29, and 9.30 predict. Explain any differences you notice.

b) Find $f^{-1}(E)$ and $g^{-1}(E)$ for $E = (-1, 1)$ and $E = [-1, 1]$. Compare your answers with what Theorems 9.26, 9.29, and 9.30 predict. Explain any differences you notice.

9.4.3. **This exercise is used in this section and in Chapter 11**. Suppose that A is open in $\mathbf{R}^n$ and $\mathbf{f} : A \to \mathbf{R}^m$. Prove that $\mathbf{f}$ is continuous on A if and only if $\mathbf{f}^{-1}(V)$ is open in $\mathbf{R}^n$ for every open subset V of $\mathbf{R}^m$.

9.4.4. Suppose that A is closed in $\mathbf{R}^n$ and $\mathbf{f} : A \to \mathbf{R}^m$. Prove that $\mathbf{f}$ is continuous on A if and only if $\mathbf{f}^{-1}(E)$ is closed in $\mathbf{R}^n$ for every closed subset E of $\mathbf{R}^m$.

9.4.5. Suppose that $E \subseteq \mathbf{R}^n$ and that $\mathbf{f} : E \to \mathbf{R}^m$.

a) Prove that $\mathbf{f}$ is continuous on E if and only if $\mathbf{f}^{-1}(B)$ is relatively closed in E for every closed subset B of $\mathbf{R}^m$.

b) Suppose that $\mathbf{f}$ is continuous on E. Prove that if V is relatively open in $\mathbf{f}(E)$, then $\mathbf{f}^{-1}(V)$ is relatively open in E, and if B is relatively closed in $\mathbf{f}(E)$, then $\mathbf{f}^{-1}(B)$ is relatively closed in E.

9.4.6. Prove that

$$f(x, y) = \begin{cases} e^{-1/|x-y|} & x \neq y \\ 0 & x = y \end{cases}$$

is continuous on $\mathbf{R}^2$.

*9.4.7. **This exercise is used in Section *9.6**. Let H be a nonempty, closed, bounded subset of $\mathbf{R}^n$.

a) Suppose that $\mathbf{f} : H \to \mathbf{R}^m$ is continuous. Prove that

$$\|\mathbf{f}\|_H := \sup_{x \in H} \|\mathbf{f}(\mathbf{x})\|$$

is finite and there exists an $\mathbf{x}_0 \in H$ such that $\|\mathbf{f}(\mathbf{x}_0)\| = \|\mathbf{f}\|_H$.

b) A sequence of functions $\mathbf{f}_k : H \to \mathbf{R}^m$ is said to converge uniformly on H to a function $\mathbf{f} : H \to \mathbf{R}^m$ if and only if for every $\varepsilon > 0$ there is an $N \in \mathbf{N}$ such that

$$k \geq N \quad \text{and} \quad \mathbf{x} \in H \quad \text{imply} \quad \|\mathbf{f}_k(\mathbf{x}) - \mathbf{f}(\mathbf{x})\| < \varepsilon.$$

Show that $\|\mathbf{f}_k - \mathbf{f}\|_H \to 0$ as $k \to \infty$ if and only if $\mathbf{f}_k \to \mathbf{f}$ uniformly on H as $k \to \infty$.

c) Prove that a sequence of functions $\mathbf{f}_k$ converges uniformly on H if and only if for every $\varepsilon > 0$ there is an $N \in \mathbf{N}$ such that

$$k, j \geq N \quad \text{implies} \quad \|\mathbf{f}_k - \mathbf{f}_j\|_H < \varepsilon.$$

9.4.8. Let $E \subset \mathbf{R}^n$ and suppose that D is *dense* in E (i.e., that $D \subseteq E$ and $\overline{D} = E$). If $\mathbf{f} : D \to \mathbf{R}^m$ is uniformly continuous on D, prove that $\mathbf{f}$ has a continuous extension to E; that is, prove that there is a continuous function $\mathbf{g} : E \to \mathbf{R}^m$ such that $\mathbf{g}(\mathbf{x}) = \mathbf{f}(\mathbf{x})$ for all $\mathbf{x} \in D$.

9.4.9. [INTERMEDIATE VALUE THEOREM]. Let E be a connected subset of $\mathbf{R}^n$. If $f : E \to \mathbf{R}$ is continuous, $f(\mathbf{a}) \neq f(\mathbf{b})$ for some $\mathbf{a}, \mathbf{b} \in E$, and y is a number which lies between $f(\mathbf{a})$ and $f(\mathbf{b})$, then prove that there is an $\mathbf{x} \in E$ such that $f(\mathbf{x}) = y$. (You may use Theorem 8.30.)

*9.4.10. **This exercise is used to prove *Corollary 11.35**.

a) A set $E \subseteq \mathbf{R}^n$ is said to be *polygonally connected* if and only if any two points $\mathbf{a}, \mathbf{b} \in E$ can be connected by a polygonal path in E; that is, there exist points $\mathbf{x}_k \in E$, $k = 1, \ldots, N$, such that $\mathbf{x}_0 = \mathbf{a}$, $\mathbf{x}_N = \mathbf{b}$ and $L(\mathbf{x}_{k-1}; \mathbf{x}_k) \subseteq E$ for $k = 1, \ldots, N$. Prove that every polygonally connected set in $\mathbf{R}^n$ is connected.

b) Let $E \subseteq \mathbf{R}^n$ be open and $\mathbf{x}_0 \in E$. Let U be the set of points $\mathbf{x} \in E$ which can be polygonally connected in E to $\mathbf{x}_0$. Prove that U is open.

c) Prove that every open connected set in $\mathbf{R}^n$ is polygonally connected.

*9.5 COMPACT SETS

This section requires no material from any other enrichment section.

In this section we give a more complete description of compact sets. Most of the results we state are trivial to prove by appealing to the hard part of Heine–Borel Theorem, specifically, that closed and bounded subsets of a Euclidean space are compact. Since this powerful result does not hold in some non-Euclidean spaces, our proofs will appeal only to the basic definition of compact sets and, hence, avoid using the Heine–Borel Theorem.

We begin by expanding our terminology concerning what we mean by a "covering."

9.35 Definition.

Let $\mathcal{V} = \{V_\alpha\}_{\alpha \in A}$ be a collection of subsets of $\mathbf{R}^n$, and suppose that $E \subseteq \mathbf{R}^n$.

i) $\mathcal{V}$ is said to *cover* E (or be a *covering* of E) if and only if

$$E \subseteq \bigcup_{\alpha \in A} V_\alpha.$$

(No assumption about V_α being open is made.)

ii) Let $\mathcal{V}$ be a covering of E. $\mathcal{V}$ is said to have a *finite* (respectively, *countable*) *subcovering* if and only if there is a finite (respectively, an at most countable) subset A_0 of A such that $\{V_\alpha\}_{\alpha \in A_0}$ covers E.

Notice that the collections of open intervals

$$\left\{\left(\frac{1}{k+1}, \frac{k}{k+1}\right)\right\}_{k \in \mathbf{N}} \quad \text{and} \quad \left\{\left(-\frac{1}{k}, \frac{k+1}{k}\right)\right\}_{k \in \mathbf{N}}$$

are open coverings of the interval $(0, 1)$. The first covering of $(0, 1)$ has no finite subcovering, but any member of the second covering covers $(0, 1)$. Thus, an open covering of an arbitrary set might not have a finite subcovering.

Our first general result about compact sets shows that every "space" contains compact sets.

9.36 Remark. *The empty set and all finite subsets of* $\mathbf{R}^n$ *are compact.*

Proof. These statements follow immediately from Definition 9.10. The empty set needs no set to cover it, and any finite set H can be covered by finitely many sets, one set for each element in H. ∎

Since the empty set and finite sets are also closed, it is natural to ask whether there is a relationship between compact sets and closed sets in general. The following three results address this question.

9.37 Remark. *A compact set is always closed.*

Proof. This result follows easily from the sequential characterization of closed sets (see the second paragraph in the proof of Theorem 9.11). ∎

Since $\{(k-1, k+1) : k \in \mathbf{N}\}$ is an open covering of the closed set $E := [1, \infty)$, the converse of Theorem 9.37 is false. The following result shows that this is not the case if E is a subset of some compact set.

9.38 Remark. *A closed subset of a compact set is compact.*

Proof. Let E be a closed subset of H, where H is compact, and suppose that $\mathcal{V} = \{V_\alpha\}_{\alpha \in A}$ is an open covering of E. Now $E^c = \mathbf{R}^n \setminus E$ is open. Thus $\mathcal{V} \cup \{E^c\}$ is an open covering of H. Since H is compact, there is a finite set $A_0 \subseteq A$ such that

$$H \subseteq E^c \cup \left(\bigcup_{\alpha \in A_0} V_\alpha \right).$$

But $E \cap E^c = \emptyset$. Therefore, E is covered by $\{V_\alpha\}_{\alpha \in A_0}$. ∎

Finally, we show that every open covering of a set in a Euclidean space has a countable subcovering.

9.39 Theorem. **[LINDELÖF].**
Let $n \in \mathbf{N}$ and let E be a subset of $\mathbf{R}^n$. If $\{V_\alpha\}_{\alpha \in A}$ is a collection of open sets and $E \subseteq \bigcup_{\alpha \in A} V_\alpha$, then there is an at most countable subset A_0 of A such that

$$E \subseteq \bigcup_{\alpha \in A_0} V_\alpha.$$

Proof. Let $\mathcal{T}$ be the collection of open balls with rational radii and rational centers (i.e., centers which belong to $\mathbf{Q}^n$). This collection is countable. Moreover, by the proof of the Borel Covering Lemma, $\mathcal{T}$ "approximates" the collection of open balls in the following sense: Given any open ball $B_r(\mathbf{x}) \subseteq \mathbf{R}^n$, there is a ball $B_\rho(\mathbf{a}) \in \mathcal{T}$ such that $\mathbf{x} \in B_\rho(\mathbf{a})$ and $B_\rho(\mathbf{a}) \subseteq B_r(\mathbf{x})$.

To prove the theorem, let $\mathbf{x} \in E$. By hypothesis, $\mathbf{x} \in V_\alpha$ for some $\alpha \in A$. Since V_α is open, there is a $r > 0$ such that $B_r(\mathbf{x}) \subset V_\alpha$. Since $\mathcal{T}$ approximates open balls, we can choose a ball $B_\mathbf{x} \in \mathcal{T}$ such that $\mathbf{x} \in B_\mathbf{x} \subseteq V_\alpha$. The collection $\mathcal{T}$ is countable and, hence, so is the subcollection

$$\{U_1, U_2, \dots\} := \{B_x : x \in E\}.$$

By the choice of the balls $B_\mathbf{x}$, for each $k \in \mathbf{N}$ there is at least one $\alpha_k \in A$ such that $U_k \subseteq V_{\alpha_k}$. Hence, by construction,

$$E \subseteq \bigcup_{\mathbf{x} \in E} B_\mathbf{x} = \bigcup_{k \in \mathbf{N}} U_k \subseteq \bigcup_{k \in \mathbf{N}} V_{\alpha_k}.$$

Thus, set $A_0 := \{\alpha_k : k \in \mathbf{N}\}$. ∎

EXERCISES

9.5.1. Identify which of the following sets are compact and which are not. If E is not compact, find the smallest compact set H (if there is one) such that $E \subset H$.

a) $\{1/k : k \in \mathbf{N}\} \cup \{0\}$
b) $\{(x, y) \in \mathbf{R}^2 : a \le x^2 + y^2 \le b\}$ for real numbers $0 < a < b$
c) $\{(x, y) \in \mathbf{R}^2 : y = \sin(1/x)$ for some $x \in (0, 1]\}$
d) $\{(x, y) \in \mathbf{R}^2 : |xy| \le 1\}$

9.5.2. Let A, B be compact subsets of $\mathbf{R}^n$. Using only Definition 9.10ii, prove that $A \cup B$ and $A \cap B$ are compact.

9.5.3. Suppose that $E \subseteq \mathbf{R}$ is compact and nonempty. Prove that $\sup E$, $\inf E \in E$.

9.5.4. Suppose that $\{V_\alpha\}_{\alpha \in A}$ is a collection of nonempty open sets in $\mathbf{R}^n$ which satisfies $V_\alpha \cap V_\beta = \emptyset$ for all $\alpha \ne \beta$ in A. Prove that A is countable. What happens to this result when *open* is omitted?

9.5.5. Prove that if V is open in $\mathbf{R}^n$, then there are open balls $B_1, B_2, \ldots$, such that

$$V = \bigcup_{j \in \mathbf{N}} B_j.$$

Prove that every open set in $\mathbf{R}$ is a countable union of open intervals.

9.5.6. Let $n \in \mathbf{N}$.

a) A subset E of $\mathbf{R}^n$ is said to be *sequentially compact* if and only if every sequence $\mathbf{x}_k$ in E has a convergent subsequence $\mathbf{x}_{k_j}$ whose limit belongs to E. Prove that every compact set is sequentially compact.
b) Prove that every sequentially compact set is closed and bounded.
c) Prove that a set $E \subset \mathbf{R}^n$ is sequentially compact if and only if it is compact.

9.5.7. Let $H \subseteq \mathbf{R}^n$. Prove that H is compact if and only if every cover $\{E_\alpha\}_{\alpha \in A}$ of H, where the E_α's are relatively open in H, has a finite subcovering.

*9.6 APPLICATIONS

This section uses no material from any prior enrichment sections.

We have seen that topological concepts (e.g., closed sets, open sets, and connected sets) are powerful theoretical tools. In this section we continue developing this theme by obtaining three independent theorems (i.e., you may cover them in any order) which further elucidate results we obtained in earlier chapters.

Our first application of topological ideas is a partial converse of Theorem 7.10. A sequence of real valued functions $\{f_k\}$ is said to be *pointwise increasing* (respectively, *pointwise decreasing*) on a subset E of $\mathbf{R}^n$ if and only if $f_k(\mathbf{x}) \le f_{k+1}(\mathbf{x})$ [respectively, $f_k(\mathbf{x}) \ge f_{k+1}(\mathbf{x})$] for all $\mathbf{x} \in E$ and $k \in \mathbf{N}$. A sequence is said to be *pointwise monotone* on E if and only if it is pointwise increasing on E or pointwise decreasing on E.

9.40 Theorem. [DINI].
Suppose that H is a compact subset of $\mathbf{R}^n$ and that $f_k : H \to \mathbf{R}$ is a pointwise monotone sequence of continuous functions. If $f_k \to f$ pointwise on H as $k \to \infty$ and if f is continuous on H, then $f_k \to f$ uniformly on H. In particular, if ϕ_k is a pointwise monotone sequence of functions continuous on an interval $[a, b]$ which converges pointwise to a continuous function, then

$$\lim_{k \to \infty} \int_a^b \phi_k(t)\, dt = \int_a^b \left(\lim_{k \to \infty} \phi_k(t) \right) dt.$$

Proof. By Theorem 7.10, we need only show that $f_k \to f$ uniformly on H. We may suppose that f_k is pointwise increasing and that $H \neq \emptyset$.

Let $\varepsilon > 0$. For each $\mathbf{x} \in H$, choose $N(\mathbf{x}) \in \mathbf{N}$ such that

$$k \geq N(\mathbf{x}) \quad \text{implies} \quad |f_k(\mathbf{x}) - f(\mathbf{x})| < \frac{\varepsilon}{3}.$$

Since f and $f_{N(\mathbf{x})}$ are continuous on H, choose an $r = r(\mathbf{x}) > 0$ such that

$$\mathbf{y} \in H \cap B_r(\mathbf{x}) \quad \text{implies} \quad |f(\mathbf{x}) - f(\mathbf{y})| < \frac{\varepsilon}{3} \quad \text{and} \quad |f_{N(\mathbf{x})}(\mathbf{x}) - f_{N(\mathbf{x})}(\mathbf{y})| < \frac{\varepsilon}{3}.$$

By the Heine–Borel Theorem, choose $\mathbf{x}_j \in H$ and $r_j = r(\mathbf{x}_j)$ such that

$$H \subset \bigcup_{j=1}^{M} B_{r_j}(\mathbf{x}_j).$$

Set $N = \max\{N(\mathbf{x}_1), \dots, N(\mathbf{x}_M)\}$, let $\mathbf{x} \in H$, and suppose that $k \geq N$. Since $\mathbf{x} \in B_{r_j}(\mathbf{x}_j)$ for some $j \in \{1, \dots, M\}$ and $k \geq N(\mathbf{x}_j)$, it follows that

$$|f(\mathbf{x}) - f_k(\mathbf{x})| = f(\mathbf{x}) - f_k(\mathbf{x}) \leq f(\mathbf{x}) - f_{N(\mathbf{x}_j)}(\mathbf{x})$$
$$\leq |f(\mathbf{x}) - f(\mathbf{x}_j)| + |f(\mathbf{x}_j) - f_{N(\mathbf{x}_j)}(\mathbf{x}_j)|$$
$$+ |f_{N(\mathbf{x}_j)}(\mathbf{x}_j) - f_{N(\mathbf{x}_j)}(\mathbf{x})|$$
$$< \frac{\varepsilon}{3} + \frac{\varepsilon}{3} + \frac{\varepsilon}{3} = \varepsilon.$$

Since this inequality holds for all $\mathbf{x} \in H$, we conclude that $f_k \to f$ uniformly on H as $k \to \infty$. $\blacksquare$

Our next application of topological ideas is a characterization of Riemann integrability of a function f by the size of the set of points of discontinuity of f. To measure the size of this set, we make the following definition. (Recall that $|I|$ denotes the length of an interval I.)

9.41 Definition.

i) A set $E \subset \mathbf{R}$ is said to be of *measure zero* if and only if for every $\varepsilon > 0$ there is a countable collection of intervals $\{I_j\}_{j \in \mathbf{N}}$ which covers E such that

$$\sum_{j=1}^{\infty} |I_j| \le \varepsilon.$$

ii) A function $f : [a, b] \rightarrow \mathbf{R}$ is said to be *almost everywhere continuous* on $[a, b]$ if and only if the set of points $x \in [a, b]$ where f is discontinuous is a set of measure zero.

Notice that, by definition, if E is of measure zero, then every subset of E is also of measure zero. Loosely speaking, a set is of measure zero if it is so sparse that it can be covered by a sequence of intervals whose total length is as small as we wish.

It is easy to see that a single point $E = \{x\}$ is a set of measure zero. Indeed, $I_1 := (x - \varepsilon/2, x + \varepsilon/2)$, $I_k := \emptyset$, for $k \ge 2$, cover E and have total length ε. Modifying this technique, we can show that any finite set is a set of measure zero (see also Remark 9.42 below). On the other hand, by the Heine–Borel Theorem, any open covering of $[a, b]$ has a finite subcovering; hence, any covering of $[a, b]$ by open intervals must have total length greater than or equal to $b - a$. In particular, a nondegenerate interval cannot be of measure zero.

The following result shows that if a set is small in the set theoretical sense, then it is small in the measure theoretical sense.

9.42 Remark. *Every at most countable set of real numbers is a set of measure zero.*

Proof. We may suppose that E is countable, say $E = \{x_1, x_2, \ldots\}$. Given $\varepsilon > 0$ and $j \in \mathbf{N}$, set

$$I_j = (x_j - \varepsilon 2^{-j-1}, x_j + \varepsilon 2^{-j-1}).$$

Then $x_j \in I_j$ and $|I_j| = \varepsilon 2^{-j}$ for $j \in \mathbf{N}$. Therefore, $E \subseteq \cup_{j=1}^{\infty} I_j$ and

$$\sum_{j=1}^{\infty} |I_j| = \varepsilon \sum_{j=1}^{\infty} \frac{1}{2^j} = \varepsilon. \qquad \blacksquare$$

The converse of Remark 9.42 is false; that is, there exist uncountable sets of measure zero (see Exercise 9.6.9 below).

The following result shows that the countable union of sets of measure zero is a set of measure zero.

9.43 Remark. *If $E_1, E_2, \ldots$ is a sequence of sets of measure zero, then*

$$E = \bigcup_{k=1}^{\infty} E_k$$

is also a set of measure zero.

Proof. Let $\varepsilon > 0$. By hypothesis, given $k \in \mathbf{N}$ we can choose a collection of intervals $\{I_j^{(k)}\}_{j \in \mathbf{N}}$ which covers E_k such that

$$\sum_{j=1}^{\infty} |I_j^{(k)}| < \frac{\varepsilon}{2^k}.$$

Then the collection $\{I_j^{(k)}\}_{k,j \in \mathbf{N}}$ is countable, covers E, and

$$\sum_{k=1}^{\infty} \sum_{j=1}^{\infty} |I_j^{(k)}| \leq \sum_{k=1}^{\infty} \frac{\varepsilon}{2^k} = \varepsilon.$$

Consequently, E is of measure zero. ∎

To facilitate our discussion of points of discontinuity, we introduce the following concepts.

9.44 Definition.

Let $[a, b]$ be a closed interval and $f : [a, b] \to \mathbf{R}$ be bounded.

i) The *oscillation* of f on an interval J which intersects $[a, b]$ is defined to be

$$\Omega_f(J) := \sup_{x,y \in J \cap [a,b]} (f(x) - f(y)).$$

ii) The *oscillation* of f at a point $t \in [a, b]$ is defined to be

$$\omega_f(t) := \lim_{h \to 0+} \Omega_f((t - h, t + h)),$$

when this limit exists.

9.45 Remark. *If $f : [a, b] \to \mathbf{R}$ is bounded, then $\omega_f(t)$ exists for all $t \in [a, b]$ and satisfies $0 \leq \omega_f(t) < \infty$.*

Proof. Fix $t \in [a, b]$ and for each interval J, set

$$M_J = \sup_{x \in J \cap [a,b]} f(x), \qquad m_J = \inf_{x \in J \cap [a,b]} f(x).$$

Since $\sup(-f(x)) = -\inf f(x)$, it is obvious that

$$\Omega_f(J) = M_J - m_J \geq 0. \tag{5}$$

Suppose for simplicity that $t \in (a, b)$, and choose h_0 so small that $(t - h_0, t + h_0) \subset (a, b)$. For each $0 < h < h_0$, set

$$F(h) = \Omega_f((t - h, t + h)).$$

By the Monotone Property of Suprema, $F(h)$ is increasing on $(0, h_0)$ and, hence, has a finite limit as $h \to 0+$. By (5), $F(h) \geq 0$. Therefore, $\omega_f(t)$ exists and is both finite and nonnegative. ∎

The next result shows that, by using the oscillation function ω_f, we can represent the set of points of discontinuity of any bounded f as a countable union.

9.46 Remark. *Let $f : [a, b] \to \mathbf{R}$ be bounded. If E represents the set of points of discontinuity of f in $[a, b]$, then*

$$E = \bigcup_{j=1}^{\infty} \left\{ t \in [a, b] : \omega_f(t) \geq \frac{1}{j} \right\}.$$

Proof. By (5), f is continuous at $t \in [a, b]$ if and only if $\omega_f(t) = 0$. Hence, t belongs to E if and only if $\omega_f(t) > 0$. Since, by the Archimedean Principle, $\omega_f(t) > 0$ if and only if $\omega_f(t) \geq 1/j$ for some $j \in \mathbf{N}$, the result follows at once. ∎

We need two technical results about the oscillation of f at a point t.

9.47 Lemma.
Let $f : [a, b] \to \mathbf{R}$ be bounded. For each $\varepsilon > 0$, the set

$$H = \{ t \in [a, b] : \omega_f(t) \geq \varepsilon \}$$

is compact.

Proof. By definition, H is bounded (by $\max\{|a|, |b|\}$). Hence, if the lemma is false, then H is not closed. Hence, there are points $t_k \in H$ such that $t_k \to t$ as $k \to \infty$ but $t \notin H$. Since $\omega_f(t) < \varepsilon$, it follows that there is an $h_0 > 0$ such that

$$\Omega_f((t - h_0, t + h_0)) < \varepsilon. \tag{6}$$

Since $t_k \to t$, choose $N \in \mathbf{N}$ so that

$$\left(t_N - \frac{h_0}{2}, t_N + \frac{h_0}{2} \right) \subset (t - h_0, t + h_0).$$

Then, by (6), $\Omega_f((t_N - h_0/2, t_N + h_0/2)) < \varepsilon$. Therefore, $\omega_f(t_N) < \varepsilon$, which contradicts the fact that $t_N \in H$.   ∎

9.48 Lemma.
Let I be a closed, bounded interval and $f : I \to \mathbf{R}$ be bounded. If $\varepsilon > 0$ and $\omega_f(t) < \varepsilon$ for all $t \in I$, then there is a $\delta > 0$ such that $\Omega_f(J) < \varepsilon$ for all closed intervals $J \subseteq I$ which satisfy $|J| < \delta$.

Proof. For each $t \in I$, choose $\delta_t > 0$ such that

$$\Omega_f((t - \delta_t, t + \delta_t)) < \varepsilon. \tag{7}$$

Since $\delta_t/2 > 0$, use the Heine–Borel Theorem to choose $t_1, \ldots, t_N$ such that

$$I \subset \bigcup_{j=1}^{N} \left(t_j - \frac{\delta_{t_j}}{2}, t_j + \frac{\delta_{t_j}}{2} \right)$$

and set

$$\delta = \min_{1 \leq j \leq N} \frac{\delta_{t_j}}{2}.$$

If $J \subseteq I$, then

$$J \cap \left(t_j - \frac{\delta_{t_j}}{2}, t_j + \frac{\delta_{t_j}}{2} \right) \neq \emptyset$$

for some $j \in \{1, \ldots, N\}$. If J also satisfies $|J| < \delta$, then it follows from $2\delta \leq \delta_{t_j}$ that $J \subseteq (t_j - \delta_{t_j}, t_j + \delta_{t_j})$. In particular, (7) implies

$$\Omega_f(J) \leq \Omega_f((t_j - \delta_{t_j}, t_j + \delta_{t_j})) < \varepsilon. \qquad ∎$$

9.49 Theorem.   [LEBESGUE].
Let $f : [a, b] \to \mathbf{R}$ be bounded. Then f is Riemann integrable on $[a, b]$ if and only if f is almost everywhere continuous on $[a, b]$. In particular, if f is bounded and has countably many points of discontinuity on $[a, b]$, then f is integrable on $[a, b]$.

Proof. Let E be the set of points of discontinuity of f in $[a, b]$. Suppose that f is integrable but E is not of measure zero. By Remarks 9.43 and 9.46, there is a $j_0 \in \mathbf{N}$ such that

$$H := \left\{ t \in [a, b] : \omega_f(t) \geq \frac{1}{j_0} \right\}$$

is not of measure zero. In particular, there is an $\varepsilon_0 > 0$ such that if $\{I_k\}_{k\in\mathbf{N}}$ is any collection of intervals which covers H, then

$$\sum_{k=1}^{\infty} |I_k| \geq \varepsilon_0. \tag{8}$$

Let $P = \{x_0, \ldots, x_n\}$ be a partition of $[a, b]$. If $(x_{k-1}, x_k) \cap H \neq \emptyset$, then, by definition, $M_k(f) - m_k(f) \geq 1/j_0$. Hence,

$$U(f, P) - L(f, P) = \sum_{k=1}^{n} (M_k(f) - m_k(f))\Delta x_k$$

$$\geq \sum_{(x_{k-1}, x_k) \cap H \neq \emptyset} (M_k(f) - m_k(f))\Delta x_k$$

$$\geq \frac{1}{j_0} \sum_{(x_{k-1}, x_k) \cap H \neq \emptyset} \Delta x_k$$

But $\{[x_{k-1}, x_k] : (x_{k-1}, x_k) \cap H \neq \emptyset\}$ is a collection of intervals which covers H. Hence, it follows from (8) that

$$U(f, P) - L(f, P) \geq \frac{\varepsilon_0}{j_0} > 0.$$

Therefore, f cannot be integrable on $[a, b]$.

Conversely, suppose that E is of measure zero. Let $M = \sup_{x\in[a,b]} f(x)$ and $m = \inf_{x\in[a,b]} f(x)$. Given $\varepsilon > 0$, choose $j_0 \in \mathbf{N}$ such that

$$\frac{M - m + b - a}{j_0} < \varepsilon.$$

Since E is of measure zero, so is

$$H = \left\{ t \in [a, b] : \omega_f(t) \geq \frac{1}{j_0} \right\}.$$

Hence, by Definition 9.41, there exists a collection of intervals which covers H, whose lengths sum to a real number less than $1/(2j_0)$. By expanding these intervals slightly, we may suppose that there exist open intervals $I_1, I_2, \ldots,$ which cover H such that

$$\sum_{v=1}^{\infty} |I_v| < \frac{1}{j_0}.$$

Hence, by Lemma 9.47, we can choose $N \in \mathbf{N}$ such that $\{I_1, I_2, \ldots, I_N\}$ covers H and

$$\sum_{v=1}^{N} |I_v| < \frac{1}{j_0}. \tag{9}$$

We must find a partition P such that $U(f, P) - L(f, P) < \varepsilon$. The endpoints of the I_v's form part of this partition. Other points will come from further division of that part of $[a, b]$ not covered by the I_v's. Indeed, let $I' \subseteq [a, b] \setminus (\cup_{v=1}^{N} I_v)$. Since the I_v's cover H, $\omega_f(t) < 1/j_0$ for all $t \in I'$. Hence, by Lemma 9.48, there is a $\delta > 0$ such that if $J \subseteq I'$ satisfies $|J| < \delta$, then $\Omega_f(J) < 1/j_0$. Subdivide $[a, b] \setminus (\cup_{v=1}^{N} I_v)$ into intervals J_ℓ, $\ell = 1, \ldots, s$, such that $|J_\ell| < \delta$. Then

$$\Omega_f(J_\ell) < \frac{1}{j_0} \tag{10}$$

for $\ell = 1, \ldots, s$.

Let $P = \{x_0, x_1, \ldots, x_n\}$ represent the collection of points x such that x is an endpoint of some I_v or of some J_ℓ. Notice that if $(x_{k-1}, x_k) \cap H \neq \emptyset$, then x_{k-1} and x_k are endpoints of some I_v, whence, by (9),

$$\sum_{(x_{k-1}, x_k) \cap H \neq \emptyset} (M_k(f) - m_k(f)) \Delta x_k \leq \frac{M - m}{j_0}.$$

On the other hand, if $(x_{k-1}, x_k) \cap H = \emptyset$, then x_{k-1} and x_k are endpoints of some J_ℓ, whence, by (10),

$$\sum_{(x_{k-1}, x_k) \cap H = \emptyset} (M_k(f) - m_k(f)) \Delta x_k \leq \frac{1}{j_0} \sum_{k=1}^{n} \Delta x_k = \frac{b - a}{j_0}.$$

Consequently,

$$U(f, P) - L(f, P) = \sum_{k=1}^{n} (M_k(f) - m_k(f)) \Delta x_k \leq \frac{M - m + b - a}{j_0} < \varepsilon.$$

We conclude that f is integrable on $[a, b]$. ∎

Recall that if $\alpha > 0$ and $f(x)$ is positive, then

$$f^\alpha(x) := e^{\alpha \log(f(x))}.$$

Suppose that f is Riemann integrable. Although Corollary 5.23 implies that f^n is integrable for each $n \in \mathbf{N}$, we have not yet investigated the integrability of noninteger powers of f (e.g., $\sqrt{f}$ and $\sqrt[3]{f}$). The following result shows that Lebesgue's Theorem answers the question of integrability for all positive powers of f, rational or irrational.

9.50 Corollary. *If* $f : [a, b] \rightarrow [0, \infty)$ *is Riemann integrable, then so is* f^α *for every* $\alpha > 0$.

In our final application, we use connectivity to characterize the graph of a continuous function.

9.51 Theorem. [CLOSED GRAPH THEOREM].
Let I be a closed interval and $f : I \rightarrow \mathbf{R}$. *Then f is continuous on I if and only if the graph of f is closed and connected in* $\mathbf{R}^2$.

Proof. For any interval $J \subseteq I$, let $\mathcal{G}(J)$ represent the graph of $y = f(x)$ for $x \in J$. Suppose that f is continuous on I. The function $x \longmapsto (x, f(x))$ is continuous from I into $\mathbf{R}^2$, and I is connected in $\mathbf{R}$. Thus $\mathcal{G}(I)$ is connected in $\mathbf{R}^2$ by Theorem 9.30. To prove that $\mathcal{G}(I)$ is closed, we shall use Theorem 9.8. Let $x_k \in I$ and $(x_k, f(x_k)) \rightarrow (x, y)$ as $k \rightarrow \infty$. Then $x_k \rightarrow x$ and $f(x_k) \rightarrow y$, as $k \rightarrow \infty$. Hence, $x \in I$ and since f is continuous, $f(x_k) \rightarrow f(x)$. In particular, the graph of f is closed.

Conversely, suppose that the graph of f is closed and connected in $\mathbf{R}^2$. We first show that f satisfies the Intermediate Value Theorem on I. Indeed, suppose to the contrary that there exist $x_1 < x_2$ in I with $f(x_1) \neq f(x_2)$ and a value y_0 between $f(x_1)$ and $f(x_2)$ such that $f(t) \neq y_0$ for all $t \in [x_1, x_2]$. Suppose for simplicity that $f(x_1) < f(x_2)$. Since $f(t) \neq y_0$ for any $t \in [x_1, x_2]$, the open sets

$$U = \{(x, y) : x < x_1\} \cup \{(x, y) : x < x_2, \ y < y_0\},$$
$$V = \{(x, y) : x > x_2\} \cup \{(x, y) : x > x_1, \ y > y_0\}$$

separate $\mathcal{G}(I)$, a contradiction. Therefore, f satisfies the Intermediate Value Theorem on I.

If f is not continuous on I, then there exist numbers $x_0 \in I$, $\varepsilon_0 > 0$, and $x_k \in I$ such that $x_k \rightarrow x_0$ and $|f(x_k) - f(x_0)| > \varepsilon_0$. By symmetry, we may suppose that $f(x_k) > f(x_0) + \varepsilon_0$ for infinitely many k's, say

$$f(x_{k_j}) > f(x_0) + \varepsilon_0 > f(x_0), \qquad j \in \mathbf{N}.$$

By the Intermediate Value Theorem, choose c_j between x_{k_j} and x_0 such that $f(c_j) = f(x_0) + \varepsilon_0$. By construction, $(c_j, f(c_j)) \rightarrow (x_0, f(x_0) + \varepsilon_0)$ and $c_j \rightarrow x_0$ as $j \rightarrow \infty$. Hence, the graph of f on I is not closed. ∎

EXERCISES

9.6.1. Suppose that $f_k : [a, b] \rightarrow [0, \infty)$ for $k \in \mathbf{N}$ and that

$$f(x) := \sum_{k=1}^{\infty} f_k(x)$$

converges pointwise on $[a, b]$. If f and f_k are continuous on $[a, b]$ for each $k \in \mathbf{N}$, prove that

$$\int_a^b \sum_{k=1}^\infty f_k(x)\, dx = \sum_{k=1}^\infty \int_a^b f_k(x)\, dx.$$

9.6.2. Let E be closed and bounded in $\mathbf{R}^n$. Suppose that $g, f_k, g_k : E \to \mathbf{R}$ are continuous on E with $g_k \geq 0$ and $f_1 \geq f_2 \ldots \geq f_k \geq 0$ for $k \in \mathbf{N}$. If $g = \sum_{k=1}^\infty g_k$ converges pointwise on E, prove that $\sum_{k=1}^\infty f_k g_k$ converges uniformly on E.

9.6.3. Suppose that $f, f_k : \mathbf{R} \to \mathbf{R}$ are continuous and nonnegative. Prove that if $f(x) \to 0$ as $x \to \pm\infty$ and $f_k \uparrow f$ everywhere on $\mathbf{R}$, then $f_k \to f$ uniformly on $\mathbf{R}$.

9.6.4. For each of the following functions, find a formula for $\omega_f(t)$.

a)
$$f(x) = \begin{cases} 1 & x \in \mathbf{Q} \\ 0 & x \notin \mathbf{Q} \end{cases}$$

b)
$$f(x) = \begin{cases} 1 & x \geq 0 \\ 0 & x < 0 \end{cases}$$

c)
$$f(x) = \begin{cases} \sin(1/x) & x \neq 0 \\ 0 & x = 0 \end{cases}$$

9.6.5. Prove that $(1 - x/k)^k \to e^{-x}$ uniformly on any closed, bounded subset of $\mathbf{R}$.

9.6.6. Show that if $f : [a, b] \to \mathbf{R}$ is integrable and $g : f([a, b]) \to \mathbf{R}$ is continuous, then $g \circ f$ is integrable on $[a, b]$. (Notice by Remark 3.34 that this result is false if g is allowed even one point of discontinuity.)

9.6.7. Using Theorem 7.10 or Theorem 9.30, prove that each of the following limits exists. Find a value for the limit in each case.

a)
$$\lim_{k \to \infty} \int_0^{\pi/2} \sin x \sqrt{\frac{2k}{4k - 3x}}\, dx$$

b)
$$\lim_{k \to \infty} \int_0^1 x^2 f\left(\frac{k}{k^2 + x}\right) dx$$

where f is continuously differentiable on $[0, 1]$ and $f'(0) > 0$.

c)
$$\lim_{k \to \infty} \int_0^1 x^3 \cos\left(\frac{\log k + x}{k + x}\right) dx$$

d)
$$\lim_{k \to \infty} \int_{-1}^{1} \left(1 + \frac{x}{k}\right)^k e^x \, dx$$

9.6.8. a) Prove that for every $\varepsilon > 0$ there is a sequence of open intervals $\{I_k\}_{k \in \mathbf{N}}$ which covers $[0, 1] \cap \mathbf{Q}$ such that

$$\sum_{k=1}^{\infty} |I_k| < \varepsilon.$$

b) Prove that if $\{I_k\}_{k \in \mathbf{N}}$ is a sequence of open intervals which covers $[0, 1]$, then there is an $N \in \mathbf{N}$ such that

$$\sum_{k=1}^{N} |I_k| \geq 1.$$

9.6.9. Let E_1 be the unit interval $[0, 1]$ with its middle third $(1/3, 2/3)$ removed (i.e., $E_1 = [0, 1/3] \cup [2/3, 1]$). Let E_2 be E_1 with its middle thirds removed; that is,

$$E_2 = [0, 1/9] \cup [2/9, 1/3] \cup [2/3, 7/9] \cup [8/9, 1].$$

Continuing in this manner, generate nested sets E_k such that each E_k is the union of 2^k closed intervals of length $1/3^k$. The *Cantor set* is the set

$$E := \bigcap_{k=1}^{\infty} E_k.$$

Assume that every point $x \in [0, 1]$ has a binary expansion and a ternary expansion; that is, there exist $a_k \in \{0, 1\}$ and $b_k \in \{0, 1, 2\}$ such that

$$x = \sum_{k=1}^{\infty} \frac{a_k}{2^k} = \sum_{k=1}^{\infty} \frac{b_k}{3^k}.$$

(e.g., if $x = 1/3$, then $a_{2k-1} = 0$, $a_{2k} = 1$ for all k and either $b_1 = 1$, $b_k = 0$ for $k > 1$ or $b_1 = 0$ and $b_k = 1$ for all $k > 1$.)

a) Prove that E is a nonempty compact set of measure zero.
b) Show that a point $x \in [0, 1]$ belongs to E if and only if x has a ternary expansion whose digits satisfy $b_k \neq 1$ for all $k \in \mathbf{N}$.
c) Define $f : E \to [0, 1]$ by

$$f\left(\sum_{k=1}^{\infty} \frac{b_k}{3^k}\right) = \sum_{k=1}^{\infty} \frac{b_k/2}{2^k}.$$

Prove that there is a countable subset E_0 of E such that f is 1–1 from $E \backslash E_0$ onto $[0, 1]$ (i.e., prove that E is uncountable).

d) Extend f from E to $[0, 1]$ by making f constant on the middle thirds $E_{k-1} \backslash E_k$. Prove that $f : [0, 1] \to [0, 1]$ is continuous and increasing.

(*Note*: The function f is almost everywhere constant on $[0, 1]$; that is, constant off a set of measure zero. Yet it begins at $f(0) = 0$ and ends at $f(1) = 1$.)

CHAPTER 10

Metric Spaces

This chapter, an alternative to Chapter 9, covers topological ideas in a metric space setting. *If you have already covered Chapter 9, skip this one and proceed directly to Chapter 11.*

10.1 INTRODUCTION

The following concept shows up in many parts of analysis.

10.1 Definition.

A *metric space* is a set X together with a function $\rho : X \times X \to \mathbf{R}$ (called the *metric* of X) which satisfies the following properties for all $x, y, z \in X$:

POSITIVE DEFINITE $\quad \rho(x, y) \geq 0$ with $\rho(x, y) = 0$ if and only if $x = y$,

SYMMETRIC $\quad \rho(x, y) = \rho(y, x)$,

TRIANGLE INEQUALITY $\quad \rho(x, y) \leq \rho(x, z) + \rho(z, y)$.

[Notice that by definition, $\rho(x, y)$ is finite valued for all $x, y \in X$.]

We are already very familiar with a whole class of metric spaces.

10.2 *EXAMPLE.*

Every Euclidean space $\mathbf{R}^n$ is a metric space with metric $\rho(\mathbf{x}, \mathbf{y}) = \|\mathbf{x} - \mathbf{y}\|$.

(We shall call this the *usual metric* on $\mathbf{R}^n$. Unless specified otherwise, we shall always use the usual metric on $\mathbf{R}^n$.)

Proof. By Theorems 1.7 and 8.6, ρ is a metric on $\mathbf{R}^n$. ∎

For the remainder of this chapter, unless otherwise noted, X and Y will denote general metric spaces with respective metrics ρ and τ. We shall develop a theory of convergence (for both sequences and functions) for arbitrary metric spaces. According to Example 10.2, this theory is valid (and will be used by us almost exclusively) on $\mathbf{R}^n$. Why, then, subject ourselves to such stark generality? Why not stick with the concrete Euclidean space case? There are at least three answers to these questions:

1) *Economy.* You will soon discover that there are many other metric spaces which crop up in analysis (e.g., all Hilbert spaces, all normed linear spaces,

and many function spaces, including the space of continuous functions on a closed, bounded interval). Our general theory of convergence in metric spaces will be valid for each of these examples, too.

2) *Visualization.* As we mentioned in Section 1.2, analysis has a strong geometric flavor. Working in an abstract metric space only makes that aspect more apparent.

3) *Simplicity.* Emphasizing the fact that $\mathbf{R}^n$ is a metric space strips $\mathbf{R}$ of all extraneous details (the field operations, the order relation, decimal expansions) so that we can focus our attention on the underlying concept (distance) which governs convergence. Mathematics frequently benefits from such abstraction. Instead of becoming more difficult, generality actually makes the proofs easier to construct.

On the other hand, $\mathbf{R}^2$ provides a good and sufficiently general model for most of the theory of abstract metric spaces (especially convergence of sequences and continuity of functions). For this reason, we often draw two-dimensional pictures to illustrate ideas and motivate proofs in an arbitrary metric space. (For example, see the proof of Remark 10.9 below.) We must not, however, mislead ourselves by believing that $\mathbf{R}^2$ provides a complete picture. Metric spaces have such simple structure that they can take on many bizarre forms. With that in mind, we introduce several more examples.

10.3 EXAMPLE.

$\mathbf{R}$ is a metric space with metric

$$\sigma(x, y) = \begin{cases} 0 & x = y \\ 1 & x \neq y. \end{cases}$$

(This metric is called the *discrete metric*.)

> **Proof.** The function σ is obviously positive definite and symmetric. To prove that σ satisfies the Triangle Inequality, we consider three cases. If $x = z$, then $\sigma(x, y) = 0 + \sigma(z, y) = \sigma(x, z) + \sigma(z, y)$. A similar equality holds if $y = z$. Finally, if $x \neq z$ and $y \neq z$, then $\sigma(x, y) \leq 1 < 2 = \sigma(x, z) + \sigma(z, y)$. ∎

Comparing Examples 10.2 and 10.3, we see that a given set can have more than one metric. Hence, to describe a particular metric space, we must specify both the set X and the metric ρ.

10.4 EXAMPLE.

If $E \subseteq X$, then E is a metric space with metric ρ. (We shall call such metric spaces E *subspaces* of X.)

> **Proof.** If the Positive Definite Property, the Symmetric Property, and the Triangle Inequality hold for all $x, y \in X$, then they hold for all $x, y \in E$. ∎

A particular example of a subspace is provided by the set of rationals in $\mathbf{R}$.

10.5 EXAMPLE.

Q is a metric space with metric $\rho(x, y) = |x - y|$.

Metric spaces are by no means confined to numbers and vectors. Here is an important metric space whose "points" are functions.

10.6 EXAMPLE.

Let $C[a, b]$ represent the collection of continuous $f : [a, b] \to \mathbf{R}$ and

$$\|f\| := \sup_{x \in [a,b]} |f(x)|.$$

Then $\rho(f, g) := \|f - g\|$ is a metric on $C[a, b]$.

Proof. By the Extreme Value Theorem, $\|f\|$ is finite for each $f \in C[a, b]$. By definition, $\|f\| \geq 0$ for all f, and $\|f\| = 0$ if and only if $f(x) = 0$ for every $x \in [a, b]$. Thus ρ is positive definite. Since ρ is obviously symmetric, it remains to verify the Triangle Inequality. But

$$\|f + g\| = \sup_{x \in [a,b]} |f(x) + g(x)| \leq \sup_{x \in [a,b]} |f(x)| + \sup_{x \in [a,b]} |g(x)| = \|f\| + \|g\|. \blacksquare$$

It is interesting to note that convergence in this metric space means uniform convergence (see Exercise 10.1.8).

There are two ways to generalize open and closed intervals to arbitrary metric spaces. One way is to use the metric directly as follows.

10.7 Definition.

Let $a \in X$ and $r > 0$. The *open ball* (in X) with *center a* and *radius r* is the set

$$B_r(a) := \{x \in X : \rho(x, a) < r\},$$

and the *closed ball* (in X) with *center a* and *radius r* is the set

$$\{x \in X : \rho(x, a) \leq r\}.$$

Notice by Theorem 1.6 that in **R** (with the usual metric), the open ball (respectively, the closed ball) centered at a of radius r is $(a - r, a + r)$ (respectively, $[a - r, a + r]$) (i.e., open balls are open intervals and closed balls are closed intervals). With respect to the discrete metric, however, balls look quite different. For example, for each $0 < r < 1$ and each a in the discrete metric space, the closed and open balls centered at a of radius r are both equal to $\{a\}$.

Another way to generalize open and closed intervals to X is to specify what *open* and *closed* mean. Notice that every point x in an open interval I is surrounded by points in I. The same property holds for complements of closed intervals. This leads us to the following definition.

10.8 Definition.

---

i) A set $V \subseteq X$ is said to be *open* if and only if for every $x \in V$ there is an $\varepsilon > 0$ such that the open ball $B_\varepsilon(x)$ is contained in V.

ii) A set $E \subseteq X$ is said to be *closed* if and only if $E^c := X \setminus E$ is open.

Our first result about these concepts shows that they are consistent as applied to balls.

10.9 Remark. *Every open ball is open, and every closed ball is closed.*

Proof. Let $B_r(a)$ be an open ball. By definition, we must prove that given $x \in B_r(a)$ there is an $\varepsilon > 0$ such that $B_\varepsilon(x) \subseteq B_r(a)$. Let $x \in B_r(a)$ and set $\varepsilon = r - \rho(x, a)$. (Look at Figure 8.5 to see why this choice of ε should work.) If $y \in B_\varepsilon(x)$, then by the Triangle Inequality, assumption, and the choice of ε,

$$\rho(y, a) \leq \rho(y, x) + \rho(x, a) < \varepsilon + \rho(x, a) = r.$$

Thus, by Definition 10.7, $y \in B_r(a)$. In particular, $B_\varepsilon(x) \subseteq B_r(a)$. Similarly, we can show that $\{x \in X : \rho(x, a) > r\}$ is also open. Hence, every closed ball is closed. ∎

Here are more examples of open sets and closed sets.

10.10 Remark. *If $a \in X$, then $X \setminus \{a\}$ is open and $\{a\}$ is closed.*

Proof. By Definition 10.8, it suffices to prove that the complement of every *singleton* $E := \{a\}$ is open. Let $x \in E^c$ and set $\varepsilon = \rho(x, a)$. Then, by Definition 10.7, $a \notin B_\varepsilon(x)$, so $B_\varepsilon(x) \subseteq E^c$. Therefore, E^c is open by Definition 10.8. ∎

Students sometimes mistakenly believe that every set is either open or closed. Some sets are neither open nor closed (like the interval $[0, 1)$). And, as the following result shows, every metric space contains two special sets which are both open and closed.

10.11 Remark. *In an arbitrary metric space, the empty set $\emptyset$ and the whole space X are both open and closed.*

Proof. Since $X = \emptyset^c$ and $\emptyset = X^c$, it suffices by Definition 10.8 to prove that $\emptyset$ and X are both open. Because the empty set contains no points, "every" point $x \in \emptyset$ satisfies $B_\varepsilon(x) \subseteq \emptyset$. (This is called the *vacuous implication*.) Therefore, $\emptyset$ is open. On the other hand, since $B_\varepsilon(x) \subseteq X$ for all $x \in X$ and all $\varepsilon > 0$, it is clear that X is open. ∎

For some metric spaces (like $\mathbf{R}^n$), these are the only two sets which are simultaneously open and closed. For other metric spaces, there are many such sets.

10.12 EXAMPLE.

Every subset of the discrete space **R** is both open and closed.

> *Proof.* It suffices to prove that every subset of **R** is open (with respect to the discrete metric). Let $E \subseteq \mathbf{R}$. By Remark 10.11, we may assume that E is nonempty. Let $a \in E$. Since $B_1(a) = \{a\}$, some open ball containing a is a subset of E. By Definition 10.8, E is open. ∎

To see how these concepts are connected with limits, we examine convergence of sequences in an arbitrary metric space. Using the analogy between the metric ρ and the absolute value, we can transfer much of the theory of limits of sequences from **R** to any metric space. Here are the basic definitions.

10.13 Definition.

Let $\{x_n\}$ be a sequence in X.

i) $\{x_n\}$ *converges* (in X) if there is a point $a \in X$ (called the *limit* of x_n) such that for every $\varepsilon > 0$ there is an $N \in \mathbf{N}$ such that

$$n \geq N \quad \text{implies} \quad \rho(x_n, a) < \varepsilon.$$

ii) $\{x_n\}$ is *Cauchy* if for every $\varepsilon > 0$ there is an $N \in \mathbf{N}$ such that

$$n, m \geq N \quad \text{implies} \quad \rho(x_n, x_m) < \varepsilon.$$

iii) $\{x_n\}$ is *bounded* if there is an $M > 0$ and a $b \in X$ such that $\rho(x_n, b) \leq M$ for all $n \in \mathbf{N}$.

Modifying the proofs in Chapter 2, by doing little more than replacing $|x - y|$ by $\rho(x, y)$, we can establish the following result.

10.14 Theorem. *Let X be a metric space.*

  i) *A sequence in X can have at most one limit.*
 ii) *If $x_n \in X$ converges to a and $\{x_{n_k}\}$ is any subsequence of $\{x_n\}$, then x_{n_k} converges to a as $k \to \infty$.*
iii) *Every convergent sequence in X is bounded.*
iv) *Every convergent sequence in X is Cauchy.*

The following result shows that, by using open sets, we can describe convergence of sequences in an arbitrary metric space without reference to the distance function. Later in this chapter, we shall use this point of view to great advantage.

10.15 Remark. *Let $x_n \in X$. Then $x_n \to a$ as $n \to \infty$ if and only if for every open set V which contains a there is an $N \in \mathbf{N}$ such that $n \geq N$ implies $x_n \in V$.*

Proof. Suppose that $x_n \to a$, and let V be an open set which contains a. By Definition 10.8, there is an $\varepsilon > 0$ such that $B_\varepsilon(a) \subseteq V$. Given this ε, use Definition 10.13 to choose an $N \in \mathbf{N}$ such that $n \geq N$ implies $x_n \in B_\varepsilon(a)$. By the choice of ε, $x_n \in V$ for all $n \geq N$.

Conversely, let $\varepsilon > 0$ and set $V = B_\varepsilon(a)$. Then V is an open set which contains a; hence, by hypothesis, there is an $N \in \mathbf{N}$ such that $n \geq N$ implies $x_n \in V$. In particular, $\rho(x_n, a) < \varepsilon$ for all $n \geq N$.   ∎

The following result, which we shall use many times, shows that convergent sequences can also be used to characterize closed sets.

10.16 Theorem. *Let $E \subseteq X$. Then E is closed if and only if the limit of every convergent sequence $x_k \in E$ satisfies*

$$\lim_{k \to \infty} x_k \in E.$$

Proof. The theorem is vacuously satisfied if E is the empty set.

Suppose that $E \neq \emptyset$ is closed but some sequence $x_n \in E$ converges to a point $x \in E^c$. Since E is closed, E^c is open. Thus, by Remark 10.15, there is an $N \in \mathbf{N}$ such that $n \geq N$ implies $x_n \in E^c$, a contradiction.

Conversely, suppose that E is a nonempty set such that every convergent sequence in E has its limit in E. If E is not closed, then, by Remark 10.11, $E \neq X$, and, by definition, E^c is nonempty and not open. Thus, there is at least one point $x \in E^c$ such that no ball $B_r(x)$ is contained in E^c. Let $x_k \in B_{1/k}(x) \cap E$ for $k = 1, 2, \ldots$. Then $x_k \in E$ and $\rho(x_k, x) < 1/k$ for all $k \in \mathbf{N}$. Now $1/k \to 0$ as $k \to \infty$, so it follows from the Squeeze Theorem (these are real sequences) that $\rho(x_k, x) \to 0$ as $k \to \infty$ (i.e., $x_k \to x$ as $k \to \infty$). Thus, by hypothesis, $x \in E$, a contradiction.   ∎

Notice that the Bolzano–Weierstrass Theorem and Cauchy's Theorem are missing from Theorem 10.14. There is a simple reason for this. As the next two remarks show, neither of these results holds in an arbitrary metric space.

10.17 Remark. *The discrete space contains bounded sequences which have no convergent subsequences.*

Proof. Let $X = \mathbf{R}$ be the discrete metric space introduced in Example 10.3. Since $\sigma(0, k) = 1$ for all $k \in \mathbf{N}$, $\{k\}$ is a bounded sequence in X. Suppose that there exist integers $k_1 < k_2 < \ldots$ and an $x \in X$ such that $k_j \to x$ as $j \to \infty$. Then there is an $N \in \mathbf{N}$ such that $\sigma(k_j, x) < 1$ for $j \geq N$ (i.e., $k_j = x$ for all $j \geq N$). This contradiction proves that $\{k\}$ has no convergent subsequences.   ∎

10.18 Remark. *The metric space $X = \mathbf{Q}$, introduced in Example 10.5, contains Cauchy sequences which do not converge.*

Proof. Choose (by the Density of Rationals) points $q_k \in \mathbf{Q}$ such that $q_k \to \sqrt{2}$. Then $\{q_k\}$ is Cauchy (by Theorem 10.14iv) but does not converge in X since $\sqrt{2} \notin X$. ∎

This leads us to the following concept.

10.19 Definition.

A metric space X is said to be *complete* if and only if every Cauchy sequence $x_n \in X$ converges to some point in X.

At this point, you should read Section 9.1 to see how these concepts play out in the concrete Euclidean space setting. Notice by Theorem 9.6 that $\mathbf{R}^n$ is complete for all $n \in \mathbf{N}$. What can be said about complete metric spaces in general?

10.20 Remark. *By Definition 10.19, a complete metric space X satisfies two properties: 1) Every Cauchy sequence in X converges; 2) the limit of every Cauchy sequence in X stays in X.*

Property 2), by Theorem 10.16, means that X is closed. Hence, it is natural to ask, Is there a simple relationship between complete subspaces and closed subsets?

10.21 Theorem. *Let X be a complete metric space and E be a subset of X. Then E (as a subspace) is complete if and only if E (as a subset) is closed.*

Proof. Suppose that E is complete and that $x_n \in E$ converges. By Theorem 10.14iv, $\{x_n\}$ is Cauchy. Since E is complete, it follows from Definition 10.19 that the limit of $\{x_n\}$ belongs to E. Thus, by Theorem 10.16, E is closed.

Conversely, suppose that E is closed and that $x_n \in E$ is Cauchy in E. Since the metrics on X and E are identical, $\{x_n\}$ is Cauchy in X. Since X is complete, it follows that $x_n \to x$, as $n \to \infty$, for some $x \in X$. But E is closed, so x must belong to E. Thus E is complete by definition. ∎

EXERCISES

10.1.1. If $a, b \in X$ and $\rho(a, b) < \varepsilon$ for all $\varepsilon > 0$, prove that $a = b$.

10.1.2. Prove that $\{x_k\}$ is bounded in X if and only if $\sup_{k \in \mathbf{N}} \rho(x_k, a) < \infty$ for all $a \in X$.

10.1.3. Let $\mathbf{R}^n$ be endowed with the usual metric and suppose that $\{\mathbf{x}_k\}$ is a sequence in $\mathbf{R}^n$ with components $x_k^{(j)}$; that is,

$$\mathbf{x}_k = \left(x_k^{(1)}, x_k^{(2)}, \ldots, x_k^{(n)} \right).$$

a) Use Remark 8.7 to prove that $\{\mathbf{x}_k\}$ is bounded in $\mathbf{R}^n$ if and only if there is a $C > 0$ such that $|x_k^{(j)}| \leq C$ for all $k \in \mathbf{N}$ and all $j \in \{1, 2, \ldots, n\}$.

b) Let $\mathbf{a} \in \mathbf{R}^n$. Prove that $\mathbf{x}_k \to \mathbf{a}$ as $n \to \infty$ if and only if $x_k^{(j)} \to a_k$, as $k \to \infty$, for every $j \in \{1, 2, \ldots, n\}$.

c) Find the limit of each of the following sequences.

$$\left(\frac{k^2 + 1}{1 - k^2}, e^{1/k} \right) \qquad \left(\sqrt{k+1} - \sqrt{k}, \frac{\sqrt{k+2}}{\sqrt[3]{k^2 - k + 1}} \right)$$

$$\left(\left(\frac{k}{k+1} \right)^k, \frac{1}{5^k}, \frac{\sqrt{4^k + 1}}{5^k - 1} \right)$$

10.1.4. a) Let $a \in X$. Prove that if $x_n = a$ for every $n \in \mathbf{N}$, then x_n converges. What does it converge to?

b) Let $X = \mathbf{R}$ with the discrete metric. Prove that $x_n \to a$ as $n \to \infty$ if and only if $x_n = a$ for large n.

10.1.5. a) Let $\{x_n\}$ and $\{y_n\}$ be sequences in X which converge to the same point. Prove that $\rho(x_n, y_n) \to 0$ as $n \to \infty$.

b) Show that the converse of part a) is false.

10.1.6. Let $\{x_n\}$ be Cauchy in X. Prove that $\{x_n\}$ converges if and only if at least one of its subsequences converges.

10.1.7. Prove that the discrete space $\mathbf{R}$ is complete.

10.1.8. a) Prove that the metric space $\mathcal{C}[a, b]$ in Example 10.6 is complete.

b) Let $\|f\|_1 := \int_a^b |f(x)| dx$ and define

$$\text{dist}(f, g) := \|f - g\|_1$$

for each pair $f, g \in \mathcal{C}[a, b]$. Prove that this distance function also makes $\mathcal{C}[a, b]$ a metric space.

c) Prove that the metric space $\mathcal{C}[a, b]$ defined in part b) is not complete.

10.1.9. a) Show that if $x \in B_r(a)$, then there is an $\varepsilon > 0$ such that the closed ball centered at x of radius ε is a subset of $B_r(a)$.

b) If $a \neq b$ are distinct points in X, prove that there is an $r > 0$ such that $B_r(a) \cap B_r(b) = \emptyset$.

c) Show that given two balls $B_r(a)$ and $B_s(b)$, and a point $x \in B_r(a) \cap B_s(b)$, there are radii c and d such that

$$B_c(x) \subseteq B_r(a) \cap B_s(b) \quad \text{and} \quad B_d(x) \supseteq B_r(a) \cup B_s(b).$$

10.1.10. a) A subset E of X is said to be *sequentially compact* if and only if every sequence $x_n \in E$ has a convergent subsequence whose limit belongs to E. Prove that every sequentially compact set is closed and bounded.

b) Prove that **R** is closed but not sequentially compact.

c) Prove that every closed bounded subset of **R** is sequentially compact.

10.1.12. Prove Theorem 10.14.

10.2 LIMITS OF FUNCTIONS

In the preceding section we used results in Chapter 2 as a model for the theory of limits of sequences in an arbitrary metric space X. In this section we use results in Chapter 3 as a model to develop a theory of limits of functions which take one metric space X to another Y.

A straightforward adaptation of Definition 3.1 leads us to guess that, in an arbitrary metric space, $f(x) \to L$ as $x \to a$ if for every $\varepsilon > 0$ there is a $\delta > 0$ such that

$$0 < \rho(x, a) < \delta \quad \text{implies} \quad \tau(f(x), L) < \varepsilon.$$

The only problem with this definition is that there may be no x which satisfies $0 < \rho(x, a) < \delta$; for example, if X is the set **N** together with the metric $\rho(x, y) = |x - y|$ and $\delta = 1$. To prevent our theory from collapsing into the vacuous case, we introduce the following idea.

10.22 Definition.

A point $a \in X$ is said to be a *cluster point* (of X) if and only if $B_\delta(a)$ contains infinitely many points for each $\delta > 0$.

For example, every point in any Euclidean space $\mathbf{R}^n$ is a cluster point (of $\mathbf{R}^n$).

Notice that any concept defined on a metric space X is automatically defined on all subsets of X. Indeed, since any subset E of X is itself a metric space (see Example 10.4 above), the definition can be applied to E as well as to X.

To be more specific, let E be a subspace of X (i.e., a nonempty subset of X). By Definition 10.7 an open ball in E has the form

$$B_r^E(a) := \{x \in E : \rho(x, a) < r\}.$$

Since the metrics on X and E are the same, it follows that

$$B_r^E(a) = B_r(a) \cap E,$$

where $B_r(a)$ is an open ball in X. A similar statement holds for closed balls. We shall call these balls *relative balls* (in E). In particular, in the subspace **Q** of Example 10.5 above, the relative open balls take on the form $B_r(a) = (a - r, a + r) \cap \mathbf{Q}$ and the relative closed balls the form $[a - r, a + r] \cap \mathbf{Q}$.

What, then, does it mean for a set E to have a cluster point? By Definition 10.22, a point $a \in X$ is a cluster point of a nonempty set $E \subseteq X$ if and only if the relative ball $E \cap B_\delta(a)$ contains infinitely many points for each $\delta > 0$.

The etymology of the term *cluster point* is obvious. A cluster point of E is a point near which E "clusters." Cluster points are also called *points of accumulation*.

Notice that, by definition, no finite set has cluster points. On the other hand, a set may have infinitely many cluster points. Indeed, by the Density of Rationals (Theorem 1.18), every point of **R** is a cluster point of **Q**.

Here are two more examples of sets and their cluster points.

10.23 EXAMPLE.

Show that 0 is the only cluster point of the set

$$ E = \left\{ \frac{1}{n} : n \in \mathbf{N} \right\}. $$

Solution. By Theorem 1.16 (the Archimedean Principle), given $\delta > 0$ there is an $N \in \mathbf{N}$ such that $1/N < \delta$. Since $n \geq N$ implies $1/n \leq 1/N$, it follows that $(-\delta, \delta) \cap E$ contains infinitely many points. Thus 0 is a cluster point of E.

On the other hand, if $x_0 \neq 0$, then choose $\delta < |x_0|$, and notice that either $x_0 - \delta > 0$ or $x_0 + \delta < 0$. Thus $(x_0 - \delta, x_0 + \delta) \cap E$ contains at most finitely many points (i.e., x_0 is not a cluster point of E). ∎

10.24 EXAMPLE.

Show that every point in the interval $[0, 1]$ is a cluster point of the open interval $(0, 1)$.

Solution. Let $x_0 \in [0, 1]$ and $\delta > 0$. Then $x_0 + \delta > 0$ and $x_0 - \delta < 1$. In particular, $(x_0 - \delta, x_0 + \delta) \cap (0, 1)$ is itself a nondegenerate interval, say (a, b). But (a, b) contains infinitely many points; for example, $(a + b)/2$, $(2a + b)/3$, $(3a + b)/4, \ldots$. Therefore, x_0 is a cluster point of $(0, 1)$. ∎

We are now prepared to define limits of functions on metric spaces.

10.25 Definition.

Let a be a cluster point of X and $f : X \setminus \{a\} \to Y$. Then $f(x)$ is said to *converge to L, as x approaches a*, if and only if for every $\varepsilon > 0$ there is a $\delta > 0$ such that

$$ 0 < \rho(x, a) < \delta \quad \text{implies} \quad \tau(f(x), L) < \varepsilon. \tag{1} $$

In this case we write $f(x) \to L$ as $x \to a$, or

$$ L = \lim_{x \to a} f(x), $$

and call L the *limit* of $f(x)$ as x approaches a.

As in Chapter 3, we can introduce an algebraic structure on functions that take X to $\mathbf{R}^m$. Given $f, g : X \to \mathbf{R}^m$ and $c \in \mathbf{R}$, define the pointwise sum, pointwise product, and scalar product by

$$(f+g)(x) := f(x)+g(x), \quad (fg)(x) := f(x)g(x), \text{ and } (cf)(x) := cf(x), \quad x \in X.$$

When $m = 1$, define the pointwise quotient by

$$(f/g)(x) := \frac{f(x)}{g(x)}, \qquad x \in X \quad \text{and} \quad g(x) \neq 0.$$

By modifying the proofs presented in Chapter 3, we can prove the following results about limits of functions on metric spaces.

10.26 Theorem. *Let a be a cluster point of X and $f, g : X \setminus \{a\} \to Y$.*

i) *If $f(x) = g(x)$ for all $x \in X \setminus \{a\}$ and $f(x)$ has a limit as $x \to a$, then $g(x)$ also has a limit as $x \to a$, and*

$$\lim_{x \to a} g(x) = \lim_{x \to a} f(x).$$

ii) [SEQUENTIAL CHARACTERIZATION OF LIMITS]. *The limit*

$$L := \lim_{x \to a} f(x)$$

exists if and only if $f(x_n) \to L$ as $n \to \infty$ for every sequence $x_n \in X \setminus \{a\}$ which converges to a as $n \to \infty$.

iii) *Suppose that $Y = \mathbf{R}^n$. If $f(x)$ and $g(x)$ have a limit as x approaches a, then so do $(f + g)(x)$, $(f \cdot g)(x)$, $(\alpha f)(x)$, and $(f/g)(x)$ [when $Y = \mathbf{R}$ and the limit of $g(x)$ is nonzero]. In fact,*

$$\lim_{x \to a} (f + g)(x) = \lim_{x \to a} f(x) + \lim_{x \to a} g(x),$$

$$\lim_{x \to a} (\alpha f)(x) = \alpha \lim_{x \to a} f(x),$$

$$\lim_{x \to a} (f \cdot g)(x) = \lim_{x \to a} f(x) \cdot \lim_{x \to a} g(x),$$

and [when $Y = \mathbf{R}$ and the limit of $g(x)$ is nonzero]

$$\lim_{x \to a} \left(\frac{f}{g}\right)(x) = \frac{\lim_{x \to a} f(x)}{\lim_{x \to a} g(x)}.$$

iv) [SQUEEZE THEOREM FOR FUNCTIONS]. *Suppose that $Y = \mathbf{R}$. If $h : X \setminus \{a\} \to \mathbf{R}$ satisfies $g(x) \leq h(x) \leq f(x)$ for all $x \in X \setminus \{a\}$, and*

$$\lim_{x \to a} f(x) = \lim_{x \to a} g(x) = L,$$

then the limit of h exists, as x → a, and

$$\lim_{x \to a} h(x) = L.$$

v) [COMPARISON THEOREM FOR FUNCTIONS]. *Suppose that $Y = \mathbf{R}$. If $f(x) \le g(x)$ for all $x \in X \setminus \{a\}$, and if f and g have a limit as x approaches a, then*

$$\lim_{x \to a} f(x) \le \lim_{x \to a} g(x).$$

At this point you should read Section 9.3 to see how these concepts play out in the concrete Euclidean space setting. Pay special attention to Theorem 9.16 and Example 9.18 (which show how to evaluate a limit in $\mathbf{R}^n$) and Examples 9.19 and 9.20 (which show how to prove that a specific limit in $\mathbf{R}^n$ does not exist).

Here is the metric space version of Definition 3.20.

10.27 Definition.

Let E be a nonempty subset of X and $f : E \to Y$.

i) f is said to be *continuous at a point $a \in E$* if and only if given $\varepsilon > 0$ there is a $\delta > 0$ such that

$$\rho(x, a) < \delta \quad \text{and} \quad x \in E \quad \text{imply} \quad \tau(f(x), f(a)) < \varepsilon.$$

ii) f is said to be *continuous on E* (notation: $f : E \to Y$ is continuous) if and only if f is continuous at every $x \in E$.

Notice that this definition is valid whether a is a cluster point or not. Modifying corresponding proofs in Chapter 3, we can prove the following results.

10.28 Theorem.   *Let E be a nonempty subset of X and $f, g : E \to Y$.*

i) *f is continuous at $a \in E$ if and only if $f(x_n) \to f(a)$, as $n \to \infty$, for all sequences $x_n \in E$ which converge to a.*

ii) *Suppose that $Y = \mathbf{R}^n$. If f, g are continuous at a point $a \in E$ (respectively, continuous on a set E), then so are $f + g$, $f \cdot g$, and αf (for any $\alpha \in \mathbf{R}$). Moreover, in the case $Y = \mathbf{R}$, f/g is continuous at $a \in E$ when $g(a) \ne 0$ [respectively, on E when $g(x) \ne 0$ for all $x \in E$].*

The following result shows that the composition of two continuous functions is continuous.

10.29 Theorem.   *Suppose that X, Y, and Z are metric spaces and that a is a cluster point of X. Suppose further that $f : X \to Y$ and $g : f(X) \to Z$. If $f(x) \to L$ as $x \to a$ and g is continuous at L, then*

$$\lim_{x \to a} (g \circ f)(x) = g \left(\lim_{x \to a} f(x) \right).$$

We shall examine the metric space analogues of the Extreme Value Theorem, the Intermediate Value Theorem, and uniform continuity in Section 10.4.

EXERCISES

10.2.1. Find all cluster points of each of the following sets.

a) $E = \mathbf{R} \setminus \mathbf{Q}$
b) $E = [a, b)$, $a, b \in \mathbf{R}$, $a < b$
c) $E = \{(-1)^n n : n \in \mathbf{N}\}$
d) $E = \{x_n : n \in \mathbf{N}\}$, where $x_n \to x$ as $n \to \infty$
e) $E = \{1, 1, 2, 1, 2, 3, 1, 2, 3, 4, \ldots\}$

10.2.2. a) A point a in a metric space X is said to be *isolated* if and only if there is an $r > 0$ so small that $B_r(a) = \{a\}$. Show that a point $a \in X$ is not a cluster point of X if and only if a is isolated.

b) Prove that the discrete space has no cluster points.

10.2.3. Prove that a is a cluster point for some $E \subseteq X$ if and only if there is a sequence $x_n \in E \setminus \{a\}$ such that $x_n \to a$ as $n \to \infty$.

10.2.4. a) Let E be a nonempty subset of X. Prove that a is a cluster point of E if and only if for each $r > 0$, $E \cap B_r(a) \setminus \{a\}$ is nonempty.

b) Prove that every bounded infinite subset of $\mathbf{R}$ has at least one cluster point.

10.2.5. Prove Theorem 10.26.
10.2.6. Prove Theorem 10.28.
10.2.7. Prove Theorem 10.29.
10.2.8. Prove that if $f_n \in \mathcal{C}[a, b]$, then $f_n \to f$ uniformly on $[a, b]$ if and only if $f_n \to f$ in the metric of $\mathcal{C}[a, b]$ (see Example 10.6).
10.2.9. Suppose that X is a metric space which satisfies the following condition.

10.30 Definition.

X is said to satisfy the *Bolzano–Weierstrass Property* if and only if every bounded sequence $x_n \in X$ has a convergent subsequence.

---

a) Prove that if E is a closed, bounded subset of X and $x_n \in E$, then there is an $a \in E$ and a subsequence x_{n_k} of x_n such that $x_{n_k} \to a$ as $k \to \infty$.
b) If E is closed and bounded in X and $f : E \to \mathbf{R}$ is continuous on E, prove that f is bounded on E.
c) Prove that under the hypotheses of part b) that there exist points x_m, $x_M \in E$ such that

$$f(x_M) = \sup_{x \in E} f(x) \quad \text{and} \quad f(x_m) = \inf_{x \in E} f(x).$$

10.3 INTERIOR, CLOSURE, AND BOUNDARY

Thus far, we have used *open* and *closed* mostly for identification. At this point, we begin to examine these concepts in more depth. Our first result shows that open sets and closed sets behave very differently with respect to unions and intersections.

10.31 Theorem. *Let X be a metric space.*

i) *If $\{V_\alpha\}_{\alpha \in A}$ is any collection of open sets in X, then*

$$\bigcup_{\alpha \in A} V_\alpha$$

is open.

ii) *If $\{V_k : k = 1, 2, \ldots, n\}$ is a finite collection of open sets in X, then*

$$\bigcap_{k=1}^{n} V_k := \bigcap_{k \in \{1,2,\ldots,n\}} V_k$$

is open.

iii) *If $\{E_\alpha\}_{\alpha \in A}$ is any collection of closed sets in X, then*

$$\bigcap_{\alpha \in A} E_\alpha$$

is closed.

iv) *If $\{E_k : k = 1, 2, \ldots, n\}$ is a finite collection of closed sets in X, then*

$$\bigcup_{k=1}^{n} E_k := \bigcup_{k \in \{1,2,\ldots,n\}} E_k$$

is closed.

v) *If V is open in X and E is closed in X, then $V \setminus E$ is open and $E \setminus V$ is closed.*

Proof. i) Let $x \in \bigcup_{\alpha \in A} V_\alpha$. Then $x \in V_\alpha$ for some $\alpha \in A$. Since V_α is open, it follows that there is an $r > 0$ such that $B_r(x) \subseteq V_\alpha$. Thus $B_r(x) \subseteq \bigcup_{\alpha \in A} V_\alpha$ (i.e., this union is open).

ii) Let $x \in \bigcap_{k=1}^{n} V_k$. Then $x \in V_k$ for $k = 1, 2, \ldots, n$. Since each V_k is open, it follows that there are numbers $r_k > 0$ such that $B_{r_k}(x) \subseteq V_k$. Let $r = \min\{r_1, \ldots, r_n\}$. Then $r > 0$ and $B_r(x) \subseteq V_k$ for all $k = 1, 2, \ldots, n$; that is, $B_r(x) \subseteq \bigcap_{k=1}^{n} V_k$. Hence, this intersection is open.

iii) By DeMorgan's Law (Theorem 1.36) and part i),

$$\left(\bigcap_{\alpha \in A} E_\alpha \right)^c = \bigcup_{\alpha \in A} E_\alpha^c$$

is open, so $\bigcap_{\alpha \in A} E_\alpha$ is closed.

iv) By DeMorgan's Law and part ii),

$$\left(\bigcup_{k=1}^{n} E_k\right)^c = \bigcap_{k=1}^{n} E_k^c$$

is open, so $\bigcup_{k=1}^{n} E_k$ is closed.

v) Since $V \setminus E = V \cap E^c$ and $E \setminus V = E \cap V^c$, the former is open by part ii), and the latter is closed by part iii). ∎

The finiteness hypothesis in Theorem 10.31 is critical, even for the case $X = \mathbf{R}$.

10.32 Remark. *Statements ii) and iv) of Theorem 10.31 are false if arbitrary collections are used in place of finite collections.*

Proof. In the metric space $X = \mathbf{R}$,

$$\bigcap_{k \in \mathbf{N}} \left(-\frac{1}{k}, \frac{1}{k}\right) = \{0\}$$

is closed and

$$\bigcup_{k \in \mathbf{N}} \left[\frac{1}{k+1}, \frac{k}{k+1}\right] = (0, 1)$$

is open. ∎

Theorem 10.31 has many applications. Our first application is that every set contains a largest open set and is contained in a smallest closed set. To facilitate our discussion, we introduce the following topological operations.

10.33 Definition.

Let E be a subset of a metric space X.

i) The *interior* of E is the set

$$E^o := \bigcup \{V : V \subseteq E \text{ and } V \text{ is open in } X\}.$$

ii) The *closure* of E is the set

$$\overline{E} := \bigcap \{B : B \supseteq E \text{ and } B \text{ is closed in } X\}.$$

Notice that every set E contains the open set $\emptyset$ and is contained in the closed set X. Hence, the sets E^o and $\overline{E}$ are well defined. Also notice, by Theorem 10.31, that the interior of a set is always open and the closure of a set is always closed.

The following result shows that E^o is the largest open set contained in E, and $\overline{E}$ is the smallest closed set which contains E.

10.34 Theorem. *Let $E \subseteq X$. Then*

i) $E^o \subseteq E \subseteq \overline{E}$,
ii) *if V is open and $V \subseteq E$, then $V \subseteq E^o$, and*
iii) *if C is closed and $C \supseteq E$, then $C \supseteq \overline{E}$.*

Proof. Since every open set V in the union defining E^o is a subset of E, it is clear that the union of these V's is a subset of E. Thus $E^o \subseteq E$. A similar argument establishes $E \subseteq \overline{E}$. This proves i).

By Definition 10.33, if V is an open subset of E, then $V \subseteq E^o$ and if C is a closed set containing E, then $\overline{E} \subseteq C$. This proves ii) and iii). ∎

In particular, the interior of a bounded interval with endpoints a and b is (a, b), and its closure is $[a, b]$. In fact, it is evident by parts ii) and iii) that $E = E^o$ if and only if E is open and $E = \overline{E}$ if and only if E is closed. We shall use this observation many times below.

The following examples illustrate the fact that the interior of a nice enough set E in $\mathbf{R}^2$ can be obtained by removing all its "edges," and the closure of E by adding all its "edges."

10.35 EXAMPLE.

Find the interior and closure of the set $E = \{(x, y) : -1 \le x \le 1 \text{ and } -|x| < y < |x|\}$.

Solution. Graph $y = |x|$ and $x = \pm 1$, and observe that E is a bow-tie-shaped region with "solid" vertical edges (see Figure 8.6). Now, by Definition 10.8, any open set in $\mathbf{R}^2$ must contain a disk around each of its points. Since E^o is the largest open set inside E, it is clear that

$$E^o = \{(x, y) : -1 < x < 1 \text{ and } -|x| < y < |x|\}.$$

Similarly,

$$\overline{E} = \{(x, y) : -1 \le x \le 1 \text{ and } -|x| \le y \le |x|\}. \qquad ∎$$

10.36 EXAMPLE.

Find the interior and closure of the set $E = B_1(-2, 0) \cup B_1(2, 0) \cup \{(x, 0) : -1 \le x \le 1\}$.

Solution. Draw a graph of this region. It turns out to be "dumbbell shaped": two open disks joined by a straight line. Thus $E^o = B_1(-2, 0) \cup B_1(2, 0)$ and

$$\overline{E} = \overline{B_1(-2, 0)} \cup \overline{B_1(2, 0)} \cup \{(x, 0) : -1 \le x \le 1\}. \qquad ∎$$

One of the most important results from one-dimensional calculus is the Fundamental Theorem of Calculus. It states that the behavior of a derivative f' on an interval $[a, b]$, as measured by the integral, is completely determined by the values of f at the endpoints of $[a, b]$. What shall we use for "endpoints" of an arbitrary set in X? Notice that the endpoints a, b are the only points which lie near both $[a, b]$ and the complement of $[a, b]$. Using this as a cue, we introduce the following concept.

10.37 Definition.

Let $E \subseteq X$. The *boundary* of E is the set

$$\partial E := \{x \in X : \text{for all } r > 0, \quad B_r(x) \cap E \neq \emptyset \text{ and } B_r(x) \cap E^c \neq \emptyset\}.$$

[We will refer to the last two conditions in the definition of ∂E by saying $B_r(x)$ *intersects* E *and* E^c.]

10.38 EXAMPLE.

Describe the boundary of the set

$$E = \{(x, y) : x^2 + y^2 \leq 9 \text{ and } (x - 1)(y + 2) > 0\}.$$

Solution. Graph the relations $x^2 + y^2 = 9$ and $(x - 1)(y + 2) = 0$ to obtain a region with solid curved edges and dotted straight edges (see Figure 8.7). By definition, then, the boundary of E is the union of these curved and straight edges (all made solid). Rather than describing ∂E analytically (which would involve solving for the intersection points of the straight lines $x = 1$, $y = -2$, and the circle $x^2 + y^2 = 9$), it is easier to describe ∂E by using set algebra.

$$\partial E = \{(x, y) : x^2 + y^2 \leq 9 \text{ and } (x - 1)(y + 2) \geq 0\}$$
$$\setminus \{(x, y) : x^2 + y^2 < 9 \text{ and } (x - 1)(y + 2) > 0\} \qquad \blacksquare$$

It turns out that set algebra can be used to describe the boundary of any set.

10.39 Theorem. *Let $E \subseteq X$. Then $\partial E = \overline{E} \setminus E^o$.*

Proof. By Definition 10.37, it suffices to show

$$x \in \overline{E} \text{ if and only if } B_r(x) \cap E \neq \emptyset \text{ for all } r > 0, \text{ and} \qquad (2)$$
$$x \notin E^o \text{ if and only if } B_r(x) \cap E^c \neq \emptyset \text{ for all } r > 0. \qquad (3)$$

We will provide the details for (2) and leave the proof of (3) as an exercise. Suppose that $x \in \overline{E}$ but $B_{r_0}(x) \cap E = \emptyset$ for some $r_0 > 0$. Then $(B_{r_0}(x))^c$ is a closed set which contains E; hence, by Theorem 10.34iii, $\overline{E} \subseteq (B_{r_0}(x))^c$. It follows that $\overline{E} \cap B_{r_0}(x) = \emptyset$ (e.g., $x \notin \overline{E}$, a contradiction).

Conversely, suppose that $x \notin \overline{E}$. Since $(\overline{E})^c$ is open, there is an $r_0 > 0$ such that $B_{r_0}(x) \subseteq (\overline{E})^c$. In particular, $\emptyset = B_{r_0}(x) \cap \overline{E} \supseteq B_{r_0}(x) \cap E$ for some $r_0 > 0$. ∎

We have introduced topological operations (interior, closure, and boundary). The following result answers the question, How do these operations interact with the set operations (union and intersection)?

10.40 Theorem. *Let $A, B \subseteq X$. Then*

i) $\qquad\qquad (A \cup B)^o \supseteq A^o \cup B^o, \quad (A \cap B)^o = A^o \cap B^o,$

ii) $\qquad\qquad \overline{A \cup B} = \overline{A} \cup \overline{B}, \qquad \overline{A \cap B} \subseteq \overline{A} \cap \overline{B},$

iii) $\partial(A \cup B) \subseteq \partial A \cup \partial B, \quad and \quad \partial(A \cap B) \subseteq (A \cap \partial B) \cup (B \cap \partial A) \cup (\partial A \cap \partial B).$

Proof. i) Since the union of two open sets is open, $A^o \cup B^o$ is an open subset of $A \cup B$. Hence, by Theorem 10.34ii, $A^o \cup B^o \subseteq (A \cup B)^o$.

Similarly, $(A \cap B)^o \supseteq A^o \cap B^o$. On the other hand, if $V \subset A \cap B$, then $V \subset A$ and $V \subset B$. Thus, $(A \cap B)^o \subseteq A^o \cap B^o$.

ii) Since $\overline{A} \cup \overline{B}$ is closed and contains $A \cup B$, it is clear that by Theorem 10.34iii), $\overline{A \cup B} \subseteq \overline{A} \cup \overline{B}$. Similarly, $\overline{A \cap B} \subseteq \overline{A} \cap \overline{B}$. To prove the reverse inequality for union, suppose that $x \notin \overline{A \cup B}$. Then there is a closed set E which contains $A \cup B$ such that $x \notin E$. Since E contains both A and B, it follows that $x \notin \overline{A}$ and $x \notin \overline{B}$. This proves part ii).

iii) Let $x \in \partial(A \cup B)$; that is, suppose that $B_r(x)$ intersects both $A \cup B$ and $(A \cup B)^c$ for all $r > 0$. Since $(A \cup B)^c = A^c \cap B^c$, it follows that $B_r(x)$ intersects both A^c and B^c for all $r > 0$. Thus, $B_r(x)$ intersects A and A^c for all $r > 0$, or $B_r(x)$ intersects B and B^c for all $r > 0$ (i.e., $x \in \partial A \cup \partial B$). This proves the first set inequality in part iii).

To prove the second set inequality, fix $x \in \partial(A \cap B)$ [i.e., suppose that $B_r(x)$ intersects $A \cap B$ and $(A \cap B)^c$ for all $r > 0$]. If $x \in (A \cap \partial B) \cup (B \cap \partial A)$, then there is nothing to prove. If $x \notin (A \cap \partial B) \cup (B \cap \partial A)$, then $x \in (A^c \cup (\partial B)^c) \cap (B^c \cup (\partial A)^c)$. Hence, it remains to prove that $x \in A^c \cup (\partial B)^c$ implies $x \in \partial A$ and $x \in B^c \cup (\partial A)^c$ implies $x \in \partial B$. By symmetry, we need only prove the first implication.

Case 1. $x \in A^c$. Since $B_r(x)$ intersects A, it follows that $x \in \partial A$.

Case 2. $x \in (\partial B)^c$. Since $B_r(x)$ intersects B, it follows that $B_r(x) \subseteq B$ for small $r > 0$. Since $B_r(x)$ also intersects $A^c \cup B^c$, it must be the case that $B_r(x)$ intersects A^c. In particular, $x \in \partial A$. ∎

EXERCISES

10.3.1. Find the interior, closure, and boundary of each of the following subsets of **R**.

a) $E = \{1/n : n \in \mathbf{N}\}$

b) $E = \bigcup_{n=1}^{\infty} \left(\dfrac{1}{n+1}, \dfrac{1}{n} \right)$

c) $E = \bigcup (-n, n)$

d) $E = \mathbf{Q}$

10.3.2. Identify which of the following sets are open, which are closed, and which are neither. Find E^o, $\overline{E}$, and ∂E and sketch E in each case.

a) $E = \{(x, y) : x^2 + 4y^2 \leq 1\}$

b) $E = \{(x, y) : x^2 - 2x + y^2 = 0\} \cup \{(x, 0) : x \in [2, 3]\}$

c) $E = \{(x, y) : y \geq x^2, \ 0 \leq y < 1\}$

d) $E = \{(x, y) : x^2 - y^2 < 1, \ -1 < y < 1\}$

10.3.3. Let $a \in X$, $s < r$,

$$V = \{x \in X : s < \rho(x, a) < r\}, \quad \text{and} \quad E = \{x \in X : s \leq \rho(x, a) \leq r\}.$$

Prove that V is open and E is closed.

10.3.4. Suppose that $A \subseteq B \subseteq X$. Prove that $\overline{A} \subseteq \overline{B}$ and $A^o \subseteq B^o$.

10.3.5. **This exercise is used in Section 10.5.** Show that if E is nonempty and closed in X and $a \notin E$, then $\inf_{x \in E} \rho(x, a) > 0$.

10.3.6. Prove (3).

10.3.7. Show that Theorem 10.40 is best possible in the following sense.

a) There exist sets A, B in $\mathbf{R}$ such that $(A \cup B)^o \neq A^o \cup B^o$.

b) There exist sets A, B in $\mathbf{R}$ such that $\overline{A \cap B} \neq \overline{A} \cap \overline{B}$.

c) There exist sets A, B in $\mathbf{R}$ such that $\partial(A \cup B) \neq \partial A \cup \partial B$ and $\partial(A \cap B) \neq (A \cap \partial B) \cup (B \cap \partial A) \cup (\partial A \cap \partial B)$.

10.3.8. **This exercise is used many times from Section 10.5 onward.** Let Y be a subspace of X.

a) Show that a set V is open in Y if and only if there is an open set U in X such that $V = U \cap Y$.

b) Show that a set E is closed in Y if and only if there is a closed set A in X such that $E = A \cap Y$.

10.3.9. Let $f : \mathbf{R} \to \mathbf{R}$. Prove that f is continuous on $\mathbf{R}$ if and only if $f^{-1}(I)$ is open in $\mathbf{R}$ for every open interval I.

10.3.10. Let V be a subset of X.

a) Prove that V is open in X if and only if there is a collection of open balls $\{B_\alpha : \alpha \in A\}$ such that

$$V = \bigcup_{\alpha \in A} B_\alpha.$$

b) What happens to this result if *open* is replaced by *closed*?

10.3.11. Let $E \subseteq X$ be closed.

a) Prove that $\partial E \subseteq E$.

b) Prove that $\partial E = E$ if and only if $E^o = \emptyset$.

c) Show that b) is false if E is not closed.

10.4 COMPACT SETS

In Chapter 3 we proved the Extreme Value Theorem for functions defined on **R**. In this section we shall extend that result to functions defined on an arbitrary metric space. To replace the hypothesis "closed, bounded interval" used in the real case, we introduce "compactness," a concept which gives us a powerful tool for extending local results to global ones (see especially Remark 10.44 and Theorems 10.52, 10.69, and 12.46).

Since compactness of E depends on how E can be "covered" by a collection of open sets, we begin by introducing the following terminology.

10.41 Definition.

Let $\mathcal{V} = \{V_\alpha\}_{\alpha \in A}$ be a collection of subsets of a metric space X and suppose that E is a subset of X.

  i) $\mathcal{V}$ is said to *cover* E (or be a *covering* of E) if and only if

$$E \subseteq \bigcup_{\alpha \in A} V_\alpha.$$

 ii) $\mathcal{V}$ is said to be an *open covering* of E if and only if $\mathcal{V}$ covers E and each V_α is open.
iii) Let $\mathcal{V}$ be a covering of E. $\mathcal{V}$ is said to have a *finite* (respectively, *countable*) *subcovering* if and only if there is a finite (respectively, countable) subset A_0 of A such that $\{V_\alpha\}_{\alpha \in A_0}$ covers E.

Notice that the collections of open intervals

$$\left\{\left(\frac{1}{k+1}, \frac{k}{k+1}\right)\right\}_{k \in \mathbf{N}} \quad \text{and} \quad \left\{\left(-\frac{1}{k}, \frac{k+1}{k}\right)\right\}_{k \in \mathbf{N}}$$

are open coverings of the interval $(0, 1)$. The first covering of $(0, 1)$ has no finite subcover, but any member of the second covering covers $(0, 1)$. Thus an open covering of an arbitrary set may or may not have a finite subcovering.

Sets that always have finite subcoverings are important enough to be given a name.

10.42 Definition.

A subset H of a metric space X is said to be *compact* if and only if every open covering of H has a finite subcover.

To get a feeling for what this definition means, we make some elementary observations concerning compact sets in general.

10.43 Remark. *The empty set and all finite subsets of a metric space are compact.*

Proof. These statements follow immediately from Definition 10.42. The empty set needs no set to cover it, and any finite set H can be covered by finitely many sets, one set for each element in H. ∎

Since the empty set and finite sets are also closed, it is natural to ask whether there is a relationship between compact sets and closed sets in general. The following three results address this question in an arbitrary metric space.

10.44 Remark. *A compact set is always closed.*

Proof. Suppose that H is compact but not closed. Then H is nonempty and (by Theorem 10.16) there is a convergent sequence $x_k \in H$ whose limit x does not belong to H. For each $y \in H$, set $r(y) := \rho(x, y)/2$. Since x does not belong to H, $r(y) > 0$; hence, each $B_{r(y)}(y)$ is open and contains y; that is, $\{B_{r(y)}(y) : y \in H\}$ is an open covering of H. Since H is compact, we can choose points y_j and radii $r_j := r(y_j)$ such that $\{B_{r_j}(y_j) : j = 1, 2, \ldots, N\}$ covers H.

Set $r := \min\{r_1, \ldots, r_N\}$. (This is a finite set of positive numbers, so r is also positive.) Since $x_k \to x$ as $k \to \infty$, $x_k \in B_r(x)$ for large k. But $x_k \in B_r(x) \cap H$ implies $x_k \in B_{r_j}(y_j)$ for some $j \in \mathbf{N}$. Therefore, it follows from the choices of r_j and r, and from the Triangle Inequality, that

$$r_j \geq \rho(x_k, y_j) \geq \rho(x, y_j) - \rho(x_k, x)$$
$$= 2r_j - \rho(x_k, x) > 2r_j - r \geq 2r_j - r_j = r_j,$$

a contradiction. ∎

The following result is a partial converse of Remark 10.44 (see also Theorem 10.50 below).

10.45 Remark. *A closed subset of a compact set is compact.*

Proof. Let E be a closed subset of H, where H is compact in X and suppose that $\mathcal{V} = \{V_\alpha\}_{\alpha \in A}$ is an open covering of E. Now $E^c = X \setminus E$ is open; hence, $\mathcal{V} \cup \{E^c\}$ is an open covering of H. Since H is compact, there is a finite set $A_0 \subseteq A$ such that

$$H \subseteq E^c \cup \left(\bigcup_{\alpha \in A_0} V_\alpha \right).$$

But $E \cap E^c = \emptyset$. Therefore, E is covered by $\{V_\alpha\}_{\alpha \in A_0}$. ∎

Here is the connection between closed, bounded sets and compact sets.

10.46 Theorem. *Let H be a subset of a metric space X. If H is compact, then H is closed and bounded.*

Proof. Suppose that H is compact. By Remark 10.44, H is closed. It is also bounded. Indeed, fix $b \in X$ and observe that $\{B_n(b) : n \in \mathbf{N}\}$ covers X. Since H is compact, it follows that

$$H \subset \bigcup_{n=1}^{N} B_n(b)$$

for some $N \in \mathbf{N}$. Since these balls are nested, we conclude that $H \subset B_N(b)$ (i.e., H is bounded). ∎

10.47 Remark. *The converse of Theorem 10.46 is false for arbitrary metric spaces.*

Proof. Let $X = \mathbf{R}$ be the discrete metric space introduced in Example 10.3. Since $\sigma(0, x) \leq 1$ for all $x \in \mathbf{R}$, every subset of X is bounded. Since $x_k \to x$ in X implies $x_k = x$ for large k, every subset of X is closed. Thus $[0, 1]$ is a closed, bounded subset of X. Since $\{x\}_{x \in [0,1]}$ is an uncountable open covering of $[0, 1]$, which has no finite subcover, we conclude that $[0, 1]$ is closed and bounded, but not compact. ∎

The problem here is that the discrete space has too many open sets. To identify a large class of metric spaces for which the converse of Theorem 10.46 DOES hold, we need a property which cuts the "number of essential" open sets down to a reasonable size.

10.48 Definition.

A metric space X is said to be *separable* if and only if it contains a countable dense subset (i.e., if and only if there is a countable set Z of X such that for every point $a \in X$ there is a sequence $x_k \in Z$ such that $x_k \to a$ as $k \to \infty$).

It is easy to see (Theorem 9.3) that all Euclidean spaces are separable. The space $C[a, b]$ is also separable (see Exercise 10.7.1). Hence, the hypothesis of separability is not an unusual requirement.

The following result makes clear what we meant above by "number of essential" open sets. It shows that every open covering of a set in a separable metric space has a countable subcovering.

10.49 Theorem. [LINDELÖF].
Let E be a subset of a separable metric space X. If $\{V_\alpha\}_{\alpha \in A}$ is a collection of open sets and $E \subseteq \cup_{\alpha \in A} V_\alpha$, then there is a countable subset $\{\alpha_1, \alpha_2, \ldots\}$ of A such that

$$E \subseteq \bigcup_{k=1}^{\infty} V_{\alpha_k}.$$

Proof. Let Z be a countable dense subset of X, and consider the collection $\mathcal{T}$ of open balls with centers in Z and rational radii. This collection is countable. Moreover, it "approximates" all other open sets in the following sense:

CLAIM: Given any open ball $B_r(x) \subset X$, there is a ball $B_q(a) \in \mathcal{T}$ such that $x \in B_q(a)$ and $B_q(a) \subseteq B_r(x)$.

PROOF OF CLAIM: Let $B_r(x) \subset X$ be given. By Definition 10.48, choose $a \in Z$ such that $\rho(x, a) < r/4$, and choose by Theorem 1.18 a rational $q \in \mathbf{Q}$ such that $r/4 < q < r/2$. Since $r/4 < q$, we have $x \in B_q(a)$. Moreover, if $y \in B_q(a)$, then

$$\rho(x, y) \le \rho(x, a) + \rho(a, y) < q + \frac{r}{4} < \frac{r}{2} + \frac{r}{4} < r.$$

Therefore, $B_q(a) \subseteq B_r(x)$. This establishes the claim.

To prove the theorem, let $x \in E$. By hypothesis, $x \in V_\alpha$ for some $\alpha \in A$. Hence, by the claim, there is a ball $B_x \in \mathcal{T}$ such that

$$x \in B_x \subseteq V_\alpha. \tag{4}$$

The collection $\mathcal{T}$ is countable; hence, so is the subcollection

$$\{U_1, U_2, \dots\} := \{B_x : x \in E\}. \tag{5}$$

By (4), for each $k \in \mathbf{N}$ there is at least one $\alpha_k \in A$ such that $U_k \subseteq V_{\alpha_k}$. Hence, by (5),

$$E \subseteq \bigcup_{x \in E} B_x = \bigcup_{k=1}^{\infty} U_k \subseteq \bigcup_{k=1}^{\infty} V_{\alpha_k}. \qquad \blacksquare$$

We are prepared to obtain a converse of Theorem 10.46. (For the definition of the Bolzano–Weierstrass Property, see Exercise 10.2.9.)

10.50 Theorem. [HEINE–BOREL].
Let X be a separable metric space which satisfies the Bolzano–Weierstrass Property and H be a subset of X. Then H is compact if and only if it is closed and bounded.

Proof. By Theorem 10.46, every compact set is closed and bounded.

Conversely, suppose to the contrary that H is closed and bounded but not compact. Let $\mathcal{V}$ be an open covering of H which has no finite subcover of H. By Lindelöf's Theorem, we may suppose that $\mathcal{V} = \{V_k\}_{k \in \mathbf{N}}$; that is,

$$H \subseteq \bigcup_{k \in \mathbf{N}} V_k. \tag{6}$$

By the choice of $\mathcal{V}$, $\cup_{j=1}^{k} V_j$ cannot contain H for any $k \in \mathbf{N}$. Thus we can choose a point

$$x_k \in H \setminus \bigcup_{j=1}^{k} V_j \tag{7}$$

for each $k \in \mathbf{N}$. Since H is bounded, the sequence x_k is bounded. Hence, by the Bolzano–Weierstrass Property, there is a subsequence x_{k_ν} which converges to some x as $\nu \to \infty$. Since H is closed, $x \in H$. Hence, by (6), $x \in V_N$ for some $N \in \mathbf{N}$. But V_N is open; hence, there is an $M \in \mathbf{N}$ such that $\nu \geq M$ implies $k_\nu > N$ and $x_{k_\nu} \in V_N$. This contradicts (7). We conclude that H is compact. ∎

Since $\mathbf{R}^n$ satisfies the hypotheses of Theorem 10.50 (see Theorems 9.3 and 9.5), it follows that a subset of a Euclidean space is compact if and only if it is closed and bounded.

We now turn our attention to uniform continuity on an arbitrary metric space.

10.51 Definition.

Let X be a metric space, E be a nonempty subset of X, and $f : E \to Y$. Then f is said to be *uniformly continuous* on E (notation: $f : E \to Y$ is uniformly continuous) if and only if given $\varepsilon > 0$ there is a $\delta > 0$ such that

$$\rho(x, a) < \delta \quad \text{and} \quad x, a \in E \quad \text{imply} \quad \tau(f(x), f(a)) < \varepsilon.$$

In the real case, we proved that uniform continuity and continuity were equivalent on closed, bounded intervals. That result, whose proof relied on the Bolzano–Weierstrass Theorem, is not true in an arbitrary metric space. If we strengthen the hypothesis from closed and bounded to compact, however, the result is valid for any metric space.

10.52 Theorem. *Suppose that E is a compact subset of X and that $f : X \to Y$. Then f is uniformly continuous on E if and only if f is continuous on E.*

Proof. If f is uniformly continuous on a set, then it is continuous whether or not the set is compact.

Conversely, suppose that f is continuous on E. Given $\varepsilon > 0$ and $a \in E$, choose $\delta(a) > 0$ such that

$$x \in B_{\delta(a)}(a) \quad \text{and} \quad x \in E \quad \text{imply} \quad \tau(f(x), f(a)) < \frac{\varepsilon}{2}.$$

Since $a \in B_\delta(a)$ for all $\delta > 0$, it is clear that $\{B_{\delta(a)/2}(a) : a \in E\}$ is an open covering of E. Since E is compact, choose finitely many points $a_j \in E$ and numbers $\delta_j := \delta(a_j)$ such that

$$E \subseteq \bigcup_{j=1}^{N} B_{\delta_j/2}(a_j). \tag{8}$$

Set $\delta := \min\{\delta_1/2, \ldots, \delta_N/2\}$.

Suppose that $x, a \in E$ with $\rho(x, a) < \delta$. By (8), x belongs to $B_{\delta_j/2}(a_j)$ for some $1 \leq j \leq N$. Hence,

$$\rho(a, a_j) \leq \rho(a, x) + \rho(x, a_j) < \frac{\delta_j}{2} + \frac{\delta_j}{2} = \delta_j;$$

that is, a also belongs to $B_{\delta_j}(a_j)$. It follows, therefore, from the choice of δ_j that

$$\tau(f(x), f(a)) \leq \tau(f(x), f(a_j)) + \tau(f(a_j), f(a)) < \frac{\varepsilon}{2} + \frac{\varepsilon}{2} = \varepsilon.$$

This proves that f is uniformly continuous on E. ∎

EXERCISES

10.4.1. Identify which of the following sets are compact and which are not. If E is not compact, find the smallest compact set H (if there is one) such that $E \subset H$.

a) $\{1/k : k \in \mathbf{N}\} \cup \{0\}$
b) $\{(x, y) \in \mathbf{R}^2 : a \leq x^2 + y^2 \leq b\}$ for real numbers $0 < a < b$
c) $\{(x, y) \in \mathbf{R}^2 : y = \sin(1/x)$ and $x \in (0, 1]\}$
d) $\{(x, y) \in \mathbf{R}^2 : |xy| \leq 1\}$

10.4.2. Let A, B be compact subsets of X. Prove that $A \cup B$ and $A \cap B$ are compact.

10.4.3. Suppose that $E \subseteq \mathbf{R}$ is compact and nonempty. Prove that $\sup E$, $\inf E \in E$.

10.4.4. Suppose that $\{V_\alpha\}_{\alpha \in A}$ is a collection of nonempty open sets in X which satisfies $V_\alpha \cap V_\beta = \emptyset$ for all $\alpha \neq \beta$ in A. Prove that if X is separable, then A is countable. What happens to this result when *open* is omitted?

10.4.5. Prove that if V is open in a separable metric space X, then there are open balls $B_1, B_2, \ldots$ such that

$$V = \bigcup_{j \in \mathbf{N}} B_j.$$

Prove that every open set in $\mathbf{R}$ is a countable union of open intervals.

10.4.6. Suppose that X is a separable metric space which satisfies the Bolzano–Weierstrass Property, that Y is a complete metric space, and that E is a bounded subset of X. Prove that a function $f : E \to Y$ is uniformly continuous on E if and only if f can be continuously extended to $\overline{E}$; that is, if and only if there exists a continuous function $g : \overline{E} \to Y$ such that $f(x) = g(x)$ for all $x \in E$.

10.4.7. Suppose that X satisfies the Bolzano–Weierstrass Property and that A and B are compact subsets of X. Prove that if $A \cap B = \emptyset$ and if

$$\text{dist}\,(A, B) := \inf\{\rho(x, y) : x \in A \quad \text{and} \quad y \in B\},$$

then dist $(A, B) > 0$. Show that even in the space $\mathbf{R}^2$, there exist subsets A and B which are closed and satisfy $A \cap B = \emptyset$, but dist$(A, B) = 0$.

10.4.8. a) Prove that *Cantor's Intersection Theorem* holds for nested compact sets in an arbitrary metric space; that is, if H_1, H_2, ... is a nested sequence of nonempty compact sets in X, then

$$\bigcap_{k=1}^{\infty} H_k \neq \emptyset.$$

    b) Prove that $(\sqrt{2}, \sqrt{3}) \cap \mathbf{Q}$ is closed and bounded but not compact in the metric space $\mathbf{Q}$ introduced in Example 10.5.

    c) Show that Cantor's Intersection Theorem does not hold in an arbitrary metric space if *compact* is replaced by *closed and bounded*.

10.4.9. Prove that the Bolzano–Weierstrass Property does not hold for $\mathcal{C}[a, b]$ and $\|f\|$ (see Example 10.6). Namely, prove that if $f_n(x) = x^n$, then $\|f_n\|$ is bounded but $\|f_{n_k} - f\|$ does not converge for any $f \in \mathcal{C}[0, 1]$ and any subsequence $\{n_k\}$.

10.4.10. Let X be a metric space.

    a) Prove that if $E \subseteq X$ is compact, then E is sequentially compact (see Exercise 10.1.10).

    b) Prove that if X is separable and satisfies the Bolzano–Weierstrass Property, then a set $E \subseteq X$ is sequentially compact if and only if it is compact.

10.5 CONNECTED SETS

We have introduced open sets (analogues of open intervals), closed sets (analogues of closed intervals), and compact sets (analogues of closed bounded intervals) in order to develop a calculus of functions of several variables, in Chapters 11 through 13, which parallels that developed for functions of a single variable in Chapters 2 through 5. Some of the earlier theory, however, depended on a property of intervals not yet discussed. For example, the proof of the Intermediate Value Theorem tacitly used the fact that an interval is connected (i.e., is unbroken and all of one piece). We shall also use connected sets in Chapter 13 to provide a sufficiently broad definition of surfaces for computational ease. Thus we introduce the following idea.

10.53 Definition.

Let X be a metric space.

i) A pair of nonempty open sets U, V in X is said to *separate* X if and only if $X = U \cup V$ and $U \cap V = \emptyset$.

ii) X is said to be *connected* if and only if X cannot be separated by any pair of open sets U, V.

Loosely speaking, a connected space is all in one piece (i.e., cannot be broken into smaller, nonempty, open pieces which do not share any common points). Indeed, we shall see that **R**, under the usual metric, is connected. On the other hand, under the discrete metric, **R** is not connected (since $(-\infty, 0]$ and $(0, \infty)$ are both "open" in the discrete space).

Recall (Example 10.4) that every subset of X is a metric space. Hence Definition 10.53 also defines what it means for a subset E of X to be connected. We can always find two subsets of an arbitrary metric space which are connected: 1) The empty set is connected, since it can never be written as the union of nonempty sets. 2) Every singleton $E = \{a\}$ is also connected since, if $E = U \cup V$, where both U and V are nonempty, then E has at least two points.

To obtain deeper results about connectivity, it is convenient to introduce the following concepts. (These concepts will also be used to study continuous functions in the next section.)

10.54 Definition.

Let X be a metric space and $E \subseteq X$.

i) A set $U \subseteq E$ is said to be *relatively open* in E if and only if there is a set V open in X such that $U = E \cap V$.

ii) A set $A \subseteq E$ is said to be *relatively closed* in E if and only if there is a set C closed in X such that $A = E \cap C$.

For example, the set E of Example 10.35 is relatively open in the subspace $Y := \{(x, y) : -1 \le x \le 1\}$ and relatively closed in the subspace $Z := \{(x, y) : -|x| < y < |x|\}$. Indeed, $V = Z$ is open in $\mathbf{R}^2$ (it contains none of its boundary), $A = Y$ is closed in $\mathbf{R}^2$ (it contains all its boundary), and $E = V \cap Y$, $E = A \cap Z$.

Recall (Exercise 10.3.8) that a subset A of E is open (respectively, closed) in the *subspace* E if and only if it is relatively open (respectively, relatively closed) in the *set* E. Thus all Definition 10.54 does is codify the "subspace topology."

By Definition 10.53, then, a set E is connected if there are no nonempty sets U, V, relatively open in E, such that $E = U \cup V$ and $U \cap V = \emptyset$. The following result, which is usually easier to use than Definition 10.53, shows that when "separating" a nonconnected set, we can use open sets instead of relatively open sets. (The converse of this result is also true, but harder to prove—see Theorem 10.57 below.)

10.55 Remark. *Let $E \subseteq X$. If there exists a pair of open sets A, B in X which separate E (i.e., if $E \subseteq A \cup B$, $A \cap B = \emptyset$, $A \cap E \neq \emptyset$, and $B \cap E \neq \emptyset$), then E is not connected.*

Proof. Set $U = A \cap E$ and $V = B \cap E$. It suffices to prove that U and V are relatively open in E and separate E. It is clear by hypothesis and the remarks above that U and V are nonempty, they are both relatively open in E, and $U \cap V = \emptyset$. It remains to prove that $E = U \cup V$. But E is a subset of $A \cup B$, so $E \subseteq U \cup V$. On the other hand, both U and V are subsets of E, so $E \supseteq U \cup V$. We conclude that $E = U \cup V$. ∎

Thus when looking for "separations" of a given set $E \subset X$, we can confine our attention to open sets in X. Here are several examples. The set **Q** is not connected since the pair $A = (-\infty, \sqrt{2})$, $B = (\sqrt{2}, \infty)$ separate **Q**. Example 10.35 is not connected since $\{(x, y) : x < 0\}$ and $\{(x, y) : x > 0\}$ are open in $\mathbf{R}^2$ (neither of them contains any of their boundary points) and separate the bow tie set E. Notice that Example 10.36 is connected in $\mathbf{R}^2$.

There is a simple description of all connected subsets of **R**.

10.56 Theorem. *A subset E of **R** is connected if and only if E is an interval.*

Proof. Let E be a connected subset of **R**. If E is empty or contains only one point, then E is a degenerate interval. Hence we may suppose that E contains at least two points.

Set $a = \inf E$ and $b = \sup E$. Notice that $-\infty \leq a < b \leq \infty$. Suppose for simplicity that $a, b \notin E$; that is, $E \subseteq (a, b)$. If $E \neq (a, b)$, then there is an $x \in (a, b)$ such that $x \notin E$. By the Approximation Property, $E \cap (a, x) \neq \emptyset$ and $E \cap (x, b) \neq \emptyset$, and, by assumption, $E \subseteq (a, x) \cup (x, b)$. Hence, E is separated by the open sets (a, x), (x, b), a contradiction.

Conversely, suppose that E is an interval which is not connected. Then there are sets U, V, relatively open in E, which separate E (i.e., $E = U \cup V$, $U \cap V = \emptyset$), and there exist points $x_1 \in U$ and $x_2 \in V$. We may suppose that $x_1 < x_2$. Since $x_1, x_2 \in E$ and E is an interval, $I_0 := [x_1, x_2] \subseteq E$. Define f on I_0 by

$$f(x) = \begin{cases} 0 & x \in U \\ 1 & x \in V. \end{cases}$$

Since $U \cap V = \emptyset$, f is well defined. We claim that f is continuous on I_0. Indeed, fix $x_0 \in [x_1, x_2]$. Since $U \cup V = E \supseteq I_0$, it is evident that $x_0 \in U$ or $x_0 \in V$. We may suppose the former. Let $y_k \in I_0$ and suppose that $y_k \to x_0$ as $k \to \infty$. Since U is relatively open, there is an $\varepsilon > 0$ such that $(x_0 - \varepsilon, x_0 + \varepsilon) \cap E \subset U$. Since $y_k \in E$ and $y_k \to x_0$, it follows that $y_k \in U$ for large k.

Hence $f(y_k) = 0 = f(x_0)$ for large k. Therefore, f is continuous at x_0 by the Sequential Characterization of Continuity.

We have proved that f is continuous on I_0. Hence, by the Intermediate Value Theorem (Theorem 3.29), f must take on the value $1/2$ somewhere on I_0. This is a contradiction, since by construction, f takes on only the values 0 or 1. $\blacksquare$

We can use this result to prove that a real function is continuous on a closed, bounded interval if and only if its graph is closed and connected (see Theorem 9.51 in the preceding chapter).

We close this section by showing that the converse of Remark 10.55 is also true. This result is optional because we do not use it elsewhere.

*10.57 Theorem. *Let $E \subseteq X$. If there exist sets U, V, relatively open in E, such that $U \cap V = \emptyset$, $E = U \cup V$, $U \neq \emptyset$, and $V \neq \emptyset$, then there is a pair of open sets A, B which separates E.*

Proof. We first show that

$$\overline{U} \cap V = \emptyset. \tag{9}$$

Indeed, since V is relatively open in E, there is a set Ω, open in X, such that $V = E \cap \Omega$. Since $U \cap V = \emptyset$, it follows that $U \subset \Omega^c$. This last set is closed in X. Therefore,

$$\overline{U} \subseteq \overline{\Omega^c} = \Omega^c;$$

that is, (9) holds.

Next, we use (9) to construct the set B. Set

$$\delta_x = \inf\{\rho(x, u) : u \in \overline{U}\}, \quad x \in V, \quad \text{and} \quad B = \bigcup_{x \in V} B_{\delta_x/2}(x).$$

Clearly, B is open in X. Since $\delta_x > 0$ for each $x \notin \overline{U}$ (see Exercise 10.3.5), B contains V; hence, $B \cap E \supseteq V$. The reverse inequality also holds since by construction $B \cap U = \emptyset$ and by hypothesis $E = U \cup V$. Therefore, $B \cap E = V$. Similarly, we can construct an open set A such that $A \cap E = U$ by setting

$$\varepsilon_y = \inf\{\rho(v, y) : v \in \overline{V}\}, \quad y \in U \quad \text{and} \quad A = \bigcup_{y \in U} B_{\varepsilon_y/2}(y).$$

To prove that the pair A, B separates E, it remains to prove that $A \cap B = \emptyset$. Suppose to the contrary that there is a point $a \in A \cap B$. Then $a \in B_{\delta_x/2}(x)$ for some $x \in V$ and $a \in B_{\varepsilon_y/2}(y)$ for some $y \in U$. We may suppose that $\delta_x \leq \varepsilon_y$. Then

$$\rho(x, y) \leq \rho(x, a) + \rho(a, y) < \frac{\delta_x}{2} + \frac{\varepsilon_y}{2} \leq \varepsilon_y.$$

Therefore, $\rho(x, y) < \inf\{\rho(v, y) : v \in \overline{V}\}$. Since $x \in V$, this is impossible. We conclude that $A \cap B = \emptyset$. ∎

EXERCISES

10.5.1. a) Let $a \leq b$ and $c \leq d$ be real numbers. Sketch a graph of the rectangle

$$[a, b] \times [c, d] := \{(x, y) : x \in [a, b], y \in [c, d]\},$$

and decide whether this set is compact or connected. Explain your answers.

b) Sketch a graph of set

$$B_1(-2, 0) \cup B_1(2, 0) \cup \{(x, 0) : -1 < x < 1\},$$

and decide whether this set is compact or connected. Explain your answers.

10.5.2. a) Sketch a graph of the set

$$\{(x, y) : x^2 + 2y^2 < 6, \ y \geq 0\},$$

and decide whether this set is relatively open or relatively closed in the subspace $\{(x, y) : y \geq 0\}$. Do the same for the subspace $\{(x, y) : x^2 + 2y^2 < 6\}$. Explain your answers.

b) Sketch a graph of set

$$\{(x, y) : x^2 + y^2 \leq 1, \ (x - 2)^2 + y^2 < 2\},$$

and decide whether this set is relatively open or relatively closed in the subspace $\overline{B_1(0, 0)}$. Do the same for the subspace $B_{\sqrt{2}}(2, 0)$. Explain your answers.

10.5.3. a) Prove that the intersection of two connected sets in $\mathbf{R}$ is connected. Show that this is false if $\mathbf{R}$ is replaced by $\mathbf{R}^2$.

b) Generalize part a) as follows. If $\{E_\alpha\}_{\alpha \in A}$ is an arbitrary collection of connected sets in $\mathbf{R}$, then

$$\bigcap_{\alpha \in A} E_\alpha$$

is also connected.

10.5.4. Prove that if $E \subseteq \mathbf{R}$ is connected, then E^o is also connected. Show that this is false if $\mathbf{R}$ is replaced by $\mathbf{R}^2$.

10.5.5. Suppose that $E \subseteq X$ is connected and that $E \subseteq A \subseteq \overline{E}$. Prove that A is connected.

10.5.6. Suppose that X and Y are metric spaces and that $f : X \to Y$. If X is compact and connected, and if to every $x \in X$ there corresponds an open ball B_x such that $x \in B_x$ and $f(y) = f(x)$ for all $y \in B_x$, prove that f is constant on X.

10.5.7. **This exercise is used in Section 10.6.** Let $H \subseteq X$. Prove that H is compact if and only if every cover $\{E_\alpha\}_{\alpha \in A}$ of H, where the E_α's are relatively open in H, has a finite subcover.

10.5.8. A set E in a metric space is called *clopen* if it is both open and closed.

    a) Prove that every metric space has at least two clopen sets.
    b) Prove that a metric space is connected if and only if it contains exactly two clopen sets.

10.5.9. Let X be a metric space. Prove that X is connected if and only if every nonempty proper subset of X has a nonempty boundary.

*10.5.10. **This exercise is used to prove *Corollary 11.35.**

    a) A set $E \subseteq \mathbf{R}^n$ is said to be *polygonally connected* if and only if any two points $\mathbf{a}, \mathbf{b} \in E$ can be connected by a polygonal path in E; that is, there exist points $\mathbf{x}_k \in E$, $k = 1, \ldots, N$, such that $\mathbf{x}_0 = \mathbf{a}$, $\mathbf{x}_N = \mathbf{b}$ and $L(\mathbf{x}_{k-1}; \mathbf{x}_k) \subseteq E$ for $k = 1, \ldots, N$. Prove that every polygonally connected set in $\mathbf{R}^n$ is connected.
    b) Let $E \subseteq \mathbf{R}^n$ be open and $\mathbf{x}_0 \in E$. Let U be the set of points $\mathbf{x} \in E$ which can be polygonally connected in E to $\mathbf{x}_0$. Prove that U is open.
    c) Prove that every open connected set in $\mathbf{R}^n$ is polygonally connected.

10.5.11. Suppose that $\{E_\alpha\}_{\alpha \in A}$ is a collection of connected sets in a metric space X such that $\bigcap_{\alpha \in A} E_\alpha \neq \emptyset$. Prove that

$$E = \bigcup_{\alpha \in A} E_\alpha$$

is connected.

10.6 CONTINUOUS FUNCTIONS

In this section we discuss the behavior of images and inverse images of open sets, closed sets, compact sets, and connected sets under continuous functions. We shall use these results many times in the sequel.

Recall that if X and Y are metric spaces (with respective metrics ρ and τ), then a function $f : X \to Y$ is continuous on X if and only if given $a \in X$ and $\varepsilon > 0$ there is a $\delta > 0$ such that $\rho(x, a) < \delta$ implies $\tau(f(x), f(a)) < \varepsilon$; that is, such that

$$B_\delta(a) \subseteq f^{-1}(B_\varepsilon(f(a))). \tag{10}$$

This observation can be used to give the following simple but powerful characterization of continuous functions which can be stated without using the metric of X (see also Exercise 10.6.3).

10.58 Theorem. *Suppose that $f : X \to Y$. Then f is continuous if and only if $f^{-1}(V)$ is open in X for every open V in Y.*

Proof. Suppose that f is continuous on X and that V is open in Y. We may suppose that $f^{-1}(V)$ is nonempty. Let $a \in f^{-1}(V)$; that is, $f(a) \in V$. Since V is open, choose $\varepsilon > 0$ such that $B_\varepsilon(f(a)) \subseteq V$. Since f is continuous at a, choose $\delta > 0$ such that (10) holds. Evidently,

$$B_\delta(a) \subseteq f^{-1}(B_\varepsilon(f(a))) \subseteq f^{-1}(V). \tag{11}$$

Since $a \in f^{-1}(V)$ was arbitrary, we have shown that every point in $f^{-1}(V)$ is interior to $f^{-1}(V)$. Thus $f^{-1}(V)$ is open.

Conversely, let $\varepsilon > 0$ and $a \in X$. The ball $V = B_\varepsilon(f(a))$ is open in Y. By hypothesis, $f^{-1}(V)$ is open. Since $a \in f^{-1}(V)$, it follows that there is a $\delta > 0$ such that $B_\delta(a) \subseteq f^{-1}(V)$. This means that if $\rho(x, a) < \delta$, then $\tau(f(x), f(a)) < \varepsilon$. Therefore, f is continuous at $a \in X$. ∎

By using the subspace (i.e., relative) topology, we see that Theorem 10.58 contains the following criterion for f to be continuous on a subset of X.

10.59 Corollary. *Let $E \subseteq X$ and $f : E \to Y$. Then f is continuous on E if and only if $f^{-1}(V) \cap E$ is relatively open in E for all open sets V in Y.*

We shall refer to Theorem 10.58 and its corollary by saying that open sets are invariant under inverse images by continuous functions. It is interesting to notice that closed sets are also invariant under inverse images by continuous functions (see Exercises 10.6.3 and 10.6.4).

It is natural to ask whether compact sets and connected sets are invariant under inverse images by continuous functions. The following examples show that, even for the metric space **R**, the answer to this question is no.

10.60 EXAMPLES.

i) If $f(x) = 1/x$ and $H = [0, 1]$, then f is continuous on $(0, \infty)$ and H is compact, but $f^{-1}(H) = [1, \infty)$ is not compact.
ii) If $f(x) = x^2$ and $E = (1, 4)$, then f is continuous on **R** and E is connected, but $f^{-1}(E) = (-2, -1) \cup (1, 2)$ is not connected.

The next two results show that compact sets and connected sets are invariant under *images*, rather than inverse images, by continuous functions.

10.61 Theorem. *If H is compact in X and $f : H \to Y$ is continuous on H, then $f(H)$ is compact in Y.*

Proof. Suppose that $\{V_\alpha\}_{\alpha \in A}$ is an open covering of $f(H)$. By Theorem 1.37,

$$H \subseteq f^{-1}(f(H)) \subseteq f^{-1}\left(\bigcup_{\alpha \in A} V_\alpha\right) = \bigcup_{\alpha \in A} f^{-1}(V_\alpha).$$

Hence, by Corollary 10.59, $\{f^{-1}(V_\alpha)\}_{\alpha \in A}$ is a covering of H whose sets are all relatively open in H. Since H is compact, there are indices $\alpha_1, \alpha_2, \ldots, \alpha_N$ such that

$$H \subseteq \bigcup_{j=1}^{N} f^{-1}(V_{\alpha_j})$$

(see Exercise 10.5.7). It follows from Theorem 1.37 that

$$f(H) \subseteq f\left(\bigcup_{j=1}^{N} f^{-1}(V_{\alpha_j})\right) = \bigcup_{j=1}^{N} (f \circ f^{-1})(V_{\alpha_j}) = \bigcup_{j=1}^{N} V_{\alpha_j}.$$

Therefore, $f(H)$ is compact. ∎

10.62 Theorem. *If E is connected in X and $f : E \to Y$ is continuous on E, then $f(E)$ is connected in Y.*

Proof. Suppose that $f(E)$ is not connected. By Definition 10.53, there exists a pair $U, V \subset Y$ of relatively open sets in $f(E)$ which separates $f(E)$. By Exercise 10.6.4, $f^{-1}(U) \cap E$ and $f^{-1}(V) \cap E$ are relatively open in E. Since $f(E) = U \cup V$, we have

$$E = (f^{-1}(U) \cap E) \cup (f^{-1}(V) \cap E).$$

Since $U \cap V = \emptyset$, we also have $f^{-1}(U) \cap f^{-1}(V) = \emptyset$. Thus $f^{-1}(U) \cap E$, $f^{-1}(V) \cap E$ is a pair of relatively open sets which separates E. Hence, by Definition 10.53, E is not connected, a contradiction. ∎

(*Note:* Theorems 10.61 and 10.62 do not hold if *compact* or *connected* are replaced by *open* or *closed*. For example, if $f(x) = x^2$ and $V = (-1, 1)$, then f is continuous on **R** and V is open, but $f(V) = [0, 1)$ is neither open nor closed.)

Suppose that f is a real function continuous on a closed, bounded interval $[a, b]$. Then the function $F(x) = (x, f(x))$ is continuous from **R** into $\mathbf{R}^2$. Since the graph of $y = f(x)$ for $x \in [a, b]$ is the image of $[a, b]$ under F, it follows from Theorems 10.61 and 10.62 that the graph of f is compact and connected. It is interesting to note that this property actually characterizes continuity of real functions (see Theorem 9.51 in the preceding chapter).

To illustrate the power of the topological point of view presented above, compare the proofs of the following theorem and Exercise 10.6.5 with those of Theorems 3.26 and 3.29.

10.63 Theorem.   [EXTREME VALUE THEOREM].
Let H be a nonempty, compact subset of X and suppose that $f : H \to \mathbf{R}$ is continuous. Then

$$M := \sup\{f(x) : x \in H\} \quad and \quad m := \inf\{f(x) : x \in H\}$$

are finite real numbers and there exist points $x_M, x_m \in H$ such that $M = f(x_M)$ and $m = f(x_m)$.

Proof. By symmetry, it suffices to prove the result for M. Since H is compact, $f(H)$ is compact. Hence, by the Theorem 10.46, $f(H)$ is closed and bounded. Since $f(H)$ is bounded, M is finite. By the Approximation Property, choose $x_k \in H$ such that $f(x_k) \to M$ as $k \to \infty$. Since $f(H)$ is closed, $M \in f(H)$. Therefore, there is an $x_M \in H$ such that $M = f(x_M)$. A similar argument shows that m is finite and attained on H. ∎

The following analogue of Theorem 4.32 will be used in Chapter 13 to examine change of parametrizations of curves and surfaces.

10.64 Theorem. *If H is a compact subset of X and $f : H \to Y$ is 1–1 and continuous, then f^{-1} is continuous on $f(H)$.*

Proof. By Exercise 10.6.4a, it suffices to show that $(f^{-1})^{-1}$ takes closed sets in X to relatively closed sets in $f(H)$. Let E be closed in X. Then $E \cap H$ is a closed subset of H, so by Remark 10.45, $E \cap H$ is compact. Hence, by Theorem 10.61, $f(E \cap H)$ is compact, in particular, closed. Since f is 1–1, $f(E \cap H) = f(E) \cap f(H)$ (see Exercise 1.5.7). Since $f(E \cap H)$ and $f(H)$ are closed, it follows that $f(E) \cap f(H)$ is relatively closed in $f(H)$. Since $(f^{-1})^{-1} = f$, we conclude that $(f^{-1})^{-1}(E) \cap f(H)$ is relatively closed in $f(H)$. ∎

If you are interested in how to use these topological ideas to study real functions further, you may read Section 9.6 in the preceding chapter now.

EXERCISES

10.6.1. Let $f(x) = \sin x$ and $g(x) = x/|x|$ if $x \neq 0$ and $g(0) = 0$.

a) Find $f(E)$ and $g(E)$ for $E = (0, \pi)$, $E = [0, \pi]$, $E = (-1, 1)$, and $E = [-1, 1]$. Compare your answers to what Theorems 10.58, 10.61, and 10.62 predict. Explain any departures from the predictions.

b) Find $f^{-1}(E)$ and $g^{-1}(E)$ for $E = (0, 1)$, $E = [0, 1]$, $E = (-1, 1)$, and $E = [-1, 1]$. Compare your answers to what Theorems 10.58, 10.61, and 10.62 predict. Explain any departures from the predictions.

10.6.2. Let $f(x) = \sqrt{x}$ and $g(x) = 1/x$ if $x \neq 0$ and $g(0) = 0$.

a) Find $f(E)$ and $g(E)$ for $E = (0, 1)$, $E = [0, 1)$, and $E = [0, 1]$. Compare your answers to what Theorems 10.58, 10.61, and 10.62 predict. Explain any departures from the predictions.

b) Find $f^{-1}(E)$ and $g^{-1}(E)$ for $E = (-1, 1)$ and $E = [-1, 1]$. Compare your answers to what Theorems 10.58, 10.61, and 10.62 predict. Explain any departures from the predictions.

10.6.3. Suppose that $f : X \rightarrow Y$. Prove that f is continuous if and only if $f^{-1}(C)$ is closed in X for every set C closed in Y.

10.6.4. Suppose that $E \subseteq X$ and that $f : E \rightarrow Y$.

a) Prove that f is continuous on E if and only if $f^{-1}(A) \cap E$ is relatively closed in E for all closed sets A in Y.

b) Suppose that f is continuous on E. Prove that if V is relatively open in $f(E)$, then $f^{-1}(V)$ is relatively open in E, and if A is relatively closed in $f(E)$, then $f^{-1}(A)$ is relatively closed in E.

10.6.5. [INTERMEDIATE VALUE THEOREM]. Let E be a connected subset of X. If $f : E \rightarrow \mathbf{R}$ is continuous, $f(a) \neq f(b)$ for some $a, b \in E$, and y is a number which lies between $f(a)$ and $f(b)$, then prove that there is an $x \in E$ such that $f(x) = y$. (You may use Theorem 10.56.)

10.6.6. Suppose that H is a nonempty compact subset of X and that Y is a Euclidean space.

a) If $f : H \rightarrow Y$ is continuous, prove that

$$\|f\|_H := \sup_{x \in H} \|f(x)\|_Y$$

is finite and there exists an $x_0 \in H$ such that $\|f(x_0)\|_Y = \|f\|_H$.

b) A sequence of functions $f_k : H \rightarrow Y$ is said to converge uniformly on H to a function $f : H \rightarrow Y$ if and only if given $\varepsilon > 0$ there is an $N \in \mathbf{N}$ such that

$$k \geq N \quad \text{and} \quad x \in H \quad \text{imply} \quad \|f_k(x) - f(x)\|_Y < \varepsilon.$$

Show that $\|f_k - f\|_H \rightarrow 0$ as $k \rightarrow \infty$ if and only if $f_k \rightarrow f$ uniformly on H as $k \rightarrow \infty$.

c) Prove that a sequence of functions f_k converges uniformly on H if and only if, given $\varepsilon > 0$, there is an $N \in \mathbf{N}$ such that

$$k, j \geq N \quad \text{implies} \quad \|f_k - f_j\|_H < \varepsilon.$$

10.6.7. Suppose that E is a compact subset of X.

a) If $f, g : E \rightarrow \mathbf{R}^n$ are uniformly continuous, prove that $f + g$ and $f \cdot g$ are uniformly continuous. Did you need compactness for both results?

b) If $g : E \rightarrow \mathbf{R}$ is continuous on E and $g(x) \neq 0$ for $x \in E$, prove that $1/g$ is a bounded function.

c) If $f, g : E \rightarrow \mathbf{R}$ are uniformly continuous on E and $g(x) \neq 0$ for $x \in E$, prove that f/g is uniformly continuous on E.

10.6.8. Suppose that $E \subseteq X$ and that $f : E \to Y$.

    a) If f is uniformly continuous on E and $x_n \in E$ is Cauchy in X, prove that $f(x_n)$ is Cauchy in Y.

    b) Suppose that D is a *dense* subspace of X (i.e., that $D \subset X$ and $\overline{D} = X$). If Y is complete and $f : D \to Y$ is uniformly continuous on D, prove that f has a continuous extension to X; that is, prove that there is a continuous function $g : X \to Y$ such that $g(x) = f(x)$ for all $x \in D$.

10.6.9. Suppose that X is connected. Prove that if there is a nonconstant, continuous function $f : X \to \mathbf{R}$, then X has uncountably many points.

*10.7 STONE–WEIERSTRASS THEOREM

This section uses the Binomial Series (Theorem 7.52). Since these results are not used elsewhere, it can be skipped.

One of the oldest questions in analysis is the following:

APPROXIMATION QUESTION. Can one use polynomials to approximate continuous functions on an interval $[a, b]$?

For well over 150 years, mathematicians thought that the way to answer this question was to use Taylor polynomials. When Cauchy showed that this approach was doomed to failure, even for the smaller space $\mathcal{C}^\infty[a, b]$ (see Remark 7.41), other avenues of investigation were opened up. The main result of this section, the Stone–Weierstrass Theorem, shows that the answer to the Approximation Question is yes in a very general sense. You will explore some of the consequences of this powerful result in the exercises. It has many other far-reaching consequences as well and is valid in a much more general setting than compact metric spaces.

For each metric space X, $\mathcal{C}(X)$ will represent the collection of continuous functions from X to $\mathbf{R}$. The Stone–Weierstrass approach to approximation requires that the collection of approximating functions be closed under pointwise addition, pointwise multiplication, and scalar multiplication. Thus we begin with the following concept.

10.65 Definition.

A set A is said to be a (real function) *algebra* in $\mathcal{C}(X)$ if and only if

  i) $\emptyset \neq A \subseteq \mathcal{C}(X)$.
  ii) If $f, g \in A$, then $f + g$ and fg both belong to A.
  iii) If $f \in A$ and $c \in \mathbf{R}$, then $cf \in A$.

We notice once and for all that every algebra of functions contains the zero function, since $f \in A$ implies $-f \in A$; hence $0 = f - f \in A$.

It is well known that the collection of one variable polynomials is an algebra in $\mathcal{C}(\mathbf{R})$. By Theorem 10.26, $\mathcal{C}(X)$ itself is an algebra in $\mathcal{C}(X)$ for any metric space X.

As you know, there are several types of convergence for function sequences. The Stone–Weierstrass Theorem is a result about the strongest of these: uniform convergence. To streamline our presentation, we introduce a metric on $\mathcal{C}(X)$ which governs uniform convergence. Namely, we define $\text{dist}(f, g) := \|f - g\|$, where *uniform norm* of an $f \in \mathcal{C}(X)$ is defined to be

$$\|f\| := \sup_{x \in X} |f(x)|.$$

If X is compact, then $\text{dist}(f, g)$ is a metric on $\mathcal{C}(X)$. Indeed, by the Extreme Value Theorem, $\text{dist}(f, g)$ is finite for each $f, g \in \mathcal{C}(X)$. And since the absolute value metric on $\mathbf{R}$ is positive definite, symmetric, and satisfies the Triangle Inequality, it is easy to verify that $\text{dist}(f, g)$ does, too.

Notice that $\|f_n - f\| \to 0$ as $n \to \infty$ if and only if for each $\varepsilon > 0$ there is an $N \in \mathbf{N}$ such that $n \geq N$ implies $|f_n(x) - f(x)| < \varepsilon$ for all $x \in X$. Thus convergence of a sequence f_n in the metric of $\mathcal{C}(X)$ is equivalent to uniform convergence of f_n on X.

10.66 Definition.

Let X be a metric space.

a) A subset A of $\mathcal{C}(X)$ is said to be (*uniformly*) *closed* if and only if for each sequence $f_n \in A$ that satisfies $\|f_n - f\| \to 0$ as $n \to \infty$, the limit function f belongs to A.

b) A subset A of $\mathcal{C}(X)$ is said to be *uniformly dense* in $\mathcal{C}(X)$ if and only if given $\varepsilon > 0$ and $f \in \mathcal{C}(X)$ there is function $g \in A$ such that $\|g - f\| < \varepsilon$.

It is easy to see that $\mathcal{C}(X)$ is uniformly closed (modify the proof of Theorem 7.9). If we can show that the polynomials are uniformly dense in $\mathcal{C}[a, b]$, then the answer to the Approximation Question is evidently yes.

It turns out that uniformly closed algebras which contain the constant functions are also closed under pointwise maxima and minima.

10.67 Lemma.
Suppose that X is a compact metric space and that A is a closed algebra in $\mathcal{C}(X)$ which contains the constant functions. For each $f, g \in \mathcal{C}(X)$ and each $x \in X$, define $(f \wedge g)(x) := \min\{f(x), g(x)\}$ and $(f \vee g)(x) := \max\{f(x), g(x)\}$. If $f, g \in A$, then $f \wedge g$ and $f \vee g$ both belong to A.

Proof. Since $f \wedge g$ and $f \vee g$ can be defined as an algebraic combination of $(f + g)$ and $|f - g|$ (see Exercise 3.1.8), and A is an algebra, it suffices to prove that $f \in A$ implies $|f| \in A$. If $\|f\| = 0$, then $|f| = 0 \in A$ and there is nothing to prove. Hence, we may suppose that $M := \|f\| > 0$.

Let $\varepsilon > 0$. Recall (see Theorem 7.52) that the Binomial Series

$$|t| = (1 - (1 - t^2))^{1/2}$$
$$= 1 - \frac{1}{2}(1 - t^2) - \frac{1}{2 \cdot 4}(1 - t^2)^2 \tag{12}$$
$$- \sum_{k=3}^{\infty} \frac{1 \cdot 3 \cdot 5 \cdots (2k - 3)}{2 \cdot 4 \cdot 6 \cdots (2k)}(1 - t^2)^k$$

converges uniformly on compact subsets of $(-\sqrt{2}, \sqrt{2})$; hence on $[-1, 1]$. In particular, there is an $N \in \mathbf{N}$ such that if $n \geq N$ and

$$P_n(t) := 1 - \frac{1}{2}(1 - t^2) - \frac{1}{2 \cdot 4}(1 - t^2)^2 - \sum_{k=3}^{n} \frac{1 \cdot 3 \cdot 5 \cdots (2k - 3)}{2 \cdot 4 \cdot 6 \cdots (2k)}(1 - t^2)^k, \tag{13}$$

then $|P_n(t) - |t|| < \varepsilon$ for all $t \in [-1, 1]$.

Fix $x \in X$ and set $g_n(x) := P_n(f(x)/M)$. Since A is an algebra which contains the constant functions, $g_n \in A$. Since $M = \|f\| > 0$ implies that $t := f(x)/M \in [-1, 1]$, we have by the choice of P that

$$\left| g_n(x) - \left| \frac{f(x)}{M} \right| \right| = |P_n(t) - |t|| < \varepsilon$$

for all $n \geq N$ and all $x \in X$. Thus $Mg_n \in A$ and $Mg_n \to |f|$ uniformly on X, as $n \to \infty$. Since A is uniformly closed, we conclude that $|f|$ belongs to A. ∎

We will identify conditions on an algebra A which guarantee that A is uniformly dense in $\mathcal{C}(X)$. Since the collection of constant functions forms an algebra in $\mathcal{C}[a, b]$ which is not uniformly dense in $\mathcal{C}[a, b]$, such an A must contain some nonconstant functions. The following concept addresses this issue.

10.68 Definition.

A subset A of $\mathcal{C}(X)$ *separates points* of X if and only if given $x, y \in X$ with $x \neq y$ there exists an $f \in A$ such that $f(x) \neq f(y)$.

We are now prepared to identify a huge class of dense algebras in $\mathcal{C}(X)$ when X is compact.

10.69 Stone–Weierstrass Theorem.   *Suppose that X is a compact metric space. If A is an algebra in $\mathcal{C}(X)$ that separates points of X and contains the constant functions, then A is uniformly dense in $\mathcal{C}(X)$.*

Proof. Fix $x, y \in X$ with $x \neq y$. Let a, b be any points in $\mathbf{R}$. Since A separates points, there is a $g \in A$ such that $g(x) \neq g(y)$. Since A is an algebra that contains the constants—in particular, contains $g(x)$ and $g(y)$—the function

$$f(t) := a \frac{g(t) - g(y)}{g(x) - g(y)} + b \frac{g(t) - g(x)}{g(y) - g(x)}$$

also belongs to A. Since $f(x) = a$ and $f(y) = b$, we have proved the following: Given $x, y \in X$,

if $x \neq y$ and $a, b \in \mathbf{R}$, then $f(x) = a$ and $f(y) = b$ for some $f \in A$. $\qquad$ (14)

Let B represent the uniform closure of A (i.e., B is the set of all functions f for which there exists a sequence $f_n \in A$ such that $\| f_n - f \| \to 0$ as $n \to \infty$). Since A is an algebra, so is B. Since B is by definition uniformly closed, it contains its pointwise maxima and minima (see Lemma 10.67).

Let $F \in \mathcal{C}(X)$ and $\varepsilon > 0$. We must show that there is a $G \in B$ such that

$$F(x) - \varepsilon < G(x) < F(x) + \varepsilon \qquad (15)$$

for all $x \in X$. We will do this a point at a time and use the compactness of X to globalize these local estimates.

Fix $x_0 \in X$. For each $y \neq x_0$, use (14) to choose an $f_y \in A \subseteq B$ such that

$$f_y(x_0) = F(x_0) \quad \text{and} \quad f_y(y) = F(y).$$

Since f_y and F are continuous, the set $V_y := \{x \in X : f_y(x) < F(x) + \varepsilon\}$ is open. Since $x_0, y \in V_y$ for all $y \in X$, it is clear that

$$X = \bigcup_{y \neq x_0} V_y.$$

Since X is compact, there exist $y_j \in X$ such that $X = \bigcup_{j=1}^{N} V_{y_j}$. Let $f_j = f_{y_j}$ for $j = 1, \ldots, N$ and set $g_{x_0} := f_1 \wedge \cdots \wedge f_N$. Then $g_{x_0} \in B$. Moreover, by construction,

$$g_{x_0}(x_0) = F(x_0) \wedge \cdots \wedge F(x_0) = F(x_0).$$

Since $x \in X$ implies $x \in V_{y_k}$ for some $k \in [1, N]$, we also have

$$g_{x_0}(x) \leq f_k(x) < F(x) + \varepsilon. \qquad (16)$$

This is essentially the right half of (15).

To finish the proof, repeat the argument by covering X by a finite collection of open sets $\{W_{x_j}\}_{j=1}^{M}$, where

$$W_{x_j} := \{x \in X : g_{x_j}(x) > F(x) - \varepsilon\}.$$

Let $g_j = g_{x_j}$ and set $G := g_1 \vee \cdots \vee g_M$. By (16), $G(x) < F(x) + \varepsilon$ for all $x \in X$. But since $x \in X$ implies that $x \in W_{x_j}$ for some j, we also have

$$G(x) \geq g_j(x) \geq F(x) - \varepsilon$$

for all $x \in X$. Therefore, (15) holds for all $x \in X$.   ∎

EXERCISES

10.7.1. a) Prove that given $f \in C[a, b]$, then there is a sequence of one-variable polynomials P_n such that $P_n \to f$ uniformly on $[a, b]$ as $n \to \infty$.

  b) Prove that the metric space $C[a, b]$ (see Example 10.6) is separable.

10.7.2. A *polynomial* on $\mathbf{R}^n$ is a function of the form

$$P(x_1, x_2, \ldots, x_n) = \sum_{j_1=0}^{N_1} \cdots \sum_{j_n=0}^{N_n} a_{j_1, \ldots, j_n} x_1^{j_1} \cdots x_n^{j_n},$$

where the $a_{j_1, \ldots, j_n}$'s are scalars and $N_1, \ldots, N_n \in \mathbf{N}$. Prove that if A is compact in $\mathbf{R}^n$ and $f \in C(A)$, then there is a sequence of polynomials P_k on $\mathbf{R}^n$ such that $P_k \to f$ uniformly on A as $k \to \infty$.

10.7.3. Let $R = [a, b] \times [c, d]$ be a rectangle in $\mathbf{R}^2$. A function f is said to have *separated variables* if

$$P(x, y) = \sum_{k=1}^{N} c_k f_k(x) g_k(y)$$

for some scalars c_k and functions $f_k \in C[a, b]$, $g_k \in C[c, d]$. Prove that given $f \in C(R)$ there is a sequence of functions with separated variables, P_n, such that $P_n \to f$ uniformly on R as $n \to \infty$.

10.7.4. Use Exercise 10.7.1 to prove that if $f \in C[a, b]$ and

$$\int_a^b f(x) x^k \, dx = 0$$

for $k = 0, 1, \ldots$, then $f(x) = 0$ for all $x \in [a, b]$.

10.7.5. Use Exercise 10.7.3 to prove that if $f \in C([a, b] \times [c, d])$, then

$$\int_a^b \left(\int_c^d f(x, y) \, dy \right) dx = \int_c^d \left(\int_a^b f(x, y) \, dx \right) dy.$$

10.7.6. Let $T = [0, 2\pi)$.

  a) Prove that the function

$$\Phi(x) := (\cos x, \sin x)$$

is 1–1 from T onto $\partial B_1(0, 0) \subseteq \mathbf{R}^2$.

b) Prove that

$$p(x, y) := \|\Phi(x) - \Phi(y)\|$$

is a metric on T.

c) Prove that a function f is continuous on (T, p) if and only if it is continuous and periodic on $[0, 2\pi]$; that is, if and only if f has an extension to $[0, 2\pi]$ which is continuous in the usual sense which also satisfies $f(0) = f(2\pi)$.

d) A function P is called a *trigonometric polynomial* if

$$P(x) = \sum_{k=0}^{N} a_k \cos(kx) + b_k \sin(kx)$$

for some scalars a_k, b_k. Prove that given $f \in \mathcal{C}(T)$ there is a sequence of trigonometric polynomials P_n such that $P_n \to f$ uniformly on $[0, 2\pi]$ as $n \to \infty$.

10.7.7. Use Exercise 10.7.6 to prove that if f is continuous and periodic on $[0, 2\pi]$ and

$$\int_0^{2\pi} f(x) \cos(kx) \, dx = \int_0^{2\pi} f(x) \sin(kx) \, dx = 0$$

for $k = 0, 1, \ldots$, then $f(x) = 0$ for all $x \in [a, b]$.

CHAPTER 11

Differentiability on $\mathbf{R}^n$

11.1 PARTIAL DERIVATIVES AND PARTIAL INTEGRALS

The most natural way to define derivatives and integrals of functions of several variables is to allow one variable to move at a time. The corresponding objects, partial derivatives and partial integrals, are the subjects of this section. Our main goal is to identify conditions under which partial derivatives, partial integrals, and evaluation of limits commute with each other (e.g., under which the limit of a partial integral is the partial integral of the limit).

We begin with some notation. The *Cartesian product* of a finite collection of sets $E_1, E_2, \ldots, E_n$ is the set of ordered n-tuples defined by

$$E_1 \times E_2 \times \cdots \times E_n := \{(x_1, x_2, \ldots, x_n) : x_j \in E_j \text{ for } j = 1, 2, \ldots, n\}.$$

Thus the Cartesian product of n subsets of $\mathbf{R}$ is a subset of $\mathbf{R}^n$. By a *rectangle* in $\mathbf{R}^n$ (or an *n-dimensional rectangle*) we mean a Cartesian product of n closed, nondegenerate, bounded intervals. An n-dimensional rectangle $H = [a_1, b_1] \times \cdots \times [a_n, b_n]$ is called an *n-dimensional cube with side s* if $|b_j - a_j| = s$ for $j = 1, \ldots, n$.

Let $f : \{x_1\} \times \cdots \times \{x_{j-1}\} \times [a, b] \times \{x_{j+1}\} \times \cdots \times \{x_n\} \to \mathbf{R}$. We shall denote the function

$$g(t) := f(x_1, \ldots, x_{j-1}, t, x_{j+1}, \ldots, x_n), \qquad t \in [a, b],$$

by $f(x_1, \ldots, x_{j-1}, \cdot, x_{j+1}, \ldots, x_n)$. If g is integrable on $[a, b]$, then the *partial integral* of f on $[a, b]$ with respect to x_j is defined by

$$\int_a^b f(x_1, \ldots, x_n) \, dx_j := \int_a^b g(t) \, dt.$$

If g is differentiable at some $t_0 \in (a, b)$, then the *partial derivative* (or *first-order partial derivative*) of f at $(x_1, \ldots, x_{j-1}, t_0, x_{j+1}, \ldots x_n)$ with respect to x_j is defined by

$$\frac{\partial f}{\partial x_j}(x_1, \ldots, x_{j-1}, t_0, x_{j+1}, \ldots, x_n) := g'(t_0).$$

We will also denote this partial derivative by $f_{x_j}(x_1, \ldots, x_{j-1}, t_0, x_{j+1}, \ldots, x_n)$. Thus the partial derivative f_{x_j} exists at a point $\mathbf{a}$ if and only if the limit

$$\frac{\partial f}{\partial x_j}(\mathbf{a}) := \lim_{h \to 0} \frac{f(\mathbf{a} + h\mathbf{e}_j) - f(\mathbf{a})}{h}$$

exists. (Some authors use f_j to denote the partial derivative f_{x_j}. To avoid confusing first-order partial derivatives with sequences and components of functions, we will not use this notation.)

We extend partial derivatives to vector-valued functions in the following way. Suppose that $\mathbf{a} = (a_1, \ldots, a_n) \in \mathbf{R}^n$ and $\mathbf{f} = (f_1, f_2, \ldots, f_m) : \{a_1\} \times \cdots \times \{a_{j-1}\} \times I \times \{a_{j+1}\} \times \cdots \times \{a_n\} \to \mathbf{R}^m$, where $j \in \{1, 2, \ldots, n\}$ is fixed and I is an open interval containing a_j. If for each $k = 1, 2, \ldots, m$ the first-order partial derivative $\partial f_k / \partial x_j$ exists at $\mathbf{a}$, then we define the *first-order partial derivative* of $\mathbf{f}$ with respect to x_j to be the vector-valued function

$$\mathbf{f}_{x_j}(\mathbf{a}) := \frac{\partial \mathbf{f}}{\partial x_j}(\mathbf{a}) := \left(\frac{\partial f_1}{\partial x_j}(\mathbf{a}), \ldots, \frac{\partial f_m}{\partial x_j}(\mathbf{a}) \right).$$

Higher-order partial derivatives are defined by iteration. For example, the *second-order partial derivative* of $\mathbf{f}$ with respect to x_j and x_k is defined by

$$\mathbf{f}_{x_j x_k} := \frac{\partial^2 \mathbf{f}}{\partial x_k \, \partial x_j} := \frac{\partial}{\partial x_k}\left(\frac{\partial \mathbf{f}}{\partial x_j} \right)$$

when it exists. Second-order partial derivatives are called *mixed* when $j \neq k$.

This brings us to the following important collection of functions.

11.1 Definition.

Let V be a nonempty, open subset of $\mathbf{R}^n$, let $\mathbf{f} : V \to \mathbf{R}^m$, and let $p \in \mathbf{N}$.

i) $\mathbf{f}$ is said to be $\mathcal{C}^p$ on V if and only if each partial derivative of $\mathbf{f}$ of order $k \leq p$ exists and is continuous on V.
ii) $\mathbf{f}$ is said to be $\mathcal{C}^\infty$ on V if and only if $\mathbf{f}$ is $\mathcal{C}^p$ on V for all $p \in \mathbf{N}$.

Clearly, if $\mathbf{f}$ is $\mathcal{C}^p$ on V and $q < p$, then $\mathbf{f}$ is $\mathcal{C}^q$ on V. By making obvious modifications in Definition 11.1 using Definition 4.6, we can also define what it means for a function to be $\mathcal{C}^p$ on a rectangle H. We shall denote the collection of functions which are $\mathcal{C}^p$ on an open set V, respectively, on a rectangle H, by $\mathcal{C}^p(V)$, respectively, by $\mathcal{C}^p(H)$.

For simplicity, we shall state all results in this section for the case $n = 2$ and $m = 1$, using x for x_1 and y for x_2. (It is too cumbersome to do otherwise.) It is clear that, with appropriate changes in notation, these results also hold for any $n, m \in \mathbf{N}$.

Since partial derivatives and partial integrals are essentially one-dimensional concepts, each one-dimensional result about derivatives and integrals contains information about partial derivatives and partial integrals. Here are three examples.

1) By the Product Rule (Theorem 4.10), if f_x and g_x exist, then

$$\frac{\partial}{\partial x}(fg) = f\frac{\partial g}{\partial x} + g\frac{\partial f}{\partial x}.$$

2) By the Mean Value Theorem (Theorem 4.15), if $f(\cdot, y)$ is continuous on $[a, b]$ and the partial derivative $f_x(\cdot, y)$ exists on (a, b), then there is a point $c \in (a, b)$ (which may depend on y as well as a and b) such that

$$f(b, y) - f(a, y) = (b - a)\frac{\partial f}{\partial x}(c, y).$$

3) By the Fundamental Theorem of Calculus (Theorem 5.28), if $f(\cdot, y)$ is continuous on $[a, b]$, then

$$\frac{\partial}{\partial x}\int_a^x f(t, y)\, dt = f(x, y),$$

and if the partial derivative $f_x(\cdot, y)$ exists and is integrable on $[a, b]$, then

$$\int_a^b \frac{\partial f}{\partial x}(x, y)\, dx = f(b, y) - f(a, y).$$

Our first result about the commutation of partial derivatives, partial integrals, and evaluation of limits deals with interchanging two first-order partial derivatives (see also Exercise 11.2.11).

11.2 Theorem. *Suppose that V is open in $\mathbf{R}^2$, that $(a, b) \in V$, and that $f : V \to \mathbf{R}$. If f is C^1 on V, and if one of the mixed second partial derivatives of f exists on V and is continuous at the point (a, b), then the other mixed second partial derivative exists at (a, b) and*

$$\frac{\partial^2 f}{\partial y\, \partial x}(a, b) = \frac{\partial^2 f}{\partial x\, \partial y}(a, b).$$

NOTE: *These hypotheses are met if $f \in C^2(V)$.*

Proof. Suppose that f_{yx} exists on V and is continuous at the point (a, b). Consider $\Delta(h, k) := f(a + h, b + k) - f(a + h, b) - f(a, b + k) + f(a, b)$, defined for $|h|, |k| < r/\sqrt{2}$, where $r > 0$ is so small that $B_r(a, b) \subset V$. Apply the Mean Value Theorem twice to choose scalars $s, t \in (0, 1)$ such that

$$\Delta(h, k) = k\frac{\partial f}{\partial y}(a + h, b + tk) - k\frac{\partial f}{\partial y}(a, b + tk) = hk\frac{\partial^2 f}{\partial x\, \partial y}(a + sh, b + tk).$$

Since this last mixed partial derivative is continuous at the point (a, b), we have

$$\lim_{k \to 0} \lim_{h \to 0} \frac{\Delta(h, k)}{hk} = \frac{\partial^2 f}{\partial x \, \partial y}(a, b). \tag{1}$$

On the other hand, the Mean Value Theorem also implies that there is a scalar $u \in (0, 1)$ such that

$$\Delta(h, k) = f(a + h, b + k) - f(a, b + k) - f(a + h, b) + f(a, b)$$

$$= h \frac{\partial f}{\partial x}(a + uh, b + k) - h \frac{\partial f}{\partial x}(a + uh, b).$$

Hence, it follows from (1) that

$$\lim_{k \to 0} \lim_{h \to 0} \frac{1}{k} \left(\frac{\partial f}{\partial x}(a + uh, b + k) - \frac{\partial f}{\partial x}(a + uh, b) \right)$$

$$= \lim_{k \to 0} \lim_{h \to 0} \frac{\Delta(h, k)}{hk} = \frac{\partial^2 f}{\partial x \, \partial y}(a, b).$$

Since f_x is continuous on $B_r(a, b)$, we can let $h = 0$ in the first expression. We conclude by definition that

$$\frac{\partial^2 f}{\partial y \, \partial x}(a, b) = \lim_{k \to 0} \frac{1}{k} \left(\frac{\partial f}{\partial x}(a, b + k) - \frac{\partial f}{\partial x}(a, b) \right) = \frac{\partial^2 f}{\partial x \, \partial y}(a, b). \qquad \blacksquare$$

We shall refer to the conclusion of Theorem 11.2 by saying the first partial derivatives of f *commute*. Thus, if f is C^2 on an open subset V of $\mathbf{R}^n$, if $\mathbf{a} \in V$, and if $j \neq k$, then

$$\frac{\partial^2 f}{\partial x_j \partial x_k}(\mathbf{a}) = \frac{\partial^2 f}{\partial x_k \partial x_j}(\mathbf{a}).$$

The following example shows that Theorem 11.2 is false if the assumption about continuity of the second-order partial derivative is dropped.

11.3 EXAMPLE.

Prove that

$$f(x, y) = \begin{cases} xy \left(\dfrac{x^2 - y^2}{x^2 + y^2} \right) & (x, y) \neq \mathbf{0} \\ 0 & (x, y) = \mathbf{0} \end{cases}$$

is C^1 on $\mathbf{R}^2$, both mixed second partial derivatives of f exist on $\mathbf{R}^2$, but the first partial derivatives of f do not commute at $(0, 0)$; that is, $f_{xy}(0, 0) \neq f_{yx}(0, 0)$.

Proof. By the one-dimensional Product and Quotient Rules,

$$\frac{\partial f}{\partial x}(x, y) = xy\frac{\partial}{\partial x}\left(\frac{x^2 - y^2}{x^2 + y^2}\right) + \frac{\partial}{\partial x}(xy)\left(\frac{x^2 - y^2}{x^2 + y^2}\right)$$

$$= xy\left(\frac{4xy^2}{(x^2 + y^2)^2}\right) + y\left(\frac{x^2 - y^2}{x^2 + y^2}\right)$$

for $(x, y) \neq (0, 0)$. Since $2|xy| \leq x^2 + y^2$, we have $|f_x(x, y)| \leq 2|y|$. Therefore, $f_x(x, y) \to 0$ as $(x, y) \to (0, 0)$. On the other hand, by definition

$$\frac{\partial f}{\partial x}(0, y) = \lim_{h \to 0} y\left(\frac{h^2 - y^2}{h^2 + y^2}\right) = -y$$

for all $y \in \mathbf{R}$; hence, $f_x(0, 0) = 0$. This proves that f_x exists and is continuous on $\mathbf{R}^2$ with value $-y$ at $(0, y)$. A similar argument shows that f_y exists and is continuous on $\mathbf{R}^2$ with value x at $(x, 0)$. It follows that the mixed second partial derivatives of f exist on $\mathbf{R}^2$, and

$$\frac{\partial^2 f}{\partial y\, \partial x}(0, 0) = -1 \neq 1 = \frac{\partial^2 f}{\partial x\, \partial y}(0, 0). \qquad \blacksquare$$

The following result shows that we can interchange a limit sign and a partial integral sign when the integrand is continuous on a rectangle.

11.4 Theorem. *Let $H = [a, b] \times [c, d]$ be a rectangle and let $f : H \to \mathbf{R}$ be continuous. If*

$$F(y) = \int_a^b f(x, y)\, dx,$$

then F is continuous on $[c, d]$; that is,

$$\lim_{\substack{y \to y_0 \\ y \in [c,d]}} \int_a^b f(x, y)\, dx = \int_a^b \lim_{\substack{y \to y_0 \\ y \in [c,d]}} f(x, y)\, dx$$

for all $y_0 \in [c, d]$.

Proof. For each $y \in [c, d]$, $f(\cdot, y)$ is continuous on $[a, b]$. Hence, by Theorem 5.10, $F(y)$ exists for $y \in [c, d]$.

Fix $y_0 \in [c, d]$ and let $\varepsilon > 0$. Since H is compact, f is uniformly continuous on H. Hence, choose $\delta > 0$ such that $\|(x, y) - (z, w)\| < \delta$ and $(x, y), (z, w) \in H$ imply

$$|f(x, y) - f(z, w)| < \frac{\varepsilon}{b - a}.$$

Since $|y - y_0| = \|(x, y) - (x, y_0)\|$, it follows that

$$|F(y) - F(y_0)| \le \int_a^b |f(x, y) - f(x, y_0)| \, dx < \varepsilon$$

for all $y \in [c, d]$ which satisfy $|y - y_0| < \delta$. We conclude that F is continuous on $[c, d]$. ∎

The following result shows that we can interchange a derivative and an integral sign when the first partial derivative of the integrand is sufficiently smooth. We will refer to this process as *differentiating under the integral sign*.

11.5 Theorem. *Let $H = [a, b] \times [c, d]$ be a rectangle in $\mathbf{R}^2$ and let $f : H \to \mathbf{R}$. Suppose that $f(\cdot, y)$ is integrable on $[a, b]$ for each $y \in [c, d]$ and that the partial derivative $f_y(x, \cdot)$ exists on $[c, d]$ for each $x \in [a, b]$. If the two-variable function $f_y(x, y)$ is continuous on H, then*

$$\frac{d}{dy} \int_a^b f(x, y) \, dx = \int_a^b \frac{\partial f}{\partial y}(x, y) \, dx$$

for all $y \in [c, d]$.

NOTE: *These hypotheses are met if $f \in \mathcal{C}^1(H)$.*

Proof. Recall that "$f_y(x, \cdot)$ exists on $[c, d]$" means $f_y(x, \cdot)$ exists on (c, d), and

$$f_y(x, c) := \lim_{h \to 0+} \frac{f(x, c + h) - f(x, c)}{h}, \quad f_y(x, d) := \lim_{h \to 0-} \frac{f(x, d + h) - f(x, d)}{h}$$

both exist (see Definition 4.6). Hence, it suffices to show that

$$\lim_{h \to 0+} \int_a^b \frac{f(x, y + h) - f(x, y)}{h} \, dx = \int_a^b \frac{\partial f}{\partial y}(x, y) \, dx$$

for $y \in [c, d)$, and that

$$\lim_{h \to 0-} \int_a^b \frac{f(x, y + h) - f(x, y)}{h} \, dx = \int_a^b \frac{\partial f}{\partial y}(x, y) \, dx$$

for $y \in (c, d]$. The arguments are similar; we provide the details only for the first identity.

Fix $x \in [a, b]$ and $y \in [c, d)$, and let $h > 0$ be so small that $y + h \in [c, d)$. Let $\varepsilon > 0$. By uniform continuity, choose a $\delta > 0$ so small that $|y - z| < \delta$ and $x \in [a, b]$ imply $|f_y(x, y) - f_y(x, z)| < \varepsilon/(b-a)$. By the Mean Value Theorem, choose a point $z(x; h)$ between y and $y + h$ such that

$$F(x, y, h) := \frac{f(x, y + h) - f(x, y)}{h} = \frac{\partial f}{\partial y}(x, z(x; h)).$$

Since $|z(x; h) - y| = z(x; h) - y \leq h$, it follows that if $0 < h < \delta$, then

$$\left| \int_a^b F(x, y, h) \, dx - \int_a^b \frac{\partial f}{\partial y}(x, y) \, dx \right| \leq \int_a^b \left| \frac{\partial f}{\partial y}(x, z(x; h)) - \frac{\partial f}{\partial y}(x, y) \right| dx < \varepsilon.$$

Therefore,

$$\frac{d}{dy} \int_a^b f(x, y) \, dx = \int_a^b \frac{\partial f}{\partial y}(x, y) \, dx. \qquad \blacksquare$$

Thus if $H = [a_1, b_1] \times \cdots \times [a_n, b_n]$ is an n-dimensional rectangle, if f is $\mathcal{C}^1$ on H, and if $k \neq j$, then

$$\frac{\partial}{\partial x_k} \int_{a_j}^{b_j} f(x_1, \ldots, x_n) \, dx_j = \int_{a_j}^{b_j} \frac{\partial f}{\partial x_k}(x_1, \ldots, x_n) \, dx_j. \qquad (2)$$

The rest of this section contains optional material which shows what happens to the results above when the improper integral is used.

We begin by borrowing a concept from the theory of infinite series.

---

*11.6 Definition.

Let $a < b$ be extended real numbers, let I be an interval in $\mathbf{R}$, and suppose that $f : (a, b) \times I \to \mathbf{R}$. The improper integral

$$\int_a^b f(x, y) \, dx$$

is said to *converge uniformly* on I if and only if $f(\cdot, y)$ is improperly integrable on (a, b) for each $y \in I$ and given $\varepsilon > 0$ there exist real numbers $A, B \in (a, b)$ such that

$$\left| \int_a^b f(x, y) \, dx - \int_\alpha^\beta f(x, y) \, dx \right| < \varepsilon$$

for all $a < \alpha < A$, $B < \beta < b$, and all $y \in I$.

For most applications, the following simple test for uniform convergence of an improper integral will be used instead of Definition 11.6 (compare with Theorem 7.15).

*11.7 Theorem.   [WEIERSTRASS M-TEST].
Suppose that $a < b$ are extended real numbers, that I is an interval in $\mathbf{R}$, that $f : (a, b) \times I \to \mathbf{R}$, and that $f(\cdot, y)$ is locally integrable on the interval (a, b) for each $y \in I$. If there is a function $M : (a, b) \to \mathbf{R}$, absolutely integrable on (a, b), such that

$$|f(x, y)| \leq M(x)$$

for all $x \in (a, b)$ and $y \in I$, then

$$\int_a^b f(x, y)\, dx$$

converges uniformly on I.

Proof. Let $\varepsilon > 0$. By hypothesis and the Comparison Test for improper integrals, $\int_a^b f(x, y)\, dx$ exists and is finite for each $y \in I$. Moreover, since $M(x)$ is improperly integrable on (a, b), there exist real numbers A, B such that $a < A < B < b$ and

$$\int_a^A M(x)\, dx + \int_B^b M(x)\, dx < \varepsilon.$$

Thus for each $a < \alpha < A < B < \beta < b$ and each $y \in I$, we have

$$\left| \int_a^b f(x, y)\, dx - \int_\alpha^\beta f(x, y)\, dx \right| \le \int_a^\alpha |f(x, y)|\, dx + \int_\beta^b |f(x, y)|\, dx$$

$$\le \int_a^A M(x)\, dx + \int_B^b M(x)\, dx < \varepsilon. \qquad \blacksquare$$

The following is an improper integral analogue of Theorem 11.4.

*11.8 Theorem. *Suppose that $a < b$ are extended real numbers, that $c < d$ are finite real numbers, and that $f : (a, b) \times [c, d] \to \mathbf{R}$ is continuous. If*

$$F(y) = \int_a^b f(x, y)\, dx$$

converges uniformly on $[c, d]$, then F is continuous on $[c, d]$; that is,

$$\lim_{\substack{y \to y_0 \\ y \in [c,d]}} \int_a^b f(x, y)\, dx = \int_a^b \lim_{\substack{y \to y_0 \\ y \in [c,d]}} f(x, y)\, dx$$

for all $y_0 \in [c, d]$.

Proof. Let $\varepsilon > 0$ and $y_0 \in [c, d]$. Choose real numbers A, B such that $a < A < B < b$ and

$$\left| F(y) - \int_A^B f(x, y)\, dx \right| < \frac{\varepsilon}{3}$$

for all $y \in [c, d]$. By Theorem 11.4, choose $\delta > 0$ such that

$$\left| \int_A^B (f(x, y) - f(x, y_0))\, dx \right| < \frac{\varepsilon}{3}$$

for all $y \in [c, d]$ which satisfy $|y - y_0| < \delta$. Then

$$|F(y) - F(y_0)| \le \left| F(y) - \int_A^B f(x, y)\, dx \right| + \left| \int_A^B (f(x, y) - f(x, y_0))\, dx \right|$$

$$+ \left| F(y_0) - \int_A^B f(x, y_0)\, dx \right|$$

$$< \frac{\varepsilon}{3} + \frac{\varepsilon}{3} + \frac{\varepsilon}{3} = \varepsilon$$

for all $y \in [c, d]$ which satisfy $|y - y_0| < \delta$. ∎

The proof of Theorem 11.5 can be modified to prove the following result.

*11.9 Theorem. *Suppose that $a < b$ are extended real numbers, that $c < d$ are finite real numbers, that $f : (a, b) \times [c, d] \to \mathbf{R}$ is continuous, and that the improper integral*

$$F(y) = \int_a^b f(x, y)\, dx$$

exists for all $y \in [c, d]$. If $f_y(x, y)$ exists and is continuous on $(a, b) \times [c, d]$ and if

$$\phi(y) = \int_a^b \frac{\partial f}{\partial y}(x, y)\, dx$$

converges uniformly on $[c, d]$, then F is differentiable on $[c, d]$ and $F'(y) = \phi(y)$; that is,

$$\frac{d}{dy} \int_a^b f(x, y)\, dx = \int_a^b \frac{\partial f}{\partial y}(x, y)\, dx$$

for all $y \in [c, d]$.

For a result about interchanging two partial integrals, see Theorem 12.31 and Exercise 12.3.10.

EXERCISES

11.1.1. Compute all mixed second-order partial derivatives of each of the following functions and verify that the mixed partial derivatives are equal.

    a) $f(x, y) = xe^y$ b) $f(x, y) = \cos(xy)$ c) $f(x, y) = \dfrac{x + y}{x^2 + 1}$

11.1.2. For each of the following functions, compute f_x and determine where it is continuous.

a)
$$f(x, y) = \begin{cases} \dfrac{x^4 + y^4}{x^2 + y^2} & (x, y) \neq (0, 0) \\ 0 & (x, y) = (0, 0) \end{cases}$$

b)
$$f(x, y) = \begin{cases} \dfrac{x^2 - y^2}{\sqrt[3]{x^2 + y^2}} & (x, y) \neq (0, 0) \\ 0 & (x, y) = (0, 0) \end{cases}$$

11.1.3. Suppose that $r > 0$, that $\mathbf{a} \in \mathbf{R}^n$, and that $\mathbf{f} : B_r(\mathbf{a}) \to \mathbf{R}^m$. If all first-order partial derivatives of $\mathbf{f}$ exist on $B_r(\mathbf{a})$ and satisfy $\mathbf{f}_{x_j}(\mathbf{x}) = \mathbf{0}$ for all $\mathbf{x} \in B_r(\mathbf{a})$ and all $j = 1, 2, \ldots, n$, prove that $\mathbf{f}$ has only one value on $B_r(\mathbf{a})$.

11.1.4. Suppose that $H = [a, b] \times [c, d]$ is a rectangle, that $f : H \to \mathbf{R}$ is continuous, and that $g : [a, b] \to \mathbf{R}$ is integrable. Prove that

$$F(y) = \int_a^b g(x) f(x, y) \, dx$$

is uniformly continuous on $[c, d]$.

11.1.5. Evaluate each of the following expressions.

a)
$$\lim_{y \to 0} \int_0^1 e^{x^3 y^2 + x} \, dx$$

b)
$$\frac{d}{dy} \int_0^1 \sin(e^x y - y^3 + \pi - e^x) \, dx \quad \text{at } y = 1$$

c)
$$\frac{\partial}{\partial x} \int_1^3 \sqrt{x^3 + y^3 + z^3 - 2} \, dz \quad \text{at } (x, y) = (1, 1)$$

11.1.6. Suppose that f is a continuous real function.

a) If $\int_0^1 f(x) \, dx = 1$, find the exact value of

$$\lim_{y \to 0} \int_0^2 f(|x - 1|) e^{x^2 y + xy^2} \, dx.$$

b) If f is C^1 on $\mathbf{R}$ and $\int_0^\pi f'(x) \sin x \, dx = e$, find the exact value of

$$e + \lim_{y \to 0} \int_0^\pi f(x) \cos(y^5 + \sqrt[3]{y} + x) \, dx.$$

c) If $\int_0^1 f(\sqrt{x}) e^x \, dx = 6$, find the exact value of

$$\frac{d}{dx} \int_0^1 f(y) e^{xy + y^2} \, dy \quad \text{at } x = 0.$$

*11.1.7. Evaluate each of the following expressions.

a)
$$\lim_{y \to 0+} \int_0^1 \frac{x \cos y}{\sqrt[3]{1-x+y}} \, dx$$

b)
$$\frac{d}{dy} \int_\pi^\infty \frac{e^{-xy} \sin x}{x} \, dx \quad \text{at } y = 1$$

*11.1.8. a) Prove that

$$\int_0^1 \frac{\cos(x^2 + y^2)}{\sqrt{x}} \, dx$$

converges uniformly on $(-\infty, \infty)$.
   b) Prove that $\int_0^\infty e^{-xy} dx$ converges uniformly on $[1, \infty)$.
   c) Prove that $\int_0^\infty y e^{-xy} dx$ exists for each $y \in [0, \infty)$ and converges uniformly on any $[a, b] \subset (0, \infty)$ but that it does not converge uniformly on $[0, 1]$.

---

*11.10 Definition.

The *Laplace transform* of a function $f : (0, \infty) \to \mathbf{R}$ is said to exist at a point $s \in (0, \infty)$ if and only if the integral

$$\mathcal{L}\{f\}(s) := \int_0^\infty e^{-st} f(t) \, dt$$

converges. (*Note*: This integral is improper at ∞ and may be improper at 0.)

---

*11.1.9. Prove that

a)
$$\mathcal{L}\{1\}(s) = \frac{1}{s}, \qquad s > 0$$

b)
$$\mathcal{L}\{t^n\}(s) = \frac{n!}{s^{n+1}}, \qquad s > 0, \; n \in \mathbf{N}$$

c)
$$\mathcal{L}\{e^{at}\}(s) = \frac{1}{s-a}, \qquad s > a, \; a \in \mathbf{R}$$

d)
$$\mathcal{L}\{\cos(bt)\}(s) = \frac{s}{s^2 + b^2}, \qquad s > 0, \; b \in \mathbf{R}$$

e)
$$\mathcal{L}\{\sin(bt)\}(s) = \frac{b}{s^2 + b^2}, \qquad s > 0, \; b \in \mathbf{R}$$

*11.1.10. Suppose that $f : (0, \infty) \to \mathbf{R}$ is continuous and bounded and that $\mathcal{L}\{f\}$ exists at some $a \in (0, \infty)$. Let

$$\phi(t) = \int_0^t e^{-au} f(u) \, du, \qquad t \in (0, \infty).$$

a) Prove that

$$\int_0^N e^{-st} f(t) \, dt = \phi(N) e^{-(s-a)N} + (s-a) \int_0^N e^{-(s-a)t} \phi(t) \, dt$$

for all $N \in \mathbf{N}$.

b) Prove that the integral $\int_0^\infty e^{-(s-a)t} \phi(t) dt$ converges uniformly on $[b, \infty)$ for any $b > a$ and

$$\int_0^\infty e^{-st} f(t) \, dt = (s-a) \int_0^\infty e^{-(s-a)t} \phi(t) \, dt, \qquad s > a.$$

c) Prove that $\mathcal{L}\{f\}$ exists, is continuous on (a, ∞), and satisfies

$$\lim_{s \to \infty} \mathcal{L}\{f\}(s) = 0.$$

d) Let $g(t) = tf(t)$ for $t \in (0, \infty)$. Prove that $\mathcal{L}\{f\}$ is differentiable on (a, ∞) and

$$\frac{d}{ds} \mathcal{L}\{f\}(s) = -\mathcal{L}\{g\}(s)$$

for all $s \in (a, \infty)$.

e) If, in addition, f' is continuous and bounded on $(0, \infty)$, prove that

$$\mathcal{L}(f')(s) = s\mathcal{L}(f)(s) - f(0)$$

for all $s \in (a, \infty)$.

*11.1.11. Using Exercises 11.1.9 and 11.1.10, find the Laplace transforms of each of the functions te^t, $t \sin \pi t$, and $t^2 \cos t$.

11.2 THE DEFINITION OF DIFFERENTIABILITY

In this section we define what it means for a vector function to be differentiable at a point. Whatever our definition, we expect two things: If $\mathbf{f}$ is differentiable at $\mathbf{a}$, then $\mathbf{f}$ will be continuous at $\mathbf{a}$, and all first-order partial derivatives of $\mathbf{f}$ will exist at $\mathbf{a}$.

Working by analogy with the one-variable case, we guess that $\mathbf{f}$ is differentiable at $\mathbf{a}$ if and only if all its first-order partial derivatives exist at $\mathbf{a}$. The following example shows that this guess is wrong even when the range of $\mathbf{f}$ is one dimensional.

11.11 *EXAMPLE*.

Prove that the first-order partial derivatives of

$$f(x, y) = \begin{cases} x + y & x = 0 \quad \text{or} \quad y = 0 \\ 1 & \text{otherwise} \end{cases}$$

exist at $(0, 0)$, but f is not continuous at $(0, 0)$.

Proof. Since

$$\lim_{x \to 0} f(x, x) = 1 \neq 0 = f(0, 0),$$

it is clear that f is not continuous at $(0, 0)$. However, the first-order partial derivatives of f DO exist since

$$f_x(0, 0) = \lim_{h \to 0} \frac{f(h, 0) - f(0, 0)}{h} = 1$$

and, similarly, $f_y(0, 0) = 1$. ∎

Even if we restrict our attention to those functions **f** which are continuous and have first-order partial derivatives, we still cannot be sure that **f** is differentiable (see Exercise 11.2.7). How, then, shall we define differentiability in $\mathbf{R}^n$?

When a mathematical analogy breaks down, it is often helpful to reformulate the problem in its original setting. For functions of one variable, we found that f is differentiable at a if and only if there is a linear function $T \in \mathcal{L}(\mathbf{R}; \mathbf{R})$ such that

$$\lim_{h \to 0} \frac{f(a + h) - f(a) - T(h)}{h} = 0$$

(see Theorem 4.3). Thus f is differentiable at $a \in \mathbf{R}$ if and only if there is a $T \in \mathcal{L}(\mathbf{R}; \mathbf{R})$ such that the function $\varepsilon(h) := f(a + h) - f(a) - T(h)$ converges to zero so fast that $\varepsilon(h)/h \to 0$ as $h \to 0$. This leads us to the following definition.

11.12 Definition.

Suppose that $\mathbf{a} \in \mathbf{R}^n$, that V is an open set containing $\mathbf{a}$, and that $\mathbf{f} : V \to \mathbf{R}^m$.

i) **f** is said to be *differentiable at* $\mathbf{a}$ if and only if there is a $\mathbf{T} \in \mathcal{L}(\mathbf{R}^n; \mathbf{R}^m)$ such that the function

$$\varepsilon(\mathbf{h}) := \mathbf{f}(\mathbf{a} + \mathbf{h}) - \mathbf{f}(\mathbf{a}) - \mathbf{T}(\mathbf{h})$$

(defined for $\|\mathbf{h}\|$ sufficiently small) satisfies $\varepsilon(\mathbf{h})/\|\mathbf{h}\| \to \mathbf{0}$ as $\mathbf{h} \to \mathbf{0}$.

ii) **f** is said to be *differentiable on a set* E if and only if E is nonempty and **f** is differentiable at every point in E.

We shall see (Theorem 11.14) that if **f** is differentiable at $\mathbf{a}$, then there is only one **T** that satisfies Definition 11.12. Its representing $m \times n$ matrix (see

Theorem 8.15) is called the *total derivative* of $\mathbf{f}$ ("total" as opposed to partial derivatives of $\mathbf{f}$) and will be denoted by $D\mathbf{f}(\mathbf{a})$.

The following result shows that if $\mathbf{f}$ is differentiable, then it is continuous. Thus differentiability, as defined in 11.12, rules out pathology such as Example 11.11.

11.13 Theorem. *If a vector function $\mathbf{f}$ is differentiable at $\mathbf{a}$, then $\mathbf{f}$ is continuous at $\mathbf{a}$.*

Proof. Suppose that $\mathbf{f}$ is differentiable at $\mathbf{a}$. Then by Definition 11.12 there exist a $\mathbf{T} \in \mathcal{L}(\mathbf{R}^n; \mathbf{R}^m)$ and a $\delta > 0$ such that $\|\mathbf{f}(\mathbf{a}+\mathbf{h}) - \mathbf{f}(\mathbf{a}) - \mathbf{T}(\mathbf{h})\| \leq \|\mathbf{h}\|$ for all $\|\mathbf{h}\| < \delta$. By the triangle inequality (Theorem 8.6iii) and the definition of the operator norm (see Theorem 8.17), it follows that

$$\|\mathbf{f}(\mathbf{a}+\mathbf{h}) - \mathbf{f}(\mathbf{a})\| \leq \|\mathbf{T}\| \, \|\mathbf{h}\| + \|\mathbf{h}\|$$

for $\|\mathbf{h}\| < \delta$. Since $\|\mathbf{T}\|$ is a finite real number, we conclude from the Squeeze Theorem that $\mathbf{f}(\mathbf{a}+\mathbf{h}) \to \mathbf{f}(\mathbf{a})$ as $\mathbf{h} \to \mathbf{0}$ (i.e., $\mathbf{f}$ is continuous at $\mathbf{a}$). ∎

If $\mathbf{f}$ is differentiable at $\mathbf{a}$, is there an easy way to compute the total derivative $D\mathbf{f}(\mathbf{a})$? The following result shows that the answer to this question is yes.

11.14 Theorem. *Let $\mathbf{f}$ be a vector function. If $\mathbf{f}$ is differentiable at $\mathbf{a}$, then all first-order partial derivatives of $\mathbf{f}$ exist at $\mathbf{a}$. Moreover, the total derivative of $\mathbf{f}$ at $\mathbf{a}$ is unique and can be computed by*

$$D\mathbf{f}(\mathbf{a}) = \left[\frac{\partial f_i}{\partial x_j}(\mathbf{a}) \right]_{m \times n} := \begin{bmatrix} \dfrac{\partial f_1}{\partial x_1}(\mathbf{a}) & \cdots & \dfrac{\partial f_1}{\partial x_n}(\mathbf{a}) \\ \vdots & \ddots & \vdots \\ \dfrac{\partial f_m}{\partial x_1}(\mathbf{a}) & \cdots & \dfrac{\partial f_m}{\partial x_n}(\mathbf{a}) \end{bmatrix}.$$

Proof. Since $\mathbf{f}$ is differentiable, we know that there is an $m \times n$ matrix $B := [b_{ij}]$ such that

$$\frac{\mathbf{f}(\mathbf{a}+\mathbf{h}) - \mathbf{f}(\mathbf{a}) - B\mathbf{h}}{\|\mathbf{h}\|} \to \mathbf{0} \qquad \text{as } \mathbf{h} \to \mathbf{0}. \tag{3}$$

Fix $1 \leq j \leq n$ and set $\mathbf{h} = t\mathbf{e}_j$ for some $t > 0$. Since $\|\mathbf{h}\| = t$, we have

$$\frac{\mathbf{f}(\mathbf{a}+\mathbf{h}) - \mathbf{f}(\mathbf{a}) - B\mathbf{h}}{\|\mathbf{h}\|} := \frac{\mathbf{f}(\mathbf{a}+t\mathbf{e}_j) - \mathbf{f}(\mathbf{a})}{t} - B\mathbf{e}_j.$$

Take the limit of this identity as $t \to 0+$, using (3) and the definition of matrix multiplication. We obtain

$$\lim_{t \to 0+} \frac{\mathbf{f}(\mathbf{a}+t\mathbf{e}_j) - \mathbf{f}(\mathbf{a})}{t} = B\mathbf{e}_j = (b_{1j}, \ldots, b_{mj}).$$

A similar argument shows that the limit of this quotient as $t \to 0-$ also exists and equals $(b_{1j}, \ldots, b_{mj})$. Since a vector function converges if and only if its components converge (see Theorem 9.16), it follows that the first-order partial derivative of each component f_i with respect to x_j exists at $\mathbf{a}$ and satisfies

$$\frac{\partial f_i}{\partial x_j}(\mathbf{a}) = b_{ij}$$

for $i = 1, 2, \ldots, m$. In particular, for each differentiable $\mathbf{f}$ and each point $\mathbf{a}$, there is only one $\mathbf{T}$ that satisfies Definition 11.12, and its representing matrix is

$$Df(\mathbf{a}) := [b_{ij}]_{m \times n} = \left[\frac{\partial f_i}{\partial x_j}(\mathbf{a}) \right]_{m \times n}. \qquad \blacksquare$$

The fact that for each function $\mathbf{f}$ differentiable at $\mathbf{a}$, there is only one matrix B which satisfies (3) will be referred to as *the uniqueness of the total derivative*.

By Theorem 11.14 and the uniqueness of the total derivative, we now have several ways to find out whether a given vector function $\mathbf{f}$ is differentiable. Indeed, $\mathbf{f}$ is differentiable at a point $\mathbf{a}$ if and only if there exists an $m \times n$ matrix B such that

$$\lim_{\mathbf{h} \to 0} \frac{\mathbf{f}(\mathbf{a} + \mathbf{h}) - \mathbf{f}(\mathbf{a}) - B\mathbf{h}}{\|\mathbf{h}\|} = \mathbf{0},$$

if and only if

$$\lim_{\mathbf{h} \to 0} \frac{\|\mathbf{f}(\mathbf{a} + \mathbf{h}) - \mathbf{f}(\mathbf{a}) - B\mathbf{h}\|}{\|\mathbf{h}\|} = 0,$$

or if and only if

$$\lim_{\mathbf{h} \to 0} \frac{\mathbf{f}(\mathbf{a} + \mathbf{h}) - \mathbf{f}(\mathbf{a}) - Df(\mathbf{a})(\mathbf{h})}{\|\mathbf{h}\|} = \mathbf{0}. \qquad (4)$$

We shall use these three descriptions interchangeably. The advantage of the first two conditions is that they can be applied without computing partial derivatives of $\mathbf{f}$ (e.g., see the proofs of Theorems 11.20 and 11.28). The last condition is more concrete but can only be used if you can compute the first-order partial derivatives of $\mathbf{f}$ at $\mathbf{a}$ (see Example 11.18). Notice that existence of these partial derivatives is NOT enough to conclude that $\mathbf{f}$ is differentiable (see Exercise 11.2.7).

Abusing the notation a bit, if all first-order partial derivatives of function $\mathbf{f}$ exist at a point $\mathbf{a}$, we shall denote the *Jacobian matrix*

$$\left[\frac{\partial f_i}{\partial x_j}(\mathbf{a}) \right]_{m \times n}$$

by $Df(\mathbf{a})$. We shall only call it the "total derivative" when $\mathbf{f}$ is differentiable at $\mathbf{a}$; that is, when $\mathbf{f}$ satisfies (4).

If $n = 1$ or $m = 1$, the Jacobian matrix $D\mathbf{f}$ is an $m \times 1$ or $1 \times n$ matrix and, hence, can be identified with a vector. Most applied mathematicians represent $D\mathbf{f}$ in these cases by different notations. For the case $n = 1$,

$$
D\mathbf{f}(a) = \begin{bmatrix} f_1'(a) \\ \vdots \\ f_m'(a) \end{bmatrix}
$$

is sometimes denoted in vector notation by

$$
\mathbf{f}'(a) := \left(f_1'(a), \ldots, f_m'(a) \right).
$$

For the case $m = 1$,

$$
D f(\mathbf{a}) = \begin{bmatrix} \dfrac{\partial f}{\partial x_1}(\mathbf{a}) & \cdots & \dfrac{\partial f}{\partial x_n}(\mathbf{a}) \end{bmatrix}
$$

is sometimes denoted in vector notation by

$$
\nabla f(\mathbf{a}) := \left(\frac{\partial f}{\partial x_1}(\mathbf{a}), \ldots, \frac{\partial f}{\partial x_n}(\mathbf{a}) \right).
$$

($\nabla \mathbf{f}$ is called the *gradient* of $\mathbf{f}$ because it identifies the direction of steepest ascent. For this connection and a relationship between gradients and directional derivatives, see Exercise 11.4.11.)

If we strengthen the conclusion of Theorem 11.14, we can obtain a reverse implication.

11.15 Theorem. *Let V be open in $\mathbf{R}^n$, let $\mathbf{a} \in V$, and suppose that $\mathbf{f} : V \to \mathbf{R}^m$. If all first-order partial derivatives of $\mathbf{f}$ exist in V and are continuous at $\mathbf{a}$, then $\mathbf{f}$ is differentiable at $\mathbf{a}$.*

NOTE: *These hypotheses are met if $\mathbf{f}$ is C^1 on V.*

Proof. Since a function converges if and only if each of its components converge (see Theorem 9.16), we may suppose that $m = 1$. By definition, then, it suffices to show that if f is real valued and has continuous first partial derivatives on V, then

$$
\lim_{\mathbf{h} \to 0} \frac{f(\mathbf{a} + \mathbf{h}) - f(\mathbf{a}) - \nabla f(\mathbf{a}) \cdot \mathbf{h}}{\|\mathbf{h}\|} = 0.
$$

Let $\mathbf{a} = (a_1, \ldots, a_n)$. Suppose that $r > 0$ is so small that $B_r(\mathbf{a}) \subset V$. Fix $\mathbf{h} = (h_1, \ldots, h_n) \neq \mathbf{0}$ in $B_r(\mathbf{0})$. By telescoping and using the one-dimensional

Mean Value Theorem, we can choose numbers c_j between a_j and $a_j + h_j$ such that

$$f(\mathbf{a} + \mathbf{h}) - f(\mathbf{a}) = f(a_1 + h_1, \ldots, a_n + h_n) - f(a_1, a_2 + h_2, \ldots, a_n + h_n)$$
$$+ \cdots + f(a_1, \ldots, a_{n-1}, a_n + h_n) - f(a_1, \ldots, a_n)$$
$$= \sum_{j=1}^{n} h_j \frac{\partial f}{\partial x_j}(a_1, \ldots, a_{j-1}, c_j, a_{j+1} + h_{j+1}, \ldots, a_n + h_n).$$

Therefore,

$$f(\mathbf{a} + \mathbf{h}) - f(\mathbf{a}) - \nabla f(\mathbf{a}) \cdot \mathbf{h} = \mathbf{h} \cdot \boldsymbol{\delta}, \tag{5}$$

where $\boldsymbol{\delta} \in \mathbf{R}^n$ is the vector with components

$$\delta_j = \frac{\partial f}{\partial x_j}(a_1, \ldots, a_{j-1}, c_j, a_{j+1} + h_{j+1}, \ldots, a_n + h_n) - \frac{\partial f}{\partial x_j}(a_1, \ldots, a_n).$$

Since the first-order partial derivatives of f are continuous at $\mathbf{a}$, $\delta_j \to 0$ for each $1 \le j \le n$ (i.e., $\|\boldsymbol{\delta}\| \to 0$ as $\mathbf{h} \to \mathbf{0}$). Moreover, by the Cauchy–Schwarz Inequality and (5),

$$0 \le \frac{|f(\mathbf{a} + \mathbf{h}) - f(\mathbf{a}) - \nabla f(\mathbf{a}) \cdot \mathbf{h}|}{\|\mathbf{h}\|} = \frac{|\mathbf{h} \cdot \boldsymbol{\delta}|}{\|\mathbf{h}\|} \le \|\boldsymbol{\delta}\|. \tag{6}$$

It follows from the Squeeze Theorem that the first quotient in (6) converges to 0 as $\mathbf{h} \to \mathbf{0}$. Thus f is differentiable at $\mathbf{a}$ by definition. ∎

If all first-order partial derivatives of a vector function $\mathbf{f}$ exist and are continuous at a point $\mathbf{a}$ (respectively, on an open set V), we shall call $\mathbf{f}$ *continuously differentiable* at $\mathbf{a}$ (respectively, on V). By Theorem 11.15, every continuously differentiable function is differentiable. In particular, every function which is $\mathcal{C}^p$ on an open set V, for some $1 \le p \le \infty$, is continuously differentiable on V.

These results suggest the following procedure to determine whether a vector function $\mathbf{f}$ is differentiable at a point $\mathbf{a}$.

1) Compute all first-order partial derivatives of $\mathbf{f}$ at $\mathbf{a}$. If one of these does not exist, then $\mathbf{f}$ is not differentiable at $\mathbf{a}$ (Theorem 11.14).
2) If all first-order partial derivatives exist and are continuous at $\mathbf{a}$, then $\mathbf{f}$ is differentiable at $\mathbf{a}$ (Theorem 11.15).
3) If the first-order partial derivatives of $\mathbf{f}$ exist but one of them fails to be continuous at $\mathbf{a}$, then use the definition of differentiability directly. By the uniqueness of the total derivative, this will involve trying to verify (3) or (4). A decision about whether this limit exists and equals zero will involve methods outlined in Section 9.3.

We close with some examples.

11.16 EXAMPLE.

Is $\mathbf{f}(x, y) = (\cos(xy), \ln x - e^y)$ differentiable at $(1, 1)$?

Solution. Since $\mathbf{f}_x = (-y \sin(xy), 1/x)$ and $\mathbf{f}_y = (-x \sin(xy), -e^y)$ both exist and are continuous at any $(x, y) \in \mathbf{R}^2$ with $x > 0$, $\mathbf{f}$ is differentiable at any such (x, y), in particular, at $(1, 1)$.   ∎

11.17 EXAMPLE.

Is

$$
f(x, y) = \begin{cases} \dfrac{y^2}{x^2 + y^2} & (x, y) \neq (0, 0) \\ 0 & (x, y) = (0, 0) \end{cases}
$$

differentiable at $(0, 0)$?

Solution. Again we begin by looking at the first-order partial derivatives of f. By the one-dimensional Quotient Rule, if $(x, y) \neq (0, 0)$, then

$$
\frac{\partial f}{\partial x}(x, y) = -\frac{2xy^2}{(x^2 + y^2)^2}.
$$

To see whether the partial derivatives exist at $(0, 0)$, we apply the definition of f_x directly:

$$
\frac{\partial f}{\partial x}(0, 0) = \lim_{h \to 0} \frac{f(h, 0) - f(0, 0)}{h} = \lim_{h \to 0} \frac{0}{h} = 0.
$$

Thus $f_x(0, 0) = 0$ DOES exist even though the formula approach above crashed. Notice, then, that we cannot rely on the rules of differentiation alone to compute partial derivatives.

What about $f_y(0, 0)$? Again, we use the definition of f_y, not the formula:

$$
\frac{\partial f}{\partial y}(0, 0) = \lim_{k \to 0} \frac{f(0, k) - f(0, 0)}{k} = \lim_{k \to 0} \frac{1}{k}.
$$

Since this last limit does not exist, $f_y(0, 0)$ does not exist. Hence f cannot be differentiable at $(0, 0)$.   ∎

Our final example shows that the converse of Theorem 11.15 is false.

11.18 EXAMPLE.

Prove that

$$
f(x, y) = \begin{cases} (x^2 + y^2) \sin \dfrac{1}{\sqrt{x^2 + y^2}} & (x, y) \neq (0, 0) \\ 0 & (x, y) = (0, 0) \end{cases}
$$

is differentiable on $\mathbf{R}^2$ but not continuously differentiable at $(0, 0)$.

Proof. If $(x, y) \neq (0, 0)$, then we can use the one-dimensional Product Rule to verify that both f_x and f_y exist and are continuous, for example,

$$f_x(x, y) = \frac{-x}{\sqrt{x^2 + y^2}} \cos \frac{1}{\sqrt{x^2 + y^2}} + 2x \sin \frac{1}{\sqrt{x^2 + y^2}}.$$

Thus f is differentiable on $\mathbf{R}^2 \setminus \{(0, 0)\}$. Since $f_x(x, 0)$ has no limit as $x \to 0$, the partial derivative f_x is not continuous at $(0, 0)$. A similar statement holds for f_y. Thus to check differentiability at $(0, 0)$ we must return to the definition. First, we compute the partial derivatives at $(0, 0)$. By definition,

$$f_x(0, 0) = \lim_{t \to 0} \frac{f(t, 0) - f(0, 0)}{t} = \lim_{t \to 0} t \sin \frac{1}{|t|} = 0,$$

and similarly, $f_y(0, 0) = 0$. Thus, both first partials exist at $(0, 0)$ and $\nabla f(0, 0) = (0, 0)$.

To prove that f is differentiable at $(0, 0)$, we must verify (4) for $\mathbf{a} = (0, 0)$. But it is clear that

$$\frac{f(h, k) - f(0, 0) - \nabla f(0, 0) \cdot (h, k)}{\|(h, k)\|} = \sqrt{h^2 + k^2} \sin \frac{1}{\sqrt{h^2 + k^2}} \to 0$$

as $(h, k) \to (0, 0)$. Thus f is differentiable at $(0, 0)$. ∎

EXERCISES

11.2.1. Suppose, for $j = 1, 2, \ldots, n$, that f_j are real functions continuously differentiable on the interval $(-1, 1)$. Prove that

$$g(\mathbf{x}) := f_1(x_1) \cdots f_n(x_n)$$

is differentiable on the cube $(-1, 1) \times (-1, 1) \times \cdots \times (-1, 1)$.

11.2.2. Suppose that $\mathbf{f}, \mathbf{g} : \mathbf{R} \to \mathbf{R}^m$ are differentiable at a and there is a $\delta > 0$ such that $\mathbf{g}(x) \neq \mathbf{0}$ for all $0 < |x - a| < \delta$. If $\mathbf{f}(a) = \mathbf{g}(a) = \mathbf{0}$ and $D\mathbf{g}(a) \neq \mathbf{0}$, prove that

$$\lim_{x \to a} \frac{\|\mathbf{f}(x)\|}{\|\mathbf{g}(x)\|} = \frac{\|D\mathbf{f}(a)\|}{\|D\mathbf{g}(a)\|}.$$

11.2.3. Prove that $f(x, y) = \sqrt{|xy|}$ is not differentiable at $(0, 0)$.

11.2.4. Prove that

$$f(x, y) = \begin{cases} \dfrac{x^2 + y^2}{\sin \sqrt{x^2 + y^2}} & 0 < \|(x, y)\| < \pi \\ 0 & (x, y) = (0, 0) \end{cases}$$

is not differentiable at $(0, 0)$.

11.2.5. Prove that

$$f(x, y) = \begin{cases} \dfrac{x^4 + y^4}{(x^2 + y^2)^\alpha} & (x, y) \neq (0, 0) \\ 0 & (x, y) = (0, 0) \end{cases}$$

is differentiable on $\mathbf{R}^2$ for all $\alpha < 3/2$.

11.2.6. Prove that if $\alpha > 1/2$, then

$$f(x, y) = \begin{cases} |xy|^\alpha \log(x^2 + y^2) & (x, y) \neq (0, 0) \\ 0 & (x, y) = (0, 0) \end{cases}$$

is differentiable at $(0, 0)$.

11.2.7. Prove that

$$f(x, y) = \begin{cases} \dfrac{x^3 - xy^2}{x^2 + y^2} & (x, y) \neq (0, 0) \\ 0 & (x, y) = (0, 0) \end{cases}$$

is continuous on $\mathbf{R}^2$ and has first-order partial derivatives everywhere on $\mathbf{R}^2$, but f is not differentiable at $(0, 0)$.

11.2.8. **This exercise is used several times in this chapter and the next.** Suppose that $\mathbf{T} \in \mathcal{L}(\mathbf{R}^n; \mathbf{R}^m)$. Prove that $\mathbf{T}$ is differentiable everywhere on $\mathbf{R}^n$ with

$$D\mathbf{T}(\mathbf{a}) = \mathbf{T} \quad \text{for} \quad \mathbf{a} \in \mathbf{R}^n.$$

11.2.9. Let $r > 0$, $f : B_r(\mathbf{0}) \to \mathbf{R}$, and suppose that there exists an $\alpha > 1$ such that $|f(\mathbf{x})| \leq \|\mathbf{x}\|^\alpha$ for all $\mathbf{x} \in B_r(\mathbf{0})$. Prove that f is differentiable at $\mathbf{0}$. What happens to this result when $\alpha = 1$?

11.2.10. Let V be open in $\mathbf{R}^n$, $\mathbf{a} \in V$, and $\mathbf{f} : V \to \mathbf{R}^m$.

---

*11.19 Definition.

If $\mathbf{u}$ is a *unit* vector in $\mathbf{R}^n$ (i.e., $\|\mathbf{u}\| = 1$), then the *directional derivative* of $\mathbf{f}$ at $\mathbf{a}$ in the direction $\mathbf{u}$ is defined by

$$D_\mathbf{u}\mathbf{f}(\mathbf{a}) := \lim_{t \to 0} \frac{\mathbf{f}(\mathbf{a} + t\mathbf{u}) - \mathbf{f}(\mathbf{a})}{t}$$

when this limit exists.

---

a) Prove that $D_\mathbf{u}\mathbf{f}(\mathbf{a})$ exists for $\mathbf{u} = \mathbf{e}_k$ if and only if $\mathbf{f}_{x_k}(\mathbf{a})$ exists, in which case

$$D_{\mathbf{e}_k}\mathbf{f}(\mathbf{a}) = \frac{\partial \mathbf{f}}{\partial x_k}(\mathbf{a}).$$

b) Show that if $\mathbf{f}$ has directional derivatives at $\mathbf{a}$ in all directions $\mathbf{u}$, then the first-order partial derivatives of $\mathbf{f}$ exist at $\mathbf{a}$. Use Example 11.11 to show that the converse of this statement is false.

c) Prove that the directional derivatives of

$$f(x, y) = \begin{cases} \dfrac{x^2 y}{x^4 + y^2} & (x, y) \neq (0, 0) \\ 0 & (x, y) = (0, 0) \end{cases}$$

exist at $(0, 0)$ in all directions $\mathbf{u}$, but f is neither continuous nor differentiable at $(0, 0)$.

11.2.11. Let $r > 0$, $(a, b) \in \mathbf{R}^2$, $f : B_r(a, b) \to \mathbf{R}$, and suppose that the first-order partial derivatives f_x and f_y exist in $B_r(a, b)$ and are differentiable at (a, b).

a) Set $\Delta(h) = f(a+h, b+h) - f(a+h, b) - f(a, b+h) + f(a, b)$ and prove for h sufficiently small that

$$\frac{\Delta(h)}{h} = f_y(a + h, b + th) - f_y(a, b) - \nabla f_y(a, b) \cdot (h, th)$$
$$- \big(f_y(a, b + th) - f_y(a, b) - \nabla f_y(a, b) \cdot (0, th) \big) + h f_{yx}(a, b)$$

for some $t \in (0, 1)$.

b) Prove that

$$\lim_{h \to 0} \frac{\Delta(h)}{h^2} = f_{yx}(a, b).$$

c) Prove that

$$\frac{\partial^2 f}{\partial x \, \partial y}(a, b) = \frac{\partial^2 f}{\partial y \partial x}(a, b).$$

11.3   DERIVATIVES, DIFFERENTIALS, AND TANGENT PLANES

In this section we begin to explore the analogy between Df and f'. First we examine how the total derivative interacts with the algebra of functions.

11.20 Theorem.   *Let $\alpha \in \mathbf{R}$, $\mathbf{a} \in \mathbf{R}^n$, and suppose that $\mathbf{f}$ and $\mathbf{g}$ are vector functions. If $\mathbf{f}$ and $\mathbf{g}$ are differentiable at $\mathbf{a}$, then $\mathbf{f} + \mathbf{g}$, $\alpha \mathbf{f}$, and $\mathbf{f} \cdot \mathbf{g}$ are all differentiable at $\mathbf{a}$. In fact,*

$$D(\mathbf{f} + \mathbf{g})(\mathbf{a}) = D\mathbf{f}(\mathbf{a}) + D\mathbf{g}(\mathbf{a}), \tag{7}$$
$$D(\alpha \mathbf{f})(\mathbf{a}) = \alpha \, D\mathbf{f}(\mathbf{a}), \tag{8}$$

and

$$D(\mathbf{f} \cdot \mathbf{g})(\mathbf{a}) = \mathbf{g}(\mathbf{a}) D\mathbf{f}(\mathbf{a}) + \mathbf{f}(\mathbf{a}) D\mathbf{g}(\mathbf{a}). \tag{9}$$

[The sums which appear on the right side of (7) and (9) represent matrix addition, and the products which appear on the right side of (9) represent matrix multiplication.]

Proof. The proofs of these rules are similar. We provide the details only for the hardest of them, (9). Let

$$\mathbf{T} = \mathbf{g}(\mathbf{a})D\mathbf{f}(\mathbf{a}) + \mathbf{f}(\mathbf{a})D\mathbf{g}(\mathbf{a}). \tag{10}$$

Since $\mathbf{g}(\mathbf{a})$ and $\mathbf{f}(\mathbf{a})$ are $1 \times m$ matrices, and $D\mathbf{f}(\mathbf{a})$ and $D\mathbf{g}(\mathbf{a})$ are $m \times n$ matrices, $\mathbf{T}$ is a $1 \times n$ matrix, the right size for the total derivative of $\mathbf{f} \cdot \mathbf{g}$. By the uniqueness of the total derivative, we need only show that

$$\lim_{\mathbf{h}\to 0} \frac{(\mathbf{f} \cdot \mathbf{g})(\mathbf{a}+\mathbf{h}) - (\mathbf{f} \cdot \mathbf{g})(\mathbf{a}) - \mathbf{T}(\mathbf{h})}{\|\mathbf{h}\|} = 0.$$

Since by (10),

$$
\begin{aligned}
(\mathbf{f} \cdot \mathbf{g})&(\mathbf{a}+\mathbf{h}) - (\mathbf{f} \cdot \mathbf{g})(\mathbf{a}) - \mathbf{T}(\mathbf{h}) \\
&= (\mathbf{f} \cdot \mathbf{g})(\mathbf{a}+\mathbf{h}) - (\mathbf{f} \cdot \mathbf{g})(\mathbf{a}) - \mathbf{g}(\mathbf{a})D\mathbf{f}(\mathbf{a})(\mathbf{h}) - \mathbf{f}(\mathbf{a})D\mathbf{g}(\mathbf{a})(\mathbf{h}) \\
&= (\mathbf{f}(\mathbf{a}+\mathbf{h}) - \mathbf{f}(\mathbf{a}) - D\mathbf{f}(\mathbf{a})(\mathbf{h})) \cdot \mathbf{g}(\mathbf{a}+\mathbf{h}) \\
&\quad + (D\mathbf{f}(\mathbf{a})(\mathbf{h})) \cdot (\mathbf{g}(\mathbf{a}+\mathbf{h}) - \mathbf{g}(\mathbf{a})) \\
&\quad + \mathbf{f}(\mathbf{a}) \cdot (\mathbf{g}(\mathbf{a}+\mathbf{h}) - \mathbf{g}(\mathbf{a}) - D\mathbf{g}(\mathbf{a})(\mathbf{h})) \\
&=: T_1(\mathbf{h}) + T_2(\mathbf{h}) + T_3(\mathbf{h}),
\end{aligned}
$$

it suffices to verify $T_j(\mathbf{h})/\|\mathbf{h}\| \to 0$ as $\mathbf{h} \to 0$ for $j = 1, 2, 3$.

Set $\varepsilon(\mathbf{h}) = \mathbf{f}(\mathbf{a}+\mathbf{h}) - \mathbf{f}(\mathbf{a}) - D\mathbf{f}(\mathbf{a})(\mathbf{h})$ and $\delta(\mathbf{h}) = \mathbf{g}(\mathbf{a}+\mathbf{h}) - \mathbf{g}(\mathbf{a}) - D\mathbf{g}(\mathbf{a})(\mathbf{h})$ for $\|\mathbf{h}\|$ sufficiently small. Since $\mathbf{f}$ and $\mathbf{g}$ are differentiable at a, we know that $\varepsilon(\mathbf{h})/\|\mathbf{h}\|$ and $\delta(\mathbf{h})/\|\mathbf{h}\|$ both converge to zero as $\mathbf{h} \to 0$.

To estimate T_1, use the Cauchy–Schwarz Inequality and the definition of ε to verify

$$|T_1(\mathbf{h})| \le \|\mathbf{f}(\mathbf{a}+\mathbf{h}) - \mathbf{f}(\mathbf{a}) - D\mathbf{f}(\mathbf{a})(\mathbf{h})\| \, \|\mathbf{g}(\mathbf{a}+\mathbf{h})\| = \|\varepsilon(\mathbf{h})\| \, \|\mathbf{g}(\mathbf{a}+\mathbf{h})\|.$$

Since $\mathbf{g}$ is continuous at $\mathbf{a}$ (Theorem 11.13) and $\varepsilon(\mathbf{h})/\|\mathbf{h}\| \to 0$ as $\mathbf{h} \to 0$, it follows that $T_1(\mathbf{h})/\|\mathbf{h}\| \to 0$ as $\mathbf{h} \to 0$. A similar argument shows that $T_3(\mathbf{h})/\|\mathbf{h}\| \to 0$ as $\mathbf{h} \to 0$.

To estimate T_2, observe by the Cauchy–Schwarz Inequality and the definition of the operator norm (see Theorem 8.17) that

$$|T_2(\mathbf{h})| \le \|D\mathbf{f}(\mathbf{a})(\mathbf{h})\| \, \|\mathbf{g}(\mathbf{a}+\mathbf{h}) - \mathbf{g}(\mathbf{a})\| \le \|D\mathbf{f}(\mathbf{a})\| \, \|\mathbf{h}\| \, \|\mathbf{g}(\mathbf{a}+\mathbf{h}) - \mathbf{g}(\mathbf{a})\|.$$

Thus $|T_2(\mathbf{h})|/\|\mathbf{h}\| \le \|D\mathbf{f}(\mathbf{a})\| \, \|\mathbf{g}(\mathbf{a}+\mathbf{h}) - \mathbf{g}(\mathbf{a})\| \to 0$ as $\mathbf{h} \to 0$. We conclude that $\mathbf{f} \cdot \mathbf{g}$ is differentiable at $\mathbf{a}$ and its total derivative is $\mathbf{T}$. ∎

Formula (7) is called the *Sum Rule*; (8) is sometimes called the *Homogeneous Rule*; and (9) is called the *Dot Product Rule*. (We note that a quotient rule also holds for real-valued functions; see Exercise 11.3.6.)

Continuing to explore the analogy between $D\mathbf{f}$ and f', let g be a real function and $\mathbf{f}$ be a real-valued vector function. We know that g is differentiable at a point a if and only if the curve $y = g(x)$ has a unique tangent line at $(a, g(a))$, in which case $g'(a)$ is the slope of that tangent line. What happens in the multidimensional case? Working by analogy, we expect that f is differentiable at a point $\mathbf{a} \in \mathbf{R}^n$ if and only if the surface $z = f(\mathbf{x})$ has a unique tangent hyperplane at the point $(\mathbf{a}, f(\mathbf{a})) := (a_1, \ldots, a_n, f(a_1, \ldots, a_n)) \in \mathbf{R}^{n+1}$. Moreover, we expect that the normal vector $\mathbf{n}$ of that tangent hyperplane is somehow related to the total derivative $\nabla f(\mathbf{a})$. We shall show that both of these expectations are correct, and that the relationship between $\mathbf{n}$ and $\nabla f(\mathbf{a})$ is a simple one [see (12) below and Exercise 11.6.9b]. Thus, for the case $m = 1$, Definition 11.12 captures both the analytic and geometric spirit of the one-dimensional derivative.

First we define what we mean by a tangent hyperplane.

11.21 Definition.

Let S be a subset of $\mathbf{R}^m$ and $\mathbf{c} \in S$. A hyperplane Π with normal $\mathbf{n}$ is said to be tangent to S at $\mathbf{c}$ if and only if Π contains $\mathbf{c}$ and

$$\lim_{k \to \infty} \mathbf{n} \cdot \frac{\mathbf{c}_k - \mathbf{c}}{\|\mathbf{c}_k - \mathbf{c}\|} = 0 \tag{11}$$

for all sequences $\mathbf{c}_k \in S \backslash \{\mathbf{c}\}$ which converge to $\mathbf{c}$.

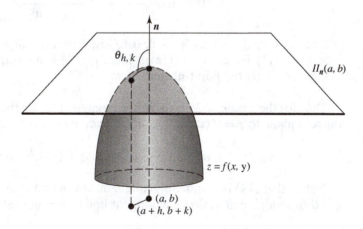

FIGURE 11.1

Definition 11.21 is illustrated for the case $n = 3$ in Figure 11.1. There S is the surface $z = f(x, y)$, $\mathbf{c} = (a, b, f(a, b))$, and $\theta_{h,k}$ represents the angle between $\mathbf{n}$ and the vector from $\mathbf{c}$ to $(a+h, b+k, f(a+h, b+k))$. Notice, by (2) in Section 8.1, that (11) is equivalent to assuming that the angle between $\mathbf{n}$ and $\mathbf{c}_k - \mathbf{c}$ converges to $\pi/2$ for all sequences $\mathbf{c}_k \in S \backslash \{\mathbf{c}\}$ which converge to $\mathbf{c}$. Hence the definition of a "tangent hyperplane" makes geometric sense.

It is easy to see that surfaces generated by differentiable vector functions have tangent hyperplanes.

11.22 Theorem. *Suppose that V is open in $\mathbf{R}^n$, that $\mathbf{a} \in V$, and that $f : V \to \mathbf{R}$. If f is differentiable at $\mathbf{a}$, then the surface*

$$S := \{(\mathbf{x}, z) \in \mathbf{R}^{n+1} : z = f(\mathbf{x}) \quad and \quad \mathbf{x} \in V\}$$

has a tangent hyperplane at $(\mathbf{a}, f(\mathbf{a}))$ with normal

$$\mathbf{n} = (\nabla f(\mathbf{a}), -1) := (f_{x_1}(\mathbf{a}), f_{x_2}(\mathbf{a}), \ldots, f_{x_n}(\mathbf{a}), -1). \tag{12}$$

An equation of this tangent hyperplane is given by

$$z = f(\mathbf{a}) + \nabla f(\mathbf{a}) \cdot (\mathbf{x} - \mathbf{a}). \tag{13}$$

Proof. Let $\mathbf{c}_k \in S$ with $\mathbf{c}_k \neq (\mathbf{a}, f(\mathbf{a}))$ and suppose that $\mathbf{c}_k \to (\mathbf{a}, f(\mathbf{a}))$ as $k \to \infty$. Then $\mathbf{c}_k = (\mathbf{a}_k, f(\mathbf{a}_k))$ for some $\mathbf{a}_k \in V$ and $\mathbf{a}_k \to \mathbf{a}$ as $k \to \infty$. For $\|\mathbf{h}\|$ small, set $\varepsilon(\mathbf{h}) = f(\mathbf{a} + \mathbf{h}) - f(\mathbf{a}) - \nabla f(\mathbf{a}) \cdot \mathbf{h}$ and define $\mathbf{n}$ by (12). Since

$$\|\mathbf{c}_k - \mathbf{c}\| = \sqrt{\|\mathbf{a}_k - \mathbf{a}\|^2 + |f(\mathbf{a}_k) - f(\mathbf{a})|^2} \geq \|\mathbf{a}_k - \mathbf{a}\|,$$

it is clear by (12) that

$$0 \leq \left| \mathbf{n} \cdot \frac{\mathbf{c}_k - \mathbf{c}}{\|\mathbf{c}_k - \mathbf{c}\|} \right| \leq \frac{|\varepsilon(\mathbf{a}_k - \mathbf{a})|}{\|\mathbf{a}_k - \mathbf{a}\|}.$$

Since $\varepsilon(\mathbf{h})/\|\mathbf{h}\| \to 0$ as $\mathbf{h} \to \mathbf{0}$, it follows from the Squeeze Theorem that $\mathbf{n}$ satisfies (11) for $\mathbf{c} := (\mathbf{a}, f(\mathbf{a}))$. Finally, (13) is an equation of this tangent hyperplane by the point-normal form. ∎

Thus for the case $n = 2$, if f is differentiable at (a, b), then an equation of that tangent plane to $z = f(x, y)$ at (a, b) is given by

$$z = f(a, b) + (x - a) f_x(a, b) + (y - b) f_y(a, b). \tag{14}$$

Notice that (14) is completely analogous to the one variable case. Namely, if g is differentiable at a, then the tangent line to $y = g(x)$ at the point $(a, g(a))$ is

$$y = g(a) + g'(a)(x - a). \tag{15}$$

It is interesting to note that the converse of Theorem 11.22 is also true (see Theorem 11.27 below).

There is another analogy between Df and f' worth mentioning. Recall that if f is a real function, then the change in $y = f(x)$ as x moves from a to $a + \Delta x$ is defined by $\Delta y = f(a + \Delta x) - f(a)$. For many concrete situations, it is convenient and useful to approximate Δy by the Leibnizian differential $dy := f'(a)\, dx$, where $dx = \Delta x$ is a small real number (see Figure 11.2). Does a similar situation prevail for functions on $\mathbf{R}^n$?

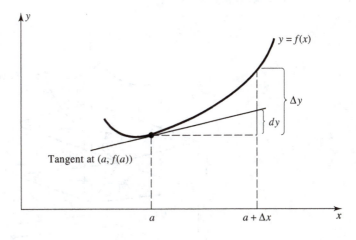

FIGURE 11.2

To answer this question, suppose that $z = f(\mathbf{x})$ is a vector function from n variables to one variable, differentiable at $\mathbf{a}$; that $\Delta z := f(\mathbf{a} + \Delta \mathbf{x}) - f(\mathbf{a})$, where $\Delta \mathbf{x} := (\Delta x_1, \dots, \Delta x_n)$; and that $d\mathbf{x} = \Delta \mathbf{x}$ is a vector with small norm. Comparing (14) and (15), we define the *first total differential* of a vector function from n variables to one variable to be

$$ dz := \nabla f(\mathbf{a}) \cdot \Delta \mathbf{x} := \sum_{j=1}^{n} \frac{\partial f}{\partial x_j}(\mathbf{a})\, dx_j. $$

Is dz a good approximation to Δz?

11.23 Remark. *Let $f : \mathbf{R}^n \rightarrow \mathbf{R}$ be differentiable at $\mathbf{a}$ and $\Delta \mathbf{x} = (\Delta x_1, \dots, \Delta x_n)$. Then*

$$ \frac{\Delta z - dz}{\|\Delta \mathbf{x}\|} \rightarrow 0 \quad \text{as} \quad \Delta \mathbf{x} \rightarrow \mathbf{0}. $$

In particular, the differential dz approximates Δz.

Proof. By definition, if f is differentiable at $\mathbf{a}$, then $\varepsilon(\mathbf{h}) := f(\mathbf{a}+\mathbf{h}) - f(\mathbf{a}) - \nabla f(\mathbf{a}) \cdot \mathbf{h}$ satisfies $\varepsilon(\mathbf{h})/\|\mathbf{h}\| \rightarrow \mathbf{0}$ as $\mathbf{h} \rightarrow \mathbf{0}$. Since $\Delta z = f(\mathbf{a}+\mathbf{h}) - f(\mathbf{a})$ for $\mathbf{h} := \Delta \mathbf{x}$ and $dz = \nabla f(\mathbf{a}) \cdot \mathbf{h}$, it follows that $(\Delta z - dz)/\|\Delta \mathbf{x}\| \rightarrow 0$ as $\Delta \mathbf{x} \rightarrow \mathbf{0}$. ∎

Figure 11.2 gives us a useful geometric interpretation of the one-dimensional differential. Is there an analogous interpretation for two-dimensional differentials? Specifically, if $z = f(x, y)$, does the total differential dz and the increment Δz play an analogous geometric role in $\mathbf{R}^3$ that dy and Δy played in $\mathbf{R}^2$?

The two-dimensional picture corresponding to Figure 11.2 involves a tangent plane and a wedge-shaped region (see Figure 11.3). Namely, let $z_0 = f(a, b)$ and consider the wedge-shaped region $\mathcal{W}$ with vertical sides parallel to the xz- and

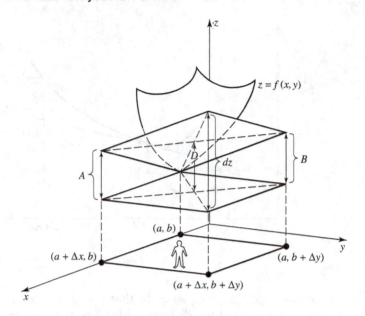

FIGURE 11.3

yz-planes whose base has vertices $\mathbf{c}_0 := (a, b, z_0)$, $\mathbf{c}_1 := (a + \Delta x, b, z_0)$, $\mathbf{c}_2 := (a, b+\Delta y, z_0)$, $\mathbf{c}_3 := (a+\Delta x, b+\Delta y, z_0)$, and whose top is tangent to $z = f(x, y)$ at $\mathbf{c}_0$. Let A represent the length of the vertical edge of $\mathcal{W}$ based at $\mathbf{c}_1$, B the length of the edge based at $\mathbf{c}_2$, and C the length of the edge based at $\mathbf{c}_3$. If dz is to play the same role in Figure 11.3 that dy plays in Figure 11.2, then it must be the case that $C = dz$. This is actually easy to verify. Since the diagonals of rectangles bisect one another, the line segment from the intersection of the diagonals in the base of $\mathcal{W}$ to the intersection of the diagonals in the top of $\mathcal{W}$ must be parallel to the z-axis. Thus the length D of this line segment can be computed two ways. On the one hand, $D = C/2$. On the other hand, $D = (A + B)/2$. Therefore, $C = A + B$. But from one-dimensional calculus, $A = f_x(a, b)\, dx$ and $B = f_y(a, b)\, dy$. Consequently,

$$C = A + B = \frac{\partial f}{\partial x}(a, b)\, dx + \frac{\partial f}{\partial y}(a, b)\, dy = dz.$$

We conclude that the first total differential of vector functions plays exactly the same role that it did for real functions.

We close this section with some optional material about tangent planes and applications of the first total differential.

Notice by Remark 11.23 that if f is differentiable at $\mathbf{a}$, then the total differential of f can be used to approximate the change of f as $\mathbf{x}$ moves from $\mathbf{a}$ to $\mathbf{a} + \mathbf{h}$ for $\|\mathbf{h}\|$ sufficiently small. This suggests that the differential can be used to approximate a function.

*11.24 EXAMPLE.

Use differentials to approximate the change of $f(x, y) = x^2y - y^3$ as (x, y) moves from $(0, 1)$ to $(0.02, 1.01)$.

Solution. Let $z = x^2y - y^3$, $a = 0$, and $b = 1$. Then $dx = 0.02$ and $dy = 0.01$. Since $dz = 2xy\, dx + (x^2 - 3y^2)\, dy$, we have

$$\Delta z \approx 0(0.02) - 3(0.01) = -0.03.$$

Note that $\Delta z = f(0.02, 1.01) - f(0, 1) = -0.029897\ldots$ is very close to -0.03. ■

*11.25 EXAMPLE.

Use differentials to approximate $(5.97)\sqrt[4]{16.03}$.

Solution. Let $z = y\sqrt[4]{x}$, $a = 16$, and $b = 6$. Then $dx = 0.03$ and $dy = -0.03$. Since

$$dz = \frac{y}{4\sqrt[4]{x^3}}dx + \sqrt[4]{x}\, dy,$$

we have

$$\Delta z \approx \frac{6(0.03)}{4\sqrt[4]{(16)^3}} + \sqrt[4]{16}(-0.03) \approx -0.054375.$$

Thus,

$$z \approx 6\sqrt[4]{16} - 0.054375 = 11.945625.$$

Note that the actual value of $5.97\sqrt[4]{16.03}$ is $11.945593\ldots$. Thus our approximation is good to three decimal places. ■

*11.26 EXAMPLE.

Find the maximum percentage error for the calculated value of the volume of a right circular cylinder if the radius can be measured with a maximum error of 3% and the altitude can be measured with a maximum error of 2%.

Solution. The volume of a right circular cylinder is $V = \pi r^2 h$, where r is the radius and h is the altitude. Hence, the differential of V is $dV = 2\pi rh\, dr + \pi r^2\, dh$. Thus

$$\frac{dV}{V} = 2\frac{dr}{r} + \frac{dh}{h}.$$

Since the percentage error of a variable x is $\Delta x/x \approx dx/x$, it follows that the maximum percentage error in calculating the volume V is approximately 8%:

$$\frac{dV}{V} = 2(\pm0.03) + (\pm0.02) = \pm0.08.$$ ■

We close this section by showing that the converse of Theorem 11.22 holds. (The proof presented here is based on Taylor [13][1].)

*11.27 Theorem. *Let V be open in $\mathbf{R}^2$, let $(a, b) \in V$, and let $f : V \to \mathbf{R}$. Then f is differentiable at (a, b) if and only if $z = f(x, y)$ has a nonvertical tangent plane Π at $\mathbf{c} := (a, b, f(a, b))$, in which case $\Pi = \Pi_{\mathbf{n}}(\mathbf{c})$ for*

$$\mathbf{n} = (f_x(a, b), f_y(a, b), -1). \tag{16}$$

Proof. If f is differentiable at (a, b), then by Theorem 11.22, $z = f(x, y)$ has a nonvertical tangent plane with normal given by (16).

Conversely, suppose that the surface $S := \{(x, y, z) : z = f(x, y)$ for $(x, y) \in V\}$ has a nonvertical tangent plane Π at $\mathbf{c}$ with normal $\mathbf{N}$ whose third component is γ. Since $\gamma \neq 0$, $\mathbf{n} := -\mathbf{N}/\gamma =: (n_1, n_2, -1)$ is also a normal of Π. Let $\mathbf{n}_{1,2} = (n_1, n_2)$, let $\mathbf{a} := (a, b)$, and let $\mathbf{h} := (h, k) \neq \mathbf{0}$ be chosen so that $\mathbf{a} + \mathbf{h} \in V$. Set

$$\varepsilon(\mathbf{h}) := \Delta z - \mathbf{n}_{1,2} \cdot \mathbf{h},$$

where $\Delta z := f(a + h, b + k) - f(a, b)$, and observe (by the uniqueness of the total derivative) that if we prove that $\varepsilon(\mathbf{h})/\|\mathbf{h}\| \to 0$ as $\mathbf{h} \to \mathbf{0}$, then f is differentiable at $\mathbf{a}$, and $\mathbf{n}_{1,2} = \nabla f(a, b)$ as required.

Set $\mathbf{c_h} := (a + h, b + k, f(a + h, b + k))$. Since Π is tangent to S at $\mathbf{c}$, and since $\mathbf{n} \cdot (\mathbf{c_h} - \mathbf{c}) = \mathbf{n}_{1,2} \cdot \mathbf{h} - \Delta z$, Definition 11.21 implies

$$Q := Q(\mathbf{h}) := \frac{\mathbf{n}_{1,2} \cdot \mathbf{h} - \Delta z}{\sqrt{\|\mathbf{h}\|^2 + (\Delta z)^2}} \to 0 \tag{17}$$

as $\mathbf{h} \to \mathbf{0}$. Since the expression Q defined in (17) is a quadratic in Δz, use the quadratic formula to solve for Δz:

$$\Delta z = \frac{-\mathbf{n}_{1,2} \cdot \mathbf{h} \pm |Q|\sqrt{(\mathbf{n}_{1,2} \cdot \mathbf{h})^2 + (1 - Q^2)\|\mathbf{h}\|^2}}{Q^2 - 1}. \tag{18}$$

It follows that

$$\varepsilon(\mathbf{h}) = \Delta z - \mathbf{n}_{1,2} \cdot \mathbf{h} = \frac{-\mathbf{n}_{1,2} \cdot \mathbf{h} \, Q^2 \pm |Q|\sqrt{(\mathbf{n}_{1,2} \cdot \mathbf{h})^2 + (1 - Q^2)\|\mathbf{h}\|^2}}{Q^2 - 1}. \tag{19}$$

To estimate $\varepsilon(\mathbf{h})$, observe by the Cauchy–Schwarz Inequality that $|\mathbf{n}_{1,2} \cdot \mathbf{h}| \leq \|\mathbf{n}_{1,2}\| \|\mathbf{h}\|$ and that $(\mathbf{n}_{1,2} \cdot \mathbf{h})^2 + (1 - Q^2)\|\mathbf{h}\|^2 \leq (\|\mathbf{n}_{1,2}\|^2 + 1 - Q) \|\mathbf{h}\|^2$. Substituting these estimates into (19), we obtain

$$\frac{|\varepsilon(\mathbf{h})|}{\|\mathbf{h}\|} \leq \frac{Q^2 \|\mathbf{n}_{1,2}\| + |Q|\sqrt{\|\mathbf{n}_{1,2}\|^2 + 1 - Q^2}}{1 - Q^2}.$$

---

[1] Angus E. Taylor, *Advanced Calculus* (Boston: Ginn and Company, 1955). Reprinted with permission of John Wiley & Sons, Inc.

Since $Q \to 0$ as $\mathbf{h} \to \mathbf{0}$, we conclude by the Squeeze Theorem that $\varepsilon(\mathbf{h})/\|\mathbf{h}\| \to 0$ as $\mathbf{h} \to \mathbf{0}$. In particular, f is differentiable at (a, b) and $(f_x(a, b), f_y(a, b), -1)$ is normal to its tangent plane there.   ∎

EXERCISES

11.3.1. For each of the following, prove that $\mathbf{f}$ and $\mathbf{g}$ are differentiable on their domains, and find formulas for $D(\mathbf{f} + \mathbf{g})(\mathbf{x})$ and $D(\mathbf{f} \cdot \mathbf{g})(\mathbf{x})$.

   a) $\mathbf{f}(x, y) = x - y, \quad \mathbf{g}(x, y) = x^2 + y^2$

   b) $\mathbf{f}(x, y) = xy, \quad \mathbf{g}(x, y) = x \sin x - \cos y$

   c) $\mathbf{f}(x, y) = (\cos(xy), x \log y), \quad \mathbf{g}(x, y) = (y, x)$

   d) $\mathbf{f}(x, y, z) = (y, x - z), \quad \mathbf{g}(x, y, z) = (xyz, y^2)$

11.3.2. For each of the following functions, find an equation of the tangent plane to $z = f(x, y)$ at $\mathbf{c}$.

   a) $f(x, y) = x^2 + y^2, \quad \mathbf{c} = (1, -1, 2)$
   b) $f(x, y) = x^3 y - xy^3, \quad \mathbf{c} = (1, 1, 0)$
   c) $f(x, y, z) = xy + \sin z, \quad \mathbf{c} = (1, 0, \pi/2, 1)$

11.3.3. Find all points on the paraboloid $z = x^2 + y^2$ (see Appendix D) where the tangent plane is parallel to the plane $x + y + z = 1$. Find equations of the corresponding tangent planes. Sketch the graphs of these functions to see that your answer agrees with your intuition.

11.3.4. Let $\mathcal{K}$ be the cone, given by $z = \sqrt{x^2 + y^2}$.

   a) Find an equation of each plane tangent to $\mathcal{K}$ which is perpendicular to the plane $x + z = 5$.
   b) Find an equation of each plane tangent to $\mathcal{K}$ which is parallel to the plane $x - y + z = 1$.

11.3.5. Prove (7) and (8) in Theorem 11.20.

11.3.6. [QUOTIENT RULE]. Suppose that $f : \mathbf{R}^n \to \mathbf{R}$ is differentiable at $\mathbf{a}$ and that $f(\mathbf{a}) \neq 0$.

   a) Show that for $\|\mathbf{h}\|$ sufficiently small, $f(\mathbf{a} + \mathbf{h}) \neq 0$.
   b) Prove that $Df(\mathbf{a})(\mathbf{h})/\|\mathbf{h}\|$ is bounded for all $\mathbf{h} \in \mathbf{R}^n \setminus \{\mathbf{0}\}$.
   c) If $T := -Df(\mathbf{a})/f^2(\mathbf{a})$, show that

$$\frac{1}{f(\mathbf{a} + \mathbf{h})} - \frac{1}{f(\mathbf{a})} - T(\mathbf{h}) = \frac{f(\mathbf{a}) - f(\mathbf{a} + \mathbf{h}) + Df(\mathbf{a})(\mathbf{h})}{f(\mathbf{a})f(\mathbf{a} + \mathbf{h})}$$
$$+ \frac{(f(\mathbf{a} + \mathbf{h}) - f(\mathbf{a}))Df(\mathbf{a})(\mathbf{h})}{f^2(\mathbf{a})f(\mathbf{a} + \mathbf{h})}$$

for $\|\mathbf{h}\|$ sufficiently small.

d) Prove that $1/f(\mathbf{x})$ is differentiable at $\mathbf{x} = \mathbf{a}$ and

$$D\left(\frac{1}{f}\right)(\mathbf{a}) = -\frac{Df(\mathbf{a})}{f^2(\mathbf{a})}.$$

e) Prove that if f and g are real-valued vector functions which are differentiable at some $\mathbf{a}$, and if $g(\mathbf{a}) \neq 0$, then

$$D\left(\frac{f}{g}\right)(\mathbf{a}) = \frac{g(\mathbf{a})Df(\mathbf{a}) - f(\mathbf{a})Dg(\mathbf{a})}{g^2(\mathbf{a})}.$$

11.3.7. [CROSS-PRODUCT RULE]. Suppose that V is open in $\mathbf{R}^n$, that $\mathbf{f}, \mathbf{g} : V \to \mathbf{R}^3$, and that $\mathbf{a} \in V$. If $\mathbf{f}$ and $\mathbf{g}$ are differentiable at $\mathbf{a}$, prove that $\mathbf{f} \times \mathbf{g}$ is differentiable at $\mathbf{a}$ and

$$D(\mathbf{f} \times \mathbf{g})(\mathbf{a})(\mathbf{y}) = \mathbf{f}(\mathbf{a}) \times (D\mathbf{g}(\mathbf{a})(\mathbf{y})) - \mathbf{g}(\mathbf{a}) \times (D\mathbf{f}(\mathbf{a})(\mathbf{y}))$$

for all $\mathbf{y} \in \mathbf{R}^n$.

*11.3.8. Compute the differential of each of the following functions.

a) $z = x^2 + y^2$ b) $z = \sin(xy)$ c) $z = \dfrac{xy}{1 + x^2 + y^2}$

*11.3.9. Let $w = x^2 y + z$. Use differentials to approximate Δw as (x, y, z) moves from $(1,2,1)$ to $(1.01, 1.98, 1.03)$. Compare your approximation with the actual value of Δw.

*11.3.10. The time T it takes for a pendulum to complete one full swing is given by

$$T = 2\pi \sqrt{\frac{L}{g}},$$

where g is the acceleration due to gravity and L is the length of the pendulum. If g can be measured with a maximum error of 1%, how accurately must L be measured (in terms of percentage error) so that the calculated value of T has a maximum error of 2%?

*11.3.11. Suppose that

$$\frac{1}{w} = \frac{1}{x} + \frac{1}{y} + \frac{1}{z},$$

where each variable x, y, z can be measured with a maximum error of $p\%$. Prove that the calculated value of w also has a maximum error of $p\%$.

11.4 THE CHAIN RULE

Here is the Chain Rule for vector functions.

11.28 Theorem.   [CHAIN RULE].
Suppose that **f** *and* **g** *are vector functions. If* **g** *is differentiable at* **a** *and* **f** *is differentiable at* **g(a)**, *then* **f** ∘ **g** *is differentiable at* **a** *and*

$$D(\mathbf{f} \circ \mathbf{g})(\mathbf{a}) = D\mathbf{f}(\mathbf{g}(\mathbf{a}))D\mathbf{g}(\mathbf{a}). \tag{20}$$

[The product $D\mathbf{f}(\mathbf{g}(\mathbf{a}))D\mathbf{g}(\mathbf{a})$ is matrix multiplication.]

Proof. To specify the dimensions, suppose that $\mathbf{a} \in \mathbf{R}^n$, $\mathbf{b} := \mathbf{g}(\mathbf{a}) \in \mathbf{R}^m$, and $\mathbf{f}(\mathbf{b}) \in \mathbf{R}^p$. Set $\mathbf{T} = D\mathbf{f}(\mathbf{g}(\mathbf{a}))D\mathbf{g}(\mathbf{a})$ and observe that $\mathbf{T}$, the product of a $p \times m$ matrix with an $m \times n$ matrix, is a $p \times n$ matrix, the right size for the total derivative of $\mathbf{f} \circ \mathbf{g}$. By the uniqueness of the total derivative, we must show that

$$\lim_{\mathbf{h} \to 0} \frac{\mathbf{f}(\mathbf{g}(\mathbf{a} + \mathbf{h})) - \mathbf{f}(\mathbf{g}(\mathbf{a})) - \mathbf{T}(\mathbf{h})}{\|\mathbf{h}\|} = \mathbf{0}. \tag{21}$$

Set

$$\varepsilon(\mathbf{h}) = \mathbf{g}(\mathbf{a} + \mathbf{h}) - \mathbf{g}(\mathbf{a}) - D\mathbf{g}(\mathbf{a})(\mathbf{h}), \tag{22}$$

and

$$\delta(\mathbf{k}) = \mathbf{f}(\mathbf{b} + \mathbf{k}) - \mathbf{f}(\mathbf{b}) - D\mathbf{f}(\mathbf{b})(\mathbf{k}) \tag{23}$$

for $\|\mathbf{h}\|$ and $\|\mathbf{k}\|$ sufficiently small. By hypothesis, $\varepsilon(\mathbf{h})/\|\mathbf{h}\| \to \mathbf{0}$ in $\mathbf{R}^m$ as $\mathbf{h} \to \mathbf{0}$ in $\mathbf{R}^n$, and $\delta(\mathbf{k})/\|\mathbf{k}\| \to \mathbf{0}$ in $\mathbf{R}^p$ as $\mathbf{k} \to \mathbf{0}$ in $\mathbf{R}^m$. Fix $\mathbf{h}$ small and set $\mathbf{k} = \mathbf{g}(\mathbf{a} + \mathbf{h}) - \mathbf{g}(\mathbf{a})$. Since (23) and (22) imply

$$\mathbf{f}(\mathbf{g}(\mathbf{a} + \mathbf{h})) - \mathbf{f}(\mathbf{g}(\mathbf{a})) = \mathbf{f}(\mathbf{b} + \mathbf{k}) - \mathbf{f}(\mathbf{b}) = D\mathbf{f}(\mathbf{b})(\mathbf{k}) + \delta(\mathbf{k})$$
$$= D\mathbf{f}(\mathbf{b})(D\mathbf{g}(\mathbf{a})(\mathbf{h}) + \varepsilon(\mathbf{h})) + \delta(\mathbf{k})$$
$$= \mathbf{T}(\mathbf{h}) + D\mathbf{f}(\mathbf{b})(\varepsilon(\mathbf{h})) + \delta(\mathbf{k}),$$

we have $\mathbf{f}(\mathbf{g}(\mathbf{a} + \mathbf{h})) - \mathbf{f}(\mathbf{g}(\mathbf{a})) - \mathbf{T}(\mathbf{h}) = D\mathbf{f}(\mathbf{b})(\varepsilon(\mathbf{h})) + \delta(\mathbf{k}) =: \mathbf{T}_1(\mathbf{h}) + \mathbf{T}_2(\mathbf{h})$. It remains to verify that $\mathbf{T}_j(\mathbf{h})/\|\mathbf{h}\| \to \mathbf{0}$ as $\mathbf{h} \to \mathbf{0}$ for $j = 1, 2$.

Since $\varepsilon(\mathbf{h})/\|\mathbf{h}\| \to \mathbf{0}$ as $\mathbf{h} \to \mathbf{0}$ and $D\mathbf{f}(\mathbf{b})(\varepsilon(\mathbf{h}))$ is matrix multiplication, it is clear that $\mathbf{T}_1(\mathbf{h})/\|\mathbf{h}\| \to D\mathbf{f}(\mathbf{b})(\mathbf{0}) = \mathbf{0}$ as $\mathbf{h} \to \mathbf{0}$. On the other hand, by (22), the triangle inequality, and the definition of the operator norm, we have

$$\|\mathbf{k}\| := \|\mathbf{g}(\mathbf{a} + \mathbf{h}) - \mathbf{g}(\mathbf{a})\| = \|D\mathbf{g}(\mathbf{a})(\mathbf{h}) + \varepsilon(\mathbf{h})\| \le \|D\mathbf{g}(\mathbf{a})\| \cdot \|\mathbf{h}\| + \|\varepsilon(\mathbf{h})\|.$$

Thus $\|\mathbf{k}\|/\|\mathbf{h}\|$ is bounded for $\|\mathbf{h}\|$ sufficiently small. Since $\mathbf{k} = \mathbf{0}$ implies $\|\mathbf{T}_2(\mathbf{h})\| = 0$, we may suppose that $\mathbf{k} \ne \mathbf{0}$. Since $\mathbf{h} \to \mathbf{0}$ implies $\mathbf{k} \to \mathbf{0}$, it follows that

$$\frac{\|\mathbf{T}_2(\mathbf{h})\|}{\|\mathbf{h}\|} = \frac{\|\mathbf{k}\|}{\|\mathbf{h}\|} \cdot \frac{\|\delta(\mathbf{k})\|}{\|\mathbf{k}\|} \to \mathbf{0}$$

as $\mathbf{h} \to \mathbf{0}$. We conclude that $\mathbf{f} \circ \mathbf{g}$ is differentiable at $\mathbf{a}$ and the derivative is $D\mathbf{f}(\mathbf{g}(\mathbf{a}))D\mathbf{g}(\mathbf{a})$. ∎

The Chain Rule can be used to compute individual partial derivatives without writing out the entire matrices $D\mathbf{f}$ and $D\mathbf{g}$. For example, suppose that $f(u_1, \ldots, u_m)$ is differentiable from $\mathbf{R}^m$ to $\mathbf{R}$, that $\mathbf{g}(x_1, \ldots, x_n)$ is differentiable from $\mathbf{R}^n$ to $\mathbf{R}^m$, and that $z = f(\mathbf{g}(x_1, \ldots, x_n))$. Since $Df = \nabla f$ and the jth column of $D\mathbf{g}$ consists of first partial derivatives, with respect to x_j, of the components $u_k := g_k(x_1, \ldots, x_n)$, it follows from the Chain Rule and the definition of matrix multiplication that

$$\frac{\partial z}{\partial x_j} = \frac{\partial f}{\partial u_1} \frac{\partial u_1}{\partial x_j} + \cdots + \frac{\partial f}{\partial u_m} \frac{\partial u_m}{\partial x_j}$$

for $j = 1, 2, \ldots, n$. Here are two concrete examples which illustrate this principle.

11.29 EXAMPLES.

i) If $F, G, H : \mathbf{R}^2 \to \mathbf{R}$ are differentiable and $z = F(x, y)$, where $x = G(r, \theta)$, and $y = H(r, \theta)$, then

$$\frac{\partial z}{\partial r} = \frac{\partial z}{\partial x} \frac{\partial x}{\partial r} + \frac{\partial z}{\partial y} \frac{\partial y}{\partial r} \quad \text{and} \quad \frac{\partial z}{\partial \theta} = \frac{\partial z}{\partial x} \frac{\partial x}{\partial \theta} + \frac{\partial z}{\partial y} \frac{\partial y}{\partial \theta}.$$

ii) If $f : \mathbf{R}^3 \to \mathbf{R}$ and $\phi, \psi, \sigma : \mathbf{R} \to \mathbf{R}$ are differentiable and $w = f(x, y, z)$, where $x = \phi(t)$, $y = \psi(t)$, and $z = \sigma(t)$, then

$$\frac{dw}{dt} = \frac{\partial w}{\partial x} \frac{dx}{dt} + \frac{\partial w}{\partial y} \frac{dy}{dt} + \frac{\partial w}{\partial z} \frac{dz}{dt}.$$

EXERCISES

11.4.1. Let $F : \mathbf{R}^3 \to \mathbf{R}$ and $f, g, h : \mathbf{R}^2 \to \mathbf{R}$ be C^2 functions. If $w = F(x, y, z)$, where $x = f(p, q)$, $y = g(p, q)$, and $z = h(p, q)$, find formulas for w_p, w_q, and w_{pp}.

11.4.2. Let $r > 0$, let $\mathbf{a} \in \mathbf{R}^n$, and suppose that $\mathbf{g} : B_r(\mathbf{a}) \to \mathbf{R}^m$ is differentiable at $\mathbf{a}$.

   a) If $f : B_r(\mathbf{g}(\mathbf{a})) \to \mathbf{R}$ is differentiable at $\mathbf{g}(\mathbf{a})$, prove that the partial derivatives of $h = f \circ \mathbf{g}$ are given by

$$\frac{\partial h}{\partial x_j}(\mathbf{a}) = \nabla f(\mathbf{g}(\mathbf{a})) \cdot \frac{\partial \mathbf{g}}{\partial x_j}(\mathbf{a})$$

   for $j = 1, 2, \ldots, n$.

   b) If $n = m$ and $\mathbf{f} : B_r(\mathbf{g}(\mathbf{a})) \to \mathbf{R}^n$ is differentiable at $\mathbf{g}(\mathbf{a})$, prove that

$$\det(D(\mathbf{f} \circ \mathbf{g})(\mathbf{a})) = \det(D\mathbf{f}(\mathbf{g}(\mathbf{a}))) \det(D\mathbf{g}(\mathbf{a})).$$

11.4.3. Suppose that $k \in \mathbf{N}$ and that $f : \mathbf{R}^n \to \mathbf{R}$ is homogeneous of order k; that is, that $f(\rho\mathbf{x}) = \rho^k f(\mathbf{x})$ for all $\mathbf{x} \in \mathbf{R}^n$ and all $\rho \in \mathbf{R}$. If f is differentiable on $\mathbf{R}^n$, prove that

$$x_1 \frac{\partial f}{\partial x_1}(\mathbf{x}) + \cdots + x_n \frac{\partial f}{\partial x_n}(\mathbf{x}) = kf(\mathbf{x})$$

for all $\mathbf{x} = (x_1, \ldots, x_n) \in \mathbf{R}^n$.

11.4.4. Let $f, g : \mathbf{R} \to \mathbf{R}$ be twice differentiable. Prove that $u(x, y) := f(xy)$ satisfies

$$x \frac{\partial u}{\partial x} - y \frac{\partial u}{\partial y} = 0,$$

and $v(x, y) := f(x - y) + g(x + y)$ satisfies the *wave equation*; that is,

$$\frac{\partial^2 v}{\partial x^2} - \frac{\partial^2 v}{\partial y^2} = 0.$$

11.4.5. Let $f, g : \mathbf{R}^2 \to \mathbf{R}$ be differentiable and satisfy the *Cauchy–Riemann equations*; that is, that

$$\frac{\partial f}{\partial x} = \frac{\partial g}{\partial y} \quad \text{and} \quad \frac{\partial f}{\partial y} = -\frac{\partial g}{\partial x}$$

hold on $\mathbf{R}^2$. If $u(r, \theta) = f(r\cos\theta, r\sin\theta)$ and $v(r, \theta) = g(r\cos\theta, r\sin\theta)$, prove that

$$\frac{\partial u}{\partial r} = \frac{1}{r}\frac{\partial v}{\partial \theta}, \qquad \frac{\partial v}{\partial r} = -\frac{1}{r}\frac{\partial u}{\partial \theta} \qquad r \neq 0.$$

11.4.6. Let $f : \mathbf{R}^2 \to \mathbf{R}$ be C^2 on $\mathbf{R}^2$ and set $u(r, \theta) = f(r\cos\theta, r\sin\theta)$. If f satisfies the *Laplace equation*; that is, if

$$\frac{\partial^2 f}{\partial x^2} + \frac{\partial^2 f}{\partial y^2} = 0,$$

prove for each $r \neq 0$ that

$$\frac{1}{r^2}\frac{\partial^2 u}{\partial \theta^2} + \frac{1}{r}\frac{\partial u}{\partial r} + \frac{\partial^2 u}{\partial r^2} = 0.$$

11.4.7. Let

$$u(x, t) = \frac{e^{-x^2/4t}}{\sqrt{4\pi t}}, \qquad t > 0, \ x \in \mathbf{R}.$$

a) Prove that u satisfies the *heat equation* (i.e., $u_{xx} - u_t = 0$ for all $t > 0$ and $x \in \mathbf{R}$).

b) If $a > 0$, prove that $u(x, t) \to 0$, as $t \to 0+$, uniformly for $x \in [a, \infty)$.

11.4.8. Let $u : \mathbf{R} \to [0, \infty)$ be differentiable. Prove that for each $(x, y, z) \neq (0, 0, 0)$,

$$F(x, y, z) := u\left(\sqrt{x^2 + y^2 + z^2} \right)$$

satisfies

$$\left(\left(\frac{\partial F}{\partial x} \right)^2 + \left(\frac{\partial F}{\partial y} \right)^2 + \left(\frac{\partial F}{\partial z} \right)^2 \right)^{1/2} = \left| u'\left(\sqrt{x^2 + y^2 + z^2} \right) \right|.$$

11.4.9. Suppose that $z = F(x, y)$ is differentiable at (a, b), that $F_y(a, b) \neq 0$, and that I is an open interval containing a. Prove that if $f : I \to \mathbf{R}$ is differentiable at a, $f(a) = b$, and $F(x, f(x)) = 0$ for all $x \in I$, then

$$\frac{df}{dx}(a) = \frac{-\dfrac{\partial F}{\partial x}(a, b)}{\dfrac{\partial F}{\partial y}(a, b)}.$$

11.4.10. Suppose that I is a nonempty, open interval and that $\mathbf{f} : I \to \mathbf{R}^m$ is differentiable on I. If $\mathbf{f}(I) \subseteq \partial B_r(\mathbf{0})$ for some fixed $r > 0$, prove that $\mathbf{f}(t)$ is orthogonal to $\mathbf{f}'(t)$ for all $t \in I$.

11.4.11. Let V be open in $\mathbf{R}^n$, $\mathbf{a} \in V$, $f : V \to \mathbf{R}$, and suppose that f is differentiable at $\mathbf{a}$.

a) Prove that the directional derivative $D_{\mathbf{u}} f(\mathbf{a})$ exists (see Exercise 11.2.10) for each $\mathbf{u} \in \mathbf{R}^n$ such that $\|\mathbf{u}\| = 1$ and $D_{\mathbf{u}} f(\mathbf{a}) = \nabla f(\mathbf{a}) \cdot \mathbf{u}$.

b) If $\nabla f(\mathbf{a}) \neq \mathbf{0}$ and θ represents the angle between $\mathbf{u}$ and $\nabla f(\mathbf{a})$, prove that $D_{\mathbf{u}} f(\mathbf{a}) = \|\nabla f(\mathbf{a})\| \cos \theta$.

c) Show that as $\mathbf{u}$ ranges over all unit vectors in $\mathbf{R}^n$, the maximum of $D_{\mathbf{u}} f(\mathbf{a})$ is $\|\nabla f(\mathbf{a})\|$, and it occurs when $\mathbf{u}$ is parallel to $\nabla f(\mathbf{a})$.

11.5 THE MEAN VALUE THEOREM AND TAYLOR'S FORMULA

Using $D\mathbf{f}$ as a replacement for f', we guess that the multidimensional analogue of the Mean Value Theorem is $\mathbf{f}(\mathbf{x}) - \mathbf{f}(\mathbf{a}) = D\mathbf{f}(\mathbf{c})(\mathbf{x} - \mathbf{a})$ for some $\mathbf{c}$ "between" $\mathbf{x}$ and $\mathbf{a}$; that is, some $\mathbf{c} \in L(\mathbf{x}; \mathbf{a})$, the line segment from $\mathbf{a}$ to $\mathbf{x}$. The following result shows that this guess is correct when $\mathbf{f}$ is real valued (see also Exercises 11.5.6 and 11.5.9).

11.30 Theorem. [MEAN VALUE THEOREM FOR REAL VALUED FUNCTIONS].
Let V be open in $\mathbf{R}^n$ and suppose that $f : V \to \mathbf{R}$ is differentiable on V. If $\mathbf{x}, \mathbf{a} \in V$ and $L(\mathbf{x}; \mathbf{a}) \subset V$, then there is a $\mathbf{c} \in L(\mathbf{x}; \mathbf{a})$ such that

$$f(\mathbf{x}) - f(\mathbf{a}) = \nabla f(\mathbf{c}) \cdot (\mathbf{x} - \mathbf{a}). \tag{24}$$

Proof. Let

$$g(t) = a + t(x - a), \qquad t \in R,$$

and notice by Exercise 11.2.8 that $g : R \to R^n$ is differentiable with $Dg(t) = x - a$ for all $t \in R$. Since $L(x; a) \subseteq V$ and V is open, choose $\delta > 0$ such that $g(t) \in V$ for all $t \in I_\delta := (-\delta, 1 + \delta)$. By the Chain Rule,

$$D(f \circ g)(t) = Df(g(t))(x - a), \qquad t \in I_\delta. \tag{25}$$

But $f \circ g : I_\delta \to R$ is a real function and f is real valued, so $D(f \circ g) = (f \circ g)'$ and $Df = \nabla f$. Hence, by the one-dimensional Mean Value Theorem and (25), there is a $t_0 \in (0, 1)$ such that

$$f(x) - f(a) = (f \circ g)(1) - (f \circ g)(0) = (f \circ g)'(t_0) = \nabla f(c) \cdot (x - a)$$

for $c = g(t_0)$. ∎

The following result shows that this result does not extend to vector-valued functions, even when the domain is one dimensional.

11.31 Remark. *The function $f(t) = (\cos t, \sin t)$ is differentiable on R and satisfies $f(2\pi) = f(0)$, but there is no $c \in R$ such that $Df(c) = (0, 0)$.*

Proof. $Df(t) = (-\sin t, \cos t)$ exists and is continuous for $t \in R$ but $(0, 0) \neq (-\sin t, \cos t)$ for $t \in R$. ∎

But any vector-valued function f can be turned into a scalar-valued function by taking the dot product of f with any vector u. Combining this observation with Theorem 11.30, we obtain the following multivariable version of the Mean Value Theorem.

11.32 Theorem.   [MEAN VALUE THEOREM FOR VECTOR VALUED FUNCTIONS].
Let V be open in R^n and suppose that $f : V \to R^m$ is differentiable on V. If $x, a \in V$ and $L(x; a) \subseteq V$, then given any $u \in R^m$ there is a $c \in L(x; a)$ such that

$$u \cdot (f(x) - f(a)) = u \cdot (Df(c)(x - a)).$$

Proof. Let $u \in R^m$. Set $h(x) := u \cdot f(x)$ and observe by the Dot Product Rule [see (9) above] that

$$Dh(x) = u \cdot D(f)(x)$$

for all $x \in V$. Since h is scalar valued, it follows from Theorem 11.30 that there is a $c \in L(x; a)$ such that

$$u \cdot (f(x) - f(a)) = h(x) - h(a) = Dh(c)(x - a) = u \cdot (Df(c)(x - a)). \qquad ∎$$

Sets which satisfy the hypothesis "$L(\mathbf{x}; \mathbf{a}) \subseteq V$" come up often enough to warrant a name.

11.33 Definition.

A subset E of $\mathbf{R}^n$ is said to be *convex* if and only if $L(\mathbf{x}; \mathbf{a}) \subseteq E$ for all $\mathbf{x}, \mathbf{a} \in E$.

Using this terminology, we see that the Mean Value Theorems above hold for any C^1 function on a convex, open set V.

It is easy to see that balls and rectangles are convex. For example, if $\mathbf{x}, \mathbf{a} \in B_r(\mathbf{b})$, then

$$\|((1 - t)\mathbf{a} + t\mathbf{x}) - \mathbf{b}\| = \|(1 - t)(\mathbf{a} - \mathbf{b}) + t(\mathbf{x} - \mathbf{b})\| < (1 - t)r + tr = r.$$

On the other hand, Figure 11.4 is an example of a nonconvex set in $\mathbf{R}^2$ (because the line segment which joins $\mathbf{a}$ to $\mathbf{b}$ contains some points outside V).

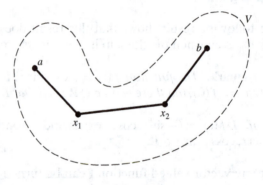

FIGURE 11.4

As in the one-dimensional case, the Mean Value Theorem is used most often to obtain information about a function from properties of its derivative. Here is a typical example.

11.34 Corollary. *Let V be an open set in $\mathbf{R}^n$, let H be a compact subset of V, and suppose that $\mathbf{f} : V \to \mathbf{R}^m$ is C^1 on V. If E is a convex subset of H, then there is a constant M (which depends on H and $\mathbf{f}$ but not on E) such that*

$$\|\mathbf{f}(\mathbf{x}) - \mathbf{f}(\mathbf{a})\| \le M\|\mathbf{x} - \mathbf{a}\|$$

for all $\mathbf{x}, \mathbf{a} \in E$.

Proof. Since H is compact and the entries of $D\mathbf{f}$ are continuous on H, we have by the Extreme Value Theorem (Theorem 9.32 or 10.63) and the proof of Theorem 8.17 that the operator norm of $D\mathbf{f}$ is bounded on H; that is, that

$$M := \sup_{\mathbf{c} \in H} \|D\mathbf{f}(\mathbf{c})\|$$

is finite. Notice that M depends only on H and $\mathbf{f}$.

Let $\mathbf{x}, \mathbf{a} \in E$ and $\mathbf{u} = \mathbf{f}(\mathbf{x}) - \mathbf{f}(\mathbf{a})$. Since E is convex, $L(\mathbf{x}; \mathbf{a}) \subseteq E$. Hence, by Theorem 11.32, there is a $\mathbf{c} \in L(\mathbf{x}; \mathbf{a})$ such that

$$\|\mathbf{f}(\mathbf{x}) - \mathbf{f}(\mathbf{a})\|^2 = \mathbf{u} \cdot (\mathbf{f}(\mathbf{x}) - \mathbf{f}(\mathbf{a})) = \mathbf{u} \cdot (D\mathbf{f}(\mathbf{c})(\mathbf{x} - \mathbf{a}))$$
$$= (\mathbf{f}(\mathbf{x}) - \mathbf{f}(\mathbf{a})) \cdot (D\mathbf{f}(\mathbf{c})(\mathbf{x} - \mathbf{a})).$$

It follows from the Cauchy–Schwarz Inequality and the definition of the operator norm that

$$\|\mathbf{f}(\mathbf{x}) - \mathbf{f}(\mathbf{a})\|^2 \le \|\mathbf{f}(\mathbf{x}) - \mathbf{f}(\mathbf{a})\| \, \|D\mathbf{f}(\mathbf{c})\| \, \|\mathbf{x} - \mathbf{a}\|.$$

If $\|\mathbf{f}(\mathbf{x}) - \mathbf{f}(\mathbf{a})\| = 0$, there is nothing to prove. Otherwise, we can divide the inequality above by $\|\mathbf{f}(\mathbf{x}) - \mathbf{f}(\mathbf{a})\|$ to obtain

$$\|\mathbf{f}(\mathbf{x}) - \mathbf{f}(\mathbf{a})\| \le \|D\mathbf{f}(\mathbf{c})\| \, \|\mathbf{x} - \mathbf{a}\| \le M \|\mathbf{x} - \mathbf{a}\|. \qquad \blacksquare$$

As the following optional result shows, for some applications of the Mean Value Theorem, the convexity hypothesis can be replaced by connectivity. (This is an analogue of the one-dimensional result: If $f' = 0$ on $[a, b]$, then f is constant on $[a, b]$.)

*11.35 Corollary. *Suppose that V is open and connected in $\mathbf{R}^n$ and that $\mathbf{f} : V \to \mathbf{R}^m$ is differentiable on V. If $D\mathbf{f}(\mathbf{c}) = O$ for all $\mathbf{c} \in V$, then $\mathbf{f}$ is constant on V.*

Proof. Fix $\mathbf{a} \in V$, and let $\mathbf{x} \in V$. Since V is open and connected, V is polygonally connected (see Exercise 9.4.10 or 10.5.10). Thus, there exist points $\mathbf{x}_0 = \mathbf{a}, \mathbf{x}_1, \ldots, \mathbf{x}_k = \mathbf{x}$ such that $L(\mathbf{x}_{j-1}; \mathbf{x}_j) \subseteq V$ for $j = 1, 2, \ldots, k$ (see Figure 11.4).

Let $\mathbf{u} = \mathbf{f}(\mathbf{x}) - \mathbf{f}(\mathbf{a})$ and choose by Theorem 11.32 points $\mathbf{c}_j \in L(\mathbf{x}_{j-1}; \mathbf{x}_j)$ such that

$$\mathbf{u} \cdot (\mathbf{f}(\mathbf{x}_j) - \mathbf{f}(\mathbf{x}_{j-1})) = \mathbf{u} \cdot (D\mathbf{f}(\mathbf{c}_j)(\mathbf{x}_j - \mathbf{x}_{j-1})) = 0$$

for $j = 1, 2, \ldots, k$. Summing over j and telescoping, we see by the choice of $\mathbf{u}$ that

$$0 = \sum_{j=1}^{k} \mathbf{u} \cdot (\mathbf{f}(\mathbf{x}_j) - \mathbf{f}(\mathbf{x}_{j-1})) = \mathbf{u} \cdot (\mathbf{f}(\mathbf{x}) - \mathbf{f}(\mathbf{a})) = \|\mathbf{f}(\mathbf{x}) - \mathbf{f}(\mathbf{a})\|^2.$$

Therefore, $\mathbf{f}(\mathbf{x}) = \mathbf{f}(\mathbf{a})$. $\qquad \blacksquare$

To obtain a multidimensional version of Taylor's Formula, we need to define higher-order differentials. Let $p \ge 1$, let V be open in $\mathbf{R}^n$, let $\mathbf{a} \in V$, and let $f : V \to \mathbf{R}$. We shall say that f has a pth-order total differential at $\mathbf{a}$ if and only if the $(p - 1)$-st order partial derivatives of f exist on V and are differentiable at $\mathbf{a}$, in which case we call

$$D^{(p)} f(\mathbf{a}; \mathbf{h}) := \sum_{i_1=1}^{n} \cdots \sum_{i_p=1}^{n} \frac{\partial^p f}{\partial x_{i_1} \dots \partial x_{i_p}}(\mathbf{a}) h_{i_1} \cdots h_{i_p}, \qquad \mathbf{h} = (h_1, \dots, h_n) \in \mathbf{R}^n$$

the *pth-order total differential* of f at $\mathbf{a}$. Notice that

$$D^{(p)} f(\mathbf{a}; \mathbf{h}) = D^{(1)}(D^{(p-1)} f)(\mathbf{a}; \mathbf{h})$$

$$= \sum_{j=1}^{n} \frac{\partial}{\partial x_j} \left(\sum_{i_1=1}^{n} \cdots \sum_{i_{p-1}=1}^{n} \frac{\partial^{p-1} f}{\partial x_{i_1} \dots \partial x_{i_{p-1}}}(\mathbf{a}) h_{i_1} \cdots h_{i_{p-1}} \right) h_j$$

for $p > 1$. Also notice that if $z = f(\mathbf{x})$, then $D^{(1)} f(\mathbf{a}; \Delta \mathbf{x})$ is the first total differential dz defined in Section 11.3, and also is the total derivative of f at $\mathbf{a}$ evaluated at $\Delta \mathbf{x}$:

$$D^{(1)} f(\mathbf{a}; \Delta \mathbf{x}) := \sum_{j=1}^{n} \frac{\partial f}{\partial x_j}(\mathbf{a}) \, \Delta x_j = \nabla f(\mathbf{a}) \cdot \Delta \mathbf{x} = Df(\mathbf{a})(\Delta \mathbf{x}).$$

For the case $n = 2$, this differential has a simple geometric interpretation (see Figure 11.3 above).

Although total differentials look messy to evaluate, when f is a sufficiently smooth function of two variables, they are relatively easy to calculate using binomial coefficients (see the next example and Exercise 11.5.2).

11.36 EXAMPLE.

Suppose that $f : V \to \mathbf{R}$ is C^2 on V. Find a formula for the second total differential of f at $(a, b) \in V$.

Solution. By definition,

$$D^{(2)} f((a, b); (h, k)) = h^2 \frac{\partial^2 f}{\partial x^2}(a, b) + hk \frac{\partial^2 f}{\partial x \, \partial y}(a, b)$$

$$+ hk \frac{\partial^2 f}{\partial y \, \partial x}(a, b) + k^2 \frac{\partial^2 f}{\partial y^2}(a, b).$$

But by Theorem 11.2, $f_{xy}(a, b) = f_{yx}(a, b)$. Therefore,

$$D^{(2)} f((a, b); (h, k)) = h^2 \frac{\partial^2 f}{\partial x^2}(a, b) + 2hk \frac{\partial^2 f}{\partial x \, \partial y}(a, b) + k^2 \frac{\partial^2 f}{\partial y^2}(a, b). \qquad \blacksquare$$

Thus the second total differential of $f(x, y) = (xy)^2$ is

$$D^{(2)} f((x, y); (h, k)) = 2y^2 h^2 + 8xyhk + 2x^2 k^2.$$

Here is a multidimensional version of Taylor's Formula.

11.37 Theorem.   [TAYLOR'S FORMULA ON $\mathbf{R}^n$].
Let $p \in \mathbf{N}$, let V be open in $\mathbf{R}^n$, let $\mathbf{x}, \mathbf{a} \in V$, and suppose that $f : V \to \mathbf{R}$. If the pth total differential of f exists on V and $L(\mathbf{x}; \mathbf{a}) \subseteq V$, then there is a point $\mathbf{c} \in L(\mathbf{x}; \mathbf{a})$ such that

$$f(\mathbf{x}) = f(\mathbf{a}) + \sum_{k=1}^{p-1} \frac{1}{k!} D^{(k)} f(\mathbf{a}; \mathbf{h}) + \frac{1}{p!} D^{(p)} f(\mathbf{c}; \mathbf{h})$$

for $\mathbf{h} := \mathbf{x} - \mathbf{a}$.

NOTE: *These hypotheses are met if V is convex and f is C^p on V.*

Proof. Let $\mathbf{h} = \mathbf{x} - \mathbf{a}$. As in the proof of Theorem 11.32, choose $\delta > 0$ so small that $\mathbf{a} + t\mathbf{h} \in V$ for $t \in I_\delta := (-\delta, 1 + \delta)$. The function $F(t) = f(\mathbf{a} + t\mathbf{h})$ is differentiable on I_δ and, by the Chain Rule,

$$F'(t) = Df(\mathbf{a} + t\mathbf{h})(\mathbf{h}) = \sum_{k=1}^{n} \frac{\partial f}{\partial x_k}(\mathbf{a} + t\mathbf{h}) \, h_k.$$

In fact, a simple induction argument can be used to verify that

$$F^{(j)}(t) = \sum_{i_1=1}^{n} \cdots \sum_{i_j=1}^{n} \frac{\partial^j f}{\partial x_{i_1} \cdots \partial x_{i_j}}(\mathbf{a} + t\mathbf{h}) \, h_{i_1} \cdots h_{i_j}$$

for $j = 1, 2, \ldots, p$. Thus

$$F^{(j)}(0) = D^{(j)} f(\mathbf{a}; \mathbf{h}) \quad \text{and} \quad F^{(p)}(t) = D^{(p)} f(\mathbf{a} + t\mathbf{h}; \mathbf{h}) \tag{26}$$

for $j = 1, \ldots, p - 1$, and $t \in I_\delta$.

We have proved that the real function F has a derivative of order p everywhere on $I_\delta \supset [0, 1]$. Therefore, by the one-dimensional Taylor Formula and (26),

$$f(\mathbf{x}) - f(\mathbf{a}) = F(1) - F(0) = \sum_{j=1}^{p-1} \frac{1}{j!} F^{(j)}(0) + \frac{1}{p!} F^{(p)}(t)$$

$$= \sum_{j=1}^{p-1} \frac{1}{j!} D^{(j)} f(\mathbf{a}; \mathbf{h}) + \frac{1}{p!} D^{(p)} f(\mathbf{a} + t\mathbf{h}; \mathbf{h})$$

for some $t \in (0, 1)$. Thus set $\mathbf{c} = \mathbf{a} + t\mathbf{h}$.   ∎

11.38 *EXAMPLE.*

Write Taylor's Formula for $f(x, y) = \cos(xy)$, $\mathbf{a} = (0, 0)$, and $p = 3$.

Solution. It is easy to verify that f_x, f_y, f_{xx}, f_{xy}, and f_{yy} are all zero at $(0, 0)$, so $D^{(1)} f((0, 0); (x, y)) = 0$ and $D^{(2)} f((0, 0); (x, y)) = 0$. Since $f_{xxx}(x, y) = y^3 \sin(xy)$, $f_{xxy} = -2y \cos(xy) + xy^2 \sin(xy)$, $f_{xyy} = -2x \cos(xy) + x^2 y \sin(xy)$, and $f_{yyy} = x^3 \sin(xy)$, Theorem 11.2 implies $D^{(3)} f((c, d); (x, y)) = (x^3 + y^3) \sin(cd) + 3(x^2 y + xy^2) \sin(cd) - 6(x + y) \cos(cd)$. Thus by Taylor's Formula, for all $(x, y) \in \mathbf{R}^2$ there is a point (c, d) on the line segment between $(0, 0)$ and (x, y) such that

$$\cos(xy) = 1 + \left(\frac{x^3 + y^3}{6} \right) \sin(cd) + \left(\frac{x^2 y + xy^2}{2} \right) \sin(cd) - (x + y) \cos(cd).$$

EXERCISES

11.5.1. a) Write out an expression in powers of $(x+1)$ and $(y-1)$ for $f(x, y) = x^2 + xy + y^2$.
  b) Write Taylor's Formula for $f(x, y) = \sqrt{x} + \sqrt{y}$, $\mathbf{a} = (1, 4)$, and $p = 3$.
  c) Write Taylor's Formula for $f(x, y) = e^{xy}$, $\mathbf{a} = (0, 0)$, and $p = 4$.

11.5.2. Suppose that $f : \mathbf{R}^2 \to \mathbf{R}$ is $\mathcal{C}^p$ on $B_r(x_0, y_0)$ for some $r > 0$. Prove that, given $(x, y) \in B_r(x_0, y_0)$, there is a point (c, d) on the line segment between (x_0, y_0) and (x, y) such that

$$f(x, y) = f(x_0, y_0) + \sum_{k=1}^{p-1} \frac{1}{k!} \left(\sum_{j=0}^{k} \binom{k}{j} (x - x_0)^j (y - y_0)^{k-j} \frac{\partial^k f}{\partial x^j \, \partial y^{k-j}} (x_0, y_0) \right)$$

$$+ \frac{1}{p!} \sum_{j=0}^{p} \binom{p}{j} (x - x_0)^j (y - y_0)^{p-j} \frac{\partial^p f}{\partial x^j \, \partial y^{p-j}} (c, d).$$

11.5.3. Suppose that $f : \mathbf{R}^n \to \mathbf{R}$ and $g : \mathbf{R}^n \to \mathbf{R}^n$ are differentiable on $\mathbf{R}^n$ and that there exist $r > 0$ and $\mathbf{a} \in \mathbf{R}^n$ such that $Dg(\mathbf{x})$ is the identity matrix, I, for all $\mathbf{x} \in B_r(\mathbf{a})$. Prove that there is a function $\mathbf{h} : B_r(\mathbf{a}) \setminus \{\mathbf{a}\} \to B_r(\mathbf{x})$ such that

$$\frac{|f(\mathbf{g}(\mathbf{x})) - f(\mathbf{g}(\mathbf{a}))|}{\|\mathbf{x} - \mathbf{a}\|} \leq \|Df((\mathbf{g} \circ \mathbf{h})(\mathbf{x}))\|$$

for all $\mathbf{x} \in B_r(\mathbf{a}) \setminus \{\mathbf{a}\}$.

11.5.4. Suppose that V is convex and open in $\mathbf{R}^n$ and that $\mathbf{f} : V \to \mathbf{R}^n$ is differentiable on V. If there exists an $\mathbf{a} \in V$ such that $D\mathbf{f}(\mathbf{x}) = D\mathbf{f}(\mathbf{a})$ for all $\mathbf{x} \in V$, prove that there exist a linear function $\mathbf{S} \in \mathcal{L}(\mathbf{R}^n; \mathbf{R}^n)$ and a vector $\mathbf{c} \in \mathbf{R}^n$ such that $\mathbf{f}(\mathbf{x}) = \mathbf{S}(\mathbf{x}) + \mathbf{c}$ for all $\mathbf{x} \in V$.

11.5.5. [INTEGRAL FORM OF TAYLOR'S FORMULA]. Let $p \in \mathbf{N}$, V be an open set in $\mathbf{R}^n$, $\mathbf{x}, \mathbf{a} \in V$, and $f : V \to \mathbf{R}$ be $\mathcal{C}^p$ on V. If $L(\mathbf{x}; \mathbf{a}) \subset V$ and $\mathbf{h} = \mathbf{x} - \mathbf{a}$, prove that

$$f(\mathbf{x}) - f(\mathbf{a}) = \sum_{k=1}^{p-1} \frac{1}{k!} D^{(k)} f(\mathbf{a}; \mathbf{h}) + \frac{1}{(p-1)!} \int_0^1 (1-t)^{p-1} D^{(p)} f(\mathbf{a} + t\mathbf{h}; \mathbf{h}) \, dt.$$

11.5.6. Let $r > 0$, $a, b \in \mathbf{R}$, $f : B_r(a, b) \to \mathbf{R}$ be differentiable, and $(x, y) \in B_r(a, b)$.

a) Let $g(t) = f(tx + (1-t)a, y) + f(a, ty + (1-t)b)$ and compute the derivative of g.

b) Prove that there are numbers c between a and x, and d between b and y such that

$$f(x, y) - f(a, b) = (x - a) f_x(c, y) + (y - b) f_y(a, d).$$

(This is Exercise 12.20 in Apostol [1].)

11.5.7. Suppose that $0 < r < 1$ and that $f : B_1(0) \to \mathbf{R}$ is continuously differentiable. If there is an $\alpha > 0$ such that $|f(\mathbf{x})| \le \|\mathbf{x}\|^\alpha$ for all $\mathbf{x} \in B_r(\mathbf{0})$, prove that there is an $M > 0$ such that $|f(\mathbf{x})| \le M\|\mathbf{x}\|$ for $\mathbf{x} \in B_r(\mathbf{0})$.

11.5.8. Suppose that V is open in $\mathbf{R}^n$, that $f : V \to \mathbf{R}$ is C^2 on V, and that $f_{x_j}(\mathbf{a}) = 0$ for some $\mathbf{a} \in H$ and all $j = 1, \dots, n$. Prove that if H is a compact convex subset of V, then there is a constant M such that

$$|f(\mathbf{x}) - f(\mathbf{a})| \le M\|\mathbf{x} - \mathbf{a}\|^2$$

for all $\mathbf{x} \in H$.

11.5.9. Let $f : \mathbf{R}^n \to \mathbf{R}$. Suppose that for each unit vector $\mathbf{u} \in \mathbf{R}^n$, the directional derivative $D_\mathbf{u} f(\mathbf{a} + t\mathbf{u})$ exists for $t \in [0, 1]$ (see Definition 11.19). Prove that

$$f(\mathbf{a} + \mathbf{u}) - f(\mathbf{a}) = D_\mathbf{u} f(\mathbf{a} + t\mathbf{u})$$

for some $t \in (0, 1)$.

11.5.10. Suppose that V is open in $\mathbf{R}^2$, that $(a, b) \in V$, and that $f : V \to \mathbf{R}$ is C^3 on V. Prove that

$$\lim_{r \to 0} \frac{4}{\pi r^2} \int_0^{2\pi} f(a + r\cos\theta, b + r\sin\theta) \cos(2\theta) \, d\theta = f_{xx}(a, b) - f_{yy}(a, b).$$

11.5.11. Suppose that V is open in $\mathbf{R}^2$, that $H = [a, b] \times [0, c] \subset V$, that $u : V \to \mathbf{R}$ is C^2 on V, and that $u(x_0, t_0) \ge 0$ for all $(x_0, t_0) \in \partial H$.

a) Show that, given $\varepsilon > 0$, there is a compact set $K \subset H^o$ such that $u(x, t) \ge -\varepsilon$ for all $(x, t) \in H \setminus K$.

b) Suppose that $u(x_1, t_1) = -\ell < 0$ for some $(x_1, t_1) \in H^o$, and choose $r > 0$ so small that $2rt_1 < \ell$. Apply part a) to $\varepsilon := \ell/2 - rt_1$ to choose the compact set K, and prove that the minimum of

$$w(x, t) := u(x, t) + r(t - t_1)$$

on H occurs at some $(x_2, t_2) \in K$.

c) Prove that if u satisfies the *heat equation* (i.e., $u_{xx} - u_t = 0$ on V), and if $u(x_0, t_0) \geq 0$ for all $(x_0, t_0) \in \partial H$, then $u(x, t) \geq 0$ for all $(x, t) \in H$.

11.5.12. a) Prove that every convex set in $\mathbf{R}^n$ is connected.
  b) Show that the converse of part a) is false.
  *c) Suppose that $f : \mathbf{R} \to \mathbf{R}$. Prove that f is convex (as a function) if and only if $E := \{(x, y) : y \geq f(x)\}$ is convex (as a set in $\mathbf{R}^2$).

11.6 THE INVERSE FUNCTION THEOREM

By the one-dimensional Inverse Function Theorem (Theorem 4.33), if $g : \mathbf{R} \to \mathbf{R}$ is 1–1 and differentiable with $g'(x_0) \neq 0$, then g^{-1} is differentiable at $y_0 = g(x_0)$ and

$$(g^{-1})'(y_0) = \frac{1}{g'(x_0)}.$$

In this section we obtain a multivariable analogue of this result (i.e., an Inverse Function Theorem for vector functions $\mathbf{f}$ from n variables to n variables). What shall we use for hypotheses? We needed g to be 1–1 so that the inverse function g^{-1} existed. For the same reason, we shall assume that $\mathbf{f}$ is 1–1. We needed $g'(x_0)$ to be nonzero so that we could divide by it. In the multidimensional case, $D\mathbf{f}(\mathbf{a})$ is a matrix; hence "divisibility" corresponds to invertibility. Since an $n \times n$ matrix is invertible if and only if it has a nonzero determinant (see Appendix C), we shall assume that the *Jacobian* of $\mathbf{f}$

$$\Delta_{\mathbf{f}}(\mathbf{a}) := \det(D\mathbf{f}(\mathbf{a})) \neq 0.$$

The word *Jacobian* is used because it was Jacobi who first recognized the importance of $\Delta_{\mathbf{f}}$ and its connection with volume (see Exercise 12.4.6).

The proof of the Inverse Function Theorem on $\mathbf{R}^n$ is not simple and lies somewhat deeper than the preceding results of this chapter. Before presenting it, we first prove two preliminary results which explore the consequences of the hypothesis $\Delta_{\mathbf{f}} \neq 0$.

11.39 Lemma.
Let V be open and nonempty in $\mathbf{R}^n$ and let $\mathbf{f} : V \to \mathbf{R}^n$ be continuous. If $\mathbf{f}$ is 1–1 and has first-order partial derivatives on V, and if $\Delta_{\mathbf{f}} \neq 0$ on V, then $\mathbf{f}^{-1}$ is continuous on $\mathbf{f}(V)$.

STRATEGY: To prove that $\mathbf{f}^{-1}$ is continuous on V, it suffices to prove (apply Exercise 9.4.3 or Theorem 10.58 to $\mathbf{f}^{-1}$) that $\mathbf{f}(W) = (\mathbf{f}^{-1})^{-1}(W)$ is open for all open $W \subseteq V$. Thus, given $\mathbf{b} \in \mathbf{f}(W)$, say $\mathbf{b} = \mathbf{f}(\mathbf{a})$ for some $\mathbf{a} \in W$, we must find a $\rho > 0$ such that $B_\rho(\mathbf{b}) \subseteq \mathbf{f}(W)$. We will actually show more: that if $\overline{B_r(\mathbf{a})} \subset W$ for some $r > 0$, then there is a $\rho > 0$ such that $B_\rho(\mathbf{b}) \subseteq \mathbf{f}(B_r(\mathbf{a}))$; that is, $\mathbf{y} \in B_\rho(\mathbf{f}(\mathbf{a}))$ implies that $\mathbf{y} = \mathbf{f}(\mathbf{c})$ for some $\mathbf{c} \in B_r(\mathbf{a})$.

Where should we look to find such a $\mathbf{c}$? To show that $\mathbf{f}(\mathbf{c}) - \mathbf{y} = \mathbf{0}$, we might first try finding a point $\mathbf{c} \in B_r(\mathbf{a})$ that minimizes $\|\mathbf{f}(\mathbf{x}) - \mathbf{y}\|$ as $\mathbf{x}$ ranges over $\overline{B_r(\mathbf{a})}$

and then try showing that the minimum value is actually zero. It is relatively easy to prove that the minimum value is zero using the hypothesis that $\Delta_{\mathbf{f}} \neq 0$. Moreover, since $\|\mathbf{f}(\mathbf{x}) - \mathbf{y}\|$ is continuous on the compact set $\overline{B_r(\mathbf{a})}$, the minimum value must be attained by some $\mathbf{c}$, but $\mathbf{c}$ might belong to the boundary of $B_r(\mathbf{a})$, not its interior. By controlling the size of ρ, we can keep the minimum value from occurring on the boundary. Here are the details.

Proof. Suppose that W is an open subset of V and let $\mathbf{b} \in \mathbf{f}(W)$. Choose $\mathbf{a} \in W$ such that $\mathbf{b} = \mathbf{f}(\mathbf{a})$. Since W is open, choose $r_0 > 0$ such that $B_{r_0}(\mathbf{a}) \subset W$, and observe for any $r \in (0, r_0)$ that $\partial B_r(\mathbf{a}) \subset \overline{B_r(\mathbf{a})} \subset W$.

Since $\mathbf{f}$ is 1–1 on W, the real-valued function

$$g(\mathbf{x}) := \|\mathbf{f}(\mathbf{x}) - \mathbf{f}(\mathbf{a})\|, \qquad \mathbf{x} \in \partial B_r(\mathbf{a}),$$

is positive on the compact set $\partial B_r(\mathbf{a})$. Since $\mathbf{f}$ is continuous on W, it follows from the Extreme Value Theorem that g attains a positive minimum on $\partial B_r(\mathbf{a})$; that is,

$$m = \inf_{\mathbf{x} \in \partial B_r(\mathbf{a})} g(\mathbf{x}) > 0.$$

Set $\rho = m/2$ and fix $\mathbf{y} \in B_\rho(\mathbf{f}(\mathbf{a}))$. To show that $\mathbf{y} \in \mathbf{f}(B_r(\mathbf{a}))$, notice that since the function $h(\mathbf{x}) := \|\mathbf{f}(\mathbf{x}) - \mathbf{y}\|$ is continuous on the compact set $\overline{B_r(\mathbf{a})}$, it also attains its minimum there. Thus there is a $\mathbf{c} \in \overline{B_r(\mathbf{a})}$ such that $h(\mathbf{c}) \le h(\mathbf{x})$ for all $\mathbf{x} \in \overline{B_r(\mathbf{a})}$.

To show that $\mathbf{c} \in B_r(\mathbf{a})$, suppose to the contrary that $\mathbf{c} \notin B_r(\mathbf{a})$; that is, that $\mathbf{c} \in \partial B_r(\mathbf{a})$. Then $\|\mathbf{f}(\mathbf{c}) - \mathbf{f}(\mathbf{a})\| \ge m = 2\rho$. Since $\mathbf{y} \in B_\rho(\mathbf{f}(\mathbf{a}))$ implies that $h(\mathbf{a}) = \|\mathbf{f}(\mathbf{a}) - \mathbf{y}\| < \rho$, the minimum of $\|h(\mathbf{x})\|$ must also be less that ρ; that is, $\rho > h(\mathbf{c})$. Therefore, it follows from the triangle inequality that

$$\rho > h(\mathbf{c}) = \|\mathbf{f}(\mathbf{c}) - \mathbf{y}\| \ge \|\mathbf{f}(\mathbf{c}) - \mathbf{f}(\mathbf{a})\| - \|\mathbf{f}(\mathbf{a}) - \mathbf{y}\| > 2\rho - \rho = \rho,$$

a contradiction. Thus $\mathbf{c} \in B_r(\mathbf{a})$.

It remains to prove that $\mathbf{y} = \mathbf{f}(\mathbf{c})$. Notice that, since $h(\mathbf{c}) \ge 0$, $h^2(\mathbf{c})$ is the minimum of h^2 on $\overline{B_r(\mathbf{a})}$. Thus, by one-dimensional calculus,

$$\frac{\partial h^2}{\partial x_k}(\mathbf{c}) = 0$$

for $k = 1, \ldots, n$. Since $h^2(\mathbf{x}) = \sum_{j=1}^{n}(f_j(\mathbf{x}) - y_j)^2$, it follows that

$$0 = \frac{1}{2}\frac{\partial h^2}{\partial x_k}(\mathbf{c}) = \sum_{j=1}^{n}(f_j(\mathbf{c}) - y_j)\frac{\partial f_j}{\partial x_k}(\mathbf{c}).$$

This is a system of n linear equations in n unknowns, $f_j(\mathbf{c}) - y_j$. Since the matrix of coefficients of this system has determinant $\Delta_{\mathbf{f}}(\mathbf{c}) \neq 0$, it follows from Cramer's Rule (see Appendix C) that this system has only the trivial solution; that is, $f_j(\mathbf{c}) - y_j = 0$ for all $j = 1, \ldots, n$. In particular, $\mathbf{y} = \mathbf{f}(\mathbf{c})$. ∎

Our second preliminary result shows that if the Jacobian of a continuously differentiable function $\mathbf{f}$ is nonzero at a point, then $\mathbf{f}$ must be 1–1 near that point. (This will provide a key step in the proof of the Inverse Function Theorem below.)

11.40 Lemma.
Let V be open in $\mathbf{R}^n$ and $\mathbf{f} : V \to \mathbf{R}^n$ be $\mathcal{C}^1$ on V. If $\Delta_{\mathbf{f}}(\mathbf{a}) \neq 0$ for some $\mathbf{a} \in V$, then there is an $r > 0$ such that $B_r(\mathbf{a}) \subset V$, $\mathbf{f}$ is 1–1 on $B_r(\mathbf{a})$, $\Delta_{\mathbf{f}}(\mathbf{x}) \neq 0$ for all $\mathbf{x} \in B_r(\mathbf{a})$, and

$$\det \left[\frac{\partial f_i}{\partial x_j}(\mathbf{c}_i) \right]_{n \times n} \neq 0$$

for all $\mathbf{c}_1, \dots \mathbf{c}_n \in B_r(\mathbf{a})$.

STRATEGY: The idea behind the proof is simple. If $\mathbf{f}$ is not 1–1 on some $B_r(\mathbf{a})$, then there exist $\mathbf{x}, \mathbf{y} \in B_r(\mathbf{a})$ such that $\mathbf{x} \neq \mathbf{y}$ and $\mathbf{f}(\mathbf{x}) = \mathbf{f}(\mathbf{y})$. Since $L(\mathbf{x}; \mathbf{y}) \subset B_r(\mathbf{a})$, we have by Theorem 11.30 (the Mean Value Theorem) that

$$0 = f_i(\mathbf{y}) - f_i(\mathbf{x}) = \sum_{k=1}^{n} \frac{\partial f_i}{\partial x_k}(\mathbf{c}_i)(y_k - x_k) \tag{27}$$

for $\mathbf{x} = (x_1, \dots, x_n)$, $\mathbf{y} = (y_1, \dots, y_n)$, $\mathbf{c}_i \in L(\mathbf{x}; \mathbf{y})$, and $i = 1, \dots, n$. Notice that (27) is a system of n linear equations in n unknowns, $(y_k - x_k)$. If we can show, for sufficiently small r, that the matrix of coefficients of (27) has nonzero determinant for any choice of $\mathbf{c}_i \in B_r(\mathbf{a})$, then by Cramer's Rule the linear system (27) has only one solution: $y_k - x_k = 0$ for $k = 1, \dots, n$. This would imply $\mathbf{x} = \mathbf{y}$, a contradiction. Here are the details.

Proof. To show that there is an $r > 0$ such that the matrix of coefficients of the linear system (27) is nonzero for all $\mathbf{c}_i \in B_r(\mathbf{a})$, let $V^{(n)} = V \times \dots \times V$ represent the n-fold Cartesian product of V with itself, and define $h : V^{(n)} \to \mathbf{R}$ by

$$h(\mathbf{x}_1, \mathbf{x}_2, \dots, \mathbf{x}_n) = \det \left[\frac{\partial f_i}{\partial x_j}(\mathbf{x}_i) \right]_{n \times n}.$$

Since the determinant of a matrix is defined using products and differences of its entries (see Appendix C), we have by hypothesis that h is continuous on $V^{(n)}$. Since $h(\mathbf{a}, \dots, \mathbf{a}) = \Delta_{\mathbf{f}}(\mathbf{a}) \neq 0$, it follows that there is an $r > 0$ such that $B_r(\mathbf{a}) \subset V$ and $h(\mathbf{c}_1, \dots, \mathbf{c}_n) \neq 0$ for $\mathbf{c}_i \in B_r(\mathbf{a})$. In particular, the determinant matrix of coefficients of the linear system (27) is nonzero for all $\mathbf{c}_i \in B_r(\mathbf{a})$, and $\Delta_{\mathbf{f}}(\mathbf{x}) = h(\mathbf{x}, \dots, \mathbf{x}) \neq 0$ for all $\mathbf{x} \in B_r(\mathbf{a})$. ∎

We now prove a multidimensional version of the Inverse Function Theorem.

11.41 Theorem. [THE INVERSE FUNCTION THEOREM].
Let V be open in $\mathbf{R}^n$ and $\mathbf{f} : V \to \mathbf{R}^n$ be $\mathcal{C}^1$ on V. If $\Delta_{\mathbf{f}}(\mathbf{a}) \neq 0$ for some $\mathbf{a} \in V$, then there exists an open set W containing $\mathbf{a}$ such that

i) **f** *is 1–1 on W,*
ii) **f**$^{-1}$ *is* $\mathcal{C}^1$ *on* **f**(W), *and*
iii) *for each* **y** $\in$ **f**(W),

$$D(\mathbf{f}^{-1})(\mathbf{y}) = [D\mathbf{f}(\mathbf{f}^{-1}(\mathbf{y}))]^{-1},$$

where $[\,]^{-1}$ *represents matrix inversion (see Theorem C.5).*

Proof. By Lemma 11.40, there is an open ball B centered at **a** such that **f** is 1–1 and $\Delta_{\mathbf{f}} \neq 0$ on B, and

$$\Delta := \det\left[\frac{\partial f_i}{\partial x_j}(\mathbf{c}_i)\right]_{n \times n} \neq 0$$

for all $\mathbf{c}_i \in B$. Let B_0 be an open ball centered at **a** which is smaller than B (i.e., the radius of B_0 is strictly less than the radius of B). Then $\overline{B_0} \subset B$, **f** is 1–1 on B_0 and, by Lemma 11.39, **f**$^{-1}$ is continuous on **f**(B_0).

Let W be any open ball centered at **a** which is smaller than B_0. Then **f** is 1–1 on W and **f**(W) is open. To show that the first partial derivatives of **f**$^{-1}$ exist and are continuous on **f**(W), fix $\mathbf{y}_0 \in$ **f**(W) and $1 \le i, k \le n$. Choose $t \in \mathbf{R} \setminus \{0\}$ so small that $\mathbf{y}_0 + t\mathbf{e}_k \in$ **f**(W), and choose $\mathbf{x}_0, \mathbf{x}_1 \in W$ such that $\mathbf{x}_0 = $ **f**$^{-1}(\mathbf{y}_0)$ and $\mathbf{x}_1 = $ **f**$^{-1}(\mathbf{y}_0 + t\mathbf{e}_k)$. Observe that for each $i = 1, 2, \dots, n$,

$$f_i(\mathbf{x}_1) - f_i(\mathbf{x}_0) = \begin{cases} t & k = i \\ 0 & k \neq i, \end{cases}$$

Hence, by Theorem 11.30 (the Mean Value Theorem), there exist points $\mathbf{c}_i \in L(\mathbf{x}_0; \mathbf{x}_1)$ such that

$$\nabla f_i(\mathbf{c}_i) \cdot \frac{\mathbf{x}_1 - \mathbf{x}_0}{t} = \frac{f_i(\mathbf{x}_1) - f_i(\mathbf{x}_0)}{t} = \begin{cases} 1 & k = i \\ 0 & k \neq i \end{cases} \quad i = 1, 2, \dots, n. \quad (28)$$

Let $x_0^{(j)}$ (respectively, $x_1^{(j)}$) denote the jth component of $\mathbf{x}_0$ (respectively, $\mathbf{x}_1$). Since (28) is a system of n linear equations in n variables $(x_1^{(j)} - x_0^{(j)})/t$ whose coefficient matrix has determinant Δ (which is nonzero by the choice of B), we see by Cramer's Rule that the solutions of (28) satisfy

$$\frac{(\mathbf{f}^{-1})_j(\mathbf{y}_0 + t\mathbf{e}_k) - (\mathbf{f}^{-1})_j(\mathbf{y}_0)}{t} := \frac{x_1^{(j)} - x_0^{(j)}}{t} = \mathcal{Q}_j(t), \quad (29)$$

where $\mathcal{Q}_j(t)$ is a quotient of determinants whose entries are 0s or 1s, or first-order partial derivatives of components of **f** evaluated at the $\mathbf{c}_i$'s. Since $t \to 0$ implies $\mathbf{x}_1 \to \mathbf{x}_0$, $\mathbf{c}_i \to \mathbf{x}_0$, and $\mathbf{y}_0 + t\mathbf{e}_k \to \mathbf{y}_0$, it follows that $\mathcal{Q}_j(t)$ converges to $\mathcal{Q}_j$, a quotient of determinants whose entries are 0s or 1s, or first-order partial derivatives of components of **f** evaluated at $\mathbf{x}_0 = $ **f**$^{-1}(\mathbf{y}_0)$. Since **f**$^{-1}$ is continuous on $\mathbf{f}(W)$, $\mathcal{Q}_j$ must be continuous at each $\mathbf{y}_0 \in \mathbf{f}(W)$. Taking the limit of (29) as $t \to 0$, we see that the first-order partial derivatives of $(\mathbf{f}^{-1})_j$ exist at $\mathbf{y}_0$ and equal $\mathcal{Q}_j$; that is, $\mathbf{f}^{-1}$ is continuously differentiable on $\mathbf{f}(W)$.

It remains to verify iii). Fix $\mathbf{y} \in \mathbf{f}(W)$, and observe, by the Chain Rule and Exercise 11.2.8, that

$$I = DI(\mathbf{y}) = D(\mathbf{f} \circ \mathbf{f}^{-1})(\mathbf{y}) = D\mathbf{f}(\mathbf{f}^{-1}(\mathbf{y}))D\mathbf{f}^{-1}(\mathbf{y}).$$

By the uniqueness of matrix inverses, we conclude that

$$D(\mathbf{f}^{-1})(\mathbf{y}) = [D\mathbf{f}(\mathbf{f}^{-1}(\mathbf{y}))]^{-1}. \qquad \blacksquare$$

Of course, the value $D\mathbf{f}^{-1}(\mathbf{y})$ is not unique because $\mathbf{f}^{-1}$ may have several branches. For example, if $f(x) = x^2$, then $f^{-1}(1) = \pm 1$, depending on whether we take the inverse of $f(x)$ near $x = 1$ or $x = -1$.

11.42 Remark. *The hypothesis "$\Delta_\mathbf{f} \neq 0$" in Lemma 11.39 can be relaxed.*

Proof. If $f(x) = x^3$, then $f : \mathbf{R} \to \mathbf{R}$ and its inverse $f^{-1}(x) = \sqrt[3]{x}$ are continuous on $\mathbf{R}$, but $\Delta_f(0) = f'(0) = 0$. $\qquad \blacksquare$

11.43 Remark. *The hypothesis "$\Delta_\mathbf{f} \neq 0$" in Theorem 11.41 cannot be relaxed. In fact, if $\mathbf{f} : B_r(\mathbf{a}) \to \mathbf{R}^n$ is differentiable at $\mathbf{a}$ and its inverse $\mathbf{f}^{-1}$ exists and is differentiable at $\mathbf{f}(\mathbf{a})$, then $\Delta_\mathbf{f}(\mathbf{a}) \neq 0$.*

Proof. Suppose to the contrary that $\mathbf{f}$ is differentiable at $\mathbf{a}$ but $\Delta_\mathbf{f}(\mathbf{a}) = 0$. By Exercise 11.2.8 and the Chain Rule,

$$I = D(\mathbf{f}^{-1} \circ \mathbf{f})(\mathbf{a}) = D(\mathbf{f}^{-1})(\mathbf{f}(\mathbf{a}))D\mathbf{f}(\mathbf{a}).$$

Taking the determinant of this identity, we have

$$1 = \Delta_{\mathbf{f}^{-1}}(\mathbf{f}(\mathbf{a}))\Delta_\mathbf{f}(\mathbf{a}) = 0,$$

a contradiction. $\qquad \blacksquare$

11.44 Remark. *The hypothesis "$\mathbf{f}$ is C^1 on V" in Theorem 11.41 cannot be relaxed, even when f is a real function.*

Proof. If $f(x) = x + 2x^2 \sin(1/x)$, $x \neq 0$, and $f(0) = 0$, then $f : \mathbf{R} \to \mathbf{R}$ is differentiable on $V := (-1, 1)$ and $f'(0) = 1 \neq 0$. However, since

$$f\left(\frac{2}{(4k-1)\pi}\right) < f\left(\frac{2}{(4k+1)\pi}\right) < f\left(\frac{2}{(4k-3)\pi}\right)$$

for $k \in \mathbf{N}$, f is not 1–1 on any open set which contains 0. Therefore, no open subset of $f(V)$ can be chosen on which f^{-1} exists. $\qquad \blacksquare$

Although Theorem 11.41 says **f** must be 1–1 on some subset W of V, it does not say that **f** is 1–1 on V.

11.45 Remark. *The set W chosen in Theorem 11.41 is, in general, a proper subset of V, even when V is connected.*

Proof. Set $\mathbf{f}(x, y) = (x^2 - y^2, xy)$ and $V = \mathbf{R}^2 \setminus \{(0, 0)\}$. Then $\Delta_{\mathbf{f}} = 2(x^2 + y^2) \neq 0$ for $(x, y) \in V$, but $\mathbf{f}(x, -y) = \mathbf{f}(-x, y)$ for all $(x, y) \in \mathbf{R}^2$. Thus **f** is not 1–1 on V. ∎

Sometimes vector functions from p variables to n variables are defined implicitly by relations on $\mathbf{R}^{n+p}$. On rare occasions, such a relation can be solved explicitly as follows.

*11.46 EXAMPLE.

If $x_0^2 + s_0^2 + t_0^2 = 1$ and $x_0 \neq 0$, prove that there exist an $r > 0$ and a function $g(s, t)$, continuously differentiable on $B_r(s_0, t_0)$, such that $x_0 = g(s_0, t_0)$ and

$$x^2 + s^2 + t^2 = 1$$

for $x = g(s, t)$ and $(s, t) \in B_r(s_0, t_0)$.

Proof. Solve $x^2 + s^2 + t^2 = 1$ for x to obtain

$$x = \pm\sqrt{1 - s^2 - t^2}.$$

Which sign shall we take? If $x_0 > 0$, set $g(s, t) = \sqrt{1 - s^2 - t^2}$. By the Chain Rule,

$$\frac{\partial g}{\partial s} = \frac{-s}{\sqrt{1 - s^2 - t^2}} \quad \text{and} \quad \frac{\partial g}{\partial t} = \frac{-t}{\sqrt{1 - s^2 - t^2}}.$$

Thus g is differentiable at any point (s, t) which lies inside the two-dimensional unit ball (i.e., which satisfies $s^2 + t^2 < 1$). Since $x_0^2 + s_0^2 + t_0^2 = 1$ and $x_0 > 0$, (s_0, t_0, x_0) lies on the boundary of the three-dimensional unit ball in stx space a distance x_0 units above the st plane (see Figure 11.5). In particular, if $r := 1 - \sqrt{1 - x_0^2}$ and $(s, t) \in B_r(s_0, t_0)$, then $s^2 + t^2 < 1$. Therefore, g is continuously differentiable on $B_r(s_0, t_0)$. If $x_0 < 0$, a similar argument works for $g(s, t) = -\sqrt{1 - s^2 - t^2}$. ∎

We cannot expect that all relations can be solved explicitly as we did in Example 11.46 above. It is most fortunate, therefore, that once we know a solution exists, we can often approximate that solution by numerical methods.

The crux of the matter, then, is which relations have solutions? In order to state a result about the existence of solutions to a relation, we introduce additional notation. Let V be an open subset of $\mathbf{R}^n$, $\mathbf{f} : V \to \mathbf{R}^m$, and $\mathbf{a} \in V$. Then

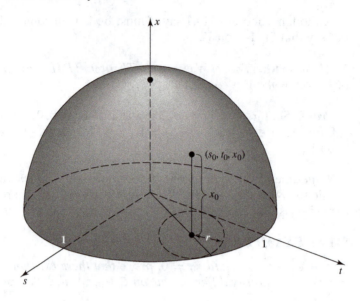

FIGURE 11.5

the *partial Jacobian* of $\mathbf{f}$ generated by a subset $\{k_1, k_2, \ldots, k_n\}$ of $\{1, 2, \ldots, m\}$ at the point $\mathbf{a}$ is the number

$$\frac{\partial(f_{k_1}, \ldots, f_{k_n})}{\partial(x_1, \ldots, x_n)}(\mathbf{a}) := \det\left[\frac{\partial f_{k_i}}{\partial x_j}(\mathbf{a})\right]_{n \times n} = \det\begin{bmatrix} \dfrac{\partial f_{k_1}}{\partial x_1}(\mathbf{a}) & \cdots & \dfrac{\partial f_{k_1}}{\partial x_n}(\mathbf{a}) \\ \vdots & \ddots & \vdots \\ \dfrac{\partial f_{k_n}}{\partial x_1}(\mathbf{a}) & \cdots & \dfrac{\partial f_{k_n}}{\partial x_n}(\mathbf{a}) \end{bmatrix}$$

provided all these partial derivatives exist. For the case $n = m$, the corresponding partial Jacobian is just the Jacobian $\Delta_{\mathbf{f}}(\mathbf{a})$. We shall use partial Jacobians again in Chapter 12 to discuss change of variables for integrals in $\mathbf{R}^n$, and in Chapter 13 to introduce differential forms of order 2.

For the next several pages, we shall represent a vector $(x_1, \ldots, x_n, t_1, \ldots, t_p)$ in $\mathbf{R}^{n+p}$ by $(\mathbf{x}, \mathbf{t})$. Here is a result about the existence of solutions to relations. It states as follows: If $\mathbf{F}$ is a $\mathcal{C}^1$ function which satisfies $\mathbf{F}(\mathbf{x}_0, \mathbf{t}_0) = \mathbf{0}$ at some point $(\mathbf{x}_0 \mathbf{t}_0)$, then the implicit relation $\mathbf{F}(\mathbf{x}, \mathbf{t}) = \mathbf{0}$ can be solved for the variables $x_1, \ldots, x_n$, when $\mathbf{t}$ is near $\mathbf{t}_0$, provided the partial Jacobian of $\mathbf{F}$ with respect to $x_1, x_2, \ldots, x_n$ (the variables we want to solve for) is not zero at $(\mathbf{x}_0, \mathbf{t}_0)$.

11.47 Theorem. [THE IMPLICIT FUNCTION THEOREM].
Suppose that V is open in $\mathbf{R}^{n+p}$, and that $\mathbf{F} = (F_1, \ldots, F_n) : V \to \mathbf{R}^n$ is $\mathcal{C}^1$ on V. Suppose further that $\mathbf{F}(\mathbf{x}_0, \mathbf{t}_0) = \mathbf{0}$ for some $(\mathbf{x}_0, \mathbf{t}_0) \in V$, where $\mathbf{x}_0 \in \mathbf{R}^n$ and $\mathbf{t}_0 \in \mathbf{R}^p$. If

$$\frac{\partial(F_1, \ldots, F_n)}{\partial(x_1, \ldots, x_n)}(\mathbf{x}_0, \mathbf{t}_0) \neq 0,$$

then there is an open set $W \subset \mathbf{R}^p$ containing $\mathbf{t}_0$ and a unique continuously differentiable function $\mathbf{g} : W \to \mathbf{R}^n$ such that $\mathbf{g}(\mathbf{t}_0) = \mathbf{x}_0$, and $\mathbf{F}(\mathbf{g}(\mathbf{t}), \mathbf{t}) = \mathbf{0}$ for all $\mathbf{t} \in W$.

STRATEGY: The idea behind the proof is simple. If $\mathbf{F}$ took its range in $\mathbf{R}^{n+p}$ instead of $\mathbf{R}^n$ and had nonzero Jacobian, then, by the Inverse Function Theorem, $\mathbf{F}^{-1}$ would exist and be differentiable on some open set. Presumably, the first n components of $\mathbf{F}^{-1}$ would solve $\mathbf{F}$ for the variables $x_1, \ldots, x_n$. Thus we should extend $\mathbf{F}$ (in the simplest possible way) to a function $\widetilde{\mathbf{F}}$ which takes its range in $\mathbf{R}^{n+p}$ and has nonzero Jacobian, and apply the Inverse Function Theorem to $\widetilde{\mathbf{F}}$. Here are the details.

Proof. For each $(\mathbf{x}, \mathbf{t}) \in V$, set

$$\widetilde{\mathbf{F}}(\mathbf{x}, \mathbf{t}) = (F_1(\mathbf{x}, \mathbf{t}), \ldots, F_n(\mathbf{x}, \mathbf{t}), t_1, \ldots, t_p). \tag{30}$$

Clearly, $\widetilde{\mathbf{F}} : V \to \mathbf{R}^{n+p}$ and

$$D\widetilde{\mathbf{F}} = \begin{bmatrix} \left[\frac{\partial F_i}{\partial x_j}\right]_{n \times n} & B \\ O_{p \times n} & I_{p \times p} \end{bmatrix},$$

where $O_{p \times n}$ represents a zero matrix, $I_{p \times p}$ represents an identity matrix, and B represents a certain $n \times p$ matrix whose entries are first-order partial derivatives of F_j's with respect to t_k's. Expanding the determinant of $D\widetilde{\mathbf{F}}$ along the bottom rows first, we see by hypothesis that

$$\Delta_{\widetilde{\mathbf{F}}}(\mathbf{x}_0, \mathbf{t}_0) = 1 \cdot \frac{\partial(F_1, \ldots, F_n)}{\partial(x_1, \ldots, x_n)}(\mathbf{x}_0, \mathbf{t}_0) \neq 0.$$

Since $\widetilde{\mathbf{F}}(\mathbf{x}_0, \mathbf{t}_0) = (\mathbf{0}, \mathbf{t}_0)$, it follows from the Inverse Function Theorem that there exist open sets Ω_1 containing $(\mathbf{x}_0, \mathbf{t}_0)$ and $\Omega_2 := \widetilde{\mathbf{F}}(\Omega_1)$ containing $(\mathbf{0}, \mathbf{t}_0)$ such that $\widetilde{\mathbf{F}}$ is 1–1 on Ω_1, and $\mathbf{G} := \widetilde{\mathbf{F}}^{-1}$ is 1–1 and continuously differentiable on Ω_2.

Let $\boldsymbol{\phi} = (G_1, \ldots, G_n)$. Since $\mathbf{G} = \widetilde{\mathbf{F}}^{-1}$ is 1–1 from Ω_2 onto Ω_1, it is evident by (30) that

$$\boldsymbol{\phi}(\widetilde{\mathbf{F}}(\mathbf{x}, \mathbf{t})) = \mathbf{x} \tag{31}$$

for all $(\mathbf{x}, \mathbf{t}) \in \Omega_1$ and

$$\widetilde{\mathbf{F}}(\boldsymbol{\phi}(\mathbf{x}, \mathbf{t}), \mathbf{t}) = (\mathbf{x}, \mathbf{t}) \tag{32}$$

for all $(\mathbf{x}, \mathbf{t}) \in \Omega_2$. Define $\mathbf{g}$ on $W := \{\mathbf{t} \in \mathbf{R}^p : (\mathbf{0}, \mathbf{t}) \in \Omega_2\}$ by $\mathbf{g}(\mathbf{t}) = \boldsymbol{\phi}(\mathbf{0}, \mathbf{t})$. Since Ω_2 is open in $\mathbf{R}^{n+p}$, W is open in $\mathbf{R}^p$. Since $\mathbf{G}$ is continuously differentiable on Ω_2 and $\boldsymbol{\phi}$ represents the first n components of $\mathbf{G}$, $\mathbf{g}$ is continuously differentiable on W. By the definition of $\mathbf{g}$, the choice of $\mathbf{x}_0$, and (31), we have

$$\mathbf{g}(\mathbf{t}_0) = \boldsymbol{\phi}(\mathbf{0}, \mathbf{t}_0) = \boldsymbol{\phi}(\widetilde{\mathbf{F}}(\mathbf{x}_0, \mathbf{t}_0)) = \mathbf{x}_0.$$

Moreover, by (30) and (32) we have $\mathbf{F}(\phi(\mathbf{x}, t), t) = \mathbf{x}$ for all $(\mathbf{x}, t) \in \Omega_2$. Specializing to the case $\mathbf{x} = \mathbf{0}$, we obtain $\mathbf{F}(\mathbf{g}(t), t) = \mathbf{0}$ for $t \in W$.

It remains to show uniqueness. But if $\mathbf{h} : W \to \mathbf{R}^n$ satisfies $\mathbf{F}(\mathbf{h}(t), t) = \mathbf{0} = \mathbf{F}(\mathbf{g}(t), t)$ [i.e., $\widetilde{\mathbf{F}}(\mathbf{h}(t), t) = (\mathbf{0}, t) = \widetilde{\mathbf{F}}(\mathbf{g}(t), t)$], then $\mathbf{g}(t) = \mathbf{h}(t)$ for all $t \in W$, since $\widetilde{\mathbf{F}}$ is 1–1 on Ω_1. ∎

Theorem 11.47 is an existence theorem. It states that a solution $\mathbf{g}$ exists without giving us any idea how to find it. Fortunately, for many applications it is not as important to be able to write an explicit formula for $\mathbf{g}$ as it is to know that $\mathbf{g}$ exists.

Here is an example for which an explicit solution is unobtainable.

11.48 EXAMPLE.

Prove that there is a function $g(s, t)$, continuously differentiable on some $B_r(1, 0)$, such that $1 = g(1, 0)$, and

$$sx^2 + tx^3 + 2\sqrt{t + s} + t^2 x^4 - x^5 \cos t - x^6 = 1$$

for $x = g(s, t)$ and $(s, t) \in B_r(1, 0)$.

Proof. If $F(x, s, t) = sx^2 + tx^3 + 2\sqrt{t + s} + t^2 x^4 - x^5 \cos t - x^6 - 1$, then $F(1, 1, 0) = 0$, and $F_x = 2sx + 3tx^2 + 4t^2 x^3 - 5x^4 \cos t - 6x^5$ is nonzero at the point $(1, 1, 0)$. Applying the Implicit Function Theorem to F, with $n = 1$, $p = 2$, $x_0 = 1$, and $(s_0, t_0) = (1, 0)$, we conclude that such a g exists. ∎

Even when an explicit solution is obtainable, it is frequently easier to apply the Implicit Function Theorem than it is to solve a relation explicitly for one or more of its variables. Indeed, consider Example 11.46 again. Let $F(x, s, t) = 1 - x^2 - s^2 - t^2$ and notice that $F_x = -2x$. Thus, by the Implicit Function Theorem, a continuously differentiable solution $x = g(s, t)$ exists for each $x_0 \neq 0$.

The following example shows that the Implicit Function Theorem can be used to prove that differentiable solutions to a system of equations exist simultaneously.

11.49 EXAMPLE.

Prove that there exist functions $u, v : \mathbf{R}^4 \to \mathbf{R}$, continuously differentiable on some ball B centered at the point $(x, y, z, w) = (2, 1, -1, -2)$, such that $u(2, 1, -1, -2) = 4$, $v(2, 1, -1, -2) = 3$, and the equations

$$u^2 + v^2 + w^2 = 29, \qquad \frac{u^2}{x^2} + \frac{v^2}{y^2} + \frac{w^2}{z^2} = 17$$

both hold for all (x, y, z, w) in B.

Proof. Set $n = 2$, $p = 4$, and

$$\mathbf{F}(u, v, x, y, z, w) = (u^2 + v^2 + w^2 - 29, u^2/x^2 + v^2/y^2 + w^2/z^2 - 17).$$

Then $\mathbf{F}(4, 3, 2, 1, -1, -2) = (0, 0)$, and

$$\frac{\partial(F_1, F_2)}{\partial(u, v)} = \det \begin{bmatrix} 2u & 2v \\ 2u/x^2 & 2v/y^2 \end{bmatrix} = 4uv \left(\frac{1}{y^2} - \frac{1}{x^2} \right).$$

This determinant is nonzero when $u = 4$, $v = 3$, $x = 2$, and $y = 1$. Therefore, such functions u, v exist by the Implicit Function Theorem.   ∎

EXERCISES

11.6.1. For each of the following functions, prove that $\mathbf{f}^{-1}$ exists and is differentiable in some nonempty, open set containing (a, b), and compute $D(\mathbf{f}^{-1})(a, b)$.

a) $\mathbf{f}(u, v) = (3u - v, 2u + 5v)$ at any $(a, b) \in \mathbf{R}^2$
b) $\mathbf{f}(u, v) = (u + v, \sin u + \cos v)$ at $(a, b) = (0, 1)$
c) $\mathbf{f}(u, v) = (uv, u^2 + v^2)$ at $(a, b) = (2, 5)$
d) $\mathbf{f}(u, v) = (u^3 - v^2, \sin u - \log v)$ at $(a, b) = (-1, 0)$

11.6.2. For each of the following functions, find out whether the given expression can be solved for z in a nonempty, open set V containing $(0, 0, 0)$. Is the solution differentiable near $(0, 0)$?

a) $xyz + \sin(x + y + z) = 0$
b) $x^2 + y^2 + z^2 + \sqrt{\sin(x^2 + y^2) + 3z + 4} = 2$
c) $xyz(2 \cos y - \cos z) + (z \cos x - x \cos y) = 0$
d) $x + y + z + g(x, y, z) = 0$, where g is any continuously differentiable function which satisfies $g(0, 0, 0) = 0$ and $g_z(0, 0, 0) > 0$

11.6.3. Prove that there exist functions $u(x, y)$, $v(x, y)$, and $w(x, y)$, and an $r > 0$ such that u, v, w are continuously differentiable and satisfy the equations

$$u^5 + xv^2 - y + w = 0$$
$$v^5 + yu^2 - x + w = 0$$
$$w^4 + y^5 - x^4 = 1$$

on $B_r(1, 1)$, and $u(1, 1) = 1$, $v(1, 1) = 1$, $w(1, 1) = -1$.

11.6.4. Find conditions on a point (x_0, y_0, u_0, v_0) such that there exist real-valued functions $u(x, y)$ and $v(x, y)$ which are continuously differentiable near (x_0, y_0) and satisfy the simultaneous equations

$$xu^2 + yv^2 + xy = 9$$
$$xv^2 + yu^2 - xy = 7.$$

Prove that the solutions satisfy $u^2 + v^2 = 16/(x + y)$.

11.6.5. Given nonzero numbers $x_0, y_0, u_0, v_0, s_0, t_0$ which satisfy the simultaneous equations

(*)
$$\begin{aligned}
u^2 + sx + ty &= 0 \\
v^2 + tx + sy &= 0 \\
2s^2 x + 2t^2 y - 1 &= 0 \\
s^2 x - t^2 y &= 0,
\end{aligned}$$

prove that there exist functions $u(x, y)$, $v(x, y)$, $s(x, y)$, $t(x, y)$, and an open ball B containing (x_0, y_0), such that u,v,s,t are continuously differentiable and satisfy (*) on B, and such that $u(x_0, y_0) = u_0$, $v(x_0, y_0) = v_0$, $s(x_0, y_0) = s_0$, and $t(x_0, y_0) = t_0$.

11.6.6. Let $E = \{(x, y) : 0 < y < x\}$ and set $\mathbf{f}(x, y) = (x+y, xy)$ for $(x, y) \in E$.

   a) Prove that $\mathbf{f}$ is 1–1 from E onto $\{(s, t) : s > 2\sqrt{t}, t > 0\}$ and find a formula for $\mathbf{f}^{-1}(s, t)$.

   b) Use the Inverse Function Theorem to compute $D(\mathbf{f}^{-1})(\mathbf{f}(x, y))$ for $(x, y) \in E$.

   c) Use the formula you obtained in part a) to compute $D(\mathbf{f}^{-1})(s, t)$ directly. Check to see that this answer agrees with the one you found in part b).

11.6.7. Suppose that V is open in $\mathbf{R}^n$, that $\mathbf{a} \in V$, and that $F : V \to \mathbf{R}$ is C^1 on V. If $F(\mathbf{a}) = 0 \neq F_{x_j}(\mathbf{a})$ and $\mathbf{u}^{(j)} := (x_1, \ldots, x_{j-1}, x_{j+1}, \ldots, x_n)$ for $j = 1, 2, \ldots, n$, prove that there exist open sets W_j containing $(a_1, \ldots, a_{j-1}, a_{j+1}, \ldots, a_n)$, an $r > 0$, and functions $g_j(\mathbf{u}^{(j)})$, C^1 on W_j, such that $F(x_1, \ldots, x_{j-1}, g_j(\mathbf{u}^{(j)}), x_{j+1}, \ldots, x_n) = 0$ on W_j and

$$\frac{\partial g_1}{\partial x_n} \frac{\partial g_2}{\partial x_1} \frac{\partial g_3}{\partial x_2} \cdots \frac{\partial g_n}{\partial x_{n-1}} = (-1)^n$$

on $B_r(\mathbf{a})$.

11.6.8. Suppose that $\mathbf{f} : \mathbf{R}^2 \to \mathbf{R}^2$ has continuous first-order partial derivatives in some ball $B_r(x_0, y_0)$, $r > 0$. Prove that if $\Delta_{\mathbf{f}}(x_0, y_0) \neq 0$, then

$$\frac{\partial f_1^{-1}}{\partial x}(f(x_0, y_0)) = \frac{\partial f_2/\partial y(x_0, y_0)}{\Delta_{\mathbf{f}}(x_0, y_0)}, \quad \frac{\partial f_1^{-1}}{\partial y}(f(x_0, y_0)) = \frac{-\partial f_1/\partial y(x_0, y_0)}{\Delta_{\mathbf{f}}(x_0, y_0)},$$

and

$$\frac{\partial f_2^{-1}}{\partial x}(f(x_0, y_0)) = \frac{-\partial f_2/\partial x(x_0, y_0)}{\Delta_{\mathbf{f}}(x_0, y_0)}, \quad \frac{\partial f_2^{-1}}{\partial y}(f(x_0, y_0)) = \frac{\partial f_1/\partial x(x_0, y_0)}{\Delta_{\mathbf{f}}(x_0, y_0)}.$$

11.6.9 . **This exercise is used in Section *11.7.** Let $F : \mathbf{R}^3 \to \mathbf{R}$ be continuously differentiable in some open set containing (a, b, c) with $F(a, b, c) = 0$ and $\nabla F(a, b, c) \neq 0$.

a) Prove that the graph of the relation $F(x, y, z) = 0$; that is, that the set $\mathcal{G} := \{(x, y, z) : F(x, y, z) = 0\}$ has a tangent plane at (a, b, c).
b) Prove that a normal of the tangent plane to $\mathcal{G}$ at (a, b, c) is given by $\nabla F(a, b, c)$.

11.6.10. Suppose that $\mathbf{f} := (u, v) : \mathbf{R} \to \mathbf{R}^2$ is C^2 and that $(x_0, y_0) = \mathbf{f}(t_0)$.

a) Prove that if $\mathbf{f}'(t_0) \neq \mathbf{0}$, then $u'(t_0)$ and $v'(t_0)$ cannot both be zero.
b) If $\mathbf{f}'(t_0) \neq \mathbf{0}$, show that either there is a C^1 function g such that $g(x_0) = t_0$ and $u(g(x)) = x$ for x near x_0, or there is a C^1 function h such that $h(y_0) = t_0$ and $v(h(y)) = y$ for y near y_0.

11.6.11. Let $\mathcal{H}$ be the hyperboloid of one sheet, given by $x^2 + y^2 - z^2 = 1$.

a) Use Exercise 11.6.9 to prove that at every point $(a, b, c) \in \mathcal{H}$, $\mathcal{H}$ has a tangent plane whose normal is given by $(-a, -b, c)$.
b) Find an equation of each plane tangent to $\mathcal{H}$ which is perpendicular to the xy-plane.
c) Find an equation of each plane tangent to $\mathcal{H}$ which is parallel to the plane $x + y - z = 1$.

*11.7 OPTIMIZATION

This section uses no material from any other enrichment section.
    In this section we discuss how to find extreme values of differentiable functions of several variables.

11.50 Definition.

Let V be open in $\mathbf{R}^n$, let $\mathbf{a} \in V$, and suppose that $f : V \to \mathbf{R}$.

i) $f(\mathbf{a})$ is called a *local minimum* of f if and only if there is an $r > 0$ such that $f(\mathbf{a}) \leq f(\mathbf{x})$ for all $\mathbf{x} \in B_r(\mathbf{a})$.
ii) $f(\mathbf{a})$ is called a *local maximum* of f if and only if there is an $r > 0$ such that $f(\mathbf{a}) \geq f(\mathbf{x})$ for all $\mathbf{x} \in B_r(\mathbf{a})$.
iii) $f(\mathbf{a})$ is called a *local extremum* of f if and only if $f(\mathbf{a})$ is a local maximum or a local minimum of f.

The following result shows that, as in the one-dimensional case, extrema of real-valued differentiable functions occur among points where the "derivative" is zero.

11.51 Remark. *If the first-order partial derivatives of f exist at $\mathbf{a}$, and $f(\mathbf{a})$ is a local extremum of f, then $\nabla f(\mathbf{a}) = \mathbf{0}$.*

Proof. The one-dimensional function $g(t) = f(a_1, \ldots, a_{j-1}, t, a_{j+1}, \ldots, a_n)$ has a local extremum at $t = a_j$ for each $j = 1, \ldots, n$. Hence, by the one-dimensional theory,

$$\frac{\partial f}{\partial x_j}(\mathbf{a}) = g'(a_j) = 0.$$ ∎

As in the one-dimensional case, $\nabla f(\mathbf{a}) = \mathbf{0}$ is necessary but not sufficient for $f(\mathbf{a})$ to be a local extremum.

11.52 Remark. *There exist continuously differentiable functions which satisfy* $\nabla f(\mathbf{a}) = \mathbf{0}$ *such that* $f(\mathbf{a})$ *is neither a local maximum nor a local minimum.*

Proof. Consider

$$f(x, y) = y^2 - x^2.$$

Since the first-order partial derivatives of f exist and are continuous every-where on $\mathbf{R}^2$, f is continuously differentiable on $\mathbf{R}^2$. Moreover, it is evident that $\nabla f(\mathbf{0}) = \mathbf{0}$, but $f(\mathbf{0})$ is not a local extremum (see Figure 11.6). ∎

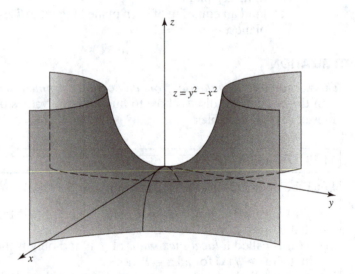

$$z = y^2 - x^2$$

FIGURE 11.6

The fact that the graph of this function resembles a saddle motivates the fol-lowing terminology.

11.53 Definition.

Let V be open in $\mathbf{R}^n$, let $\mathbf{a} \in V$, and let $f : V \to \mathbf{R}$ be differentiable at $\mathbf{a}$. Then $\mathbf{a}$ is called a *saddle point* of f if $\nabla f(\mathbf{a}) = \mathbf{0}$ and there is a $r_0 > 0$ such that given any $0 < \rho < r_0$ there are points $\mathbf{x}, \mathbf{y} \in B_\rho(\mathbf{a})$ which satisfy $f(\mathbf{x}) < f(\mathbf{a}) < f(\mathbf{y})$.

By the Extreme Value Theorem, if f is continuous on a compact set H, then it *attains* its maximum and minimum on H; that is, there exist points $\mathbf{a}, \mathbf{b} \in H$ such that

$$f(\mathbf{a}) = \sup_{\mathbf{x} \in H} f(\mathbf{x}) \quad \text{and} \quad f(\mathbf{b}) = \inf_{\mathbf{x} \in H} f(\mathbf{x}).$$

When f is a function of two variables, these points can sometimes be found by combining Remark 11.51 with one-dimensional techniques.

11.54 EXAMPLE.

Find the maximum and minimum of $f(x, y) = x^2 - x + y^2 - 2y$ on $H = \overline{B_2(0, 0)}$.

Solution. If $\nabla f(x, y) = (0, 0)$, then $(x, y) = (1/2, 1)$. Since this point belongs to H, it might be an extremum of f on H. Set it aside. (If it did not belong to H, we would discard it.)

By polar coordinates, $\partial H = \{(x, y) = (2\cos\theta, 2\sin\theta) : \theta \in [0, 2\pi]\}$ is essentially a one-dimensional set. Thus to find the extrema of f on ∂H, we must optimize $h(\theta) := f(2\cos\theta, 2\sin\theta) = 2(2 - \cos\theta - 2\sin\theta)$ on $[0, 2\pi]$. Since $h'(\theta) = 0$ implies $\tan\theta = 2$, the critical points of h are $\theta = \arctan 2 \approx 1.10715$ and $\theta = \arctan 2 + \pi \approx 4.24874$. This provides two more candidates for extrema of f on H : $(x, y) \approx (0.4472, 0.8944)$ and $(x, y) \approx (-0.4472, -0.8944)$. Finally, the endpoints of $[0, 2\pi]$ provide a fourth candidate: $(x, y) = (2, 0)$.

Evaluating f at these four points, we see that the maximum of f on H is $f(-0.4472, -0.8944) \approx 3.236$ and the minimum of f on H is $f(1/2, 1) = -1.25$. [The values $f(0.4472, 0.8944) \approx -1.236$ and $f(2, 0) = 2$ are neither maxima nor minima of f on H.] ∎

Using the second-order total differential $D^{(2)}f$ introduced in Section 11.5, we can obtain a multidimensional analogue of the Second Derivative Test. First, we prove a technical result.

11.55 Lemma.
Let V be open in $\mathbf{R}^n$, $\mathbf{a} \in V$, and $f : V \to \mathbf{R}$. If all second-order partial derivatives of f exist at $\mathbf{a}$ and $D^{(2)}f(\mathbf{a}; \mathbf{h}) > 0$ for all $\mathbf{h} \neq \mathbf{0}$, then there is an $m > 0$ such that

$$D^{(2)}f(\mathbf{a}; \mathbf{x}) \geq m\|\mathbf{x}\|^2 \tag{33}$$

for all $\mathbf{x} \in \mathbf{R}^n$.

Proof. Set $H = \{\mathbf{x} \in \mathbf{R}^n : \|\mathbf{x}\| = 1\}$ and consider the function

$$g(\mathbf{x}) := D^{(2)}f(\mathbf{a}; \mathbf{x}) := \sum_{j=1}^{n}\sum_{k=1}^{n} \frac{\partial^2 f}{\partial x_k \, \partial x_j}(\mathbf{a}) \, x_j x_k, \qquad \mathbf{x} \in \mathbf{R}^n.$$

By hypothesis, g is continuous and positive on $\mathbf{R}^n \setminus \{\mathbf{0}\}$ and, hence, on H. Since H is compact, it follows from the Extreme Value Theorem that g has a positive minimum m on H.

Clearly, (33) holds for $\mathbf{x} = \mathbf{0}$. If $\mathbf{x} \neq \mathbf{0}$, then $\mathbf{x}/\|\mathbf{x}\| \in H$, and it follows from the choice of g and m that

$$D^{(2)} f(\mathbf{a}; \mathbf{x}) = \frac{g(\mathbf{x})}{\|\mathbf{x}\|^2} \|\mathbf{x}\|^2 = g\left(\frac{\mathbf{x}}{\|\mathbf{x}\|}\right) \|\mathbf{x}\|^2 \geq m\|\mathbf{x}\|^2.$$

We conclude that (33) holds for all $\mathbf{x} \in \mathbf{R}^n$. ∎

11.56 Theorem. [THE SECOND DERIVATIVE TEST].
Let V be open in $\mathbf{R}^n$, $\mathbf{a} \in V$, and suppose that $f : V \to \mathbf{R}$ satisfies $\nabla f(\mathbf{a}) = \mathbf{0}$. Suppose further that the second-order total differential of f exists on V and is continuous at $\mathbf{a}$.

i) *If $D^{(2)} f(\mathbf{a}; \mathbf{h}) > 0$ for all $\mathbf{h} \neq \mathbf{0}$, then $f(\mathbf{a})$ is a local minimum of f.*
ii) *If $D^{(2)} f(\mathbf{a}; \mathbf{h}) < 0$ for all $\mathbf{h} \neq \mathbf{0}$, then $f(\mathbf{a})$ is a local maximum of f.*
iii) *If $D^{(2)} f(\mathbf{a}; \mathbf{h})$ takes on both positive and negative values for $\mathbf{h} \in \mathbf{R}^n$, then $\mathbf{a}$ is a saddle point of f.*

Proof. Choose $r > 0$ such that $B_r(\mathbf{a}) \subset V$, and suppose for a moment that there is a function $\varepsilon : B_r(\mathbf{0}) \to \mathbf{R}$ such that $\varepsilon(\mathbf{h}) \to 0$ as $\mathbf{h} \to \mathbf{0}$ and

$$f(\mathbf{a} + \mathbf{h}) - f(\mathbf{a}) = \frac{1}{2} D^{(2)} f(\mathbf{a}; \mathbf{h}) + \|\mathbf{h}\|^2 \varepsilon(\mathbf{h}) \tag{34}$$

for $\|\mathbf{h}\|$ sufficiently small. If $D^{(2)} f(\mathbf{a}; \mathbf{h}) > 0$ for $\mathbf{h} \neq \mathbf{0}$, then (33) and (34) imply

$$f(\mathbf{a} + \mathbf{h}) - f(\mathbf{a}) \geq \left(\frac{m}{2} + \varepsilon(\mathbf{h})\right) \|\mathbf{h}\|^2$$

for $\|\mathbf{h}\|$ sufficiently small. Since $m > 0$ and $\varepsilon(\mathbf{h}) \to 0$ as $\mathbf{h} \to \mathbf{0}$, it follows that $f(\mathbf{a} + \mathbf{h}) - f(\mathbf{a}) > 0$ for $\|\mathbf{h}\|$ sufficiently small; that is, $f(\mathbf{a})$ is a local minimum. Similarly, if $D^{(2)} f(\mathbf{a}; \mathbf{h}) < 0$ for $\mathbf{h} \neq \mathbf{0}$, then $f(\mathbf{a})$ is a local maximum. This proves parts i) and ii).

To prove part iii), fix $\mathbf{h} \in \mathbf{R}^n$ and notice that (34) implies

$$f(\mathbf{a} + t\mathbf{h}) - f(\mathbf{a}) = t^2 \left(\frac{1}{2} D^{(2)} f(\mathbf{a}; \mathbf{h}) + \|\mathbf{h}\|^2 \varepsilon(t\mathbf{h})\right)$$

for $t \in \mathbf{R}$. Since $\varepsilon(t\mathbf{h}) \to 0$ as $t \to 0$, it follows that $f(\mathbf{a} + t\mathbf{h}) - f(\mathbf{a})$ takes on the same sign as $D^{(2)} f(\mathbf{a}; \mathbf{h})$ for t small. In particular, if $D^{(2)} f(\mathbf{a}; \mathbf{h})$ takes on both positive and negative values as $\mathbf{h}$ varies, then $\mathbf{a}$ is a saddle point.

It remains to find a function $\varepsilon : B_r(\mathbf{0}) \to \mathbf{R}$ such that $\varepsilon(\mathbf{h}) \to 0$ as $\mathbf{h} \to \mathbf{0}$, and (34) holds for all $\|\mathbf{h}\|$ sufficiently small. Set $\varepsilon(\mathbf{0}) = 0$ and

$$\varepsilon(\mathbf{h}) = \frac{f(\mathbf{a} + \mathbf{h}) - f(\mathbf{a}) - \frac{1}{2} D^{(2)} f(\mathbf{a}; \mathbf{h})}{\|\mathbf{h}\|^2}, \qquad \mathbf{h} \in B_r(\mathbf{0}), \ \mathbf{h} \neq \mathbf{0}.$$

By the definition of $\varepsilon(\mathbf{h})$, (34) holds for $\mathbf{h} \in B_r(\mathbf{0})$. Does $\varepsilon(\mathbf{h}) \to 0$ as $\mathbf{h} \to \mathbf{0}$? Fix $\mathbf{h} = (h_1, h_2, \ldots, h_n) \in B_r(\mathbf{0})$. Since $\nabla f(\mathbf{a}) = \mathbf{0}$, Taylor's Formula implies

$$f(\mathbf{a} + \mathbf{h}) - f(\mathbf{a}) = \frac{1}{2} D^{(2)} f(\mathbf{c}; \mathbf{h})$$

for some $\mathbf{c} \in L(\mathbf{a}; \mathbf{a} + \mathbf{h})$; that is,

$$f(\mathbf{a} + \mathbf{h}) - f(\mathbf{a}) - \frac{1}{2} D^{(2)} f(\mathbf{a}; \mathbf{h}) = \frac{1}{2} \left(D^{(2)} f(\mathbf{c}; \mathbf{h}) - D^{(2)} f(\mathbf{a}; \mathbf{h}) \right)$$
$$= \frac{1}{2} \sum_{j=1}^{n} \sum_{k=1}^{n} \left(\frac{\partial^2 f}{\partial x_j \, \partial x_k}(\mathbf{c}) - \frac{\partial^2 f}{\partial x_j \, \partial x_k}(\mathbf{a}) \right) h_j h_k.$$

Since $|h_j h_k| \le \|\mathbf{h}\|^2$ and the second-order partial derivatives of f are continuous at $\mathbf{a}$, it follows that

$$0 \le |\varepsilon(\mathbf{h})| \le \frac{1}{2} \left(\sum_{j=1}^{n} \sum_{k=1}^{n} \left| \frac{\partial^2 f}{\partial x_j \, \partial x_k}(\mathbf{c}) - \frac{\partial^2 f}{\partial x_j \, \partial x_k}(\mathbf{a}) \right| \right) \to 0$$

as $\mathbf{h} \to \mathbf{0}$. We conclude by the Squeeze Theorem that $\varepsilon(\mathbf{h}) \to 0$ as $\mathbf{h} \to \mathbf{0}$. ∎

The following result shows that the strict inequalities in Theorem 11.56 cannot be relaxed.

11.57 Remark. *If* $D^{(2)} f(\mathbf{a}; \mathbf{h}) \ge 0$, *then* $f(\mathbf{a})$ *can be a local minimum or* $\mathbf{a}$ *can be a saddle point.*

Proof. $f(0, 0)$ is a local minimum of $f(x, y) = x^4 + y^2$, and $(0, 0)$ is a saddle point of $f(x, y) = x^3 + y^2$. ∎

In practice, it is not easy to determine the sign of $D^{(2)} f(\mathbf{a}; \mathbf{h})$. For the case $n = 2$, the second total differential $D^{(2)} f(\mathbf{a}; \mathbf{h})$ is a *quadratic form* (i.e., has the form $Ah^2 + 2Bhk + Ck^2$). The following result shows that the sign of a quadratic form is determined completely by the *discriminant* $D = AC - B^2$.

11.58 Lemma.
Let $A, B, C \in \mathbf{R}$, $D = AC - B^2$, and $\phi(h, k) = Ah^2 + 2Bhk + Ck^2$.

i) *If* $D > 0$, *then* A *and* $\phi(h, k)$ *have the same sign for all* $(h, k) \neq (0, 0)$.
ii) *If* $D < 0$, *then* $\phi(h, k)$ *takes on both positive and negative values as* (h, k) *varies over* $\mathbf{R}^2$.

Proof.

i) Suppose that $D > 0$. Then $A \neq 0$ and $A\phi(h, k)$ is a sum of two squares:

$$A\phi(h, k) = A^2 h^2 + 2ABhk + ACk^2 = (Ah + Bk)^2 + Dk^2.$$

Since $A \neq 0 \neq D$, at least one of these squares is positive for each $(h, k) \neq (0, 0)$. It follows that A and $\phi(h, k)$ have the same sign for all $(h, k) \neq (0, 0)$.

ii) Suppose that $D < 0$. Then either $A \neq 0$ or $B \neq 0$.

If $A \neq 0$, then $A\phi(h, k)$ is a difference of two squares:

$$A\phi(h, k) = (Ah + Bk - \sqrt{|D|}\, k)(Ah + Bk + \sqrt{|D|}\, k).$$

The lines $Ah + Bk - \sqrt{|D|}k = 0$ and $Ah + Bk + \sqrt{|D|}k = 0$ divide the hk-plane into four open regions (see Figure 11.7). Since $A\phi(h, k)$ is positive on two of these regions and negative on the other two, it follows that $\phi(h, k)$ takes on both positive and negative values as (h, k) varies over $\mathbf{R}^2$.

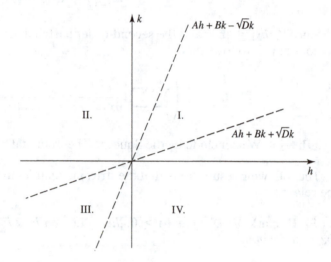

FIGURE 11.7

If $A = 0$ and $B \neq 0$, then

$$\phi(h, k) = 2Bhk + Ck^2 = (2Bh + Ck)k.$$

Since $B \neq 0$, the lines $2Bh + Ck = 0$ and $k = 0$ divide the hk-plane into four open regions. As before, $\phi(h, k)$ takes on both positive and negative values as (h, k) varies over $\mathbf{R}^2$. ∎

This result leads us to the following simple test for extrema and saddle points.

11.59 Theorem. *Let V be open in $\mathbf{R}^2$, $(a, b) \in V$, and suppose that $f : V \to \mathbf{R}$ satisfies $\nabla f(a, b) = \mathbf{0}$. Suppose further that the second-order total differential of f exists on V and is continuous at (a, b), and set*

$$D = f_{xx}(a, b) f_{yy}(a, b) - f_{xy}^2(a, b).$$

i)   *If $D > 0$ and $f_{xx}(a, b) > 0$, then $f(a, b)$ is a local minimum.*
ii) *If $D > 0$ and $f_{xx}(a, b) < 0$, then $f(a, b)$ is a local maximum.*
iii) *If $D < 0$, then (a, b) is a saddle point.*

Proof. Set $A = f_{xx}(a, b)$, $B = f_{xy}(a, b)$, and $C = f_{yy}(a, b)$. Apply Theorem 11.56 and Lemma 11.58.   ∎

(For a discriminant which works for functions on $\mathbf{R}^n$, see Colley [3], p. 250.)

11.60 Remark. *If the discriminant $D = 0$, $f(a, b)$ may be a local maximum, a local minimum, or (a, b) may be a saddle point.*

Proof. The function $f(x, y) = x^2$ has zero discriminant at $(a, b) = (0, 0)$, and $0 = f(0, 0)$ is a local minimum for f. On the other hand, $f(x, y) = x^3$ has zero discriminant at $(a, b) = (0, 0)$, and $(0, 0)$ is a saddle point for f.   ∎

In practice, we often wish to optimize a function subject to certain constraints. (For example, we do not simply want to build the cheapest shipping container, but the cheapest shipping container which will fit in a standard railway car and will not fall apart after several trips.)

11.61 Definition.

Let V be open in $\mathbf{R}^n$, $\mathbf{a} \in V$, and $f, g_j : V \to \mathbf{R}$ for $j = 1, 2, \ldots, m$.

i) $f(\mathbf{a})$ is called a *local minimum of f subject to the constraints* $g_j(\mathbf{a}) = 0$, $j = 1, \ldots, m$, if and only if there is a $\rho > 0$ such that $\mathbf{x} \in B_\rho(\mathbf{a})$ and $g_j(\mathbf{x}) = 0$ for all $j = 1, \ldots, m$ imply $f(\mathbf{x}) \geq f(\mathbf{a})$.
ii) $f(\mathbf{a})$ is called a *local maximum of f subject to the constraints* $g_j(\mathbf{a}) = 0$, $j = 1, \ldots, m$, if and only if there is a $\rho > 0$ such that $\mathbf{x} \in B_\rho(\mathbf{a})$ and $g_j(\mathbf{x}) = 0$ for all $j = 1, \ldots, m$ imply $f(\mathbf{x}) \leq f(\mathbf{a})$.

The following example serves two purposes: to illustrate Definition 11.61 and to motivate Theorem 11.63.

*11.62 EXAMPLE.

Find all points on the ellipsoid $x^2 + 2y^2 + 3z^2 = 1$ (see Appendix D) which lie closest to or farthest from the origin.

Solution. We must optimize the distance formula $\sqrt{x^2 + y^2 + z^2}$; equivalently, we must optimize the function $f(x, y, z) = x^2 + y^2 + z^2$ subject to the constraint $g(x, y, z) = x^2 + 2y^2 + 3z^2 - 1 = 0$. Using g to eliminate the variable x in f, we see that f takes on the form

$$\phi(y, z) = 1 - y^2 - 2z^2.$$

Solving $\nabla\phi(y,z) = (0,0)$, we obtain $(y,z) = (0,0)$ (i.e., $x^2 = 1$). Thus, elimination of x leads to the points $(\pm 1, 0, 0)$. Similarly, elimination of y leads to $(0, \pm 1/\sqrt{2}, 0)$, and elimination of z leads to $(0, 0, \pm 1/\sqrt{3})$. Checking the distance formula, we see that the maximum distance is 1, which occurs at the points $(\pm 1, 0, 0)$, and the minimum distance is $1/\sqrt{3}$, which occurs at the points $(0, 0, \pm 1/\sqrt{3})$. (The points $(0, \pm 1/\sqrt{2}, 0)$ are saddle points, that is, correspond neither to a maximum nor to a minimum.) ∎

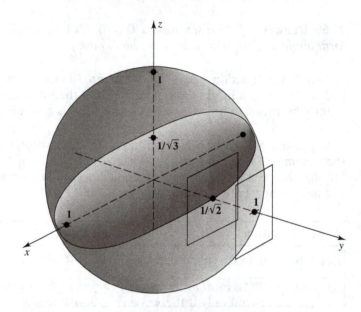

FIGURE 11.8

Optimizing a function subject to constraints, as above, by eliminating one or more of the variables is called the *direct method*. There is another, more geometric, method for solving Example 11.62. Notice that the points on the ellipsoid $g(x, y, z) = x^2 + 2y^2 + 3z^2 - 1 = 0$ which are closest to and farthest from the origin occur at points where the tangent planes of the ellipsoid $g(x, y, z) = 0$ and the sphere $f(x, y, z) = 1$ are parallel (see Figure 11.8). Recall that two nonzero vectors $\mathbf{a}$ and $\mathbf{b}$ are parallel if and only if $\mathbf{a} + \lambda\mathbf{b} = \mathbf{0}$ for some scalar $\lambda \neq 0$. Since normal vectors of the tangent planes of $f(x, y, z) = 1$ and $g(x, y, z) = 0$ are ∇f and ∇g (see Exercise 11.6.9b), it follows that extremal points (x, y, z) of $f(x, y, z)$ subject to the constraints $g(x, y, z) = 0$ must satisfy

$$\nabla f(x, y, z) + \lambda \nabla g(x, y, z) = \mathbf{0} \tag{35}$$

for some $\lambda \neq 0$. For the case at hand, (35) implies $(2x, 2y, 2z) + \lambda(2x, 4y, 6z) = (0, 0, 0)$. Combining this equation with the constraint $g(x, y, z) = 0$, we have four equations in four unknowns:

$$x(\lambda + 1), \qquad y(2\lambda + 1) = 0, \qquad z(3\lambda + 1) = 0, \quad \text{and} \quad x^2 + 2y^2 + 3z^2 = 1.$$

Solving these equations, we obtain three pairs of solutions: $(\pm 1, 0, 0)$ (when $\lambda = -1$), $(0, \pm 1/\sqrt{2}, 0)$ (when $\lambda = -1/2$), and $(0, 0, \pm 1/\sqrt{3})$ (when $\lambda = -1/3$). Hence, we obtain the same solutions with the geometric method as we did with the direct method.

The following result shows that the geometric method is valid when the functions have nothing to do with spheres and ellipsoids, even when several constraints are used. This is fortunate since the direct method cannot be used unless the constraints are relatively simple.

11.63 Theorem.   [LAGRANGE MULTIPLIERS].
Let $m < n$, V be open in $\mathbf{R}^n$, and $f, g_j : V \to \mathbf{R}$ be C^1 on V for $j = 1, 2, \ldots, m$. Suppose that there is an $\mathbf{a} \in V$ such that

$$\frac{\partial(g_1, \ldots, g_m)}{\partial(x_1, \ldots, x_m)}(\mathbf{a}) \neq 0.$$

If $f(\mathbf{a})$ is a local extremum of f subject to the constraints $g_k(\mathbf{a}) = 0$, $k = 1, \ldots, m$, then there exist scalars $\lambda_1, \lambda_2, \ldots, \lambda_m$ such that

$$\nabla f(\mathbf{a}) + \sum_{k=1}^{m} \lambda_k \nabla g_k(\mathbf{a}) = \mathbf{0}. \tag{36}$$

Proof. Equation (36) is a system of n equations in m unknowns, $\lambda_1, \lambda_2, \ldots, \lambda_m$:

$$\sum_{k=1}^{m} \lambda_k \frac{\partial g_k}{\partial x_j}(\mathbf{a}) = -\frac{\partial f}{\partial x_j}(\mathbf{a}), \qquad j = 1, 2, \ldots, n. \tag{37}$$

The first m of these equations forms a system of m linear equations in m variables whose matrix of coefficients has a nonzero determinant and, hence, this system uniquely determines the λ_k's. What remains to be seen is that because $f(\mathbf{a})$ is a local extremum subject to the constraints $g_k(\mathbf{a}) = 0$, these same λ_k's also satisfy (37) for $j = m + 1, \ldots, n$. This is a question about implicit functions.

Let $p = n - m$. As in the proof of the Implicit Function Theorem, write vectors in $\mathbf{R}^{m+p}$ in the form $\mathbf{x} = (\mathbf{y}, \mathbf{t}) = (y_1, \ldots, y_m, t_1, \ldots, t_p)$. We must show that

$$0 = \frac{\partial f}{\partial t_\ell}(\mathbf{a}) + \sum_{k=1}^{m} \lambda_k \frac{\partial g_k}{\partial t_\ell}(\mathbf{a}) \tag{38}$$

for $\ell = 1, \ldots, p$.

Let $\mathbf{g} = (g_1, \ldots, g_m)$, and choose $\mathbf{y}_0 \in \mathbf{R}^m$, $\mathbf{t}_0 \in \mathbf{R}^p$ such that $\mathbf{a} = (\mathbf{y}_0, \mathbf{t}_0)$. By hypothesis, $\mathbf{g}(\mathbf{y}_0, \mathbf{t}_0) = \mathbf{0}$ and the partial Jacobian of $\mathbf{g}$ (with respect to the variables y_j) is nonzero at $(\mathbf{y}_0, \mathbf{t}_0)$. Hence, by the Implicit Function Theorem, there is an open set $W \subset \mathbf{R}^p$ which contains $\mathbf{t}_0$, and a function $\mathbf{h} : W \to \mathbf{R}^m$ such that $\mathbf{h}$ is continuously differentiable on W, $\mathbf{h}(\mathbf{t}_0) = \mathbf{y}_0$, and

$$\mathbf{g}(\mathbf{h}(\mathbf{t}), \mathbf{t}) = \mathbf{0}, \qquad \mathbf{t} \in W. \tag{39}$$

For each $\mathbf{t} \in W$ and $k = 1, \ldots, m$, set

$$G_k(\mathbf{t}) = g_k(\mathbf{h}(\mathbf{t}), \mathbf{t}) \quad \text{and} \quad \mathbf{F}(\mathbf{t}) = \mathbf{f}(\mathbf{h}(\mathbf{t}), \mathbf{t}).$$

We shall use the functions $G_1, \ldots, G_m$ and $\mathbf{F}$ to verify (38) for $\ell = 1, \ldots, p$. Fix such an ℓ. By (39), each G_k is identically zero on W and, hence, has derivative zero there. Since $\mathbf{t}_0 \in W$ and $(\mathbf{h}(\mathbf{t}_0), \mathbf{t}_0) = (\mathbf{y}_0, \mathbf{t}_0) = \mathbf{a}$, it follows from the Chain Rule that

$$O = DG_k(\mathbf{t}_0) = \begin{bmatrix} \dfrac{\partial g_k}{\partial x_1}(\mathbf{a}) & \cdots & \dfrac{\partial g_k}{\partial x_n}(\mathbf{a}) \end{bmatrix} \begin{bmatrix} \dfrac{\partial h_1}{\partial t_1}(\mathbf{t}_0) & \cdots & \dfrac{\partial h_1}{\partial t_p}(\mathbf{t}_0) \\ \vdots & \ddots & \vdots \\ \dfrac{\partial h_m}{\partial t_1}(\mathbf{t}_0) & \cdots & \dfrac{\partial h_m}{\partial t_p}(\mathbf{t}_0) \\ 1 & \cdots & 0 \\ \vdots & \ddots & \vdots \\ 0 & \cdots & 1 \end{bmatrix}.$$

Hence, the ℓth component of $DG_k(\mathbf{t}_0)$ is given by

$$0 = \sum_{j=1}^{m} \frac{\partial g_k}{\partial x_j}(\mathbf{a}) \frac{\partial h_j}{\partial t_\ell}(\mathbf{t}_0) + \frac{\partial g_k}{\partial t_\ell}(\mathbf{a}) \tag{40}$$

for $k = 1, 2, \ldots, m$. Multiplying (40) by λ_k and adding, we obtain

$$0 = \sum_{k=1}^{m} \sum_{j=1}^{m} \lambda_k \frac{\partial g_k}{\partial x_j}(\mathbf{a}) \frac{\partial h_j}{\partial t_\ell}(\mathbf{t}_0) + \sum_{k=1}^{m} \lambda_k \frac{\partial g_k}{\partial t_\ell}(\mathbf{a})$$

$$= \sum_{j=1}^{m} \left(\sum_{k=1}^{m} \lambda_k \frac{\partial g_k}{\partial x_j}(\mathbf{a}) \right) \frac{\partial h_j}{\partial t_\ell}(\mathbf{t}_0) + \sum_{k=1}^{m} \lambda_k \frac{\partial g_k}{\partial t_\ell}(\mathbf{a}).$$

Hence, it follows from (37) that

$$0 = -\sum_{j=1}^{m} \frac{\partial f}{\partial x_j}(\mathbf{a}) \frac{\partial h_j}{\partial t_\ell}(\mathbf{t}_0) + \sum_{k=1}^{m} \lambda_k \frac{\partial g_k}{\partial t_\ell}(\mathbf{a}). \tag{41}$$

Suppose that $f(\mathbf{a})$ is a local maximum subject to the constraints $\mathbf{g}(\mathbf{a}) = \mathbf{0}$. Set $E_0 = \{\mathbf{x} \in V : \mathbf{g}(\mathbf{x}) = \mathbf{0}\}$, and choose an n-dimensional open ball $B(\mathbf{a})$ such that

$$\mathbf{x} \in B(\mathbf{a}) \cap E_0 \quad \text{implies} \quad f(\mathbf{x}) \le f(\mathbf{a}). \tag{42}$$

Since $\mathbf{h}$ is continuous, choose a p-dimensional open ball $B(\mathbf{t}_0)$ such that $\mathbf{t} \in B(\mathbf{t}_0)$ implies $(\mathbf{h}(\mathbf{t}), \mathbf{t}) \in B(\mathbf{a})$. By (42), $F(\mathbf{t}_0)$ is a local maximum of F on $B(\mathbf{t}_0)$. Hence, $\nabla F(\mathbf{t}_0) = \mathbf{0}$. Applying the Chain Rule as above, we obtain

$$0 = \sum_{j=1}^{m} \frac{\partial f}{\partial x_j}(\mathbf{a}) \frac{\partial h_j}{\partial t_\ell}(\mathbf{t}_0) + \frac{\partial f}{\partial t_\ell}(\mathbf{a}) \tag{43}$$

[compare with (40)]. Adding (43) and (41), we conclude that

$$0 = \frac{\partial f}{\partial t_\ell}(\mathbf{a}) + \sum_{k=1}^{m} \lambda_k \frac{\partial g_k}{\partial t_\ell}(\mathbf{a}). \quad \blacksquare$$

11.64 EXAMPLE.

Find all extrema of $x^2 + y^2 + z^2$ subject to the constraints $x - y = 1$ and $y^2 - z^2 = 1$.

Solution. Let $f(x, y, z) = x^2 + y^2 + z^2$, $g(x, y, z) = x - y - 1$, and $h(x, y, z) = y^2 - z^2 - 1$. Then (36) takes on the form $\nabla f + \lambda \nabla g + \mu \nabla h = \mathbf{0}$; that is,

$$(2x, 2y, 2z) + \lambda(1, -1, 0) + \mu(0, 2y, -2z) = (0, 0, 0).$$

In particular, $2x + \lambda = 0$, $2y + 2\mu y - \lambda = 0$, and $2z - 2\mu z = 0$. From this last equation, either $\mu = 1$ or $z = 0$.

If $\mu = 1$, then $\lambda = 4y$. Since $2x + \lambda = 0$, we find that $x = -2y$. From $g = 0$ we obtain $-3y = 1$ (i.e., $y = -1/3$). Substituting this into $h = 0$, we obtain $z^2 = -8/9$, a contradiction.

If $z = 0$, then from $h = 0$ we obtain $y = \pm 1$. Since $g = 0$, we obtain $x = 2$ when $y = 1$, and $x = 0$ when $y = -1$. Thus, the only candidates for extrema of f subject to the constraints $g = 0 = h$ are $f(2, 1, 0) = 5$ and $f(0, -1, 0) = 1$. To decide whether these are maxima, minima, or neither, look at the problem from a geometric point of view. The problem requires us to find points on the intersection of the plane $x - y = 1$ and the hyperbolic cylinder $y^2 - z^2 = 1$ which lie closest to the origin. Evidently, both of these points correspond to local minima, and there is no maximum (see Figure 11.9). In particular, the minimum of $x^2 + y^2 + z^2$ subject to the given constraints is 1, attained at the point $(0, -1, 0)$. $\blacksquare$

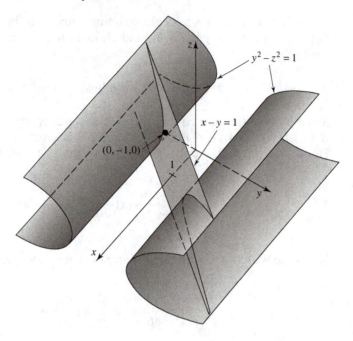

FIGURE 11.9

EXERCISES

11.7.1. Find all local extrema of each of the following functions.

a) $f(x, y) = x^2 - xy + y^3 - y$
b) $f(x, y) = \sin x + \cos y$
c) $f(x, y, z) = e^{x+y} \cos z$
d) $f(x, y) = ax^2 + bxy + cy^2$, where $a \neq 0$ and $b^2 - 4ac \neq 0$

11.7.2. For each of the following, find the maximum and minimum of f on H.

a) $f(x, y) = x^2 + 2x - y^2$ and $H = \{(x, y) : x^2 + 4y^2 \leq 4\}$
b) $f(x, y) = x^2 + 2xy + 3y^2$, and H is the region bounded by the triangle with vertices $(1,0)$, $(1,2)$, $(3,0)$
c) $f(x, y) = x^3 + 3xy - y^3$, and $H = [-1, 1] \times [-1, 1]$

11.7.3. For each of the following, use Lagrange multipliers to find all extrema of f subject to the given constraints

a) $f(x, y) = x + y^2$ and $x^2 + y^2 = 4$
b) $f(x, y) = x^2 - 4xy + 4y^2$ and $x^2 + y^2 = 1$
c) $f(x, y, z) = xy$, $x^2 + y^2 + z^2 = 1$ and $x + y + z = 0$
d) $f(x, y, z, w) = 3x + y + w$, $3x^2 + y + 4z^3 = 1$ and $-x^3 + 3z^4 + w = 0$

11.7.4. Suppose that $f : \mathbf{R}^n \to \mathbf{R}^m$ is differentiable at $\mathbf{a}$, and that $g : \mathbf{R}^m \to \mathbf{R}$ is differentiable at $\mathbf{b} = f(\mathbf{a})$. Prove that if $g(\mathbf{b})$ is a local extremum of g, then $\nabla(g \circ f)(\mathbf{a}) = \mathbf{0}$.

11.7.5. Suppose that V is open in $\mathbf{R}^2$, that $(a, b) \in V$, and that $f : V \to \mathbf{R}$ has second-order total differential on V with $f_x(a, b) = f_y(a, b) = 0$. If the second-order partial derivatives of f are continuous at (a, b) and exactly two of the three numbers $f_{xx}(a, b)$, $f_{xy}(a, b)$, and $f_{yy}(a, b)$ are zero, prove that (a, b) is a saddle point if $f_{xy}(a, b) \neq 0$.

11.7.6. Suppose that V is an open set in $\mathbf{R}^n$, that $\mathbf{a} \in V$, and that $f : V \to \mathbf{R}$ is C^2 on V. If $f(\mathbf{a})$ is a local minimum of f, prove that $D^{(2)} f(\mathbf{a}; \mathbf{h}) \geq 0$ for all $\mathbf{h} \in \mathbf{R}^n$.

11.7.7. Let a, b, c, D, E be real numbers with $c \neq 0$.

a) If $DE > 0$, find all extrema of $ax + by + cz$ subject to the constraint $z = Dx^2 + Ey^2$. Prove that a maximum occurs when $cD < 0$ and a minimum when $cD > 0$.

b) What can you say when $DE < 0$?

11.7.8. [IMPLICIT METHOD].

a) Suppose that $f, g : \mathbf{R}^3 \to \mathbf{R}$ are differentiable at a point (a, b, c), and that $f(a, b, c)$ is an extremum of f subject to the constraint $g(x, y, z) = k$, where k is a constant. Prove that

$$\frac{\partial f}{\partial x}(a, b, c)\frac{\partial g}{\partial z}(a, b, c) - \frac{\partial f}{\partial z}(a, b, c)\frac{\partial g}{\partial x}(a, b, c) = 0$$

and

$$\frac{\partial f}{\partial y}(a, b, c)\frac{\partial g}{\partial z}(a, b, c) - \frac{\partial f}{\partial z}(a, b, c)\frac{\partial g}{\partial y}(a, b, c) = 0.$$

b) Use part a) to find all extrema of $f(x, y, z) = 4xy + 2xz + 2yz$ subject to the constraint $xyz = 16$.

11.7.9 . **This exercise is used in Section *14.4.**

a) Let $p > 1$. Find all extrema of $f(\mathbf{x}) = \sum_{k=1}^n x_k^2$ subject to the constraint $\sum_{k=1}^n |x_k|^p = 1$.

b) Prove that

$$\frac{1}{n^{(2-p)/(2p)}} \left(\sum_{k=1}^n |x_k|^p \right)^{1/p} \leq \left(\sum_{k=1}^n x_k^2 \right)^{1/2} \leq \left(\sum_{k=1}^n |x_k|^p \right)^{1/p}$$

for all $x_1, \ldots, x_n \in \mathbf{R}$, $n \in \mathbf{N}$, and $1 \leq p \leq 2$.

11.7.10. [LEAST SQUARES OR LINEAR REGRESSION].

Suppose that points $\mathbf{x}$ and $\mathbf{y}$ are fixed in $\mathbf{R}^n$ and set

$$d_0 := n \sum_{k=1}^n x_k^2 - \left(\sum_{k=1}^n x_k \right)^2.$$

a) Prove that if

$$F(a, b) := \sum_{k=1}^{n} (y_k - (ax_k + b))^2$$

for $(a, b) \in \mathbf{R}^2$, then the system

$$\frac{\partial F}{\partial a}(a, b) = 0 = \frac{\partial F}{\partial b}(a, b)$$

is solved by

$$a_0 := \frac{n \sum_{k=1}^{n} x_k y_k - \sum_{k=1}^{n} x_k \sum_{k=1}^{n} y_k}{d_0}$$

and

$$b_0 := \frac{\sum_{k=1}^{n} x_k^2 \sum_{k=1}^{n} y_k - \sum_{k=1}^{n} x_k \sum_{k=1}^{n} x_k y_k}{d_0}.$$

b) Prove that if a_0 and b_0 are given by part a), then the straight line whose graph is closest to the points $(x_1, y_1), \ldots, (x_n, y_n)$—that is, such that

$$\sum_{k=1}^{n} (y_k - (mx_k + b))^2$$

is minimized—is the line $\lambda(x) = a_0 x + b_0$.

CHAPTER 12

Integration on $\mathbf{R}^n$

12.1 JORDAN REGIONS

In this section we define grids and use them to identify special subsets of $\mathbf{R}^n$, called *Jordan regions*, which have a well-defined volume. In the next section, when we define integrals of multivariable functions on Jordan regions, grids will play the role that partitions did in the one-variable case.

Throughout this chapter, R will represent a nondegenerate n-dimensional rectangle; that is,

$$R = [a_1, b_1] \times \cdots \times [a_n, b_n] := \{\mathbf{x} \in \mathbf{R}^n : x_j \in [a_j, b_j] \text{ for } j = 1, \ldots, n\}, \quad (1)$$

where $a_j < b_j$ for $j = 1, 2, \ldots, n$. A *grid* on R is a collection of n-dimensional rectangles $\mathcal{G} = \{R_1, \ldots, R_p\}$ obtained by subdividing the sides of R; that is, for each $j = 1, \ldots, n$ there are integers $v_j \in \mathbf{N}$ and partitions $\mathcal{P}_j = \mathcal{P}_j(\mathcal{G}) = \{x_k^{(j)} : k = 1, \ldots, v_j\}$ of $[a_j, b_j]$ such that $\mathcal{G}$ is the collection of rectangles of the form $I_1 \times \ldots \times I_n$, where each $I_j = [x_{k-1}^{(j)}, x_k^{(j)}]$ for some $k = 1, \ldots, v_j$ (see Figure 12.1). A grid $\mathcal{G}$ is said to be *finer* than a grid $\mathcal{H}$ if and only if each partition $\mathcal{P}_j(\mathcal{G})$ is finer than the corresponding partition $\mathcal{P}_j(\mathcal{H})$, $j = 1, \ldots, n$. Notice that given two grids $\mathcal{G}$ and $\mathcal{H}$, there is a grid $\mathcal{I}$ which is finer than both $\mathcal{G}$ and $\mathcal{H}$. [Such a grid can be constructed by taking $\mathcal{P}_j(\mathcal{I}) = \mathcal{P}_j(\mathcal{G}) \cup \mathcal{P}_j(\mathcal{H})$ for $j = 1, \ldots, n$.]

If R is an n-dimensional rectangle of the form (1), then the *volume* of R is defined to be

$$|R| = (b_1 - a_1) \ldots (b_n - a_n).$$

(When $n = 1$, we shall call $|R|$ the *length* of R, and when $n = 2$, we shall call $|R|$ the *area* of R.) Notice that given $\varepsilon > 0$ there exists a rectangle R^* such that $R \subset (R^*)^o$ and $|R^*| = |R| + \varepsilon$. Indeed, since $b_j - a_j + 2\delta \to b_j - a_j$ as $\delta \to 0$, we can choose $\delta > 0$ so small that $R^* := [a_1 - \delta, b_1 + \delta] \times \cdots \times [a_n - \delta, b_n + \delta]$ satisfies $|R^*| = |R| + \varepsilon$.

We want to define the integral of a multivariable function on a variety of sets; for example, the integral of a function of two variables on rectangles, disks, triangles, ellipses, and the integral of a function of three variables on balls, cones, ellipsoids, pyramids, and so on. One property these regions all have in common is that they have a well-defined "area" or "volume."

How shall we define the volume of a general set E? Let R be a rectangle which contains E. If E is simple enough, we should be able to get a good approximation for the volume of E by choosing a sufficiently fine grid $\mathcal{G}$ on R and adding up the collective volumes of all rectangles in $\mathcal{G}$ which intersect $\overline{E}$.

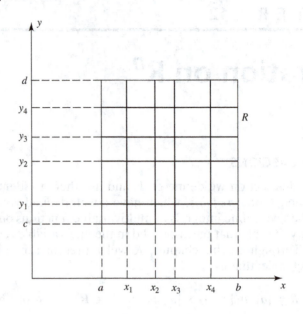

FIGURE 12.1

Accordingly, we define the *outer sums* of E with respect to a grid $\mathcal{G}$ on a rectangle R by

$$V(E; \mathcal{G}) := \sum_{R_j \cap \overline{E} \neq \emptyset} |R_j|,$$

where the *empty sum* is by definition zero. Notice that since the empty sum is defined to be zero, $V(\emptyset; \mathcal{G}) = 0$ for all grids $\mathcal{G}$.

Figure 12.2 illustrates an outer sum for a particular set E and grid $\mathcal{G}$. The rectangles which intersect $\overline{E}$ have been shaded; those which cover ∂E are darker than those which are contained in E^o. Notice, even for this crude grid, that the shaded region is a fair approximation to the "volume" of E.

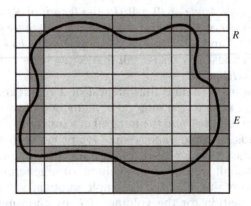

FIGURE 12.2

The following result shows that as the grids get finer, the outer sum approximations to the volume of E get better.

12.1 Remark. *Let R be an n-dimensional rectangle.*

i) *Let E be a subset of R, and let $\mathcal{G}$, $\mathcal{H}$ be grids on R. If $\mathcal{G}$ is finer than $\mathcal{H}$, then*

$$V(E; \mathcal{G}) \leq V(E; \mathcal{H}).$$

ii) *If A and B are subsets of R and $A \subseteq B$, then*

$$V(A; \mathcal{G}) \leq V(B; \mathcal{G}).$$

Proof. i) Since $\mathcal{G}$ is finer than $\mathcal{H}$, each $Q \in \mathcal{H}$ is a finite union of R_j's in $\mathcal{G}$. If $Q \cap \overline{E} \neq \emptyset$, then some of the R_j's in Q intersect $\overline{E}$ and others might not (see Figure 12.3, where the darker lines represent the grid $\mathcal{H}$, the lighter lines represent $\mathcal{G} \setminus \mathcal{H}$, and the R_j's which intersect $\overline{E}$ are shaded). Let $\mathcal{I}_1 = \{R \in \mathcal{G} : R \cap \overline{E} \neq \emptyset\}$ and $\mathcal{I}_2 = \{R \in \mathcal{G} \setminus \mathcal{I}_1 : R \subseteq Q$ for some $Q \in \mathcal{H}$ with $Q \cap \overline{E} \neq \emptyset\}$. Then

$$V(E; \mathcal{H}) = \sum_{R \in \mathcal{I}_1} |R| + \sum_{R \in \mathcal{I}_2} |R| \geq \sum_{R \in \mathcal{I}_1} |R| = V(E; \mathcal{G}).$$

ii) If $A \subseteq B$, then $\overline{A} \subseteq \overline{B}$ (see Exercise 8.4.3). Thus, every rectangle which appears in the sum $V(A; \mathcal{G})$ also appears in the sum $V(B; \mathcal{G})$. Since all $|R_j|$'s are nonnegative, it follows that $V(A; \mathcal{G}) \leq V(B; \mathcal{G})$. ∎

In view of this, we guess that the volume of a set E can be computed by taking the infimum of all outer sums of E. Unfortunately, this guess is wrong unless some restriction is made on the set E. To see why a restriction is necessary, notice that any reasonable definition of volume should satisfy the following

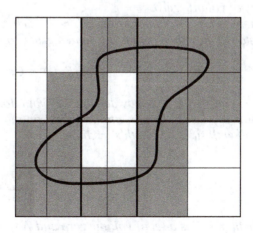

FIGURE 12.3

property: If $E = A \cup B$, where $B = E \setminus A$, then the volume of E equals the sum of the volumes of A and B. The following example shows that this property does not hold if A is too fractured.

12.2 EXAMPLE.

If $R = [0, 1] \times [0, 1]$, $A = \{(x, y) : x, y \in \mathbf{Q} \cap [0, 1]\}$, and $B = R \setminus A$, then $V(A; \mathcal{G}) + V(B; \mathcal{G}) = 2V(R; \mathcal{G})$ no matter how fine $\mathcal{G}$ is.

Proof. Let $\mathcal{G} = \{R_1, \ldots, R_p\}$ be a grid on R. Since each R_j is nondegenerate, it is clear by the Density of Rationals (Theorem 1.18) that $R_j \cap A \neq \emptyset$ for all $j \in [1, p]$. Hence $V(A; \mathcal{G}) = |R| = 1$. Similarly, the Density of the Irrationals (Exercise 1.3.3) implies $V(B; \mathcal{G}) = |R| = 1$. ∎

The real problem with A is that its boundary, $\partial A := \overline{A} \setminus A^o = R$, is too big. To avoid this type of pathology, we will restrict our attention to "Jordan" regions; that is to sets whose boundaries are small in the following sense (see Definition 12.5 and the darkly shaded rectangles in Figure 12.2).

12.3 Definition.

A subset E of $\mathbf{R}^n$ is said to be *of volume zero* if and only if given $\varepsilon > 0$ there is rectangle $R \supseteq E$ and a grid $\mathcal{G} = \{R_1, \ldots, R_p\}$ on R such that $V(E; \mathcal{G}) < \varepsilon$.

Recall that E is covered by $\{Q_k\}_{k=1}^q$ means that $E \subseteq \bigcup_{k=1}^q Q_k$. By Definition 12.3, E is of volume zero if and only if it is covered by rectangles (from a grid $\mathcal{G}$) whose total volume is as small as one wishes. The next result contains two other (highly useful) descriptions of sets of volume zero.

12.4 Theorem. *For every subset E of $\mathbf{R}^n$, the following three conditions are equivalent.*

i) *E is of volume zero.*

ii) *There is an absolute constant $C > 0$ such that for each $\varepsilon > 0$ there is a rectangle R, which contains E, and a grid $\mathcal{G}$ on R such that*

$$V(E; \mathcal{G}) < C\varepsilon.$$

(The constant $C > 0$ may depend on E but does not depend on ε or $\mathcal{G}$.)

iii) *For every $\varepsilon > 0$ there is a finite collection of cubes Q_k of the same size; that is, all with sides of length s, such that*

$$E \subset \bigcup_{k=1}^q Q_k \quad \text{and} \quad \sum_{k=1}^q |Q_k| < \varepsilon.$$

In particular, if E is a set of volume zero and $A \subseteq E$, then both A and ∂A are sets of volume zero.

Proof. *i) implies ii).* If $V(E; \mathcal{G}) < \varepsilon$, then $V(E; \mathcal{G}) < C\varepsilon$ for $C = 1$.

ii) implies iii). Let $\varepsilon > 0$ and set $\eta = \varepsilon/(2C)$. By hypothesis, there exists a grid $\mathcal{G}$ such that if $\{R_1, \dots, R_p\}$ represents all rectangles in $\mathcal{G}$ which intersect $\overline{E}$, then

$$\overline{E} \subset \bigcup_{j=1}^{p} R_j \quad \text{and} \quad \sum_{j=1}^{p} |R_j| < C\eta.$$

By increasing the size of the R_j's slightly, we may suppose that the sides of each R_j have rational lengths, and $\sum_{j=1}^{p} |R_j| < 2C\eta = \varepsilon$. (These rectangles no longer form a grid because they may have some overlap, but they still cover $\overline{E}$ and hence E itself.)

The lengths of the sides of the R_j's have a common denominator, say d. By using a grid fine enough, we can divide each R_j into cubes $Q_k^{(j)}$, for $k = 1, 2, \dots, \nu_j$ and some choice of $\nu_j \in \mathbf{N}$, such that each $Q_k^{(j)}$ has sides of common length $s = 1/d$. Since $|R_j| = \sum_{k=1}^{\nu_j} |Q_k^{(j)}|$, it follows that

$$\sum_{j=1}^{p} \sum_{k=1}^{\nu_j} |Q_k^{(j)}| = \sum_{j=1}^{p} |R_j| < \varepsilon.$$

iii) implies i). Suppose that E can be covered by finitely many cubes

$$Q_k = [a_1^{(k)}, b_1^{(k)}] \times \cdots \times [a_n^{(k)}, b_n^{(k)}]$$

whose volumes sum to a number less than ε. Let R be a rectangle which contains the union of the Q_k's. For each $j = 1, 2, \dots, n$, the endpoints $\{a_j^{(1)}, b_j^{(1)}, \dots, a_j^{(q)}, b_j^{(q)}\}$ can be arranged in increasing order to form a partition of the jth side of R. Thus there is a grid $\mathcal{G} = \{R_1, \dots, R_p\}$ so fine that each Q_k is a union of the R_j's (see Figure 12.4). Thus $V(E; \mathcal{G}) \leq \sum_{k=1}^{q} |Q_k| < \varepsilon$; that is, E is a set of volume zero by definition. This completes the proof that conditions i), ii), and iii) are equivalent.

Finally, suppose that $A \subseteq E$ and $\mathcal{G}$ is a grid on some rectangle that contains E. Since $\partial A \subseteq \overline{A} \subseteq \overline{E}$, it is clear that $V(\partial A, \mathcal{G}) \leq V(A, \mathcal{G}) \leq V(E, \mathcal{G})$. Thus if E is a set of volume zero, then so are A and ∂A.   ∎

We are now prepared to define volume.

12.5 Definition.

A subset E of $\mathbf{R}^n$ is called a *Jordan region* if and only if $E \subseteq R$ for some n-dimensional rectangle R and ∂E is of volume zero, in which case we define the *volume* (or *Jordan content*) of E by

$$\text{Vol}(E) := \inf_{\mathcal{G}} V(E; \mathcal{G}) := \inf\{V(E; \mathcal{G}) : \mathcal{G} \text{ ranges over all grids on } R\}.$$

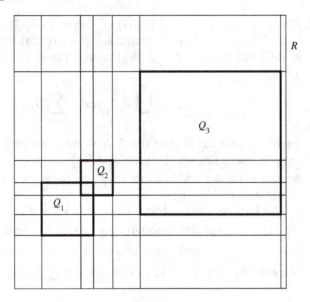

FIGURE 12.4

We sometimes shall call Vol(E) *length* when $n = 1$ and *area* when $n = 2$. Notice, then, that the empty set is of length, area, and volume zero.

By Theorem 12.4, it is clear that a set A is of volume zero if and only if Vol(A) = 0. Since rectangles are bounded, it is also clear by Definition 12.5 that every Jordan region is bounded. The converse of this last statement is false. The set A in Example 12.2 is bounded but not a Jordan region.

Notice that by Theorem 12.4 and Definition 12.5, a set E is a Jordan region if and only if its boundary can be covered by cubes whose total volume is as small as one wishes. We shall use this observation many times in the sequel.

Before we continue, we need to show that Vol(E) does not depend on the rectangle R chosen to generate the grids $\mathcal{G}$. To this end, let R and Q be rectangles which contain E. Since the intersection of two rectangles is a rectangle, we may suppose that $E \subseteq Q \subset R$. Since $Q \subset R$, it is easy to see that

$$\inf_{\mathcal{H} \text{ on } Q} V(E; \mathcal{H}) \leq \inf_{\mathcal{G} \text{ on } R} V(E; \mathcal{G}).$$

On the other hand, given $\varepsilon > 0$, choose, for each grid $\mathcal{H}$ on Q, a rectangle Q^* such that $Q \subset (Q^*)^o$ and $|Q^*| = |Q| + \varepsilon$. Let $\mathcal{H}_0$ be the grid formed by adding the endpoints of Q^* and R to $\mathcal{H}$; that is, if

$$Q^* = [\alpha_1, \beta_1] \times \cdots \times [\alpha_n, \beta_n] \text{ and } R = [\gamma_1, \delta_1] \times \cdots \times [\gamma_n, \delta_n],$$

then $\mathcal{P}_j(\mathcal{H}_0) = \mathcal{P}_j(\mathcal{H}) \cup \{\alpha_j, \beta_j, \gamma_j, \delta_j\}$. Then $\mathcal{G}_0 := \mathcal{H}_0 \cap R$ is a grid on R whose rectangles which intersect $\overline{E}$ are either part of $\mathcal{H}$ to begin with or the thin ones formed by adding the endpoints of Q^*. Hence,

$$V(E; \mathcal{H}) + \varepsilon \geq V(E; \mathcal{G}_0) \geq \inf_{\mathcal{G} \text{ on } R} V(E; \mathcal{G}).$$

It follows that

$$\inf_{\mathcal{H} \text{ on } Q} V(E; \mathcal{H}) \leq \inf_{\mathcal{G} \text{ on } R} V(E; \mathcal{G}) \leq \inf_{\mathcal{H} \text{ on } Q} V(E; \mathcal{H}) + \varepsilon,$$

for every $\varepsilon > 0$. By letting $\varepsilon \to 0$, we verify that the definition of volume does not depend on the rectangle R.

Next, we show that the two definitions of the volume of a rectangle (length times width times etc. versus the infimum of outer sums) agree.

12.6 Remark. *If R is an n-dimensional rectangle, then R is a Jordan region in $\mathbf{R}^n$ and*

$$\text{Vol}(R) = |R|.$$

Proof. Let $\varepsilon > 0$ and suppose that

$$R = [a_1, b_1] \times \cdots \times [a_n, b_n].$$

Since $b_j - a_j - 2\delta \to b_j - a_j$ as $\delta \to 0$, we can choose $\delta > 0$ so small that if

$$Q = [a_1 + \delta, b_1 - \delta] \times \cdots \times [a_n + \delta, b_n - \delta],$$

then $|R| - |Q| < \varepsilon$.

Let $\mathcal{G} := \{H_1, \ldots, H_q\}$ be the grid on R determined by

$$\mathcal{P}_j(\mathcal{G}) = \{a_j, a_j + \delta, b_j - \delta, b_j\}.$$

Then it is clear that an $H_j \in \mathcal{G}$ intersects ∂R if and only if $H_j \neq Q$. Hence,

$$V(\partial R; \mathcal{G}) := \sum_{H_j \cap \partial R \neq \emptyset} |H_j| = |R| - |Q| < \varepsilon.$$

This proves that R is a Jordan region.

To compute the volume of R using Definition 12.5, let $\mathcal{G} = \{R_1, \ldots, R_p\}$ be any grid on R. Since $R_j \cap R \neq \emptyset$ for all $R_j \in \mathcal{G}$, it follows from definition that $V(R; \mathcal{G}) = |R|$. Taking the infimum of this identity over all grids $\mathcal{G}$ on R, we find that $\text{Vol}(R) = |R|$. ∎

In general, it is not easy to decide whether or not a given set is a Jordan region. Topology alone cannot resolve this problem since there are open sets in $\mathbf{R}^n$ which are not Jordan regions (see Spivak [12], p. 56). In practice, however, it is usually easy to show that a specific set E is a Jordan region by applying Theorem 12.4 to ∂E. Here is a typical example.

12.7 Theorem. *If E_1 and E_2 are Jordan regions, then $E_1 \cup E_2$ is a Jordan region and*

$$\text{Vol}(E_1 \cup E_2) \leq \text{Vol}(E_1) + \text{Vol}(E_2).$$

Proof. We begin by proving that $E_1 \cup E_2$ is a Jordan region. Since E_1 and E_2 are Jordan regions, use Theorem 12.4 to choose cubes $\{S_j\}$ which cover ∂E_1 (respectively, cubes $\{Q_k\}$ which cover ∂E_2) such that

$$\sum_{j=1}^{p} |S_j| < \frac{\varepsilon}{2} \quad \text{and} \quad \sum_{k=1}^{p} |Q_k| < \frac{\varepsilon}{2}.$$

But by Theorem 8.37 or 10.40, $\partial(E_1 \cup E_2) \subseteq \partial E_1 \cup \partial E_2$. Thus $\{S_j\} \cup \{Q_k\}$ is a collection of cubes which covers $\partial(E_1 \cup E_2)$ whose volumes sum to a number less than ε. Hence by Theorem 12.4 and Definition 12.5, $E_1 \cup E_2$ is a Jordan region.

To estimate the volume of $E_1 \cup E_2$, let $\mathcal{G}$ be a grid on a rectangle which contains $E_1 \cup E_2$. If R_j intersects $\overline{E_1 \cup E_2}$, then by Theorem 8.37 (or 10.40) R_j intersects $\overline{E}_1$ or $\overline{E}_2$ (or both). Hence, $V(E_1 \cup E_2; \mathcal{G}) \leq V(E_1; \mathcal{G}) + V(E_2; \mathcal{G})$. Taking the infimum of this inequality over all grids $\mathcal{G}$, we obtain

$$\text{Vol}(E_1 \cup E_2) \leq \text{Vol}(E_1) + \text{Vol}(E_2). \qquad \blacksquare$$

By iterating this result, we see that the collection of Jordan regions is closed under finite unions. This is also the case for intersections and set differences (see Exercise 12.1.6b).

Theorem 12.4 can also be used to show that spheres, ellipsoids, and, in fact, all "projectable regions" (just about anything you can draw) are Jordan regions (see Exercise 12.1.4 and Theorem 12.39).

To evaluate integrals of multivariable functions over unions of sets, we introduce the following concept.

12.8 Definition.

Let $\mathcal{E} := \{E_\ell\}_{\ell \in \mathbf{N}}$ be a collection of subsets of $\mathbf{R}^n$.

i) $\mathcal{E}$ is said to be *nonoverlapping* if and only if $E_j \cap E_k$ is of volume zero for $j \neq k$.

ii) $\mathcal{E}$ is said to be *pairwise disjoint* if and only if $E_j \cap E_k = \emptyset$ for $j \neq k$.

Notice that since $\emptyset$ is of volume zero, every collection of pairwise disjoint sets is nonoverlapping. (The converse of this statement is false—see Exercise 12.1.3 below.)

In order to prove a change-of-variables formula in $\mathbf{R}^n$ in Section 12.4, we need to identify conditions under which a C^1 function preserves Jordan regions (see Theorem 12.10). Since Jordan regions are sets whose boundaries are of volume zero, we first prove a result about functions which take sets of volume zero to sets of volume zero.

12.9 Lemma.
Suppose that V is a bounded, open set in $\mathbf{R}^n$ and that $\phi : V \to \mathbf{R}^n$ is 1–1 and C^1 on V with $\Delta_\phi \neq 0$. If E is of volume zero and $\overline{E} \subset V$, then $\phi(E)$ is of volume

zero. In particular, if $\{E_k\}_{k\in\mathbf{N}}$ is a nonoverlapping collection of sets in $\mathbf{R}^n$ with $\overline{E}_k \subset V$ for all $k \in \mathbf{N}$, then $\{\phi(E_k)\}_{k\in\mathbf{N}}$ is a nonoverlapping collection of sets in $\mathbf{R}^n$.

Proof. Since $\overline{E} \subset V$, for each $\mathbf{x} \in \overline{E}$ there is an $r(\mathbf{x}) > 0$ such that $\overline{B_{r(\mathbf{x})}(\mathbf{x})} \subset V$. Since

$$\overline{E} \subseteq \bigcup_{\mathbf{x}\in\overline{E}} B_{r(\mathbf{x})}(\mathbf{x}),$$

it follows from the Heine–Borel Theorem that there exist finitely many $\mathbf{x}_k \in \overline{E}$ such that the bounded open set

$$U := \bigcup_{k=1}^{N} B_{r(\mathbf{x}_k)}(\mathbf{x}_k)$$

contains $\overline{E}$. Set $H := \overline{U}$. Since U is bounded, H is compact. Moreover, the construction guarantees that $\overline{E} \subset H^o \subset H \subset V$. Thus by Corollary 11.34, there exists an $M > 0$ which depends only on H, ϕ, and n such that

$$\|\phi(\mathbf{x}) - \phi(\mathbf{y})\| \leq M\|\mathbf{x} - \mathbf{y}\|, \qquad \mathbf{x}, \mathbf{y} \in Q, \tag{2}$$

for all cubes $Q \subseteq H$.

Let $\epsilon > 0$ and set $C := (2A)^n$, where $A := M\sqrt{n}$. Since E is of volume zero and $\overline{E} \subset H^o$, use Theorem 12.4 to choose cubes $Q_1, \ldots, Q_q$, all the same size with sides of length s, such that $Q_j \subset H$,

$$E \subset \bigcup_{j=1}^{q} Q_j, \quad \text{and} \quad \sum_{j=1}^{q} |Q_j| < \frac{\epsilon}{C}. \tag{3}$$

Fix j and fix $\mathbf{x}_0 \in Q_j$. By (2) and Remark 8.7,

$$\|\phi(\mathbf{x}_0) - \phi(\mathbf{x})\| \leq Ms\sqrt{n} = sA$$

for all $\mathbf{x} \in Q_j$. Thus $\phi(Q_j)$ is contained in the cube $\widetilde{Q}_j := I_1 \times \cdots \times I_n$, where

$$I_k := [\phi_k(\mathbf{x}_0) - sA, \phi_k(\mathbf{x}_0) + sA].$$

In particular, it follows from the left side of (3) and Theorem 1.37 that

$$\phi(E) \subset \phi\left(\bigcup_{j=1}^{q} Q_j\right) = \bigcup_{j=1}^{q} \phi(Q_j) \subseteq \bigcup_{j=1}^{q} \widetilde{Q}_j.$$

But the sides of $\tilde{Q}_j$ are of length $2sA$, so $|\tilde{Q}_j| = (2sA)^n = Cs^n$. In particular, the right side of (3) implies that

$$\sum_{j=1}^{q} |\tilde{Q}_j| = \sum_{j=1}^{q} Cs^n = C \sum_{j=1}^{q} |Q_j| < C\frac{\varepsilon}{C} = \varepsilon.$$

Therefore, $\text{Vol}(\phi(E)) = 0$ by Theorem 12.4.

Finally, by what we just proved, if $E_k \cap E_j$ is of volume zero, then so is $\phi(E_k \cap E_j)$. But by Exercise 1.5.7 (since ϕ is 1–1),

$$\phi(E_k) \cap \phi(E_j) = \phi(E_k \cap E_j).$$

Thus $\{\phi(E_k)\}$ is nonoverlapping when $\{E_k\}$ is nonoverlapping. ∎

12.10 Theorem. *Suppose that V is a bounded, open set in $\mathbf{R}^n$ and that $\phi : V \to \mathbf{R}^n$ is 1–1 and C^1 on V with $\Delta_\phi \neq 0$. If E is a Jordan region and $\overline{E} \subset V$, then $\phi(E)$ is a Jordan region.*

Proof. By Definition 12.5, Theorem 12.4, and Lemma 12.9, it suffices to prove that $\partial(\phi(E)) \subseteq \phi(\partial E)$. By Lemma 11.39, the set $\phi(E^o)$ is open, and by Theorem 9.29 (or 10.61), the set $\phi(\overline{E})$ is closed. It follows from Theorem 8.32 (or 10.34) that $\phi(E^o) \subseteq (\phi(E))^o$ and $\phi(\overline{E}) \supseteq \overline{\phi(E)}$. Therefore,

$$\partial(\phi(E)) = \overline{\phi(E)} \setminus (\phi(E))^o \subseteq \phi(\overline{E}) \setminus \phi(E^o) = \phi(\overline{E} \setminus E^o) = \phi(\partial E). \quad ∎$$

We close this section with some optional results which will not be used elsewhere. They show that the volume of a set can also be approximated from below using inner sums.

We introduced outer sums (analogues of upper sums) and defined the volume of a Jordan region as the infimum of all outer sums. In order to calculate the volume of a specific set, it is sometimes convenient to have inner sums (analogues of the lower sums we used to define integrals in Chapter 5). Given $E \subset \mathbf{R}^n$, a subset of some n-dimensional rectangle R, and $\mathcal{G} = \{R_j : j = 1, \ldots, p\}$, a grid on R, the *inner sums* of E with respect to $\mathcal{G}$ are defined by

$$v(E; \mathcal{G}) := \sum_{R_j \subset E^o} |R_j|,$$

where the empty sum is again interpreted to be zero. Thus $v(E; \mathcal{G}) = 0$ for all grids $\mathcal{G}$ and all sets E satisfying $E^o = \emptyset$.

Inner and outer sums can be used to define inner and outer volume of ANY bounded set, in the same way that upper and lower sums were used to define upper and lower integrals of any bounded function (see Definition 12.13 below). If $\mathcal{G}$ is fine enough and E is Jordan, the inner sum of a Jordan region E with respect to $\mathcal{G}$ should approximate $\text{Vol}(E)$; just as $V(E; \mathcal{G})$ overestimated $\text{Vol}(E)$, each $v(E; \mathcal{G})$ underestimates Vol(E). [In Figure 12.2, the underestimate $v(E; \mathcal{G})$ is represented by the lightly shaded rectangles. You might refine the grid there and revisualize the inner and outer sums to illustrate that these estimates get better as the grid gets finer.]

Since $v(E; \mathcal{H})$ is either zero or a sum of nonnegative terms, it is clear that $v(E; \mathcal{H}) \geq 0$ for all grids $\mathcal{H}$. If we combine this observation with the proof of Remark 12.1i, we can also establish the following result.

12.11 Remark. *Let R be an n-dimensional rectangle, let E be a subset of R, and let $\mathcal{G}, \mathcal{H}$ be grids on R. If $\mathcal{G}$ is finer than $\mathcal{H}$, then*

$$0 \leq v(E; \mathcal{H}) \leq v(E; \mathcal{G}) \leq V(E; \mathcal{G}) \leq V(E; \mathcal{H}).$$

This leads us to the following fundamental principle.

12.12 Remark. *Let R be an n-dimensional rectangle and E be a subset of R. If $\mathcal{G}$ and $\mathcal{H}$ are grids on R, then*

$$0 \leq v(E; \mathcal{G}) \leq V(E; \mathcal{H}).$$

Proof. Let $\mathcal{I}$ be a grid finer than both $\mathcal{G}$ and $\mathcal{H}$. By Remark 12.11,

$$0 \leq v(E; \mathcal{G}) \leq v(E; \mathcal{I}) \leq V(E; \mathcal{I}) \leq V(E; \mathcal{H}). \qquad \blacksquare$$

Using the sums $v(E; \mathcal{G})$ and $V(E; \mathcal{G})$, we can define inner and outer volume of any bounded set E.

12.13 Definition.

Let E be a bounded subset of $\mathbf{R}^n$ and let R be an n-dimensional rectangle which satisfies $E \subseteq R$. The *inner volume* of E is defined by

$$\underline{\text{Vol}}(E) := \sup\{v(E; \mathcal{G}) : \mathcal{G} \text{ ranges over all grids on } R\},$$

and the *outer volume* of E is defined by

$$\overline{\text{Vol}}(E) := \inf\{V(E; \mathcal{G}) : \mathcal{G} \text{ ranges over all grids on } R\}.$$

As before, we can show that this definition is independent of the rectangle R chosen to generate the grids $\mathcal{G}$.

When E is a Jordan region, the outer and inner volume of E is precisely the volume of E.

12.14 Theorem. *Let E be a bounded subset of $\mathbf{R}^n$. Then E is a Jordan region if and only if $\overline{\text{Vol}}(E) = \underline{\text{Vol}}(E)$.*

Proof. Let $E \subset \mathbf{R}^n$ and suppose that R is a rectangle which contains E. We shall show that for all grids $\mathcal{G}$ on R,

$$V(E; \mathcal{G}) - v(E; \mathcal{G}) = V(\partial E; \mathcal{G}). \tag{4}$$

If $E^o = \emptyset$, then $\partial E = \overline{E}$ and (4) is obvious. Otherwise, suppose that $R_j \in \mathcal{G}$ is a rectangle which appears in the sum represented by the left side of (4); that is, R_j intersects $\overline{E}$ but R_j is not a subset of E^o. If R_j does not appear in the sum represented by the right side of (4), then $R_j \cap \partial E = \emptyset$. It follows that the pair E^o, $(\mathbf{R}^n \backslash E)^o$ separates R_j, a contradiction since all rectangles are connected (see Remark 9.34). Therefore, every rectangle which appears in the sum represented by the left side of (4) also appears in the sum represented by the right side; that is,

$$V(E; \mathcal{G}) - v(E; \mathcal{G}) \le V(\partial E; \mathcal{G}).$$

On the other hand, suppose that $R_j \in \mathcal{G}$ is a rectangle which appears in the sum represented by the right side of (4); that is, $R_j \cap \overline{\partial E} \ne \emptyset$. Recall from Theorems 8.24 and 8.36 (or 10.39 and 10.31) that $\partial E = \overline{E} \backslash E^o$ is closed, so $R_j \cap \partial E \ne \emptyset$. It follows that R_j intersects $\overline{E}$ but R_j is not a subset of E^o. Thus every rectangle which appears in the sum represented by the right side of (4) also appears in the sum represented by the left side; that is, $V(E; \mathcal{G}) - v(E; \mathcal{G}) \ge V(\partial E; \mathcal{G})$. This proves (4).

To prove the theorem, suppose that E is a Jordan region. By definition, $\mathrm{Vol}(\partial E) = 0$. Since by (4), $V(\partial E; \mathcal{G}) = V(E; \mathcal{G}) - v(E; \mathcal{G}) \ge \overline{\mathrm{Vol}}(E) - \underline{\mathrm{Vol}}(E)$, it follows (by taking the infimum of this last inequality over all grids $\mathcal{G}$) that

$$0 = \mathrm{Vol}(\partial E) \ge \overline{\mathrm{Vol}}(E) - \underline{\mathrm{Vol}}(E) \ge 0. \tag{5}$$

Thus $\overline{\mathrm{Vol}}(E) = \underline{\mathrm{Vol}}(E)$.

Conversely, suppose that $\overline{\mathrm{Vol}}(E) = \underline{\mathrm{Vol}}(E)$. By the Approximation Property, given $\varepsilon > 0$, there exist grids $\mathcal{H}_1$ and $\mathcal{H}_2$ such that

$$\overline{\mathrm{Vol}}(E) + \varepsilon > V(E; \mathcal{H}_1) \quad \text{and} \quad \underline{\mathrm{Vol}}(E) - \varepsilon < v(E; \mathcal{H}_2).$$

If $\mathcal{G}$ is a grid on R which is finer than both $\mathcal{H}_1$ and $\mathcal{H}_2$, it follows from Remark 12.11 that

$$\overline{\mathrm{Vol}}(E) + \varepsilon > V(E; \mathcal{G}) \quad \text{and} \quad \underline{\mathrm{Vol}}(E) - \varepsilon < v(E; \mathcal{G}).$$

Subtracting these inequalities, we see by (4) that

$$0 \le V(\partial E; \mathcal{G}) = V(E; \mathcal{G}) - v(E; \mathcal{G}) < \overline{\mathrm{Vol}}(E) - \underline{\mathrm{Vol}}(E) + 2\varepsilon = 2\varepsilon.$$

Hence E is a Jordan region by definition. ∎

EXERCISES

12.1.1. a) For $m = 1, 2, 3$, let $\mathcal{G}_m$ be the grid on $[0, 1] \times [0, 1]$ generated by

$$\mathcal{P}_j(\mathcal{G}_m) = \{k/2^m : k = 0, 1, \ldots, 2^m\},$$

where $j = 1, 2$. For each of the following sets, compute $V(E; \mathcal{G}_m)$.
  $\alpha)$ $E = \{(x, y) \in [0, 1] \times [0, 1] : x = 0 \text{ or } y = 0\}$.
  $\beta)$ $E = \{(x, y) \in [0, 1] \times [0, 1] : y \leq x\}$.
  $\gamma)$ $E = \{(x, y) \in [0, 1] \times [0, 1] : (2x - 1)^2 + (2y - 1)^2 \leq 1\}$.
  b) For each E in part a), compute $v(E; \mathcal{G}_m)$.

12.1.2. a) Prove that every finite subset of $\mathbf{R}^n$ is a Jordan region of volume zero.
  b) Show that, even in $\mathbf{R}^2$, part a) is not true if *finite* is replaced by *countable*.
  c) By an interval in $\mathbf{R}^2$ we mean a set of the form

$$\{(x, c) : a \leq x \leq b\} \quad \text{or} \quad \{(c, y) : a \leq y \leq b\}$$

for some $a, b, c \in \mathbf{R}$. Prove that every interval in $\mathbf{R}^2$ is a Jordan region.

12.1.3. Prove that every grid is a nonoverlapping collection of Jordan regions.

12.1.4. a) Prove that the boundary of an open ball $B_r(\mathbf{a})$ is given by

$$\partial B_r(\mathbf{a}) = \{\mathbf{x} : \|\mathbf{x} - \mathbf{a}\| = r\}.$$

  b) Prove that $B_r(\mathbf{a})$ is a Jordan region for all $\mathbf{a} \in \mathbf{R}^n$ and all $r \geq 0$.

12.1.5. Let E be a Jordan region in $\mathbf{R}^n$.

  a) Prove that E^o and $\overline{E}$ are Jordan regions.
  b) | **This exercise is used in Section 12.2.** Prove that $\text{Vol}(E^o) = \text{Vol}(\overline{E}) = \text{Vol}(E)$.
  c) Prove that $\text{Vol}(E) > 0$ if and only if $E^o \neq \emptyset$.
  d) Let $f : [a, b] \to \mathbf{R}$ be continuous on $[a, b]$. Prove that the graph of $y = f(x)$, $x \in [a, b]$, is a Jordan region in $\mathbf{R}^2$.
  e) Does part d) hold if *continuous* is replaced by *integrable*? How about *bounded*?

12.1.6. **This exercise is used in Section *12.5.** Suppose that E_1, E_2 are Jordan regions in $\mathbf{R}^n$.

  a) Prove that if $E_1 \subseteq E_2$, then

$$\text{Vol}(E_1) \leq \text{Vol}(E_2).$$

  b) Prove that $E_1 \cap E_2$ and $E_1 \setminus E_2$ are Jordan regions.
  c) Prove that if E_1, E_2 are nonoverlapping, then

$$\text{Vol}(E_1 \cup E_2) = \text{Vol}(E_1) + \text{Vol}(E_2).$$

d) If $E_2 \subseteq E_1$, prove that

$$\text{Vol}(E_1 \setminus E_2) = \text{Vol}(E_1) - \text{Vol}(E_2).$$

e) Prove that

$$\text{Vol}(E_1 \cup E_2) = \text{Vol}(E_1) + \text{Vol}(E_2) - \text{Vol}(E_1 \cap E_2).$$

$\boxed{\textbf{12.1.7}}$. **This exercise is used in Section *12.6.** Let $E \subset \mathbf{R}^n$. The *translation* of E by an $\mathbf{x} \in \mathbf{R}^n$ is the set $\mathbf{x} + E = \{\mathbf{y} \in \mathbf{R}^n : \mathbf{y} = \mathbf{x} + \mathbf{z} \text{ for some } \mathbf{z} \in E\}$, and the *dilation* of E by a scalar $\alpha > 0$ is the set $\alpha E = \{\mathbf{y} \in \mathbf{R}^n : \mathbf{y} = \alpha \mathbf{z} \text{ for some } \mathbf{z} \in E\}$.

a) Prove that E is a Jordan region if and only if $\mathbf{x} + E$ is a Jordan region, in which case $\text{Vol}(\mathbf{x} + E) = \text{Vol}(E)$.

b) Prove that E is a Jordan region if and only if αE is a Jordan region, in which case $\text{Vol}(\alpha E) = \alpha^n \text{Vol}(E)$.

12.1.8. A set $E \subset \mathbf{R}^n$ is said to be of *measure zero* if and only if given $\varepsilon > 0$ there is a sequence of rectangles $R_1, R_2, \dots$ which covers E such that $\sum_{k=1}^{\infty} |R_k| < \varepsilon$.

a) Prove that if $E \subset \mathbf{R}^n$ is of volume zero, then E is of measure zero.

b) Prove that if $E \subset \mathbf{R}^n$ is at most countable, then E is of measure zero.

c) Prove that there is a set $E \subset \mathbf{R}^2$ of measure zero which does not have zero area and, in fact, is not even a Jordan region.

*12.1.9. Show that if $E \subset \mathbf{R}^n$ is bounded and has only finitely many cluster points, then E is a Jordan region.

12.2 RIEMANN INTEGRATION ON JORDAN REGIONS

By analogy with the one-variable case, the integral of a nonnegative function f over a Jordan region E should be the volume of the set $\{(\mathbf{x}, t) : \mathbf{x} \in E, 0 \le t \le f(\mathbf{x})\}$. We should be able to approximate this volume by using $(n + 1)$-dimensional rectangles whose heights approximate $t = f(\mathbf{x})$ and whose bases belong to some grid on E (see Figure 12.5). This leads us to the following definition (compare with Definition 5.13).

12.15 Definition.

Let E be a Jordan region in $\mathbf{R}^n$, let $f : E \to \mathbf{R}$ be a bounded function, let R be an n-dimensional rectangle such that $E \subseteq R$, and let $\mathcal{G} = \{R_1, \dots, R_p\}$ be a grid on R. Extend f to $\mathbf{R}^n$ by setting $f(\mathbf{x}) = 0$ for $\mathbf{x} \in \mathbf{R}^n \setminus E$.

i) The *upper sum* of f on E with respect to $\mathcal{G}$ is

$$U(f, \mathcal{G}) := \sum_{R_j \cap E \ne \emptyset} M_j |R_j|,$$

where $M_j = \sup_{\mathbf{x} \in R_j} f(\mathbf{x})$.

---

ii) The *lower sum* of f on E with respect to $\mathcal{G}$ is

$$L(f, \mathcal{G}) := \sum_{R_j \cap E \neq \emptyset} m_j |R_j|,$$

where $m_j = \inf_{\mathbf{x} \in R_j} f(\mathbf{x})$.

iii) The *upper* and *lower integrals* of f on E are defined by

$$(L) \int_E f(\mathbf{x}) \, d\mathbf{x} := (L) \int_E f \, dV := \sup_{\mathcal{G}} L(f, \mathcal{G})$$

and

$$(U) \int_E f(\mathbf{x}) \, d\mathbf{x} := (U) \int_E f \, dV := \inf_{\mathcal{G}} U(f, \mathcal{G}),$$

where the supremum and infimum are taken over all grids $\mathcal{G}$ on R.

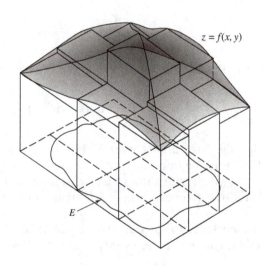

$z = f(x, y)$

E

FIGURE 12.5

Using the fact that $f(\mathbf{x}) = 0$ when $\mathbf{x} \in R \setminus E$, we can modify the proofs of Remarks 5.7, 5.8, and 5.14 to establish the following result.

12.16 Remark. *Let E be a nonempty Jordan region in $\mathbf{R}^n$, let $f : E \to \mathbf{R}$ be bounded, and let R be a rectangle which contains E.*

i) *If $\mathcal{G}$ and $\mathcal{H}$ are grids on R, then*

$$L(f, \mathcal{G}) \leq U(f, \mathcal{H}).$$

ii) *The upper and lower integrals of f over E always exist, do not depend on the choice of R, and satisfy*

$$(L) \int_E f \, dV \le (U) \int_E f \, dV. \tag{6}$$

12.17 Definition.

A real-valued bounded function f defined on a Jordan region E is said to be (*Riemann*) *integrable* on E if and only if for every $\varepsilon > 0$ there is a grid $\mathcal{G}$ such that

$$U(f, \mathcal{G}) - L(f, \mathcal{G}) < \varepsilon.$$

By modifying the proof of Theorem 5.15, we can establish the following result.

12.18 Remark. *Let E be a Jordan region in $\mathbf{R}^n$ and suppose that $f : E \to \mathbf{R}$ is bounded. Then f is integrable on E if and only if*

$$(L) \int_E f \, dV = (U) \int_E f \, dV. \tag{7}$$

When f is integrable on E, we denote the common value in (7) by

$$\int_E f(\mathbf{x}) \, d\mathbf{x} \quad \text{or} \quad \int_E f \, dV$$

and call it the *integral* of f over E. For $n = 2$ (respectively, $n = 3$) we shall frequently denote the integral $\int_E f \, dV$ by $\iint_E f \, dA$ (respectively, by $\iiint_E f \, dV$).

The following result shows that evaluation of Riemann integrals over Jordan regions reduces to evaluation of Riemann integrals over rectangles.

12.19 Remark. *Let E be a Jordan region in $\mathbf{R}^n$, let R be an n-dimensional rectangle which contains E, and suppose that $f : E \to \mathbf{R}$ is integrable on E. If*

$$g(\mathbf{x}) = \begin{cases} f(\mathbf{x}) & \mathbf{x} \in E \\ 0 & \mathbf{x} \notin E, \end{cases}$$

then g is integrable on R and

$$\int_E f \, dV = \int_R g \, dV. \tag{8}$$

Proof. By Definition 12.15, the upper and lower sums of f and g are identical; hence, they have the same upper and lower integrals. It follows from Remark 12.18 that they have the same integrals. ∎

This last proof worked because we defined the upper and lower integrals of a function f on E by extending f to be zero off E. We did this to be sure that $U(f; \mathcal{G})$ was an overestimate of the integral of f and $L(f; \mathcal{G})$ was an underestimate. Unfortunately, the abrupt change from f to 0 at the boundary of E introduces additional complications. The next result shows that since the boundary of E is of volume zero, we can ignore what happens at the boundary.

12.20 Theorem. *Let E be a Jordan region and suppose that $f : E \to \mathbf{R}$ is bounded. Then given $\varepsilon > 0$ there is a grid $\mathcal{G}_0$ such that if $\mathcal{G} := \{R_1, \ldots, R_p\}$ is any grid finer than $\mathcal{G}_0$ and M_j, m_j are defined as in Definition 12.15, then*

$$\left| (U) \int_E f(\mathbf{x})\, dx - \sum_{R_j \subset E^o} M_j\, |R_j| \right| < \varepsilon$$

and

$$\left| (L) \int_E f(\mathbf{x})\, dx - \sum_{R_j \subset E^o} m_j\, |R_j| \right| < \varepsilon.$$

Proof. Let $\varepsilon > 0$ and choose $M > 0$ such that $|f(\mathbf{x})| \le M$ for all $\mathbf{x} \in E$. Since $\mathrm{Vol}(\partial E) = 0$, we can choose a grid $\mathcal{H}_1$ such that $V(\partial E; \mathcal{H}_1) < \varepsilon/(2M)$. Moreover, by the Approximation Property of Infima, we can choose a grid $\mathcal{H}_2$ such that

$$(U) \int_E f(\mathbf{x})\, dx \le U(f; \mathcal{H}_2) < (U) \int_E f(\mathbf{x})\, dx + \frac{\varepsilon}{2}.$$

Let $\mathcal{G}_0$ be a grid finer than both $\mathcal{H}_1$ and $\mathcal{H}_2$, and suppose that $\mathcal{G} = \{R_1, \ldots, R_p\}$ is finer than $\mathcal{G}_0$. Since each R_j is connected, it is easy to see that if R_j intersects E but R_j is not a subset of E^o, then R_j intersects ∂E. [Indeed, if $R_j \cap \overline{\partial E} = \emptyset$, then the pair E^o, $(\mathbf{R}^n \backslash E)^o$ separates R_j, a contradiction since all rectangles are connected—see Remark 9.34.] Since $\mathcal{G}$ is finer than $\mathcal{H}_1$ and $\mathcal{H}_2$, it follows that

$$\left| (U) \int_E f(\mathbf{x})\, dx - \sum_{R_j \subset E^o} M_j\, |R_j| \right| \le \frac{\varepsilon}{2} + \left| U(f; \mathcal{G}) - \sum_{R_j \subset E^o} M_j\, |R_j| \right|$$

$$\le \frac{\varepsilon}{2} + \sum_{R_j \cap \partial E \ne \emptyset} |M_j|\, |R_j|$$

$$\le \frac{\varepsilon}{2} + M\, V(\partial E; \mathcal{G}) < \varepsilon.$$

A similar proof establishes the inequality involving lower sums and lower integrals.   ∎

It follows that if $E^o = \emptyset$, then the upper and lower integrals of any bounded f are zero; that is, $\int_E f\, dV = 0$.

Can we avoid worrying about the boundary by redefining the numbers M_j and m_j in Definition 12.15? For example, why not just define $M_j = \sup_{\mathbf{x} \in R_j \cap E} f(\mathbf{x})$? This approach will not work because the infimum of these upper sums will not equal the integral of f. For example, suppose that $\mathcal{G}_0 = \{[0, 1] \times [0, 1]\}$ and that $\mathcal{G} = \{R_1, R_2, R_3, R_4\}$, where the R_j's are formed by bisecting the sides of $\mathcal{G}_0$; that is, each R_j is exactly one-fourth of the unit square. Let $E = R_1$ and suppose that $f = 1$ on R_1^o, but $f = -1$ otherwise. If M_j is defined as above, then $U(f, \mathcal{G}_0) = 1$ but $U(f, \mathcal{G}) = -1/2$, which is LESS than $\int_E f \, dV = 1/4$. Evidently, in order to define f on E by looking at grids on a rectangle which contains E, we must extend f to be zero off E.

Our first application of Theorem 12.20 is an analogue of Theorem 5.10.

12.21 Theorem. *If E is a closed Jordan region in $\mathbf{R}^n$ and $f : E \to \mathbf{R}$ is continuous on E, then f is integrable on E.*

Proof. Since by hypothesis E is closed and bounded, f is bounded on E (apply the Extreme Value Theorem and the Heine–Borel Theorem). To show that f is integrable on E, let $\varepsilon > 0$ and R be a rectangle which contains E. By Theorem 12.20, there is a grid $\mathcal{G}_0$ on R such that if $\mathcal{G} = \{R_1, \dots, R_p\}$ is any grid which is finer than $\mathcal{G}_0$, then

$$\left| (U) \int_E f \, dV - (L) \int_E f \, dV - \sum_{R_j \subset E^o} (M_j - m_j) \, |R_j| \right| < \varepsilon. \tag{9}$$

Since f is uniformly continuous on E, choose $\delta > 0$ such that

$$\|\mathbf{x} - \mathbf{y}\| < \delta \quad \text{and} \quad \mathbf{x}, \mathbf{y} \in E \quad \text{imply} \quad |f(\mathbf{x}) - f(\mathbf{y})| < \varepsilon.$$

Make $\mathcal{G}$ finer by insisting that for each $R_j \in \mathcal{G}$, $\|\mathbf{x} - \mathbf{y}\| < \delta$ when $\mathbf{x}, \mathbf{y} \in R_j$. Then the choice of δ implies that $M_j - m_j < \varepsilon$ for all j which satisfy $R_j \subset E$. Hence it follows from Remark 12.16 and (9) that

$$0 \le (U) \int_E f \, dV - (L) \int_E f \, dV < \varepsilon + \sum_{R_j \subset E^o} (M_j - m_j)|R_j|$$

$$< \varepsilon + \varepsilon V(E; \mathcal{G}) \le \varepsilon(1 + |R|).$$

Since $\varepsilon > 0$ was arbitrary, we conclude that $(U) \int_E f \, dV = (L) \int_E f \, dV$ (i.e., f is integrable on E). $\blacksquare$

The proof of Theorem 12.21 shows that the hypothesis that E be closed can be weakened if we insist that f be uniformly continuous on E. (See also Exercises 9.4.8 and 12.2.11.)

The following result shows that the volume of a Jordan region can be computed by integration.

12.22 Theorem. *If E is a Jordan region, then*

$$\text{Vol}(E) = \int_E 1 \, d\mathbf{x}.$$

Proof. By Exercise 12.1.5b, we may suppose that E is closed. Let R be a rectangle containing E and $\mathcal{G} = \{R_1, \ldots, R_p\}$ be a grid on R. Define $f(\mathbf{x}) = 1$ for $\mathbf{x} \in E$ and $f(\mathbf{x}) = 0$ for $\mathbf{x} \notin E$, and notice by Theorem 12.21 that f is integrable on E. Since $R_j \cap E \neq \emptyset$ implies $R_j \cap \overline{E} \neq \emptyset$, and $M_j(f) = 1$ when $R_j \cap E \neq \emptyset$, it is clear, by the definition of upper sums and outer sums, that $U(f, \mathcal{G}) \leq V(E; \mathcal{G})$. Taking the infimum of this inequality over all grids $\mathcal{G}$, and applying Theorem 12.21 together with Definitions 12.15 and 12.5, we have

$$\int_E 1 \, d\mathbf{x} = \inf_{\mathcal{G}} U(f, \mathcal{G}) \leq \inf_{\mathcal{G}} V(E; \mathcal{G}) = \text{Vol}(E).$$

On the other hand, since $\text{Vol}(\partial E) = 0$, given $\varepsilon > 0$ we can choose $\mathcal{G}$ so that $V(\partial E; \mathcal{G}) < \varepsilon$. Since $m_j(f) = 0$ when $R_j \cap E^c \neq \emptyset$, and $m_j(f) = 1$ when $R_j \subseteq E$, it follows that

$$\int_E 1 \, d\mathbf{x} \geq L(f; \mathcal{G}) = \sum_{R_j \cap E \neq \emptyset} m_j |R_j|$$

$$= \sum_{R_j \subseteq E} |R_j| \geq \sum_{R_j \cap \overline{E} \neq \emptyset} |R_j| - \sum_{R_j \cap \partial E \neq \emptyset} |R_j|$$

$$= V(E; \mathcal{G}) - V(\partial E; \mathcal{G}) \geq \text{Vol}(E) - \varepsilon.$$

Since $\varepsilon > 0$ was arbitrary, it follows that $\int_E 1 \, d\mathbf{x} \geq \text{Vol}(E)$.   ∎

As in the one-dimensional case, the integral of a sum of functions over a union of regions can be broken into simpler pieces.

12.23 Theorem.   [LINEAR PROPERTIES].
Let E be a Jordan region in $\mathbf{R}^n$, let $f, g : E \to \mathbf{R}$, and let α be a scalar.

i) *If f, g are integrable on E, then so are αf and $f + g$. In fact,*

$$\int_E \alpha f \, dV = \alpha \int_E f \, dV \tag{10}$$

and

$$\int_E (f + g) \, d\mathbf{x} = \int_E f \, dV + \int_E g \, dV. \tag{11}$$

ii) *If $E_1, E_2 \subseteq E$ are nonoverlapping Jordan regions and f is integrable on both E_1 and E_2, then f is integrable on $E_1 \cup E_2$ and*

$$\int_{E_1 \cup E_2} f \, dV = \int_{E_1} f \, dV + \int_{E_2} f \, dV. \tag{12}$$

Proof. We suppose for simplicity that $\alpha > 0$. Let $\varepsilon > 0$ and choose a grid $\mathcal{G}$ such that

$$U(f, \mathcal{G}) - \varepsilon < \int_E f \, dV < L(f, \mathcal{G}) + \varepsilon. \tag{13}$$

Notice that $U(\alpha f, \mathcal{G}) = \alpha U(f, \mathcal{G})$ and $L(\alpha f, \mathcal{G}) = \alpha L(f, \mathcal{G})$. Multiplying (13) by α, we obtain

$$U(\alpha f, \mathcal{G}) - \alpha \varepsilon < \alpha \int_E f \, dV < L(\alpha f, \mathcal{G}) + \alpha \varepsilon.$$

In particular,

$$\inf_{\mathcal{G}} U(\alpha f, \mathcal{G}) < \alpha \int_E f \, dV + \alpha \varepsilon$$

and

$$\sup_{\mathcal{G}} L(\alpha f, \mathcal{G}) > \alpha \int_E f \, dV - \alpha \varepsilon.$$

Taking the limit of these inequalities as $\varepsilon \to 0$, we conclude that

$$\inf_{\mathcal{G}} U(\alpha f, \mathcal{G}) \le \alpha \int_E f \, dV \le \sup_{\mathcal{G}} L(\alpha f, \mathcal{G}).$$

This proves (10).

To prove (11), choose a grid $\mathcal{G}$ such that

$$U(f, \mathcal{G}) - \varepsilon < \int_E f \, dV < L(f, \mathcal{G}) + \varepsilon$$

and

$$U(g, \mathcal{G}) - \varepsilon < \int_E g \, dV < L(g, \mathcal{G}) + \varepsilon.$$

Adding these inequalities, we have

$$U(f, \mathcal{G}) + U(g, \mathcal{G}) - 2\varepsilon < \int_E f \, dV + \int_E g \, dV < L(f, \mathcal{G}) + L(g, \mathcal{G}) + 2\varepsilon.$$

By definition, $U(f + g, \mathcal{G}) \le U(f, \mathcal{G}) + U(g, \mathcal{G})$ and $L(f + g, \mathcal{G}) \ge L(f, \mathcal{G}) + L(g, \mathcal{G})$. Therefore,

$$U(f + g, \mathcal{G}) - 2\varepsilon < \int_E f \, dV + \int_E g \, dV < L(f + g, \mathcal{G}) + 2\varepsilon;$$

that is,

$$\inf_{\mathcal{G}} U(f + g, \mathcal{G}) \le \int_E f \, dV + \int_E g \, dV \le \sup_{\mathcal{G}} L(f + g, \mathcal{G}).$$

This proves (11). ∎

To prove (12), let $\varepsilon > 0$ and apply Theorem 12.20 three times to choose a grid $\mathcal{G}_0$ so that if $\mathcal{G} = \{R_1, \ldots, R_p\}$ is finer than $\mathcal{G}_0$, then

$$\left| \int_{E_i} f \, dV - \sum_{R_j \subset E_i^o} M_j |R_j| \right| < \varepsilon \tag{14}$$

for $i = 1, 2$, and

$$\left| (U) \int_{E_1 \cup E_2} f \, dV - \sum_{R_j \subset (E_1 \cup E_2)^o} M_j |R_j| \right| < \varepsilon. \tag{15}$$

Since E_1 and E_2 are nonoverlapping, we may also assume that

$$V(E_1 \cap E_2; \mathcal{G}) < \varepsilon. \tag{16}$$

Let $M = \max\{|M_1|, \ldots, |M_p|\}$. Since each R_j is connected and $E_1^o \cap E_2^o = \emptyset$, it is easy to see that each $R_j \subset (E_1 \cup E_2)^o$ satisfies one and only one of the following three conditions: i) $R_j \subset E_1^o$; ii) $R_j \subset E_2^o$; or iii) $R_j \cap \overline{E_1 \cap E_2} \neq \emptyset$. Hence, it follows from (15), (16), and (14) that

$$(U) \int_{E_1 \cup E_2} f \, dV < \varepsilon + \sum_{R_j \subset (E_1 \cup E_2)^o} M_j |R_j|$$

$$\leq \varepsilon + \sum_{R_j \subset E_1^o} M_j |R_j| + \sum_{R_j \subset E_2^o} M_j |R_j| + M \, V(E_1 \cap E_2; \mathcal{G})$$

$$< 3\varepsilon + \int_{E_1} f \, dV + \int_{E_2} f \, dV + M\varepsilon.$$

Since $\varepsilon > 0$ was arbitrary, we obtain

$$(U) \int_{E_1 \cup E_2} f \, dV \leq \int_{E_1} f \, dV + \int_{E_2} f \, dV.$$

A similar argument establishes

$$(L) \int_{E_1 \cup E_2} f \, dV \geq \int_{E_1} f \, dV + \int_{E_2} f \, dV.$$

Thus (12) holds. ∎

The following result shows that the value of an integral remains the same when the integrand is changed on a set of volume zero (compare with Exercise 5.1.6).

12.24 Theorem. *Let E be a Jordan region in $\mathbf{R}^n$ and let $f, g : E \to \mathbf{R}$ be bounded functions.*

i) *If E_0 is of volume zero, then g is integrable on E_0 and*

$$\int_{E_0} g \, dV = 0.$$

ii) *If f is integrable on E, if E_0 is a subset of E of volume zero, and if $g(x) = f(x)$ for all $\mathbf{x} \in E \setminus E_0$, then g is integrable on E and*

$$\int_E g \, dV = \int_E f \, dV.$$

Proof. i) If $E_0^o \neq \emptyset$, then E_0 contains a ball, hence a nondegenerate rectangle, so $\mathrm{Vol}(E) > 0$, a contradiction. Since $E_0^o = \emptyset$, it follows from Theorem 12.20 that

$$(U) \int_{E_0} g \, dV = (L) \int_{E_0} g \, dV = 0.$$

ii) Since $f = g$ on $E \setminus E_0$, it follows from the proof of Theorem 12.23ii and part i) above that

$$\int_E f \, dV = \int_{E \setminus E_0} f \, dV + \int_{E_0} f \, dV$$

$$= (U) \int_{E \setminus E_0} g \, dV + (U) \int_{E_0} g \, dV \geq (U) \int_E g \, dV.$$

Similarly, $\int_E f \, dV \leq (L) \int_E g \, dV$. ∎

This suggests a way to define the integral of f on a Jordan region E when f is not defined on all of E. Indeed, if f is defined on $E \setminus E_0$, where E_0 is of volume zero, and if the function

$$g(\mathbf{x}) := \begin{cases} f(\mathbf{x}) & \mathbf{x} \in E \setminus E_0 \\ 0 & \mathbf{x} \in E_0 \end{cases}$$

is integrable on E, then define

$$\int_E f \, dV := \int_E g \, dV.$$

For example,

$$\int_0^2 \frac{x^2 - 1}{x - 1} \, dx = \int_0^2 (x + 1) \, dx = 4.$$

Henceforth, the phrase "$f : E \to \mathbf{R}$ is integrable" includes the possibility that f may not be defined on a subset of E of volume zero.

The following result is a multidimensional analogue of Theorems 5.21 and 5.22.

12.25 Theorem.   [COMPARISON THEOREM FOR MULTIPLE INTEGRALS].
Let E be a Jordan region in $\mathbf{R}^n$ and suppose that $f, g : E \to \mathbf{R}$ are integrable on E.

i) *If $f(\mathbf{x}) \le g(\mathbf{x})$ for $\mathbf{x} \in E$, then*

$$\int_E f \, dV \le \int_E g \, dV.$$

ii) *If m, M are scalars which satisfy $m \le f(\mathbf{x}) \le M$ for $\mathbf{x} \in E$, then*

$$m \, \mathrm{Vol}(E) \le \int_E f \, dV \le M \, \mathrm{Vol}(E).$$

iii) *The function $|f|$ is integrable on E and*

$$\left| \int_E f \, dV \right| \le \int_E |f| \, dV. \tag{17}$$

Proof. i) If $f \le g$ on E, then $L(f, \mathcal{G}) \le L(g, \mathcal{G})$ for any grid $\mathcal{G}$. Taking the supremum of this inequality over all grids $\mathcal{G}$ verifies part i).

ii) By Theorem 12.22, (10), and part i),

$$m \, \mathrm{Vol}(E) = \int_E m \, d\mathbf{x} \le \int_E f \, dV \le \int_E M \, d\mathbf{x} = M \, \mathrm{Vol}(E).$$

iii) Let $\varepsilon > 0$ and choose by Definition 12.17 a grid $\mathcal{G} = \{R_1, \ldots, R_p\}$ such that

$$U(f, \mathcal{G}) - L(f, \mathcal{G}) < \varepsilon. \tag{18}$$

By repeating the argument which verified (10) in Theorem 5.22, we have

$$\sup_{\mathbf{x} \in R_j} |f(\mathbf{x})| - \inf_{\mathbf{x} \in R_j} |f(\mathbf{x})| \le \sup_{\mathbf{x} \in R_j} f(\mathbf{x}) - \inf_{\mathbf{x} \in R_j} f(\mathbf{x}).$$

Hence, it follows from (18) that

$$U(|f|, \mathcal{G}) - L(|f|, \mathcal{G}) \le U(f, \mathcal{G}) - L(f, \mathcal{G}) < \varepsilon.$$

Thus $|f|$ is integrable on E. Since $-|f| \le f \le |f|$, we conclude by part i) that

$$-\int_E |f|\, dV \le \int_E f\, dV \le \int_E |f|\, dV. \qquad \blacksquare$$

12.26 Theorem. [MEAN VALUE THEOREM FOR MULTIPLE INTEGRALS].

Let E be a Jordan region in $\mathbf{R}^n$ and let $f, g : E \to \mathbf{R}$ be integrable on E with $g(\mathbf{x}) \ge 0$ for all $\mathbf{x} \in E$.

i) *There is a number c satisfying*

$$\inf_{\mathbf{x} \in E} f(\mathbf{x}) \le c \le \sup_{\mathbf{x} \in E} f(\mathbf{x}) \qquad (19)$$

such that

$$c\int_E g\, dV = \int_E fg\, dV. \qquad (20)$$

ii) *There is a number c satisfying (19) such that*

$$c \operatorname{Vol}(E) = \int_E f\, dV.$$

Proof. i) By hypothesis, the product fg is integrable on E (see Exercise 12.2.8). Let $m = \inf_{\mathbf{x} \in E} f(\mathbf{x})$ and $M = \sup_{\mathbf{x} \in E} f(\mathbf{x})$. Since $g \ge 0$, Theorem 12.25 implies that

$$m\int_E g\, dV \le \int_E fg\, dV \le M\int_E g\, dV. \qquad (21)$$

If $\int_E g\, dV = 0$, then $\int_E f(\mathbf{x})g\, dV = 0$ by (21), so (20) holds for any c. If $\int_E g\, dV \ne 0$, then (20) holds for

$$c = \frac{\int_E fg\, dV}{\int_E g\, dV}.$$

ii) Apply part i) to $g(\mathbf{x}) = 1$. $\qquad \blacksquare$

We close this section with some optional material which generalizes a concept introduced in Section 9.6.

---

*12.27 Definition.

A set $E \subset \mathbf{R}^n$ is said to be of *measure zero* if and only if for every $\varepsilon > 0$ there is a countable collection of rectangles $\{R_j\}_{j \in \mathbf{N}}$ such that

$$E \subset \bigcup_{j=1}^{\infty} R_j \quad \text{and} \quad \sum_{j=1}^{\infty} |R_j| < \varepsilon.$$

---

*12.28 Remark. *If $E_1,\ E_2, \ldots$ is a sequence of subsets of $\mathbf{R}^n$ and each E_k is of measure zero, then*

$$E = \bigcup_{k=1}^{\infty} E_k$$

is also of measure zero.

Proof. Let $\varepsilon > 0$. For each $k \in \mathbf{N}$, choose a collection of rectangles $\{R_j^{(k)}\}_{j \in \mathbf{N}}$ which covers E_k such that

$$\sum_{j=1}^{\infty} |R_j^{(k)}| < \frac{\varepsilon}{2^k}.$$

Clearly, the collection $\{R_j^{(k)}\}_{j,k \in \mathbf{N}}$ is countable, covers E, and

$$\sum_{k=1}^{\infty} \sum_{j=1}^{\infty} |R_j^{(k)}| \leq \sum_{k=1}^{\infty} \frac{\varepsilon}{2^k} = \varepsilon.$$

Consequently, E is of measure zero. ∎

Every singleton $E = \{\mathbf{a}\}$ in $\mathbf{R}^n$ is of measure zero. In fact, by comparing Definition 12.27 with Theorem 12.4, it is clear that every set of volume zero is a set of measure zero. The converse of this statement is false. Indeed, for each $a \in \mathbf{R}$ the set $\{(a, y) : y \in [0, 1]\}$ is of volume zero and, hence, is of measure zero. Thus, by Remark 12.28, $E := \mathbf{Q} \times [0, 1]$ is a set of measure zero. On the other hand, it is clear that $\underline{\text{Vol}}(E) = 0 < 1 \leq \overline{\text{Vol}}(E)$, so E is not a set of volume zero; in fact, E is not even a Jordan region.

An analogue of Lebesgue's Theorem holds for multiple integrals.

*12.29 Theorem. *Let E be a Jordan region and let $f : E \to \mathbf{R}$ be bounded.*

i) *f is Riemann integrable on E if and only if the set of points of discontinuity of f on E is of measure zero.*

ii) *Suppose that V is an open set in $\mathbf{R}^n$ such that $\overline{E} \subset V$, and that $\phi : V \to \mathbf{R}^n$ is 1–1 and ϕ^{-1} is $\mathcal{C}^1$ on $\phi(V)$ with $\Delta_{\phi^{-1}} \neq 0$. If f is integrable on $\phi(E)$, then $f \circ \phi$ is integrable on E.*

Proof. i) This part can be verified by modifying the proof of Theorem 9.49 (see Spivak [12], p. 53).

ii) By part i) and Theorem 12.10, it suffices to show that the set of points of discontinuity of $f \circ \phi$ on E is a set of measure zero. Let $\varepsilon > 0$. Since f is integrable on $\phi(E)$, its set of points of discontinuity, D, can be covered by cubes Q_k such that $\sum_{k=1}^{\infty} |Q_k| < \varepsilon$. Set $\psi = \phi^{-1}$ and apply (2), with ψ in place of ϕ, to choose an absolute constant C and cubes Q_k^{ψ} such that $\psi(Q_k) \subseteq Q_k^{\psi}$ and $|Q_k^{\psi}| \leq C|Q_k|$. Then $\{Q_k^{\psi}\}$ covers $\psi(D) = \phi^{-1}(D)$ and

$$\sum_{k=1}^{\infty} |Q_k^{\psi}| \leq C \sum_{k=1}^{\infty} |Q_k| < C\varepsilon.$$

Hence, $\phi^{-1}(D) := \psi(D)$ is a set of measure zero. But since D is the set of points of discontinuity of f on $\phi(E)$, $\phi^{-1}(D)$ is the set of points of discontinuity of $f \circ \phi$ on E. Hence $f \circ \phi$ is Riemann integrable by part i). $\blacksquare$

EXERCISES

12.2.1. Using Exercise 1.4.4a, compute the upper and lower sums $U(f, \mathcal{G}_m)$, $L(f, \mathcal{G}_m)$ for $m \in \mathbf{N}$, where $f(x, y) = xy$ and $\mathcal{G}_m$ is determined by

$$\mathcal{P}_j(\mathcal{G}_m) = \{k/2^m : k = 0, 1, \ldots, 2^m\}$$

for $j = 1, 2$. Prove that

$$\lim_{m \to \infty} U(f, \mathcal{G}_m) - L(f, \mathcal{G}_m) = 0.$$

12.2.2. Let E be a Jordan region in $\mathbf{R}^n$ with $E \subseteq [0, 1] \times \cdots \times [0, 1]$. If f, g are integrable on E with

$$\int_E f \, dV = 1 \quad \text{and} \quad \int_E g \, dV = -1,$$

and if $g(\mathbf{x}) \leq f(\mathbf{x})$ for all $\mathbf{x} \in E$, prove that for each $j \in \{1, 2, \ldots, n\}$ there is a $0 \leq t_j \leq 2$ such that

$$\int_E x_j^2 (f(\mathbf{x}) - g(\mathbf{x})) \, d\mathbf{x} = t_j.$$

12.2.3. **This exercise is used in Sections 12.4, 13.5, and 13.6.** Let E be an open Jordan region in $\mathbf{R}^n$ and $\mathbf{x}_0 \in E$. If $f : E \to \mathbf{R}$ is integrable on E and continuous at $\mathbf{x}_0$, prove that

$$\lim_{r \to 0+} \frac{1}{\mathrm{Vol}(B_r(\mathbf{x}_0))} \int_{B_r(\mathbf{x}_0)} f \, dV = f(\mathbf{x}_0).$$

12.2.4. a) Suppose that E is a Jordan region in $\mathbf{R}^n$ and that $f_k : E \to \mathbf{R}$ are integrable on E for $k \in \mathbf{N}$. If $f_k \to f$ uniformly on E as $k \to \infty$, prove that f is integrable on E and

$$\lim_{k \to \infty} \int_E f_k(\mathbf{x}) \, d\mathbf{x} = \int_E f(\mathbf{x}) \, d\mathbf{x}.$$

b) Prove that

$$\lim_{k \to \infty} \iint_E \cos(x/k) e^{y/k} \, dA$$

exists, and find its value for any Jordan region E in $\mathbf{R}^2$.

12.2.5. If $E_0 \subset E$ are Jordan regions in $\mathbf{R}^n$ and $f : E \to \mathbf{R}$ is integrable on E, prove that f is integrable on E_0.

12.2.6. Let H be a closed, connected, nonempty Jordan region and suppose that $f : H \to \mathbf{R}$ is continuous. If $g : H \to \mathbf{R}$ is integrable and nonnegative on H, prove that there is an $\mathbf{x}_0 \in H$ such that

$$f(\mathbf{x}_0) \int_H g(\mathbf{x}) \, d\mathbf{x} = \int_H f(\mathbf{x}) g(\mathbf{x}) \, d\mathbf{x}.$$

12.2.7. Suppose that $Q := \{(x, y) \in \mathbf{R} : x > 0 \text{ and } y > 0\}$ and that f is a continuous function on $\mathbf{R}^2$ whose first-order partial derivative satisfies $|f_x| \le 1$. If

$$F(x, y) := \frac{1}{x^3} \iint_{B_x(0,0)} (f(u, y) - f(v, y)) \, d(u, v)$$

for $(x, y) \in Q$, prove that F is bounded on Q.
[Hint: You may use polar coordinates to change variables in F.]

12.2.8. Suppose that E is a Jordan region in $\mathbf{R}^n$ and that $f, g : E \to \mathbf{R}$ is integrable on E.

a) Modifying the proof of Corollary 5.23, prove that fg is integrable on E.

b) Prove that $f \vee g$ and $f \wedge g$ are integrable on E (see Exercise 3.1.8).

12.2.9. Suppose that V is open in $\mathbf{R}^n$ and that $f : V \to \mathbf{R}$ is continuous. Prove that if

$$\int_E f \, dV = 0$$

for all nonempty Jordan regions $E \subset V$, then $f = 0$ on V.

12.2.10. Suppose that E is a Jordan region and that $f : E \to \mathbf{R}$ is integrable.

a) If $f(E) \subseteq H$, for some compact set H, and $\phi : H \to \mathbf{R}$ is continuous, prove that $\phi \circ f$ is integrable on E.

*b) Show that part a) is false if ϕ has even one point of discontinuity.

12.2.11. Prove the following special case of Theorem 12.29i. Suppose that E and E_0 are a Jordan regions in $\mathbf{R}^n$, and that $f : E \to \mathbf{R}$ is bounded. If f is continuous on $E \setminus E_0$, then f is integrable on E.

12.3 ITERATED INTEGRALS

If $f(x_1, \ldots, x_k, \ldots, x_j, \ldots, x_n)$ is defined for $x_k \in [c, d]$ and $x_j \in [a, b]$, $j \neq k$, then we shall call

$$\int_c^d \int_a^b f(x_1, \ldots, x_n) \, dx_j \, dx_k := \int_c^d \left(\int_a^b f(x_1, \ldots, x_n) \, dx_j \right) dx_k$$

an *iterated integral*, when the integrals on the right side exist. In a similar way, we define higher-order iterated integrals.

In the preceding section we defined the Riemann integral of a multivariable function but developed no practical way to evaluate it. In this section we show that, for a large collection of Jordan regions E, integrals over E can be evaluated using iterated integrals.

For simplicity, we begin with the two-dimensional case. Recall that if $\phi : [a, b] \to \mathbf{R}$ is bounded, then the upper Riemann integral, $(U) \int_a^b \phi(x) \, dx$, and the lower Riemann integral, $(L) \int_a^b \phi(x) \, dx$, both exist and are finite.

12.30 Lemma.
Let $R = [a, b] \times [c, d]$ be a two-dimensional rectangle and suppose that $f : R \to \mathbf{R}$ is bounded. If $f(x, \cdot)$ is integrable on $[c, d]$ for each $x \in [a, b]$, then

$$(L) \iint_R f \, dA \le (L) \int_a^b \left(\int_c^d f(x, y) dy \right) dx$$

$$\le (U) \int_a^b \left(\int_c^d f(x, y) dy \right) dx \le (U) \iint_R f \, dA. \tag{22}$$

Proof. Let $R_{ij} = [x_{i-1}, x_i] \times [y_{j-1}, y_j]$, where $\{x_0, \ldots, x_k\}$ is a partition of $[a, b]$ and $\{y_0, \ldots, y_\ell\}$ is a partition of $[c, d]$. Then $\mathcal{G} = \{R_{ij} : i = 1, 2, \ldots, k, j = 1, 2 \ldots, \ell\}$ is a grid on R.

Let $\varepsilon > 0$, choose $\mathcal{G}$ so that

$$U(f, \mathcal{G}) - \varepsilon < (U) \iint_R f \, dA, \tag{23}$$

and set

$$M_{ij} = \sup_{(x, y) \in R_{ij}} f(x, y). \tag{24}$$

Since $(U) \int_a^b \phi(x)\, dx = \sum_{i=1}^k (U) \int_{x_{i-1}}^{x_i} \phi(x)\, dx$ and

$$(U) \int_a^b (\phi(x) + \psi(x))\, dx \le (U) \int_a^b \phi(x)\, dx + (U) \int_a^b \psi(x)\, dx$$

for any bounded functions ϕ and ψ defined on $[a, b]$ (see Exercise 5.1.7), we can write

$$(U) \int_a^b \left(\int_c^d f(x, y) dy \right) dx = \sum_{i=1}^k (U) \int_{x_{i-1}}^{x_i} \left(\sum_{j=1}^{\ell} \int_{y_{j-1}}^{y_j} f(x, y) dy \right) dx$$

$$\le \sum_{i=1}^k \sum_{j=1}^{\ell} (U) \int_{x_{i-1}}^{x_i} \left(\int_{y_{j-1}}^{y_j} f(x, y) dy \right) dx$$

$$\le \sum_{i=1}^k \sum_{j=1}^{\ell} M_{ij} (x_i - x_{i-1})(y_j - y_{j-1}) = U(f, \mathcal{G}).$$

It follows from (23) that

$$(U) \int_a^b \left(\int_c^d f(x, y) dy \right) dx < (U) \iint_R f\, dA + \varepsilon.$$

Taking the limit of this inequality as $\varepsilon \to 0$, we obtain

$$(U) \int_a^b \left(\int_c^d f(x, y)\, dy \right) dx \le (U) \iint_R f\, dA.$$

Similarly,

$$(L) \int_a^b \left(\int_c^d f(x, y)\, dy \right) dx \ge (L) \iint_R f\, dA. \qquad \blacksquare$$

We are now prepared to show that, under reasonable conditions, a double integral over a rectangle reduces to an iterated integral.

12.31 Theorem. [FUBINI'S THEOREM].
*Let $R = [a, b] \times [c, d]$ be a two-dimensional rectangle and let $f : R \to \mathbf{R}$.
Suppose that $f(x, \cdot)$ is integrable on $[c, d]$ for each $x \in [a, b]$, that $f(\cdot, y)$ is integrable on $[a, b]$ for each $y \in [c, d]$, and that f is integrable on R (as a function of two variables). Then*

$$\iint_R f\, dA = \int_a^b \int_c^d f(x, y)\, dy\, dx = \int_c^d \int_a^b f(x, y)\, dx\, dy. \qquad (25)$$

NOTE: *These hypotheses hold if f is continuous on the rectangle $[a, b] \times [c, d]$.*

Proof. For each $x \in [a, b]$, set $g(x) = \int_c^d f(x, y)\, dy$. Since f is integrable on R, Lemma 12.30 implies that

$$\iint_R f\, dA = (U) \int_a^b g(x)\, dx = (L) \int_a^b g(x)\, dx.$$

Hence, g is integrable on $[a, b]$ and the first identity in (25) holds. Reversing the roles of x and y, we obtain

$$\iint_R f\, dA = \int_c^d \int_a^b f(x, y)\, dx\, dy.$$

Hence, the second identity in (25) holds. ∎

The second identity in Fubini's Theorem is as important as the first. It tells us that, under certain conditions, the order of integration in an iterated integral can be reversed. Frequently, one of these iterated integrals is easier to evaluate than the other.

12.32 EXAMPLE.

Find

$$\int_0^1 \int_0^1 y^3 e^{xy^2}\, dy\, dx.$$

Solution. This iterated integral looks tough to integrate. However, if we change the order of integration, using Fubini's Theorem, and substitute $u = y^2$, we obtain

$$\int_0^1 \int_0^1 y^3 e^{xy^2}\, dx\, dy = \int_0^1 y(e^{y^2} - 1)\, dy = \frac{e - 2}{2}. \quad ∎$$

The following three remarks show that the hypotheses of Fubini's Theorem cannot be relaxed. First, we show that existence of both iterated integrals is not enough to apply Fubini. You must also verify that f is integrable in the two-dimensional sense.

12.33 Remark. *There exists a function $f : \mathbf{R}^2 \to \mathbf{R}$ such that $f(x, \cdot)$ and $f(\cdot, y)$ are both integrable on $[0, 1]$, but the iterated integrals are not equal.*

Proof. Set

$$f(x, y) = \begin{cases} 2^{2n} & (x, y) \in [2^{-n}, 2^{-n+1}) \times [2^{-n}, 2^{-n+1}), \ n \in \mathbf{N}, \\ -2^{2n+1} & (x, y) \in [2^{-n-1}, 2^{-n}) \times [2^{-n}, 2^{-n+1}), \ n \in \mathbf{N}, \\ 0 & \text{otherwise.} \end{cases}$$

Notice that for each fixed $y_0 \in [0, 1)$, $f(x, y_0)$ takes on only two nonzero values and is integrable on $[0, 1)$ in x. For example, if $y_0 \in [2^{-n}, 2^{-n+1})$, then $f(x, y_0) = 2^{2n}$ for $x \in [2^{-n}, 2^{-n+1})$, and $f(x, y_0) = -2^{2n+1}$ for $x \in [2^{-n-1}, 2^{-n})$; hence, $f(x, y_0)$ is bounded on $[0, 1)$, and

$$\int_0^1 f(x, y_0) \, dx = \int_{2^{-n}}^{2^{-n+1}} 2^{2n} \, dx - \int_{2^{-n-1}}^{2^{-n}} 2^{2n+1} \, dx = 2^n - 2^n = 0. \tag{26}$$

The same is true for $f(x_0, y)$ when $x_0 \in [0, 1/2)$, but when $x_0 \in [1/2, 1)$, $f(x_0, y)$ takes on only one nonzero value — namely, $f(x_0, y) = 4$ when $y \in [1/2, 1)$, and equals zero otherwise. It follows that

$$\int_0^1 \int_0^1 f(x, y) \, dy \, dx = \int_{1/2}^1 \int_{1/2}^1 4 \, dy \, dx = 1.$$

On the other hand, by (26) we have

$$\int_0^1 \int_0^1 f(x, y) \, dx \, dy = 0.$$

Thus the iterated integrals of f are not equal. ∎

The function in Remark 12.33 is not bounded. The following example shows that even when f is bounded, existence of the iterated integrals is not enough to conclude that f is integrable in the two-dimensional sense.

12.34 Remark. *There exists a bounded function $f : \mathbf{R}^2 \to \mathbf{R}$ such that $f(x, \cdot)$ and $f(\cdot, y)$ are both integrable on $[0, 1]$, but f is not integrable on $[0, 1] \times [0, 1]$.*

Proof. Set

$$f(x, y) = \begin{cases} 1 & (x, y) = \left(\dfrac{p}{2^n}, \dfrac{q}{2^n} \right), \quad 0 < p, q < 2^n, n \in \mathbf{N}, \\ 0 & \text{otherwise.} \end{cases}$$

Notice that if $x_0 = p/2^n$, then $f(x_0, y) = 1$ only when $y = q/2^n$ for some $q = 1, 2, \ldots, 2^n - 1$. Hence, for each fixed $x_0 \in [0, 1]$, $f(x_0, y) = 0$ except for finitely many y's. It follows from Exercise 5.1.6 that

$$\int_0^1 f(x, y) \, dy = 0$$

for all $x \in [0, 1]$. A similar statement holds for the dx integral. Consequently,

$$\int_0^1 \int_0^1 f(x, y) \, dy \, dx = \int_0^1 \int_0^1 f(x, y) \, dx \, dy = 0.$$

To see that the double integral of f does not exist, let $R_j := [a, b] \times [c, d]$ be a nondegenerate rectangle in $[0, 1] \times [0, 1]$. It is easy to verify that $[a, b]$ and $[c, d]$ both contain irrational points and points of the form $p/2^n$ (just use density of irrationals, and repeat the proof of Theorem 1.18 with 2^n in place of n). Thus if $\mathcal{G} = \{R_j\}$ is a grid on $[0, 1] \times [0, 1]$, then $M_j(f) = 1$ and $m_j(f) = 0$ for all j, and $U(f, \mathcal{G}) - L(f, \mathcal{G}) = 1 - 0 = 1$. Hence, f is not integrable on $[0, 1] \times [0, 1]$. ∎

Thus we cannot assume that a function of several variables is integrable just because its iterated integrals exist and are equal. (See also Exercises 12.3.5 and 12.3.9.)

The next result is starred because it uses Lebesgue's characterization of Riemann integrability (see Theorems 9.49 and 12.29i).

*12.35 Remark. *There exists a function $f : \mathbf{R}^2 \to \mathbf{R}$ such that f is integrable on $[0, 1] \times [0, 1]$, $f(\cdot, y)$ is integrable on $[0, 1]$ for all $y \in [0, 1]$, but $f(x, \cdot)$ is not integrable on $[0, 1]$ for infinitely many $x \in [0, 1]$.*

Proof. Let

$$f(x, y) = \begin{cases} 0 & \text{when } x = 0 \text{ or when } x \text{ or } y \text{ is irrational} \\ 1/q & \text{when } x, y \in \mathbf{Q} \text{ and } x = p/q \text{ is in reduced form.} \end{cases}$$

By the argument of Example 3.33, the function f is continuous and zero on the set $([0, 1] \setminus \mathbf{Q}) \times [0, 1]$. Hence, by Lebesgue's Theorem, f is integrable on the square $R = [0, 1] \times [0, 1]$. By computing its lower sums, we find that $\iint_R f \, dA = 0$.

Similarly, for each $y \in [0, 1]$, $f(\cdot, y)$ is integrable on $[0, 1]$ with $\int_0^1 f(x, y) \, dx = 0$. Thus

$$\int_0^1 \left(\int_0^1 f(x, y) \, dx \right) dy = \iint_R f \, dA = 0.$$

On the other hand, since for each nonzero $x \in \mathbf{Q}$ the function $f(x, \cdot)$ is nowhere continuous, it cannot be integrable on $[0, 1]$. Therefore, the other iterated integral in Fubini's Theorem does not exist. ∎

Fubini's Theorem shows us how to evaluate a double integral over a rectangle by means of iterated integrals. The following result shows that the integral of a continuous function over a rectangle in $\mathbf{R}^n$ can be evaluated using n partial integrals.

12.36 Lemma.
Let $R = [a_1, b_1] \times \cdots \times [a_n, b_n]$ be an n-dimensional rectangle and let $f : R \to \mathbf{R}$ be integrable on R. If, for each $\mathbf{x} := (x_1, \ldots, x_{n-1}) \in R_n := [a_1, b_1] \times \cdots \times [a_{n-1}, b_{n-1}]$, the function $f(\mathbf{x}, \cdot)$ is integrable on $[a_n, b_n]$, then

$$\int_{a_n}^{b_n} f(\mathbf{x}, t) \, dt$$

is integrable on R_n, and

$$\int_R f(\mathbf{x}, t) \, d(\mathbf{x}, t) = \int_{R_n} \int_{a_n}^{b_n} f(\mathbf{x}, t) \, dt \, d\mathbf{x}. \tag{27}$$

Proof. By repeating the argument of Lemma 12.30, we have

$$(L) \int_R f(\mathbf{x}, t) \, d(\mathbf{x}, t) \le (L) \int_{R_n} \int_{a_n}^{b_n} f(\mathbf{x}, t) \, dt \, d\mathbf{x}$$

$$\le (U) \int_{R_n} \int_{a_n}^{b_n} f(\mathbf{x}, t) \, dt \, d\mathbf{x}$$

$$\le (U) \int_R f(\mathbf{x}, t) \, d(\mathbf{x}, t)$$

for any bounded f. Since f is integrable on R, it follows that (27) holds. ∎

Using this result in conjunction with Remark 12.19, we can evaluate integrals over a large collection of nonrectangular Jordan regions. To this end, we shall call a nonempty set $E \subset \mathbf{R}^n$ a *projectable region* if and only if there is a closed Jordan region $H \subset \mathbf{R}^{n-1}$, an index $j \in \{1, \dots, n\}$, and continuous functions $\phi, \psi : H \to \mathbf{R}$ such that

$$E = \{(x_1, \dots, x_n) \in \mathbf{R}^n : (x_1, \dots, \widehat{x}_j, \dots, x_n) \in H$$
$$\text{and} \quad \phi(x_1, \dots, \widehat{x}_j, \dots, x_n) \le x_j \le \psi(x_1, \dots, \widehat{x}_j, \dots, x_n)\}.$$

[The notation $\widehat{x}_j$ means the variable x_j is missing; hence, $(x_1, \dots, \widehat{x}_j, \dots, x_n)$ is a point in $\mathbf{R}^{n-1}$.] In this case, we say that E is *generated* by j, H, ϕ, and ψ.

We are more specific for regions in $\mathbf{R}^2$ and $\mathbf{R}^3$. A set $E \subset \mathbf{R}^2$ is called a *region of type I* if and only if $E = \{(x, y) : x \in [a, b], \ \phi(x) \le y \le \psi(x)\}$ and a *region of type II* if and only if $E = \{(x, y) : y \in [a, b], \ \phi(y) \le x \le \psi(y)\}$, where $\phi, \psi : [a, b] \to \mathbf{R}$ are continuous functions. Similarly, a set $E \subset \mathbf{R}^3$ is called a *region of type I* if and only if $E = \{(x, y, z) : (x, y) \in H, \ \phi(x, y) \le z \le \psi(x, y)\}$, a *region of type II* if and only if $E = \{(x, y, z) : (x, z) \in H, \ \phi(x, z) \le y \le \psi(x, z)\}$, and a *region of type III* if and only if $E = \{(x, y, z) : (y, z) \in H, \ \phi(y, z) \le x \le \psi(y, z)\}$, where $\phi, \psi : H \to \mathbf{R}$ are continuous functions and H is a closed Jordan region in $\mathbf{R}^2$.

12.37 EXAMPLE.

Prove that the set E in $\mathbf{R}^2$ bounded by $y = x$ and $y = x^2$ is a region of types I and II.

Proof. The set E can be described by

$$\{(x, y) : x^2 \leq y \leq x,\ x \in [0, 1]\} \quad \text{or} \quad \{(x, y) : y \leq x \leq \sqrt{y},\ y \in [0, 1]\}$$

(see Figure 12.6). ∎

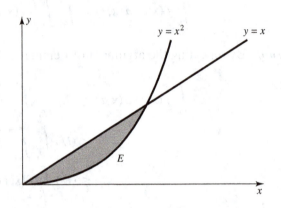

FIGURE 12.6

12.38 EXAMPLE.

Prove that the set E of points (x, y, z) which satisfy $4x^2 + y^2 + z^2 \leq 1$ is a region of types I, II, and III.

Proof. The set E, an ellipsoid, can be described by

$$E = \{(x, y, z) : -\sqrt{1 - 4x^2 - y^2} \leq z \leq \sqrt{1 - 4x^2 - y^2},\ (x, y) \in H\},$$

where $H = \{(x, y) : 4x^2 + y^2 \leq 1\}$. A similar argument shows that E is of types II and III. ∎

Before we show how to evaluate multiple integrals over projectable regions, we introduce additional terminology. For each $k = 1, \ldots, n$ the set

$$\Pi_k = \{\mathbf{x} \in \mathbf{R}^n : x_k = 0\}$$

will be called a *coordinate hyperplane*. Given a set $E \subseteq \mathbf{R}^n$, the *projection* of E onto the coordinate hyperplane Π_k is the set E_k of points $(x_1, \ldots, x_{k-1}, 0, x_{k+1}, \ldots, x_n)$ such that $(x_1, \ldots, x_k, \ldots, x_n) \in E$ for some $x_k \in \mathbf{R}$. For example, in $\mathbf{R}^3$ the coordinate hyperplane Π_1 corresponds to the yz-plane, and the projection of the three-dimensional ball $B_r(x_0, y_0, z_0)$ onto Π_1 is essentially the two-dimensional ball $B_r(y_0, z_0)$ (see Figure 12.7).

The following result shows that multiple integrals over most projectable regions can be evaluated using iterated integrals.

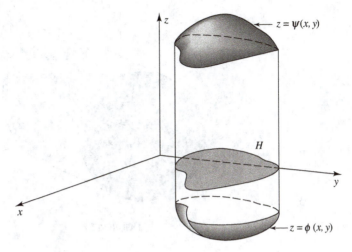

FIGURE 12.8

It follows from Theorem 12.4 that B is of volume zero. A similar argument shows that T is of volume zero.

To estimate the volume of S, set

$$M = \sup_{\mathbf{x} \in H} \psi(\mathbf{x}) \quad \text{and} \quad m = \inf_{\mathbf{x} \in H} \phi(\mathbf{x}).$$

Since H is a Jordan region, choose a grid $\{Q_1, \ldots, Q_p\}$ in $\mathbf{R}^{n-1}$ such that

$$\sum_{Q_k \cap \partial H \neq \emptyset} |Q_k| < \varepsilon.$$

Set $R_k = Q_k \times [m, M]$ and observe that $\mathcal{G} := \{R_1, \ldots, R_p\}$ is a grid in $\mathbf{R}^n$, and

$$V(S; \mathcal{G}) \leq \sum_{k=1}^{p} |R_k| < (M - m)\varepsilon.$$

Hence it follows from Theorem 12.4 that S is of volume zero. We conclude that ∂E is of volume zero (i.e., E is a Jordan region).

To prove (28), let $R = [a_1, b_1] \times \cdots \times [a_n, b_n]$ be an n-dimensional rectangle which contains E, and define g on R by $g(\mathbf{x}, t) = f(\mathbf{x}, t)$ when $(\mathbf{x}, t) \in E$, and $g(\mathbf{x}, t) = 0$ otherwise. By Remark 12.19 and Lemma 12.36,

$$\int_E f(\mathbf{x}, t) \, d(\mathbf{x}, t) = \int_{a_1}^{b_1} \cdots \int_{a_n}^{b_n} g(x_1, \ldots, x_n) \, dx_n \ldots dx_1$$

$$= \int_H \left(\int_{a_n}^{b_n} g(\mathbf{x}, t) \, dt \right) dx.$$

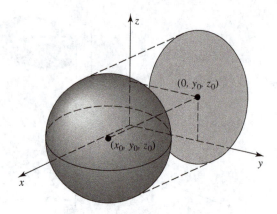

FIGURE 12.7

12.39 Theorem. *Let E be a projectable region in $\mathbf{R}^n$ generated by j, H, ϕ, and ψ. Then E is a Jordan region in $\mathbf{R}^n$. Moreover, if $f : E \to \mathbf{R}$ is continuous on E, then*

$$\int_E f(\mathbf{x})\, d\mathbf{x} = \int_H \left(\int_{\phi(x_1,\dots\,\widehat{x}_j,\dots\,x_n)}^{\psi(x_1,\dots,\widehat{x}_j,\dots\,x_n)} f(x_1,\dots,x_n)\, dx_j \right) d(x_1,\dots,\widehat{x}_j,\dots,x_n).$$

$$(28)$$

Proof. By symmetry, we may suppose that $j = n$. Thus

$$E = \{(\mathbf{x}, t) : \mathbf{x} = (x_1,\dots,x_{n-1}) \in H \quad \text{and} \quad \phi(\mathbf{x}) \le t \le \psi(\mathbf{x})\}.$$

To show that E is a Jordan region, we must show that the volume of ∂E is zero. Now ∂E is made up of "lower-dimensional pieces," a bottom $B = \{(\mathbf{x}, t) : \mathbf{x} \in H$ and $t = \phi(\mathbf{x})\}$, a top $T = \{(\mathbf{x}, t) : \mathbf{x} \in H$ and $t = \psi(\mathbf{x})\}$, and a side $S = \{(\mathbf{x}, t) : \mathbf{x} \in \partial H$ and $\phi(\mathbf{x}) \le t \le \psi(\mathbf{x})\}$. (Figure 12.8 illustrates the situation for the case $n = 3$.) Hence, we must show that B, T, and S are of volume zero.

To estimate the volume of B, notice that since H is compact, ϕ is uniformly continuous on H. Thus, given $\varepsilon > 0$, there is a $\delta > 0$ such that

$$\mathbf{x}, \mathbf{y} \in H \quad \text{and} \quad \|\mathbf{x} - \mathbf{y}\| < \delta \quad \text{imply} \quad |\phi(\mathbf{x}) - \phi(\mathbf{y})| < \varepsilon. \qquad (29)$$

Since H is bounded, H is contained in some $(n-1)$-dimensional cube Q. Divide Q into subcubes $Q_1, \dots, Q_p$ such that $\mathbf{x}, \mathbf{y} \in Q_k$ implies $\|\mathbf{x} - \mathbf{y}\| < \delta$, and let $R_k = Q_k \times [\phi(\mathbf{a}_k) - 2\varepsilon, \phi(\mathbf{a}_k) + 2\varepsilon]$ for some $\mathbf{a}_k \in Q_k$, $k = 1, 2, \dots, p$. Then $\mathcal{G} := \{R_1, \dots, R_p\}$ is grid in $\mathbf{R}^n$, and, by (29),

$$V(B; \mathcal{G}) \le \sum_{k=1}^{p} |R_k| = 4\varepsilon \sum_{k=1}^{p} |Q_k| = 4\varepsilon |Q|.$$

But for each $\mathbf{x} = (x_1, \ldots, x_{n-1}) \in H$, we have

$$g(\mathbf{x}, t) = \begin{cases} f(\mathbf{x}, t) & \phi(\mathbf{x}) \le t \le \psi(\mathbf{x}) \\ 0 & \text{otherwise.} \end{cases}$$

Therefore,

$$\int_{a_n}^{b_n} g(\mathbf{x}, t) \, dt = \int_{\phi(\mathbf{x})}^{\psi(\mathbf{x})} f(\mathbf{x}, t) \, dt. \qquad \blacksquare$$

Although we have stated Theorem 12.39 for continuous f, the result is evidently true whenever Lemma 12.36 applies; for example, if f is integrable on E and $f(\mathbf{x}, \cdot)$ is integrable on $[a_n, b_n]$ for each fixed $\mathbf{x} \in H$.

If the set H is itself projectable, then Theorem 12.39 can be applied again to H. Thus if E is nice enough, an integral over E can be evaluated using n partial integrals. We close this section with several examples which illustrate this principle for the cases $n = 2$ and $n = 3$.

12.40 EXAMPLE.

Find the integral of $f(x, y, z) = x$ over the region E bounded by $z = 1 - x - y$, $x = 0$, $y = 0$, and $z = 0$.

Solution. The surfaces $z = 0$ and $z = 1 - x - y$ intersect when $y = 1 - x$. The projection E_3 is bounded by the curves $x = 0$, $y = 0$, and $y = 1 - x$. These last two curves intersect when $x = 1$. Thus E is a region of type I: $E = \{(x, y, z) : 0 \le x \le 1, \ 0 \le y \le 1 - x, \ 0 \le z \le 1 - x - y\}$ (see Figure 12.9). It follows that

$$\iiint_E f \, dV = \int_0^1 \int_0^{1-x} \int_0^{1-x-y} x \, dz \, dy \, dx$$

$$= \int_0^1 \int_0^{1-x} (x - x^2 - xy) \, dy \, dx$$

$$= \frac{1}{2} \int_0^1 (x - 2x^2 + x^3) \, dx = \frac{1}{24}. \qquad \blacksquare$$

12.41 EXAMPLE.

Find the integral of $f(x, y, z) = x^2$ over the region E bounded by $|x| = 1$, $z = x^2 - y^2$, where $z \ge 0$.

Solution. The surfaces $z = 0$ and $z = x^2 - y^2$ intersect when $x^2 - y^2 = 0$ (i.e., $y = \pm x$). The curves $y = \pm x$ and $|x| = 1$ intersect when $x = \pm 1$. Thus the region E is of type I:

$$E = \{(x, y, z) : -1 \le x \le 1, \ -|x| \le y \le |x|, \ 0 \le z \le x^2 - y^2\}$$

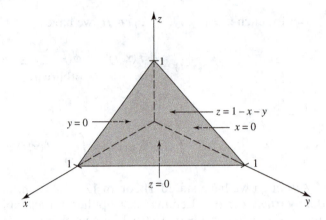

FIGURE 12.9

(see Figure 12.10). It follows that

$$
\iiint_E f \, dV = \int_{-1}^{1} \int_{-|x|}^{|x|} \int_{0}^{x^2-y^2} x^2 \, dz \, dy \, dx
$$

$$
= \int_{-1}^{1} \int_{-|x|}^{|x|} (x^2 - y^2)x^2 \, dy \, dx
$$

$$
= 4 \int_{0}^{1} \int_{0}^{x} (x^2 - y^2)x^2 dy \, dx = \frac{8}{3} \int_{0}^{1} x^5 \, dx = \frac{4}{9}. \qquad \blacksquare
$$

Although Theorem 12.39 can be used in conjunction with Theorem 12.23 to handle the case when E is a finite union of projectable subregions, we can sometimes avoid breaking E into subregions by changing our point of view. Here is a typical example.

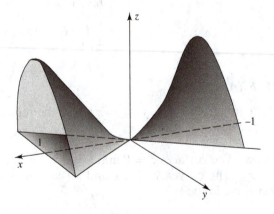

FIGURE 12.10

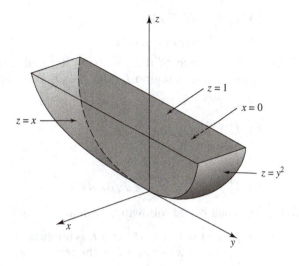

FIGURE 12.11

12.42 EXAMPLE.

Find the integral of $f(x, y, z) = x - z$ over the region bounded by $z = y^2$, $z = 1$, $z = x$, and $x = 0$.

Solution. The region E is a union of two regions of type I (see Figure 12.11, where the "back" of E is that portion of the plane $x = 0$ which is bounded by the parabola $z = y^2$, $x = 0$ here represented by a dashed line). Therefore, we must use two integrals if we integrate dz first: the integral where z varies between y^2 and 1, and the integral where z varies from x to 1. It looks complicated to set up. The solution is simpler if we integrate dx first. Indeed, E is a single region of type III since

$$E = \{(x, y, z) : -1 \leq y \leq 1,\ y^2 \leq z \leq 1,\ 0 \leq x \leq z\}.$$

Thus,

$$\iiint_E f\, dV = \int_{-1}^{1} \int_{y^2}^{1} \int_{0}^{z} (x - z)\, dx\, dz\, dy$$

$$= -\frac{1}{2} \int_{-1}^{1} \int_{y^2}^{1} z^2\, dz\, dy = \frac{1}{6} \int_{-1}^{1} (y^6 - 1)\, dy = -\frac{2}{7}. \qquad \blacksquare$$

EXERCISES

12.3.1. Evaluate each of the following iterated integrals.

a) $\int_0^1 \int_0^1 (x + y)\, dx\, dy$

b) $\int_0^3 \int_0^1 \sqrt{xy+x}\, dx\, dy$

c) $\int_0^\pi \int_0^\pi y\cos(xy)\, dy\, dx$

12.3.2. Evaluate each of the following iterated integrals. Write each as an integral over a region E, and sketch E in each case.

a) $\int_0^1 \int_x^{x^2+1} (x+1)\, dy\, dx$

b) $\int_0^1 \int_y^1 \sin(x^2)\, dx\, dy$

c) $\int_0^1 \int_{\sqrt{y}}^1 \int_0^{x^2+y^2} dz\, dx\, dy$

d) $\int_0^1 \int_{\sqrt{y}}^1 \int_{x^3}^1 \sqrt{x^3+z}\, dz\, dx\, dy$

12.3.3. For each of the following, evaluate $\int_E f\, dV$.

a) $f(x,y) = (1+x^2)^{-1}$ and E is bounded by $x=1$, $y=0$, and $y=x^3$.

b) $f(x,y) = x+y$ and E is the triangle with vertices $(0,0)$, $(0,1)$, and $(2,0)$.

c) $f(x,y) = x^2 e^{xy}$ and E is the triangle with vertices $(0,0)$, $(1,0)$, and $(1,1)$.

d) $f(x,y,z) = x$ and E is the set of points (x,y,z) such that $0 \le z \le 1-x^2$, $0 \le y \le x^2+z^2$, and $x \ge 0$.

12.3.4. Compute the volume of each of the following regions.

a) E is bounded by the surfaces $x+y+z=3$, $z=0$, and $x^2+y^2=1$.

b) E lies under the plane $z=x+y$ and over the region in the xy-plane bounded by the curves $x=\sqrt{y/2}$, $x=2\sqrt{y}$, $x+y=3$.

c) E is bounded by $z=y^2$, $x=y^2+z^2$, $x=0$, $z=1$.

d) E is bounded by $y=x^3$, $x=z^2$, $z=x^2$, and $y=0$.

12.3.5. a) Verify that the hypotheses of Fubini's Theorem hold when f is continuous on R.

b) Modify the proof of Remark 12.33 to show that Fubini's Theorem might not hold for a nonintegrable f, even if $f(x,y)$ is continuous in each variable separately; that is, if $f(x,\cdot)$ is continuous for each $x \in [a,b]$ and $f(\cdot,y)$ is continuous for each $y \in [c,d]$.

12.3.6. a) Suppose that f_k is integrable on $[a_k, b_k]$ for $k=1,\ldots,n$, and set $R = [a_1, b_1] \times \cdots \times [a_n, b_n]$. Prove that

$$\int_R f_1(x_1)\ldots f_n(x_n)\, d(x_1,\ldots,x_n)$$

$$= \left(\int_{a_1}^{b_1} f_1(x_1)\, dx_1\right) \ldots \left(\int_{a_n}^{b_n} f_n(x_n)\, dx_n\right).$$

b) If $Q = [0,1]^n$ and $\mathbf{y} := (1,1,\ldots,1)$, prove that

$$\int_Q e^{-\mathbf{x}\cdot\mathbf{y}}\, d\mathbf{x} = \left(\frac{e-1}{e}\right)^n.$$

12.3.7. The greatest integer in a real number x is the integer $[x] := n$ which
satisfies $n \le x < n + 1$. An interval $[a, b]$ is called **Z**-asymmetric if
$b + a \ne [b] + [a] + 1$.

a) Suppose that R is a two-dimensional **Z**-asymmetric rectangle (i.e.,
that both of its sides are **Z**-asymmetric). If $\psi(x, y) := (x - [x] - 1/2)(y - [y] - 1/2)$, prove that $\iint_R \psi \, dA = 0$ if and only if at least
one side of R has integer length.

b) Suppose that R is tiled by rectangles $R_1 \dots, R_N$ (i.e., that the R_j's
are **Z**-asymmetric, nonoverlapping, and that $R = \cup_{j=1}^N R_j$). Prove
that if each R_j has at least one side of integer length and R is
Z-asymmetric, then R has at least one side of integer length.

12.3.8. Let E be a nonempty Jordan region in $\mathbf{R}^2$ and $f : E \to [0, \infty)$ be
integrable on E. Prove that the volume of $\Omega = \{(x, y, z) : (x, y) \in E, 0 \le z \le f(x, y)\}$ (as given by Definition 12.5) satisfies

$$\text{Vol}(\Omega) = \iint_E f \, dA.$$

12.3.9. Let $R = [a, b] \times [c, d]$ be a two-dimensional rectangle and $f : R \to \mathbf{R}$
be bounded.

a) Prove that

$$(L) \iint_R f \, dA \le (L) \int_a^b \left((X) \int_c^d f(x, y) \, dy \right) dx$$

$$\le (U) \int_a^b \left((X) \int_c^d f(x, y) \, dy \right) dx$$

$$\le (U) \iint_R f \, dA$$

for $X = U$ or $X = L$.

b) Prove that if f is integrable on R, then

$$\iint_R f \, dA = \int_a^b \left((L) \int_c^d f(x, y) \, dy \right) dx$$

$$= \int_a^b \left((U) \int_c^d f(x, y) \, dy \right) dx.$$

c) Compute the two iterated integrals in part b) for

$$f(x, y) = \begin{cases} 1 & y \in \mathbf{Q} \\ x & y \notin \mathbf{Q} \end{cases}$$

and $R = [0, 1] \times [0, 1]$. Prove that f is not integrable on R.

*12.3.10. [FUBINI'S THEOREM FOR IMPROPER INTEGRALS]. If $a < b$ are extended real numbers, $c < d$ are finite real numbers, $f : (a, b) \times [c, d] \to \mathbf{R}$ is continuous, and

$$F(y) = \int_a^b f(x, y)\, dx$$

converges uniformly on $[c, d]$, prove that

$$\int_c^d f(x, y)\, dy$$

is improperly integrable on (a, b) and

$$\int_c^d \int_a^b f(x, y)\, dx\, dy = \int_a^b \int_c^d f(x, y)\, dy\, dx.$$

12.4 CHANGE OF VARIABLES

Recall (Theorem 5.34) that if $\phi : [a, b] \to \mathbf{R}$ is continuously differentiable and $\phi' \neq 0$ on $[a, b]$, then

$$\int_{\phi([a,b])} f(t)\, dt = \int_{[a,b]} f(\phi(x))\, |\phi'(x)|\, dx$$

for all f integrable on $\phi([a, b])$. We shall generalize this result to functions of several variables; namely, we shall identify conditions under which

$$\int_{\boldsymbol{\phi}(E)} f(\mathbf{u})\, d\mathbf{u} = \int_E f(\boldsymbol{\phi}(\mathbf{x}))|\Delta_{\boldsymbol{\phi}}(\mathbf{x})|\, d\mathbf{x} \qquad (30)$$

holds. (At this point you may wish to read the discussion following the proof of Theorem 12.46 below to see that $\Delta_{\boldsymbol{\phi}}$ takes on a familiar form when $\boldsymbol{\phi}$ is the change from polar to rectangular coordinates.)

It takes six or seven hypotheses to establish (30). These hypotheses fall into two categories:

1) *Hypotheses made so the change of variables is possible.* Since the one-dimensional result required ϕ to be continuously differentiable and $\phi' \neq 0$ (which together imply that ϕ is 1–1), we expect the corresponding hypotheses for (30) to be as follows: $\boldsymbol{\phi}$ is 1–1, continuously differentiable, and $\Delta_{\boldsymbol{\phi}} \neq 0$.

2) *Hypotheses made so the integrals in (30) exist.* There are four of these: E is a Jordan region, $\boldsymbol{\phi}(E)$ is a Jordan region, f is integrable on $\boldsymbol{\phi}(E)$, and $f \circ \boldsymbol{\phi}|\Delta_{\boldsymbol{\phi}}|$ is integrable on E. In practice, only the first and third of the hypotheses in category 2) need be verified. Indeed, if $\boldsymbol{\phi}$ satisfies all hypotheses in category 1) and E is Jordan, then $\boldsymbol{\phi}(E)$ is Jordan by Theorem 12.10, and, when f is integrable on $\boldsymbol{\phi}(E)$, $f \circ \boldsymbol{\phi}|\Delta_{\boldsymbol{\phi}}|$ is integrable on E (see Theorem 12.29ii and Exercise 12.2.8a). Moreover, the remaining hypotheses in category 2) can usually be verified by inspection. The reason for this is twofold. E is frequently projectable, hence a Jordan region, and most functions are

continuous (or nearly so), hence integrable on E. Therefore, the crucial hypotheses for (30) are those in category 1), namely, that ϕ be 1–1, continuously differentiable, and $\Delta_\phi \neq 0$.

To give an outline of a proof of (30), we introduce the following terminology. A function f is said to satisfy a certain property $\mathcal{P}$ "locally" on a set E if and only if given $\mathbf{a} \in E$ there is an open set W containing $\mathbf{a}$ such that f satisfies $\mathcal{P}$ on $W \cap E$. f is said to satisfy the property $\mathcal{P}$ "globally" on E if and only if f satisfies $\mathcal{P}$ for all points in E. To prove (30), we first obtain several preliminary results which culminate in a "local" change-of-variables formula (see Lemma 12.45) and then use this to obtain a "global" change-of-variables formula for functions ϕ which are $\mathcal{C}^1$ on an open set which contains E (see Theorem 12.46). Throughout this discussion, we assume that Δ_ϕ is never zero. In Section 12.5, we work much harder to show that the condition "$\Delta_\phi \neq 0$" can be relaxed on a set of volume zero (see Theorem 12.65).

Since every Jordan region can be approximated by rectangles, and every integrable function is almost continuous, hence locally nearly constant, we should consider (30) first in the case when $\phi(E)$ is a rectangle and f is identically 1; that is, we should prove that

$$|R| = \int_{\phi^{-1}(R)} |\Delta_\phi(\mathbf{x})| \, d\mathbf{x}. \tag{31}$$

Our first preliminary result shows that this case is a step in the right direction.

12.43 Lemma.
Let W be open in $\mathbf{R}^n$, let $\phi : W \to \mathbf{R}^n$ be 1–1 and continuously differentiable on W with $\Delta_\phi \neq 0$ on W, and suppose that ϕ^{-1} is continuously differentiable on $\phi(W)$ with $\Delta_{\phi^{-1}} \neq 0$ on $\phi(W)$. Suppose further that (31) holds for every n-dimensional rectangle $R \subset \phi(W)$. If E is a Jordan region with $\overline{E} \subset W$, if f is integrable on $\phi(E)$, and if $f \circ \phi$ is integrable on E, then

$$\int_{\phi(E)} f(\mathbf{u}) \, d\mathbf{u} = \int_E (f \circ \phi)(\mathbf{x})|\Delta_\phi(\mathbf{x})| \, d\mathbf{x}.$$

Proof. We may suppose that W is nonempty. Let E be a fixed Jordan region which satisfies $\overline{E} \subset W$ and suppose that f is integrable on $\phi(E)$. Set $f^+ = (|f| + f)/2$ and $f^- = (|f| - f)/2$. Then f^+ and f^- are both nonnegative and integrable on $\phi(E)$, and $f = f^+ - f^-$ (see Exercises 3.1.7 and 5.2.2). Since the integral of a difference is the difference of the integrals, it suffices to prove the lemma for the case when $f \geq 0$.

Let $\varepsilon > 0$. Since f is integrable on $\phi(E)$, choose a grid $\mathcal{G} = \{R_1, \ldots, R_p\}$ such that

$$\int_{\phi(E)} f(\mathbf{u}) \, d\mathbf{u} > U(f, \mathcal{G}) - \varepsilon := \sum_{R_j \cap \phi(E) \neq \emptyset} M_j |R_j| - \varepsilon, \tag{32}$$

where $M_j := \sup f(R_j) = f(\phi(\phi^{-1}(R_j)))$. Moreover, since $\overline{\phi(E)} = \phi(\overline{E}) \subset \phi(W)$, we may suppose, by refining $\mathcal{G}$ if necessary, that $R_j \cap \phi(E) \neq \emptyset$ implies

$R_j \subset \boldsymbol{\phi}(W)$. Hence, by Lemma 12.9, $\{\boldsymbol{\phi}^{-1}(R_j)\}_{R_j \cap \boldsymbol{\phi}(E) \neq \emptyset}$ is a nonoverlapping collection of Jordan regions whose union satisfies

$$\Omega_1 := \bigcup_{R_j \cap \boldsymbol{\phi}(E) \neq \emptyset} \boldsymbol{\phi}^{-1}(R_j) \supseteq \boldsymbol{\phi}^{-1}(\boldsymbol{\phi}(E)) = E.$$

Hence, (32), (31), and Theorems 12.25 and 12.23 imply that

$$\int_{\boldsymbol{\phi}(E)} f(\mathbf{u}) \, d\mathbf{u} \geq \sum_{R_j \cap \boldsymbol{\phi}(E) \neq \emptyset} M_j |R_j| - \varepsilon$$

$$= \sum_{R_j \cap \boldsymbol{\phi}(E) \neq \emptyset} M_j \int_{\boldsymbol{\phi}^{-1}(R_j)} |\Delta_{\boldsymbol{\phi}}(\mathbf{x})| \, d\mathbf{x} - \varepsilon$$

$$\geq \sum_{R_j \cap \boldsymbol{\phi}(E) \neq \emptyset} \int_{\boldsymbol{\phi}^{-1}(R_j)} f(\boldsymbol{\phi}(\mathbf{x})) |\Delta_{\boldsymbol{\phi}}(\mathbf{x})| \, d\mathbf{x} - \varepsilon$$

$$= \int_{\Omega_1} f(\boldsymbol{\phi}(\mathbf{x})) |\Delta_{\boldsymbol{\phi}}(\mathbf{x})| \, d\mathbf{x} - \varepsilon$$

$$\geq \int_E f(\boldsymbol{\phi}(\mathbf{x})) |\Delta_{\boldsymbol{\phi}}(\mathbf{x})| \, d\mathbf{x} - \varepsilon.$$

(For this last step, we used the fact that $f \geq 0$.) Since $\varepsilon > 0$ was arbitrary, we obtain

$$\int_{\boldsymbol{\phi}(E)} f(\mathbf{u}) \, d\mathbf{u} \geq \int_E f(\boldsymbol{\phi}(\mathbf{x})) |\Delta_{\boldsymbol{\phi}}(\mathbf{x})| \, d\mathbf{x}.$$

On the other hand, by Theorem 12.20 there is a grid $\mathcal{H} = \{Q_1, \dots, Q_p\}$ such that

$$\int_{\boldsymbol{\phi}(E)} f(\mathbf{u}) \, d\mathbf{u} \leq \sum_{Q_j \subset (\boldsymbol{\phi}(E))^o} m_j |Q_j| + \varepsilon,$$

where $m_j := \inf f(Q_j) = f(\boldsymbol{\phi}(\boldsymbol{\phi}^{-1}(Q_j)))$. Repeating the steps above with

$$\Omega_2 := \bigcup_{Q_j \subset (\boldsymbol{\phi}(E))^o} \boldsymbol{\phi}^{-1}(Q_j) \subseteq \boldsymbol{\phi}^{-1}(\boldsymbol{\phi}(E)) = E$$

in place of Ω_1, we see that

$$\int_{\boldsymbol{\phi}(E)} f(\mathbf{u}) \, d\mathbf{u} \leq \sum_{Q_j \subset (\boldsymbol{\phi}(E))^o} m_j |Q_j| + \varepsilon$$

$$\leq \int_{\Omega_2} f(\boldsymbol{\phi}(\mathbf{x})) |\Delta_{\boldsymbol{\phi}}(\mathbf{x})| \, d\mathbf{x} + \varepsilon$$

$$\leq \int_E f(\boldsymbol{\phi}(\mathbf{x})) |\Delta_{\boldsymbol{\phi}}(\mathbf{x})| \, d\mathbf{x} + \varepsilon.$$

We conclude that

$$\int_{\phi(E)} f(\mathbf{u})\, d\mathbf{u} = \int_E f(\phi(\mathbf{x}))|\Delta_\phi(\mathbf{x})|\, d\mathbf{x}. \qquad \blacksquare$$

Next, we show that (31) holds locally near points $\mathbf{a}$ when $\Delta_\phi(\mathbf{a}) \neq 0$ and ϕ is 1–1 and C^1.

12.44 Lemma.
Let V be open in $\mathbf{R}^n$, and $\phi : V \to \mathbf{R}^n$ be 1–1 and continuously differentiable on V. If $\Delta_\phi(\mathbf{a}) \neq 0$ for some $\mathbf{a} \in V$, then there exists an open rectangle W such that $\mathbf{a} \in W \subset V$, Δ_ϕ is nonzero on W, ϕ^{-1} is C^1 and its Jacobian is nonzero on $\phi(W)$, and such that if R is an n-dimensional rectangle contained in $\phi(W)$, then $\phi^{-1}(R)$ is Jordan and (31) holds.

Proof. The proof is by induction on n. If $n = 1$ and $\phi'(a) \neq 0$, then ϕ' is nonzero on some open interval I containing a. Hence, by Theorem 5.34, (31) holds for "rectangles" (i.e., intervals) in $\phi(I)$.

Suppose that (31) holds on $\mathbf{R}^{n-1}$, for some $n > 1$. Let $\phi : V \to \mathbf{R}^n$ be 1–1 and C^1 on V with $\Delta_\phi(\mathbf{a}) \neq 0$. Expanding the determinant of $A := D\phi(\mathbf{a})$ along the first row, it is clear that at least one of its minors, say A_{1j}, has nonzero determinant; that is, that $\psi(\mathbf{x}) := (\phi_2(\mathbf{x}), \ldots, \phi_j(\mathbf{x}), x_j, \ldots, \phi_n(\mathbf{x}))$ also has nonzero Jacobian at $\mathbf{x} = \mathbf{a}$. Since ϕ and ψ are both C^1 on V, it follows from the Inverse Function Theorem that there is an open set $W \subset V$, containing $\mathbf{a}$, such that ϕ and ψ are 1–1 on W, Δ_ϕ and Δ_ψ are nonzero on W, and ϕ^{-1} is 1–1, C^1 and $\Delta_{\phi^{-1}} \neq 0$ on $\phi(W)$. By making W smaller, if necessary, we may suppose that W is an open rectangle; that is, that there exist open intervals I_j such that $W = I_1 \times \cdots \times I_n$.

Assume for simplicity that $j = 1$; that is, that $\psi(\mathbf{x}) = (x_1, \phi_2(\mathbf{x}), \ldots, \phi_n(\mathbf{x}))$. For each $\mathbf{x} = (x_1, \ldots, x_n) \in \psi(W)$ set $\sigma(\mathbf{x}) = (\phi_1(\psi^{-1}(\mathbf{x})), x_2, \ldots, x_n)$. It is clear that $\phi = \sigma \circ \psi$, hence by the Chain Rule, $\Delta_\phi(\mathbf{x}) = \Delta_\sigma(\psi(\mathbf{x}))\Delta_\psi(\mathbf{x})$. In particular, by the choice of W,

$$\Delta_\psi(\mathbf{x}) \neq 0 \quad \text{and} \quad \Delta_\sigma(\psi(\mathbf{x})) \neq 0 \quad \text{for all } \mathbf{x} \in W. \qquad (33)$$

To show that the inductive hypothesis can be used on ψ, fix $t \in I_1$. Set $W_0 = I_2 \times \cdots \times I_n$ and $\phi^t(\mathbf{y}) = (\phi_2(t, \mathbf{y}), \ldots, \phi_n(t, \mathbf{y}))$ for each $\mathbf{y} \in W_0$. Then $\phi^t : W_0 \to \mathbf{R}^{n-1}$ is 1–1 and C^1 on W_0, and, by (33), $\Delta_{\phi^t}(\mathbf{y}) = \Delta_\psi(t, \mathbf{y}) \neq 0$ for all $\mathbf{y} \in W_0$. It follows from the inductive hypothesis that if Q_0 is an $(n-1)$-dimensional rectangle which satisfies $Q_0 \subset \phi^t(W_0)$, then $(\phi^t)^{-1}(Q_0)$ is Jordan and

$$|Q_0| = \int_{(\phi^t)^{-1}(Q_0)} |\Delta_{\phi^t}(\mathbf{y})|\, d\mathbf{y}. \qquad (34)$$

(W_0, hence, W, may have gotten smaller again.)

Let $Q = I_0 \times Q_0$ be any n-dimensional rectangle in $\psi(W)$ and integrate (34) with respect to t over I_0 to verify

$$|Q| = |I_0| \cdot |Q_0| = \int_{I_0} \int_{(\phi^t)^{-1}(Q_0)} |\Delta_{\phi^t}(\mathbf{y})| \, d\mathbf{y} \, dt.$$

But the first component of ψ satisfies $\psi_1(t, \mathbf{y}) = t$ for all $\mathbf{y} \in W$, so $\Delta_{\phi^t} = \Delta_\psi$ and $\psi^{-1}(Q)$ is the union of the "t-sections" $(\phi^t)^{-1}(Q_0)$ as t ranges over I_0. Hence, we can continue the identity above as follows:

$$|Q| = \int_{I_0} \int_{(\phi^t)^{-1}(Q_0)} |\Delta_{\phi^t}(\mathbf{y})| \, d\mathbf{y} = \int_{\psi^{-1}(Q)} |\Delta_\psi(\mathbf{u})| \, d\mathbf{u}.$$

In particular, it follows from Lemma 10.43 that

$$\int_{\psi(E)} g(\mathbf{u}) \, d\mathbf{u} = \int_E g(\psi(\mathbf{x}))|\Delta_\psi(\mathbf{x})| \, d\mathbf{x} \tag{35}$$

for all Jordan regions E which satisfy $\overline{E} \subset W$, provided g is integrable on $\psi(W)$ and $g \circ \psi$ is integrable on E.

Similarly, we can use the inductive hypothesis to prove that (31) holds for σ in place of ϕ for all n-dimensional rectangles R contained in $\phi(W)$. Hence, for each such rectangle R, we have by (35)—with $E = \psi^{-1}(\sigma^{-1}(R)) \equiv \phi^{-1}(R)$ and $g = |\Delta_\sigma|$—and the Chain Rule that

$$|R| = \int_{\sigma^{-1}(R)} |\Delta_\sigma(\mathbf{u})| \, d\mathbf{u}$$

$$= \int_{\psi^{-1}(\sigma^{-1}(R))} |\Delta_\sigma(\psi(\mathbf{x}))| \, |\Delta_\psi(\mathbf{x})| \, d\mathbf{x}$$

$$= \int_{\phi^{-1}(R)} |\Delta_\phi(\mathbf{x})| \, d\mathbf{x}. \qquad \blacksquare$$

By combining Lemmas 12.43 and 12.44, we obtain the following local version of the change-of-variables formula we want.

12.45 Lemma.
Suppose that V is open in $\mathbf{R}^n$, that $\mathbf{a} \in V$, and that $\phi : V \to \mathbf{R}^n$ is continuously differentiable on V. If $\Delta_\phi(\mathbf{a}) \neq 0$, then there exists an open rectangle $W \subset V$ containing $\mathbf{a}$ such that if E is Jordan with $\overline{E} \subset W$, if $f \circ \phi$ is integrable on E, and if f is integrable on $\phi(E)$, then

$$\int_{\phi(E)} f(\mathbf{u}) \, d\mathbf{u} = \int_E f(\phi(\mathbf{x}))|\Delta_\phi(\mathbf{x})| \, d\mathbf{x}.$$

This local change-of-variables formula contains the following global result.

12.46 Theorem. *Suppose that V is open in $\mathbf{R}^n$ and that $\boldsymbol{\phi} : V \to \mathbf{R}^n$ is 1–1 and continuously differentiable on V. If $\Delta_\phi \neq 0$ on V, if E is a Jordan region with $\overline{E} \subset V$, if $f \circ \boldsymbol{\phi}$ is integrable on E, and if f is integrable on $\boldsymbol{\phi}(E)$, then*

$$\int_{\boldsymbol{\phi}(E)} f(\mathbf{u}) \, d\mathbf{u} = \int_E f(\boldsymbol{\phi}(\mathbf{x})) |\Delta_\phi(\mathbf{x})| \, d\mathbf{x}. \tag{38}$$

Proof. Let $f : \boldsymbol{\phi}(E) \to \mathbf{R}$ be integrable, and set $H := \overline{E}$. By Lemma 12.45, given $\mathbf{a} \in H$ there is an open rectangle $W_\mathbf{a}$ such that $\mathbf{a} \in W_\mathbf{a} \subset V$ and

$$\int_{\boldsymbol{\phi}(E_i)} f(\mathbf{u}) \, d\mathbf{u} = \int_{E_i} f(\boldsymbol{\phi}(\mathbf{x})) |\Delta_\phi(\mathbf{x})| \, d\mathbf{x} \tag{39}$$

for every Jordan region E_i which satisfies $\overline{E_i} \subset W_\mathbf{a}$. Let $Q_\mathbf{a}$ be an open rectangle which satisfies $\mathbf{a} \in Q_\mathbf{a} \subset \overline{Q_\mathbf{a}} \subset W_\mathbf{a}$. Then for each $\mathbf{a} \in H$ there is an $r(\mathbf{a}) > 0$ such that $B_{r(\mathbf{a})}(\mathbf{a}) \subset Q_\mathbf{a}$. Since the Jordan region E is bounded, H is compact by the Heine–Borel Theorem. Thus there exist finitely many $\mathbf{a}_j$ such that H is covered by $B_{r(\mathbf{a}_j)}(\mathbf{a}_j)$, $j = 1, 2, \ldots, N$. Hence the open rectangles $Q_j := Q_{\mathbf{a}_j}$ satisfy

$$H \subset \bigcup_{j=1}^{N} Q_j.$$

Let R be a huge rectangle which contains H and $\mathcal{G} = \{R_1, \ldots, R_p\}$ be a grid on R so fine that each rectangle in $\mathcal{G}$ which intersects H is a subset of some $\overline{Q}_j$. (This is possible since there are only finitely many Q_j's; just use the endpoints of the Q_j's to generate $\mathcal{G}$.) Let $E_i = R_i \cap E$. Then $\overline{E_i} \subseteq R_i \cap H \subseteq \overline{Q}_j \subset W_{\mathbf{a}_j}$ for some $j \in \{1, \ldots, N\}$; that is, (39) holds. Moreover, the collection $\{E_1, \ldots, E_p\}$ is a nonoverlapping family of nonempty Jordan regions whose union is E; hence, by Theorem 1.37 and Lemma 12.9, the collection $\{\boldsymbol{\phi}(E_i) : i = 1, \ldots, p\}$ is a nonoverlapping family of nonempty Jordan regions whose union is $\boldsymbol{\phi}(E)$. It follows from Theorem 12.23 and (39) that

$$\int_{\boldsymbol{\phi}(E)} f(\mathbf{u}) \, d\mathbf{u} = \sum_{i=1}^{p} \int_{\boldsymbol{\phi}(E_i)} f(\mathbf{u}) \, d\mathbf{u}$$

$$= \sum_{i=1}^{p} \int_{E_i} f(\boldsymbol{\phi}(\mathbf{x})) |\Delta_\phi(\mathbf{x})| \, d\mathbf{x} = \int_E f(\boldsymbol{\phi}(\mathbf{x})) |\Delta_\phi(\mathbf{x})| \, d\mathbf{x}. \quad \blacksquare$$

Again, we note that in Theorem 12.46 the hypothesis that $f \circ \boldsymbol{\phi}$ be integrable is superfluous—see Theorem 12.29ii.

To see how all this works out in practice, we begin with a familiar change of variables in $\mathbf{R}^2$. Recall that *polar coordinates* in $\mathbf{R}^2$ have the form

$$x = r \cos \theta, \qquad y = r \sin \theta,$$

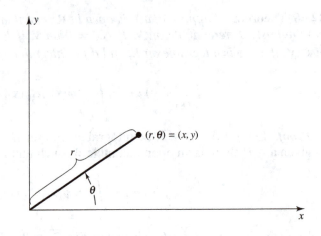

FIGURE 12.12

where $r = \|(x, y)\|$ and θ is the angle measured counterclockwise from the positive x axis to the line segment $L((0, 0); (x, y))$ (see Figure 12.12). Set $\phi(r, \theta) = (r \cos \theta, r \sin \theta)$ and observe that

$$\Delta_\phi = \det \begin{bmatrix} \cos \theta & -r \sin \theta \\ \sin \theta & r \cos \theta \end{bmatrix} = r(\cos^2 \theta + \sin^2 \theta) = r. \qquad (40)$$

Thus we abbreviate the change-of-variables formula from polar coordinates to rectangular coordinates by $dx\, dy = r\, dr\, d\theta$.

Although ϕ is not 1–1 [e.g., $\phi(0, \theta) = (0, 0)$ for all $\theta \in \mathbf{R}$] and its Jacobian is not nonzero, this does not prevent us from applying Theorem 12.46 (i.e., changing variables from polar coordinates to rectangular coordinates and vice versa). Indeed, since ϕ is 1–1 on $\Omega := \{(r, \theta) : r > 0, 0 \leq \theta < 2\pi\}$ and its Jacobian is nonzero off the set $Z := \{(r, \theta) : r = 0\}$, we can apply Theorem 12.46 to $E \cap \{(r, \theta) : r > 0\}$ and let $r \downarrow 0$. Since the end result is the same as if we applied Theorem 12.46 directly without this intermediate step, we shall do so below without any further comments. This works in part because the set Z where the hypotheses of category 1 fail (see the discussion following (30) above) is a set of volume zero (see Theorem 12.66 below).

The next two examples show that polar coordinates can be used to evaluate integrals which cannot be computed easily using rectangular coordinates.

12.47 EXAMPLE.

Find the volume of the region E bounded by $z = x^2 + y^2$, $x^2 + y^2 = 4$, and $z = 0$.

Solution. Clearly, E lies under the function $f(x, y) = x^2 + y^2$ over the region $B = B_2(0, 0)$ (see Figure 12.13). Using polar coordinates, we obtain

$$\text{Vol}(E) = \iint_B (x^2 + y^2)\, dA = \int_0^{2\pi} \int_0^2 r^3\, dr\, d\theta = 8\pi. \qquad \blacksquare$$

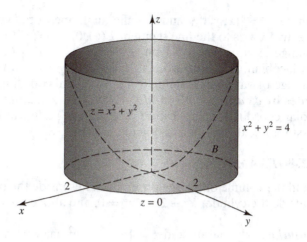

FIGURE 12.13

12.48 EXAMPLE.

Evaluate

$$\iint_E \frac{x^2 + y^2}{x} \, dA,$$

where $E = \{(x, y) : a^2 \le x^2 + y^2 \le 1 \text{ and } 0 \le y \le x\}$ for some $0 < a < 1$.

Solution. Changing to polar coordinates, we see that

$$\iint_E \frac{x^2 + y^2}{x} \, dA = \int_0^{\pi/4} \int_a^1 \frac{r^3}{r \cos \theta} \, dr \, d\theta = \frac{1 - a^3}{3} \int_0^{\pi/4} \sec \theta \, d\theta.$$

To integrate $\sec \theta$, multiply and divide by $\sec \theta + \tan \theta$. Using the change of variables $u = \sec \theta + \tan \theta$, we obtain

$$\int_0^{\pi/4} \sec \theta \, d\theta = \int_0^{\pi/4} \frac{\sec \theta \tan \theta + \sec^2 \theta}{\sec \theta + \tan \theta} \, d\theta$$

$$= \int_1^{1+\sqrt{2}} \frac{du}{u} = \log(1 + \sqrt{2}).$$

Consequently,

$$\iint_E \frac{x^2 + y^2}{x} \, dA = \frac{(1 - a^3) \log(1 + \sqrt{2})}{3}. \qquad \blacksquare$$

Recall that *cylindrical coordinates* in $\mathbf{R}^3$ have the form

$$x = r \cos \theta, \qquad y = r \sin \theta, \qquad z = z,$$

where $r = \|(x, y, 0)\|$ and θ is the angle measured counterclockwise from the positive x axis to the line segment $L((0, 0, 0); (x, y, 0))$. It is easy to see that this change of variables is 1–1 on $\Omega := \{(r, \theta, z) : r > 0, 0 \le \theta < 2\pi, z \in \mathbf{R}\}$, and its Jacobian, r, is nonzero off $Z := \{(r, \theta, z) : r = 0\}$. We shall abbreviate the change-of-variables formula from cylindrical coordinates to rectangular coordinates by $dx\,dy\,dz = r\,dz\,dr\,d\theta$. (Note that Z is a set of volume zero. As with polar coordinates, application of Theorem 12.46 can be justified by applying it first for $r > 0$, and then taking the limit as $r \downarrow 0$.)

12.49 EXAMPLE.

Find the volume of the region E which lies inside the paraboloid $x^2 + y^2 + z = 4$, outside the cylinder $x^2 - 2x + y^2 = 0$, and above the plane $z = 0$.

Solution. The paraboloid $z = 4 - x^2 - y^2$ has vertex $(0, 0, 4)$ and opens downward about the z-axis. The cylinder $x^2 - 2x + y^2 = (x - 1)^2 + y^2 - 1 = 0$ has base centered at $(1, 0)$ with radius 1. Hence, the projection E_3 lies inside the circle $x^2 + y^2 = 4$ and outside the circle $x^2 + y^2 = 2x$ (see Figure 12.14). This last circle can be described in polar coordinates by $r^2 = 2r \cos \theta$, that is, $r = 2 \cos \theta$. Thus

$$\mathrm{Vol}(E) = \iiint_E 1\,dV = \iint_{E_3} \int_0^{4-r^2} dz\,dA$$

$$= \int_{-\pi/2}^{\pi/2} \int_{2\cos\theta}^{2} (4 - r^2) r\,dr\,d\theta + \int_{\pi/2}^{3\pi/2} \int_0^2 (4 - r^2) r\,dr\,d\theta = \frac{11\pi}{2}. \quad\blacksquare$$

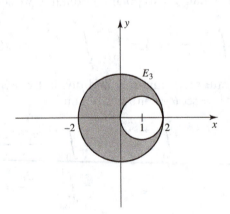

FIGURE 12.14

Recall that *spherical coordinates* in $\mathbf{R}^3$ have the form

$$x = \rho \sin \varphi \cos \theta, \qquad y = \rho \sin \varphi \sin \theta, \qquad z = \rho \cos \varphi,$$

where $\rho = \|(x, y, z)\|$, θ is the angle measured counterclockwise from the positive x axis to the line segment $L((0, 0, 0); (x, y, 0))$, and φ is the angle measured

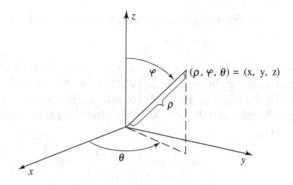

FIGURE 12.15

from the positive z-axis to the vector (x, y, z) (see Figure 12.15). Notice that this change of variables is 1–1 on $\{(\rho, \varphi, \theta) : \rho > 0, 0 < \varphi < \pi, 0 \leq \theta < 2\pi\}$ and its Jacobian, $\rho^2 \sin \varphi$ (see Exercise 12.4.8a), is nonzero off $Z := \{(\rho, \varphi, \theta) : \varphi = 0, \pi, \rho = 0\}$, a Jordan region of volume zero. Hence, application of Theorem 12.46 can justified by applying it first for $\rho > 0$ and $0 < \varphi < \pi$, and then taking the limit as $\rho, \varphi \downarrow 0$ and $\varphi \uparrow \pi$. Since the end result is the same as applying Theorem 12.46 directly to any projectable region in $\mathbf{R}^3$, we shall do so, without further comments, when changing variables to or from spherical coordinates. We shall abbreviate the change-of-variables formula from spherical coordinates to rectangular coordinates by $dx\, dy\, dz = \rho^2 \sin \varphi\, d\rho\, d\varphi\, d\theta$. (For spherical coordinates in $\mathbf{R}^n$, see the proof of Theorem 12.70.)

12.50 EXAMPLE.

Find

$$\iiint_Q x\, dV,$$

where $Q = B_3(0, 0, 0) \setminus B_2(0, 0, 0)$.

Solution. Using spherical coordinates, we have

$$\iiint_Q x\, dV = \int_0^{2\pi} \int_0^{\pi} \int_2^3 \rho \sin \varphi \cos \theta (\rho^2 \sin \varphi)\, d\rho\, d\varphi\, d\theta = 0. \qquad \blacksquare$$

Theorem 12.46 can be used for other changes of variables besides polar, cylindrical, and spherical coordinates.

12.51 EXAMPLE.

Find

$$\iint_E \sin(x + y) \cos(2x - y)\, dA,$$

where E is the region bounded by $y = 2x - 1$, $y = 2x + 3$, $y = -x$, and $y = -x + 1$.

Solution. Let $\phi(x, y) = (2x - y, x + y)$ and observe that the integral in question looks like the right side of (38) except the Jacobian is missing. By Cramer's Rule, for each fixed $u, v \in \mathbf{R}$, the system $u = 2x - y$, $v = x + y$ has a unique solution in x, y. Hence, ϕ is 1–1 on $\mathbf{R}^2$. It is obviously continuously differentiable, and its Jacobian,

$$\Delta_\phi(x, y) = \frac{\partial(u, v)}{\partial(x, y)} = \det \begin{bmatrix} 2 & -1 \\ 1 & 1 \end{bmatrix} = 3,$$

is a nonzero constant. Hence, we can make adjustments to the integral in question so that it is precisely the right side of (38):

$$\iint_E \sin(x + y) \cos(2x - y) \, dA = \frac{1}{3} \iint_E f \circ \phi(x, y) \Delta_\phi(x, y) \, d(x, y),$$

where $f(u, v) = \cos u \sin v$. It remains to compute the left side of (38) (i.e., to find what happens to E under ϕ).

Notice that $y = 2x - 1$ implies $u = 1$, $y = 2x + 3$ implies $u = -3$, $y = -x$ implies $v = 0$, and $y = -x + 1$ implies $v = 1$. Thus $\phi(E) = [-3, 1] \times [0, 1]$. Applying Theorem 12.46 and the preliminary step taken above, we find

$$\iint_E \sin(x + y) \cos(2x - y) \, dA = \frac{1}{3} \int_0^1 \int_{-3}^1 \sin v \cos u \, du \, dv$$

$$= \frac{1}{3}(\sin(1) + \sin(3))(1 - \cos(1)). \qquad \blacksquare$$

EXERCISES

12.4.1. Evaluate each of the following integrals.

a)
$$\int_0^2 \int_0^{\sqrt{4-x^2}} \sin(x^2 + y^2) \, dy \, dx$$

b)
$$\int_0^1 \int_0^x \sqrt[3]{(2y - y^2)^2} \, dy \, dx$$

c)
$$\int_a^b \int_0^x \sqrt{x^2 + y^2} \, dy \, dx, \qquad 0 \le a < b$$

12.4.2. For each of the following, find $\iint_E f \, dA$.

a) $f(x, y) = \cos(3x^2 + y^2)$ and E is the set of points satisfying $x^2 + y^2/3 \le 1$.

b) $f(x, y) = y\sqrt{x - 2y}$ and E is bounded by the triangle with vertices $(0, 0)$, $(4, 0)$, and $(4, 2)$.

12.4.3. For each of the following, find $\iiint_E f \, dV$.

a) $f(x, y, z) = z^2$ and E is the set of points satisfying $x^2 + y^2 + z^2 \le 6$ and $z \ge x^2 + y^2$.

b) $f(x, y, z) = e^z$ and E is the set of points satisfying $x^2 + y^2 + z^2 \le 9$, $x^2 + y^2 \le 1$, and $z \ge 0$.

c) $f(x, y, z) = (x - y)z$ and E is the set of points satisfying $x^2 + y^2 + z^2 \le 4$, $z \ge \sqrt{x^2 + y^2}$, and $x \ge 0$.

12.4.4. a) Prove that the volume bounded by the ellipsoid

$$\frac{x^2}{a^2} + \frac{y^2}{b^2} + \frac{z^2}{c^2} = 1$$

is $4\pi abc / 3$.

b) Let a, b, c, d be positive numbers and $r^2 < d^2/(b^2 + c^2)$. Find the volume of the region bounded by $y^2 + z^2 = r^2$, $x = 0$, and $ax + by + cz = d$.

c) Show that for any $a \ge 0$, the volume of the region bounded by the cylinders $x^2 + z^2 = a^2$ and $y^2 + z^2 = a^2$ is $16a^3/3$.

12.4.5. a) Compute $\iint_E \sqrt{x - y}\sqrt{x + 2y} \, dA$, where E is the parallelogram with vertices $(0, 0)$, $(2/3, -1/3)$, $(1, 0)$, $(1/3, 1/3)$.

b) Compute $\iint_E \sqrt[3]{2x^2 - 5xy - 3y^2} \, dA$, where E is the parallelogram bounded by the lines $y = x/3$, $y = (x-1)/3$, $y = -2x$, $y = 1-2x$.

c) Find

$$\iint_E e^{(y-x)/(y+x)} \, dA,$$

where E is the trapezoid with vertices $(1, 1)$, $(2, 2)$, $(2, 0)$, $(4, 0)$.

d) Given $\int_0^1 (1 - x)f(x) \, dx = 5$, find

$$\int_0^1 \int_0^x f(x - y) \, dy \, dx.$$

12.4.6. Suppose that V is nonempty and open in $\mathbf{R}^n$ and that $\mathbf{f} : V \to \mathbf{R}^n$ is continuously differentiable with $\Delta_{\mathbf{f}} \ne 0$ on V. Prove that

$$\lim_{r \to 0+} \frac{\text{Vol}(\mathbf{f}(B_r(\mathbf{x}_0)))}{\text{Vol}(B_r(\mathbf{x}_0))} = |\Delta_{\mathbf{f}}(\mathbf{x}_0)|$$

for every $\mathbf{x}_0 \in V$.

12.4.7. Show that Vol is *rotation invariant* in $\mathbf{R}^2$; that is, if ϕ is a rotation on $\mathbf{R}^2$ (see Exercise 8.2.9) and E is a Jordan region in $\mathbf{R}^2$, then

$$\text{Vol}(\phi(E)) = \text{Vol}(E).$$

12.4.8. a) Compute the Jacobian of the change of variables from spherical coordinates to rectangular coordinates.

b) Assuming that Vol is translation and rotation invariant (see Exercises 12.1.7 and 12.4.7), verify the following classical formulas: the volume of a sphere of radius r is $\frac{4}{3}\pi r^3$, and the volume of a right circular cone of altitude h and radius r is $\pi r^2 h/3$.

12.4.9. Let $\mathbf{v}_j = (v_{j1}, \ldots, v_{jn}) \in \mathbf{R}^n$, $j = 1, \ldots, n$, be fixed. The *parallelepiped* determined by the vectors $\mathbf{v}_j$ is the set

$$\mathcal{P}(\mathbf{v}_1, \ldots, \mathbf{v}_n) := \{t_1\mathbf{v}_1 + \cdots + t_n\mathbf{v}_n : t_j \in [0, 1]\},$$

and the determinant of the $\mathbf{v}_j$'s is the number

$$\det(\mathbf{v}_1, \ldots, \mathbf{v}_n) := \det\left[v_{jk}\right]_{n \times n}.$$

Prove that

$$\mathrm{Vol}(\mathcal{P}(\mathbf{v}_1, \ldots, \mathbf{v}_n)) = |\det(\mathbf{v}_1, \ldots, \mathbf{v}_n)|.$$

Check this formula for $n = 2$ and $n = 3$ to see that it agrees with the classical formulas for the area of a parallelogram and the volume of a parallelepiped.

12.4.10. **This exercise is used in Section *12.6.**

a) Prove that the improper integral $\int_0^\infty e^{-x^2}\, dx$ converges to a finite real number.

b) Prove that if I is the value of the integral in part a), then

$$I^2 = \lim_{N \to \infty} \int_0^{\pi/2} \int_0^N e^{-r^2} r\, dr\, d\theta.$$

c) Show that

$$\int_0^\infty e^{-x^2}\, dx = \frac{\sqrt{\pi}}{2}.$$

d) Let Q_k represent the n-dimensional cube $[-k, k] \times \cdots \times [-k, k]$. Find

$$\lim_{k \to \infty} \int_{Q_k} e^{-\|\mathbf{x}\|^2}\, d\mathbf{x}.$$

12.4.11. Let $H \subset V \subset \mathbf{R}^n$, with H convex and V open, and suppose that $\phi : V \to \mathbf{R}^n$ is $\mathcal{C}^1$.

a) Show that if E is a closed subset of H^o and

$$\epsilon_\mathbf{h}(\mathbf{x}) := \phi(\mathbf{x} + \mathbf{h}) - \phi(\mathbf{x}) - D\phi(\mathbf{x})(\mathbf{h}), \quad \text{for } \mathbf{x} \in V \text{ and } \mathbf{h} \text{ small},$$

then $\epsilon_\mathbf{h}(\mathbf{x})/\|\mathbf{h}\| \to 0$ uniformly on E, as $\mathbf{h} \to 0$.

b) Show that if R is a closed rectangle in H^o and $S := (D\phi(\mathbf{x}))^{-1}$ exists for some $\mathbf{x} \in R$, then given $\varepsilon > 0$ there are constants $\delta > 0$ and $M > 0$ and a function $T(\mathbf{x}, \mathbf{y})$ such that

$$S \circ \phi(\mathbf{x}) - S \circ \phi(\mathbf{y}) = \mathbf{x} - \mathbf{y} + T(\mathbf{x}, \mathbf{y})$$

for $\mathbf{x}, \mathbf{y} \in R$, and $\|T(\mathbf{x}, \mathbf{y})\| \leq M\varepsilon$ when $\|\mathbf{x} - \mathbf{y}\| < \delta$.

c) Use parts a) and b) to prove that if Δ_ϕ is nonzero on V, $\mathbf{x} \in H^o$, and ε is sufficiently small, then there exist numbers $C_\varepsilon > 0$, which depend only on H, ϕ, n, and ε, and a $\delta > 0$ such that $C_\varepsilon \to 1$ as $\varepsilon \to 0$ and $\mathrm{Vol}(S \circ \phi(Q)) \leq C_\varepsilon |Q|$ for all cubes $Q \subset H$ which contain $\mathbf{x}$ and satisfy $\mathrm{Vol}(Q) < \delta$.

d) Use part c) and Exercise 12.4.9 to prove that if Δ_ϕ is nonzero on V and $\mathbf{x} \in H^o$, then given any sequence of cubes Q_j which satisfy $\mathbf{x} \in Q_j$ and $\mathrm{Vol}(Q_j) \to 0$ as $j \to \infty$, it is also the case that $\mathrm{Vol}(\phi(Q_j))/|Q_j| \to |\Delta_\phi(\mathbf{x})|$ as $j \to \infty$.

*12.5 PARTITIONS OF UNITY

This section uses results from Section 9.5.

In this section we show that a smooth function can be broken into a sum of smooth functions, each of which is zero except on a small set, and use this to prove a global change-of-variables formula when the Jacobian is nonzero off a set of volume zero. This same technique can be used to prove the Fundamental Theorem of Calculus on manifolds (see [12], for example).

12.52 Definition.

Let $f : \mathbf{R}^n \to \mathbf{R}$.

i) The *support* of f is the closure of the set of points at which f is nonzero; that is,

$$\mathrm{spt}\, f := \overline{\{\mathbf{x} \in \mathbf{R}^n : f(\mathbf{x}) \neq 0\}}.$$

ii) A function f is said to have *compact support* if and only if $\mathrm{spt}\, f$ is a compact set.

12.53 EXAMPLE.

If

$$f(x) = \begin{cases} 1 & x \in \mathbf{Q} \\ 0 & x \notin \mathbf{Q}, \end{cases}$$

then $\mathrm{spt}\, f = \mathbf{R}$.

12.54 EXAMPLE.

If

$$f(x) = \begin{cases} 1 & x \in (0, 1) \\ 2 & x \in (1, 2) \\ 0 & \text{otherwise,} \end{cases}$$

then spt $f = [0, 2]$.

Since the support of a function is always closed, a function f on $\mathbf{R}^n$ has compact support if and only if spt f is bounded (see the Heine–Borel Theorem).

The following result shows that if two functions have compact support, then so does their sum (see also Exercises 12.5.1 and 12.5.2).

12.55 Remark. *If $f, g : \mathbf{R}^n \to \mathbf{R}$, then*

$$\text{spt } (f + g) \subseteq \text{spt } f \cup \text{spt } g.$$

Proof. If $(f + g)(\mathbf{x}) \neq 0$, then $f(\mathbf{x}) \neq 0$ or $g(\mathbf{x}) \neq 0$. Thus

$$\{\mathbf{x} \in \mathbf{R}^n : (f + g)(\mathbf{x}) \neq 0\} \subseteq \{\mathbf{x} \in \mathbf{R}^n : f(\mathbf{x}) \neq 0\} \cup \{\mathbf{x} \in \mathbf{R}^n : g(\mathbf{x}) \neq 0\}.$$

Since the closure of a union equals the union of its closures (see Theorem 8.37 or 10.40), it follows that spt $(f + g) \subseteq$ spt $f \cup$ spt g. ∎

Let $p \in \mathbf{N}$ or $p = \infty$. The symbol $\mathcal{C}_c^p(\mathbf{R}^n)$ will denote the collection of functions $f : \mathbf{R}^n \to \mathbf{R}$ which are $\mathcal{C}^p$ on $\mathbf{R}^n$ and have compact support. In particular, it follows from Remark 12.55 that if $f_j \in \mathcal{C}_c^p(\mathbf{R}^n)$ for $j = 1, \ldots, N$, then

$$\sum_{j=1}^{N} f_j \in \mathcal{C}_c^p(\mathbf{R}^n).$$

We will use this observation several times below.

If f is analytic (a condition stronger than $\mathcal{C}^\infty$) and has compact support, then f is identically zero (see Exercise 12.5.3). Thus it is not at all obvious that $\mathcal{C}_c^\infty(\mathbf{R}^n)$ contains anything but the zero function. Nevertheless, we shall show that $\mathcal{C}_c^\infty(\mathbf{R}^n)$ not only contains nonzero functions but has enough functions to "approximate" any compact set (see Theorem 12.58 and Exercise 12.5.6).

First, we deal with the one-dimensional case.

12.56 Lemma.
For every $a < b$ there is a function $\phi \in \mathcal{C}_c^\infty(\mathbf{R})$ such that $\phi(t) > 0$ for $t \in (a, b)$ and $\phi(t) = 0$ for $t \notin (a, b)$.

Proof. The function

$$f(t) = \begin{cases} e^{-1/t^2} & t \neq 0 \\ 0 & t = 0 \end{cases}$$

belongs to $\mathcal{C}^\infty(\mathbf{R})$ and $f^{(j)}(0) = 0$ for all $j \in \mathbf{N}$ (see Exercise 4.4.7). Hence,

$$\phi(t) = \begin{cases} e^{-1/(t-a)^2} e^{-1/(t-b)^2} & t \in (a, b) \\ 0 & \text{otherwise} \end{cases}$$

belongs to $\mathcal{C}^\infty(\mathbf{R})$, satisfies $\phi(t) > 0$ for $t \in (a, b)$, and spt $\phi = [a, b]$. ∎

Next, we show that there exists a nonzero $\mathcal{C}^\infty$ function which is constant everywhere except on a small interval.

12.57 Lemma.
For each $\delta > 0$ there is a function $\psi \in \mathcal{C}^\infty(\mathbf{R})$ such that $0 \leq \psi \leq 1$ on $\mathbf{R}$, $\psi(t) = 0$ for $t \leq 0$, and $\psi(t) = 1$ for $t > \delta$.

Proof. By Lemma 12.56, choose $\phi \in \mathcal{C}_c^\infty(\mathbf{R})$ such that $\phi(t) > 0$ for $t \in (0, \delta)$ and $\phi(t) = 0$ for $t \notin (0, \delta)$. Set

$$\psi(t) = \frac{\int_0^t \phi(u)\, du}{\int_0^\delta \phi(u)\, du}.$$

By the Fundamental Theorem of Calculus, $\psi \in \mathcal{C}^\infty(\mathbf{R})$, by construction $0 \leq \psi \leq 1$, and

$$\psi(t) = \begin{cases} 0 & t \leq 0 \\ 1 & t > \delta. \end{cases}$$ ∎

Finally, we use these one-dimensional $\mathcal{C}^\infty$ functions to construct nonzero functions in $\mathcal{C}_c^\infty(\mathbf{R}^n)$.

12.58 Theorem.   [$\mathcal{C}^\infty$ **VERSION OF URYSOHN'S LEMMA**].
Let H be compact and nonempty, let V be open in $\mathbf{R}^n$, and let $H \subset V$. Then there is an $h \in \mathcal{C}_c^\infty(\mathbf{R}^n)$ such that $0 \leq h(\mathbf{x}) \leq 1$ for all $\mathbf{x} \in \mathbf{R}^n$, $h(\mathbf{x}) = 1$ for all $\mathbf{x} \in H$, and spt $h \subset V$.

Proof. Let $\phi \in \mathcal{C}_c^\infty(\mathbf{R})$ satisfy $\phi(t) > 0$ for $t \in (-1, 1)$ and $\phi(t) = 0$ for $t \notin (-1, 1)$. For each $\varepsilon > 0$ and each $\mathbf{x} \in \mathbf{R}^n$, let $Q_\varepsilon(\mathbf{x})$ represent the n-dimensional cube

$$Q_\varepsilon(\mathbf{x}) = \{\mathbf{y} \in \mathbf{R}^n : |y_j - x_j| \leq \varepsilon \text{ for all } j = 1, \ldots, n\}.$$

Set

$$g_\varepsilon(\mathbf{y}) = \phi\left(\frac{y_1}{\varepsilon}\right) \cdots \phi\left(\frac{y_n}{\varepsilon}\right), \tag{44}$$

and observe by Theorem 4.10 (the Product Rule) that g_ε is $\mathcal{C}^\infty$ on $\mathbf{R}^n$. By construction, $g_\varepsilon(\mathbf{y}) \geq 0$ on $\mathbf{R}^n$, $g_\varepsilon(\mathbf{y}) > 0$ for all $\mathbf{y}$ in the open ball $B_\varepsilon(\mathbf{0})$, and the support of g_ε is a subset of the cube $Q_\varepsilon(\mathbf{0})$. In particular, $g_\varepsilon \in \mathcal{C}_c^\infty(\mathbf{R}^n)$.

We will use sums of translates of these g_ε's to construct a $\mathcal{C}^\infty$ function, supported on V, which is strictly positive on H. It is here that the compactness of H enters in a crucial way.

For each $\mathbf{x} \in H$, choose $\varepsilon := \varepsilon(\mathbf{x}) > 0$ such that $Q_\varepsilon(\mathbf{x}) \subset V$. Set

$$h_{\mathbf{x}}(\mathbf{y}) = g_\varepsilon(\mathbf{y} - \mathbf{x}), \qquad \mathbf{y} \in \mathbf{R}^n,$$

and notice that $h_{\mathbf{x}} \geq 0$ on $\mathbf{R}^n$, $h_{\mathbf{x}}(\mathbf{y}) > 0$ for all $\mathbf{y} \in B_\varepsilon(\mathbf{x})$, $h_{\mathbf{x}}(\mathbf{y}) = 0$ for all $\mathbf{y} \notin Q_\varepsilon(\mathbf{x})$, and $h_{\mathbf{x}} \in \mathcal{C}_c^\infty(\mathbf{R}^n)$. Since H is compact and

$$H \subset \bigcup_{\mathbf{x} \in H} B_\varepsilon(\mathbf{x}),$$

choose points $\mathbf{x}_j \in H$ and positive numbers $\varepsilon_j = \varepsilon(\mathbf{x}_j)$, $j = 1, \ldots, N$, such that

$$H \subset B_{\varepsilon_1}(\mathbf{x}_1) \cup \cdots \cup B_{\varepsilon_N}(\mathbf{x}_N).$$

Set $Q = Q_{\varepsilon_1}(\mathbf{x}_1) \cup \ldots \cup Q_{\varepsilon_N}(\mathbf{x}_N)$ and $f = h_{\mathbf{x}_1} + \ldots + h_{\mathbf{x}_N}$. Clearly, Q is compact, $Q \subset V$, and f is $\mathcal{C}^\infty$ on $\mathbf{R}^n$. If $\mathbf{x} \notin Q$, then $\mathbf{x} \notin Q_{\varepsilon_j}(\mathbf{x}_j)$ for all j; hence, $f(\mathbf{x}) = 0$. Thus spt $f \subseteq Q$. If $\mathbf{x} \in H$, then $\mathbf{x} \in B_{\varepsilon_j}(\mathbf{x}_j)$ for some j; hence, $f(\mathbf{x}) > 0$. It remains to flatten f so that it is identically 1 on H. This is where Lemma 12.57 comes in.

Since $f > 0$ on the compact set H, f has a nonzero minimum on H. Thus there is a $\delta > 0$ such that $f(\mathbf{x}) > \delta$ for $\mathbf{x} \in H$. By Lemma 12.57, choose $\psi \in \mathcal{C}^\infty(\mathbf{R})$ such that $\psi(t) = 0$ when $t \leq 0$, and $\psi(t) = 1$ when $t > \delta$. Set $h = \psi \circ f$. Clearly, $h \in \mathcal{C}_c^\infty(\mathbf{R}^n)$, spt $h \subseteq Q \subset V$, and, since $f > \delta$ on H, $h = 1$ on H. Finally, since $0 \leq \psi \leq 1$, the same is true of h. $\blacksquare$

This result leads directly to a decomposition theorem for $\mathcal{C}^\infty$ functions.

12.59 Theorem. [$\mathcal{C}^\infty$ PARTITIONS OF UNITY].
Let $\Omega \subset \mathbf{R}^n$ be nonempty and let $\{V_\alpha\}_{\alpha \in A}$ be an open covering of Ω. Then there exist functions $\phi_j \in \mathcal{C}_c^\infty(\mathbf{R}^n)$ and indices $\alpha_j \in A$, $j \in \mathbf{N}$, such that the following properties hold.

i)
$$\phi_j \geq 0 \text{ for all } j \in \mathbf{N}.$$

ii)
$$\text{spt } \phi_j \subset V_{\alpha_j} \text{ for all } j \in \mathbf{N}.$$

iii)
$$\sum_{j=1}^{\infty} \phi_j(\mathbf{x}) = 1 \text{ for all } \mathbf{x} \in \Omega.$$

iv) *If H is a nonempty compact subset of Ω, then there is a nonempty open set $W \supset H$ and an integer N such that $\phi_j(\mathbf{x}) = 0$ for all $j \geq N$ and $\mathbf{x} \in W$. In particular,*

$$\sum_{j=1}^{N} \phi_j(\mathbf{x}) = 1 \text{ for all } \mathbf{x} \in W.$$

Proof. For each $\mathbf{x} \in \Omega$, choose a bounded open set $W(\mathbf{x})$ and an index $\alpha \in A$ such that

$$\mathbf{x} \in W(\mathbf{x}) \subset \overline{W}(\mathbf{x}) \subset V_\alpha.$$

Then $\mathcal{W} = \{W(\mathbf{x}) : \mathbf{x} \in \Omega\}$ is an open covering of Ω and, by Lindelöf's Theorem, we may suppose that $\mathcal{W}$ is countable; that is, $\mathcal{W} = \{W_j\}_{j \in \mathbf{N}}$.

By construction, given $j \in \mathbf{N}$, there is an index $\alpha_j \in A$ such that

$$W_j \subset \overline{W}_j \subset V_{\alpha_j}.$$

Choose by Theorem 12.58 functions $h_j \in C_c^{\infty}(\mathbf{R}^n)$ such that $0 \leq h_j \leq 1$ on $\mathbf{R}^n$, $h_j = 1$ on $\overline{W}_j$, and spt $h_j \subset V_{\alpha_j}$ for $j \in \mathbf{N}$. Set $\phi_1 = h_1$ and for $j > 1$, set

$$\phi_j = (1 - h_1) \dots (1 - h_{j-1}) h_j.$$

Then $\phi_j \geq 0$ on $\mathbf{R}^n$, and $\phi_j \in C_c^{\infty}(\mathbf{R}^n)$ with spt $\phi_j \subseteq$ spt $h_j \subset V_{\alpha_j}$ for $j \in \mathbf{N}$. This proves parts i) and ii).

An easy induction argument establishes

$$\sum_{j=1}^{k} \phi_j = 1 - (1 - h_1) \dots (1 - h_k)$$

for $k \in \mathbf{N}$. If $\mathbf{x} \in \Omega$, then $\mathbf{x} \in W_{j_0}$ for some j_0 so $1 - h_{j_0}(\mathbf{x}) = 0$. Thus

$$\sum_{j=1}^{k} \phi_j(\mathbf{x}) = 1 - 0 = 1$$

for $k \geq j_0$. If H is a compact subset of Ω, then $H \subset W_1 \cup \dots \cup W_N$ for some $N \in \mathbf{N}$. If $W = W_1 \cup \dots \cup W_N$, then $\mathbf{x} \in W$ implies $h_k(\mathbf{x}) = 1$ for some $1 \leq k \leq N$; that is, $\phi_j(\mathbf{x}) = 0$ for all $j > N$. Hence,

$$\sum_{j=1}^{N} \phi_j(\mathbf{x}) = \sum_{j=1}^{\infty} \phi_j(\mathbf{x}) = 1$$

for all $\mathbf{x} \in W$. ∎

A sequence of functions $\{\phi_j\}_{j \in \mathbf{N}}$ is called a (C^0) *partition of unity on Ω subordinate to* a covering $\{V_\alpha\}_{\alpha \in A}$ if and only if Ω and all the V_α's are open and nonempty, the ϕ_j's are all continuous with compact support and satisfy statements i) through iv) of Theorem 12.59. By a C^p *partition of unity* on Ω we shall mean a partition of unity on Ω whose functions ϕ_j are also C^p on Ω. By Theorem 12.59, given any open covering $\mathcal{V}$ of any nonempty set $\Omega \subseteq \mathbf{R}^n$ and any extended real number $p \geq 0$, there exists a C^p partition of unity on Ω subordinate to $\mathcal{V}$.

C^p partitions of unity can be used to decompose a function f into a sum of functions f_j which have small support and are as smooth as f. For example, let f be defined on a set Ω, $\{\phi_j\}_{j \in \mathbf{N}}$ be a C^p partition of unity on Ω subordinate to a covering $\{V_j\}_{j \in \mathbf{N}}$, and $f_j = f\phi_j$. Then

$$f(\mathbf{x}) = f(\mathbf{x}) \sum_{j=1}^{\infty} \phi_j(\mathbf{x}) = \sum_{j=1}^{\infty} f(\mathbf{x})\phi_j(\mathbf{x}) = \sum_{j=1}^{\infty} f_j(\mathbf{x})$$

for all $\mathbf{x} \in \Omega$. If f is continuous on Ω and $p \geq 0$, then each f_j is continuous on Ω; if f is continuously differentiable on Ω and $p \geq 1$, then each f_j is continuously differentiable on Ω. Thus, f can be written as a sum of functions f_j which are as smooth as f. This allows us to pass from local results to global ones; for example, if we know that a certain property holds on small open sets in Ω, then we can show that a similar property holds on all of Ω by using a partition of unity subordinate to a covering of Ω which consists of small open sets.

To illustrate the power of this point of view, we now show that the integral can be extended from Jordan regions to open bounded sets, even though such sets are not always Jordan regions. This extension is a multidimensional version of the improper integral. (The proofs Theorems 12.63 and 12.64 are based on Spivak [12].[1])

STRATEGY: The idea behind this extension is fairly simple. Let V be a bounded open set and let f be *locally integrable* on V; that is, $f : V \rightarrow \mathbf{R}$ is integrable on every closed Jordan region $H \subset V$. For each $\mathbf{x} \in V$, choose an open Jordan region $V(\mathbf{x})$ so small that $\mathbf{x} \in V(\mathbf{x}) \subset V$. [e.g., $V(\mathbf{x})$ could be an open ball.] Then $\{V(\mathbf{x})\}_{\mathbf{x} \in V}$ is an open covering of V, and by Lindelöf's Theorem it has a countable subcover, say $\mathcal{V} = \{V_j\}_{j \in \mathbf{N}}$. Let $\{\phi_j\}_{j \in \mathbf{N}}$ be a partition of unity on V subordinate to $\mathcal{V}$. Since f is locally integrable on V, each $f\phi_j$ is integrable. Since $f = \sum_{j=1}^{\infty} f\phi_j$, it seems reasonable to define

$$\int_V f(\mathbf{x}) \, d\mathbf{x} = \sum_{j=1}^{\infty} \int_{V_j} f(\mathbf{x})\phi_j(\mathbf{x}) \, d\mathbf{x}.$$

Before we can proceed, we must answer two questions: Does this series converge? And if it does, will its value change when the partition of unity changes? The next two results answer these questions.

---

[1]M. Spivak, *Calculus on Manifolds*, (New York: W. A. Benjamin, Inc., 1965). Reprinted with permission of Addison-Wesley Publishing Company.

12.60 Lemma.

Let V be a bounded open set in $\mathbf{R}^n$ and let $\mathcal{V} = \{V_j\}_{j \in \mathbf{N}}$ be a sequence of nonempty open Jordan regions in V which satisfies

$$V = \bigcup_{j=1}^{\infty} V_j.$$

Suppose that $f : V \to \mathbf{R}$ is bounded on V and integrable on each V_j. If $\{\phi_j\}_{j \in \mathbf{N}}$ is any partition of unity on V subordinate to the covering $\mathcal{V}$, then

$$\sum_{j=1}^{\infty} \int_{V_j} \phi_j(\mathbf{x}) f(\mathbf{x}) \, d\mathbf{x} \tag{45}$$

converges absolutely.

Proof. Let R be an n-dimensional rectangle containing V and $M = \sup_{\mathbf{x} \in V} |f(\mathbf{x})|$. Since ϕ_j is supported on V_j, the function $\phi_j f$ is integrable on V_j. Moreover, if $E = \cup_{j=1}^{N} V_j$ we have

$$\sum_{j=1}^{N} \left| \int_{V_j} \phi_j(\mathbf{x}) f(\mathbf{x}) \, d\mathbf{x} \right| \le \sum_{j=1}^{N} \int_{E} |\phi_j(\mathbf{x}) f(\mathbf{x})| \, d\mathbf{x}$$

$$= \int_{E} \sum_{j=1}^{N} |\phi_j(\mathbf{x}) f(\mathbf{x})| \, d\mathbf{x}$$

$$\le M \int_{E} \sum_{j=1}^{N} |\phi_j(\mathbf{x})| \, d\mathbf{x} \le M \, \mathrm{Vol}(E) \le M|R| < \infty.$$

Therefore, the series in (45) converges absolutely. ∎

The value of the series in (45) depends neither on the partition of unity chosen nor the covering $\mathcal{V}$.

12.61 Lemma.

Let V be a bounded, nonempty, open set in $\mathbf{R}^n$. Suppose that $\mathcal{V} = \{V_j\}_{j \in \mathbf{N}}$ and $\mathcal{W} = \{W_k\}_{k \in \mathbf{N}}$ are sequences of nonempty open Jordan regions in $\mathbf{R}^n$ such that

$$V = \bigcup_{j=1}^{\infty} V_j = \bigcup_{k=1}^{\infty} W_k.$$

Suppose further that $f : V \to \mathbf{R}$ is bounded and locally integrable on V. If $\{\phi_j\}_{j \in \mathbf{N}}$ is a partition of unity on V subordinate to $\mathcal{V}$ and $\{\psi_k\}_{k \in \mathbf{N}}$ is a partition of unity on V subordinate to $\mathcal{W}$, then

$$\sum_{j=1}^{\infty} \int_{V_j} \phi_j(\mathbf{x}) f(\mathbf{x}) \, d\mathbf{x} = \sum_{k=1}^{\infty} \int_{W_k} \psi_k(\mathbf{x}) f(\mathbf{x}) \, d\mathbf{x}. \tag{46}$$

Proof. By Lemma 12.60, both sums in (46) converge absolutely. By Exercise 12.5.5, $\{\phi_j \psi_k\}_{j,k \in \mathbf{N}}$ is a partition of unity on V subordinate to the covering $\{V_j \cap W_k\}_{j,k \in \mathbf{N}}$. Thus

$$\sum_{j=1}^{\infty} \sum_{k=1}^{\infty} \int_V \phi_j(\mathbf{x}) \psi_k(\mathbf{x}) f(\mathbf{x}) \, d\mathbf{x}$$

also converges absolutely. Fix $j \in \mathbf{N}$. Since $\operatorname{spt} \phi_j$ is compact, choose $N \in \mathbf{N}$ so large that $\psi_k(\mathbf{x}) = 0$ for $k > N$ and $\mathbf{x} \in \operatorname{spt} \phi_j$. Hence,

$$\int_{V_j} \phi_j(\mathbf{x}) f(\mathbf{x}) \, d\mathbf{x} = \int_{V_j} \phi_j(\mathbf{x}) \sum_{k=1}^{N} \psi_k(\mathbf{x}) f(\mathbf{x}) \, d\mathbf{x}$$

$$= \sum_{k=1}^{N} \int_{V_j \cap W_k} \phi_j(\mathbf{x}) \psi_k(\mathbf{x}) f(\mathbf{x}) \, d\mathbf{x}$$

$$= \sum_{k=1}^{\infty} \int_{V_j \cap W_k} \phi_j(\mathbf{x}) \psi_k(\mathbf{x}) f(\mathbf{x}) \, d\mathbf{x}.$$

Thus

$$\sum_{j=1}^{\infty} \int_{V_j} \phi_j(\mathbf{x}) f(\mathbf{x}) \, d\mathbf{x} = \sum_{j=1}^{\infty} \sum_{k=1}^{\infty} \int_{V_j \cap W_k} \phi_j(\mathbf{x}) \psi_k(\mathbf{x}) f(\mathbf{x}) \, d\mathbf{x}.$$

Reversing the roles of j and k, we also have

$$\sum_{k=1}^{\infty} \int_{W_k} \psi_k(\mathbf{x}) f(\mathbf{x}) \, d\mathbf{x} = \sum_{k=1}^{\infty} \sum_{j=1}^{\infty} \int_{V_j \cap W_k} \phi_j(\mathbf{x}) \psi_k(\mathbf{x}) f(\mathbf{x}) \, d\mathbf{x}.$$

Since these series are absolutely convergent, we may reverse the order of summation in the last double series. ∎

Using Lemma 12.61, we define the integral of a locally integrable function f over a bounded open set V as follows.

12.62 Definition.

Let V be a bounded, nonempty, open set in $\mathbf{R}^n$ and let $f : V \to \mathbf{R}$ be bounded and locally integrable on V. The *integral* of f on V is defined to be

$$I_V(f) := \sum_{j=1}^{\infty} \int_{V_j} \phi_j(\mathbf{x}) f(\mathbf{x}) \, d\mathbf{x},$$

where $\{\phi_j\}_{j\in\mathbf{N}}$ is any partition of unity on V subordinate to an open covering $\mathcal{V} = \{V_j\}_{j\in\mathbf{N}}$ such that each V_j is a nonempty Jordan region and

$$V = \bigcup_{j=1}^{\infty} V_j.$$

The following result shows that this definition agrees with the old one when V is a Jordan region. Thus, we shall use the notation $\int_V f(\mathbf{x}) \, d\mathbf{x}$ for $I_V(f)$.

12.63 Theorem. *If E is a nonempty, open Jordan region in $\mathbf{R}^n$ and $f : E \to \mathbf{R}$ is integrable on E, then*

$$\int_E f(\mathbf{x}) \, d\mathbf{x} = I_E(f).$$

Proof. Let $\varepsilon > 0$. Since E is a Jordan region, choose a grid $\mathcal{G} = \{Q_1, \dots, Q_p\}$ of some n-dimensional rectangle $R \supset E$ such that

$$\sum_{Q_\ell \cap \partial E \neq \emptyset} |Q_\ell| < \varepsilon. \tag{47}$$

Let

$$H = \bigcup_{Q_\ell \subset E} Q_\ell.$$

Clearly, H is compact and by (47), $\mathrm{Vol}(E \backslash H) < \varepsilon$ (see Exercise 12.1.6d).

Set $M = \sup_{\mathbf{x}\in E} |f(\mathbf{x})|$. Let $\{R_j\}_{j\in\mathbf{N}}$ be a sequence of rectangles such that $R_j \subset E$ and $E = \bigcup_{j=1}^{\infty} R_j^o$, and let $\{\phi_j\}_{j\in\mathbf{N}}$ be a partition of unity on E subordinate to $\mathcal{V} = \{R_j^o\}_{j\in\mathbf{N}}$. Since H is compact, choose $N_1 \in \mathbf{N}$ such that $\phi_j(\mathbf{x}) = 0$ for $j > N_1$ and $\mathbf{x} \in H$. Then, for any $N \geq N_1$, we have

$$\left| \int_E f(\mathbf{x})\, d\mathbf{x} - \sum_{j=1}^{N} \int_{R_j} \phi_j(\mathbf{x}) f(\mathbf{x})\, d\mathbf{x} \right| = \left| \int_E f(\mathbf{x})\, d\mathbf{x} - \sum_{j=1}^{N} \int_E \phi_j(\mathbf{x}) f(\mathbf{x})\, d\mathbf{x} \right|$$

$$\leq \int_E \left| f(\mathbf{x}) - \sum_{j=1}^{N} \phi_j(\mathbf{x}) f(\mathbf{x}) \right| d\mathbf{x}$$

$$\leq M \int_E \left| 1 - \sum_{j=1}^{N} \phi_j(\mathbf{x}) \right| d\mathbf{x}$$

$$\leq M \operatorname{Vol}(E \setminus H) < M\varepsilon.$$

We conclude that $I_E(f)$ exists and equals $\int_E f(\mathbf{x})\, d\mathbf{x}$. ∎

We now prove a change-of-variables formula valid for all open bounded sets.

12.64 Theorem. *Suppose that V is a bounded, nonempty, open set in $\mathbf{R}^n$, that $\phi : V \to \mathbf{R}^n$ is 1–1 and continuously differentiable on V, and that $\phi(V)$ is bounded. If $\Delta_\phi \neq 0$ on V, then*

$$\int_{\phi(V)} f(\mathbf{u})\, d\mathbf{u} = \int_V f(\phi(\mathbf{x})) |\Delta_\phi(\mathbf{x})|\, d\mathbf{x},$$

for all bounded $f : \phi(V) \to \mathbf{R}$, provided f is locally integrable on $\phi(V)$.

Proof. For each $\mathbf{a} \in V$, choose by Theorem 12.45 an open rectangle $W(\mathbf{a})$ such that $\overline{W}(\mathbf{a}) \subset V$ and

$$\int_{\phi(W(\mathbf{a}))} f(\mathbf{u})\, d\mathbf{u} = \int_{W(\mathbf{a})} f(\phi(\mathbf{x})) |\Delta_\phi(\mathbf{x})|\, d\mathbf{x}. \tag{48}$$

Set $\mathcal{W} = \{W(\mathbf{a})\}_{\mathbf{a} \in V}$. Then $\mathcal{W}$ is an open covering of V. By Lindelöf's Theorem, we may assume that $\mathcal{W} = \{W_j\}_{j \in \mathbf{N}}$. Let $\{\phi_j\}_{j \in \mathbf{N}}$ be a partition of unity on V subordinate to $\mathcal{W}$; that is, a sequence of C^∞ functions such that

$$\operatorname{spt} \phi_j \subset W_j \subset V, \quad j \in \mathbf{N}, \quad \text{and} \quad \sum_{j=1}^{\infty} \phi_j(\mathbf{x}) = 1$$

for all $\mathbf{x} \in V$. By Theorem 12.10, each $\phi(W_j)$ is a Jordan region. By Theorem 11.39, each $\phi(W_j)$ is open. And by Exercise 12.5.4, $\{\phi_j \circ \phi^{-1}\}_{j \in \mathbf{N}}$ is a partition of unity on $\phi(V)$ subordinate to the open covering $\{\phi(W_j)\}_{j \in \mathbf{N}}$. Hence, by Definition 12.62 and (48),

$$\int_{\phi(V)} f(\mathbf{u}) \, d\mathbf{u} = \sum_{j=1}^{\infty} \int_{\phi(W_j)} (\phi_j \circ \phi^{-1})(\mathbf{u}) f(\mathbf{u}) \, d\mathbf{u}$$

$$= \sum_{j=1}^{\infty} \int_{W_j} \phi_j(\mathbf{x}) f(\phi(\mathbf{x})) |\Delta_\phi(\mathbf{x})| \, dx$$

$$= \int_V f(\phi(\mathbf{x})) |\Delta_\phi(\mathbf{x})| \, dx. \qquad \blacksquare$$

Finally, we are prepared to prove a change-of-variables formula for functions whose Jacobians are zero on a set of volume zero.

12.65 Theorem. [CHANGE OF VARIABLES FOR MULTIPLE INTEGRALS].
Suppose that E is a Jordan region in $\mathbf{R}^n$ and that $\phi : E \to \mathbf{R}^n$ is 1–1 and continuous on E, and $\mathcal{C}^1$ on E^o. Suppose further that there exists a closed subset Z of E such that $Vol(Z) = Vol(\phi(Z)) = 0$. If $\phi(\partial E)$ is of volume zero and if $\Delta_\phi(\mathbf{x}) \neq 0$ for all $\mathbf{x} \in E^o \setminus Z$, then

$$\int_{\phi(E)} f(\mathbf{u}) \, d\mathbf{u} = \int_E f(\phi(\mathbf{x})) |\Delta_\phi(\mathbf{x})| \, dx$$

provided f is integrable on $\phi(E)$.

Proof. By the proof of Lemma 12.9, $\partial(\phi(E))$ is a subset of $\phi(E)$. Since this last set is of volume zero, it follows that $\phi(E)$ is a Jordan region.

Set $V := E^o \setminus Z$ and $E_0 := (E \setminus E^o) \cup Z$. Then V is a bounded open set, E_0 is of volume zero, $E = V \cup E_0$, and $V \cap E_0 = \emptyset$.

Since ϕ is 1–1 on E, $\phi(E) = \phi(V) \cup \phi(E_0)$ is a disjoint decomposition of the Jordan region $\phi(E)$. Moreover, by Lemma 11.39, the set $\phi(V)$ is bounded and open. Since

$$\phi(E_0) = \phi(E \setminus E^o) \cup \phi(Z) \subseteq \phi(\partial E) \cup \phi(Z),$$

a set of volume zero, $\phi(E_0)$ is also a set of volume zero. We conclude by Theorems 12.23, 12.24, and 12.65 that

$$\int_{\phi(E)} f(\mathbf{u}) \, d\mathbf{u} = \int_{\phi(V)} f(\mathbf{u}) \, d\mathbf{u}$$

$$= \int_V f(\phi(\mathbf{x})) |\Delta_\phi(\mathbf{x})| \, dx = \int_E f(\phi(\mathbf{x})) |\Delta_\phi(\mathbf{x})| \, dx. \qquad \blacksquare$$

We close by noting that, as general as it is, even this result can be improved. If something called the Lebesgue integral is used instead of the Riemann integral, the condition that $\Delta_\phi \neq 0$ can be dropped altogether (see Spivak [12], p. 72).

EXERCISES

12.5.1. If $f, g : \mathbf{R}^n \to \mathbf{R}$, prove that $\mathrm{spt}(fg) \subseteq \mathrm{spt}\, f \cap \mathrm{spt}\, g$.

12.5.2. Prove that if $f, g \in \mathcal{C}_c^\infty(\mathbf{R}^n)$, then so are fg and αf for any scalar α.

*12.5.3. Prove that if f is analytic on $\mathbf{R}$ and $f(x_0) \neq 0$ for some $x_0 \in \mathbf{R}$, then $f \notin C_c^\infty(\mathbf{R})$.

12.5.4. Suppose that V is a bounded, nonempty, open set in $\mathbf{R}^n$ and that $\phi : V \to \mathbf{R}^n$ is 1–1 and continuously differentiable on V with $\Delta_\phi \neq 0$ on V. Let $\mathcal{W} = \{W_j\}_{j \in \mathbf{N}}$ be an open covering of V and $\{\phi_j\}_{j \in \mathbf{N}}$ be a C^p partition of unity on V subordinate to $\mathcal{W}$, where $p \geq 1$. Prove that $\{\phi_j \circ \phi^{-1}\}_{j \in \mathbf{N}}$ is a C^1 partition of unity on $\phi(V)$ subordinate to the open covering $\{\phi(W_j)\}_{j \in \mathbf{N}}$.

12.5.5. Let V be open in $\mathbf{R}^n$ and $\mathcal{V} = \{V_j\}_{j \in \mathbf{N}}$, $\mathcal{W} = \{W_k\}_{k \in \mathbf{N}}$ be coverings of V. If $\{\phi_j\}_{j \in \mathbf{N}}$ is a C^p partition of unity on V subordinate to $\mathcal{V}$ and $\{\psi_k\}_{k \in \mathbf{N}}$ is a C^p partition of unity on V subordinate to $\mathcal{W}$, prove that $\{\phi_j \psi_k\}_{j,k \in \mathbf{N}}$ is a C^p partition of unity on V subordinate to the covering $\{V_j \cap W_k\}_{j,k \in \mathbf{N}}$.

12.5.6. Show that, given any compact Jordan region $H \subset \mathbf{R}^n$, there is a sequence of C^∞ functions ϕ_j such that

$$\lim_{j \to \infty} \int_{\mathbf{R}^n} \phi_j \, dV = \mathrm{Vol}(H)$$

*12.6 THE GAMMA FUNCTION AND VOLUME

The last result of this section uses Dini's Theorem from Section 9.6.
In this section we introduce the gamma function and use it to find a formula for the volume of any n-dimensional ball and an asymptotic estimate of $n!$.

Recall that if $f : (0, \infty) \to \mathbf{R}$ is locally integrable on $(0, \infty)$, then

$$\int_0^\infty f(t) \, dt = \lim_{\substack{x \to 0+ \\ y \to \infty}} \int_x^y f(t) \, dt.$$

In particular, it is easy to check that $\int_0^\infty e^{-\alpha t} \, dt$ is finite for all $\alpha > 0$.

12.66 Definition.

The *gamma function* is defined by

$$\Gamma(x) = \int_0^\infty t^{x-1} e^{-t} \, dt, \qquad x \in (0, \infty),$$

when this (improper) integral converges.

By definition,

$$\Gamma(1) = \int_0^\infty e^{-t} \, dt = 1,$$

and

$$\Gamma(1/2) = \int_0^\infty t^{-1/2} e^{-t} \, dt = 2 \int_0^\infty e^{-u^2} \, du = \sqrt{\pi}.$$

(We used the change of variables $t = u^2$ and Exercise 12.4.10.) It turns out that $\Gamma(x)$ is defined for all $x \in (0, \infty)$.

12.67 Theorem. *For each $x \in (0, \infty)$, $\Gamma(x)$ exists and is finite, $\Gamma(x+1) = x\Gamma(x)$, and $\Gamma(n) = (n-1)!$ for $n \in \mathbf{N}$.*

Proof. Write

$$\Gamma(x) = \int_0^1 t^{x-1}e^{-t}\, dt + \int_1^\infty t^{x-1}e^{-t}\, dt =: I_1 + I_2.$$

By l'Hôpital's Rule,

$$\lim_{t\to\infty} e^{-t/2}t^y = 0$$

for all $y \in \mathbf{R}$. Hence, $e^{-t}t^{x-1} \le e^{-t/2}$ for t large and it follows, from Theorem 5.43 (the Comparison Theorem), that I_2 is finite for all $x \in \mathbf{R}$.

To show that I_1 is finite for $x > 0$, suppose first that $x \ge 1$. Then $t^{x-1} \le 1$ for all $t \in [0, 1]$ and

$$I_1 = \int_0^1 t^{x-1}e^{-t}\, dt \le \int_0^1 e^{-t}\, dt = 1 - \frac{1}{e} < \infty.$$

Therefore, $\Gamma(x)$ is finite for all $x \ge 1$. Next, suppose that $0 < x < 1$. Then $x + 1 \ge 1$, so $\Gamma(x + 1)$ is finite. Integration by parts yields

$$\Gamma(x) = \int_0^\infty t^{x-1}e^{-t}\, dt = \frac{t^x e^{-t}}{x}\Big|_{t=0}^\infty + \frac{1}{x}\int_0^\infty t^x e^{-t}\, dt = \frac{1}{x}\Gamma(x+1).$$

Therefore, $\Gamma(x)$ is finite when $0 < x < 1$.

This argument also verifies $x\Gamma(x) = \Gamma(x+1)$ for $x \in (0, \infty)$. Since $\Gamma(1) = 1$, it follows that $\Gamma(n) = (n-1)!$ for all $n \in \mathbf{N}$. ∎

The gamma function can be used to evaluate certain integrals which cannot be evaluated by using elementary techniques of integration.

12.68 Theorem. *If $x, y \in (0, \infty)$, then*

i)
$$\int_0^1 v^{y-1}(1-v)^{x-1}\, dv = \frac{\Gamma(x)\Gamma(y)}{\Gamma(x+y)},$$

and

ii)
$$\int_0^{\pi/2} \cos^{2x-1}\varphi \sin^{2y-1}\varphi\, d\varphi = \frac{\Gamma(x)\Gamma(y)}{2\Gamma(x+y)}.$$

In particular,

iii)
$$\int_0^\pi \sin^{k-2}\varphi\, d\varphi = \frac{\Gamma((k-1)/2)\Gamma(1/2)}{\Gamma(k/2)}$$

holds for all integers $k > 2$.

Proof. To prove part i), make the change of variables $v = u/(1+u)$ and write

$$\int_0^1 v^{y-1}(1-v)^{x-1}\,dv = \int_0^\infty \left(\frac{u}{1+u}\right)^{y-1}\left(1 - \frac{u}{1+u}\right)^{x-1}\frac{du}{(1+u)^2}$$
$$= \int_0^\infty u^{y-1}\left(\frac{1}{1+u}\right)^{x+y}\,du.$$

It follows from two more changes of variables $[s = t/(1+u)$ and $w = su]$ and Fubini's Theorem that

$$\Gamma(x+y)\int_0^1 v^{y-1}(1-v)^{x-1}\,dv$$

$$= \int_0^\infty \int_0^\infty u^{y-1}\left(\frac{1}{1+u}\right)^{x+y} t^{x+y-1}e^{-t}\,dt\,du$$

$$= \int_0^\infty \int_0^\infty u^{y-1}s^{x+y-1}e^{-s(u+1)}\,ds\,du$$

$$= \int_0^\infty s^{x-1}e^{-s}\left(\int_0^\infty u^{y-1}s^y e^{-su}\,du\right)ds$$

$$= \int_0^\infty s^{x-1}e^{-s}\left(\int_0^\infty w^{y-1}e^{-w}\,dw\right)ds = \Gamma(x)\Gamma(y).$$

To prove part ii) use the change of variables $v = \sin^2\varphi$ and part i) to verify

$$\int_0^{\pi/2} \cos^{2x-1}\varphi \, \sin^{2y-1}\varphi \, d\varphi = \frac{1}{2}\int_0^1 v^{y-1}(1-v)^{x-1}\,dv = \frac{\Gamma(x)\Gamma(y)}{2\Gamma(x+y)}.$$

Specializing to the case $y = (k-1)/2$ and $x = 1/2$, we obtain part iii). ∎

The connection between the gamma function and volume is contained in the following result.

12.69 Theorem. *If $r > 0$ and $\mathbf{a} \in \mathbf{R}^n$, then*

$$\text{Vol}(B_r(\mathbf{a})) = \frac{2r^n \pi^{n/2}}{n\Gamma(n/2)}.$$

Proof. By translation invariance (see Exercise 12.1.7) and Theorem 12.22, $\text{Vol}(B_r(\mathbf{a})) = \int_B 1\,d\mathbf{x}$ for $B = B_r(\mathbf{0})$. We suppose for simplicity that $n \geq 2$, and we introduce a change of variables in $\mathbf{R}^n$ analogous to spherical coordinates. Namely, let

$$x_1 = \rho \cos\varphi_1, \quad x_2 = \rho \sin\varphi_1 \cos\varphi_2, \quad x_3 = \rho \sin\varphi_1 \sin\varphi_2 \cos\varphi_3, \quad \ldots,$$
$$x_{n-1} = \rho \sin\varphi_1 \ldots \sin\varphi_{n-2}\cos\theta, \quad \text{and} \quad x_n = \rho \sin\varphi_1 \ldots \sin\varphi_{n-2}\sin\theta,$$

where $0 \leq \rho \leq r$, $0 \leq \theta \leq 2\pi$, and $0 \leq \varphi_j \leq \pi$ for $j = 1, \ldots, n-2$. An easy induction argument shows that this change of variables has Jacobian

$$\Delta := \rho^{n-1} \sin^{n-2} \varphi_1 \sin^{n-3} \varphi_2 \ldots \sin^2 \varphi_{n-3} \sin \varphi_{n-2}. \tag{49}$$

Hence, by Theorems 12.65 (or Theorem 12.46 and a limiting argument) and 12.68iii,

$$\mathrm{Vol}(B_r(\mathbf{a})) = \int_B 1 \, d\mathbf{x}$$

$$= \int_0^r \int_0^\pi \cdots \int_0^\pi \int_0^{2\pi} \rho^{n-1} \sin^{n-2} \varphi_1 \ldots \sin \varphi_{n-2} \, d\theta \, d\varphi_1 \ldots d\varphi_{n-2} \, d\rho$$

$$= \frac{2\pi r^n}{n} \left(\int_0^\pi \sin^{n-2} \varphi \, d\varphi \right) \cdots \left(\int_0^\pi \sin \varphi \, d\varphi \right)$$

$$= \frac{2\pi r^n}{n} \cdot \frac{\Gamma((n-1)/2)\Gamma(1/2)}{\Gamma(n/2)} \cdot \frac{\Gamma((n-2)/2)\Gamma(1/2)}{\Gamma((n-1)/2)} \cdots \frac{\Gamma(1)\Gamma(1/2)}{\Gamma(3/2)}.$$

Canceling all superfluous factors and substituting the value $\sqrt{\pi}$ for $\Gamma(1/2)$, we conclude that

$$\mathrm{Vol}(B_r(\mathbf{a})) = \frac{2\pi r^n}{n} \left(\frac{\Gamma^{n-2}(1/2)}{\Gamma(n/2)} \right) = \frac{2r^n \pi^{n/2}}{n\Gamma(n/2)}. \qquad \blacksquare$$

This formula agrees with what we already know. For $n = 1$ we have

$$\mathrm{Vol}(B_r(0)) = \frac{2r\pi^{1/2}}{\Gamma(1/2)} = 2r,$$

for $n = 2$ we have

$$\mathrm{Vol}(B_r(0, 0)) = \frac{2r^2\pi}{2\Gamma(1)} = \pi r^2,$$

and for $n = 3$ we have

$$\mathrm{Vol}(B_r(0, 0, 0)) = \frac{2r^3\pi^{3/2}}{3\Gamma(3/2)} = \frac{2r^3\pi^{3/2}}{(3/2)\Gamma(1/2)} = \frac{4}{3}\pi r^3.$$

We close this section with an asymptotic estimate of $n!$. First, we obtain an integral representation for $n!/(n^{n+1/2}e^{-n})$.

12.70 Lemma.
If $\phi(x) = x - \log x - 1$, $x > 0$, then

$$\frac{n!}{n^{n+1/2}e^{-n}} = \int_{-\sqrt{n}}^{\infty} e^{-n\phi(1+t/\sqrt{n})} \, dt.$$

Proof. By Definition 12.66 and Theorem 12.67, we can write

$$n! = \Gamma(n+1) = \int_0^\infty x^n e^{-x}\, dx.$$

Making two changes of variables (first $x = ny$, then $y = 1 + t/\sqrt{n}$), we conclude that

$$\frac{n!}{n^{n+1/2}e^{-n}} = \frac{1}{\sqrt{n}} \int_0^\infty \left(\frac{x}{n}\right)^n e^{-x+n}\, dx$$

$$= \sqrt{n} \int_0^\infty y^n e^{-n(y-1)}\, dy$$

$$= \sqrt{n} \int_0^\infty e^{-n\phi(y)}\, dy = \int_{-\sqrt{n}}^\infty e^{-n\phi(1+t/\sqrt{n})}\, dt. \quad \blacksquare$$

Next, we derive some inequalities which will be used, in conjunction with Dini's Theorem, to evaluate the limit of the integral which appears in Lemma 12.70.

12.71 Lemma.
If $\phi(x) = x - \log x - 1$, $x > 0$, then

$$(x-1)\phi'(x) - 2\phi(x) > 0, \quad \text{for } 0 < x < 1,$$

and

$$(x-1)\phi'(x) - 2\phi(x) < 0, \quad \text{for } x > 1.$$

Moreover, there is an absolute constant $M > 0$ such that

$$\phi(x) \geq M(x-1)^2, \quad \text{for } 0 < x < 2, \tag{50}$$

and

$$\phi(x) \geq M(x-1), \quad \text{for } x \geq 2. \tag{51}$$

Proof. Let $\psi(x) = 2\log x - x + 1/x$ and observe that $(x-1)\phi'(x) - 2\phi(x) = \psi(x)$. Since $\psi'(x) = -(x-1)^2/x^2 < 0$ for all $x \neq 1$, ψ is decreasing on $(0, \infty)$. Since $\psi(1) = 0$, it follows that $\psi > 0$ on $(0, 1)$ and $\psi < 0$ on $(1, \infty)$. This proves the first pair of inequalities.

To prove the second pair of inequalities, observe first that, by Taylor's Formula,

$$\phi(x) = \phi(1) + \phi'(1)(x-1) + \phi''(c)\frac{(x-1)^2}{2!} = \frac{(x-1)^2}{2c^2}$$

for some c between x and 1. Thus $\phi(x) \geq (x-1)^2/8$ for all $0 < x < 2$. Next, observe, since $\phi(x) > 0$ for $x > 1$ and $\phi(x)/(x-1) \to 1$ as $x \to \infty$, that

$\phi(x)/(x-1)$ has a positive minimum, say m, on $[2, \infty)$. Thus (50) and (51) hold for $M := \min\{m, 1/8\}$. ∎

Our final preliminary result evaluates the limit of the integral which appears in Lemma 12.70.

12.72 Lemma.
If $\phi(x) = x - \log x - 1$ for $x > 0$, and $F_n(t) = e^{-n\phi(1+t/\sqrt{n})}$ for $n \in \mathbf{N}$ and $t > -\sqrt{n}$, then

$$\lim_{n \to \infty} \int_{-\sqrt{n}}^{\infty} F_n(t) \, dt = \int_{-\infty}^{\infty} e^{-t^2/2} \, dt.$$

STRATEGY: The idea behind the proof is simple. By l'Hôpital's Rule,

$$\lim_{n \to \infty} n\phi \left(1 + \frac{t}{\sqrt{n}}\right) = \lim_{n \to \infty} \frac{t}{2} \frac{\phi'(1 + t/\sqrt{n})}{1/\sqrt{n}} = \frac{t^2}{2} \lim_{n \to \infty} \phi'' \left(1 + \frac{t}{\sqrt{n}}\right) = \frac{t^2}{2},$$

so $F_n(t) \to e^{-t^2/2}$, as $n \to \infty$, for every $t \in \mathbf{R}$. Thus $\int_{-\sqrt{n}}^{\infty} F_n(t) \, dt$ should converge to $\int_{-\infty}^{\infty} e^{-t^2/2} \, dt$ as $n \to \infty$. To prove this, we break the integral over $(-\sqrt{n}, \infty)$ into three pieces: one over $(-\sqrt{n}, -\sqrt{a})$, one over $(\sqrt{a}, \infty)$, and one over $(-\sqrt{a}, \sqrt{a})$. Since $e^{-t^2/2}$ is integrable on $(-\infty, \infty)$, the first two integrals should be small for a sufficiently large. Once a is fixed, we shall use Dini's Theorem on the third integral. Here are the details.

Proof. Let $\varepsilon > 0$ and observe that

$$\left| \int_{-\sqrt{n}}^{\infty} F_n(t) \, dt - \int_{-\sqrt{n}}^{\infty} e^{-t^2/2} \, dt \right|$$

$$\leq I_1 + I_2 + I_3 + I_4$$

$$:= \left| \int_{-\sqrt{a}}^{\sqrt{a}} \left(F_n(t) - e^{-t^2/2} \right) dt \right| + \int_{|t| \geq \sqrt{a}} e^{-t^2/2} \, dt$$

$$\int_{\sqrt{a}}^{\infty} |F_n(t)| \, dt + \int_{-\sqrt{n}}^{-\sqrt{a}} |F_n(t)| \, dt$$

for any $a > 0$ and $n \in \mathbf{N}$, provided $n > a$. Hence, it suffices to prove that $|I_j| \leq \varepsilon/4$ for $j = 1, 2, 3, 4$, and n, a sufficiently large.

Let M be the constant given in Lemma 12.71, and choose $a > 0$ so large that

$$\int_{|t| \geq \sqrt{a}} e^{-Mt^2} \, dt < \frac{\varepsilon}{4}, \qquad \int_{\sqrt{a}}^{\infty} e^{-Mt} \, dt < \frac{\varepsilon}{4}, \tag{52}$$

and

$$\int_{|t| \geq \sqrt{a}} e^{-t^2/2} \, dt < \frac{\varepsilon}{4}. \tag{53}$$

By (53), $|I_2| < \varepsilon/4$.

To estimate $|I_j|$ for $j \neq 2$, fix $t > -\sqrt{a}$ and consider the function $G(x) = e^{-x\phi(1+t/\sqrt{x})}$, $x > 0$. By the Product Rule,

$$G'(x) = e^{-x\phi(1+t/\sqrt{x})}\left(\frac{t}{2\sqrt{x}}\phi'\left(1+\frac{t}{\sqrt{x}}\right) - \phi\left(1+\frac{t}{\sqrt{x}}\right)\right)$$

$$= \frac{e^{-x\phi(y)}}{2}((y-1)\phi'(y) - 2\phi(y)),$$

where $y = 1 + t/\sqrt{x}$. Thus by Lemma 12.71, $G'(x) > 0$ for $x > a$, $-\sqrt{a} < t < 0$, and $G'(x) < 0$ for $x > 0$, $t > 0$. It follows that for each $t \in (-\sqrt{a}, 0)$, $F_n(t) \uparrow e^{-t^2/2}$ as $n \to \infty$, and for each $t \in (0, \infty)$, $F_n(t) \downarrow e^{-t^2/2}$ as $n \to \infty$. Hence, by Dini's Theorem (Theorem 9.40),

$$\int_{-\sqrt{a}}^{\sqrt{a}} F_n(t)\,dt \to \int_{-\sqrt{a}}^{\sqrt{a}} e^{-t^2/2}\,dt$$

as $n \to \infty$. Thus, we can choose an $N \in \mathbf{N}$ so large that $n \geq N$ implies $|I_1| < \varepsilon$. It remains to estimate $|I_j|$ for $j = 3, 4$.

To this end, let $n > \max\{N, a\}$. By (50) and (51),

$$n\phi\left(1+\frac{t}{\sqrt{n}}\right) \geq nM\frac{t^2}{n} = Mt^2 \quad \text{for } -\sqrt{n} < t < \sqrt{n},$$

and

$$n\phi\left(1+\frac{t}{\sqrt{n}}\right) \geq nM\frac{t}{\sqrt{n}} \geq Mt \quad \text{for } t \geq \sqrt{n}.$$

Since $n > a$, it follows that

$$|I_3| + |I_4| = \int_{\sqrt{a}}^{\infty} |F_n(t)|\,dt + \int_{-\sqrt{n}}^{-\sqrt{a}} |F_n(t)|\,dt$$

$$\leq \int_{\sqrt{a} \leq |t| \leq \sqrt{n}} e^{-Mt^2}\,dt + \int_{\sqrt{n}}^{\infty} e^{-Mt}\,dt$$

$$< \int_{|t| \geq \sqrt{a}} e^{-Mt^2}\,dt + \int_{\sqrt{a}}^{\infty} e^{-Mt}\,dt.$$

We conclude by (52) that $|I_3| + |I_4| < \varepsilon/2$ as required. ∎

12.73 Theorem. [STIRLING'S FORMULA].
For $n \in \mathbf{N}$ sufficiently large, $n! \approx \sqrt{2\pi}(n^{n+1/2})e^{-n}$; that is,

$$\lim_{n\to\infty} \frac{n!}{\sqrt{2\pi}(n^{n+1/2})e^{-n}} = 1.$$

Proof. By Exercise 12.4.10 and the change of variables $t = \sqrt{2}u$, we have

$$\int_{-\infty}^{\infty} e^{-t^2/2} \, dt = \sqrt{2} \int_{-\infty}^{\infty} e^{-u^2} \, du = 2\sqrt{2} \int_{0}^{\infty} e^{-u^2} \, du = \sqrt{2\pi}.$$

Therefore, it follows from Lemmas 12.70 and 12.72 that

$$\lim_{n\to\infty} \frac{n!}{\sqrt{2\pi}\,n^{n+1/2}e^{-n}} = \lim_{n\to\infty} \frac{1}{\sqrt{2\pi}} \int_{-\sqrt{n}}^{\infty} e^{-n\phi(1+t/\sqrt{n})} \, dt$$

$$= \frac{1}{\sqrt{2\pi}} \int_{-\infty}^{\infty} e^{-t^2/2} \, dt = 1. \qquad \blacksquare$$

EXERCISES

12.6.1. Show that

$$\int_{0}^{\infty} t^2 e^{-t^2} \, dt = \frac{\sqrt{\pi}}{4}.$$

12.6.2. Show that

$$\int_{0}^{1} \frac{dx}{\sqrt{-\log x}} = \sqrt{\pi}.$$

12.6.3. Show that

$$\int_{-\infty}^{\infty} e^{\pi t - e^t} \, dt = \Gamma(\pi).$$

12.6.4. Show that the volume of a four-dimensional ball of radius r is $\pi^2 r^4/2$, and the volume of a five-dimensional ball of radius r is $8\pi^2 r^5/15$.

12.6.5. Verify (49).

12.6.6. Suppose that $n > 2$ and define an n-dimensional ellipsoid by

$$E = \left\{ (x_1, \ldots, x_n) : \frac{x_1^2}{a_1^2} + \frac{x_2^2}{a_2^2} + \cdots + \frac{x_n^2}{a_n^2} \le 1 \right\}.$$

Prove that

$$\mathrm{Vol}(E) = \frac{2a_1 \ldots a_n \pi^{n/2}}{n\Gamma(n/2)}.$$

12.6.7. Suppose that $n > 2$ and define an n-dimensional cone by

$$C = \{ (x_1, \ldots, x_n) : (h/r)\sqrt{x_2^2 + \cdots + x_n^2} \le x_1 \le h \}.$$

Prove that

$$\mathrm{Vol}(C) = \frac{2hr^{n-1}\pi^{(n-1)/2}}{n(n-1)\Gamma((n-1)/2)}.$$

12.6.8. Find the value of

$$\int_{B_r(0)} x_k^2 \, d(x_1, \ldots, x_n)$$

for each $k \in \mathbf{N}$.

12.6.9. If $f : B_1(0) \to \mathbf{R}$ is differentiable with $f(0) = 0$ and $\|\nabla f(\mathbf{x})\| \le 1$ for $\mathbf{x} \in B_1(0)$, prove that the following exists and equals 0.

$$\lim_{k \to \infty} \int_{B_1(0)} |f(\mathbf{x})|^k \, d\mathbf{x}$$

12.6.10. a) Prove that Γ is differentiable on $(0, \infty)$ with

$$\Gamma'(x) = \int_0^\infty e^{-t} t^{x-1} \log t \, dt.$$

*b) Prove that Γ is $\mathcal{C}^\infty$ and convex on $(0, \infty)$.

CHAPTER 13

Fundamental Theorems of Vector Calculus

This chapter is more descriptive and less rigorous than its predecessors. Our goal is to lay a practical foundation for calculus on manifolds.

13.1 CURVES

For a non-mathematician, a curve is a smooth line which bends, without corners; a one-dimensional object with length but no breadth. Of course, this definition is too imprecise. It is also too restrictive. Our concept of a curve will include not only "smooth" objects such as the graphs of the function $y = x^2$ and the relation $x^2 + y^2 = 1$, but also objects with "corners" such as the graph of $y = |x|$.

Recall that if $I \subseteq \mathbf{R}$ and $\boldsymbol{\phi} : I \to \mathbf{R}^m$, then the image of I under $\boldsymbol{\phi}$ is the set

$$\boldsymbol{\phi}(I) = \{\mathbf{x} \in \mathbf{R}^m : \mathbf{x} = \boldsymbol{\phi}(t) \text{ for some } t \in I\}.$$

Also recall that, given $\mathbf{a}, \mathbf{b} \in \mathbf{R}^m$ with $\mathbf{b} \neq 0$, the image of $\mathbf{R}$ under $\boldsymbol{\phi}(t) := \mathbf{a} + t\mathbf{b}$ is the straight line through $\mathbf{a}$ in the direction of $\mathbf{b}$. This is the simplest type of curve in $\mathbf{R}^m$.

A naive attempt to define a general curve in $\mathbf{R}^m$ is to insist that it simply be the image of an interval under some continuous function $\boldsymbol{\phi} : \mathbf{R} \to \mathbf{R}^m$. It turns out that this definition is too broad. There are continuous functions (called "space-filling curves") which take the unit interval $[0, 1]$ onto the unit square $[0, 1] \times [0, 1]$ (see Boas [2]). One way to fix this definition is to use homeomorphisms (i.e., continuous functions whose inverses are also continuous). Since we are interested primarily in the differential structure of curves, we take a different approach, using differentiable functions to define curves (see Definition 13.1 below).

We begin by extending the definition of partial differentiation to include functions defined on nonopen domains. Let $m, n, p \in \mathbf{N}$, and E be a nonempty subset of $\mathbf{R}^n$. A function $\mathbf{f} : E \to \mathbf{R}^m$ is said to be $\mathcal{C}^p$ (on E) if and only if there is an open set $V \supseteq E$ and a function $\mathbf{g} : V \to \mathbf{R}^m$ whose partial derivatives of orders $j \leq p$ exist and are continuous on V such that $\mathbf{f}(\mathbf{x}) = \mathbf{g}(\mathbf{x})$ for all $\mathbf{x} \in E$. In this case we define the *partial derivatives* of $\mathbf{f}$ to be equal to the partial derivatives of $\mathbf{g}$; for example, $\partial f_j / \partial x_k(\mathbf{x}) = \partial g_j / \partial x_k(\mathbf{x})$ for $k = 1, 2, \ldots, n$, $j = 1, 2, \ldots, m$, and $\mathbf{x} \in E$. A function $\mathbf{f} : E \to \mathbf{R}^m$ is said to be $\mathcal{C}^\infty$ (on E) if and only if $\mathbf{f}$ is $\mathcal{C}^p$ on E for all $p \in \mathbf{N}$. Notice that this agrees with Definition 4.6 when $n = 1$. Also notice that the Mean Value Theorem and the Inverse Function Theorem hold for functions in $C^1(E)$.

Henceforth, p will denote an element of **N** or the extended real number ∞.

13.1 Definition.

A subset C of $\mathbf{R}^m$ is called a C^p *curve* (in $\mathbf{R}^m$) if and only if there is a nondegenerate interval I (bounded or unbounded) and a C^p function $\boldsymbol{\phi} : I \to \mathbf{R}^m$ such that $\boldsymbol{\phi}$ is 1–1 on I^o and $C = \boldsymbol{\phi}(I)$. In this case, the pair $(\boldsymbol{\phi}, I)$ is called a *parametrization* of C, and C is called the *trace* of $(\boldsymbol{\phi}, I)$. The equations

$$x_j = \phi_j(t), \qquad t \in J, \quad j = 1, \ldots, m,$$

are called the *parametric equations* of C induced by the parametrization $(\boldsymbol{\phi}, I)$.

Thus the straight line through **a** in the direction of **b** is a C^∞ curve with parametrization $\boldsymbol{\phi}(t) := \mathbf{a} + t\mathbf{b}$, $I = \mathbf{R}$.

For most applications, we must assume more about curves.

13.2 Definition.

A C^p curve is called an *arc* if and only if it has a parametrization $(\boldsymbol{\phi}, I)$ where $I = [a, b]$ for some $a, b \in \mathbf{R}$. In this case, we shall call $\boldsymbol{\phi}(a)$ and $\boldsymbol{\phi}(b)$ the *endpoints* of C. An arc is said to be *closed* if and only if $\boldsymbol{\phi}(a) = \boldsymbol{\phi}(b)$.

Thus $L(\mathbf{a}; \mathbf{b})$, the line segment from **a** to **b**, is an arc. The circle $x^2 + y^2 = a^2$ is an example of a closed arc (see Example 13.4 below).

A closed arc is said to be *simple* if and only if it does not intersect itself except possibly at its endpoints. Simple closed arcs are also called *Jordan curves* because of the Jordan Curve Theorem. This theorem states that every simple closed arc C in $\mathbf{R}^2$ separates $\mathbf{R}^2$ into two pieces, a bounded connected set E and an unbounded connected set Ω, where $\partial E = \partial \Omega = C$. However, as W. F. Osgood[1] discovered, the set E is not necessarily a Jordan region.

Before we start developing a theory of curves, we look at several additional examples to see how broad Definitions 13.1 and 13.2 really are. First, we show that curves, as defined in Definition 13.1, include graphs of C^p real functions.

13.3 EXAMPLE.

Let I be an interval and let $f : I \to \mathbf{R}$ be a C^p function. Prove that the graph of $y = f(x)$ on I is a C^p curve in $\mathbf{R}^2$.

---

[1]"A Jordan Curve of Positive Area," *Transactions of the American Mathematical Society*, vol. 4 (1903), pp. 107–112.

Proof. Let $\phi(t) = (t, f(t))$. Then ϕ is C^p and 1–1 on I, and $\phi(I)$ is the graph of $y = f(x)$ as x varies over I. [We shall call (ϕ, I) the *trivial parametrization* of $y = f(x)$.] ■

By an *explicit curve* we mean a curve of the form $\phi(I)$, where either $\phi(t) = (t, f(t))$ or $\phi(t) = (f(t), t)$ for some C^p function $f : I \to \mathbf{R}$. Notice, then, that an explicit curve is a set of points (x, y) which satisfy $y = f(x)$ [respectively, $x = f(y)$] for some real C^p function f.

We have just proved that every explicit curve is a curve in $\mathbf{R}^2$. The following result shows that the converse of this statement is false.

13.4 EXAMPLE.

Prove that the circle $x^2 + y^2 = a^2$ is a C^∞ Jordan curve in $\mathbf{R}^2$.

Proof. This circle can be described in polar coordinates by $r = a$ (i.e., in rectangular coordinates by $x = a \cos\theta$, $y = a \sin\theta$). Set $\phi(t) = (a \cos t, a \sin t)$ and $I = [0, 2\pi]$. Then ϕ is C^∞, 1–1 on $[0, 2\pi)$, and $\phi(I)$ is the set of points $(x, y) \in \mathbf{R}^2$ such that $x^2 + y^2 = a^2$. [The trace of this parametrization is sketched in Figure 13.1. The arrow shows the direction the point $\phi(t)$ moves as t gets larger. For example, $\phi(0) = (a, 0)$ and $\phi(\pi/2) = (0, a)$.] ■

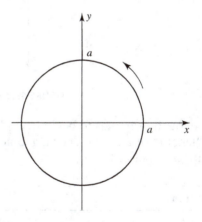

FIGURE 13.1

Recall that the graph of a C^p function on an interval is "smooth" (i.e., has a tangent line at each of its points). The following example shows that this is not the case for a general C^p curve.

13.5 EXAMPLE.

Let $\phi(t) = (\cos^3 t, \sin^3 t)$ and $I = [0, 2\pi]$. Show that (ϕ, I) is a parametrization of a C^∞ Jordan curve in $\mathbf{R}^2$ which has "corners." (This curve is called an *astroid*.)

Proof. Clearly, ϕ is $\mathcal{C}^\infty$ on I and 1–1 on $[0, 2\pi)$. Let $x = \cos^3 t$ and $y = \sin^3 t$ and observe by a double-angle formula that

$$x^2 + y^2 = \frac{3}{4} \cos^2(2t) + \frac{1}{4}.$$

Hence, $\sqrt{x^2 + y^2}$ varies from a maximum of 1 (attained when $t = 0, \pi/2$, π, $3\pi/2$, 2π) to a minimum of 1/2 (attained when $t = \pi/4, 3\pi/4, 5\pi/4, 7\pi/4$). Since I is connected and ϕ is differentiable, hence continuous, the set $\phi(I)$ must also be connected. Plotting a few points, we see that $\phi(I)$ is a four-cornered star, starting at $(1, 0)$ and moving in a counterclockwise direction from $\partial B_1(0, 0)$ to $\partial B_{1/2}(0, 0)$ and back again (see Figure 13.2). As t runs from 0 to 2π, this curve makes one complete circuit. ∎

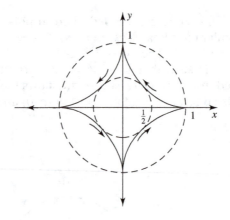

FIGURE 13.2

We have enough examples to begin to explore the theory of curves. First, we define the "length" of a curve. (For a geometric justification of this definition, see Theorem 13.17 below.)

13.6 Definition.

Let C be a $\mathcal{C}^p$ arc and (ϕ, I) be one of its parametrizations. The *arc length* of C, as measured by (ϕ, I), is defined to be

$$L(C) := \int_I \|\phi'(t)\| \, dt.$$

For example, let (ϕ, I) be the parametrization of the circle of radius a given by Example 13.4. Since $\|\phi'(t)\| = a$ for all $t \in [0, 2\pi]$, it is easy to check that $L(C) = 2\pi a$, exactly what it should be. This also demonstrates why we insisted that parametrizations be 1–1 on the interior of their domains. If ϕ were not 1–1 on $(0, 2\pi)$, some part of the circle might be traced more than once, giving an inflated value of its arc length.

Since ϕ is C^p on a closed, bounded interval I, then $\|\phi'\|$ is integrable on I by Theorem 12.21. Hence, $L(C)$ is finite for any parametrization of a C^p arc C. This is not necessarily the case if C is merely the continuous image of an interval (the space-filling curve is continuous but its length is infinite) or if C is the image of an open interval (see Exercise 13.1.4).

When C is an explicit curve, say $y = f(x)$ on $[a, b]$, and (ϕ, I) is the trivial parametrization, Definition 13.6 becomes

$$L(C) = \int_a^b \sqrt{1 + (f'(x))^2}\, dx.$$

This agrees with the formula for arc length introduced in elementary calculus texts.

Before we continue, it is important to realize that every curve has many different parametrizations. For example, the line segment $\{(x, y) \in \mathbf{R}^2 : y = x, 0 < x \le 1\}$ is the trace of $\phi(t) = (t, t)$ on $(0, 1]$, of $\psi(t) = (t/2, t/2)$ on $(0, 2]$, and of $\sigma(t) = (1/t, 1/t)$ on $[1, \infty)$. Although these functions trace the same line segment, each of them traces it differently. The function ψ traces the line "twice as slowly" as ϕ, and σ traces the line "backward" from ϕ. Therefore, a parametrization (ϕ, I) of C is a specific way of tracing out the points in C.

At this point, it is natural to ask, Does the value of arc length, $L(C)$, remain the same if we use different parametrizations of C? To answer this question, we begin by showing that any two parametrizations of the same arc are related by a one-dimensional change of variables τ.

13.7 Remark. *Let I, J be closed bounded intervals and let $\phi : I \to \mathbf{R}^m$ be 1–1 and continuous. Then $\phi(I) = \psi(J)$ for some continuous $\psi : J \to \mathbf{R}^m$ if and only if there is a continuous function τ from J onto I such that $\psi = \phi \circ \tau$.*

Proof. Since I is closed and bounded and ϕ is 1–1 and continuous on I, ϕ^{-1} is continuous from $\phi(I)$ onto I (see Theorem 9.33 or 10.64). Since $\psi(J) = \phi(I)$, it follows that $\tau := \phi^{-1} \circ \psi$ is continuous from J onto I.

Conversely, if τ is any continuous function from J onto I, then $\psi = \phi \circ \tau$ is continuous from J onto $\phi(I)$; that is, $\psi(J) = \phi(I)$. ∎

Thus if (ϕ, I) and (ψ, J) are C^p parametrizations of the same arc and ϕ is 1–1, then there is a continuous function $\tau : J \to I$, called the *transition* from J to I, such that $\psi = \phi \circ \tau$, or, equivalently, $\tau = \phi^{-1} \circ \psi$. It follows that if the transition is differentiable, then by the Chain Rule,

$$\psi'(u) = \phi'(\tau(u))\tau'(u), \qquad u \in J. \tag{1}$$

We are prepared to prove that the definition of arc length does not depend on the parametrization chosen provided the transition has a nonzero derivative.

13.8 Remark. *If (ϕ, I) and (ψ, J) are C^p parametrizations of the same arc, if $\psi = \phi \circ \tau$, where τ takes J onto I and satisfies $\tau'(u) \neq 0$ for all $u \in J$, then*

$$\int_I \|\phi'(t)\| \, dt = \int_J \|\psi'(u)\| \, du.$$

Proof. By hypothesis, $\tau(J) = I$. Hence, it follows from (1) and the Change-of-Variables Formula (Theorem 12.46) that

$$\int_I \|\phi'(t)\| \, dt = \int_{\tau(J)} \|\phi'(t)\| \, dt = \int_J \|\phi'(\tau(u))\| \, |\tau'(u)| \, du = \int_J \|\psi'(u)\| \, du. \quad \blacksquare$$

We note that the condition $\tau' \neq 0$ can be relaxed at finitely many points in J (see Exercise 13.1.8).

One productive way to think about different parametrizations of a curve C is to interpret $\phi(t)$ as the position of a particle moving along C at time t. Different parametrizations of C represent different flight plans, some faster, some slower, some forward, and some backward, but all tracing out the same set of points.

13.9 Remark. *Let (ϕ, I) be a parametrization of a C^p curve, and let $x_0 = \phi(t_0)$ for some $t_0 \in I^\circ$. If $\phi(t)$ represents the position of a moving particle at time t, then $\|\phi'(t_0)\|$ is the speed of that particle at position x_0 and, when $\phi'(t_0) \neq 0$, $\phi'(t_0)$ is a vector which points in the direction of flight at x_0.*

Proof. Let $t_0 \in I^\circ$ and notice that, for each sufficiently small $h > 0$, the quotient

$$\frac{\phi(t_0 + h) - \phi(t_0)}{h}$$

is a vector which points in the direction of flight along the curve C (see Figure 13.3). To calculate the speed of the particle, define the *natural parameter* of the curve $C := \phi(I)$ by

$$s := \ell(t) := \int_a^t \|\phi'(u)\| \, du, \qquad t \in [a, b]. \tag{2}$$

By the Fundamental Theorem of Calculus, $ds/dt = \ell'(t) = \|\phi'(t)\|$. Thus, the change of arc length s with respect to time at t_0 (i.e., the speed of the particle at x_0) is precisely $\|\phi'(t_0)\|$. $\quad \blacksquare$

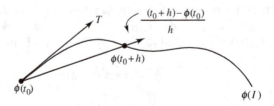

FIGURE 13.3

By elementary calculus, every explicit C^p curve is "smooth" (i.e., has a tangent line at each of its points). The astroid (Example 13.5) shows that a general C^p curve does not have to be smooth at every point.

Is there an easy way to recognize when a general C^p curve has a tangent line (in the sense of Definition 11.21) at a given point on its trace? To answer this question, let $(\phi,\ I)$ be the parametrization of the astroid given in Example 13.5, and notice that $\phi'(t) = 0$ when $t = 0, \pi/2, \pi, 3\pi/2, 2\pi$; that is, exactly at the points where the astroid $\phi(I)$ fails to have a tangent line. (Notice that if we use the flight plan analogy, this condition makes much sense. It is impossible to draw a curve at a corner without pausing to make the direction change, that is, without making the velocity of the drawing device zero.)

Could the answer to our question be this simple? Does a curve with parametrization $(\phi,\ I)$ have a tangent line at each point where $\phi' \neq 0$?

13.10 Remark. *If $(\phi,\ I)$ is a parametrization of a C^p curve C in $\mathbf{R}^2$, and $\phi'(t_0) \neq 0$ for some $t_0 \in I^\circ$, then C has a tangent line at $(x_0, y_0) := \phi(t_0)$.*

STRATEGY: By elementary calculus, the graph of every differentiable function has a tangent line at each of its points. The curve C is given by $x = \phi_1(t)$, $y = \phi_2(t)$. If we could solve the first equation for t, then by the second equation C is an explicit curve: $y = \phi_2 \circ \phi_1^{-1}(x)$. Thus we must decide: Is ϕ_1^{-1} differentiable? This looks like a job for the Implicit Function Theorem.

Proof. Let (ϕ_1, ϕ_2) represent the components of ϕ. Since $\phi'(t_0) \neq 0$, we may suppose that $\phi_1'(t_0) \neq 0$. Set $F(x, t) = \phi_1(t) - x$ and observe by the Implicit Function Theorem that there is an open interval J_0 containing x_0 and a continuously differentiable function $g : J_0 \to I$ such that $\phi_1(g(x)) = x$ for all $x \in J_0$ and $g(x_0) = t_0$. Thus the graph of $y = f(x) := \phi_2 \circ g(x)$, $x \in J_0$, coincides with the trace of ϕ on $g(J_0)$; that is, near $(x_0,\ y_0)$. It follows that C has a tangent line at $(x_0,\ y_0)$. ∎

Accordingly, we make the following definition.

13.11 Definition.

Let $(\phi,\ I)$ be a parametrization of a C^p curve C.

i) $(\phi,\ I)$ is said to be *smooth* at $t_0 \in I$ if and only if $\phi'(t_0) \neq 0$.

ii) $(\phi,\ I)$ is called *smooth* if and only if it is smooth at each point of I, in which case we call ϕ' the *tangent vector* of C induced by $(\phi,\ I)$.

iii) A curve is called *smooth* if and only if it has a smooth parametrization, unless it is a closed arc, in which case we also insist that one of its smooth parametrizations $(\psi,\ [c, d])$ satisfies $\psi'(c) = \psi'(d)$.

By definition, then, if a curve C has a smooth parametrization, then C is smooth, but even smooth arcs can have non-smooth parametrizations.

13.12 Remark. *Every smooth arc has nonsmooth parametrizations.*

Proof. Let $(\phi, [a, b])$ be a smooth parametrization of a smooth arc C. We may suppose (by a preliminary change of variables) that $0 \in (a, b)$. Then $\psi(t) := \phi(t^3)$, $J = (\sqrt[3]{a}, \sqrt[3]{b})$ is a parametrization of C. It is NOT smooth, however, since $\psi'(t) = \phi'(t^3) \cdot 3t^2 = 0$ when $t = 0$. ∎

This raises another question: When does a change in parametrization preserve smoothness? To answer this question, suppose that (ϕ, I) and (ψ, J) are parametrizations of the same curve, with ϕ 1–1 and (ϕ, I) smooth. If the transition τ, from J to I, is differentiable, then, by (1), (ψ, J) is smooth if and only if $\tau'(u) \neq 0$ for all $u \in J$. This leads us to the following definition.

13.13 Definition.

Two C^p parametrizations (ϕ, I), (ψ, J) are said to be *smoothly equivalent* if and only if they are smooth parametrizations of the same curve, and there is a C^p function τ, called the *transition* from J to I, such that $\psi = \phi \circ \tau$, $\tau(J) = I$, and $\tau'(u) \neq 0$ for all $u \in J$.

Thus, by Remark 13.8, the arc length of a curve is the same under smoothly equivalent parametrizations.

Notice that since τ' is continuous and nonzero, either τ' is positive on J or τ' is negative on J. Hence, by Theorem 4.17i, a transition τ between two smoothly equivalent parametrizations is always 1–1.

The following integral can be interpreted as the mass of a wire on a curve with density g (see Appendix E).

13.14 Definition.

Let C be a smooth arc in $\mathbf{R}^m$ with parametrization (ϕ, I), and let $g : C \to \mathbf{R}$ be continuous. Then the *line integral* of g on C is

$$\int_C g \, ds := \int_I g(\phi(t)) \|\phi'(t)\| \, dt. \tag{3}$$

For an explicit curve C given by $y = f(x)$, $x \in [a, b]$, this integral becomes

$$\int_C g \, ds = \int_a^b g(x, f(x)) \sqrt{1 + |f'(x)|^2} \, dx.$$

We note that by Definition 13.6, the line integral (3) equals the arc length of C when $g = 1$. This explains the notation ds. Indeed, the parameter s represents arc length [see (2) above], so, by the Fundamental Theorem of Calculus, $ds/dt = \|\phi'(t)\|$. Hence, the Leibnizian differential of s satisfies $ds = \|\phi'(t)\| \, dt$. We also note that the line integral of a function g on a curve is the same under smoothly equivalent parametrizations (see Exercise 13.1.8).

Since a line integral is a one-dimensional integral, it can often be evaluated by the techniques discussed in Chapter 5.

13.15 EXAMPLE.

Find $\int_C g \, ds$, where $g(x, y) = 2x + y$, $C = \phi(I)$, $\phi(t) = (\cos t, \sin t)$, and $I = [0, \pi/2]$.

Solution. Since $\|\phi'(t)\| = \|(-\sin t, \cos t)\| = 1$, we have

$$\int_C g \, ds = \int_0^{\pi/2} (2\cos t + \sin t) \, dt = 3.$$ ∎

For even the simplest applications, we must have a theory rich enough to handle curves, like the boundary of the unit square $\partial([0, 1] \times [0, 1])$, which are not smooth but a union of smooth pieces. Consequently, we extend the theory developed above to finite unions of smooth curves as follows.

A subset C of $\mathbf{R}^m$ is called a *piecewise smooth curve* (respectively, a *piecewise smooth arc*) if and only if $C = \cup_{j=1}^N C_j$, where each C_j is a smooth curve (respectively, arc) and for each $j \neq k$ either C_j and C_k are disjoint or they intersect at a single point. Thus a piecewise smooth curve might consist of several disjoint smooth pieces, like the boundary of an annulus $0 < a^2 < x^2 + y^2 < b^2$, or several connected pieces with corners, like the boundary of the perforated square $([0, 3] \times [0, 3]) \backslash ([1, 2] \times [1, 2])$.

Let $C = \cup_{j=1}^N C_j$ be a piecewise smooth curve. By a *parametrization* of C we mean a collection of smooth parametrizations (ϕ_j, I_j) of C_j. Two parametrizations $\cup_{j=1}^N(\phi_j, I_j)$ and $\cup_{j=1}^N(\psi_j, J_j)$ of C are said to be *smoothly equivalent* if and only if (ϕ_j, I_j) and (ψ_j, J_j) are smoothly equivalent for each $j \in \{1, \dots, N\}$. Finally, if C is a piecewise smooth arc, then the *arc length* of C is defined by

$$L(C) := \sum_{j=1}^N L(C_j),$$

and the *line integral on C* of a continuous function $g : C \to \mathbf{R}$ is defined by

$$\int_C g \, ds = \sum_{j=1}^N \int_{C_j} g \, ds.$$

13.16 EXAMPLE.

Parametrize the boundary C of the unit square $[0, 1] \times [0, 1]$ and compute $\int_C g \, ds$, where $g(x, y) = x^2 + y^3$.

Solution. C has four smooth pieces which can be parametrized by

$$\boldsymbol{\phi}_1(t) = (t, 0), \quad \boldsymbol{\phi}_2(t) = (1, t), \quad \boldsymbol{\phi}_3(t) = (t, 1), \quad \boldsymbol{\phi}_4(t) = (0, t),$$

for $t \in [0, 1]$. Since $\|\boldsymbol{\phi}'_j(t)\| = 1$, we have, by definition,

$$\int_C g \, ds = \int_0^1 t^2 \, dt + \int_0^1 (1 + t^3) \, dt + \int_0^1 (t^2 + 1) \, dt + \int_0^1 t^3 \, dt = \frac{19}{6}. \quad \blacksquare$$

We close this section with a geometric justification of Definition 13.6 which will not be used elsewhere.

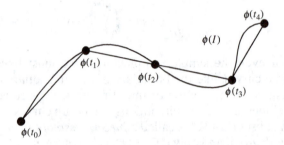

FIGURE 13.4

The arc length of some non-C^p curves can be defined by using line segments for approximation (see Figure 13.4). Namely, we say that a curve C with parametrization $(\boldsymbol{\phi}, I)$ is *rectifiable* if and only if

$$\|C\| := \sup \left\{ \sum_{j=1}^k \|\boldsymbol{\phi}(t_j) - \boldsymbol{\phi}(t_{j-1})\| : \{t_0, t_1, \ldots, t_k\} \text{ is a partition of } I \right\}$$

is finite, in which case we call $\|C\|$ the *arc length* of C.

The following result shows that every C^p arc is rectifiable, and the two definitions we have given for arc length agree.

*13.17 Theorem. *If C is a C^p arc, then $\|C\|$ is finite, and $L(C) = \|C\|$.*

STRATEGY: The idea behind the proof is simple. By the Mean Value Theorem, each term $\|\boldsymbol{\phi}(t_j) - \boldsymbol{\phi}(t_{j-1})\|$ which appears in the definition of $\|C\|$ is approximately $\|\boldsymbol{\phi}'(t_j)\|(t_j - t_{j-1})$, a term of a Riemann sum of the integral of $\|\boldsymbol{\phi}'(t)\|$. Thus, we begin by controlling the size of $\|\boldsymbol{\phi}'(t_j)\|$.

Proof. Let $\varepsilon > 0$, write $\boldsymbol{\phi} = (\phi_1, \phi_2, \ldots, \phi_m)$, and set

$$F(x_1, \ldots, x_m) = \left(\sum_{\ell=1}^m |\phi'_\ell(x_\ell)|^2 \right)^{1/2}$$

for $(x_1, \ldots, x_m)$ in the cube $I^m := I \times \cdots \times I$. By hypothesis, F is continuous on I^m, and I^m is evidently closed and bounded. Thus, F is uniformly continuous on I^m; that is, there is a $\delta > 0$ such that

$$\mathbf{x}, \mathbf{y} \in I^m \quad \text{and} \quad \|\mathbf{x} - \mathbf{y}\| < \delta \quad \text{imply} \quad |F(\mathbf{x}) - F(\mathbf{y})| < \frac{\varepsilon}{2|I|}.$$

Let $\mathcal{P} = \{u_0, \ldots, u_N\}$ be any partition of I. By Theorem 5.18, choose a partition $\mathcal{P}_0 = \{t_0, t_1, \ldots, t_k\}$ of I finer than $\mathcal{P}$ such that $\|\mathcal{P}_0\| < \delta/\sqrt{m}$ and

$$\int_I \|\boldsymbol{\phi}'(t)\| \, dt - \frac{\varepsilon}{2} < \sum_{j=1}^{k} \|\boldsymbol{\phi}'(t_j)\|(t_j - t_{j-1}) < \int_I \|\boldsymbol{\phi}'(t)\| \, dt + \frac{\varepsilon}{2}.$$

Fix $\ell \in \{1, \ldots, m\}$ and $j \in \{1, \ldots, k\}$. By Theorem 4.15ii (the one-dimensional Mean Value Theorem), choose a point $c_j(\ell) \in [t_{j-1}, t_j]$ such that

$$\phi_\ell(t_j) - \phi_\ell(t_{j-1}) = \phi'_\ell(c_j(\ell))(t_j - t_{j-1}).$$

Since $\|\mathcal{P}_0\| < \delta/\sqrt{m}$, we have $|F(t_j, \ldots, t_j) - F(c_j(1), \ldots, c_j(m))| < \varepsilon/(2|I|)$. Since $\boldsymbol{\phi}'(t) = (\phi'_1(t), \ldots, \phi'_m(t))$, we also have $F(t_j, \ldots, t_j) = \|\boldsymbol{\phi}'(t_j)\|$ and

$$F(c_j(1), \ldots, c_j(m))(t_j - t_{j-1}) = \left(\sum_{\ell=1}^{m} |\phi'_\ell(c_j(\ell))|^2 \right)^{1/2} (t_j - t_{j-1})$$

$$= \|\boldsymbol{\phi}(t_j) - \boldsymbol{\phi}(t_{j-1})\|.$$

It follows that

$$\sum_{j=1}^{k} \|\boldsymbol{\phi}'(t_j)\|(t_j - t_{j-1}) - \frac{\varepsilon}{2} < \sum_{j=1}^{k} \|\boldsymbol{\phi}(t_j) - \boldsymbol{\phi}(t_{j-1})\| < \sum_{j=1}^{k} \|\boldsymbol{\phi}'(t_j)\|(t_j - t_{j-1}) + \frac{\varepsilon}{2}.$$

Combining this double inequality with the preceding one, we obtain

$$\int_I \|\boldsymbol{\phi}'(t)\| \, dt - \varepsilon < \sum_{j=1}^{k} \|\boldsymbol{\phi}(t_j) - \boldsymbol{\phi}(t_{j-1})\| < \int_I \|\boldsymbol{\phi}'(t)\| \, dt + \varepsilon.$$

Using the left-hand inequality and the definition of $\|C\|$, we have

$$L(C) - \varepsilon = \int_I \|\boldsymbol{\phi}'(t)\| \, dt - \varepsilon < \sum_{j=1}^{k} \|\boldsymbol{\phi}(t_j) - \boldsymbol{\phi}(t_{j-1})\| \leq \|C\|.$$

It follows from Definition 13.6 that $L(C) \leq \|C\|$. On the other hand, since $\mathcal{P}_0 = \{t_0, t_1, \ldots, t_k\}$ is finer than $\mathcal{P}$, it follows from the right-hand inequality that

$$\sum_{i=1}^{N} \|\boldsymbol{\phi}(u_i) - \boldsymbol{\phi}(u_{i-1})\| \leq \sum_{j=1}^{k} \|\boldsymbol{\phi}(t_j) - \boldsymbol{\phi}(t_{j-1})\| < \int_I \|\boldsymbol{\phi}'(t)\| \, dt + \varepsilon.$$

Taking the supremum over all partitions $\{u_0, \ldots, u_N\}$ of I, we have

$$\|C\| \leq \int_I \|\boldsymbol{\phi}'(t)\| \, dt + \varepsilon;$$

that is, $\|C\| \leq L(C)$. ∎

EXERCISES

13.1.1. Let $\boldsymbol{\psi}(t) = (a \sin t, a \cos t)$, $\boldsymbol{\sigma}(t) = (a \cos 2t, a \sin 2t)$, $I = [0, 2\pi)$, and $J = [0, \pi)$. Sketch the traces of $(\boldsymbol{\psi}, I)$ and $(\boldsymbol{\sigma}, J)$. Note the "direction of flight" and the "speed" of each parametrization. Compare these parametrizations with the one given in Example 13.4.

13.1.2. Let $\mathbf{a}, \mathbf{b} \in \mathbf{R}^m$, $\mathbf{b} \neq 0$, and set $\boldsymbol{\phi}(t) = \mathbf{a} + t\mathbf{b}$. Show that $C = \boldsymbol{\phi}(\mathbf{R})$ is a smooth unbounded curve which contains $\mathbf{a}$ and $\mathbf{a} + \mathbf{b}$. Prove that the angle between $\boldsymbol{\phi}(t_1) - \boldsymbol{\phi}(0)$ and $\boldsymbol{\phi}(t_2) - \boldsymbol{\phi}(0)$ for any $t_1, t_2 \neq 0$ is 0 or π.

13.1.3. Let I be an interval and $f : I \to \mathbf{R}$ be continuously differentiable with

$$|f(\theta)|^2 + |f'(\theta)|^2 \neq 0$$

for all $\theta \in I$. Prove that the graph of $r = f(\theta)$ (in polar coordinates) is a smooth C^1 curve in $\mathbf{R}^2$.

*13.1.4. Show that the curve $y = \sin(1/x)$, $0 < x \leq 1$, is not rectifiable. Thus show that Theorem 13.17 can be false if C is not an arc.

13.1.5. Sketch the trace and compute the arc length of each of the following.

a) $\boldsymbol{\phi}(t) = (e^t \sin t, e^t \cos t, e^t)$, $t \in [0, 2\pi]$
b) $y^3 = x^2$ from $(-1, 1)$ to $(1,1)$
c) $\boldsymbol{\phi}(t) = (t^2, t^2, t^2)$, $t \in [0, 2]$
d) The astroid of Example 13.5

13.1.6. For each of the following, find a (piecewise) smooth parametrization of C and compute $\int_C g \, ds$.

a) C is the curve $y = \sqrt{9 - x^2}$, $x \geq 0$, and $g(x, y) = xy^2$.
b) C is the portion of the ellipse $x^2/a^2 + y^2/b^2 = 1$, $a, b > 0$, which lies in the first quadrant and $g(x, y) = xy$.
c) C is the intersection of the surfaces $x^2 + z^2 = 4$ and $y = x^2$, and $g(x, y, z) = \sqrt{1 + yz^2}$.
d) C is the triangle with vertices $(0, 0, 0)$, $(1, 0, 0)$, and $(0, 2, 0)$, and $g(x, y, z) = x + y + z^3$.

13.1.7. Let C be a smooth arc and $g_k : C \to \mathbf{R}$ be continuous for $n \in \mathbf{N}$.

a) If $g_k \to g$ uniformly on C, prove that $\int_C g_k \, ds \to \int_C g \, ds$ as $k \to \infty$.

*b) Suppose that $\{g_k\}$ is pointwise monotone and that $g_k \to g$ pointwise on C as $k \to \infty$. If g is continuous on $\phi(I)$, prove that $\int_C g_k \, ds \to \int_C g \, ds$ as $k \to \infty$.

13.1.8. Suppose that (ϕ, I) is a parametrization of a smooth arc in $\mathbf{R}^m$, and that $\tau : J \to \mathbf{R}$ is a C^1 function, 1–1 from J onto I. If $\tau'(u) \neq 0$ for all but finitely many $u \in J$, $\psi = \phi \circ \tau$, and $g : \phi(I) \to \mathbf{R}$ is continuous, prove that

$$\int_I g(\phi(t))\|\phi'(t)\| \, dt = \int_J g(\psi(u))\|\psi'(u)\| \, du.$$

13.1.9. [THE FOLIUM OF DESCARTES]. Let C be the piecewise smooth curve $\phi(I_1 \cup I_2)$, where $I_1 = (-\infty, -1)$, $I_2 = (-1, \infty)$, and

$$\phi(t) = \left(\frac{3t}{1+t^3}, \frac{3t^2}{1+t^3} \right).$$

Show that if $(x, y) = \phi(t)$, then $x^3 + y^3 = 3xy$. Sketch C.

13.1.10. The *absolute curvature* of a smooth curve with parametrization (ψ, I) at a point $\mathbf{x}_0 = \psi(t_0)$ is the number

$$\kappa(\mathbf{x}_0) = \lim_{t \to t_0} \frac{\theta(t)}{\ell(t)},$$

when this limit exists, where $\theta(t)$ is the angle between $\psi'(t)$ and $\psi'(t_0)$, and $\ell(t)$ is the arc length of $\psi(I)$ from $\psi(t)$ to $\psi(t_0)$. [Thus κ measures how rapidly $\theta(t)$ changes with respect to arc length.]

a) Given $\mathbf{a}, \mathbf{b} \in \mathbf{R}^n$, $\mathbf{b} \neq 0$, prove that the absolute curvature of the line Λ from $\mathbf{a}$ to $\mathbf{b}$ is zero at each point $\mathbf{x}_0$ on Λ.

b) Prove that the absolute curvature of the circle of radius r is $1/r$ at each point $\mathbf{x}_0$ on C.

13.1.11. Let C be a smooth C^2 arc with parametrization $(\phi, [a, b])$, and suppose that $s = \ell(t)$ is given by (2). The *natural parametrization* of C is the pair $(v, [0, L])$, where

$$v(s) = (\phi \circ \ell^{-1})(s) \quad \text{and} \quad L = L(C).$$

a) Prove that $\|v'(s)\| = 1$ for all $s \in [0, L]$ and the arc length of a subcurve $(v, [c, d])$ of C is $d - c$. (This is why $(v, [0, L])$ is called the natural parametrization.)

b) Show that $v'(s)$ and $v''(s)$ are orthogonal for each $s \in [0, L]$.

c) Prove that the absolute curvature (see Exercise 13.1.10) of $(v, [0, L])$ at $\mathbf{x}_0 = v(s_0)$ is $\kappa(\mathbf{x}_0) = \|v''(s_0)\|$.

d) Show that if $\mathbf{x}_0 = \boldsymbol{\phi}(t_0) = \mathbf{v}(s_0)$ and $m = 3$, then

$$\kappa(\mathbf{x}_0) = \|\mathbf{v}'(s_0) \times \mathbf{v}''(s_0)\| = \frac{\|\boldsymbol{\phi}'(t_0) \times \boldsymbol{\phi}''(t_0)\|}{\|\boldsymbol{\phi}'(t_0)\|^3}.$$

e) Prove that the absolute curvature of an explicit $\mathcal{C}^p$ curve $y = f(x)$ at (x_0, y_0) under the trivial parametrization is

$$\kappa = \frac{|y''(x_0)|}{(1 + (y'(x_0))^2)^{3/2}}.$$

13.2 ORIENTED CURVES

Every parametrization $(\boldsymbol{\phi}, I)$ of a smooth curve C determines a "direction of flight" along C; that is, determines the direction $\boldsymbol{\phi}(t)$ moves as t increases on I, equivalently, the direction in which the tangent vector $\boldsymbol{\phi}'(t)$ points. This direction is called the *orientation* of C *induced by* $(\boldsymbol{\phi}, I)$. (The arrows in Figures 13.1 and 13.2 above represent the orientation of the given parametrization.)

If C is smooth and $(\boldsymbol{\phi}, I)$ is one of its smooth parametrizations, then the *unit tangent vector* of C at $\mathbf{x}_0 = \boldsymbol{\phi}(t_0)$ is defined by

$$\mathbf{T}(\mathbf{x}_0) := \frac{\boldsymbol{\phi}'(t_0)}{\|\boldsymbol{\phi}'(t_0)\|}.$$

Suppose that $(\boldsymbol{\phi}, I)$ and $(\boldsymbol{\psi}, J)$ are smoothly equivalent parametrizations of the same curve with transition τ. Since τ' is continuous and nonzero, either $\tau'(u) > 0$ for all $u \in J$ or $\tau'(u) < 0$ for all $u \in J$. In the first case, the vectors $\boldsymbol{\phi}'(\tau(u))$ and $\boldsymbol{\psi}'(u)$ point in the same direction [see (1) in Section 13.1]; hence, these parametrizations determine the same orientation and same unit tangent. In the second case, the vectors $\boldsymbol{\phi}'(\tau(u))$ and $\boldsymbol{\psi}'(u)$ point in opposite directions and, hence, determine different orientations and opposite unit tangents. Accordingly, we make the following definition.

13.18 Definition.

Two parametrizations $(\boldsymbol{\phi}, I)$ and $(\boldsymbol{\psi}, J)$ are said to be *orientation equivalent* if and only if they are smoothly equivalent and the transition τ from J to I satisfies $\tau'(u) > 0$ for all $u \in J$.

In practice, a curve and its orientation are often described geometrically. The reader must provide a parametrization which traces the curve in the prescribed orientation. Here are two typical examples.

13.19 *EXAMPLE.*

Find a smooth parametrization of the curve C in $\mathbf{R}^3$, oriented in the clockwise direction when viewed from high up on the positive z-axis, formed by intersecting the surfaces $x^2 + 5y^2 = 5$ and $z = x^2$.

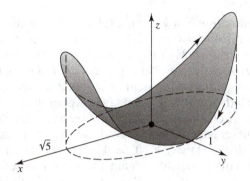

FIGURE 13.5

Solution. The elliptical cylinder $x^2 + 5y^2 = 5$ intersects the parabolic cylinder $z = x^2$ to form a "sagging ellipse" (the shaded region in Figure 13.5 represents that part of $z = x^2$ which lies inside the cylinder $x^2 + 5y^2 = 5$). Using $x = \sqrt{5}\sin t$, $y = \cos t$ to incorporate clockwise motion around the ellipse $x^2 + 5y^2 = 5$, we see that $z = x^2 = 5\sin^2 t$. Thus a smooth parametrization of C with clockwise orientation is $\phi(t) = (\sqrt{5}\sin t, \cos t, 5\sin^2 t)$ on $I = [0, 2\pi]$. ∎

13.20 EXAMPLE.

Find a smooth parametrization of the curve C in $\mathbf{R}^3$, oriented from right to left when viewed from far out the line $y = x$ in the xy-plane, formed by intersecting the surfaces $z = x^2 - y^2$ and $x + y = 1$.

Solution. The saddle surface $z = x^2 - y^2$ intersects the plane $x + y = 1$ to form a curve which cuts across the surface. Using $x = t$ as a parameter to incorporate right to left orientation, we see that $y = 1 - t$ and $z = t^2 - (1-t)^2 = 2t - 1$. Thus a smooth parametrization of C is $\phi(t) = (t, 1 - t, 2t - 1)$ on $I = \mathbf{R}$. In particular, C is a line in the direction $(1, -1, 2)$ passing through the point $(0, 1, -1)$. ∎

The following integral arises naturally in the study of fluids, electricity, and magnetism (e.g., see the discussion that follows this definition).

13.21 Definition.

Let C be a smooth arc in $\mathbf{R}^m$ with unit tangent $\mathbf{T}$, and let (ϕ, I) be a smooth parametrization of C. If $\mathbf{F} : C \to \mathbf{R}^m$ is continuous, then the *oriented line integral* of $\mathbf{F}$ *along C* is

$$\int_C \mathbf{F} \cdot \mathbf{T}\, ds := \int_C \mathbf{F} \cdot d\phi := \int_I \mathbf{F}(\phi(t)) \cdot \phi'(t)\, dt. \qquad (4)$$

The notation $\mathbf{F} \cdot d\boldsymbol{\phi}$ is self-explanatory. The notation $\mathbf{F} \cdot \mathbf{T} \, ds$ is consistent with equation (3) in Section 13.1. Indeed, $\mathbf{T} = \boldsymbol{\phi}'(t)/\|\boldsymbol{\phi}'(t)\|$ and $ds = \|\boldsymbol{\phi}'(t)\| \, dt$, so in the expression $\mathbf{F} \cdot \mathbf{T} \, ds$, the scalars $\|\boldsymbol{\phi}'(t)\|$ cancel each other out.

What does this number represent? If $\mathbf{F}$ represents the flow of a fluid, then $\mathbf{F} \cdot \mathbf{T}$ is the tangential component of $\mathbf{F}$; that is, a measure of fluid flow in the direction to which the tangent $\mathbf{T}$ points (see Appendix E). For example, suppose that C is the unit circle oriented in the counterclockwise direction and $\mathbf{F}(x, y) = (-y, x)$. The unit tangent to C at a point (x, y) is $(-y, x)$, so $\mathbf{F}$ points in the same direction that $\mathbf{T}$ does. Hence, $\mathbf{F} \cdot \mathbf{T} = 1$ is an indication that the fluid is flowing "with the tangent" rather than against it. On the other hand, if $\mathbf{G}(x, y) = (y, -x)$ and $\mathbf{H}(x, y) = (x, y)$, then $\mathbf{G} \cdot \mathbf{T} = -1$ because the fluid is flowing against the tangent, and $\mathbf{H} \cdot \mathbf{T} = 0$ because the fluid is flowing orthogonally to $\mathbf{T}$ (e.g., neither with nor against it). Therefore, the integral of $\mathbf{F} \cdot \mathbf{T} \, ds$ over C is a measure of the circulation of $\mathbf{F}$ around C in the direction of the tangent vector. If this integral is positive, it means that the net flow of the fluid is with $\mathbf{T}$ rather than against $\mathbf{T}$.

Since an oriented line integral is a one-dimensional integral, it can often be evaluated by techniques introduced in Chapter 5. Here is a typical example.

13.22 EXAMPLE.

Describe the trace of $\boldsymbol{\phi}(t) = (\cos t, \sin t, t)$, $t \in I = [0, 4\pi]$, and compute

$$\int_C \mathbf{F} \cdot \mathbf{T} \, ds,$$

where $\mathbf{F}(x, y, z) = (1, \cos z, xy)$ and $C = \boldsymbol{\phi}(I)$.

Solution. Let $(x, y, z) = \boldsymbol{\phi}(t)$. Since $x^2 + y^2 = 1$, the trace of $\boldsymbol{\phi}$ lies on the cylinder $x^2 + y^2 = 1$, $0 \le z \le 4\pi$. As t increases, the point (x, y) winds around the unit circle $x^2 + y^2 = 1$ in a counterclockwise direction. Thus the trace of $\boldsymbol{\phi}$ is a spiral (called the *circular helix*) which winds around the cylinder $x^2 + y^2 = 1$ (see Figure 13.6). As t runs from 0 to 4π, this spiral winds around the cylinder twice, and z runs from 0 to 4π. Since $\boldsymbol{\phi}'(t) = (-\sin t, \cos t, 1)$, we have

$$\int_C \mathbf{F} \cdot \mathbf{T} \, ds = \int_0^{4\pi} (1, \cos t, \cos t \sin t) \cdot (-\sin t, \cos t, 1) \, dt$$

$$= \int_0^{4\pi} (-\sin t + \cos^2 t + \sin t \cos t) \, dt = 2\pi. \qquad \blacksquare$$

The following result shows that, unlike the line integral $\int_C g \, ds$, the oriented line integral $\int_C \mathbf{F} \cdot \mathbf{T} \, ds$ can give different values for different smoothly equivalent parametrizations of the same curve.

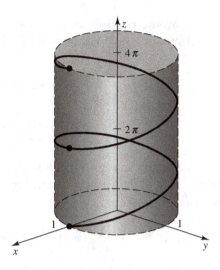

FIGURE 13.6

13.23 Remark. *If* $(\boldsymbol{\phi}, I)$ *and* $(\boldsymbol{\psi}, J)$ *are smoothly equivalent but not orientation equivalent, then*

$$\int_I \mathbf{F}(\boldsymbol{\phi}(t)) \cdot \boldsymbol{\phi}'(t) \, dt = -\int_J \mathbf{F}(\boldsymbol{\psi}(u)) \cdot \boldsymbol{\psi}'(u) \, du.$$

Proof. Let τ be the transition from J to I. Since τ' is continuous and nonzero, it is either positive on J or negative on J. Since $(\boldsymbol{\phi}, I)$ and $(\boldsymbol{\psi}, J)$ are not orientation equivalent, it follows that τ' is negative on J; that is, $|\tau'(u)| = -\tau'(u)$ for $u \in J$. Combining this observation with the Change-of-Variables Formula (Theorem 12.46) and (1) in Section 13.1, we conclude that

$$\int_I \mathbf{F}(\boldsymbol{\phi}(t)) \cdot \boldsymbol{\phi}'(t) \, dt = \int_J \mathbf{F}(\boldsymbol{\phi}(\tau(u)) \cdot \boldsymbol{\phi}'(\tau(u)) \, |\tau'(u)| \, du$$

$$= -\int_J \mathbf{F}(\boldsymbol{\psi}(u)) \cdot \boldsymbol{\psi}'(u) \, du. \qquad \blacksquare$$

By the same method, we can show that the oriented integral (4) gives identical values for orientation equivalent parametrizations of the same curve (see Exercise 13.2.5). Therefore, to evaluate an oriented integral over a curve C whose orientation has been described geometrically, we can use any smooth parametrization of C and adjust the sign of the integral to reflect the prescribed orientation. Here is a typical example.

13.24 EXAMPLE.

Find

$$\int_C \mathbf{F} \cdot \mathbf{T} \, ds,$$

where $\mathbf{F}(x, y) = (y, xy)$ and C is the unit circle $x^2 + y^2 = 1$ oriented in the clockwise direction.

Solution. The parametrization $\boldsymbol{\phi}(t) = (\cos t, \sin t)$, $t \in [0, 2\pi]$, of C has counterclockwise orientation (see Example 13.4). Thus, by Remark 13.23,

$$\int_C \mathbf{F} \cdot \mathbf{T} \, ds = -\int_0^{2\pi} (\sin t, \sin t \cos t) \cdot (-\sin t, \cos t) \, dt$$

$$= \int_0^{2\pi} (\sin^2 t - \sin t \cos^2 t) \, dt = \pi. \qquad \blacksquare$$

There is another way to represent the oriented integral (4) which uses differential notation. Recall that if $x_j = \phi_j(t)$, then $dx_j = \phi_j'(t) \, dt$. Hence, formally, $\mathbf{F}(\boldsymbol{\phi}(t)) \cdot \boldsymbol{\phi}'(t) \, dt$ looks like

$$(F_1(\boldsymbol{\phi}(t))\phi_1'(t) + \cdots + F_m(\boldsymbol{\phi}(t))\phi_m'(t)) \, dt = F_1 \, dx_1 + \cdots + F_m \, dx_m.$$

This last expression is called a *differential form of degree* 1 on $\mathbf{R}^m$ (more briefly, a *1-form*) and the functions F_j are called its *coefficients*. A 1-form is said to be *continuous* on a set E if and only if each of its coefficients is continuous on E. The *oriented integral* of a continuous 1-form on a smooth arc C in $\mathbf{R}^m$ is defined by

$$\int_C F_1 \, dx_1 + \cdots + F_m \, dx_m := \int_C \mathbf{F} \cdot \mathbf{T} \, ds,$$

where $\mathbf{F} = (F_1, \dots, F_m)$.

The following example illustrates the fact that differential forms provide a shorthand for the way an oriented line integral is computed (so we can avoid parametrizing explicit curves).

13.25 EXAMPLE.

Find

$$\int_C y \, dx + \cos x \, dy,$$

where C is the explicit curve $y = x^2 + \sin x$ oriented from $(0, 0)$ to (π, π^2).

Solution. Since $y = x^2 + \sin x$ and $dy = (2x + \cos x) \, dx$, we have

$$\int_C y \, dx + \cos x \, dy = \int_0^\pi (x^2 + \sin x) \, dx + \int_0^\pi \cos x \, (2x + \cos x) \, dx$$

$$= \frac{\pi^3}{3} + \frac{\pi}{2} - 2. \qquad \blacksquare$$

Let $C = \bigcup_{j=1}^N C_j$ be a piecewise smooth arc in $\mathbf{R}^m$ (see the discussion preceding Example 13.16) and $\mathbf{T}_j$ be a unit tangent vector for C_j. If $\mathbf{F} : C \to \mathbf{R}^m$ is continuous, then the *oriented line integral* of **F** *along* C induced by the tangents $\mathbf{T}_j$ is defined to be

$$\int_C \mathbf{F} \cdot \mathbf{T} \, ds = \sum_{j=1}^{N} \int_{C_j} \mathbf{F} \cdot \mathbf{T}_j \, ds.$$

If ω is a 1-form continuous on C, then the *oriented integral* of ω along C is defined to be

$$\int_C \omega = \sum_{j=1}^{N} \int_{C_j} \omega.$$

13.26 EXAMPLE.

Find

$$\int_C xy \, dx + (x^2 + y^2) \, dy,$$

where C is the boundary of $Q = [0, 1] \times [0, 1]$ oriented in the counterclockwise direction.

Solution. The boundary $C = \partial Q$ consists of four smooth pieces (see Figure 13.7): C_1 (which lies in the line $x = 0$), C_2 (in $y = 0$), C_3 (in $x = 1$), and C_4 (in $y = 1$). For C_1, let $x = 0$ and y run from 1 to 0 (to maintain counterclockwise orientation on C). Then

$$\int_{C_1} xy \, dx + (x^2 + y^2) \, dy = \int_1^0 y^2 \, dy = -\frac{1}{3}.$$

Similarly, the integrals over C_2, C_3, and C_4 are 0, 4/3, and $-1/2$. Hence,

$$\int_C \mathbf{F} \cdot \mathbf{T} \, ds = -\frac{1}{3} + 0 + \frac{4}{3} - \frac{1}{2} = \frac{1}{2}. \qquad \blacksquare$$

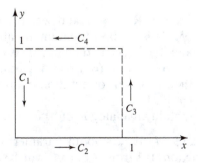

FIGURE 13.7

EXERCISES

13.2.1. For each of the following, sketch the trace of (ϕ, R), describe its orientation, and verify that it is a subset of the surface S.

a) $\phi(t) = (3t, 3\sin t, \cos t)$, $S = \{(x, y, z) : y^2 + 9z^2 = 9\}$
b) $\phi(t) = (t^2, t^3, t^2)$, $S = \{(x, y, z) : z = x\}$
c) $\phi(t) = (t, t^2, \sin t)$, $S = \{(x, y, z) : y = x^2\}$
d) $\phi(t) = (\cos t, \sin t, \cos t)$, $S = \{(x, y, z) : y^2 + z^2 = 1\}$
e) $\phi(t) = (\sin t, \sin t, t)$, $S = \{(x, y, z) : y = x\}$

13.2.2. For each of the following, find a (piecewise) smooth parametrization of C and compute $\int_C \mathbf{F} \cdot \mathbf{T} \, ds$.

a) C is the curve $y = x^2$ from $(1, 1)$ to $(3, 9)$, and $\mathbf{F}(x, y) = (xy, y - x)$.
b) C is the intersection of the elliptical cylinder $y^2 + 2z^2 = 1$ with the plane $x = -1$, oriented in the counterclockwise direction when viewed from far out the positive x-axis, and $\mathbf{F}(x, y, z) = (\sqrt{x^3 + y^3 + 5}, z, x^2)$.
c) C is the intersection of the bent plane $y = |x|$ with the elliptical cylinder $x^2 + 3z^2 = 1$, oriented in the clockwise direction when viewed from far out the positive y-axis, and $\mathbf{F}(x, y, z) = (z, -z, x + y)$.

13.2.3. For each of the following, compute $\int_C \omega$.

a) C is the polygonal path consisting of the line segment from $(1,1)$ to $(2,1)$ followed by the line segment from $(2,1)$ to $(2,3)$, and $\omega = y \, dx + x \, dy$.
b) C is the intersection of $z = x^2 + y^2$ and $x^2 + y^2 + z^2 = 1$, oriented in the counterclockwise direction when viewed from high up the positive z-axis, and $\omega = dx + (x + y) \, dy + (x^2 + xy + y^2) \, dz$.
c) C is the boundary of the rectangle $R = [a, b] \times [c, d]$, oriented in the counterclockwise direction, and $\omega = xy \, dx + (x + y) \, dy$.
d) C is the intersection of $y = x$ and $y = z^2$, $0 \le z \le 1$, oriented from left to right when viewed from far out the y-axis, and $\omega = \sqrt{x} \, dx + \cos y \, dy - dz$.

13.2.4. a) Let $c \in \mathbf{R}$, $\delta > 0$, and set $\tau(u) = \delta u + c$ for $u \in \mathbf{R}$. Prove that if (ϕ, I) is a smooth parametrization of some curve, if $J = \tau^{-1}(I)$, and if $\psi = \phi \circ \tau$, then (ψ, J) is orientation equivalent to (ϕ, I).
b) Prove that if (ϕ, I) is a parametrization of some smooth arc, then it has an orientation equivalent parametrization of the form $(\psi, [0, 1])$.
c) Obtain an analogue of b) for piecewise smooth curves.

13.2.5. Let (ϕ, I) be a smooth parametrization of some arc and τ be a C^1 function, 1–1 from J onto I, which satisfies $\tau'(u) > 0$ for all but finitely many $u \in J$. If $\psi = \phi \circ \tau$, prove that

$$\int_I \mathbf{F}(\boldsymbol{\phi}(t)) \cdot \boldsymbol{\phi}'(t)\, dt = \int_J \mathbf{F}(\boldsymbol{\psi}(u)) \cdot \boldsymbol{\psi}'(u)\, du$$

for any continuous $\mathbf{F} : \boldsymbol{\phi}(I) \to \mathbf{R}^m$.

13.2.6. **This exercise is used in Section 13.5.** Let $f : [a, b] \to \mathbf{R}$ be $\mathcal{C}^1$ on $[a, b]$ with $f'(t) \neq 0$ for $t \in [a, b]$. Prove that the explicit curve $x = f^{-1}(y)$, as y runs from $f(a)$ to $f(b)$, is orientation equivalent to the explicit curve $y = f(x)$, as x runs from a to b.

13.2.7. Let $V \neq \emptyset$ be open in $\mathbf{R}^2$. A function $\mathbf{F} : V \to \mathbf{R}^2$ is said to be *conservative* on V if and only if there is a function $f : V \to \mathbf{R}$ such that $\mathbf{F} = \nabla f$ on V. Let $(x, y) \in V$ and let $\mathbf{F} = (P, Q) : V \to \mathbf{R}^2$ be continuous on V.

a) Suppose that $C(x)$ is a horizontal line segment terminating at (x, y); that is, a line segment of the form $L((x_1, y); (x, y))$, oriented from (x_1, y) to (x, y). If $C(x)$ is a subset of V, prove that

$$\frac{\partial}{\partial x} \int_{C(x)} \mathbf{F} \cdot \mathbf{T}\, ds = P(x, y).$$

Make and prove a similar statement for $\partial/\partial y$ and vertical line segments in V terminating at (x, y).

b) Let $(x_0, y_0) \in V$. Prove that

$$(*) \qquad\qquad \int_C \mathbf{F} \cdot \mathbf{T}\, ds = 0$$

for all closed piecewise smooth curves $C \subset V$ if and only if for all $(x, y) \in V$, the integrals

$$f(x, y) := \int_{C(x,y)} \mathbf{F} \cdot \mathbf{T}\, ds$$

give the same value for all piecewise smooth curves $C(x, y)$ which start at (x_0, y_0), end at (x, y), and stay inside V.

c) Prove that $\mathbf{F}$ is conservative on V if and only if $(*)$ holds for all closed piecewise smooth curves C which are subsets of V.

d) Prove that if $\mathbf{F}$ is $\mathcal{C}^1$ and satisfies $(*)$ for all closed piecewise smooth curves C which are subsets of V, then

$$\frac{\partial P}{\partial y} = \frac{\partial Q}{\partial x}.$$

Note: If V is nice enough, the converse of this statement also holds (see Exercise 13.6.8).

*13.2.8. Suppose that $f : [0, 1] \to \mathbf{R}$ is increasing and continuously differentiable on $[0, 1]$. Let $\mathbf{T}$ be the right triangle whose vertices are $(0, f(0))$,

$(1, f(0))$, and $(1, f(1))$. If c represents the hypotenuse of **T**, a and b represent the legs of **T**, and L represents the arc length of the explicit curve $y = f(x)$, $x \in [0, 1]$, prove that $c \leq L \leq a + b$.

13.3 SURFACES

In this section we define surfaces and unoriented surface integrals, concepts which are two-dimensional analogues of arcs and the line integrals discussed in Section 13.1. Recall that a smooth arc is parametrized on a closed, bounded interval. On what shall we parametrize a smooth surface? Evidently, we need to use some type of closed, bounded set in $\mathbf{R}^2$. Although we could use rectangles, such a restriction would be awkward when dealing with explicit surfaces with curved projections, or with surfaces described by cylindrical or spherical coordinates. It is much more efficient to build greater generality into the definition of surface, using *two-dimensional regions* instead of rectangles (i.e., using sets of the following type for $m = 2$).

13.27 Definition.

An *m-dimensional region* is a set $E \subset \mathbf{R}^m$ such that $E = \overline{V}$ for some nonempty, open, connected Jordan region V in $\mathbf{R}^m$.

Notice that every closed, bounded interval is a one-dimensional region, every two-dimensional rectangle and the closure of every two-dimensional ball or ellipse is a two-dimensional region, and every three-dimensional rectangle and the closure of every three-dimensional ball or ellipsoid is a three-dimensional region.

13.28 Definition.

A subset S of $\mathbf{R}^3$ is called a C^p *surface* (in $\mathbf{R}^3$) if and only if there is a pair $(\boldsymbol{\phi}, E)$ such that E is a two-dimensional region, $\boldsymbol{\phi} : E \to \mathbf{R}^3$ is C^p on E and 1–1 on E^o, and $S = \boldsymbol{\phi}(E)$. In this case we call $(\boldsymbol{\phi}, E)$ a *parametrization* of S, S the *trace* of $(\boldsymbol{\phi}, E)$, and the equations

$$x = \phi_1(u, v), \quad y = \phi_2(u, v), \quad z = \phi_3(u, v), \qquad (u, v) \in E$$

the *parametric equations* of S induced by $(\boldsymbol{\phi}, E)$.

Earlier, we called the graph of a function $z = f(x, y)$ a surface. The following result shows that this designation is compatible with Definition 13.28 when f is C^p.

13.29 EXAMPLE.

Let E be a two-dimensional region and let $f : E \to \mathbf{R}$ be a C^p function. Prove that the graph of $z = f(x, y)$ is a C^p surface.

Proof. If $\phi(u, v) = (u, v, f(u, v))$, then ϕ is C^p and 1–1 on E, and $\phi(E)$ is the graph of $z = f(x, y)$. [This is called the *trivial parametrization* of $z = f(x, y)$.] ∎

In a similar way we define trivial parametrizations of surfaces of the form $x = f(y, z)$ and $y = f(x, z)$. For example, the trivial parametrization of the surface $x = f(y, z)$, $(y, z) \in E$, is given by (ϕ, E), where $\phi(u, v) = (f(u, v), u, v)$. By an *explicit surface over E* we shall mean a surface of the form $x = f(y, z)$, $y = f(x, z)$, or $z = f(x, y)$, where $f : E \to \mathbf{R}$ is a C^p function and E is a two-dimensional region. By the proof of Example 13.29, every explicit surface is a C^p surface.

The next four examples, which provide model parametrizations for certain kinds of surfaces, show that not every surface is an explicit surface.

13.30 EXAMPLE.

Show that the truncated cylinder $x^2 + y^2 = 1$, $0 \leq z \leq 2$, is a C^∞ surface.

Proof. Let $\phi(u, v) = (\cos u, \sin u, v)$ and $E = [0, 2\pi] \times [0, 2]$, and notice that ϕ is 1–1 on E^o and C^∞ on E. The corresponding parametric equations are $x = \cos u$, $y = \sin u$, $z = v$. Clearly, $x^2 + y^2 = 1$. Thus $\phi(E)$ is a subset of the cylinder $x^2 + y^2 = 1$, $0 \leq z \leq 2$. Since E is connected, so is $\phi(E)$. To see that $\phi(E)$ is the entire cylinder, look at the images of horizontal line segments in E. The image of the line segment $v = v_0$ is a circle lying in the plane $z = v_0$, centered at $(0, 0, v_0)$, of radius 1 (see Figure 13.8). Thus, as v_0 ranges from 0 to 2, the images of horizontal lines $v = v_0$ cover the entire cylinder $x^2 + y^2 = 1$, $0 \leq z \leq 2$. ∎

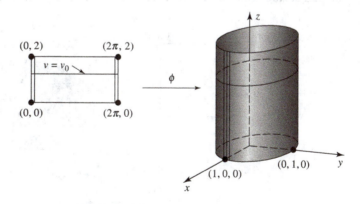

FIGURE 13.8

13.31 EXAMPLE.

Show that the sphere $x^2 + y^2 + z^2 = a^2$ is a C^∞ surface.

Proof. Let $\phi(u, v) = (a \cos u \cos v, a \sin u \cos v, a \sin v)$ and $E = [0, 2\pi] \times [-\pi/2, \pi/2]$. Clearly, ϕ is C^∞ on E. The corresponding parametric equations are $x = a \cos u \cos v$, $y = a \sin u \cos v$, $z = a \sin v$. Since $x^2 + y^2 = a^2 \cos^2 v$, we have $x^2 + y^2 + z^2 = a^2$. Thus $\phi(E)$ is a subset of the sphere centered at the origin of radius a. The image of the horizontal line segment $v = v_0$ is a circle, lying in the plane $z = a \sin v_0$, centered at $(0, 0, a \sin v_0)$ of radius $a \cos v_0$ (see Figure 13.9). The image of the top edge (respectively, bottom edge) of E (i.e., of the horizontal line $v = \pi/2$) (respectively, $v = -\pi/2$) is the north pole $(0, 0, a)$ [respectively, the south pole $(0, 0, -a)$]. Thus, as v_0 ranges from $-\pi/2$ to $\pi/2$, the images of horizontal lines $v = v_0$ cover the entire sphere $x^2 + y^2 + z^2 = a^2$. ∎

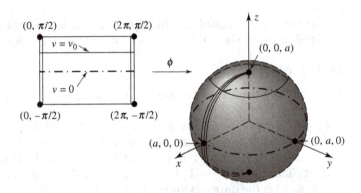

FIGURE 13.9

Let C represent the circle in the xz-plane centered at $(a, 0, 0)$ of radius b, where $a > b$. The *torus* centered at the origin with radii $a > b$ is the donut-shaped surface obtained by revolving C about the z axis (see Figure 13.10).

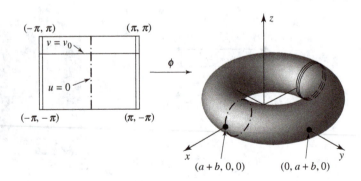

FIGURE 13.10

13.32 EXAMPLE.

Show that the torus centered at the origin with radii $a > b$ is a C^∞ surface.

Proof. Let $\phi(u, v) = ((a + b \cos v) \cos u, (a + b \cos v) \sin u, b \sin v)$ and $E = [-\pi, \pi] \times [-\pi, \pi]$, and notice that ϕ is 1–1 on E^o and C^∞ on E. The image of

$u = 0$ is a circle in the xz-plane centered at $(a, 0, 0)$ of radius b. The images of horizontal lines $v = v_0$ are circles, parallel to the xy-plane, centered at $(0, 0, b \sin v_0)$ of radius $(a + b \cos v_0)$. The image of the lines $v = \pm \pi$ is a circle in the xy-plane centered at $(0, 0, 0)$ of radius $a - b$. Thus $\phi(E)$ covers the entire torus. ∎

13.33 EXAMPLE.

Let $b > 0$. Show that the truncated cone $z = \sqrt{x^2 + y^2}$, $0 \le z \le b$, is a C^∞ surface.

Proof. Let $(x, y, z) = \phi(u, v) = (v \cos u, \, v \sin u, \, v)$ and $E = [0, 2\pi] \times [0, b]$, and notice that ϕ is 1–1 on E^o and C^∞ on E. Clearly, $x^2 + y^2 = z^2$ and $0 \le z \le b$. Thus $\phi(E)$ is a subset of the given cone. The image of a horizontal line $v = v_0$, $0 < v_0 \le b$, is a circle in the plane $z = v_0$ centered at $(0, 0, v_0)$ of radius v_0 (see Figure 13.11). Thus $\phi(E)$ is the cone $z = \sqrt{x^2 + y^2}$, $0 \le z \le b$. Notice that the image of the line $v = 0$ is the vertex $(0, 0, 0)$. ∎

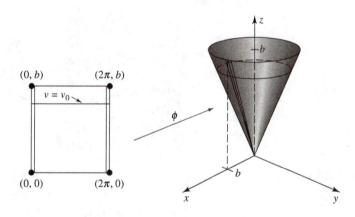

FIGURE 13.11

Let S be a C^p surface with parametrization (ϕ, E), and suppose that $(u_0, v_0) \in E^o$. If $\phi = (\phi_1, \phi_2, \phi_3)$, then by the Implicit Function Theorem (see the proof of Remark 13.10), we can show that if at least one of the partial Jacobians is nonzero at (u_0, v_0)—that is, if

$$\Delta_{\phi_i, \phi_j}(u_0, v_0) := \frac{\partial(\phi_i, \phi_j)}{\partial(u, v)}(u_0, v_0) \ne 0 \tag{5}$$

for some $i \ne j$—then there is a C^p explicit surface (ψ, B) such that $(x_0, y_0, z_0) := \phi(u_0, v_0) \in \psi(B)$ and $\psi(B) \subset \phi(E)$. Since differentiable explicit surfaces have tangent planes (see Theorem 11.22), it follows that if (5) is satisfied for some $i \ne j$ and $(x_0, y_0, z_0) = \phi(u_0, v_0)$, then S has a tangent plane at (x_0, y_0, z_0).

The following result shows how to use a parametrization of a surface to compute a normal to its tangent plane.

13.34 Remark. *Let S be a C^p surface, let (ϕ, E) be one of its parametrizations, and set $\phi =: (\phi_1, \phi_2, \phi_3)$. If (5) holds at some $(u_0, v_0) \in E^o$ and some $i \neq j$, then a normal to the tangent plane of S at $(x_0, y_0, z_0) = \phi(u_0, v_0)$ is given by*

$$(\phi_u \times \phi_v)(u_0, v_0) := \left(\Delta_{\phi_2,\phi_3}(u_0, v_0), \Delta_{\phi_3,\phi_1}(u_0, v_0), \Delta_{\phi_1,\phi_2}(u_0, v_0)\right). \quad (6)$$

Proof. Let Π be the tangent plane to S at $\phi(u_0, v_0)$. To compute a normal to Π we need only find two vectors which lie in Π. But $\phi_u(u_0, v_0)$ is tangent to the curve $\phi(u, v_0)$ and $\phi_v(u_0, v_0)$ is tangent to the curve $\phi(u_0, v)$ (see Figure 13.3). Hence, $\phi_u(u_0, v_0)$ and $\phi_v(u_0, v_0)$ both lie in Π (see Figure 13.12). Therefore, a normal to Π at (x_0, y_0, z_0) is given by the cross product

$$\phi_u(u_0, v_0) \times \phi_v(u_0, v_0) = (\Delta_{\phi_2,\phi_3}(u_0, v_0), \Delta_{\phi_3,\phi_1}(u_0, v_0), \Delta_{\phi_1,\phi_2}(u_0, v_0)). \quad \blacksquare$$

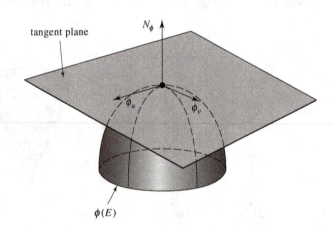

FIGURE 13.12

If (ϕ, E) is a parametrization of a C^1 surface S, we shall use the notation

$$N_\phi(u, v) := \phi_u(u, v) \times \phi_v(u, v), \quad (u, v) \in E,$$

to represent the vector (6). When (5) holds for some $i \neq j$, we shall call $N_\phi(u_0, v_0)$ *the normal induced by ϕ* on S. It is easy to check that if $z = f(x, y)$ is an explicit surface and ϕ is its trivial parametrization, then $N_\phi = (-f_x, -f_y, 1)$. This is equivalent to the normal we used for explicit surfaces before (see Theorem 11.22).

Normal vectors play the same role for surfaces that tangent vectors played for curves. (For example, we shall use normal vectors to define area of surfaces, smooth surfaces, and orientation of surfaces.) Indeed, many of the concepts for curves can be brought over to surfaces by replacing ϕ' by N_ϕ. For example, compare the following definition with Definition 13.11.

13.35 Definition.

Suppose that (ϕ, E) is a parametrization of a C^p surface.

  i) (ϕ, E) is said to be *smooth* at a point $(u_0, v_0) \in E$ if and only if $N_\phi(u_0, v_0) \neq 0$ [equivalently, if and only if $\|N_\phi(u_0, v_0)\| > 0$].
 ii) (ϕ, E) is said to be *smooth* if and only if it is smooth at each point in E.
iii) (ϕ, E) is said to be smooth off a set $E_0 \subset E$ if and only if (ϕ, E) is smooth at each point in $E \setminus E_0$.

Notice that the trivial parametrization of an explicit surface is always smooth.

Analogous to the situation for curves, a surface with a smooth parametrization must have a tangent plane at each of its points (see Exercise 13.3.7). On the other hand, a surface with tangent planes at each point can have nonsmooth parametrizations. For example, the parametrization ϕ of the sphere given in Example 13.31 satisfies

$$\|N_\phi\| = \|(a^2 \cos u \cos^2 v, a^2 \sin u \cos^2 v, a^2 \sin v \cos v)\| = a^2 |\cos v|$$

and, hence, is not smooth when $v = \pm\pi/2$. (This happens because this parametrization takes the lines $v = \pm\pi/2$ to the north and south pole, and, hence, is not 1–1 there.)

We shall call a surface S *smooth* if and only if for each point $x_0 \in S$ there is a parametrization (ϕ, E) of S which is smooth at (u_0, v_0), where $x_0 = \phi(u_0, v_0)$. Other authors call a surface smooth only when it has a smooth parametrization. This definition is inadequate for most "closed" surfaces (i.e., surfaces which are the boundary of some three-dimensional region) because those surfaces have no (globally) smooth parametrizations. (See, e.g., discussion of the parametrization of the sphere in the preceding paragraph. The sphere IS smooth by our definition, however, since we can find other parametrizations which are "smooth" at the north and south poles, for example, the trivial parametrizations of each hemisphere.) This is typical. Every surface smooth by our definition is a union of surfaces with smooth parametrizations—see Exercise 13.4.7.

The following result shows what happens to the normal vector N_ϕ under a change of parameter.

13.36 Theorem. *Let (ϕ, E) and (ψ, B) be parametrizations of the same C^p surface. If τ is a C^1 function which takes B into E such that $\psi = \phi \circ \tau$, then*

$$N_\psi(u, v) = \Delta_\tau(u, v) N_\phi(\tau(u, v))$$

for each $u, v \in B$.

Proof. Let $\phi = (\phi_1, \phi_2, \phi_3)$ and $\psi = (\psi_1, \psi_2, \psi_3)$. By Remark 13.34,

$$N_\psi = (\Delta_{(\psi_2, \psi_3)}, \Delta_{(\psi_3, \psi_1)}, \Delta_{(\psi_1, \psi_2)}).$$

Since, by hypothesis, $(\psi_i, \psi_j) = (\phi_i, \phi_j) \circ \tau$ for $i, j = 1, 2, 3$, it follows from the Chain Rule that

$$\Delta_{(\psi_i, \psi_j)}(u, v) = \Delta_\tau(u, v)\Delta_{(\phi_i, \phi_j)}(\tau(u, v))$$

for any $u, v \in B$. Therefore, $N_\psi = \Delta_\tau \cdot (N_{\phi \circ \tau})$ on B. ∎

This leads us to the following definition (compare with Definition 13.13).

13.37 Definition.

Two C^p parametrizations (ϕ, E), (ψ, B) are said to be *smoothly equivalent* if and only if they are smooth parametrizations of the same surface and there is a C^p function τ, which takes B onto E, such that $\psi = \phi \circ \tau$ and $\Delta_\tau(u, v) \neq 0$ for all $(u, v) \in B$. The function τ is called the *transition* from B to E.

Analogous to Definitions 13.6 and 13.14, we define surface area and the surface integral as follows.

13.38 Definition.

Let S be a smooth C^p surface and (ϕ, E) be one of its parametrizations.

i) The *surface area* of S is defined to be

$$\sigma(S) := \int_E \|N_\phi(u, v)\| \, d(u, v).$$

ii) If $g : S \to \mathbf{R}$ is continuous, then the *surface integral* of g on S is defined to be

$$\iint_S g \, d\sigma := \int_E g(\phi(u, v)) \, \|N_\phi(u, v)\| \, d(u, v). \tag{7}$$

The surface integral (7) can be interpreted as the mass of a membrane with shape $\phi(E)$ and density g (see Appendix E). For an explicit C^p surface S given by $z = f(x, y)$, $(x, y) \in E$, this integral looks like

$$\iint_S g \, d\sigma = \int_E g(x, y)\sqrt{1 + f_x^2(x, y) + f_y^2(x, y)} \, d(x, y). \tag{8}$$

It can be argued on heuristic grounds that this is the right definition for surface area (see Appendix E). In fact, we could have defined the surface area of S by approximating it with planar regions, as we defined $\|C\|$ below Example 13.16, by approximating it by line segments (see Price [10], p. 360). This approach, however, works only under suitable restrictions. Indeed, even when using triangular regions to approximate a bounded cylinder, the total area of the approximating regions may become infinite (see Spivak [12], p. 130). We prefer Definition 13.38i because it is both direct and easy to use.

Notice that by Theorem 12.24, (7) makes sense when the normal $N_\phi(u, v)$ is undefined on a set of area zero. Thus the surface integral can be defined for some nonsmooth surfaces (e.g., for cones).

It is easy to see that surface area and the surface integral are invariant under smoothly equivalent parametrizations, even when the condition $\Delta_\tau \neq 0$ is relaxed on a closed set of area zero (see Exercise 13.3.5). It is also easy to see that if a surface S is a subset of $\mathbf{R}^2$, then its surface area, as defined by Definition 13.38, is the same as the area of S, as defined by Definition 12.3 (see Exercise 13.3.4).

To compute a surface integral, we must find a suitable parametrization of the given surface and apply Definition 13.38.

13.39 EXAMPLE.

Find $\iint_S g \, d\sigma$, where S is the hemisphere $z = \sqrt{a^2 - x^2 - y^2}$ and $g(x, y, z) = \sqrt{z}$.

Solution. Let ϕ be the function defined in Example 13.31 and $E = [0, 2\pi] \times [0, \pi/2]$. Then (ϕ, E) is a parametrization of the hemisphere S and $\|N_\phi\| = a^2 \cos v$. Therefore,

$$\iint_S g \, d\sigma = \iint_{E_0} a^2 \cos v \sqrt{a \sin v} \, du \, dv = 2\pi a^{5/2} \int_0^{\pi/2} \cos v \sqrt{\sin v} \, dv = \frac{4\pi}{3} a^{5/2}.$$ ∎

Continuity of g is assumed in Definition 13.38 only so that the integral on the right-hand side of (7) makes sense. If one of the iterated integrals is a convergent improper integral, we can extend the definition of the surface integral in the obvious way. Using this observation, we now offer a second solution to Example 13.39 using the trivial parametrization.

ALTERNATE SOLUTION. The explicit surface $z = \sqrt{a^2 - x^2 - y^2}$ has normal $N = (-z_x, -z_y, 1) = (x/z, y/z, 1)$. [This normal does not exist on $\partial B_a(0, 0)$, but since $\partial B_a(0, 0)$ is of area zero, we can ignore it when integrating over $B_a(0, 0)$.] Notice that on S, $\|N\| = a/z$. Thus, by (8) and polar coordinates,

$$\iint_S g \, d\sigma = \int_{B_a(0,0)} \frac{a\sqrt{z}}{z} \, d(x, y) = a \int_0^{2\pi} \int_0^a r(a^2 - r^2)^{-1/4} \, dr \, d\theta = \frac{4\pi}{3} a^{5/2}.$$

[The inner integral (with respect to r) is an improper integral.]   ∎

For even the simplest applications, we must have a theory rich enough to handle surfaces, like the boundary of the unit cube $\partial([0, 1] \times [0, 1] \times [0, 1])$, which are not smooth but a union of smooth pieces. Consequently, we shall extend the theory developed above to finite unions of smooth surfaces. This expanded theory will be introduced using informal geometric descriptions instead of formal statements. For now, these vague descriptions will suffice because the concrete surfaces which arise in practice are easy to visualize. (Spivak [12] contains a rigorous and more mathematically satisfying treatment of these ideas.)

Before describing piecewise smooth surfaces, we must distinguish between interior points (points which lie "inside" a surface) and boundary points (points which lie on the "edge" of a surface). To illustrate the difference, consider the truncated cylinder S parametrized by $(\boldsymbol{\phi}, E)$ in Example 13.30. A point $(x, y, z) \in S$ lies inside S if $0 < z < 2$, and on its edge if $z = 0$ or $z = 2$. (Look at Figure 13.8 to see why this terminology is appropriate.) Naively, we might guess that (x, y, z) lies on the edge of $\boldsymbol{\phi}(E)$ if and only if $(x, y, z) \notin \boldsymbol{\phi}(E^o)$. This guess is incorrect, even for the cylinder; for example, $(1, 0, 1) = \boldsymbol{\phi}(0, 1)$ does not belong to $\boldsymbol{\phi}(E^o)$ but does not belong to an edge of the cylinder either. (Instead, it lies on a "seam" of S.) Evidently, to define the interior and boundary of a general surface S, we must describe them geometrically. We cannot define the interior and boundary of a surface by using a particular parametrization $(\boldsymbol{\phi}, E)$.

Accordingly, let S be a C^p surface in $\mathbf{R}^3$. Imagine yourself standing on a point $(x, y, z) \in S$. We shall say that (x, y, z) is *interior* to S if you are surrounded on all sides by points in S (i.e., if you take a sufficiently small step in any direction you remain on S). We shall denote the set of interior points of a surface S by $\mathrm{Int}(S)$ and shall define the (*manifold*) *boundary* of a surface S by $\partial S := S \setminus \mathrm{Int}(S)$.

We have used the same notation to denote the boundary of a surface as we did to denote the boundary of a set (see Definition 8.34 or 10.37) even though these concepts are not the same. We made this choice because it homogenizes the statements of all the fundamental theorems of multidimensional calculus. To avoid ambiguity, we shall henceforth refer to the boundary of a region E (i.e., to $\overline{E} \setminus E^o$) as the *topological boundary* of E. No confusion will arise because the only boundary we use in connection with surfaces is the manifold boundary, and the only boundary we use in connection with m-dimensional regions is the topological boundary.

A surface S is said to be *closed* if and only if $\partial S = \emptyset$. For example, if $a > 0$, then the sphere $x^2 + y^2 + z^2 = a^2$ is closed, but the hemisphere $z = \sqrt{a^2 - x^2 - y^2}$ (respectively, the truncated paraboloid $z = x^2 + y^2$, $0 \leq z \leq 1$) is not closed, since its boundary is $x^2 + y^2 = a^2$, $z = 0$ (respectively, $x^2 + y^2 = 1$, $z = 1$).

By the Jordan Curve Theorem, a closed arc C divides $\mathbf{R}^2$ into two or more disjoint connected sets, the bounded components "surrounded" by C and the unbounded component which lies "outside" C. This is not the case for closed surfaces. Indeed, there are closed smooth surfaces (the Klein bottle is one example) which surround no points and, hence, do not divide $\mathbf{R}^3$ into disjoint sets (see Hocking and Young [4], p. 237).

A set $S \subset \mathbf{R}^3$ will be called a *piecewise smooth surface* if and only if $S = \cup_{j=1}^{N} S_j$, where each $S_j = (\boldsymbol{\phi}_j, E_j)$ is a smooth surface and for each $j \neq k$ either $S_j \cap S_k$ is empty, or a portion of the boundary of S_j is matched to a portion of the boundary of S_k. Thus a piecewise smooth surface might consist of several disjoint, smooth surfaces, like the topological boundary of the *corona* $0 < a \leq \|(x, y, z)\| \leq b$, or several connected pieces with ridges, like the concentric boxes $\partial[([0, 3] \times [0, 3] \times [0, 3]) \setminus ([1, 2] \times [1, 2] \times [1, 2])]$. We make the further restriction that the intersection of any three S_j's is empty, or a finite set. This prevents a piecewise smooth surface from doubling back on itself more than once along any given edge.

Let $S = \cup_{j=1}^{N} S_j$ be a piecewise smooth surface. By a *parametrization* of S we mean a collection of smooth parametrizations (ϕ_j, E_j) of S_j. Two parametrizations (ϕ_j, E_j), (ψ_j, B_j) are said to be *smoothly equivalent* if and only if (ϕ_j, E_j) is smoothly equivalent to (ψ_j, B_j) for $j = 1, \ldots, N$. The *boundary*, ∂S, of S is defined to be the union of all points which belong to the closure of an unmatched portion of ∂S_j. (e.g., the boundary of the box formed by removing the face $z = 1$ from the unit cube $[0, 1] \times [0, 1] \times [0, 1]$ is the unit square in the plane $z = 1$, and the boundary of the union of $x^2 + y^2 = 1$, $-3 \le z \le 0$, and $z = \sqrt{1 - x^2 - y^2}$ is the unit circle in the plane $z = -3$.) The *surface area* of S is defined by

$$\sigma(S) = \sum_{j=1}^{N} \sigma(S_j)$$

and the *surface integral* of a real-valued function g continuous on S is defined by

$$\iint_S g \, d\sigma = \sum_{j=1}^{N} \iint_{S_j} g \, d\sigma.$$

13.40 EXAMPLE.

Let S be the tetrahedron formed by taking the topological boundary of the region bounded by $x = 0$, $y = 0$, $z = 0$, and $x + y + z = 1$. Find a piecewise smooth parametrization S and compute $\iint_S g \, d\sigma$, where $g(x, y, z) = x + y^2 + z^3$.

Solution. The tetrahedron has four faces which can be parametrized by $\phi_1(u, v) = (u, v, 0)$, $\phi_2(u, v) = (0, u, v)$, $\phi_3(u, v) = (u, 0, v)$, $\phi_4(u, v) = (u, v, 1 - u - v)$, where (u, v) belongs to E, the triangular region with vertices $(0, 0)$, $(1, 0)$, and $(0, 1)$. Since $\|N_{\phi_j}\| = 1$ for $j = 1, 2, 3$ and $\|N_{\phi_4}\| = \sqrt{3}$, we have

$$\iint_S g \, d\sigma = \int_0^1 \int_0^{1-u} (u + v^2) \, dv \, du + \int_0^1 \int_0^{1-u} (u^2 + v^3) \, dv \, du$$

$$+ \int_0^1 \int_0^{1-u} (u + v^3) \, dv \, du$$

$$+ \sqrt{3} \int_0^1 \int_0^{1-u} (u + v^2 + (1 - u - v)^3)) \, dv \, du$$

$$= \int_0^1 \int_0^{1-u} ((2 + \sqrt{3})u + u^2 + (1 + \sqrt{3})v^2 + 2v^3$$

$$+ \sqrt{3}(1 - u - v)^3) \, dv \, du$$

$$= \int_0^1 \left((2 + \sqrt{3})u - (1 + \sqrt{3})u^2 - u^3 + \frac{1 + \sqrt{3}}{3}(1 - u)^3 \right.$$
$$\left. + \frac{2 + \sqrt{3}}{4}(1 - u)^4 \right) du$$
$$= \frac{3}{10}(2 + \sqrt{3}). \qquad \blacksquare$$

EXERCISES

13.3.1. For each of the following, find the surface area of S.

a) S is the conical shell given by $z = \sqrt{x^2 + y^2}$, where $a \le z \le b$.

b) S is the sphere given in Example 13.31.

c) S is the torus given in Example 13.32.

13.3.2. For each of the following, find a (piecewise) smooth parametrization of S and of ∂S, and compute $\iint_S g \, d\sigma$.

a) S is the portion of the surface $z = x^2 - y^2$ which lies above the xy-plane and between the planes $x = 1$ and $x = -1$, and $g(x, y, z) = \sqrt{1 + 4x^2 + 4y^2}$.

b) S is the surface $y = x^3$, $0 \le y \le 8$, $0 \le z \le 4$, and $g(x, y, z) = x^3 z$.

c) S is the portion of the hemisphere $z = \sqrt{9 - x^2 - y^2}$ which lies outside the cylinder $2x^2 + 2y^2 = 9$, and $g(x, y, z) = x + y + z$.

13.3.3. Find a parametrization (ϕ, E) of the ellipsoid

$$\frac{x^2}{a^2} + \frac{y^2}{b^2} + \frac{z^2}{c^2} = 1$$

which is smooth off the topological boundary ∂E.

13.3.4. a) Suppose that E is a two-dimensional region and that $S = \{(x, y, z) \in \mathbf{R}^3 : (x, y) \in E \text{ and } z = 0\}$. Prove that

$$\text{Area } (E) = \iint_S d\sigma$$

and that

$$\iint_S g \, d\sigma = \int_E g(x, y, 0) \, d(x, y)$$

for each continuous $g : E \to \mathbf{R}$.

b) Let $f : [a, b] \to \mathbf{R}$ be a C^p function, let C be the curve in $\mathbf{R}^2$ determined by $z = f(x)$, $a \le x \le b$, and let S be the surface in $\mathbf{R}^3$ determined by $z = f(x)$, $a \le x \le b$, $c \le y \le d$. Show that $\sigma(S) = (d - c)L(C)$.

c) Let $f : [a, b] \to \mathbf{R}$ be a C^p function and let S be the surface obtained by revolving the curve $y = f(x)$, $a \le x \le b$, about the x-axis. Prove that the surface area of S is

$$\sigma(S) = 2\pi \int_a^b |f(x)|\sqrt{1 + |f'(x)|^2}\, dx.$$

13.3.5. Suppose that $\psi(B)$ and $\phi(E)$ are $\mathcal{C}^p$ surfaces and that $\psi = \phi \circ \tau$, where τ is a $\mathcal{C}^1$ function from B onto Z.

a) If (ψ, B) and (ϕ, E) are smooth and τ is 1–1 with $\Delta_\tau \neq 0$ on B, prove that

$$\iint_E g(\phi(u, v))\|N_\phi(u, v)\|\, du\, dv = \iint_B g(\psi(s, t))\|N_\psi(s, t)\|\, ds\, dt$$

for all continuous $g : \phi(E) \to \mathbf{R}$.

*b) If Z is a closed subset of B of area zero such that (ψ, B) is smooth off Z, τ is 1–1, and $\Delta_\tau \neq 0$ on $B^o \backslash Z$, prove that

$$\iint_E g(\phi(u, v))\|N_\phi(u, v)\|\, du\, dv = \iint_B g(\psi(s, t))\|N_\psi(s, t)\|\, ds\, dt$$

for all continuous $g : \phi(E) \to \mathbf{R}$.

13.3.6. Suppose that $f : B_3(0, 0) \to \mathbf{R}$ is differentiable with $\|\nabla f(x, y)\| \leq 1$ for all $(x, y) \in B_3(0, 0)$. Prove that if S is the paraboloid $2z = x^2 + y^2$, $0 \leq z \leq 4$, then

$$\iint_S |f(x, y) - f(0, 0)|\, d\sigma \leq 40\pi.$$

13.3.7. Suppose that $\phi(E)$ is a $\mathcal{C}^p$ surface and that $(x_0, y_0, z_0) = \phi(u_0, v_0)$, where $(u_0, v_0) \in E^o$. If $N_\phi(u_0, v_0) \neq 0$, prove that $\phi(E)$ has a tangent plane at (x_0, y_0, z_0).

13.3.8. Let $\psi(B)$ be a smooth surface. Set $E = \|\psi_u\|$, $F = \psi_u \cdot \psi_v$, and $G = \|\psi_v\|$. Prove that the surface area of S is $\int_B \sqrt{E^2 G^2 - F^2}\, d(u, v)$.

13.3.9. Suppose that S is a $\mathcal{C}^1$ surface with parametrization (ϕ, E) which is smooth at $(x_0, y_0, z_0) = \phi(u_0, v_0)$. Let (ψ, I) be a parametrization of a $\mathcal{C}^1$ curve in E which passes through the point (u_0, v_0) [i.e., there is a $t_0 \in I$ such that $\psi(t_0) = (u_0, v_0)$]. Prove that $(\phi \circ \psi)'(t_0) \cdot (\phi_u \times \phi_v)(u_0, v_0) = 0$.

13.4 ORIENTED SURFACES

Recall that a smooth curve $\phi(I)$ is oriented by using the tangent vector $\phi'(t)$ to choose a "positive direction." Analogously, a smooth surface $S = \phi(E)$ will be oriented by using the normal vector N_ϕ to choose a "positive side." Since smooth surfaces are by definition connected, such a choice will be possible if S has two, and only two, sides.

A new complication arises here. There are smooth surfaces which have only one side. (The following example of such a surface can be made out of paper by taking a long narrow strip by the narrow edges, twisting it once, and gluing the narrow edges together.)

13.41 *EXAMPLE.* [THE MÖBIUS STRIP].

Sketch the trace of (ϕ, E), where $\phi(u, v) = ((2 + v \sin(u/2)) \cos u, (2 + v \sin(u/2)) \sin u, v \cos(u/2))$ and $E = [-\pi, \pi] \times [-1, 1]$.

Solution. The image of the horizontal line $v = 0$ under ϕ is $(2 \cos u, 2 \sin u, 0)$ (i.e., the circle in the xy-plane centered at the origin of radius 2). The image of each vertical line $u = u_0$ is a line segment in $\mathbf{R}^3$ which rotates through space as u_0 increases. For example, the image of $u = 0$ is $(2, 0, v)$, $-1 \le v \le 1$, and the image of $u = \pm\pi$ is the seam $S_0 := (-2 \mp v, 0, 0)$, $-1 \le v \le 1$; that is, the set of points $\{(x, 0, 0) : -3 \le x \le -1\}$. Thus the trace of (ϕ, E) is given in Figure 13.13.   ∎

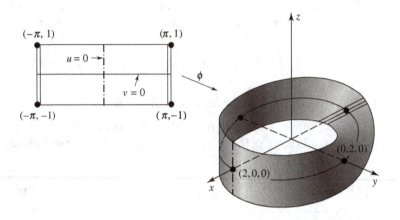

FIGURE 13.13

To avoid such anomalies, we introduce the following concepts. The *unit normal* of a smooth surface S, at a point (x_0, y_0, z_0) on S, induced by one of its parametrizations (ϕ, E) is the vector $\mathbf{n}(x_0, y_0, z_0) = N_\phi(u_0, v_0)/\|N_\phi(u_0, v_0)\|$, where $\phi(u_0, v_0) = (x_0, y_0, z_0)$. Evidently, the unit normal $\mathbf{n}$ is well defined only when

$$\frac{N_\phi(u_0, v_0)}{\|N_\phi(u_0, v_0)\|} = \frac{N_\phi(u_1, v_1)}{\|N_\phi(u_1, v_1)\|} \ne \mathbf{0}$$

for all $(u_j, v_j) \in E$ which satisfy $\phi(u_j, v_j) = (x_0, y_0, z_0)$ for $j = 0, 1$. This will surely be the case if ϕ is 1–1 and smooth on E. If ϕ fails to be 1–1 on E, however, the unit normal $\mathbf{n}$ might not be well defined, even though (ϕ, E) is smooth on E [see the Möbius strip above where $\phi(\pi, v) = \phi(-\pi, v)$ but $N_\phi(\pi, v) = -N_\phi(-\pi, v)$ for all v].

A smooth surface S is said to be *orientable* if and only if it has a smooth parametrization (ϕ, E) which *induces* an unambiguous unit normal $\mathbf{n}$ on S that varies continuously over S; that is, if $\phi(u_0, v_0) = \phi(u_1, v_1)$, then $N_\phi(u_0, v_0)$ points in the same direction as $N_\phi(u_1, v_1)$, and if (u_2, v_2) is near (u_0, v_0), then $N_\phi(u_2, v_2)$ points in approximately the same direction as $N_\phi(u_0, v_0)$. (A formal definition of orientable can be found in Spivak [12].) If S is orientable, then its unit normal can be used to choose a "positive" side (the side from which $\mathbf{n}$ points).

Henceforth, by a *parametrization of an orientable surface S* we mean a smooth $(\boldsymbol{\phi}, E)$ which induces an unambiguous unit normal on S.

13.42 Definition.

Two parametrizations $(\boldsymbol{\phi}, E)$ and $(\boldsymbol{\psi}, B)$ are said to be *orientation equivalent* if and only if they are parametrizations of the same orientable surface, smoothly equivalent with transition $\boldsymbol{\tau}$, and $\Delta_{\boldsymbol{\tau}}(u, v) > 0$ for all $(u, v) \in B$.

By Theorem 13.36, if $(\boldsymbol{\phi}, E)$ and $(\boldsymbol{\psi}, B)$ are orientation equivalent, then the normal vectors they generate point in the same direction. Thus the positive side chosen by $(\boldsymbol{\phi}, E)$ is the same as the positive side chosen by $(\boldsymbol{\psi}, B)$.

Oriented surface integrals can be defined using the unit normal in the same way that oriented line integrals were defined using the unit tangent (compare the following definition with Definition 13.21).

13.43 Definition.

Let S be a smooth orientable surface with unit normal $\mathbf{n}$ determined by a parametrization $(\boldsymbol{\phi}, E)$. If $\mathbf{F} : S \rightarrow \mathbf{R}^3$ is continuous, then the *oriented surface integral* of $\mathbf{F}$ on S is

$$\iint_S \mathbf{F} \cdot \mathbf{n} \, d\sigma := \int_E (\mathbf{F} \circ \boldsymbol{\phi})(u, v) \cdot N_{\boldsymbol{\phi}}(u, v) \, d(u, v).$$

The notation of the left-most integral is consistent with the notation in (7) since $\mathbf{n} = N_{\boldsymbol{\phi}}/\|N_{\boldsymbol{\phi}}\|$ and $d\sigma = \|N_{\boldsymbol{\phi}}\| \, d(u, v)$.

Notice that the trivial parametrization always induces an unambiguous normal on an explicit surface. In fact, if $S = \{(x, y, z) : z = f(x, y), (x, y) \in E\}$, Definition 13.43 takes the form

$$\iint_S \mathbf{F} \cdot \mathbf{n} \, d\sigma = \int_E \mathbf{F}(x, y, f(x, y)) \cdot (-f_x, -f_y, 1) \, d(x, y). \tag{9}$$

Things are not so simple for smooth surfaces which are the boundary of a three-dimensional region (like the sphere) and for surfaces which are not smooth (like the cone), because their parametrizations have at least one point where the normal is zero and, hence, the unit normal cannot be defined. Nevertheless, as was the case for the oriented line integral, the oriented surface integral can be defined when the normal fails to exist on some set of area zero (see Exercise 13.4.4). We need to be careful, however, with the definition of *orientable*. If the collection of nonsmooth points cuts across the entire surface (like the peak of a pup tent or the edge of a pyramid), we have difficulty defining what it means to have a "continuously varying" normal. We shall address this problem for piecewise smooth surfaces at the end of this section. In the meantime, notice that we can define what it means for a surface $S = \phi(E)$ to be orientable if the set of *singularities* [i.e., the set of $(x, y, z) \in \mathbf{R}^3$ such that $(x, y, z) = \phi(u, v)$ for some $(u, v) \in E$ which satisfies $N_\phi(u, v) = 0$] is finite. In particular, the standard parametrizations of spheres and cones can be used in Definition 13.43.

What does an oriented surface integral represent? If $\mathbf{F}$ represents the flow of an incompressible fluid at points on a surface S, then $\mathbf{F} \cdot \mathbf{n}$ represents the normal component of $\mathbf{F}$ (i.e., the amount of fluid which flows in the direction of $\mathbf{n}$) (see Appendix E). Thus the integral of $\mathbf{F} \cdot \mathbf{n} \, d\sigma$ on S, a measure of the flow of the fluid across the surface S in the direction of $\mathbf{n}$, is sometimes called the *flux* of $\mathbf{F}$ across S. In particular, we should not be surprised when many of these integrals turn out to be zero.

It is easy to see that the integral of $\mathbf{F} \cdot \mathbf{n} \, d\sigma$ on a surface S does not change when orientation equivalent parametrizations are used (see Exercise 13.4.4). The following result shows that a change of orientation changes the value of the oriented surface integral by a minus sign.

13.44 Remark. *If (ϕ, E) and (ψ, B) are smoothly equivalent but not orientation equivalent, then*

$$\int_E \mathbf{F}(\phi(u, v)) \cdot N_\phi(u, v) \, d(u, v) = - \int_B \mathbf{F}(\psi(s, t)) \cdot N_\psi(s, t) \, d(s, t).$$

Proof. Let τ be the transition from B to E. Since Δ_τ is continuous and nonzero on the connected set B, and (ϕ, E) and (ψ, B) are not orientation equivalent, we have $\Delta_\tau < 0$ on B. Hence, it follows from Theorem 13.36 and Theorem 12.46 (the Change-of-Variables Formula) that

$$\int_B \mathbf{F}(\psi(s, t)) \cdot N_\psi(s, t) \, d(s, t) = - \int_B |\Delta_\tau(s, t)| (\mathbf{F} \circ \phi \circ \tau)(s, t) \cdot (N_\phi \circ \tau)(s, t) \, d(s, t)$$

$$= - \int_{\tau(B)} \mathbf{F}(\phi(u, v)) \cdot N_\phi(u, v) \, d(u, v)$$

$$= - \int_E \mathbf{F}(\phi(u, v)) \cdot N_\phi(u, v) \, d(u, v). \qquad \blacksquare$$

Therefore, when evaluating an oriented integral on a surface S whose orientation has been described geometrically, we can use any smooth parametrization of S and adjust the sign of the integral to reflect the prescribed orientation. Here is a typical example.

13.45 EXAMPLE.

Find the value of $\iint_S \mathbf{F} \cdot \mathbf{n} \, d\sigma$, where $\mathbf{F}(x, y, z) = (xy, x - y, z)$, S is the planar region $x + y + z = 1$, $(x, y) \in [0, 1] \times [0, 1]$, and $\mathbf{n}$ is the downward–pointing normal.

Solution. The usual normal $(1, 1, 1)$ of the plane $x + y + z = 1$ points upward rather than downward. Thus, by Remark 13.44,

$$\iint_S \mathbf{F} \cdot \mathbf{n} \, d\sigma = -\int_0^1 \int_0^1 (xy, x - y, 1 - x - y) \cdot (1, 1, 1) \, dx \, dy = -\frac{1}{4}. \quad \blacksquare$$

It is convenient to have a "differential" version of oriented surface integrals. To see how to define differentials of degree 2, let $S = \phi(E)$ be a smooth orientable surface and $x = \phi_1(u, v)$, $y = \phi_2(u, v)$, $z = \phi_3(u, v)$. By Remark 13.34,

$$N_\phi = \left(\frac{\partial(y, z)}{\partial(u, v)}, \frac{\partial(z, x)}{\partial(u, v)}, \frac{\partial(x, y)}{\partial(u, v)} \right).$$

Therefore, the oriented surface integral of a function $\mathbf{F} = (P, Q, R) : \phi(E) \to \mathbf{R}^3$ has the form

$$\int_E \left(P \frac{\partial(y, z)}{\partial(u, v)} + Q \frac{\partial(z, x)}{\partial(u, v)} + R \frac{\partial(x, y)}{\partial(u, v)} \right) d(u, v)$$

$$=: \iint_S P \, dy \, dz + Q \, dz \, dx + R \, dx \, dy;$$

that is, we should define differentials of degree 2 by

$$dy \, dz := \frac{\partial(y, z)}{\partial(u, v)} \, d(u, v), \quad dz \, dx := \frac{\partial(z, x)}{\partial(u, v)} \, d(u, v), \quad \text{and} \quad dx \, dy := \frac{\partial(x, y)}{\partial(u, v)} \, d(u, v).$$

[These are two-dimensional analogues of the differential $dy = f'(x) \, dx$.] By a 2-*form* (or a *differential form of degree* 2) on a set $\Omega \subset \mathbf{R}^3$ we mean an expression of the form

$$P \, dy \, dz + Q \, dz \, dx + R \, dx \, dy,$$

where $P, Q, R : \Omega \to \mathbf{R}$. A 2-form is said to be continuous on Ω if and only if its *coefficients* P, Q, R are continuous on Ω. The *oriented integral* of a continuous 2-form on a smooth surface S oriented with a unit normal $\mathbf{n}$ is defined by

$$\iint_S P \, dy \, dz + Q \, dz \, dx + R \, dx \, dy = \iint_S (P, Q, R) \cdot \mathbf{n} \, d\sigma.$$

Differential forms of degree 1 were formal devices used in certain computations (e.g., to compute an oriented line integral or to estimate the increment of a function). Similarly, differential forms of degree 2 are formal devices which will be used in certain computations (e.g., to compute an oriented surface integral). They can also be used to unify the three fundamental theorems of vector calculus presented in the next two sections (see Spivak [12]). There is a less formal way to introduce differentials in which the differential dx can be interpreted as the derivative of the projection operator $(x, y, z) \longmapsto x$ (see Spivak [12], p. 89).

In general, the boundary of a surface is a curve. Since the boundary of the Möbius strip is a simple closed curve, the boundary of a surface may be orientable even when the surface is not.

Suppose that S is an oriented surface and that ∂S is a piecewise smooth curve. The orientation of S can be used to induce an orientation on ∂S in the following way. Imagine yourself standing close to ∂S on the positive side of S. The direction of positive flow on ∂S moves from right to left (i.e., as you walk around the boundary on the positive side of S in the direction of positive flow, the surface lies on your left). This orientation of ∂S is called the *positive orientation*, the *right-hand orientation*, or the orientation on ∂S *induced* by the orientation of S. When S is a subset of $\mathbf{R}^2$ (i.e., of the xy-plane), we shall say that ∂S is oriented *positively* if it carries the orientation induced by the upward pointing normal on S (i.e., the normal which points toward the upper half space $z \geq 0$). Thus if S is a bounded subset of $\mathbf{R}^2$ whose boundary is a connected piecewise smooth closed curve, then the usual orientation on S induces a counterclockwise orientation on ∂S when viewed from high up on the positive z-axis. This is not the case, however, when E has interior "holes." For example, if $E = \{(x, y) : a^2 < x^2 + y^2 < b^2\}$ for some $a > 0$, then the positive orientation is counterclockwise on $\{(x, y) : x^2 + y^2 = b^2\}$ but clockwise on $\{(x, y) : x^2 + y^2 = a^2\}$. This informal geometric description is sufficient to identify the induced orientation in most concrete situations. Here is a typical example.

13.46 EXAMPLE.

Let S be the truncated paraboloid $z = x^2 + y^2$, $0 \leq z \leq 4$, with outward-pointing normal. Parametrize ∂S with positive orientation.

Solution. The boundary of S is the circle $x^2 + y^2 = 4$ which lies in the $z = 4$ plane. The positive orientation is clockwise when viewed from high up the z-axis. Therefore, a parametrization of ∂S is given by $\boldsymbol{\phi}(t) = (2 \sin t, 2 \cos t, 4)$, $t \in [0, 2\pi]$. ∎

How do we extend these ideas to piecewise smooth surfaces? If $S = \cup S_j$, it is not enough to assume that each S_j is orientable, because the Möbius strip is the union of two orientable surfaces, namely $\boldsymbol{\phi}(E_1)$ and $\boldsymbol{\phi}(E_2)$, where $\boldsymbol{\phi}$ is given by Example 13.41 and $E_k = [\pi(k - 2), \pi(k - 1)] \times [-1, 1]$, $k = 1, 2$. We shall say that a piecewise smooth surface $S = \cup S_j$ is *orientable* if and only if one can use the normals $\pm N_{\phi_j}$ to generate a unit normal $\mathbf{n}_j$ on each piece S_j which identifies the "positive side" in a consistent way (e.g., all normals on one connected piece point outward and all normals on another connected piece point outward). If $S = \cup_{j=1}^{N} S_j$ is orientable, then the *oriented surface integral* of a continuous function $\mathbf{F} : S \to^3 \mathbf{R}$ is defined to be

$$\iint_S \mathbf{F} \cdot \mathbf{n} \, d\sigma = \sum_{j=1}^{N} \iint_{S_j} \mathbf{F} \cdot \mathbf{n}_j \, d\sigma.$$

The following three examples provide further explanation of these ideas.

13.47 EXAMPLE.

Evaluate

$$\iint_S \mathbf{F} \cdot \mathbf{n} \, d\sigma,$$

where S is the topological boundary of the solid bounded by the cylinder $x^2 + y^2 = 1$ and the planes $z = 0$, $z = 2$; $\mathbf{n}$ is the outward-pointing normal; and $\mathbf{F}(x, y, z) = (x, 0, y)$.

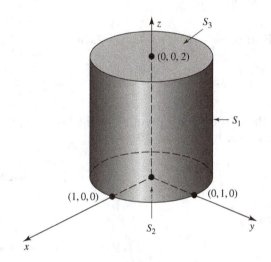

FIGURE 13.14

Solution. This surface has three smooth pieces: a vertical side S_1, a bottom S_2, and a top S_3 (see Figure 13.14). Parametrize S_1 by $\boldsymbol{\phi}(u, v) = (\cos u, \sin u, v)$, where $E = [0, 2\pi] \times [0, 2]$. Thus $N_{\boldsymbol{\phi}} = (\cos u, \sin u, 0)$ and

$$\iint_{S_1} \mathbf{F} \cdot \mathbf{n} \, d\sigma = \int_0^2 \int_0^{2\pi} \cos^2 u \, du \, dv = 2\pi.$$

Since the outward-pointing unit normal to S_2 is $\mathbf{n} = (0, 0, -1)$, we see by Exercise 13.3.4a that

$$\iint_{S_2} \mathbf{F} \cdot \mathbf{n} \, d\sigma = -\int_{B_1(0,0)} y \, d(x, y) = -\int_0^{2\pi} \int_0^1 r^2 \sin \theta \, dr \, d\theta = 0.$$

Similarly, the integral on S_3 is also zero. Therefore,

$$\iint_S \mathbf{F} \cdot \mathbf{n} \, d\sigma = 2\pi + 0 + 0 = 2\pi. \qquad \blacksquare$$

13.48 *EXAMPLE.*

Find $\iint_S \mathbf{F} \cdot \mathbf{n} \, d\sigma$, where $\mathbf{F}(x, y, z) = (x + z^2, x, z)$, S is the topological boundary of the solid bounded by the paraboloid $z = x^2 + y^2$ and the plane $z = 1$, and $\mathbf{n}$ is the outward-pointing normal.

Solution. The surface S has two smooth pieces: the paraboloid S_1 given by $z = x^2 + y^2$, $0 \le z \le 1$, and the disk S_2 given by $x^2 + y^2 \le 1$, $z = 1$. The trivial parametrization of S_1 is $\boldsymbol{\phi}(u, v) = (u, v, u^2 + v^2)$, $(u, v) \in B_1(0, 0)$. Note that $N_{\boldsymbol{\phi}} = (-2u, -2v, 1)$ points inward (the wrong way). Thus, by Remark 13.44 and polar coordinates,

$$\iint_{S_1} \mathbf{F} \cdot \mathbf{n} \, d\sigma = -\int_{B_1(0,0)} (-2u^2 - 2u(u^2 + v^2)^2 - 2uv + (u^2 + v^2)) \, d(u, v)$$

$$= \int_0^1 \int_0^{2\pi} (2r^2 \cos^2 \theta + 2r^5 \cos \theta + 2r^2 \cos \theta \sin \theta - r^2) r \, d\theta \, dr = 0.$$

Since the unit outward-pointing normal of S_2 is $\mathbf{n} = (0, 0, 1)$ and $\mathbf{F} \cdot \mathbf{n} = z = 1$ on S_2, we see by Exercise 13.3.4a that

$$\iint_{S_2} \mathbf{F} \cdot \mathbf{n} \, d\sigma = \int_{B_1(0,0)} d(x, y) = \text{Area}(B_1(0, 0)) = \pi.$$

Therefore,

$$\iint_S \mathbf{F} \cdot \mathbf{n} \, d\sigma = 0 + \pi = \pi. \qquad \blacksquare$$

13.49 *EXAMPLE.*

Compute $\iint_S \mathbf{F} \cdot \mathbf{n} \, d\sigma$, where $\mathbf{F}(x, y, z) = (x, y, z)$; S is the topological boundary of the solid bounded by the hyperboloid of one sheet $x^2 + y^2 = (z^2 + 1)^2$ and the planes $z = -1$, $z = \sqrt{3}$; and $\mathbf{n}$ is the outward-pointing normal to S.

Solution. The surface S has three smooth pieces: a top S_1, a side S_2, and a bottom S_3 (see Figure 13.15). Using $\mathbf{n} = (0, 0, 1)$ for S_1, we have

$$\iint_{S_1} \mathbf{F} \cdot \mathbf{n} \, d\sigma = \int_{B_2(0,0)} \sqrt{3} \, d(x, y) = 4\sqrt{3}\pi.$$

Similarly,

$$\iint_{S_3} \mathbf{F} \cdot \mathbf{n} \, d\sigma = 2\pi.$$

To integrate $\mathbf{F} \cdot \mathbf{n}$ on S_2, let $z = u$ and note that $x^2 + y^2 = (1 + u^2)^2$. Thus $\boldsymbol{\phi}(u, v) = ((1 + u^2)\cos v, (1 + u^2)\sin v, u)$, $(u, v) \in [-1, \sqrt{3}] \times [0, 2\pi]$, is a parametrization of S_2. Since $N_{\boldsymbol{\phi}} = (-(1 + u^2)\cos v, -(1 + u^2)\sin v, 2u(1 + u^2))$ points inward and

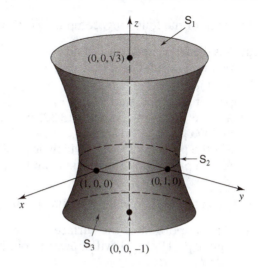

FIGURE 13.15

$$\mathbf{F} \cdot N_\phi = ((1+u^2)\cos v, \, (1+u^2)\sin v, \, u) \cdot$$
$$\cdot \, (-(1+u^2)\cos v, \, -(1+u^2)\sin v, \, 2u(1+u^2))$$
$$= -(1+u^2)^2 + 2u^2(1+u^2) = u^4 - 1,$$

we have

$$\iint_{S_2} \mathbf{F} \cdot \mathbf{n} \, d\sigma = -\int_{-1}^{\sqrt{3}} \int_0^{2\pi} (u^4 - 1) \, dv \, du$$

$$= 2\pi \int_{-1}^{\sqrt{3}} (1 - u^4) \, du = \frac{8\pi}{5}(1 - \sqrt{3}).$$

Therefore,

$$\iint_S \mathbf{F} \cdot \mathbf{n} \, d\sigma = 4\sqrt{3}\pi + 2\pi + \frac{8\pi}{5}(1 - \sqrt{3}) = \frac{6\pi}{5}(3 + 2\sqrt{3}). \qquad \blacksquare$$

EXERCISES

13.4.1. For each of the following, find a (piecewise) smooth parametrization of ∂S which agrees with the induced orientation, and compute $\int_{\partial S} \mathbf{F} \cdot \mathbf{T} \, ds$.

a) S is the truncated paraboloid $y = 9 - x^2 - z^2$, $y \geq 0$, with outward-pointing normal, and $\mathbf{F}(x, y, z) = (x^2 y, \, y^2 x, \, x + y + z)$.

b) S is the portion of the plane $x + 2y + z = 1$ which lies in the first octant, with normal which points away from the origin, and $\mathbf{F}(x, y, z) = (x - y, \, y - x, \, xz^2)$.

c) S is the truncated paraboloid $z = x^2 + y^2$, $1 \leq z \leq 4$, with outward-pointing normal, and $\mathbf{F}(x, y, z) = (5y + \cos z, \, 4x - \sin z, \, 3x \cos z + 2y \sin z)$.

13.4.2. For each of the following, compute $\iint_S \mathbf{F} \cdot \mathbf{n} \, d\sigma$.

a) S is the truncated paraboloid $z = x^2 + y^2$, $0 \le z \le 1$, $\mathbf{n}$ is the outward-pointing normal, and $\mathbf{F}(x, y, z) = (x, y, z)$.

b) S is the truncated half-cylinder $z = \sqrt{4 - y^2}$, $0 \le x \le 1$, $\mathbf{n}$ is outward-pointing normal, and $\mathbf{F}(x, y, z) = (x^2 + y^2, yz, z^2)$.

c) S is the torus in Example 13.32, $\mathbf{n}$ is the outward-pointing normal, and $\mathbf{F}(x, y, z) = (y, -x, z)$.

d) S is the portion of $z = x^2$ which lies inside the cylinder $x^2 + y^2 = 1$, $\mathbf{n}$ is the upward-pointing normal, and $\mathbf{F}(x, y, z) = (y^2 z, \cos(2 + \log(2 - x^2 - y^2)), x^2 z)$.

13.4.3. For each of the following, compute $\iint_S \omega$.

a) S is the portion of the surface $z = x^4 + y^2$ which lies over the unit square $[0, 1] \times [0, 1]$, with upward-pointing normal, and $\omega = x \, dy \, dz + y \, dz \, dx + z \, dx \, dy$.

b) S is the upper hemisphere $z = \sqrt{a^2 - x^2 - y^2}$, with outward-pointing normal, and $\omega = x \, dy \, dz + y \, dz \, dx$.

c) S is the spherical cap $z = \sqrt{a^2 - x^2 - y^2}$ which lies inside the cylinder $x^2 + y^2 = b^2$, $0 < b < a$, with upward-pointing normal, and $\omega = xz \, dy \, dz + dz \, dx + z \, dx \, dy$.

d) S is the truncated cone $z = 2\sqrt{x^2 + y^2}$, $0 \le z \le 2$, with normal which points away from the z-axis, and $\omega = x \, dy \, dz + ydz \, dx + z^2 \, dx \, dy$.

13.4.4. Suppose that $\psi(B)$ and $\phi(E)$ are C^p surfaces and that $\psi = \phi \circ \tau$, where τ is a C^1 function from B onto E.

a) If (ψ, B) and (ϕ, E) are smooth, and τ is 1–1 with $\Delta_\tau > 0$ on B, prove for all continuous $\mathbf{F} : \phi(E) \to \mathbf{R}^3$ that

$$\int_E \mathbf{F}(\phi(u, v)) \cdot N_\phi(u, v) \, d(u, v) = \int_B \mathbf{F}(\psi(s, t)) \cdot N_\psi(s, t) \, d(s, t).$$

*b) Suppose that Z is a closed subset of B of area zero, that (ψ, B) is smooth off Z, and that τ is 1–1 with $\Delta_\tau > 0$ on $B^o \backslash Z$. Prove for all continuous $\mathbf{F} : \phi(E) \to \mathbf{R}^3$ that

$$\int_E \mathbf{F}(\phi(u, v)) \cdot N_\phi(u, v) \, d(u, v) = \int_B \mathbf{F}(\psi(s, t)) \cdot N_\psi(s, t) \, d(s, t).$$

13.4.5. Let E be the solid tetrahedron bounded by $x = 0$, $y = 0$, $z = 0$, and $x + y + z = 1$, and suppose that its topological boundary, $T = \partial E$, is oriented with outward-pointing normal. Prove for all C^1 functions $P, Q, R : E \to \mathbf{R}$ that

$$\iint_{\partial E} P \, dy \, dz + Q \, dz \, dx + R \, dx \, dy = \iiint_E (P_x + Q_y + R_z) \, dV.$$

13.4.6. Let **T** be the topological boundary of the tetrahedron in Exercise 13.4.5, with outward-pointing normal, and S be the surface obtained by taking away the slanted face from **T** (i.e., S has three triangular faces, one each in the planes $x = 0$, $y = 0$, $z = 0$). If ∂S is oriented positively, prove for all C^1 functions $P, Q, R : S \to \mathbf{R}$ that

$$\int_{\partial S} P \, dx + Q \, dy + R \, dz = \iint_S (R_y - Q_z) \, dy \, dz + (P_z - R_x) \, dz \, dx$$
$$+ (Q_x - P_y) \, dx \, dy.$$

13.4.7. Suppose that S is a smooth surface.

a) Show that there exist smooth parametrizations $(\boldsymbol{\phi}_j, E_j)$ of portions of S such that $S = \cup_{j=1}^N \boldsymbol{\phi}_j(E_j)$.

b) Show that there exist nonoverlapping surfaces S_j with smooth parametrizations such that $S = \cup_{j=1}^N S_j$. What happens if S is orientable?

13.5 THEOREMS OF GREEN AND GAUSS

Recall by the Fundamental Theorem of Calculus that if f is a C^1 function, then

$$f(b) - f(a) = \int_a^b f'(t) \, dt.$$

Thus the integral of the derivative f' on $[a, b]$ is completely determined by the values f takes on the topological boundary $\{a, b\}$ of $[a, b]$.

In the next two sections we shall obtain analogues of this theorem for functions $\mathbf{F} : \Omega \to \mathbf{R}^m$, where Ω is a surface or an m-dimensional region, $m = 2$ or 3. Namely, we shall show that the integral of a "derivative" of $\mathbf{F}$ on Ω is completely determined by the values $\mathbf{F}$ takes on the "boundary" of Ω. Which "derivative" and "boundary" we use depends on whether Ω is a surface or an m-dimensional region and whether $m = 2$ or 3.

Our first fundamental theorem applies to two-dimensional regions in the plane.

13.50 Theorem. [GREEN'S THEOREM].
Let E be a two-dimensional region whose topological boundary ∂E is a piecewise smooth C^1 curve oriented positively. If $P, Q : E \to \mathbf{R}$ are C^1 and $\mathbf{F} = (P, Q)$, then

$$\int_{\partial E} \mathbf{F} \cdot \mathbf{T} \, ds = \iint_E \left(\frac{\partial Q}{\partial x} - \frac{\partial P}{\partial y} \right) dA.$$

PROOF FOR SPECIAL REGIONS. We will prove Green's Theorem when E is a finite union of nonoverlapping regions each of which is of types I and II. For a proof of Green's Theorem as stated, see Spivak [12].

Suppose first that E is a single region of types I and II. Write the integral on the left in differential notation,

$$\int_{\partial E} P\,dx + Q\,dy = \int_{\partial E} P\,dx + \int_{\partial E} Q\,dy =: I_1 + I_2.$$

We evaluate I_1 first. Since E is of type I, choose continuous functions $f, g :$ $[a, b] \to \mathbf{R}$ such that

$$E = \{(x, y) \in \mathbf{R}^2 : a \leq x \leq b,\ f(x) \leq y \leq g(x)\}.$$

Thus ∂E has a top $y = g(x)$, a bottom $y = f(x)$, and (possibly) one or two vertical sides (see Figure 13.16).

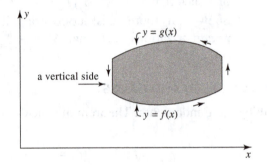

FIGURE 13.16

Since the positive orientation is counterclockwise, the trivial parametrization of the top is $y = g(x)$, where x runs from b to a, and of the bottom is $y = f(x)$, where x runs from a to b. Since $dx = 0$ on any vertical curve, the contribution of the vertical sides to I_1 is zero. Thus it follows from Definition 13.21 and the one-dimensional Fundamental Theorem of Calculus that

$$I_1 = \int_{\partial E} P\,dx = \int_a^b P(x, f(x))\,dx + \int_b^a P(x, g(x))\,dx$$

$$= -\int_a^b (P(x, g(x)) - P(x, f(x))\,dx$$

$$= -\int_a^b \int_{f(x)}^{g(x)} \frac{\partial P}{\partial y}(x, y)\,dy\,dx = -\iint_E \frac{\partial P}{\partial y}\,dA.$$

Since E is of type II, a similar argument establishes

$$I_2 = \int_{\partial E} Q\,dy = \iint_E \frac{\partial Q}{\partial x}\,dA.$$

[Here, we have changed parametrizations of ∂E, for example, replaced $y = f(x)$ by $x = f^{-1}(y)$. The value of the oriented integral does not change because these

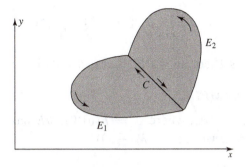

FIGURE 13.17

parametrizations are orientation equivalent—see Exercise 13.2.6.] Adding I_1 and I_2 completes the proof when E is of type I and II.

Since disjoint regions have disjoint boundaries, it remains to verify Green's Theorem for two-dimensional regions which can be divided into a finite number of regions each of which is of types I and II (see Theorem 12.23). By induction, it is enough to examine a region E which can be divided into two contiguous regions (see Figure 13.17). Notice that although E is not of type II, the regions E_1 and E_2 are both of types I and II. Applying Theorem 13.50 to each piece, we find

$$\iint_E \left(\frac{\partial Q}{\partial x} - \frac{\partial P}{\partial y} \right) dA = \iint_{E_1} \left(\frac{\partial Q}{\partial x} - \frac{\partial P}{\partial y} \right) dA + \iint_{E_2} \left(\frac{\partial Q}{\partial x} - \frac{\partial P}{\partial y} \right) dA$$

$$= \int_{\partial E_1} \mathbf{F} \cdot \mathbf{T} \, ds + \int_{\partial E_2} \mathbf{F} \cdot \mathbf{T} \, ds$$

$$= \int_{\partial E} \mathbf{F} \cdot \mathbf{T} \, ds + \int_{C \cap \partial E_1} \mathbf{F} \cdot \mathbf{T} \, ds + \int_{C \cap \partial E_2} \mathbf{F} \cdot \mathbf{T} \, ds,$$

where C is the common border between E_1 and E_2. Since ∂E_1 and ∂E_2 are oriented in the counterclockwise direction, the orientation of $C \cap \partial E_1$ is different from the orientation of $C \cap \partial E_2$. Since a change of orientation changes the sign of the integral, the integrals along C drop out. The end result is the integral of $\mathbf{F} \cdot \mathbf{T} \, ds$ on ∂E, as promised. ∎

Green's Theorem is often used to avoid tedious parametrizations.

13.51 EXAMPLE.

Find $\int_{\partial E} \mathbf{F} \cdot \mathbf{T} \, ds$, where $E = [0, 2] \times [1, 3]$, ∂E has the counterclockwise orientation, and $\mathbf{F}(x, y) = (xy, x^2 + y^2)$.

Solution. Since ∂E has four sides, direct evaluation requires four separate parametrizations. However, by Green's Theorem,

$$\int_{\partial E} \mathbf{F} \cdot \mathbf{T} \, ds = \int_0^2 \int_1^3 (2x - x) \, dy \, dx = 4. \qquad \blacksquare$$

Green's Theorem is also used to avoid difficult integrals.

13.52 EXAMPLE.

Find $\int_{\partial E} \mathbf{F} \cdot \mathbf{T} ds$, where $E = B_1(0, 0)$, ∂E has the clockwise orientation, and $\mathbf{F} = (xy^2, \arctan(\log(y + 3)) - x))$.

Solution. The second component of $\mathbf{F}$ looks tough to integrate. However, by Green's Theorem,

$$\int_{\partial E} \mathbf{F} \cdot \mathbf{T} \, ds = - \iint_{B_1(0,0)} (-1 - 2xy) \, dx \, dy$$

$$= \int_0^{2\pi} \int_0^1 (1 + 2r^2 \cos \theta \sin \theta) r \, dr \, d\theta = \pi.$$

(*Note*: The minus sign appears because ∂E is oriented in the clockwise direction.) $\qquad \blacksquare$

By Green's Theorem, the "derivative" used to obtain a fundamental theorem of calculus for two-dimensional regions in $\mathbf{R}^2$ is $Q_x - P_y$. Here are the "derivatives" which will be used when Ω is a surface in $\mathbf{R}^3$ or a three-dimensional region.

13.53 Definition.

Let E be a subset of $\mathbf{R}^3$ and let $\mathbf{F} = (P, Q, R) : E \to \mathbf{R}^3$ be C^1 on E. The *curl* of $\mathbf{F}$ is

$$\text{curl } \mathbf{F} = \left(\frac{\partial R}{\partial y} - \frac{\partial Q}{\partial z}, \frac{\partial P}{\partial z} - \frac{\partial R}{\partial x}, \frac{\partial Q}{\partial x} - \frac{\partial P}{\partial y} \right),$$

and the *divergence* of $\mathbf{F}$ is

$$\text{div } \mathbf{F} = \frac{\partial P}{\partial x} + \frac{\partial Q}{\partial y} + \frac{\partial R}{\partial z}.$$

Notice that if $\mathbf{F} = (P, Q, 0)$, where P and Q are as in Green's Theorem, then curl $\mathbf{F} \cdot \mathbf{k} = Q_x - P_y$ is the derivative used for Green's Theorem.

These derivatives take on a more easily remembered form by using the notation

$$\nabla = \left(\frac{\partial}{\partial x}, \frac{\partial}{\partial y}, \frac{\partial}{\partial z} \right).$$

Indeed, curl $\mathbf{F} = \nabla \times \mathbf{F}$ and div $\mathbf{F} = \nabla \cdot \mathbf{F}$.

If E is a three-dimensional region whose topological boundary is a piecewise smooth orientable surface, then the *positive orientation* on ∂E is determined by the unit normal which points away from E^o. If E is convex, this means that $\mathbf{n}$ points outward. This is not the case, however, when E has interior "bubbles." For example, if $E = \{\mathbf{x} : a \le \|\mathbf{x}\| \le b\}$ for some $a > 0$, then $\mathbf{n}$ points away from the origin on $\{\mathbf{x} : \|\mathbf{x}\| = b\}$ but toward the origin on $\{\mathbf{x} : \|\mathbf{x}\| = a\}$.

Our next fundamental theorem applies to the case when Ω is a three-dimensional region. This result is also called the *Gauss's Theorem*.

13.54 Theorem.   [THE DIVERGENCE THEOREM].
Let E be a three-dimensional region whose topological boundary ∂E is a piecewise smooth C^1 surface oriented positively. If $\mathbf{F} : E \to \mathbf{R}^3$ is C^1 on E, then

$$\iint_{\partial E} \mathbf{F} \cdot \mathbf{n} \, d\sigma = \iiint_E \operatorname{div} \mathbf{F} \, dV.$$

PROOF FOR SPECIAL REGIONS. We will prove the Divergence Theorem when E is a finite union of nonoverlapping three-dimensional regions each of which is of types I, II, and III. For a proof of the Divergence Theorem as stated, see Spivak [12].

Suppose first that E is a single region of types I, II, and III. Let $\mathbf{F} = (P, Q, R)$ and write the surface integral in differential form:

$$\iint_{\partial E} \mathbf{F} \cdot \mathbf{n} \, d\sigma = \iint_{\partial E} P \, dy \, dz + \iint_{\partial E} Q \, dz \, dx + \iint_{\partial E} R \, dx \, dy =: I_1 + I_2 + I_3.$$

We evaluate I_3 first.

Since E is of type I, there exist a two-dimensional region $B \subset \mathbf{R}^2$ and continuous functions $f, g : B \to \mathbf{R}$ such that

$$E = \{(x, y, z) \in \mathbf{R}^3 : (x, y) \in B, \ f(x, y) \le z \le g(x, y)\}.$$

Thus ∂E has a top $z = g(x, y)$, a bottom $z = f(x, y)$, and (possibly) a vertical side (see Figure 13.18). Any normal to ∂E on the vertical side is parallel to the xy-plane. Since $dx \, dy$ is the third component of a normal to ∂E, it must be zero on the vertical portion. Therefore, I_3 can be evaluated by integrating over the top and bottom of ∂E. Notice that, by hypothesis, the unit normal on the bottom portion points downward and the unit normal on the top portion points upward. By using trivial parametrizations and Theorem 5.28 (the Fundamental Theorem of Calculus), we obtain

$$I_3 = \iint_{\partial E} R \, dx \, dy = \int_B (R(x, y, g(x, y)) - R(x, y, f(x, y))) \, d(x, y)$$

$$= \int_B \int_{f(x,y)}^{g(x,y)} \frac{\partial R}{\partial z}(x, y, z) \, dz \, d(x, y) = \iiint_E \frac{\partial R}{\partial z} \, dV.$$

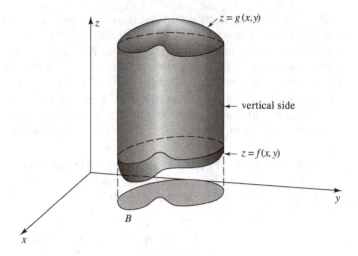

FIGURE 13.18

Similarly, since E is of type II,

$$I_2 = \iiint_E \frac{\partial Q}{\partial y} \, dV,$$

and since E is of type III,

$$I_1 = \iiint_E \frac{\partial P}{\partial x} \, dV.$$

Adding $I_1 + I_2 + I_3$ verifies the theorem.

Since disjoint regions have disjoint boundaries, it remains to verify the Divergence Theorem for three-dimensional regions which can be divided into a finite number of regions each of which is of types I, II, and III (see Theorem 12.23). But if $E = E_1 \cup E_2$ share a common boundary, then

$$\iiint_E \text{div } \mathbf{F} \, dV = \iiint_{E_1} \text{div } \mathbf{F} \, dV + \iiint_{E_2} \text{div } \mathbf{F} \, dV$$

$$= \iint_{\partial E} \mathbf{F} \cdot \mathbf{n} \, d\sigma + \iint_{S \cap \partial E_1} \mathbf{F} \cdot \mathbf{n} \, d\sigma + \iint_{S \cap \partial E_2} \mathbf{F} \cdot \mathbf{n} \, d\sigma,$$

where S is the common surface between E_1 and E_2. Since E_1 and E_2 have outward-pointing normals, the orientation of $S \cap \partial E_1$ is different from the orientation of $S \cap \partial E_2$, and the integrals over S cancel each other out. ∎

The next two examples show that, like Green's Theorem, the Divergence Theorem can be used to avoid difficult integrals and tedious parametrizations.

13.55 EXAMPLE.

Use Theorem 13.54 to evaluate $\iint_S \mathbf{F} \cdot \mathbf{n}\, d\sigma$, where S is the topological boundary of the solid $E = \{(x, y, z) : x^2 + y^2 \le z \le 1\}$, $\mathbf{n}$ is the outward-pointing normal, and $\mathbf{F}(x, y, z) = (2x + z^2, x^5 + z^7, \cos(x^2) + \sin(y^3) - z^2)$.

Solution. Since div $\mathbf{F} = 2 - 2z$, it follows from the Divergence Theorem that

$$\iint_S \mathbf{F} \cdot \mathbf{n}\, d\sigma = \iiint_E (2 - 2z)\, dV = 2 \int_0^{2\pi} \int_0^1 \int_{r^2}^1 (1 - z)r\; dz\, dr\, d\theta = \frac{\pi}{3}. \quad \blacksquare$$

13.56 EXAMPLE.

Evaluate $\iint_{\partial Q} \mathbf{F} \cdot \mathbf{n}\, d\sigma$, where Q is the unit cube $[0, 1] \times [0, 1] \times [0, 1]$, $\mathbf{n}$ is the outward-pointing normal, and $\mathbf{F}(x, y, z) = (2x - z, x^2 y, -xz^2)$.

Solution. Since ∂Q has six sides, direct evaluation of this integral requires six separate integrals. However, by the Divergence Theorem,

$$\iint_{\partial Q} \mathbf{F} \cdot \mathbf{n}\, d\sigma = \int_0^1 \int_0^1 \int_0^1 (2 + x^2 - 2xz)\, dx\, dy\, dz = \frac{11}{6}. \quad \blacksquare$$

These definitions and results take on new meaning when examined in the context of fluid flow. When $\mathbf{F}$ represents the flow of an incompressible fluid near a point $\mathbf{a}$, curl $\mathbf{F}(\mathbf{a})$ measures the tendency of the fluid to swirl in a counterclockwise direction about $\mathbf{a}$ (see Exercise 13.6.6), and div $\mathbf{F}(\mathbf{a})$ measures the tendency of the fluid to spread out from $\mathbf{a}$ (see Exercise 13.5.7). (This explains the etymology of the words *curl* and *divergence*.) For example, if $\mathbf{F}(x, y, z) = (x, y, z)$, then the fluid is not swirling at all, but spreading straight out from the origin. Accordingly, curl $\mathbf{F} = 0$ and div $\mathbf{F} = 3$. On the other hand, if $G(x, y, z) = (y, -x, 0)$, then the fluid is swirling around in a circular motion about the origin. Accordingly, curl $G = (0, 0, -1)$ but div $G = 0$. Note the minus sign in the component of curl G. This fluid swirls about the origin in a clockwise direction, so runs against counterclockwise motion.

When the fluid flows over a two-dimensional region $E \subset \mathbf{R}^2$, the integral of $\mathbf{F} \cdot \mathbf{T}\, ds$ over C represents the circulation of the fluid around C in the direction of $\mathbf{T}$ (see the comments following Definition 13.21). Thus Green's Theorem tells us that the circulation of a fluid around ∂E in the direction of the tangent is determined by how strongly the fluid swirls inside E. When $\mathbf{F}$ represents the flow of an incompressible fluid through a three-dimensional region $E \subset \mathbf{R}^3$ and $S = \partial E$, the integral $\iint_S \mathbf{F} \cdot \mathbf{n}$ represents the flux of the fluid across the surface S (see the comments following Definition 13.43). Thus the Divergence Theorem tells us that the flux of the fluid across $S = \partial E$ is determined by how strongly the fluid is spreading out inside E.

We close this section by admitting that the interpretations of curl and divergence given above are imperfect at best. For example, the vector field $\mathbf{F}(x, y, z) = (0, z, 0)$ has curl $(-1, 0, 0)$. Here the fluid is shearing in layers with flow parallel to the xy-plane in the direction of the positive y-axis when $z > 0$. Although the fluid is not swirling, it does tend to rotate a stick placed in the fluid parallel to the z-axis [e.g., the line segment $\{(0, 1, z) : 0 \le z \le 1\}$] because more force is applied to the top than the bottom. This tendency toward rotation is reflected by the value of the curl. (Notice that the rotation is clockwise and the curl has a negative first component.)

EXERCISES

13.5.1. For each of the following, evaluate $\int_C \mathbf{F} \cdot \mathbf{T}\, ds$.

    a) C is the topological boundary of the two-dimensional region in the first quadrant bounded by $x = 0$, $y = 0$, and $y = \sqrt{4 - x^2}$, oriented in the counterclockwise direction, and $\mathbf{F}(x, y) = (\sin(\sqrt{x^3 - x^2}), xy)$.

    b) C is the perimeter of the rectangle with vertices $(0, 0)$, $(2, 0)$, $(0, 3)$, $(2, 3)$, oriented in the counterclockwise direction, and $\mathbf{F}(x, y) = (e^y, \log(x + 1))$.

    c) $C = C_1 \cup C_2$, where $C_1 = \partial B_1(0, 0)$ oriented in the counterclockwise direction, $C_2 = \partial B_2(0, 0)$ oriented in the clockwise direction, and $\mathbf{F}(x, y) = (f(x^2 + y^2), xy^2)$, where f is a C^1 function on $[1, 2]$.

13.5.2. For each of the following, evaluate $\int_C \omega$.

    a) C is the topological boundary of the rectangle $[a, b] \times [c, d]$, oriented in the counterclockwise direction, and $\omega = (f(x) + y)\, dx + xy\, dy$, where $f : [0, 1] \to \mathbf{R}$ is any continuous function.

    b) C is the topological boundary of the two-dimensional region bounded by $y = x^2$ and $y = x$, oriented in the clockwise direction, and $\omega = yf(x)\, dx + (x^2 + y^2)\, dy$, where $f : [0, 1] \to \mathbf{R}$ is C^1 and satisfies $\int_0^1 xf(x)\, dx = \int_0^1 x^2 f(x)\, dx$.

    c) C is the topological boundary of a two-dimensional region E which satisfies the hypotheses of Green's Theorem, oriented positively, and $\omega = e^x \sin y\, dy - e^x \cos y\, dx$.

13.5.3. For each of the following, evaluate $\iint_S \mathbf{F} \cdot \mathbf{n}\, d\sigma$, where $\mathbf{n}$ is the outward-pointing normal.

    a) S is the topological boundary of the rectangle $[0, 1] \times [0, 2] \times [0, 3]$ and $\mathbf{F}(x, y, z) = (x + e^z, y + e^z, e^z)$.

    b) S is the truncated cylinder $x^2 + y^2 = 1$, $0 \le z \le 1$ together with the disks $x^2 + y^2 \le 1$, $z = 0, 1$, and $\mathbf{F}(x, y, z) = (x^2, y^2, z^2)$.

    c) S is the topological boundary of E, where $E \subset \mathbf{R}^3$ is bounded by $z = 2 - x^2$, $z = x^2$, $y = 0$, $z = y$, and $\mathbf{F}(x, y, z) = (x + f(y, z), y + g(x, z), z + h(x, y))$ and $f, g, h : \mathbf{R}^2 \to \mathbf{R}$ are C^1.

    d) S is the ellipsoid $x^2/a^2 + y^2/b^2 + z^2/c^2 = 1$ and $\mathbf{F}(x, y, z) = (x|y|, y|z|, z|x|)$.

13.5.4. For each of the following, find $\iint_S \omega$, where $\mathbf{n}$ is the outward-pointing normal.

a) S is the topological boundary of the three-dimensional region enclosed by $y = x^2$, $z = 0$, $z = 1$, $y = 4$, and $\omega = xyz \, dy \, dz + (x^2 + y^2 + z^2) \, dz \, dx + (x + y + z) \, dx \, dy$.

b) S is the truncated hyperboloid of one sheet $x^2 - y^2 + z^2 = 1$, $0 \le y \le 1$, together with the disks $x^2 + z^2 \le 1$, $y = 0$, and $x^2 + z^2 \le 2$, $y = 1$, and $\omega = xy|z| \, dy \, dz + x^2|z| \, dz \, dx + (x^3 + y^3) \, dx \, dy$.

c) S is the topological boundary of E, where $E \subset \mathbf{R}^3$ is bounded by the surfaces $x^2 + y + z^2 = 4$ and $4x + y + 2z = 5$, and $\omega = (x + y^2 + z^2) \, dy \, dz + (x^2 + y + z^2) \, dz \, dx + (x^2 + y^2 + z) \, dx \, dy$.

13.5.5. a) Prove that if E is a Jordan region whose topological boundary is a piecewise smooth curve oriented in the counterclockwise direction, then

$$\text{Area } (E) = \frac{1}{2} \int_{\partial E} x \, dy - y \, dx.$$

b) Find the area enclosed by the loop in the Folium of Descartes; that is, by

$$\phi(t) = \left(\frac{3t}{1 + t^3}, \frac{3t^2}{1 + t^3} \right), \qquad t \in [0, \infty).$$

c) Find an analogue of part a) for the volume of a Jordan region E in $\mathbf{R}^3$.

d) Compute the volume of the torus with radii $a > b$ (see Example 13.32).

13.5.6. a) Show that Green's Theorem does not hold if continuity of P, Q is relaxed at one point in E. [*Hint*: Consider $P = y/(x^2 + y^2)$, $Q = -x/(x^2 + y^2)$, and $E = B_1(0, 0)$.]

b) Show that the Divergence Theorem does not hold if continuity of $\mathbf{F}$ is relaxed at one point in E.

13.5.7. **This exercise is used in Section 13.6.** Suppose that V is a nonempty, open set in $\mathbf{R}^3$ and that $\mathbf{F} : V \to \mathbf{R}^3$ is C^1. Prove that

$$\text{div } \mathbf{F}(\mathbf{x}_0) = \lim_{r \to 0+} \frac{1}{\text{Vol}(B_r(\mathbf{x}_0))} \iint_{\partial B_r(\mathbf{x}_0)} \mathbf{F} \cdot \mathbf{n} \, d\sigma$$

for each $\mathbf{x}_0 \in V$, where $\mathbf{n}$ is the outward-pointing normal of $B_r(\mathbf{x}_0)$.

13.5.8. Let $\mathbf{F}, \mathbf{G} : \mathbf{R}^3 \to \mathbf{R}^3$ and $f : \mathbf{R}^3 \to \mathbf{R}$ be differentiable. Prove the following analogues of the Sum and Product Rules for the "derivatives" curl and divergence.

a) $\qquad\qquad \nabla \times (\mathbf{F} + \mathbf{G}) = (\nabla \times \mathbf{F}) + (\nabla \times \mathbf{G})$

b) $\qquad\qquad \nabla \times (f\mathbf{F}) = f(\nabla \times \mathbf{F}) + (\nabla f \times \mathbf{F})$

c) $$\nabla \cdot (f\mathbf{F}) = \nabla f \cdot \mathbf{F} + f \cdot (\nabla \cdot \mathbf{F})$$

d) $$\nabla \cdot (\mathbf{F} + \mathbf{G}) = \nabla \cdot \mathbf{F} + \nabla \cdot \mathbf{G}$$

e) $$\nabla \cdot (\mathbf{F} \times \mathbf{G}) = (\nabla \times \mathbf{F}) \cdot \mathbf{G} - (\nabla \times \mathbf{G}) \cdot \mathbf{F}$$

13.5.9. **This exercise is used in Section 13.6.** Let $E \subset \mathbf{R}^3$. Recall that the *gradient* of a $\mathcal{C}^1$ function $f : E \to \mathbf{R}$ is defined by

$$\text{grad } f := \nabla f := (f_x, f_y, f_z).$$

a) Prove that if f is $\mathcal{C}^2$ at $\mathbf{x}_0$, then curl grad $f(\mathbf{x}_0) = 0$.
b) If $\mathbf{F} : E \to \mathbf{R}^3$ is $\mathcal{C}^1$ on E and $\mathcal{C}^2$ at $\mathbf{x}_0 \in E$, prove that div curl $\mathbf{F}(\mathbf{x}_0) = 0$.
c) Suppose that E satisfies the hypotheses of the Divergence Theorem and that $f : E \to \mathbf{R}$ is a $\mathcal{C}^2$ function which is harmonic on E (see Exercise 13.5.10d). If $\mathbf{F} = \text{grad } f$ on E, prove that

$$\iint_{\partial E} f\mathbf{F} \cdot \mathbf{n} \, d\sigma = \iiint_E \|\mathbf{F}\|^2 \, dV.$$

13.5.10. Let E be a set in $\mathbf{R}^m$. For each $u : E \to \mathbf{R}$ which has second-order partial derivatives on E, *Laplace's equation* is defined by

$$\Delta u := \sum_{j=1}^{m} \frac{\partial^2 u}{\partial x_j^2}.$$

a) Show that if u is $\mathcal{C}^2$ on E, then $\Delta u = \nabla \cdot (\nabla u)$ on E.
b) [GREEN'S FIRST IDENTITY]. Show that if $E \subset \mathbf{R}^3$ satisfies the hypotheses of the Divergence Theorem, then

$$\iiint_E (u\Delta v + \nabla u \cdot \nabla v) \, dV = \iint_{\partial E} u\nabla v \cdot \mathbf{n} \, d\sigma$$

for all $\mathcal{C}^2$ functions $u, v : E \to \mathbf{R}$.
c) [GREEN'S SECOND IDENTITY]. Show that if $E \subset \mathbf{R}^3$ satisfies the hypotheses of the Divergence Theorem, then

$$\iiint_E (u\Delta v - v\Delta u) \, dV = \iint_{\partial E} (u\nabla v - v\nabla u) \cdot \mathbf{n} \, d\sigma$$

for all $\mathcal{C}^2$ functions $u, v : E \to \mathbf{R}$.
d) A function $u : E \to \mathbf{R}$ is said to be *harmonic* on E if and only if u is $\mathcal{C}^2$ on E and $\Delta u(\mathbf{x}) = 0$ for all $\mathbf{x} \in E$. Suppose that E is a nonempty open region in $\mathbf{R}^3$ which satisfies the hypotheses of the Divergence Theorem. If u is harmonic on E, u is continuous on $\overline{E}$, and $u = 0$ on ∂E, prove that $u = 0$ on $\overline{E}$.

e) Suppose that V is open and nonempty in $\mathbf{R}^2$, that u is C^2 on V, and that u is continuous on $\overline{V}$. Prove that u is harmonic on V if and only if

$$\int_{\partial E} (u_x \, dy - u_y \, dx) = 0$$

for all two-dimensional regions $E \subset V$ which satisfy the hypotheses of Green's Theorem.

13.6 STOKES'S THEOREM

Our final fundamental theorem applies to surfaces in $\mathbf{R}^3$ whose boundaries are curves.

13.57 Theorem.   [STOKES'S THEOREM].
Let S be an oriented, piecewise smooth C^2 surface in $\mathbf{R}^3$ with unit normal $\mathbf{n}$. If the boundary ∂S is a piecewise smooth C^1 curve oriented positively and $\mathbf{F}$: $S \to \mathbf{R}^3$ is C^1, then

$$\int_{\partial S} \mathbf{F} \cdot \mathbf{T} \, ds = \iint_S \operatorname{curl} \mathbf{F} \cdot \mathbf{n} \, d\sigma.$$

PROOF FOR SPECIAL REGIONS. We will prove Stokes's Theorem when E is a finite union of nonoverlapping explicit C^2 surfaces which lie over "Green's regions" (i.e., two-dimensional regions which satisfy the hypotheses of Green's Theorem). For a proof of Stokes's Theorem as stated, see Spivak [12].

Suppose first that S is a single explicit C^2 surface which lies over a "Green's region" E. Let $\mathbf{F} = (P, Q, R)$ be C^1 on S and write the line integral in differential notation:

$$\int_{\partial S} \mathbf{F} \cdot \mathbf{T} \, ds = \int_{\partial S} P \, dx + Q \, dy + R \, dz.$$

Without loss of generality, suppose that S is determined by $z = f(x, y)$, $(x, y) \in E$, where $f : E \to \mathbf{R}$ is a C^2 function, and that S is oriented with the upward-pointing normal. Thus $\mathbf{n} = N/\|N\|$, where $N = (-f_x, -f_y, 1)$.

Let $(g(t), h(t))$, $t \in [a, b]$, be a piecewise smooth parametrization of ∂E oriented in the counterclockwise direction. Then

$$\boldsymbol{\phi}(t) = (g(t), h(t), f(g(t), h(t))), \qquad t \in [a, b],$$

is a piecewise smooth parametrization of ∂S which is oriented positively (see Figure 13.19). If $x = g(t)$, $y = h(t)$, and $z = f(g(t), h(t))$, then $dx = g'(t) \, dt$, $dy = h'(t) \, dt$, and

$$dz = \frac{\partial z}{\partial x} \, dx + \frac{\partial z}{\partial y} \, dy.$$

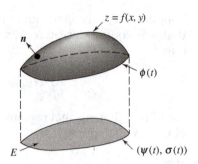

FIGURE 13.19

Thus, by definition,

$$\int_{\partial S} P\, dx + Q\, dy + R\, dz = \int_{\partial E} (P + R\frac{\partial z}{\partial x})\, dx + (Q + R\frac{\partial z}{\partial y})\, dy. \qquad (10)$$

We shall apply Green's Theorem to this last integral. By the Chain Rule and the Product Rule,

$$\frac{\partial}{\partial x}\left(Q + R\frac{\partial z}{\partial y}\right) = \frac{\partial Q}{\partial x} + \frac{\partial Q}{\partial z}\frac{\partial z}{\partial x} + \frac{\partial R}{\partial x}\frac{\partial z}{\partial y} + \frac{\partial R}{\partial z}\frac{\partial z}{\partial x}\frac{\partial z}{\partial y} + R\frac{\partial^2 z}{\partial x\, \partial y}$$

and

$$\frac{\partial}{\partial y}\left(P + R\frac{\partial z}{\partial x}\right) = \frac{\partial P}{\partial y} + \frac{\partial P}{\partial z}\frac{\partial z}{\partial y} + \frac{\partial R}{\partial y}\frac{\partial z}{\partial x} + \frac{\partial R}{\partial z}\frac{\partial z}{\partial y}\frac{\partial z}{\partial x} + R\frac{\partial^2 z}{\partial y\, \partial x}.$$

Since $z = f(x, y)$ is C^2, the mixed second-order partial derivatives above are equal. Therefore,

$$\frac{\partial}{\partial x}\left(Q + R\frac{\partial z}{\partial y}\right) - \frac{\partial}{\partial y}\left(P + R\frac{\partial z}{\partial x}\right)$$

$$= \left(\frac{\partial R}{\partial y} - \frac{\partial Q}{\partial z}\right)\left(-\frac{\partial z}{\partial x}\right) + \left(\frac{\partial P}{\partial z} - \frac{\partial R}{\partial x}\right)\left(-\frac{\partial z}{\partial y}\right) + \left(\frac{\partial Q}{\partial x} - \frac{\partial P}{\partial y}\right)$$

$$= \text{curl } \mathbf{F} \cdot N.$$

Hence, it follows from (10), Green's Theorem, and (9) that

$$\int_{\partial S} \mathbf{F} \cdot \mathbf{T}\, ds = \int_E \text{curl } \mathbf{F} \cdot N\, d(x, y) = \iint_S \text{curl } \mathbf{F} \cdot \mathbf{n}\, d\sigma.$$

Stokes's Theorem for finite unions of pairwise disjoint explicit C^2 surfaces which lie over Green's regions follows immediately. If the surfaces are contiguous, the common boundaries (as in the proofs of Green and Gauss) appear twice, each time in a different orientation and, hence, cancel each other out. ∎

Stokes's Theorem can be used to replace complicated line integrals by simple surface integrals.

13.58 EXAMPLE.

Compute $\int_C \mathbf{F} \cdot \mathbf{T}\, ds$, where C is the circle $x^2 + z^2 = 1$, $y = 0$, oriented in the counterclockwise direction when viewed from far out on the y-axis, and $\mathbf{F}(x, y, z) = (x^2 z + \sqrt{x^3 + x^2 + 2}, xy, xy + \sqrt{z^3 + z^2 + 2})$.

Solution. Since curl $\mathbf{F} = (x, x^2 - y, y)$, using Stokes's Theorem is considerably easier than trying to integrate $\mathbf{F} \cdot \mathbf{T}\, ds$ directly. Let S be the disk $x^2 + z^2 \leq 1$, $y = 0$, and notice that $\partial S = C$. Since C is oriented in the counterclockwise direction, the normal to S must point toward the positive y-axis [i.e., $\mathbf{n} = (0, 1, 0)$]. Thus curl $\mathbf{F} \cdot \mathbf{n} = x^2 - y = x^2$ on S, and Stokes's Theorem implies

$$\int_C \mathbf{F} \cdot \mathbf{T}\, ds = \iint_S x^2\, dA = \int_0^{2\pi} \int_0^1 r^3 \cos^2 \theta\, dr\, d\theta = \frac{\pi}{4}. \qquad \blacksquare$$

In Example 13.58, we could have chosen any surface S whose boundary is C. Thus Stokes's Theorem can also be used to replace complicated surface integrals by simpler ones.

13.59 EXAMPLE.

Find $\iint_S \text{curl}\, \mathbf{F} \cdot \mathbf{n}\, d\sigma$, where S is the semiellipsoid $9x^2 + 4y^2 + 36z^2 = 36$, $z \geq 0$, $\mathbf{n}$ is the upward-pointing normal, and

$$\mathbf{F}(x, y, z) = (\cos x \sin z + xy, x^3, e^{x^2 + z^2} - e^{y^2 + z^2} + \tan(xy)).$$

Solution. Let $C = \partial S$. The integral of curl $\mathbf{F} \cdot \mathbf{n}\, d\sigma$ over S and the integral of $\mathbf{F} \cdot \mathbf{T}\, ds$ over C are both complicated. But, by Stokes's Theorem, the integral of $\mathbf{F} \cdot \mathbf{T}\, ds$ over C is the same as the integral of curl $\mathbf{F} \cdot \mathbf{n}\, d\sigma$ over any oriented C^2 surface E satisfying $\partial E = C$. Let E be the two-dimensional region $9x^2 + 4y^2 \leq 36$. On E, $\mathbf{n} = (0, 0, 1)$. Thus we only need the third component of curl $\mathbf{F}$:

$$(\text{curl}\, \mathbf{F})_3 := \frac{\partial}{\partial x}(x^3) - \frac{\partial}{\partial y}(\cos x \sin z + xy) = 3x^2 - x.$$

Therefore,

$$\iint_S \text{curl}\, \mathbf{F} \cdot \mathbf{n}\, d\sigma = \int_E (3x^2 - x)\, d(x, y).$$

Let $x = 2r \cos \theta$ and $y = 3r \sin \theta$. By a change of variables,

$$\int_E (3x^2 - x)\, d(x, y) = \int_0^{2\pi} \int_0^1 (12r^2 \cos^2 \theta - 2r \cos \theta) 6r\, dr\, d\theta = 18\pi. \qquad \blacksquare$$

Stokes's Theorem can also be used to replace complicated surface integrals by simple line integrals.

13.60 EXAMPLE.

Let S be the union of the truncated paraboloid $z = x^2 + y^2$, $0 \le z \le 1$, and the truncated cylinder $x^2 + y^2 = 1$, $1 \le z \le 3$. Compute

$$\iint_S \mathbf{F} \cdot \mathbf{n}\, d\sigma,$$

where $\mathbf{n}$ is the outward-pointing normal and $\mathbf{F}(x, y, z) = (x + z^2, 0, -z - 3)$.

Solution. The boundary of S is $x^2 + y^2 = 1$, $z = 3$. To use Stokes's Theorem, we must find a function $\mathbf{G} = (P, Q, R) : S \to \mathbf{R}^3$ such that $\operatorname{curl} \mathbf{G} = \mathbf{F}$; that is, such that

$$\frac{\partial R}{\partial y} - \frac{\partial Q}{\partial z} = x + z^2, \tag{11}$$

$$\frac{\partial P}{\partial z} - \frac{\partial R}{\partial x} = 0, \tag{12}$$

and

$$\frac{\partial Q}{\partial x} - \frac{\partial P}{\partial y} = -z - 3. \tag{13}$$

Starting with (11), set

$$\frac{\partial Q}{\partial z} = -x \quad \text{and} \quad \frac{\partial R}{\partial y} = z^2. \tag{14}$$

The left side of (14) implies $Q = -xz + g(x, y)$ for some $g : \mathbf{R}^2 \to \mathbf{R}$. Similarly, the right side of (14) leads to $R = z^2 y + h(x, z)$ for some $h : \mathbf{R}^2 \to \mathbf{R}$. Thus $Q_x = -z + g_x$ will solve (13) if we set $g = 0$ and $P_y = 3$; that is, $P = 3y + \sigma(x, z)$ for some $\sigma : \mathbf{R}^2 \to \mathbf{R}$. Hence, $P_z - R_x = \sigma_z - h_x$ will satisfy (12) if $\sigma = h = 0$. Therefore, $P = 3y$, $Q = -xz$ and $R = yz^2$; that is, $\mathbf{G} = (3y, -xz, yz^2)$.

Parametrize ∂S by $\boldsymbol{\phi}(t) = (\sin t, \cos t, 3)$, $t \in [0, 2\pi]$, and observe that

$$(\mathbf{G} \circ \boldsymbol{\phi}) \cdot \boldsymbol{\phi}' = (3\cos t, -3\sin t, 9\cos t) \cdot (\cos t, -\sin t, 0) = 3\cos^2 t + 3\sin^2 t = 3.$$

Consequently, Stokes's Theorem implies

$$\iint_S \mathbf{F} \cdot \mathbf{n}\, d\sigma = \iint_S \operatorname{curl} \mathbf{G} \cdot \mathbf{n}\, d\sigma = \int_{\partial S} \mathbf{G} \cdot \mathbf{T}\, ds = \int_0^{2\pi} 3\, dt = 6\pi. \quad \blacksquare$$

The solution to Example 13.60 involved finding a function $\mathbf{G}$ which satisfied $\operatorname{curl} \mathbf{G} = \mathbf{F}$. This function is not unique. Indeed, we could have begun with

$$\frac{\partial Q}{\partial z} = -z^2 \quad \text{and} \quad \frac{\partial R}{\partial y} = x$$

instead of (14). This leads to a different solution:

$$\widetilde{\mathbf{G}}(x, y, z) = (zy, -(3x + z^3/3), xy).$$

The technique used to solve Example 13.60, however, is perfectly valid. Indeed, by Stokes's Theorem the value of the oriented line integral of $\mathbf{G} \cdot \mathbf{T}$ will be the same for all C^1 functions $\mathbf{G}$ which satisfy $\operatorname{curl} \mathbf{G} = \mathbf{F}$.

This technique works only when the system of partial differential equations $\operatorname{curl} \mathbf{G} = \mathbf{F}$ has a solution $\mathbf{G}$. To avoid searching for something which does not exist, we must be able to discern beforehand whether such a solution exists. To discover how to do this, suppose that $\mathbf{G}$ is a C^2 function which satisfies $\operatorname{curl} \mathbf{G} = \mathbf{F}$ on some set E. Then $\operatorname{div} \mathbf{F} = 0$ on E by Exercise 13.5.9b. Thus the condition $\operatorname{div} \mathbf{F} = 0$ is necessary for existence of a solution $\mathbf{G}$ to $\operatorname{curl} \mathbf{G} = \mathbf{F}$. The following result shows that if E is nice enough, this condition is also sufficient.

13.61 Theorem. *Let Ω be a ball or a rectangle with nonempty interior, and suppose that $\mathbf{F} : \Omega \to \mathbf{R}^3$ is C^1 on Ω. Then the following three statements are equivalent.*

i) *There is a C^2 function $\mathbf{G} : \Omega \to \mathbf{R}^3$ such that $\operatorname{curl} \mathbf{G} = \mathbf{F}$ on Ω.*
ii) *If E and $S = \partial E$ satisfy the hypotheses of the Divergence Theorem and $E \subset \Omega$, then*

$$\iint_S \mathbf{F} \cdot \mathbf{n} \, d\sigma = 0. \tag{15}$$

iii) *The identity $\operatorname{div} \mathbf{F} = 0$ holds everywhere on Ω.*

Proof. If i) holds, then $\operatorname{div} \mathbf{F} = \operatorname{div}(\operatorname{curl} \mathbf{G}) = 0$ since the first-order partial derivatives of $\mathbf{G}$ commute. Thus (15) holds by the Divergence Theorem. (This works for any set Ω.)

If ii) holds, then by the Divergence Theorem and Exercise 13.5.7,

$$\operatorname{div} \mathbf{F}(\mathbf{x}_0) = \lim_{r \to 0+} \frac{1}{\operatorname{Vol}(B_r(\mathbf{x}_0))} \iiint_{B_r(\mathbf{x}_0)} \operatorname{div} \mathbf{F} \, dV$$

$$= \lim_{r \to 0+} \frac{1}{\operatorname{Vol}(B_r(\mathbf{x}_0))} \iint_{\partial B_r(\mathbf{x}_0)} \mathbf{F} \cdot \mathbf{n} \, d\sigma = 0$$

for each $\mathbf{x}_0 \in \Omega^o$. Since $\operatorname{div} \mathbf{F}$ is continuous on Ω, it follows that $\operatorname{div} \mathbf{F} = 0$ everywhere on Ω. (This works for any three-dimensional region Ω.)

Finally, suppose that iii) holds. Let $\mathbf{F} = (p, q, r)$ and suppose for simplicity that $\mathbf{G} = (0, Q, R)$. If $\operatorname{curl} \mathbf{G} = \mathbf{F}$, then

$$R_y - Q_z = p, \quad -R_x = q, \quad Q_x = r. \tag{16}$$

If Ω is a ball, let (x_0, y_0, z_0) be its center; if Ω is a rectangle, let (x_0, y_0, z_0) be any point in Ω. Then given any $(x, y, z) \in \Omega$, the line segment from (x_0, y, z)

to (x, y, z) is a subset of Ω. Hence we can integrate the last two identities in (16) from x_0 to x, obtaining

$$R = -\int_{x_0}^{x} q(u, y, z)\, du + g(y, z) \quad \text{and} \quad Q = \int_{x_0}^{x} r(u, y, z)\, du + h(y, z)$$

for some $g, h : \mathbf{R}^2 \to \mathbf{R}$. Differentiating under the integral sign (Theorem 11.5), and using condition iii), the first identity becomes

$$p = R_y - Q_z = -\int_{x_0}^{x} (q_y(u, y, z) + r_z(u, y, z))\, du + g_y - h_z$$

$$= \int_{x_0}^{x} p_x(u, y, z)\, du + g_y - h_z = p(x, y, z) - p(x_0, y, z) + g_y - h_z.$$

Thus (16) can be solved by $g_y = p(x_0, y, z)$ and $h = 0$; that is,

$$Q = \int_{x_0}^{x} r(u, y, z)\, du \quad \text{and} \quad R = \int_{y_0}^{y} p(x_0, v, z)\, dv - \int_{x_0}^{x} q(u, y, z)\, du. \quad \blacksquare$$

We notice that Theorem 13.61 holds for any three-dimensional region Ω which satisfies the following property: There is a point $(x_0, y_0, z_0) \in \Omega$ such that the line segments $L((x_0, y, z); (x, y, z))$ and $L((x_0, y_0, z); (x_0, y, z))$ are both subsets of Ω for all $(x, y, z) \in \Omega$. However, as the following result shows, Theorem 13.61 is false without some restriction on Ω.

13.62 Remark. *Let* $\Omega = B_1(0, 0, 0) \setminus \{(0, 0, 0)\}$ *and*

$$\mathbf{F}(x, y, z) = \left(\frac{x}{w^{3/2}}, \frac{y}{w^{3/2}}, \frac{z}{w^{3/2}} \right),$$

where $w = w(x, y, z) = x^2 + y^2 + z^2$. *Then* div $\mathbf{F} = 0$ *on* Ω, *but there is no* $\mathbf{G}$ *which satisfies* curl $\mathbf{G} = \mathbf{F}$.

Proof. By definition,

$$\text{div } \mathbf{F} = \frac{-2x^2 + y^2 + z^2}{w^{5/2}} + \frac{x^2 - 2y^2 + z^2}{w^{5/2}} + \frac{x^2 + y^2 - 2z^2}{w^{5/2}} = 0.$$

Let S represent the unit sphere $\partial B_1(0, 0, 0)$ oriented with the outward-pointing normal, and suppose that there is a $\mathbf{G}$ such that curl $\mathbf{G} = \mathbf{F}$. On the one hand, since $\mathbf{F} = (x, y, z) = \mathbf{n}$ on S implies $\mathbf{F} \cdot \mathbf{n} = x^2 + y^2 + z^2 = 1$, we have

$$\iint_S \mathbf{F} \cdot \mathbf{n}\, d\sigma = \iint_S 1\, dA = \sigma(S) = 4\pi. \tag{17}$$

On the other hand, dividing S into the upper hemisphere S_1 and the lower hemisphere S_2, we have by Stokes's Theorem that

$$
\iint_S \mathbf{F} \cdot \mathbf{n} \, d\sigma = \iint_{S_1} \mathbf{F} \cdot \mathbf{n} \, d\sigma + \iint_{S_2} \mathbf{F} \cdot \mathbf{n} \, d\sigma
$$

$$
= \int_{\partial S_1} \mathbf{G} \cdot \mathbf{T}_1 \, ds + \int_{\partial S_2} \mathbf{G} \cdot \mathbf{T}_2 \, ds = 0.
$$

(18)

This last step follows from the fact that $\partial S_1 = \partial S_2$ and $T_1 = -T_2$. Since (17) and (18) are incompatible, we conclude that there is no $\mathbf{G}$ which satisfies curl $\mathbf{G} = \mathbf{F}$.   ∎

EXERCISES

13.6.1. For each of the following, evaluate $\int_C \mathbf{F} \cdot \mathbf{T} \, ds$.

a) C is the curve formed by intersecting the cylinder $x^2 + y^2 = 1$ with $z = -x$, oriented in the counterclockwise direction when viewed from high on the positive z-axis, and $\mathbf{F}(x, y, z) = (xy^2, 0, xyz)$.

b) C is the intersection of the cubic cylinder $z = y^3$ and the circular cylinder $x^2 + y^2 = 3$, oriented in the clockwise direction when viewed from high up the positive z-axis, and $\mathbf{F}(x, y, z) = (e^x + z, xy, ze^y)$.

13.6.2. For each of the following, evaluate $\iint_S \text{curl} \, \mathbf{F} \cdot \mathbf{n} \, d\sigma$.

a) S is the "bottomless" surface in the upper half-space $z \geq 0$ bounded by $y = x^2$, $z = 1 - y$, $\mathbf{n}$ is the outward-pointing normal, and $\mathbf{F}(x, y, z) = (x \sin z^3, y \cos z^3, x^3 + y^3 + z^3)$.

b) S is the truncated paraboloid $z = 3 - x^2 - y^2$, $z \geq 0$, $\mathbf{n}$ is the outward-pointing normal, and $\mathbf{F}(x, y, z) = (y, xyz, y)$.

c) S is the hemisphere $z = \sqrt{10 - x^2 - y^2}$, $\mathbf{n}$ is the inward-pointing normal, and $\mathbf{F}(x, y, z) = (x, x, x^2 y^3 \log(z + 1))$.

d) S is the "bottomless" tetrahedron in the upper half-space $z \geq 0$ bounded by $x = 0$, $y = 0$, $x + 2y + 3z = 1$, $z \geq 0$, $\mathbf{n}$ is the outward-pointing normal, and $\mathbf{F}(x, y, z) = (xy, yz, xz)$.

13.6.3. For each of the following, evaluate $\iint_S \mathbf{F} \cdot \mathbf{n} \, d\sigma$ using Stokes's Theorem or the Divergence Theorem.

a) S is the sphere $x^2 + y^2 + z^2 = 1$, $\mathbf{n}$ is the outward-pointing normal, and $\mathbf{F}(x, y, z) = (xz^2, x^2 y - z^3, 2xy + y^2 z)$.

b) S is the portion of the plane $z = y$ which lies inside the ball $B_1(0)$, $\mathbf{n}$ is the upward-pointing normal, and $\mathbf{F}(x, y, z) = (xy, xz, -yz)$.

c) S is the truncated cone $y = 2\sqrt{x^2 + z^2}$, $2 \leq y \leq 4$, $\mathbf{n}$ is the outward-pointing normal, and $\mathbf{F}(x, y, z) = (x, -2y, z)$.

d) S is a union of truncated paraboloids $z = 4 - x^2 - y^2$, $0 \leq z \leq 4$, and $z = x^2 + y^2 - 4$, $-4 \leq z \leq 0$, $\mathbf{n}$ is the outward-pointing normal, and

$$
\mathbf{F}(x, y, z) = (x + y^2 + \sin z, x + y^2 + \cos z, \cos x + \sin y + z).
$$

e) S is the union of three surfaces $z = x^2 + y^2$ $(0 \le z \le 2)$, $2 = x^2 + y^2$ $(2 \le z \le 5)$, and $z = 7 - x^2 - y^2$ $(5 \le z \le 6)$, $\mathbf{n}$ is the outward-pointing normal, and $\mathbf{F}(x, y, z) = (2y, 2z, 1)$.

13.6.4. For each of the following, evaluate $\int_S \omega$ using Stokes's Theorem or the Divergence Theorem.

a) S is the topological boundary of cylindrical solid $y^2 + z^2 \le 9$, $0 \le x \le 2$, with outward-pointing normal, and $\omega = xy \, dy \, dz + (x^2 - z^2) \, dz \, dx + xz \, dx \, dy$.

b) S is the truncated cylinder $x^2 + z^2 = 8$, $0 \le y \le 1$, with outward-pointing normal, and $\omega = (x - 2z) \, dy \, dz - y \, dz \, dx$.

c) S is the topological boundary of $R = [0, \pi/2] \times [0, 1] \times [0, 3]$, with outward-pointing normal, and $\omega = e^y \cos x \, dy \, dz + x^2 z \, dz \, dx + (x + y + z) \, dx \, dy$.

d) S is the intersection of the elliptic cylindrical solid $2x^2 + z^2 \le 1$ and the plane $x = y$, with normal which points toward the positive x-axis, and $\omega = x \, dy \, dz - y \, dz \, dx + \sin y \, dx \, dy$.

13.6.5. Prove that Green's Theorem is a corollary of Stokes's Theorem.

13.6.6. Let Π be a plane in $\mathbf{R}^3$ with unit normal $\mathbf{n}$ and $\mathbf{x}_0 \in \Pi$. For each $r > 0$, let S_r be the disk in Π centered at $\mathbf{x}_0$ of radius r [i.e., $S_r = B_r(\mathbf{x}_0) \cap \Pi$]. Prove that if $\mathbf{F} : B_1(\mathbf{x}_0) \to \mathbf{R}^3$ is C^1 and ∂S_r carries the orientation induced by $\mathbf{n}$, then

$$\operatorname{curl} \mathbf{F}(\mathbf{x}_0) \cdot \mathbf{n} = \lim_{r \to 0+} \frac{1}{\sigma(S_r)} \int_{\partial S_r} \mathbf{F} \cdot \mathbf{T} \, ds.$$

13.6.7. Let S be an orientable surface with unit normal $\mathbf{n}$ and nonempty boundary ∂S which satisfies the hypotheses of Stokes's Theorem.

a) Suppose that $\mathbf{F} : S \to \mathbf{R}^3 \backslash \{0\}$ is C^1, that ∂S is smooth, and that $\mathbf{T}$ is the unit tangent vector on ∂S induced by $\mathbf{n}$. If the angle between $\mathbf{T}(\mathbf{x}_0)$ and $\mathbf{F}(\mathbf{x}_0)$ is never obtuse for any $\mathbf{x}_0 \in \partial S$, and $\iint_S \operatorname{curl} \mathbf{F} \cdot \mathbf{n} \, d\sigma = 0$, prove that $\mathbf{T}(\mathbf{x}_0)$ and $\mathbf{F}(\mathbf{x}_0)$ are orthogonal for all $\mathbf{x}_0 \in \partial S$.

b) If $\mathbf{F}, \mathbf{F}_k : S \to \mathbf{R}^3$ are C^1 and $\mathbf{F}_k \to \mathbf{F}$ uniformly on ∂S, prove that

$$\lim_{k \to \infty} \iint_S \operatorname{curl} \mathbf{F}_k \cdot \mathbf{n} \, d\sigma = \iint_S \operatorname{curl} \mathbf{F} \cdot \mathbf{n} \, d\sigma.$$

13.6.8. Suppose that E is a two-dimensional region such that if $(x, y) \in E$, then the line segments from $(0, 0)$ to $(x, 0)$ and from $(x, 0)$ to (x, y) are both subsets of E. If $\mathbf{F} : E \to \mathbf{R}^2$ is C^1, prove that the following three statements are equivalent.

a) $\mathbf{F} = \nabla f$ on E for some $f : E \to \mathbf{R}$.

b) $\mathbf{F} = (P, Q)$ is *exact* (i.e., $Q_x = P_y$ on E).

c) $\int_C \mathbf{F} \cdot \mathbf{T} \, ds = 0$ for all piecewise smooth curves $C = \partial \Omega$ oriented counterclockwise, where Ω is a two-dimensional region which satisfies the hypotheses of Green's Theorem, and $\Omega \subset E$.

13.6.9. Let Ω be a three-dimensional region and $\mathbf{F} : \Omega \to \mathbf{R}^3$ be C^1 on Ω. Suppose further that, for each $(x, y, z) \in \Omega$, both the line segments $L((x, y, 0); (x, y, z))$ and $L((x, 0, 0); (x, y, 0))$ are subsets of Ω. Prove that the following statements are equivalent.

a) There is a C^2 function $\mathbf{G} : \Omega \to \mathbf{R}^3$ such that $\operatorname{curl} \mathbf{G} = \mathbf{F}$ on Ω.

b) If $\mathbf{F}$, E, and $S = \partial E$ satisfy the hypotheses of the Divergence Theorem and $E \subset \Omega$, then

$$\iint_S \mathbf{F} \cdot \mathbf{n} \, d\sigma = 0.$$

c) The identity $\operatorname{div} \mathbf{F} = 0$ holds everywhere on Ω.

13.6.10. Suppose that E satisfies the hypotheses of the Divergence Theorem and that S satisfies the hypotheses of Stokes's Theorem.

a) If $f : S \to \mathbf{R}$ is a C^2 function and $\mathbf{F} = \operatorname{grad} f$ on S, prove that

$$\iint_{\partial S} (f\mathbf{F}) \cdot \mathbf{T} \, ds = 0.$$

b) If $\mathbf{G} : E \to \mathbf{R}^3$ is a C^2 function and $\mathbf{F} = \operatorname{curl} \mathbf{G}$ on E, prove that

$$\iint_{\partial E} (f\mathbf{F}) \cdot \mathbf{n} \, d\sigma = \iiint_E \operatorname{grad} f \cdot \mathbf{F} \, dV.$$

Note: You may wish to use Exercises 13.5.8 and 13.5.9.

13.6.11. Let $\mathbf{F}$ be C^1 and exact on $\mathbf{R}^2 \backslash \{(0, 0)\}$ (see Exercise 13.6.8b).

a) Suppose that C_1 and C_2 are disjoint smooth simple curves, oriented in the counterclockwise direction, and that E is a two-dimensional region whose topological boundary ∂E is the union of C_1 and C_2. (*Note*: This means that E has a hole with one of the C_j's as the outer boundary and the other as the inner boundary.) If $(0, 0) \notin E$, prove that

$$\int_{C_1} \mathbf{F} \cdot \mathbf{T} \, ds = \int_{C_2} \mathbf{F} \cdot \mathbf{T} \, ds.$$

b) Suppose that E is a two-dimensional region which satisfies $(0, 0) \in E^o$. If ∂E is a smooth simple curve oriented in the counterclockwise direction, and

$$\mathbf{F}(x, y) = \left(\frac{-y}{x^2 + y^2}, \frac{x}{x^2 + y^2} \right),$$

compute $\int_{\partial E} \mathbf{F} \cdot \mathbf{T} \, ds$.

c) State and prove an analogue of part a) for functions $\mathbf{F} : \mathbf{R}^3 \backslash \{(0, 0, 0)\}$, three-dimensional regions, and smooth surfaces.

CHAPTER 14

Fourier Series

*14.1 INTRODUCTION

This section uses no material from any other enrichment section.

In Chapter 7 we studied power series and their partial sums, *classical* polynomials. In this chapter, we shall study the following objects.

14.1 Definition.

Let $a_k, b_k \in \mathbf{R}$ and let N be a nonnegative integer.

i) A *trigonometric series* is a series of the form

$$\frac{a_0}{2} + \sum_{k=1}^{\infty}(a_k \cos kx + b_k \sin kx).$$

ii) A *trigonometric polynomial* of order N is a function $P : \mathbf{R} \to \mathbf{R}$ of the form

$$P(x) = \frac{a_0}{2} + \sum_{k=1}^{N}(a_k \cos kx + b_k \sin kx).$$

[Here, $\cos kx$ is shorthand for $\cos(kx)$, and $\sin kx$ is shorthand for $\sin(kx)$.]

Calculus was invented with the tacit assumption that power series provided a unified function theory; that is, every function has a power series expansion (see Kline [5]). When Cauchy showed that this assumption was false (see Remark 7.41), mathematicians began to wonder whether some other type of series would provide a unified function theory. Euler (respectively, Fourier) had shown that the position of a vibrating string (respectively, the temperature along a metal rod) can be represented by trigonometric series. Thus, it was natural to ask, Does every function have a trigonometric series expansion? In this chapter we shall examine this question, and the following calculation will help to answer it.

584

14.2 Lemma. [ORTHOGONALITY].
Let k, j be nonnegative integers. Then

i)
$$\int_{-\pi}^{\pi} \cos kx \cos jx \, dx = \begin{cases} 2\pi & k = j = 0 \\ \pi & k = j \neq 0 \\ 0 & k \neq j \end{cases}$$

ii)
$$\int_{-\pi}^{\pi} \sin kx \sin jx \, dx = \begin{cases} \pi & k = j \neq 0 \\ 0 & k \neq j \end{cases}$$

and

iii)
$$\int_{-\pi}^{\pi} \sin kx \cos jx \, dx = 0.$$

Proof. Let

$$I = \int_{-\pi}^{\pi} \cos kx \cos jx \, dx.$$

If $k = j = 0$, then $I = \int_{-\pi}^{\pi} dx = 2\pi$. If $k = j \neq 0$, then by a half-angle formula and elementary integration, we have

$$I = \int_{-\pi}^{\pi} \cos^2 kx \, dx = \frac{1}{2} \int_{-\pi}^{\pi} (1 + \cos 2kx) \, dx = \pi.$$

And if $k \neq j$, then by a sum-angle formula and elementary integration, we have

$$I = \frac{1}{2} \int_{-\pi}^{\pi} (\cos(k + j)x + \cos(k - j)x) \, dx = 0.$$

This proves part i). Similar arguments prove parts ii) and iii).   ∎

Notice that the question concerning representation of functions by trigonometric series has a built-in limitation. A function $f : \mathbf{R} \to \mathbf{R}$ is said to be *periodic* (of period 2π) if and only if $f(x + 2\pi) = f(x)$ for all $x \in \mathbf{R}$. Since $\cos kx$ and $\sin kx$ are periodic, it is clear that every trigonometric polynomial is periodic. Therefore, any function which is the pointwise or uniform limit of a trigonometric series must also be periodic. For this reason, we will usually restrict our attention to the interval $[-\pi, \pi]$ and assume that $f(-\pi) = f(\pi)$.

The following definition, which introduces a special type of trigonometric series, plays a crucial role in the representation of periodic functions by trigonometric series.

14.3 Definition.

Let f be integrable on $[-\pi, \pi]$ and let N be a nonnegative integer.

i) The *Fourier coefficients* of f are the numbers

$$a_k(f) = \frac{1}{\pi} \int_{-\pi}^{\pi} f(x) \cos kx \, dx, \qquad k = 0, 1, \ldots,$$

and

$$b_k(f) = \frac{1}{\pi} \int_{-\pi}^{\pi} f(x) \sin kx \, dx, \qquad k = 1, 2, \ldots.$$

ii) The *Fourier series* of f is the trigonometric series

$$(Sf)(x) = \frac{a_0(f)}{2} + \sum_{k=1}^{\infty} (a_k(f) \cos kx + b_k(f) \sin kx).$$

iii) The *partial sum* of Sf of order N is the trigonometric polynomial defined, for each $x \in \mathbf{R}$, by $(S_0 f)(x) = a_0(f)/2$ if $N = 0$, and

$$(S_N f)(x) = \frac{a_0(f)}{2} + \sum_{k=1}^{N} (a_k(f) \cos kx + b_k(f) \sin kx)$$

if $N \in \mathbf{N}$.

The following result shows why Fourier series play such an important role in the representation of periodic functions by trigonometric series.

14.4 Theorem. [FOURIER].
If a trigonometric series

$$S := \frac{a_0}{2} + \sum_{k=1}^{\infty} (a_k \cos kx + b_k \sin kx)$$

converges uniformly on $\mathbf{R}$ *to a function* f, *then* S *is the Fourier series of* f; *that is,* $a_k = a_k(f)$ *for* $k = 0, 1, \ldots$, *and* $b_k = b_k(f)$ *for* $k = 1, 2, \ldots$.

Proof. Fix an integer $k \geq 0$. Since

$$f(x) = \frac{a_0}{2} + \sum_{j=1}^{\infty} (a_j \cos jx + b_j \sin jx)$$

converges uniformly and $\cos kx$ is bounded,

$$f(x) \cos kx = \frac{a_0}{2} \cos kx + \sum_{j=1}^{\infty} (a_j \cos jx \cos kx + b_j \sin jx \cos kx) \qquad (1)$$

also converges uniformly. Since f is the uniform limit of continuous functions, f is continuous and, hence, integrable on $[-\pi, \pi]$. Integrating (1) term by term and using orthogonality, we obtain

$$\begin{aligned}
a_k(f) &= \frac{1}{\pi} \int_{-\pi}^{\pi} f(x) \cos kx \, dx \\
&= \frac{a_0}{2\pi} \int_{-\pi}^{\pi} \cos kx \, dx + \sum_{j=1}^{\infty} \left(\frac{a_j}{\pi} \int_{-\pi}^{\pi} \cos kx \cos jx \, dx \right. \\
&\quad \left. + \frac{b_j}{\pi} \int_{-\pi}^{\pi} \cos kx \sin jx \, dx \right) \\
&= a_k.
\end{aligned}$$

A similar argument establishes $b_k(f) = b_k$. ■

There are two central questions in the study of trigonometric series.

THE CONVERGENCE QUESTION. *Given a function $f : \mathbf{R} \to \mathbf{R}$, periodic on $\mathbf{R}$ and integrable on $[-\pi, \pi]$, does the Fourier series of f converge to f?*

THE UNIQUENESS QUESTION. *If a trigonometric series S converges to some function f integrable on $[-\pi, \pi]$, is S the Fourier series of f?*

We shall answer these questions for pointwise and uniform convergence when f is continuous and of bounded variation. We notice in passing that, by Theorem 14.4, the answer to the Uniqueness Question is yes if uniform convergence is used.

The following special trigonometric polynomials arise naturally in connection with the Convergence Question (see Exercise 14.1.2).

14.5 Definition.

Let N be a nonnegative integer.

i) The *Dirichlet kernel* of order N is the function defined, for each $x \in \mathbf{R}$, by $D_0(x) = 1/2$ if $N = 0$, and

$$D_N(x) = \frac{1}{2} + \sum_{k=1}^{N} \cos kx$$

if $N \in \mathbf{N}$.

14.5 Definition. (Continued)

---

ii) The *Fejér kernel* of order N is the function defined, for each $x \in \mathbf{R}$, by $K_0(x) = 1/2$ if $N = 0$, and

$$K_N(x) = \frac{1}{2} + \sum_{k=1}^{N} \left(1 - \frac{k}{N+1} \right) \cos kx \qquad (2)$$

if $N \in \mathbf{N}$.

The following result shows that there is a simple relationship between Fejér kernels and Dirichlet kernels.

14.6 Remark. *If N is a nonnegative integer, then*

$$K_N(x) = \frac{D_0(x) + \cdots + D_N(x)}{N+1}$$

for all $x \in \mathbf{R}$.

Proof. The identity is trivial if $N = 0$. To prove the identity for $N \in \mathbf{N}$, fix $x \in \mathbf{R}$. By definition,

$$K_N(x) = \frac{1}{N+1} \left(\frac{N+1}{2} + \sum_{k=1}^{N} (N-k+1) \cos kx \right)$$

$$= \frac{1}{N+1} \left(\frac{1}{2} + \frac{N}{2} + \sum_{k=1}^{N} \sum_{j=k}^{N} 1 \cdot \cos kx \right)$$

$$= \frac{1}{N+1} \left(\frac{1}{2} + \sum_{j=1}^{N} \left(\frac{1}{2} + \sum_{k=1}^{j} \cos kx \right) \right) = \frac{D_0(x) + \cdots + D_N(x)}{N+1}. \qquad \blacksquare$$

The next result shows that Dirichlet and Fejér kernels can be represented by quotients of trigonometric functions.

14.7 Theorem. *If $x \in \mathbf{R}$ cannot be written in the form $2k\pi$ for any $k \in \mathbf{Z}$, then*

$$D_N(x) = \frac{\sin\left(N + \frac{1}{2}\right)x}{2\sin\frac{x}{2}} \qquad (3)$$

and

$$K_N(x) = \frac{2}{N+1} \left(\frac{\sin\left(\frac{N+1}{2}\right)x}{2\sin\frac{x}{2}} \right)^2 \tag{4}$$

for $N = 0, 1, \ldots$.

Proof. The formulas are trivial for $N = 0$. Fix $N \in \mathbf{N}$. Applying a sum-angle formula and telescoping, we have

$$\left(D_N(x) - \frac{1}{2} \right) \sin\frac{x}{2} = \sum_{k=1}^{N} \cos kx \sin\frac{x}{2}$$

$$= \frac{1}{2} \sum_{k=1}^{N} \left(\sin\left(k + \frac{1}{2}\right)x - \sin\left(k - \frac{1}{2}\right)x \right)$$

$$= \frac{1}{2} \left(\sin\left(N + \frac{1}{2}\right)x - \sin\frac{x}{2} \right).$$

Solving this equation for $D_N(x)$ verifies (3).
  Let $k \in \mathbf{N}$. By (3) and another sum-angle formula,

$$D_k(x) \sin^2 \frac{x}{2} = \frac{1}{2} \sin\frac{x}{2} \sin\left(k + \frac{1}{2}\right)x = \frac{1}{4}(\cos kx - \cos(k+1)x).$$

This identity also holds for $k = 0$. Applying Remark 14.6 and telescoping, we have

$$(N+1)K_N(x) \sin^2 \frac{x}{2} = \sum_{k=0}^{N} D_k(x) \sin^2 \frac{x}{2}$$

$$= \frac{1}{4} \sum_{k=0}^{N} (\cos kx - \cos(k+1)x)$$

$$= \frac{1}{4}(1 - \cos(N+1)x) = \frac{1}{2} \sin^2 \left(\frac{N+1}{2} \right)x.$$

Solving this equation for $K_N(x)$ verifies (4). ∎

  These identities will be used in the next section to obtain a partial answer to the Convergence Question.
  The next two examples illustrate the general principle that the Fourier coefficients of many common functions can be computed using integration by parts.

14.8 *EXAMPLE.*

Prove that the Fourier series of $f(x) = x$ is

$$2\sum_{k=1}^{\infty} \frac{(-1)^{k+1}}{k} \sin kx.$$

Proof. Since $x \cos kx$ is odd and $x \sin kx$ is even, we see that $a_k(f) = 0$ for $k = 0, 1, \ldots,$ and

$$b_k(f) = \frac{2}{\pi} \int_0^{\pi} x \sin kx \, dx$$

for $k = 1, 2, \ldots$. Integrating by parts, we conclude that

$$b_k(f) = \frac{2}{\pi} \left(-\frac{x \cos kx}{k} \Big|_0^{\pi} + \frac{1}{k} \int_0^{\pi} \cos kx \, dx \right) = \frac{2(-1)^{k+1}}{k}. \qquad \blacksquare$$

14.9 *EXAMPLE.*

Prove that the Fourier series of $f(x) = |x|$ is

$$\frac{\pi}{2} - \frac{4}{\pi} \sum_{k=1}^{\infty} \frac{\cos(2k-1)x}{(2k-1)^2}.$$

Proof. Since $|x| \cos kx$ is even and $|x| \sin kx$ is odd, we see that $b_k(f) = 0$ for $k = 1, 2 \ldots,$ and

$$a_k(f) = \frac{2}{\pi} \int_0^{\pi} x \cos kx \, dx$$

for $k = 0, 1, \ldots$. If $k = 0$, then

$$a_k(f) = \frac{2}{\pi} \left(\frac{\pi^2}{2} \right) = \pi;$$

that is, $a_0(f)/2 = \pi/2$. If $k > 0$, then integration by parts yields

$$a_k(f) = \frac{2}{\pi k^2} (\cos k\pi - 1) = \begin{cases} 0 & \text{if } k \text{ is even,} \\ -\dfrac{4}{\pi k^2} & \text{if } k \text{ is odd.} \end{cases} \qquad \blacksquare$$

EXERCISES

14.1.1. Compute the Fourier series of x^2 and of $\cos^2 x$.

14.1.2. Prove that if $f : \mathbf{R} \to \mathbf{R}$ is integrable on $[-\pi, \pi]$, then

$$(S_N f)(x) = \frac{1}{\pi} \int_{-\pi}^{\pi} f(t) D_N(x-t) \, dt$$

for all $x \in [-\pi, \pi]$ and $N \in \mathbf{N}$.

14.1.3. Show that if f, g are integrable on $[-\pi, \pi]$ and $\alpha \in \mathbf{R}$, then

$$a_k(f + g) = a_k(f) + a_k(g), \quad a_k(\alpha f) = \alpha a_k(f), \qquad k = 0, 1, \ldots,$$

and

$$b_k(f + g) = b_k(f) + b_k(g), \quad b_k(\alpha f) = \alpha b_k(f), \qquad k = 1, 2, \ldots.$$

14.1.4. Suppose that $f : \mathbf{R} \to \mathbf{R}$ is differentiable and periodic and that f' is integrable on $[-\pi, \pi]$. Prove that

$$a_k(f') = kb_k(f) \quad \text{and} \quad b_k(f') = -ka_k(f), \qquad k \in \mathbf{N}.$$

14.1.5. Suppose that $f_N : [-\pi, \pi] \to \mathbf{R}$ are integrable and that $f_N \to f$ uniformly on $[-\pi, \pi]$ as $N \to \infty$.

a) Prove that $a_k(f_N) \to a_k(f)$ and $b_k(f_N) \to b_k(f)$, as $N \to \infty$, uniformly in k.

b) Show that part a) holds under the weaker hypothesis

$$\lim_{N \to \infty} \int_{-\pi}^{\pi} |f(x) - f_N(x)| \, dx = 0.$$

14.1.6. Let

$$f(x) = \begin{cases} \dfrac{x}{|x|} & x \neq 0 \\ 0 & x = 0. \end{cases}$$

a) Compute the Fourier coefficients of f.

b) Prove that

$$(S_{2N} f)(x) = \frac{2}{\pi} \int_0^x \frac{\sin 2Nt}{\sin t} \, dt$$

for $x \in [-\pi, \pi]$ and $N \in \mathbf{N}$.

*c) [GIBBS'S PHENOMENON]. Prove that

$$\lim_{N \to \infty} (S_{2N} f) \left(\frac{\pi}{2N} \right) = \frac{2}{\pi} \int_0^\pi \frac{\sin t}{t} \, dt \approx 1.179.$$

*14.2 SUMMABILITY OF FOURIER SERIES

This section uses material from Section 14.1.

The Convergence Question posed in Section 14.1 is very difficult to answer, even for continuous functions. In this section we replace it with an easier question and show that the answer to this question is yes. Namely, we shall show that the Fourier series of any continuous periodic function f is uniformly summable to f. By summable, we mean the following concept.

14.10 Definition.

A series $\sum_{k=0}^{\infty} a_k$ with partial sums $s_N = \sum_{k=0}^{N} a_k$ is said to be *Cesàro summable* to L if and only if its *Cesàro means*

$$\sigma_N := \frac{s_0 + \cdots + s_N}{N + 1}$$

converge to L as $N \to \infty$.

The following result shows that summability is a generalization of convergence.

14.11 Remark. *If $\sum_{k=0}^{\infty} a_k$ converges to a finite number L, then it is Cesàro summable to L.*

Proof. Let $\varepsilon > 0$. Choose $N_1 \in \mathbf{N}$ such that $k \geq N_1$ implies $|s_k - L| < \varepsilon/2$. Use the Archimedean Principle to choose $N_2 \in \mathbf{N}$ such that $N_2 > N_1$ and

$$\sum_{k=0}^{N_1} |s_k - L| < \frac{\varepsilon N_2}{2}.$$

If $N > N_2$, then

$$|\sigma_N - L| \leq \frac{1}{N+1} \sum_{k=0}^{N_1} |s_k - L| + \frac{1}{N+1} \sum_{k=N_1+1}^{N} |s_k - L|$$

$$\leq \frac{\varepsilon N_2}{2(N+1)} + \frac{\varepsilon}{2}\left(\frac{N - N_1}{N+1}\right) < \frac{\varepsilon}{2} + \frac{\varepsilon}{2} = \varepsilon. \qquad \blacksquare$$

The converse of Remark 14.11 is false. Indeed, although the series $\sum_{k=0}^{\infty} (-1)^k$ does not converge, its Cesàro means satisfy

$$\sigma_N = \begin{cases} \dfrac{N+2}{2(N+1)} & N \text{ is even} \\[2mm] \dfrac{1}{2} & N \text{ is odd,} \end{cases}$$

whence $\sigma_N \to 1/2$ as $N \to \infty$.

It is easier to show that a series is Cesàro summable than to show that it converges. Thus the following question is easier to answer than the Convergence Question.

THE SUMMABILITY QUESTION. *Given a function $f : \mathbf{R} \to \mathbf{R}$, periodic on $\mathbf{R}$ and integrable on $[-\pi, \pi]$, is Sf Cesàro summable to f?*

The Cesàro means of a Fourier series Sf are denoted by

$$(\sigma_N f)(x) := \frac{(S_0 f)(x) + \cdots + (S_N f)(x)}{N + 1},$$

$N = 0, 1, \dots$. The following result shows that the Cesàro means of a Fourier series can always be represented by an integral equation. This is important because it allows us to estimate the remainder $\sigma_N f - f$, using techniques of integration.

14.12 Lemma.
Let $f : \mathbf{R} \to \mathbf{R}$ be periodic on $\mathbf{R}$ and integrable on $[-\pi, \pi]$. Then

$$(\sigma_N f)(x) = \frac{1}{\pi} \int_{-\pi}^{\pi} f(x - t) K_N(t)\, dt$$

for all $N = 0, 1, \dots$, and all $x \in \mathbf{R}$.

Proof. Fix $j, N \in \mathbf{N}$ and $x \in \mathbf{R}$. By definition and a sum-angle formula,

$$a_j(f) \cos jx + b_j(f) \sin jx$$

$$= \frac{1}{\pi} \int_{-\pi}^{\pi} f(u) \cos ju \cos jx\, du + \frac{1}{\pi} \int_{-\pi}^{\pi} f(u) \sin ju \sin jx\, du$$

$$= \frac{1}{\pi} \int_{-\pi}^{\pi} f(u)(\cos ju \cos jx + \sin ju \sin jx)\, du$$

$$= \frac{1}{\pi} \int_{-\pi}^{\pi} f(u) \cos j(x - u)\, du.$$

Summing this identity over integers $j = 1, 2, \dots, k$ and adding $a_0(f)/2$, we have

$$(S_k f)(x) = \frac{a_0(f)}{2} + \sum_{j=1}^{k} (a_j(f) \cos jx + b_j(f) \sin jx)$$

$$= \frac{1}{\pi} \int_{-\pi}^{\pi} f(u) \left(\frac{1}{2} + \sum_{j=1}^{k} \cos j(x - u) \right) du$$

$$= \frac{1}{\pi} \int_{-\pi}^{\pi} f(u) D_k(x - u)\, du$$

for $k = 0, 1, \dots$. Making the change of variables $t = x - u$ and using the fact that both f and D_k are periodic, we obtain

$$(S_k f)(x) = \frac{1}{\pi} \int_{-\pi}^{\pi} f(x - t) D_k(t)\, dt, \qquad k = 0, 1, \dots.$$

We conclude by Remark 14.11 that

$$(\sigma_N f)(x) = \frac{1}{N+1} \sum_{k=0}^{N} (S_k f)(x)$$

$$= \frac{1}{N+1} \sum_{k=0}^{N} \frac{1}{\pi} \int_{-\pi}^{\pi} f(x-t) D_k(t)\, dt = \frac{1}{\pi} \int_{-\pi}^{\pi} f(x-t) K_N(t)\, dt. \qquad \blacksquare$$

To answer the Summability Question we need to know more about Fejér kernels. The following result shows that Fejér kernels satisfy some very nice properties.

14.13 Lemma.
For each nonnegative integer N,

$$K_N(t) \ge 0 \quad \text{for all } t \in \mathbf{R}, \qquad (5)$$

and

$$\frac{1}{\pi} \int_{-\pi}^{\pi} K_N(t)\, dt = 1. \qquad (6)$$

Moreover, for each $0 < \delta < \pi$,

$$\lim_{N\to\infty} \int_{\delta}^{\pi} |K_N(t)|\, dt = 0. \qquad (7)$$

Proof. Fix $N \ge 0$. If $t = 2j\pi$ for some $j \in \mathbf{Z}$, then $D_k(t) = k + 1/2 \ge 0$ for all $k \ge 0$, whence $K_N(t) \ge 0$. If $t \ne 2j\pi$ for any $j \in \mathbf{Z}$, then, by Theorem 14.7,

$$K_N(t) = \frac{2}{N+1} \left(\frac{\sin\left(\frac{N+1}{2}\right)t}{2\sin\frac{t}{2}} \right)^2 \ge 0.$$

This proves (5). By Definition 14.5 and orthogonality,

$$\int_{-\pi}^{\pi} K_N(t)\, dt = \int_{-\pi}^{\pi} \left(\frac{1}{2} + \sum_{k=1}^{N} \left(1 - \frac{k}{N+1}\right) \cos kt \right) dt = \pi.$$

This proves (6).
To prove (7), fix $0 < \delta < \pi$ and observe that $\sin t/2 \ge \sin \delta/2$ for $t \in [\delta, \pi]$. Hence, it follows from Theorem 14.7 that

$$\int_{\delta}^{\pi} |K_N(t)|\, dt \le \frac{2}{N+1} \int_{\delta}^{\pi} \left(\frac{\sin\left(\frac{N+1}{2}\right)t}{2\sin\frac{\delta}{2}} \right)^2 dt \le \frac{\pi}{2(N+1)\sin^2\frac{\delta}{2}}.$$

Since δ is fixed, this last expression tends to 0 as $N \to \infty$. $\blacksquare$

Using these properties, we can answer the Summability Question for continuous functions (see also Exercises 14.2.6 and 14.2.8).

14.14 Theorem.   [FEJÉR].
Suppose that $f : \mathbf{R} \to \mathbf{R}$ is periodic on $\mathbf{R}$ and integrable on $[-\pi, \pi]$.

i) *If*

$$L = \lim_{h \to 0} \frac{f(x_0 + h) + f(x_0 - h)}{2}$$

exists for some $x_0 \in \mathbf{R}$, then $(\sigma_N f)(x_0) \to L$ as $N \to \infty$.
ii) *If f is continuous on some closed interval I, then $\sigma_N f \to f$ uniformly on I as $N \to \infty$.*

Proof. Since f is periodic, we may suppose that $x_0 \in [-\pi, \pi]$. Fix $N \in \mathbf{N}$. By (6), Lemma 14.12, and a change of variables,

$$
\begin{aligned}
(\sigma_N f)(x_0) - L &= \frac{1}{\pi} \int_{-\pi}^{\pi} K_N(t)(f(x_0 - t) - L)\, dt \\
&= \frac{2}{\pi} \int_0^{\pi} K_N(t) \left(\frac{f(x_0 + t) + f(x_0 - t)}{2} - L \right) dt \qquad (8) \\
&=: \frac{2}{\pi} \int_0^{\pi} K_N(t) F(x_0, t)\, dt.
\end{aligned}
$$

Let $\varepsilon > 0$ and choose $0 < \delta < \pi$ such that $|t| < \delta$ implies $|F(x_0, t)| < \varepsilon/3$. By (5) and (6) we have

$$\frac{2}{\pi} \int_0^{\delta} K_N(t)|F(x_0, t)|\, dt < \frac{2\varepsilon}{3\pi} \int_0^{\delta} K_N(t)\, dt \leq \frac{2\varepsilon}{3}. \qquad (9)$$

On the other hand, choose by (7) an $N_1 \in \mathbf{N}$ such that $N \geq N_1$ implies $\int_{\delta}^{\pi} K_N(t)\, dt < \varepsilon/3M$, where $M := \sup_{x \in \mathbf{R}} |F(x)|$. Then

$$\frac{2}{\pi} \int_{\delta}^{\pi} K_N(t)|F(x_0, t)|\, dt \leq M \int_{\delta}^{\pi} K_N(t)\, dt < \frac{\varepsilon}{3},$$

and it follows from (8) and (9) that

$$|(\sigma_N f)(x_0) - L| \leq \frac{2}{\pi} \int_0^{\delta} K_N(t)|F(x_0, t)|\, dt + \frac{2}{\pi} \int_{\delta}^{\pi} K_N(t)|F(x_0, t)|\, dt < \varepsilon \quad (10)$$

for all $N \geq N_1$. This proves part i).

To prove part ii), suppose that f is continuous on some closed interval I. Since f is periodic, we may suppose that $I \subseteq [-\pi, \pi]$. Thus I is closed and bounded, and f is uniformly continuous on I. Repeating the estimates above, we see that (10) holds uniformly for all $x_0 \in I$. ∎

14.15 Corollary. *If $f : \mathbf{R} \to \mathbf{R}$ is continuous and periodic, then $\sigma_N f$ converges to f uniformly on $\mathbf{R}$ as $N \to \infty$.*

14.16 Corollary. [COMPLETENESS]. *If $f : \mathbf{R} \to \mathbf{R}$ is continuous and periodic, and $a_{k-1}(f) = b_k(f) = 0$ for $k \in \mathbf{N}$, then $f(x) = 0$ for all $x \in \mathbf{R}$.*

Proof. By hypothesis, $(\sigma_N f)(x) = 0$ for all $N \in \mathbf{N}$ and $x \in \mathbf{R}$. Hence, by Corollary 14.15, $f(x) = \lim_{N \to \infty}(\sigma_N f)(x) = 0$ for all $x \in \mathbf{R}$. ∎

14.17 Corollary. *Let $f : \mathbf{R} \to \mathbf{R}$ be continuous and periodic. Then there is a sequence of trigonometric polynomials $T_1, T_2, \ldots$, such that $T_N \to f$ uniformly on $\mathbf{R}$.*

Proof. Set $T_N = \sigma_N f$ for $N \in \mathbf{N}$, and apply Corollary 14.15. ∎

EXERCISES

14.2.1. Let $E \subseteq \mathbf{R}$ and suppose that $f, f_k : \mathbf{R} \to \mathbf{R}$ are bounded functions. Prove that if $\sum_{k=0}^{\infty} f_k(x)$ converges to $f(x)$ uniformly on E, then

$$\sigma_N(x) := \sum_{k=0}^{N} \left(1 - \frac{k}{N+1}\right) f_k(x)$$

converges to $f(x)$ uniformly on E as $N \to \infty$.

14.2.2. If $f : \mathbf{R} \to \mathbf{R}$ is periodic on $\mathbf{R}$ and integrable on $[-\pi, \pi]$, prove that the Cesàro means of Sf are uniformly bounded; that is, there is an $M > 0$ such that

$$|(\sigma_N f)(x)| \le M$$

for all $x \in \mathbf{R}$ and $N \in \mathbf{N}$.

14.2.3. Let

$$S = \frac{a_0}{2} + \sum_{k=1}^{\infty}(a_k \cos kx + b_k \sin kx)$$

be a trigonometric series and set

$$\sigma_N(x) = \frac{a_0}{2} + \sum_{k=1}^{N} \left(1 - \frac{k}{N+1}\right)(a_k \cos kx + b_k \sin kx)$$

for $x \in \mathbf{R}$ and $N \in \mathbf{N}$. Prove that S is the Fourier series of some continuous periodic function $f : \mathbf{R} \to \mathbf{R}$ if and only if σ_N converges uniformly on $\mathbf{R}$, as $N \to \infty$.

14.2.4. Let f be integrable on $[-\pi, \pi]$ and $L \in \mathbf{R}$.

a) Prove that if $(\sigma_N f)(x_0) \to L$ as $N \to \infty$ and if $(Sf)(x_0)$ converges, then $(S_N f)(x_0) \to L$.

b) Prove that

$$\sin \sqrt{2}\pi + \sum_{k=1}^{\infty} \frac{4(-1)^k \sin \sqrt{2}\pi}{2 - k^2} \cos kx$$

converges to $\sqrt{2}\pi \cos \sqrt{2}x$ uniformly on compact subsets of $(0, 2\pi)$.

14.2.5. Suppose that $f : [a, b] \to \mathbf{R}$ is continuous and that

$$\int_a^b x^n f(x)\, dx = 0$$

for all integers $n \geq 0$.

a) Evaluate $\int_a^b P(x)f(x)\, dx$ for any polynomial P on $\mathbf{R}$.
b) Prove that $\int_a^b |f(x)|^2\, dx = 0$.
c) Show that $f(x) = 0$ for all $x \in [a, b]$.

14.2.6. [SUMMABILITY KERNELS]. Let $\phi_N : \mathbf{R} \to \mathbf{R}$ be a sequence of continuous, periodic functions on $\mathbf{R}$ which satisfy

$$\int_0^{2\pi} \phi_N(t)\, dt = 1 \quad \text{and} \quad \int_0^{2\pi} |\phi_N(t)|\, dt \leq M < \infty$$

for all $N \in \mathbf{N}$, and

$$\lim_{N \to \infty} \int_\delta^{2\pi - \delta} |\phi_N(t)|\, dt = 0$$

for each $0 < \delta < 2\pi$. Suppose that $f : \mathbf{R} \to \mathbf{R}$ is continuous and periodic. Prove that

$$\lim_{N \to \infty} \int_0^{2\pi} f(x - t)\phi_N(t)\, dt = f(x)$$

uniformly for $x \in \mathbf{R}$.

14.2.7. Let $[a, b]$ be a nondegenerate, closed, bounded interval.

a) Prove that given any polynomial P on $\mathbf{R}$ and any $\varepsilon > 0$, there is a polynomial Q on $\mathbf{R}$, with rational coefficients, such that $|P(x) - Q(x)| < \varepsilon$ for all $x \in [a, b]$.
*b) Prove that the space $\mathcal{C}[a, b]$ (see Example 10.6) is separable.

*14.2.8. A sequence of functions $f_N : \mathbf{R} \to \mathbf{R}$ is said to converge *almost everywhere* to a function f if and only if there is a set E of measure zero such that $f_N(x) \to f(x)$, as $N \to \infty$, for every $x \in \mathbf{R} \setminus E$. Suppose that $f : \mathbf{R} \to \mathbf{R}$ is also periodic. Prove that if f is Riemann integrable on $[-\pi, \pi]$, then $\sigma_N f \to f$ almost everywhere as $N \to \infty$.

*14.3   GROWTH OF FOURIER COEFFICIENTS

This section uses material from Sections 5.5 and 14.2.

By Theorem 14.14, a continuous periodic function f is completely determined by its Fourier coefficients. In this section we ask to what extent smoothness of f affects the growth of these coefficients.

We begin with a computational result.

14.18 Lemma.
If $f : \mathbf{R} \to \mathbf{R}$ is integrable on $[-\pi, \pi]$ and N is a nonnegative integer, then

$$\frac{1}{\pi} \int_{-\pi}^{\pi} f(x)(S_N f)(x) \, dx = \frac{|a_0(f)|^2}{2} + \sum_{k=1}^{N} \left(|a_k(f)|^2 + |b_k(f)|^2 \right) \tag{11}$$
$$= \frac{1}{\pi} \int_{-\pi}^{\pi} |(S_N f)(x)|^2 \, dx.$$

Proof. Fix $N \geq 0$. Since f and $S_N f$ are integrable on $[-\pi, \pi]$, both integrals in (11) exist. By definition and orthogonality,

$$\frac{1}{\pi} \int_{-\pi}^{\pi} f(x) \frac{a_0(f)}{2} \, dx = \frac{|a_0(f)|^2}{2} = \frac{1}{\pi} \int_{-\pi}^{\pi} (S_N f)(x) \frac{a_0(f)}{2} \, dx.$$

Similarly,

$$\frac{1}{\pi} \int_{-\pi}^{\pi} f(x) a_k(f) \cos kx \, dx = |a_k(f)|^2 = \frac{1}{\pi} \int_{-\pi}^{\pi} (S_N f)(x) a_k(f) \cos kx \, dx$$

and

$$\frac{1}{\pi} \int_{-\pi}^{\pi} f(x) b_k(f) \sin kx \, dx = |b_k(f)|^2 = \frac{1}{\pi} \int_{-\pi}^{\pi} (S_N f)(x) b_k(f) \sin kx \, dx$$

for $k \in \mathbf{N}$. Adding these identities for $k = 0, \ldots, N$ verifies (11). ∎

Next, we use this result to identify a growth condition satisfied by the Fourier coefficients of any Riemann integrable function.

14.19 Theorem.   [BESSEL'S INEQUALITY].
If $f : \mathbf{R} \to \mathbf{R}$ is (Riemann) integrable on $[-\pi, \pi]$, then $\sum_{k=1}^{\infty} |a_k(f)|^2$ and $\sum_{k=1}^{\infty} |b_k(f)|^2$ are convergent series. In fact,

$$\frac{|a_0(f)|^2}{2} + \sum_{k=1}^{\infty} \left(|a_k(f)|^2 + |b_k(f)|^2 \right) \leq \frac{1}{\pi} \int_{-\pi}^{\pi} |f(x)|^2 \, dx. \tag{12}$$

Proof. Fix $N \in \mathbf{N}$. By Lemma 14.18,

$$0 \le \frac{1}{\pi} \int_{-\pi}^{\pi} |f(x) - (S_N f)(x)|^2 \, dx$$

$$= \frac{1}{\pi} \int_{-\pi}^{\pi} |f(x)|^2 \, dx - \frac{2}{\pi} \int_{-\pi}^{\pi} f(x)(S_N f)(x) \, dx + \frac{1}{\pi} \int_{-\pi}^{\pi} |(S_N f)(x)|^2 \, dx$$

$$= \frac{1}{\pi} \int_{-\pi}^{\pi} |f(x)|^2 \, dx - \left(\frac{|a_0(f)|^2}{2} + \sum_{k=1}^{N} \left(|a_k(f)|^2 + |b_k(f)|^2 \right) \right).$$

Therefore,

$$\frac{|a_0(f)|^2}{2} + \sum_{k=1}^{N} \left(|a_k(f)|^2 + |b_k(f)|^2 \right) \le \frac{1}{\pi} \int_{-\pi}^{\pi} |f(x)|^2 \, dx$$

for all $N \in \mathbf{N}$. Taking the limit of this inequality as $N \to \infty$ verifies (12). Since $|f|^2$ is Riemann integrable when f is, it follows that both $\sum_{k=1}^{\infty} |a_k(f)|^2$ and $\sum_{k=1}^{\infty} |b_k(f)|^2$ are convergent series. ∎

14.20 Corollary. [RIEMANN–LEBESGUE LEMMA].   *If f is integrable on* $[-\pi, \pi]$, *then*

$$\lim_{k \to \infty} a_k(f) = \lim_{k \to \infty} b_k(f) = 0.$$

Proof. Since the terms of a convergent series converge to zero, it follows from Bessel's Inequality that $a_k(f)$ and $b_k(f)$ converge to zero as $k \to \infty$. ∎

Our next major result shows that Bessel's Inequality is actually an identity when f is continuous and periodic. First, we show that the partial sums of the Fourier series of a function f are the best approximations to f in the following sense.

14.21 Lemma.
Let $N \in \mathbf{N}$. If f is (Riemann) integrable on $[-\pi, \pi]$ and

$$T_N = \frac{c_0}{2} + \sum_{k=1}^{N} (c_k \cos kx + d_k \sin kx)$$

is any trigonometric polynomial of degree N, then

$$\int_{-\pi}^{\pi} |f(x) - (S_N f)(x)|^2 \, dx \le \int_{-\pi}^{\pi} |f(x) - T_N(x)|^2 \, dx.$$

Proof. Notice by (11) that

$$\int_{-\pi}^{\pi} |f(x) - T_N(x)|^2 \, dx$$

$$= \int_{-\pi}^{\pi} |f(x) - (S_N f)(x) + (S_N f)(x) - T_N(x)|^2 \, dx$$

$$= \int_{-\pi}^{\pi} |f(x) - (S_N f)(x)|^2 \, dx$$

$$+ 2 \int_{-\pi}^{\pi} (f(x) - (S_N f)(x))((S_N f)(x) - T_N(x)) \, dx$$

$$+ \int_{-\pi}^{\pi} |(S_N f)(x) - T_N(x)|^2 \, dx$$

$$\geq \int_{-\pi}^{\pi} |f(x) - (S_N f)(x)|^2 \, dx + 2 \int_{-\pi}^{\pi} ((S_N f)(x)T_N(x) - f(x)T_N(x)) \, dx.$$

This last term is zero since, by orthogonality,

$$\frac{1}{\pi} \int_{-\pi}^{\pi} ((S_N f)(x)T_N(x) - f(x)T_N(x)) \, dx$$

$$= \frac{a_0(f)c_0}{4} + \sum_{k=1}^{N}(a_k(f)c_k + b_k(f)d_k)$$

$$- \frac{c_0}{2\pi} \int_{-\pi}^{\pi} f(x) \, dx - \sum_{j=1}^{N} \frac{c_j}{\pi} \int_{-\pi}^{\pi} f(x) \cos jx \, dx$$

$$- \sum_{j=1}^{N} \frac{d_j}{\pi} \int_{-\pi}^{\pi} f(x) \sin jx \, dx$$

$$= \frac{a_0(f)c_0}{4} + \sum_{k=1}^{N}(a_k(f)c_k + b_k(f)d_k)$$

$$- \left(\frac{a_0(f)c_0}{4} + \sum_{k=1}^{N} a_k(f)c_k + b_k(f)d_k \right)$$

$$= 0.$$

Consequently,

$$\int_{-\pi}^{\pi} |f(x) - T_N(x)|^2 \, dx \geq \int_{-\pi}^{\pi} |f(x) - (S_N f)(x)|^2 \, dx. \qquad \blacksquare$$

14.22 Theorem.   [PARSEVAL'S IDENTITY].
If $f : \mathbf{R} \to \mathbf{R}$ is periodic and continuous, then

$$\frac{|a_0(f)|^2}{2} + \sum_{k=1}^{\infty} \left(|a_k(f)|^2 + |b_k(f)|^2 \right) = \frac{1}{\pi} \int_{-\pi}^{\pi} |f(x)|^2 \, dx. \tag{13}$$

Proof. By Bessel's Inequality, we need only show that the left side of (13) is greater than or equal to the right side of (13). Since f is continuous and periodic, $\sigma_N f \to f$ uniformly on $\mathbf{R}$ as $N \to \infty$ by Fejér's Theorem. Hence, it follows from Lemmas 14.18 and 14.21 that

$$\frac{1}{\pi} \int_{-\pi}^{\pi} |f(x)|^2 \, dx - \frac{|a_0(f)|^2}{2} - \sum_{k=1}^{N} \left(|a_k(f)|^2 + |b_k(f)|^2 \right)$$

$$= \frac{1}{\pi} \int_{-\pi}^{\pi} |f(x) - (S_N f)(x)|^2 \, dx \le \frac{1}{\pi} \int_{-\pi}^{\pi} |f(x) - (\sigma_N f)(x)|^2 \, dx \to 0$$

as $N \to \infty$. In particular,

$$\frac{1}{\pi} \int_{-\pi}^{\pi} |f(x)|^2 \, dx \le \frac{|a_0(f)|^2}{2} + \sum_{k=1}^{\infty} \left(|a_k(f)|^2 + |b_k(f)|^2 \right). \qquad \blacksquare$$

The Riemann–Lebesgue Lemma can be improved if f is smooth and periodic. In fact, the following result shows that the smoother f is, the more rapidly its Fourier coefficients converge to zero.

14.23 Theorem.   *Let $f : \mathbf{R} \to \mathbf{R}$ and $j \in \mathbf{N}$. If $f^{(j)}$ exists and is integrable on $[-\pi, \pi]$ and $f^{(\ell)}$ is periodic for each $0 \le \ell < j$, then*

$$\lim_{k \to \infty} k^j a_k(f) = \lim_{k \to \infty} k^j b_k(f) = 0. \tag{14}$$

Proof. Fix $k \in \mathbf{N}$. Since f is periodic, integration by parts yields

$$a_k(f') = \frac{1}{\pi} \int_{-\pi}^{\pi} f'(x) \cos kx \, dx = \frac{k}{\pi} \int_{-\pi}^{\pi} f(x) \sin kx \, dx = k b_k(f).$$

Similarly, $b_k(f') = -k a_k(f)$; hence $a_k(f'') = k b_k(f') = -k^2 a_k(f)$. Iterating, we obtain

$$|a_k(f^{(j)})| = \begin{cases} |k^j a_k(f)| & \text{when } j \text{ is even,} \\ |k^j b_k(f)| & \text{when } j \text{ is odd.} \end{cases}$$

A similar identity holds for $|b_k(f^{(j)})|$. Since the Riemann–Lebesgue Lemma implies $a_k(f^{(j)})$ and $b_k(f^{(j)}) \to 0$ as $k \to \infty$, it follows that $k^j a_k(f) \to 0$ and $k^j b_k(f) \to 0$ as $k \to \infty$. $\qquad \blacksquare$

This result shows that if f is continuously differentiable and periodic, then $ka_k(f)$ and $kb_k(f)$ both converge to zero as $k \to \infty$. Recall that if f is continuously differentiable on $[-\pi, \pi]$, then f is of bounded variation (see Remark 5.51). Thus it is natural to ask, How rapidly do $ka_k(f)$ and $kb_k(f)$ grow when f is a function of bounded variation? To answer this question, let $\{x_0, x_1, \ldots, x_n\}$ be a partition of $[-\pi, \pi]$. Using Riemann sums, the Mean Value Theorem, Abel's Formula, and $\sin kx_0 = \sin kx_n = 0$, we can convince ourselves that

$$\pi a_k(f) = \int_{-\pi}^{\pi} f(x) \cos kx \, dx \approx \sum_{j=1}^{n} f(x_j) \cos kx_j (x_j - x_{j-1})$$

$$\approx \frac{1}{k} \sum_{j=1}^{n} f(x_j)(\sin kx_j - \sin kx_{j-1})$$

$$= \frac{1}{k} \sum_{j=1}^{n-1} (f(x_j) - f(x_{j+1})) \sin kx_j.$$

Since the absolute value of this last sum is bounded by Var f, we guess that $k|a_k(f)| \le \text{Var } f/\pi$.

To prove that our guess is correct, suppose for a moment that f is increasing, periodic, and differentiable on $[-\pi, \pi]$, and $\phi(x) = \sin kx$. Then, by Definition 14.3, periodicity, integration by parts, and the Fundamental Theorem of Calculus, we can estimate the Fourier coefficients of f as follows:

$$\pi k|a_k(f)| = \left| \int_{-\pi}^{\pi} f(x)\phi'(x) \, dx \right|$$

$$= \left| f(x)\phi(x) \Big|_{-\pi}^{\pi} - \int_{-\pi}^{\pi} f'(x)\phi(x) \, dx \right|$$

$$= \left| \int_{-\pi}^{\pi} f'(x)\phi(x) \, dx \right| \le \int_{-\pi}^{\pi} f'(x) \, dx$$

$$= \sum_{j=1}^{n} \int_{x_{j-1}}^{x_j} f'(x) \, dx = \sum_{j=1}^{n} f(x_j) - f(x_{j-1}) \le \text{Var } f.$$

The following result shows that this estimate is valid even when f is neither differentiable nor increasing.

14.24 Lemma.
Suppose that f and ϕ are periodic, where f is of bounded variation on $[-\pi, \pi]$ and ϕ is continuously differentiable on $[-\pi, \pi]$. If $M := \sup_{x \in [-\pi,\pi]} |\phi(x)|$, then

$$\left| \int_{-\pi}^{\pi} f(x)\phi'(x) \, dx \right| \le M \text{ Var } f. \tag{15}$$

Proof. Since f is of bounded variation and ϕ' is continuous on $[-\pi, \pi]$, the product $f\phi'$ is integrable on $[-\pi, \pi]$ (see Corollary 5.23 and the comments following Corollary 5.57).

Let $\varepsilon > 0$ and set $C = \sup_{x \in [-\pi, \pi]} |f(x)|$. Since ϕ' is uniformly continuous and $f\phi'$ is integrable on $[-\pi, \pi]$, choose a partition $P = \{x_0, x_1, \ldots, x_{2n}\}$ of $[-\pi, \pi]$ such that

$$w, c \in [x_{j-1}, x_j] \quad \text{implies} \quad |\phi'(w) - \phi'(c)| < \frac{\varepsilon}{4\pi C} \tag{16}$$

and

$$\left| \sum_{j=1}^{2n} f(w_j)\phi'(w_j)(x_j - x_{j-1}) - \int_{-\pi}^{\pi} f(x)\phi'(x)\, dx \right| < \frac{\varepsilon}{2} \tag{17}$$

for any choice of $w_j \in [x_{j-1}, x_j]$.

Set

$$A := \sum_{j=1}^{2n} f(w_j)(\phi(x_j) - \phi(x_{j-1})),$$

where $w_j = x_j$ when j is even, $w_j = x_{j-1}$ when j is odd. By the Mean Value Theorem, choose $c_j \in [x_{j-1}, x_j]$ such that $\phi(x_j) - \phi(x_{j-1}) = \phi'(c_j)(x_j - x_{j-1})$. Then

$$A = \sum_{j=1}^{2n} f(w_j)\phi'(c_j)(x_j - x_{j-1}).$$

Hence, it follows from (17) and (16) that

$$\left| A - \int_{-\pi}^{\pi} f(x)\phi'(x)\, dx \right|$$

$$\leq \left| \sum_{j=1}^{2n} f(w_j)\phi'(c_j)(x_j - x_{j-1}) - \sum_{j=1}^{2n} f(w_j)\phi'(w_j)(x_j - x_{j-1}) \right|$$

$$+ \left| \sum_{j=1}^{2n} f(w_j)\phi'(w_j)(x_j - x_{j-1}) - \int_{-\pi}^{\pi} f(x)\phi'(x)\, dx \right|$$

$$< \sum_{j=1}^{2n} |f(w_j)|\, |\phi'(c_j) - \phi'(w_j)|(x_j - x_{j-1}) + \frac{\varepsilon}{2}$$

$$\leq \frac{\varepsilon}{4\pi} \sum_{j=1}^{2n} (x_j - x_{j-1}) + \frac{\varepsilon}{2} = \frac{\varepsilon}{2} + \frac{\varepsilon}{2} = \varepsilon.$$

Combining this observation with the Triangle Inequality, we obtain

$$\left| \int_{-\pi}^{\pi} f(x)\phi'(x)\,dx \right| \leq |A| + \varepsilon. \tag{18}$$

On the other hand, by the choice of the w_j's,

$$A = \sum_{j=1}^{n} f(x_{2j-2})(\phi(x_{2j-1}) - \phi(x_{2j-2})) + \sum_{j=1}^{n} f(x_{2j})(\phi(x_{2j}) - \phi(x_{2j-1}))$$

$$= \sum_{j=1}^{n} \phi(x_{2j-1})(f(x_{2j-2}) - f(x_{2j}))$$

$$+ \sum_{j=1}^{n} (f(x_{2j})\phi(x_{2j}) - f(x_{2j-2})\phi(x_{2j-2})).$$

Since f and ϕ are periodic, this last sum telescopes to 0. Therefore,

$$|A| = \left| \sum_{j=1}^{n} \phi(x_{2j-1})(f(x_{2j-2}) - f(x_{2j})) \right|$$

$$\leq \sum_{j=1}^{n} |\phi(x_{2j-1})| \, |f(x_{2j-2}) - f(x_{2j})| \leq M \operatorname{Var} f.$$

This, together with (18), proves that

$$\left| \int_{-\pi}^{\pi} f(x)\phi'(x)\,dx \right| \leq M \operatorname{Var} f + \varepsilon.$$

Taking the limit of this inequality as $\varepsilon \to 0$, we conclude that (15) holds. ∎

We now estimate the rate of growth of Fourier coefficients of functions of bounded variation.

14.25 Theorem. *If* $f : \mathbf{R} \to \mathbf{R}$ *is periodic and of bounded variation on* $[-\pi, \pi]$, *then*

$$|ka_k(f)| \leq \frac{\operatorname{Var} f}{\pi} \quad and \quad |kb_k(f)| \leq \frac{\operatorname{Var} f}{\pi}$$

for $k \in \mathbf{N}$.

Proof. Fix $k \in \mathbf{N}$ and set $\phi(x) = \sin kx$. Then ϕ is periodic and $\phi'(x) = k \cos kx$ is continuously differentiable on $[0, 2\pi]$. Hence, it follows from Lemma 14.24 that

$$|ka_k(f)| = \left| \frac{1}{\pi} \int_{-\pi}^{\pi} f(x)k \cos kx \, dx \right| = \left| \frac{1}{\pi} \int_{-\pi}^{\pi} f(x)\phi'(x) \, dx \right| \le \frac{\text{Var } f}{\pi}.$$

A similar argument proves that $|kb_k(f)| \le \text{Var } f/\pi$.   ∎

EXERCISES

14.3.1. If f is integrable on $[-\pi, \pi]$ and $\alpha \in \mathbf{R}$, prove that

$$\lim_{k \to \infty} \int_{-\pi}^{\pi} f(x) \sin(k + \alpha)x \, dx = 0.$$

14.3.2. Prove that there is no continuous function whose Fourier coefficients satisfy $|a_k(f)| \ge 1/\sqrt{k}$ for $k \in \mathbf{N}$.

14.3.3. Prove that if $f : \mathbf{R} \to \mathbf{R}$ belongs to $C^2(\mathbf{R})$ and f, f' are both periodic, then Sf converges to f uniformly and absolutely on $\mathbf{R}$. (See also Exercise 14.4.5.)

14.3.4. If $f : \mathbf{R} \to \mathbf{R}$ belongs to $C^\infty(\mathbf{R})$ and $f^{(j)}$ is periodic for all $j \ge 0$, prove that Sf is term-by-term differentiable on $\mathbf{R}$. In fact, show that

$$\frac{d^j f}{dx^j}(x) = \sum_{k=1}^{\infty} \frac{d^j}{dx^j}(a_k(f) \cos kx + b_k(f) \sin kx)$$

uniformly for all $j \in \mathbf{N}$.

14.3.5. Suppose that $f : \mathbf{R} \to \mathbf{R}$ is periodic on $\mathbf{R}$, integrable on $[-\pi, \pi]$, and that $a_k(f) \ge 0$ for $k = 0, 1, \ldots$.

a) Prove that $(S_k f)(0) \ge (S_j f)(0)$ for all $k \ge j \ge 0$.
b) Prove that $S_N f(0) \le 2\sigma_{2N} f(0)$ for $N \in \mathbf{N}$.
c) Prove that $\sum_{k=1}^{\infty} |a_k(f)| < \infty$.
d) Suppose that f is also even. Prove that f must be continuous and Sf converges uniformly and absolutely on $\mathbf{R}$.

14.3.6. Suppose that $f : \mathbf{R} \to \mathbf{R}$ is continuous and periodic. The *modulus of continuity* of f is defined by

$$\omega(f, \delta) = \sup_{\substack{t \in [0, 2\pi] \\ |h| \le \delta}} |f(t + h) - f(t)|.$$

a) Show that

$$a_k(f) = \frac{1}{2\pi} \int_{-\pi}^{\pi} \left(f(u) - f\left(u + \frac{\pi}{k}\right) \right) \cos ku \, du$$

for $k \in \mathbf{N}$.

b) Prove that

$$|a_k(f)| \leq \omega\left(f, \frac{\pi}{k}\right) \quad \text{and} \quad |b_k(f)| \leq \omega\left(f, \frac{\pi}{k}\right)$$

for $k \in \mathbf{N}$.

c) Use part b) to give a different proof the Riemann–Lebesgue Lemma in the special case when f is periodic and continuous.

14.3.7. a) Compute the Fourier coefficients of $f(x) = x$.

b) Prove that

$$\sum_{k=1}^{n} \frac{1}{k^2} = \frac{\pi^2}{6}.$$

*14.4 CONVERGENCE OF FOURIER SERIES

This section uses material from Sections 5.5, 14.2, and 14.3.

We shall prove that, under certain conditions, a summable series must also be convergent. Such results, called *Tauberian theorems*, will be used to obtain a partial answer to the Convergence Question posed in Section 14.1 and further results concerning the growth of Fourier coefficients.

The following result was the first Tauberian theorem discovered.

14.26 Theorem. [TAUBER].
Let $a_k \geq 0$ and $L \in \mathbf{R}$. If $\sum_{k=0}^{\infty} a_k$ is Cesàro summable to L, then

$$\sum_{k=0}^{\infty} a_k = L.$$

Proof. By Remark 14.11, it suffices to prove that $\sum_{k=0}^{\infty} a_k < \infty$. Suppose to the contrary that $\sum_{k=0}^{\infty} a_k = \infty$. Then, given $M > 0$, there is an $n_0 \in \mathbf{N}$ such that $n \geq n_0$ implies $s_n := \sum_{k=0}^{n} a_k \geq M$. Let $N > n_0$. Then

$$\sigma_N := \frac{s_0 + s_1 + \cdots + s_{n_0}}{N+1} + \frac{s_{n_0+1} + \cdots + s_N}{N+1} \geq 0 + \frac{N - n_0}{N+1}M.$$

Taking the limit of this last inequality as $N \to \infty$, we obtain $L \geq M$ for all $M > 0$. We conclude that $L = \infty$, a contradiction. ∎

This result can be used to improve the Riemann–Lebesgue Lemma for certain types of functions.

14.27 Corollary. *Let $f : \mathbf{R} \to \mathbf{R}$ be periodic on $\mathbf{R}$ and integrable on $[-\pi, \pi]$. If $a_k(f) = 0$ and $b_k(f) \geq 0$ for $k \in \mathbf{N}$, then*

$$\sum_{k=1}^{\infty} \frac{b_k(f)}{k} < \infty.$$

Proof. By considering $g = f - a_0(f)$, we may suppose that $a_0(f) = 0$. Let

$$F(x) = \int_0^x f(t)\,dt.$$

By Theorem 5.26, F is continuous on **R**. Since $a_0(f) = 0$, F is also periodic. Hence, by Fejér's Theorem, $(\sigma_N F)(0) \to F(0) = 0$ as $N \to \infty$. Integrating by parts, we obtain

$$a_k(F) = \frac{b_k(f)}{k} \geq 0 \quad \text{and} \quad b_k(F) = -\frac{a_k(f)}{k} = 0.$$

It follows that $\sum_{k=1}^{\infty} b_k(f)/k$ is Cesàro summable [to $-a_0(F)/2$] and has nonnegative terms. We conclude by Tauber's Theorem that $\sum_{k=1}^{\infty} b_k(f)/k$ converges. ∎

We are now in a position to see that the converse of the Riemann–Lebesgue Lemma is false. Indeed, if

$$\sum_{k=2}^{\infty} \frac{\sin kx}{\log k}$$

were the Fourier series of some integrable function, then, by Corollary 14.27,

$$\sum_{k=2}^{\infty} \frac{1}{k \log k}$$

would converge, a contradiction of the Integral Test.

The following result is one of the deepest Tauberian theorems.

14.28 Theorem.   [HARDY].
Let $E \subseteq \mathbf{R}$ and suppose that $f_k : E \to \mathbf{R}$ is a sequence of functions which satisfies

$$|k f_k(x)| \leq M \tag{19}$$

for all $x \in E$, all $k \in \mathbf{N}$, and some $M > 0$. If $\sum_{k=0}^{\infty} f_k$ is uniformly Cesàro summable to f on E, then $\sum_{k=0}^{\infty} f_k$ converges uniformly to f on E.

Proof. Fix $x \in E$ and suppose without loss of generality that $M \geq 1$. For each $n = 0, 1, \ldots$, set

$$s_n(x) = \sum_{k=0}^{n} f_k(x), \qquad \sigma_n(x) = \frac{s_0(x) + \cdots + s_n(x)}{n+1},$$

and consider the delayed averages

$$\sigma_{n,k}(x) := \frac{s_n(x) + \cdots + s_{n+k}(x)}{k+1}$$

defined for $n, k \geq 0$.

Let $0 < \varepsilon < 1$. For each $n \in \mathbf{N}$, choose $k = k(n) \in \mathbf{N}$ such that $k + 1 \leq n\varepsilon/(2M) < k + 2$. Then

$$\frac{n-1}{k+1} < \frac{n}{k+1} < \frac{2M}{\varepsilon} < \infty. \tag{20}$$

Moreover, since

$$\sigma_{n,k}(x) - s_n(x) = \frac{(s_n(x) - s_n(x)) + \cdots + (s_{n+k}(x) - s_n(x))}{k+1}$$

$$= \sum_{j=n}^{n+k} \left(1 - \frac{j-n}{k+1}\right) f_j(x),$$

it follows from (19) and the choice of $k = k(n)$ that

$$|\sigma_{n,k}(x) - s_n(x)| \leq \sum_{j=n+1}^{n+k} |f_j(x)| \leq M \sum_{j=n+1}^{n+k} \frac{1}{j} < \frac{M(k+1)}{n+1} < \frac{\varepsilon}{2}. \tag{21}$$

Since $\sigma_n \to f$ uniformly on E, choose $N \in \mathbf{N}$ such that

$$n \geq N \quad \text{and} \quad x \in E \quad \text{imply} \quad |\sigma_n(x) - f(x)| < \frac{\varepsilon^2}{12M}. \tag{22}$$

Since

$$\sigma_{n,k}(x) = \left(1 + \frac{n-1}{k+1}\right)\sigma_{n+k} - \left(\frac{n-1}{k+1}\right)\sigma_{n-1},$$

it follows from (20), (21), and (22) that

$$|s_n(x) - f(x)| \leq |s_n(x) - \sigma_{n,k}(x)| + |\sigma_{n,k} - f(x)|$$

$$< \frac{\varepsilon}{2} + \left(1 + \frac{n-1}{k+1}\right)|\sigma_{n+k}(x) - f(x)|$$

$$+ \left(\frac{n-1}{k+1}\right)|\sigma_{n-1}(x) - f(x)|$$

$$< \frac{\varepsilon}{2} + \left(1 + \frac{2M}{\varepsilon}\right)\left(\frac{\varepsilon^2}{12M}\right) + \frac{2M}{\varepsilon}\left(\frac{\varepsilon^2}{12M}\right)$$

$$= \frac{\varepsilon}{2} + \frac{\varepsilon^2}{12M} + \frac{\varepsilon}{3} < \frac{\varepsilon}{2} + \frac{\varepsilon}{12} + \frac{\varepsilon}{3} < \varepsilon$$

for any $n > N$ and $x \in E$. We conclude that $s_n \to f$ uniformly on E as $n \to \infty$. ∎

We are prepared to answer the Convergence Question posed in Section 14.1 for piecewise continuous functions of bounded variation.

14.29 Theorem. [DIRICHLET–JORDAN].
If $f : \mathbf{R} \to \mathbf{R}$ is periodic on $\mathbf{R}$ and of bounded variation on $[-\pi, \pi]$, then

$$\lim_{N \to \infty} (S_N f)(x) = \frac{f(x+) + f(x-)}{2}$$

for every $x \in \mathbf{R}$. If f is also continuous on some closed interval I, then

$$\lim_{N \to \infty} S_N f = f$$

uniformly on I.

Proof. Since f is periodic and of bounded variation, the one-sided limits $f(x+)$ and $f(x-)$ exist for each $x \in \mathbf{R}$, and f is Riemann integrable on $[-\pi, \pi]$ (see the comments which follow the proof of Corollary 5.57). Hence, by Fejér's Theorem, both conclusions hold if S_N is replaced by σ_N. Since Theorem 14.25 implies

$$|k a_k(f) \cos kx| \quad \text{and} \quad |k b_k(f) \cos kx| \le \frac{\text{Var } f}{\pi}$$

for $k \in \mathbf{N}$, it follows from Hardy's Theorem that both conclusions hold as stated. ∎

We close this section with an application of Fourier series to an extremal problem. We will show that among all smooth simple closed curves in $\mathbf{R}^2$ with a given arc length, the largest area is enclosed by a circle. (The proof presented here comes from Marsden [7].)

14.30 Theorem.   [THE ISOPERIMETRIC PROBLEM].
Let E be a region in $\mathbf{R}^2$ whose topological boundary $C = \partial E$ is a smooth closed simple curve of length 2π. If $A = Area(E)$, then $A \le \pi$. Moreover, $A = \pi$ if and only if $E = B_1(a, b)$ for some $a, b \in \mathbf{R}$.

Proof. Let $(v, [0, 2\pi])$ be the natural parametrization of C; that is, $\|v'(s)\| = 1$ for all $s \in [0, 2\pi]$. Set

$$a = \frac{1}{2\pi} \int_0^{2\pi} v_1(s)\, ds, \qquad b = \frac{1}{2\pi} \int_0^{2\pi} v_2(s)\, ds,$$
$$P(s) = v_1(s) - a, \quad Q(s) = v_2(s) - b, \quad \text{and} \quad \phi(s) = (P(s), Q(s))$$

for $s \in [0, 2\pi]$, where $(v_1, v_2) := v$. Clearly, $(\phi, [0, 2\pi])$ is a smooth parametrization of $\partial E - (a, b)$ whose trace is a smooth closed simple curve with arc length 2π which encloses a region with area A. Moreover,

$$|P'(s)|^2 + |Q'(s)|^2 = 1, \tag{23}$$

$$\frac{1}{2\pi}\int_0^{2\pi} P(s)\,ds = 0, \qquad \frac{1}{2\pi}\int_0^{2\pi} Q(s)\,ds = 0, \tag{24}$$

and, by Green's Theorem,

$$A = \iint_E dA = \int_{\partial E} x\,dy = \int_0^{2\pi} P(s)Q'(s)\,ds. \tag{25}$$

Let a_k, b_k (respectively, c_k, d_k) represent the Fourier coefficients of P (respectively, Q). Since $(\phi, [0, 2\pi])$ is smooth and closed, P and Q are continuously differentiable and periodic. By (24) and the Dirichlet–Jordan Theorem,

$$P(s) = \sum_{k=1}^{\infty}(a_k \cos ks + b_k \sin ks), \qquad Q(s) = \sum_{k=1}^{\infty}(c_k \cos ks + d_k \sin ks), \tag{26}$$

$$P'(s) = \sum_{k=1}^{\infty}(kb_k \cos ks - ka_k \sin ks), \quad \text{and} \quad Q'(s) = \sum_{k=1}^{\infty}(kd_k \cos ks - kc_k \sin ks) \tag{27}$$

uniformly on $[0, 2\pi]$. Hence, by (23) and Parseval's Identity,

$$2\pi = \int_0^{2\pi}(|P'(s)|^2 + |Q'(s)|^2)\,ds = \pi\sum_{k=1}^{\infty}k^2(a_k^2 + b_k^2 + c_k^2 + d_k^2).$$

Moreover, by (25) and orthogonality

$$A = \int_0^{2\pi} P(s)Q'(s)\,ds = \pi\sum_{k=1}^{\infty}k(a_kd_k - b_kc_k).$$

It follows that

$$\pi - A = \frac{\pi}{2}\sum_{k=2}^{\infty}(k^2 - k)(a_k^2 + b_k^2 + c_k^2 + d_k^2) + \frac{\pi}{2}\sum_{k=1}^{\infty}k((a_k - d_k)^2 + (c_k + b_k)^2) \geq 0.$$

In particular, $A \leq \pi$ and $A = \pi$ if and only if $a_1 = d_1$, $c_1 = -b_1$, and $a_k = b_k = c_k = d_k = 0$ for $k \geq 2$.

Suppose that $A = \pi$. Then $P(s) = a_1 \cos s + b_1 \sin s$ and $Q(s) = -b_1 \cos s + a_1 \sin s = -P(s + \frac{\pi}{2})$. Thus $P'(s) = -Q(s)$ and $Q'(s) = -P''(s) = P(s)$ for all $s \in [0, 2\pi]$. It follows from (23) that $\phi([0, 2\pi])$ is a subset of $\partial B_1(0, 0)$. Since $\phi(0) = \phi(2\pi)$, we must have $\phi([0, 2\pi]) = \partial B_1(0, 0)$. Therefore, C is the boundary of the disk $E = B_1(a, b)$. ∎

EXERCISES

14.4.1. Suppose that f is continuous and of bounded variation on $[-\pi, \pi]$. Prove that $S_N f \to f$ pointwise on $(-\pi, \pi)$ and uniformly on any $[a, b] \subset (-\pi, \pi)$.

14.4.2. a) Prove that

$$x = 2 \sum_{k=1}^{\infty} \frac{(-1)^{k+1}}{k} \sin kx$$

pointwise on $(-\pi, \pi)$ and uniformly on any $[a, b] \subset (-\pi, \pi)$.

b) Prove that

$$|x| = \frac{\pi}{2} - \frac{4}{\pi} \sum_{k=1}^{\infty} \frac{\cos(2k-1)x}{(2k-1)^2}$$

uniformly on $[-\pi, \pi]$.

c) Find a value for

$$\sum_{k=1}^{\infty} \frac{1}{(2k-1)^2}.$$

14.4.3. Prove that if f is continuous, odd, and periodic, then $\sum_{k=1}^{\infty} b_k(f)/k$ converges.

14.4.4. Let $L \in \mathbf{R}$. A series $\sum_{k=0}^{\infty} a_k$ is said to be *Abel summable* to L if and only if

$$\lim_{r \to 1-} \sum_{k=0}^{\infty} a_k r^k = L.$$

a) Let $S_k = \sum_{j=0}^{k} a_k$. Prove that

$$\sum_{k=0}^{\infty} a_k r^k = (1-r) \sum_{k=0}^{\infty} S_k r^k = (1-r)^2 \sum_{k=0}^{\infty} (k+1) \sigma_k r^k,$$

provided any one of these series converges for all $0 < r < 1$.

b) Prove that if $\sum_{k=0}^{\infty} a_k$ is Cesàro summable to L, then it is Abel summable to L.

c) Prove that if f is continuous, periodic, and of bounded variation on $\mathbf{R}$, then Sf is Abel summable to f uniformly on $\mathbf{R}$.

d) Show that if $a_k \geq 0$ and $\sum_{k=0}^{\infty} a_k$ is Abel summable to L, then $\sum_{k=0}^{\infty} a_k$ converges to L.

14.4.5. [BERNSTEIN]. Let $f : \mathbf{R} \to \mathbf{R}$ be periodic and $\alpha > 0$. Suppose that f is Lipschitz of order α; that is, there is a constant $M > 0$ such that

$$|f(x+h) - f(x)| \leq M|h|^{\alpha}$$

for all $x, h \in \mathbf{R}$.

a) Prove that

$$\frac{1}{\pi} \int_{-\pi}^{\pi} |f(x+h) - f(x-h)|^2 \, dx = 4 \sum_{k=1}^{\infty} (a_k^2(f) + b_k^2(f)) \sin^2 kh$$

holds for each $h \in \mathbf{R}$.

b) If $h = \pi/2^{n+1}$, prove that $\sin^2 kh \geq 1/2$ for all $k \in [2^{n-1}, 2^n]$.

c) Combine parts a) and b) to prove that

$$\left\{ \sum_{k=2^{n-1}}^{2^n-1} (a_k^2(f) + b_k^2(f)) \right\}^{1/2} \leq M^2 \left(\frac{\pi}{2^{n+1}} \right)^{2\alpha}$$

for $n = 1, 2, 3, \ldots$.

d) Assuming

$$\sum_{k=2^{n-1}}^{2^n-1} (|a_k(f)| + |b_k(f)|) \leq 2^{n/2} \left(\sum_{k=2^{n-1}}^{2^n-1} (a_k^2(f) + b_k^2(f)) \right)^{1/2}$$

(see Exercise 11.7.9), prove that if f is Lipschitz of order α for some $\alpha > 1/2$, then Sf converges absolutely and uniformly on $\mathbf{R}$.

e) Prove that if $f : \mathbf{R} \to \mathbf{R}$ is periodic and continuously differentiable, then Sf converges absolutely and uniformly on $\mathbf{R}$.

*14.4.6. Suppose that $f : \mathbf{R} \to \mathbf{R}$ is periodic and of bounded variation on $[-\pi, \pi]$. Prove that $S_N f \to f$ almost everywhere as $N \to \infty$ (see Exercise 14.2.8).

*14.5 UNIQUENESS

This section uses material from Section 14.4.

In this section we examine the Uniqueness Question posed in Section 14.1. We begin with the following generalization of the second derivative.

14.31 Definition.

Let $x_0 \in \mathbf{R}$ and let I be an open interval containing x_0. A function $F : I \to \mathbf{R}$ is said to have a *second symmetric derivative* at x_0 if and only if

$$D_2 F(x_0) = \lim_{h \to 0+} \frac{F(x_0 + 2h) + F(x_0 - 2h) - 2F(x_0)}{4h^2}$$

exists.

14.32 Remark. *Let $x_0 \in \mathbf{R}$ and let I be an open interval containing x_0. If F is differentiable on I and $F''(x_0)$ exists, then F has a second symmetric derivative at x_0 and $D_2 F(x_0) = F''(x_0)$.*

Proof. Set $G(t) = F(x_0 + 2t) + F(x_0 - 2t)$ for $t \in I$ and $H(t) = 4t^2$ and fix $t \in I$. By Theorem 4.15 (the Generalized Mean Value Theorem),

$$\frac{F(x_0 + 2t) + F(x_0 - 2t) - 2F(x_0)}{4t^2} = \frac{G(t) - G(0)}{H(t) - H(0)} = \frac{G'(c)}{H'(c)}$$

$$= \frac{F'(x_0 + 2c) - F'(x_0 - 2c)}{4c}$$

for some c between 0 and t. Since $c \to 0$ as $t \to 0$, it follows that

$$D_2 F(x_0) = \lim_{c \to 0} \frac{F'(x_0 + 2c) - F'(x_0 - 2c)}{4c}$$

$$= \frac{1}{2} \lim_{c \to 0} \left(\frac{F'(x_0 + 2c) - F'(x_0)}{2c} + \frac{F'(x_0) - F'(x_0 - 2c)}{2c} \right)$$

$$= \frac{1}{2}(F''(x_0) + F''(x_0)) = F''(x_0). \qquad \blacksquare$$

The converse of Remark 14.32 is false. Indeed, if

$$F(x) = \begin{cases} 1 & x > 0 \\ 0 & x = 0 \\ -1 & x < 0, \end{cases}$$

then $D_2 F(0) = 0$ but $F''(0)$ does not exist.

The following result reinforces further the analogy between the second derivative and the second symmetric derivative (see also Exercises 14.5.1 and 14.5.5).

14.33 Lemma.
Let $[a, b]$ be a closed bounded interval. If $F : [a, b] \to \mathbf{R}$ is continuous on $[a, b]$ and $D_2 F(x) = 0$ for all $x \in (a, b)$, then F is linear on $[a, b]$; that is, there exist constants m, γ such that $F(x) = mx + \gamma$ for all $x \in [a, b]$.

Proof. Let $\varepsilon > 0$. By hypothesis,

$$\phi(x) := F(x) - F(a) - \left(\frac{F(b) - F(a)}{b - a} \right)(x - a) + \varepsilon(x - a)(x - b)$$

is continuous on $[a, b]$, and, by Remark 14.32,

$$D_2 \phi(x) = D_2 F(x) + 2\varepsilon = 2\varepsilon \qquad (29)$$

for $x \in (a, b)$.

We claim that $\phi(x) \leq 0$ for $x \in [a, b]$. Clearly, $\phi(a) = \phi(b) = 0$. If $\phi(x) > 0$ for some $x \in (a, b)$, then ϕ attains its maximum at some $x_0 \in (a, b)$. By Exercise 14.5.1, $D_2 \phi(x_0) \leq 0$; hence, by (29), $2\varepsilon \leq 0$, a contradiction. This proves the claim.

Fix $x \in [a, b]$. We have shown that

$$F(x) - F(a) - \left(\frac{F(b) - F(a)}{b - a} \right)(x - a) \leq \varepsilon(x - a)(b - x).$$

A similar argument establishes that

$$F(x) - F(a) - \left(\frac{F(b) - F(a)}{b - a} \right)(x - a) \geq -\varepsilon(x - a)(b - x).$$

Therefore,

$$\left| F(x) - F(a) - \left(\frac{F(b) - F(a)}{b - a} \right)(x - a) \right| \leq \varepsilon(x - a)(b - x) \leq \varepsilon(b - a)^2.$$

Taking the limit of this inequality as $\varepsilon \to 0$, we conclude that

$$F(x) = F(a) + \left(\frac{F(b) - F(a)}{b - a} \right)(x - a)$$

for all $x \in [a, b]$; that is, F is linear on $[a, b]$. ∎

14.34 Definition.

The *second formal integral* of a trigonometric series,

$$S = \frac{a_0}{2} + \sum_{k=1}^{\infty} (a_k \cos kx + b_k \sin kx),$$

is the function

$$F(x) = \frac{a_0}{4} x^2 - \sum_{k=1}^{\infty} \frac{1}{k^2} (a_k \cos kx + b_k \sin kx).$$

By the Weierstrass M-Test, if the coefficients of S are bounded, then the second formal integral of S converges uniformly on $\mathbf{R}$. In particular, the second formal integral always exists when the coefficients of S converge to zero.

Notice that the second formal integral of a trigonometric series S is the result of integrating S twice term by term. Hence, it is not unreasonable to expect that two derivatives of the second formal integral F might recapture the original series S. Although this statement is not quite correct, the following result shows that there is a simple connection between the limit of the series S and the second *symmetric* derivative of F.

14.35 Theorem.   [RIEMANN].
Suppose that

$$S = \frac{a_0}{2} + \sum_{k=1}^{\infty} (a_k \cos kx + b_k \sin kx)$$

is a trigonometric series whose coefficients a_k, $b_k \to 0$ as $k \to \infty$ and let F be the second formal integral of S. If $S(x_0)$ converges to L for some $x_0 \in \mathbf{R}$, then $D_2 F(x_0) = L$.

Proof. Let F_N denote the partial sums of F. After several applications of Theorem B.3, we observe that

$$\lim_{h \to 0} \frac{F_N(x_0 + 2h) + F_N(x_0 - 2h) - 2F_N(x_0)}{4h^2}$$

$$= \lim_{h \to 0} \left(\frac{a_0}{2} + \sum_{k=1}^{N} (a_k \cos kx_0 + b_k \sin kx_0) \left(\frac{\sin kh}{kh} \right)^2 \right)$$

$$= \frac{a_0}{2} + \sum_{k=1}^{N} (a_k \cos kx_0 + b_k \sin kx_0)$$

holds for any $N \in \mathbf{N}$. Therefore, it suffices to show that given $\varepsilon > 0$ there is an $N \in \mathbf{N}$ such that

$$|R_N| := \left| \sum_{k=N+1}^{\infty} (a_k \cos kx_0 + b_k \sin kx_0) \left(\frac{\sin kh}{kh} \right)^2 \right| < \varepsilon \tag{30}$$

for all $|h| \leq 1$.
 Let

$$A_k = \sum_{j=k+1}^{\infty} (a_j \cos jx_0 + b_j \sin jx_0) \quad \text{and} \quad B_k = \left(\frac{\sin kh}{kh} \right)^2$$

for $k \in \mathbf{N}$. Since $A_n \to 0$ as $n \to \infty$, we have by Abel's Formula that

$$R_N := \lim_{n \to \infty} \sum_{k=N+1}^{n} (A_{k-1} - A_k) B_k$$

$$= \lim_{n \to \infty} \left((A_N - A_n) B_n - \sum_{k=N+1}^{n-1} (A_N - A_k)(B_{k+1} - B_k) \right) \tag{31}$$

$$= A_N B_{N+1} + \sum_{k=N+1}^{\infty} A_k (B_{k+1} - B_k).$$

Moreover, by the Fundamental Theorem of Calculus,

$$|B_{k+1} - B_k| = \left| \int_{kh}^{(k+1)h} \frac{d}{dt} \left(\frac{\sin t}{t} \right)^2 dt \right|. \tag{32}$$

Since

$$\frac{d}{dt} \left(\frac{\sin t}{t} \right)^2 = \frac{2 \sin t}{t} \left(\frac{t \cos t - \sin t}{t^2} \right)$$

is bounded near $t = 0$ and is bounded by $2(t + 1)/t^3 < 2/t^2$ for $t \geq 2$, it is clear that the improper integral

$$C = \int_0^\infty \left| \frac{d}{dt} \left(\frac{\sin t}{t} \right)^2 \right| dt$$

converges. Since $\{B_k\}$ is bounded and $A_N \to 0$ as $N \to \infty$, we can choose an $N \in \mathbf{N}$ such that

$$|A_N B_{N+1}| < \frac{\varepsilon}{2} \quad \text{and} \quad k \geq N \quad \text{implies} \quad |A_k| < \frac{\varepsilon}{2C}. \tag{33}$$

It follows from (32) that

$$\sum_{k=N+1}^\infty A_k (B_{k+1} - B_k) \leq \frac{\varepsilon}{2C} \sum_{k=N+1}^\infty \left| \int_{kh}^{(k+1)h} \frac{d}{dt} \left(\frac{\sin t}{t} \right)^2 dt \right|$$

$$\leq \frac{\varepsilon}{2C} \int_0^\infty \left| \frac{d}{dt} \left(\frac{\sin t}{t} \right)^2 \right| dt = \frac{\varepsilon}{2}.$$

Combining this inequality with (31) and (33), we conclude that $|R_N| < \varepsilon$. ∎

The following result shows that the hypotheses of Riemann's Theorem are satisfied by any trigonometric series which converges pointwise on a nondegenerate interval.

14.36 Theorem. [THE CANTOR–LEBESGUE LEMMA].
If

$$S = \frac{a_0}{2} + \sum_{k=1}^\infty (a_k \cos kx + b_k \sin kx)$$

is a trigonometric series which converges pointwise on a nondegenerate interval [a, b], *then its coefficients satisfy* a_k, $b_k \to 0$ *as* $k \to \infty$.

Proof. Set $\rho_0 = a_0/2$ and $\rho_k^2 = a_k^2 + b_k^2$ for $k \in \mathbf{N}$. If the result is false, then there is a $\delta > 0$ such that $\rho_k > \delta$ for infinitely many $k \in \mathbf{N}$.

Set $\theta_0 = 0$ and for each $k \in \mathbf{N}$ define $\theta_k \in \mathbf{R}$ so that $a_k = \rho_k \cos k\theta_k$, $b_k = \rho_k \sin k\theta_k$. By a sum-angle formula,

$$\frac{a_0}{2} + \sum_{k=1}^{n}(a_k \cos kx + b_k \sin kx) = \sum_{k=0}^{n} \rho_k \cos k(x - \theta_k)$$

for each $x \in \mathbf{R}$ and $n \in \mathbf{N}$. Since S converges on $[a, b]$, it follows that

$$\lim_{k \to \infty} \rho_k \cos k(x - \theta_k) = 0 \tag{34}$$

for all $x \in [a, b]$.

Set $I_0 = [a, b]$ and $k_0 = 1$. Fix $j \geq 0$ and suppose that a closed interval $I_j \subseteq I_0$ and an integer $k_j > k_0$ have been chosen. Choose $k_{j+1} > k_j$ such that $k_{j+1}|I_j| > 2\pi$ and $\rho_{k_{j+1}} > \delta$. Clearly, $k_{j+1}(x - \theta_{k_{j+1}})$ runs over an interval of length $> 2\pi$ as x runs over I_j. Hence, we can choose a closed interval $I_{j+1} \subseteq I_j$ such that

$$x \in I_{j+1} \quad \text{implies} \quad \cos k_{j+1}(x - \theta_{k_{j+1}}) \geq \frac{1}{2}.$$

By induction, then, there exist integers $1 < k_1 < k_2 < \ldots$ and a nested sequence of closed intervals $I_0 \supseteq I_1 \supseteq \ldots$ such that

$$\rho_{k_j} \cos k_j (x - \theta_{k_j}) \geq \frac{\delta}{2} \tag{35}$$

for $x \in I_j$, $j \in \mathbf{N}$. By the Nested Interval Property, there is an $x \in I_j$ for all $j \in \mathbf{N}$. This x must satisfy (35) for all $j \in \mathbf{N}$ and must belong to $[a, b]$ by construction. Since this contradicts (34), we conclude that $\rho_k \to 0$ as $k \to \infty$. ∎

We are now prepared to answer the Uniqueness Question for continuous functions of bounded variation.

14.37 Theorem. [Cantor].
Suppose that

$$S = \frac{a_0}{2} + \sum_{k=1}^{\infty}(a_k \cos kx + b_k \sin kx)$$

converges pointwise on $[-\pi, \pi]$ to a function f which is periodic and continuous on $\mathbf{R}$, and of bounded variation on $[-\pi, \pi]$. Then S is the Fourier series of f; that is, $a_k = a_k(f)$ for $k = 0, 1, \ldots$, and $b_k = b_k(f)$ for $k = 1, 2, \ldots$.

Proof. Suppose first that $f(x) = 0$ for all $x \in \mathbf{R}$. By the Cantor–Lebesgue Lemma, the coefficients a_k, b_k tend to zero as $k \to \infty$. Thus the second formal integral F of S is continuous on $\mathbf{R}$ and by Riemann's Theorem has a second symmetric derivative which satisfies $D_2 F(x) = 0$ for $x \in \mathbf{R}$. It follows that F is linear on $\mathbf{R}$; that is, there exist numbers m and γ such that

$$mx + \gamma = \frac{a_0}{4} x^2 - \sum_{k=1}^{\infty} \frac{1}{k^2} (a_k \cos kx + b_k \sin kx)$$

for $x \in \mathbf{R}$. Since the series in this expression is periodic, it must be the case that $m = a_0 = 0$; that is,

$$\gamma + \sum_{k=1}^{\infty} \frac{1}{k^2} (a_k \cos kx + b_k \sin kx) = 0$$

for all $x \in \mathbf{R}$. Since this series converges uniformly, it follows from Theorem 14.4 that $\gamma = 0$ and $a_k = b_k = 0$ for $k \in \mathbf{N}$. This proves the theorem when $f = 0$.

If f is periodic, continuous, and of bounded variation on $[-\pi, \pi]$, then $S_N f \to f$ uniformly on $\mathbf{R}$ by Theorem 14.29. Hence, the series $S - Sf$ converges pointwise on $\mathbf{R}$ to zero. It follows from the case already considered that $a_k - a_k(f) = 0$ for $k = 0, 1, \ldots,$ and $b_k - b_k(f) = 0$ for $k = 1, 2, \ldots$. ∎

EXERCISES

14.5.1. Suppose that $F : \mathbf{R} \to \mathbf{R}$ has a second symmetric derivative at some x_0. Prove that if $F(x_0)$ is a local maximum, then $D_2 F(x_0) \leq 0$, and if $F(x_0)$ is a local minimum, then $D_2 F(x_0) \geq 0$.

14.5.2. Prove that if the coefficients of a trigonometric series are bounded, then its second formal integral converges uniformly on $\mathbf{R}$.

14.5.3. Prove that if $f : \mathbf{R} \to \mathbf{R}$ is periodic, then there exists at most one trigonometric series which converges to f pointwise on $\mathbf{R}$.

14.5.4. Suppose that $f : \mathbf{R} \to \mathbf{R}$ is periodic, piecewise continuous, and of bounded variation on $\mathbf{R}$. Prove that if S is a trigonometric series which converges to $(f(x+) + f(x-))/2$ for all $x \in \mathbf{R}$, then S is the Fourier series of f.

*14.5.5. Suppose that $F : (a, b) \to \mathbf{R}$ is continuous and $D_2 F(x) > 0$ for all $x \in (a, b)$. Prove that F is convex on (a, b).

Appendices

A. ALGEBRAIC LAWS

In this appendix we derive several consequences of the ordered field axioms (i.e., Postulates 1 and 2 in Section 1.2) and show that by Postulate 3, if $t > 0$, then $\sqrt{t} \in \mathbf{R}$.

A.1 Theorem. *Let $x, a \in \mathbf{R}$.*

i) *If $a = x + a$, then $x = 0$.*

ii) *If $a = x \cdot a$ and $a \neq 0$, then $x = 1$.*

> **Proof.** i) Since the additive inverse of a exists, we can add $-a$ to the equation $a = x + a$. Using the Associative Property and the fact that 0 is the additive identity, we obtain
>
> $$0 = a + (-a) = (x + a) + (-a) = x + (a + (-a)) = x + 0 = x.$$
>
> ii) Since the multiplicative inverse of a exists, we can multiply $a = x \cdot a$ by a^{-1}. Using the Associative Property and the fact that 1 is the multiplicative identity, we obtain
>
> $$1 = a \cdot a^{-1} = (x \cdot a) \cdot a^{-1} = x \cdot (a \cdot a^{-1}) = x \cdot 1 = x. \qquad \blacksquare$$

Theorem A.1 shows that the additive and multiplicative identities are unique. The following result shows that additive and multiplicative inverses are also unique. Thus *unique* can be dropped from the statements in Postulate 1.

A.2 Theorem.

i) *If $a, b \in \mathbf{R}$ and $a + b = 0$, then $b = -a$.*

ii) *If $a, b \in \mathbf{R}$ and $ab = 1$, then $b = a^{-1}$.*

> **Proof.** i) By hypothesis and the Associative Property,
>
> $$-a = -a + (a + b) = (-a + a) + b = 0 + b = b.$$
>
> ii) Since $1 \neq 0$, $a \neq 0$. Thus it follows from hypothesis and the Associative Property that
>
> $$a^{-1} = a^{-1}(ab) = (a^{-1}a)b = 1 \cdot b = b. \qquad \blacksquare$$

A.3 Theorem. *For all $a, b \in \mathbf{R}$, $0 \cdot a = 0$, $-a = (-1) \cdot a$, $-(-a) = a$, $(-1)^2 = 1$, and $-(a - b) = b - a$.*

Proof. Since 1 is the multiplicative identity and 0 is the additive identity, it follows from the Distributive Property that

$$a + 0 \cdot a = 1 \cdot a + 0 \cdot a = (1 + 0) \cdot a = 1 \cdot a = a.$$

Hence, by Theorem A.1, $0 \cdot a = 0$. Similarly,

$$a + (-1) \cdot a = (1 + (-1)) \cdot a = 0 \cdot a = 0.$$

Since additive inverses are unique, it follows that $(-1) \cdot a = -a$. Since $-a + a = a + (-a) = 0$, a similar argument proves that $-(-a) = a$. Substituting $a = -1$, we have

$$(-1)(-1) = -(-1) = 1.$$

Finally, for any $a, b \in \mathbf{R}$, we also have

$$-(a - b) = (-1)(a - b) = (-1)a + (-1)(-b) = -a + b = b - a. \qquad \blacksquare$$

A.4 Theorem. *Let $a, b, c \in \mathbf{R}$.*
i) *If $a \cdot b = 0$, then $a = 0$ or $b = 0$.*
ii) *If $a \cdot b = a \cdot c$ and $a \neq 0$, then $b = c$.*

Proof. i) If $a = 0$, we are done. If $a \neq 0$, then multiplying the identity $0 = a \cdot b$ by a^{-1}, we have

$$0 = a^{-1} \cdot 0 = a^{-1} \cdot (a \cdot b) = (a^{-1} \cdot a) \cdot b = 1 \cdot b = b.$$

ii) If $a \cdot b = a \cdot c$, then by Theorem A.3 we have

$$a \cdot (b - c) = a \cdot (b + (-1)c) = a \cdot b + (-1)a \cdot c = a \cdot b - a \cdot c = 0.$$

Since $a \neq 0$, it follows from part i) that $b - c = 0$ (i.e., $b = c$). $\qquad \blacksquare$

A subset E of $\mathbf{R}$ is called *inductive* if

$$1 \in E$$

and

$$\text{for every } x \in E, \ x + 1 \text{ also belongs } to \ E. \tag{1}$$

Notice by Postulate 1, $\mathbf{R}$ is an inductive set.

Define $\mathbf{N}$ to be the set of elements which belong to ALL inductive sets, and set $\mathbf{Z} := \{k \in \mathbf{R} : k \in \mathbf{N}, -k \in \mathbf{N}, \text{ or } k = 0\}$. Notice that $\mathbf{N}$ is the smallest inductive set (i.e., $\mathbf{N}$ is inductive), and $\mathbf{N} \subseteq E$ for any inductive set E. Indeed, since 1 belongs to all inductive sets, $1 \in \mathbf{N}$. If $k \in \mathbf{N}$ and E is any inductive set, then $k \in E$. Since E is inductive, it follows that $k + 1 \in E$ for all inductive sets E, so $k + 1 \in \mathbf{N}$. Finally, if $k \in \mathbf{N}$ and E is inductive, then by definition, $k \in E$. Thus $\mathbf{N} \subseteq E$.

Since **N** is the smallest inductive set, it automatically satisfies the Principle of Mathematical Induction. Indeed, if $E \subseteq \mathbf{N}$ satisfies "$1 \in E$" and "$k \in E$ implies $k + 1 \in E$," then E is an inductive subset of the smallest inductive set **N** and, hence, must be equal to **N**.

It is now fairly easy to prove that **N** and **Z**, as defined above, satisfies the assumptions made in Remark 1.1. We begin with an easy consequence of the Trichotomy Property.

A.5 *Lemma.*

 Given $n \in \mathbf{Z}$, one and only one of the following statements holds: $n \in \mathbf{N}$, $-n \in \mathbf{N}$, or $n = 0$.

Proof. Since $(0, \infty)$ is an inductive set, all elements of **N** are positive. By the definition of **Z**, given $n \in \mathbf{Z}$, one of the following statements holds: $n \in \mathbf{N}$, $-n \in \mathbf{N}$, or $n = 0$. It follows that either $n > 0$, $n < 0$, or $n = 0$. Since the Trichotomy Property implies that only one of these conditions can hold for a given n, the lemma is proved. ∎

Next, we show that **N** and **Z** satisfy the assumptions made in Remark 1.1ii and iii.

A.6 Theorem.

 i) *If $n \in \mathbf{Z}$, then $n \in \mathbf{N}$ if and only if $n \geq 1$.*
 ii) *No $n \in \mathbf{Z}$ satisfies $0 < n < 1$.*

Proof. i) Since $[1, \infty)$ is an inductive set, $\mathbf{N} \subseteq [1, \infty)$. Thus every $n \in \mathbf{N}$ satisfies $n \geq 1$.

Conversely, if $n \in \mathbf{Z}$ and $n \geq 1$, then $n \neq 0$ so by Lemma A.5, either $n \in \mathbf{N}$ or $-n \in \mathbf{N}$. But if $-n \in \mathbf{N}$, then by what we've already proved, $-n \geq 1$ so by the Second Multiplicative Property and Example 1.2, $n \leq -1 < 1$, a contradiction of the Trichotomy Property. We conclude that $n \in \mathbf{N}$.

ii) If $n \in \mathbf{Z}$ and $n > 0$, then by Lemma A.5, $n \in \mathbf{N}$. Hence by part i), $n \geq 1$ so it cannot satisfy $n < 1$. ∎

We need two more preliminary results before we prove that **Z** is closed under addition, subtraction, and multiplication.

A.7 *Lemma.*

 If $n \in \mathbf{Z}$, then $n \pm 1 \in \mathbf{Z}$.

Proof. Fix $n \in \mathbf{Z}$. By Lemma A.5, $n = 0$, or $n \in \mathbf{N}$, or $-n \in \mathbf{N}$.

If $n = 0$, then $n \pm 1 = \pm 1$ belong to **Z** by definition.

If $n \in \mathbf{N}$, then $n + 1 \in \mathbf{N} \subset \mathbf{Z}$ because **N** is an inductive set. On the other hand, since $A = \{k \in \mathbf{N} : k - 1 \in \mathbf{Z}\}$ is obviously an inductive set, we have $A = \mathbf{N}$ (i.e., $n \in \mathbf{N}$ implies $n - 1 \in \mathbf{Z}$).

Finally, if $-n \in \mathbf{N}$, then by what we just proved, $-n \mp 1 \in \mathbf{Z}$, so by definition, $n \pm 1 = -(-n \mp 1) \in \mathbf{Z}$.   ∎

Next, we prove that $\mathbf{N}$ is closed under addition and multiplication.

A.8 Theorem.   *If $n, m \in \mathbf{N}$, then $n + m$ and nm both belong to $\mathbf{N}$.*

Proof. Fix $n \in \mathbf{N}$, and consider the set $A := \{m \in \mathbf{N} : n + m \in \mathbf{N}\}$. Since $\mathbf{N}$ is an inductive set, $1 \in A$. If $m \in A$ for some $m \geq 1$, then $n + m \in \mathbf{N}$. Since $\mathbf{N}$ is inductive, it follows that $n + (m + 1) = (n + m) + 1 \in \mathbf{N}$ (i.e., $m + 1 \in A$). Thus, by induction, $A = \mathbf{N}$ and closure holds for addition.

For multiplication, consider $B = \{m \in \mathbf{N} : nm \in \mathbf{N}\}$. Clearly, $1 \in B$. If some $m \in B$ (i.e., $mn \in \mathbf{N}$), then $n(m + 1) = nm + n$ also belongs to $\mathbf{N}$ since we have already proved that $\mathbf{N}$ is closed under addition. Thus $m \in B$ implies $m + 1 \in B$. By induction $B = \mathbf{N}$.   ∎

We are now prepared to prove that $\mathbf{Z}$ is closed under addition, subtraction, and multiplication.

A.9 Theorem.   *If $n, m \in \mathbf{Z}$, then $n + m$, $n - m$, and nm all belong to $\mathbf{Z}$.*

Proof. Fix $n \in \mathbf{Z}$. By Lemma A.5, $n = 0$, or $n \in \mathbf{N}$, or $-n \in \mathbf{N}$.

Suppose that $n = 0$. Then for all $m \in \mathbf{Z}$, $n \pm m = \pm m \in \mathbf{Z}$ and $nm = 0 \in \mathbf{Z}$.

Suppose that $n \in \mathbf{N}$ and $m \in \mathbf{Z}$. By Lemma A.5, either $m \in \mathbf{N}$, $m = 0$, or $-m \in \mathbf{N}$. In the first two cases, we have by Lemma A.8 or a triviality that $n + m$ and nm both belong to $\mathbf{N}$, hence also to $\mathbf{Z}$. If $-m \in \mathbf{N}$, then consider the set $A = \{k \in \mathbf{N} : n - k \in \mathbf{Z}\}$. By Lemma A.8, $1 \in A$. If $k \in A$, then $n - (k + 1) = (n - k) - 1 \in \mathbf{Z}$ by the Inductive Hypothesis and Lemma A.8 again. By induction, then, $A = \mathbf{N}$ (i.e., $n - k \in \mathbf{Z}$ for all $k \in \mathbf{N}$). In particular, $n - m \in \mathbf{Z}$ as promised. A similar argument shows that $nm \in \mathbf{Z}$.

Finally, if $-n \in \mathbf{N}$, then by what we just proved, $(-n) \pm m$ and $(-n)m \in \mathbf{Z}$ for all $m \in \mathbf{Z}$. Since $\mathbf{Z}$ is closed under multiplication by -1, we conclude that $n \pm m$ and nm belong to $\mathbf{Z}$.   ∎

Our next result shows that the Complete Ordered Field axioms (Postulates 1, 2, and 3) guarantee that every positive real number has a square root. The same method can be used to show that if $t > 0$ and $n \in \mathbf{N}$, then $\sqrt[n]{t}$ exists. Thus $t^{m/n} := \sqrt[n]{t^m}$ can be defined using only the Complete Ordered Field axioms.

A.10 Theorem.   *If $t > 0$, then $\sqrt{t}$ exists; that is, there is a number $b \in (0, \infty)$ which satisfies $b^2 = t$.*

Proof. Our proof has two steps.

Step 1. *A perfect square can always be made a little bigger or a little smaller; namely, given $a < b^2 < c$ there exist real numbers x and y such that $x < b < y$ and*

$$a < x^2 < b^2 < y^2 < c.$$

To prove this, let $\varepsilon = \min \{0.5, (c - b^2)/(2b + 1)\}$ and set $y = b + \varepsilon$. Since $b > 0$ and $b^2 < c$, the number ε is positive. Thus $b < y$. Since $\varepsilon < 1$, we also have $\varepsilon^2 < \varepsilon$. Therefore,

$$y^2 = b^2 + 2\varepsilon b + \varepsilon^2 < b^2 + 2\varepsilon b + \varepsilon = b^2 + (2b + 1)\varepsilon.$$

Since $\varepsilon \leq (c - b^2)/(2b + 1)$ implies $(2b + 1)\varepsilon \leq c - b^2$, it follows that

$$y^2 < b^2 + (2b + 1)\varepsilon \leq b^2 + (c - b^2) = c.$$

We have proved that $b^2 < y^2 < c$.

A similar, but simpler, proof establishes the existence of x. First, set $x = b - \varepsilon$, where $\varepsilon = (b^2 - a)/2b$, and observe by hypothesis that $\varepsilon > 0$. Hence $x < b$. Next, notice that

$$x^2 = b^2 - 2\varepsilon b + \varepsilon^2 > b^2 - 2\varepsilon b + 0 = b^2 - 2\varepsilon b.$$

Since $2\varepsilon b = b^2 - a$, it follows that $x^2 > b^2 - (b^2 - a) = a$.

Step 2. *If $E = \{x \in \mathbf{R} : x > 0 \text{ and } x^2 < t\}$, then $b = \sup E$ exists and is the square root of t.*

To show that E has a supremum, notice first that by the Archimedean Principle, there is an $n \in \mathbf{N}$ such that $n > 1/t$. Thus E is nonempty since $x = 1/n \in E$. Next, observe that E is bounded above by $\max\{t, 1\}$. Indeed, if $t < 1$ and $x \in E$ satisfies $x \geq 1$, then $t > x^2 \geq 1^2 = 1$, a contradiction. Thus E is bounded above by 1 when $t < 1$. On the other hand, if $t \geq 1$ and $x \in E$ satisfies $x > t$, then $t > x^2 > t^2 \geq t$, a contradiction. Thus E is bounded above by t when $t \geq 1$.

We have proved that E is nonempty and bounded above. Thus by the Completeness Axiom, $b := \sup E$ exists and is finite. It remains to verify $b^2 = t$.

Suppose to the contrary that $b^2 \neq t$. By the Trichotomy Property, either $b^2 < t$ or $b^2 > t$. We shall show that both these assumptions lead to contradictions.

Case 1. $b^2 < t$. Apply Step 1 with $c = t$ to choose a $y > b$ such that $y^2 < t$. Since $y^2 < t$, $y \in E$. Since $b = \sup E$, it follows that $y \leq b$. This contradicts the choice that $y > b$.

Case 2. $b^2 > t$. Apply Step 1 with $a = t$ to choose an $x < b$ such that $t < x^2$. Let $w \in E$ (i.e., $w > 0$ and $w^2 < t$). Since $t < x^2$, the Transitive Property implies $w^2 < x^2$. This last inequality implies that $w < x$. Indeed, if $w \geq x$, then by squaring both sides we have $w^2 \geq x^2$, a contradiction.

We have proved that $w < x$ for all $w \in E$ (i.e., x is an upper bound of E). Since b is the supremum of E, we conclude that $b \leq x$, a contradiction of the choice that $x < b$. ∎

We close this section by proving that, under mild assumptions, the Axiom of Induction and the Well-Ordering Principle are equivalent.

A.11 Theorem. *Suppose that the Ordered Field Axioms hold and that*

$$n \in \mathbf{N} \quad and \quad n \neq 1 \quad imply \ n - 1 \in \mathbf{N}. \tag{2}$$

Then the Axiom of Induction holds if and only if the Well-Ordering Principle holds.

Proof. By the proof of Theorem 1.11 [which used (2) at a crucial spot], the Well-Ordering Principle implies the Axiom of Induction.

Conversely, if the Axiom of Induction holds, then the elements of **N** belong to every inductive set. This is the "definition" of **N** we made following (1) above. It follows that all results above are valid. (We will use Theorem A.6.)

Suppose that E is a nonempty subset of **N**, and consider the set

$$A := \{x \in \mathbf{N} : x \leq e \quad \text{for all} \quad e \in E\}.$$

A is nonempty since $1 \in A$. A is not the whole set **N** since if $e_0 \in E$, then $e_0 + 1$ cannot belong to A. Hence, by the Axiom of Induction, A cannot be an inductive set. In particular, there is an $x \in A$ such that $x + 1 \notin A$.

We claim that this x is a least element of E; that is, x is a lower bound of E and $x \in E$. That x is a lower bound of E is obvious, since by construction, $x \in A$ implies $x \leq e$ for all $e \in E$. On the other hand, if $x \notin E$, then $x < e$ for all $e \in E$. Hence, by Theorem A.6ii, $x + 1 \leq e$ for all $e \in E$ (i.e., $x + 1 \in A$), a contradiction of the choice of x. ∎

B. TRIGONOMETRY

In this appendix we derive some trigonometric identities by using elementary geometry and algebra.

Let (x, y) be a point on the unit circle $x^2 + y^2 = 1$ and θ be the angle measured counterclockwise from the positive x-axis to the line segment from $(0, 0)$ to (x, y) (see Figure B.1a). [We shall refer to (x, y) as the point *determined* by the angle θ.] Define

$$\sin \theta = y, \quad \cos \theta = x, \quad \text{and} \quad \tan \theta = \frac{y}{x}.$$

By the Law of Similar Triangles, given a right triangle with base angle θ, altitude a, base b, and hypotenuse h (see Figure B.1b), $\sin \theta = a/h$, $\cos \theta = b/h$, and $\tan \theta = a/b = \sin \theta / \cos \theta$.

B.1 Theorem. *Given a circle $C : x^2 + y^2 = r^2$ of radius r, let $s(\theta)$ represent the length of the arc on C swept out by θ, and $A(\theta)$ represent the area of the angular sector swept out by θ (see Figure B.1a). If the angle θ is measured in radians (not degrees), then*

$$s(\theta) = r\theta \quad and \quad A(\theta) = \frac{r^2 \theta}{2}.$$

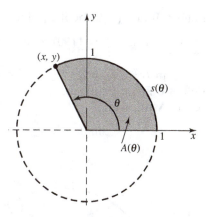

FIGURE B.1a

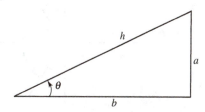

FIGURE B.1b

Proof. Since there are 2π radians in a complete circle and the circumference of a circle of radius r is $2\pi r$, we have

$$\frac{s(\theta)}{2\pi r} = \frac{\theta}{2\pi};$$

that is, $s(\theta) = r\theta$. Similarly, since the area of a circle is πr^2, we have

$$\frac{A(\theta)}{\pi r^2} = \frac{\theta}{2\pi};$$

that is, $A(\theta) = r^2\theta/2$. $\blacksquare$

B.2 Theorem.

i) $\sin(0) = 0$ *and* $\cos(0) = 1$.

ii) *For any* $\theta \in \mathbf{R}$, $|\sin\theta| \le 1$, $|\cos\theta| \le 1$, $\sin(-\theta) = -\sin\theta$, $\cos(-\theta) = \cos\theta$, *and* $\sin^2\theta + \cos^2\theta = 1$.

iii) *If* θ *is measured in radians, then* $\sin(\pi/2) = 1$, $\cos(\pi/2) = 0$, $\sin(\theta + 2\pi) = \sin\theta$, *and* $\cos(\theta + 2\pi) = \cos\theta$. *Moreover, if* $0 < \theta < \pi/2$, *then* $0 < \theta\cos\theta < \sin\theta < \theta$.

iv) *If* $\theta \in \mathbf{R}$ *is measured in radians, then* $|\sin\theta| \le |\theta|$.

Proof. Let $\theta \in \mathbf{R}$ and (x, y) be the point on the unit circle determined by θ.

i) If $\theta = 0$, then $(x, y) = (1, 0)$ (see Figure B.1a). Hence, $\sin(0) = 0$ and $\cos(0) = 1$.

ii) Clearly, $|\sin \theta| = |y| = \sqrt{y^2} \le \sqrt{x^2 + y^2} = 1$, and, similarly, $|\cos \theta| \le 1$. By definition (see Figure B.2a), $\sin(-\theta) = -y = -\sin \theta$ and $\cos(-\theta) = x = \cos \theta$. Moreover,

$$\sin^2 \theta + \cos^2 \theta = x^2 + y^2 = 1.$$

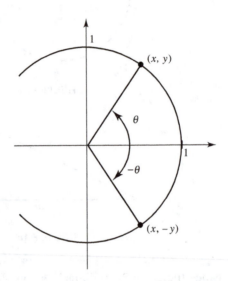

FIGURE B.2a

iii) If $\theta = \pi/2$, then $(x, y) = (0, 1)$, so $\sin(\pi/2) = 1$ and $\cos(\pi/2) = 0$. Fix $\theta \in (0, \pi/2)$ and consider Figure B.2b. Since $\sin \theta$ is the altitude of triangle ABC and the shortest distance between two points is a straight line, we have by Theorem B.1 that

$$\sin \theta < s(\theta) = \theta.$$

On the other hand, the triangle ABC is a proper subset of the angular sector swept out by θ, which is a proper subset of the triangle ABD. Hence,

$$\text{Area}\,(ABC) < A(\theta) < \text{Area}\,(ABD).$$

Since the area of a triangle is one-half the product of its base and its altitude, it follows from Theorem B.1 that

$$\frac{\sin \theta}{2} < \frac{\theta}{2} < \frac{\tan \theta}{2}. \tag{3}$$

But $0 < \cos \theta < 1$ for all $\theta \in (0, \pi/2)$. Multiplying (3) by $2 \cos \theta$, we conclude that

$$\sin \theta \cos \theta < \theta \cos \theta < \sin \theta. \tag{4}$$

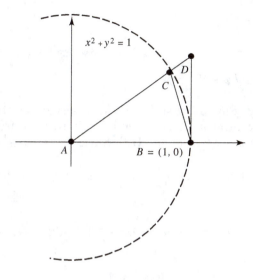

FIGURE B.2b

iv) By part iii), $|\sin\theta| = \sin\theta \le \theta = |\theta|$ for all $0 \le \theta \le \pi/2$. Since $\sin(-\theta) = -\sin\theta$, it follows that $|\sin\theta| \le |\theta|$ for all $\theta \in [-\pi/2, \pi/2]$. But if $\theta \notin [-\pi/2, \pi/2]$, then $|\sin\theta| \le 1 < \pi/2 < |\theta|$. Therefore, $|\sin\theta| \le |\theta|$ for all $\theta \in \mathbf{R}$. ∎

The next result shows how to compute the sine and cosine of a sum of angles.

B.3 Theorem.
i) [SUM-ANGLE FORMULAS]. *If* $\theta, \varphi \in \mathbf{R}$, *then*

$$\cos(\theta \pm \varphi) = \cos\theta \cos\varphi \mp \sin\theta \sin\varphi,$$

and

$$\sin(\theta \pm \varphi) = \sin\theta \cos\varphi \pm \cos\theta \sin\varphi.$$

ii) [DOUBLE-ANGLE FORMULAS]. *If* $\theta \in \mathbf{R}$, *then*

$$\cos^2\theta = \frac{1 + \cos(2\theta)}{2},$$
$$\sin^2\theta = \frac{1 - \cos(2\theta)}{2},$$

and

$$\cos\theta = 1 - 2\sin^2(\theta/2).$$

iii) [SHIFT FORMULAS]. *If* φ *is measured in radians, then*

$$\sin\varphi = \cos\left(\frac{\pi}{2} - \varphi\right),$$

and

$$\cos\varphi = \sin\left(\frac{\pi}{2} - \varphi\right)$$

for all $\varphi \in \mathbf{R}$.

Proof. Suppose first that $\theta > \varphi$. Consider the chord A cut from the unit circle by a central angle $\theta - \varphi$, and the chord B cut from the unit circle by a central angle $\varphi - \theta$ (see Figure B.3). Since $\sin^2\theta + \cos^2\theta = 1$, we have

$$A^2 = (\cos\theta - \cos\varphi)^2 + (\sin\theta - \sin\varphi)^2 = 2 - 2(\cos\theta\cos\varphi + \sin\theta\sin\varphi), \quad (5)$$

and

$$B^2 = (\cos(\theta - \varphi) - 1)^2 + (\sin(\theta - \varphi))^2 = 2 - 2\cos(\theta - \varphi).$$

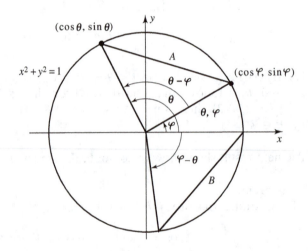

FIGURE B.3

Since $|\theta - \varphi| = |\varphi - \theta|$, the lengths of these chords must be equal. Thus

$$\cos(\theta - \varphi) = \cos\theta\cos\varphi + \sin\theta\sin\varphi \qquad (6)$$

for $\theta < \varphi$. A similar argument establishes (6) for $\varphi < \theta$. Since (6) is trivial when $\theta = \varphi$, we have proved that (6) holds for all θ and φ. Combining the identities $\sin(-\theta) = -\sin\theta$ and $\cos(-\theta) = \cos\theta$ with (6), we obtain

$$\cos(\theta + \varphi) = \cos(\theta - (-\varphi)) = \cos\theta\cos\varphi - \sin\theta\sin\varphi.$$

This and (6) verify the first identity in part i).

Applying this identity to $\theta = \pi/2$, we see by Theorem B.2ii that

$$\cos\left(\frac{\pi}{2} - \varphi\right) = \cos\left(\frac{\pi}{2}\right)\cos\varphi + \sin\left(\frac{\pi}{2}\right)\sin\varphi = \sin\varphi;$$

that is, the first identity in part iii) holds. Combining the first identities in parts i) and iii), we obtain

$$
\begin{aligned}
\sin(\theta \pm \varphi) &= \cos\left(\left(\frac{\pi}{2} - \theta\right) \mp \varphi\right) \\
&= \cos\left(\frac{\pi}{2} - \theta\right)\cos(-\varphi) \mp \sin\left(\frac{\pi}{2} - \theta\right)\sin(-\varphi) \\
&= \sin\theta\cos\varphi \pm \cos\theta\sin\varphi.
\end{aligned}
$$

This proves the second identity in part i). Specializing to the case $\theta = \pi/2$, we obtain $\sin(\pi/2 - \varphi) = \cos\varphi$. Thus parts i) and iii) have been proved.

To establish part ii), notice by part i) and Theorem B.2 that

$$
\cos(2\theta) = \cos(\theta + \theta) = \cos^2\theta - \sin^2\theta = 2\cos^2\theta - 1.
$$

Hence, $\cos^2\theta = (1 + \cos(2\theta))/2$. Similar arguments establish the rest of part ii). ∎

We close this section with the Law of Cosines, a generalization of the Pythagorean Theorem.

B.4 Theorem.   [LAW OF COSINES].
If T is a triangle with sides of length a, b, c, and θ is the angle opposite the side of length c, then

$$
c^2 = a^2 + b^2 - 2ab\cos\theta.
$$

Proof. Suppose without loss of generality that θ is acute, and rotate T so b is its base. Let h be the altitude of T, and notice that h cuts a right triangle out of T whose sides are a and h and the angle opposite h is θ. By the definition of $\sin\theta$ and $\cos\theta$, $h = a\sin\theta$ and the length d of the base of this right triangle is $d = b - a\cos\theta$. Substituting these values into the equation $c^2 = h^2 + d^2$ (which follows directly from the Pythagorean Theorem), we obtain

$$
\begin{aligned}
c^2 &= (a\sin\theta)^2 + (b - a\cos\theta)^2 \\
&= a^2\sin^2\theta + a^2\cos^2\theta + b^2 - 2ab\cos\theta = a^2 + b^2 - 2ab\cos\theta. \quad\blacksquare
\end{aligned}
$$

C. MATRICES AND DETERMINANTS

In this appendix we prove several elementary results about matrices and determinants. We assume the student is familiar with the concept of row and column reduction to canonical form.

Recall that an $m \times n$ matrix B is a rectangular array which has m rows and n columns:

$$B = [b_{ij}]_{m \times n} = \begin{bmatrix} b_{11} & b_{12} & \cdots & b_{1n} \\ b_{21} & b_{22} & \cdots & b_{2n} \\ \vdots & \vdots & \ddots & \vdots \\ b_{m1} & b_{m2} & \cdots & b_{mn} \end{bmatrix}.$$

The notation b_{ij} indicates the *entry* in the ith row and jth column. We shall call B *real* if all its entries b_{ij} belong to $\mathbf{R}$.

C.1 Definition.

Let $B = [b_{ij}]_{m \times n}$ and $C = [c_{k\ell}]_{p \times q}$ be real matrices.

i) B and C are said to be *equal* if $m = p$, $n = q$, and $b_{ij} = c_{ij}$ for $i = 1, 2, \ldots, m$, and $j = 1, 2, \ldots, n$.

ii) The $m \times n$ *zero matrix* is the matrix $O = O_{m \times n} = [b_{ij}]_{m \times n}$, where $b_{ij} = 0$ for $i = 1, \ldots, m$, and $j = 1, \ldots, n$.

iii) The $n \times n$ *identity matrix* is the matrix $I = I_{n \times n} = [b_{ij}]_{n \times n}$, where $b_{ii} = 1$ for $i = 1, \ldots, n$, and $b_{ij} = 0$ for $i \neq j$, $i, j = 1, \ldots, n$.

iv) The *product* of a matrix B and a scalar α is defined by

$$\alpha B = [\alpha b_{ij}]_{m \times n}.$$

v) The *negative* of a matrix B is defined by $-B = (-1)B$.

vi) When $m = p$ and $n = q$, the *sum* of B and C is defined by

$$B + C = [b_{ij} + c_{ij}]_{m \times n}.$$

vii) When $n = p$, the *product* of B and C is defined by

$$BC = \left[\sum_{\nu=1}^{n} b_{i\nu} c_{\nu j} \right]_{m \times q}.$$

C.2 EXAMPLE.

Compute $B + C$, $3B$, $-C$, BC, and CB, where

$$B = \begin{bmatrix} 1 & 0 \\ 2 & 3 \end{bmatrix} \quad \text{and} \quad C = \begin{bmatrix} -1 & 1 \\ -2 & 0 \end{bmatrix}.$$

SOLUTION. By definition,

$$B + C = \begin{bmatrix} 0 & 1 \\ 0 & 3 \end{bmatrix}, \qquad 3B = \begin{bmatrix} 3 & 0 \\ 6 & 9 \end{bmatrix},$$

$$-C = \begin{bmatrix} 1 & -1 \\ 2 & 0 \end{bmatrix}, \qquad BC = \begin{bmatrix} -1 & 1 \\ -8 & 2 \end{bmatrix}, \quad \text{and} \quad CB = \begin{bmatrix} 1 & 3 \\ -2 & 0 \end{bmatrix}. \qquad \blacksquare$$

These operations do not satisfy all the usual laws of algebra. (For example, the last two computations show that matrix multiplication is not commutative.) Here is a list of algebraic laws satisfied by real matrices.

C.3 Theorem. *Let $A = [a_{ij}]$, $B = [b_{ij}]$, and $C = [c_{ij}]$ be real matrices and α, β be scalars.*

   i) $(\alpha + \beta)C = \alpha C + \beta C$.
  ii) *If $B + C$ is defined, then $\alpha(B + C) = \alpha B + \alpha C$, and $B + C = C + B$.*
 iii) *If BC is defined, then $\alpha(BC) = (\alpha B)C = B(\alpha C)$.*
 iv) *If AB and AC are defined, then $A(B + C) = AB + AC$. If BA and CA are defined, then $(B + C)A = BA + CA$.*
  v) *If $A + B$ and $B + C$ are defined, then $(A + B) + C = A + (B + C)$. If AB and BC are defined, then $(AB)C = A(BC)$.*
 vi) *If B is an $m \times n$ matrix, then*

$$B + O_{m \times n} = B, \qquad B - B = O_{m \times n},$$
$$BO_{n \times q} = O_{m \times q}, \qquad O_{p \times m} B = O_{p \times n}, \quad \text{and} \quad 0B = O_{m \times n}.$$

vii) *If B is an $n \times n$ matrix, then*

$$I_{n \times n} B = B I_{n \times n} = B.$$

Proof. By definition,

$$(\alpha + \beta)C = [(\alpha + \beta)c_{ij}] = [\alpha c_{ij} + \beta c_{ij}] = \alpha C + \beta C,$$

and

$$\alpha(B + C) = \alpha[b_{ij} + c_{ij}] = [\alpha(b_{ij} + c_{ij})] = [\alpha b_{ij}] + [\alpha c_{ij}] = \alpha B + \alpha C.$$

A similar argument establishes $B + C = C + B$.

Let B be an $m \times n$ matrix and C be an $n \times q$ matrix. By definition,

$$\alpha(BC) = \left[\alpha \sum_{\nu=1}^{n} b_{i\nu} c_{\nu j} \right] = \left[\sum_{\nu=1}^{n} (\alpha b_{i\nu}) c_{\nu j} \right] = (\alpha B)C.$$

A similar argument establishes $\alpha(BC) = B(\alpha C)$.

Let A be an $m \times n$ matrix and B, C be $n \times q$ matrices. By definition,

$$A(B + C) = \left[\sum_{v=1}^{n} a_{iv}(b_{vj} + c_{vj}) \right] = \left[\sum_{v=1}^{n} a_{iv}b_{vj} + \sum_{v=1}^{n} a_{iv}c_{vj} \right] = AB + AC.$$

A similar argument establishes $(B + C)A = BA + CA$.

Let A be an $m \times n$ matrix, B be an $n \times p$ matrix, and C be a $p \times q$ matrix. By definition,

$$(AB)C = \left[\sum_{v=1}^{n} a_{iv}b_{vj} \right] [c_{jk}]$$

$$= \left[\sum_{j=1}^{p} \left(\sum_{v=1}^{n} a_{iv}b_{vj} \right) c_{jk} \right]$$

$$= \left[\sum_{v=1}^{n} \left(a_{iv} \sum_{j=1}^{p} b_{vj}c_{jk} \right) \right]$$

$$= A \left[\sum_{j=1}^{p} b_{vj}c_{jk} \right] = A(BC).$$

A similar argument establishes $(A + B) + C = A + (B + C)$.

By definition,

$$B + O_{m \times n} = [b_{ij} + 0] = [b_{ij}] = B, \qquad B - B = [b_{ij} - b_{ij}] = O_{m \times n},$$

$$BO_{n \times q} = \left[\sum_{v=1}^{n} b_{iv} \cdot 0 \right] = O_{m \times q}, \qquad O_{p \times m}B = \left[\sum_{v=1}^{m} 0 \cdot b_{vj} \right] = O_{p \times n},$$

and $0 \cdot B = [0 \cdot b_{ij}] = O_{m \times n}$. And, since $I = [\delta_{ij}]$, where

$$\delta_{ij} = \begin{cases} 1 & i = j, \\ 0 & i \neq j, \end{cases}$$

we have $I_{n \times n}B = \left[\sum_{v=1}^{n} \delta_{iv}b_{vj} \right] = [b_{ij}] = B = BI_{n \times n}$. ∎

A square matrix is a matrix with as many rows as columns. Clearly, if B and C are square real matrices of the same size, then both $B + C$ and BC are defined. This gives room for more algebraic structure. An $n \times n$ real matrix B is said to be *invertible* if and only if there is an $n \times n$ matrix B^{-1}, called the *inverse* of B, which satisfies

$$BB^{-1} = B^{-1}B = I.$$

The following result shows that matrix inverses are unique.

C.4 Theorem. *Let A, B be $n \times n$ real matrices. If B is invertible and $BA = I$, then $B^{-1} = A$.*

Proof. By Theorem C.3 and definition,

$$B^{-1} = B^{-1}I = B^{-1}(BA) = (B^{-1}B)A = IA = A. \qquad \blacksquare$$

If $B = [b_{ij}]_{n \times n}$ is square, recall that the *minor matrix* B_{ij} of B is the $(n-1) \times (n-1)$ matrix obtained by removing the ith row and the jth column from B. For example, if

$$B = \begin{bmatrix} 1 & 2 & 3 \\ -1 & -2 & -3 \\ 4 & 5 & 6 \end{bmatrix},$$

then

$$B_{21} = \begin{bmatrix} 2 & 3 \\ 5 & 6 \end{bmatrix}.$$

Minor matrices can be used to define an operation on square real matrices (the determinant) which makes invertible matrices easy to identify (see Theorem C.6 below).

The *determinant* can be defined recursively as follows. Let B be an $n \times n$ real matrix.

i) If $n = 1$, then the determinant of B is defined by $\det[b] = b$.

ii) If $n = 2$, then the determinant of B is defined by

$$\det \begin{bmatrix} a & b \\ c & d \end{bmatrix} = ad - bc.$$

iii) If $n > 2$, then the determinant of B is defined recursively by

$$\det[b_{ij}]_{n \times n} = b_{11} \det B_{11} - b_{12} \det B_{12} + \cdots + (-1)^{n-1} b_{1n} \det B_{1n},$$

where B_{1j} are minor matrices of B.

The following result shows what an elementary column operation does to the determinant of a matrix.

C.5 Theorem. *Let $B = [b_{ij}]$ and $C = [c_{ij}]$ be $n \times n$ real matrices, $n \geq 2$.*

i) *If C is obtained from B by interchanging two columns, then $\det C = -\det B$.*

ii) *If C is obtained from B by multiplying one column of B by a scalar α, then $\det C = \alpha \det B$.*

iii) *If C is obtained from B by multiplying one column of B by a scalar and adding it to another column of B, then $\det C = \det B$.*

Proof. Since

$$\det \begin{bmatrix} a & b \\ c & d \end{bmatrix} = ad - bc = -(bc - ad) = -\det \begin{bmatrix} b & a \\ d & c \end{bmatrix},$$

part i) holds for 2×2 matrices. Suppose that part i) holds for $(n-1) \times (n-1)$ matrices. Suppose further that there are indices $j_0 < j_1$ such that $b_{ij_0} = c_{ij_1}$ and $b_{ij_1} = c_{ij_0}$ for $i = 1, \ldots, n$. By the inductive hypothesis, $\det C_{1j} = -\det B_{1j}$ for $j \neq j_0$ and $j \neq j_1$,

$$\det C_{1j_0} = (-1)^{j_1-j_0-1} \det B_{1j_1}, \quad \text{and} \quad \det C_{1j_1} = (-1)^{j_0-j_1+1} \det B_{1j_0}.$$

Hence, by definition,

$$\det C = c_{11} \det C_{11} - c_{12} \det C_{12} + \cdots + (-1)^{n-1} c_{1n} \det C_{1n}$$
$$= -b_{11} \det B_{11} + b_{12} \det B_{12} + \cdots - (-1)^{n-1} b_{1n} \det B_{1n} = -\det B.$$

Thus i) holds for all $n \in \mathbf{N}$. Similar arguments establish parts ii) and iii). ∎

In the same way we can show that Theorem C.5 holds if *column* is replaced by *row*. It follows that we can compute the determinant of a real matrix by expanding along any row or any column, with an appropriate adjustment of signs. For example, to expand along the ith row, interchange the ith row with the first row, expand along the new first row, and use Theorem C.5 to relate everything back to B. In particular, we see that

$$\det[b_{ij}]_{n \times n} = (-1)^{i+1} b_{i1} \det B_{i1} + (-1)^{i+2} b_{i2} \det B_{i2} + \cdots + (-1)^{i+n} b_{in} \det B_{in}.$$

The numbers $(-1)^{i+j} \det B_{ij}$ are called the *cofactors* of b_{ij} in $\det B$

The operations in Theorem C.5 are called *elementary column operations*. They can be simulated by matrix multiplication. Indeed, an *elementary matrix* is a matrix obtained from the identity matrix by a single elementary column operation. Thus elementary matrices fall into three categories: $E(i \leftrightarrow j)$, the matrix obtained by interchanging the ith and jth columns of I; $E(\alpha i)$, the matrix obtained by multiplying the ith column of I by $\alpha \neq 0$; and $E(\alpha i + j)$, the matrix obtained by multiplying the ith column of I by $\alpha \neq 0$ and adding it to the jth column. Notice that an elementary column operation on B can be obtained by multiplying B by an elementary matrix; for example, $E(i \leftrightarrow j)B$ is the matrix obtained by interchanging the ith and jth columns of B.

These observations can be used to show that the determinant is multiplicative.

C.6 Theorem. *If B, C are $n \times n$ real matrices, then*

$$\det(BC) = \det B \det C.$$

Moreover, B is invertible if and only if $\det(B) \neq 0$.

Proof. It is easy to check that

$$\det(E(i \leftrightarrow j)) = -1, \quad \det(E(\alpha i)) = \alpha, \quad \text{and} \quad \det(E(\alpha i + j)) = 1.$$

Hence, by Theorem C.5,

$$\det(EA) = \det E \det A \tag{7}$$

holds for any $n \times n$ matrix A and any $n \times n$ elementary matrix E.

The matrix B can be reduced, by a sequence of elementary column operations, to a matrix V, where $V = I$ if B is invertible, and V has at least one zero column if B is not invertible (see Noble and Daniel [9], p85). It follows that there exist elementary matrices $E_1, \ldots, E_p$ such that $A = E_1 \ldots E_p V$. Hence, by (7),

$$\det(B) = \det(E_1 \ldots E_p V)$$
$$= \det(E_1) \det(E_2 \ldots E_p V) = \cdots = \det(E_1 \ldots E_p)\det(V).$$

In particular, B is invertible if and only if $\det B \neq 0$.

Suppose that B is invertible. Then $V = I$ and by (7),

$$\det(BC) = \det(E_1 \ldots E_p)\det(VC) = \det B \det C.$$

If B is not invertible, then BC is not invertible either (see Noble and Daniel [9], p204). Hence, $\det(BC) = 0$ and we have

$$\det(BC) = 0 = \det B \det C. \qquad \blacksquare$$

The *transpose* of a matrix $B = [b_{ij}]$ is the matrix B^T obtained from B by making the ith row of B the ith column of B^T; that is, the $(i \times j)$th entry of B^T is b_{ji}. The *adjoint* of an $n \times n$ matrix B is the transpose of the matrix of cofactors of B; that is,

$$\mathrm{adj}(B) = [(-1)^{i+j} \det B_{ij}]^T.$$

The adjoint can be used to give an explicit formula for the inverse of an invertible matrix.

C.7 Theorem. *Suppose that B is a square real matrix. If B is invertible, then*

$$B^{-1} = \frac{1}{\det B} \mathrm{adj}(B). \tag{8}$$

Proof. Set $[c_{ij}] = B \, \mathrm{adj}\,(B)$. By definition,

$$c_{ij} = (-1)^{1+j} b_{i1} B_{j1} + \cdots + (-1)^{n+j} b_{in} B_{jn}.$$

If $i = j$, then c_{ij} is an expansion of the determinant of B along the ith row of B (i.e., $c_{ii} = \det B$). If $i \neq j$, then c_{ij} is a determinant of a matrix with two identical rows so c_{ij} is zero. It follows that

$$B \, \mathrm{adj}(B) = \det B \cdot I.$$

We conclude by Theorem C.4 that (8) holds.   ∎

In particular,

$$\begin{bmatrix} a & b \\ c & d \end{bmatrix}^{-1} = \frac{1}{ad - bc} \begin{bmatrix} d & -b \\ -c & a \end{bmatrix}. \tag{9}$$

The following result shows how the determinant can be used to solve systems of linear equations. (This result is of great theoretical interest but of little practical use because it requires lots of storage to use on a computer. Most packaged routines which solve systems of linear equations use methods more efficient than Cramer's Rule; e.g., Gaussian elimination.)

C.8 Theorem.   [CRAMER'S RULE].
Let $c_1, c_2, \ldots, c_n \in \mathbf{R}$ and $B = [b_{ij}]_{n \times n}$ be a square real matrix. The system

$$\begin{aligned} b_{11}x_1 + b_{12}x_2 + \cdots + b_{1n}x_n &= c_1 \\ b_{21}x_1 + b_{22}x_2 + \cdots + b_{2n}x_n &= c_2 \end{aligned} \tag{10}$$

$$\vdots \qquad\qquad \vdots$$

$$b_{n1}x_1 + b_{n2}x_2 + \cdots + b_{nn}x_n = c_n$$

of n linear equations in n unknowns has a unique solution if and only if the matrix B has a nonzero determinant, in which case

$$x_j = \frac{\det C(j)}{\det B},$$

where $C(j)$ is obtained from B by replacing the jth column of B by the column matrix $[c_1 \ldots c_n]^T$. In particular, if $c_j = 0$ for all j and $\det B \neq 0$, then the system (10) has only the trivial solution $x_j = 0$ for $j = 1, 2, \ldots, n$.

Proof. The system (10) is equivalent to the matrix equation

$$BX = C,$$

where $B = [b_{ij}]$, $X = [x_1 \ldots x_n]^T$, and $C = [c_1 \ldots c_n]^T$. If $\det B \neq 0$, then by Theorem C.7,

$$X = B^{-1}C = \frac{1}{\det B} \operatorname{adj}(B)C.$$

By definition, $\operatorname{adj}(B)C$ is a column matrix whose jth "row" is the number

$$(-1)^{1+j}c_1 \det B_{1j} + (-1)^{2+j}c_2 \det B_{2j} + \cdots + (-1)^{n+j}c_n \det B_{nj} = \det C(j).$$

[We expanded the determinant of $C(j)$ along the jth column.] Thus $x_j = \det C(j)$.

Conversely, if $BX = C$ has a unique solution, B can be row reduced to I. Thus B is invertible (i.e., $\det B \neq 0$).   ∎

D. QUADRIC SURFACES

A *quadric surface* is a surface which is the graph of a relation in $\mathbf{R}^3$ of the form

$$Ax^2 + By^2 + Cz^2 + Dx + Ey + Fz + Gxy + Hyz + Izx = J,$$

where $A, B, \ldots, J \in \mathbf{R}$ and not all A, B, C, G, H, I are zero. We shall only consider the cases when $G = H = I = 0$. These include the following special types.

1. The *ellipsoid*, the graph of

$$Ax^2 + By^2 + Cz^2 = 1,$$

   where A, B, C are all positive

2. The *hyperboloid of one sheet*, the graph of

$$Ax^2 + By^2 + Cz^2 = 1,$$

   where two of A, B, C are positive and the other is negative.

3. The *hyperboloid of two sheets*, the graph of

$$Ax^2 + By^2 + Cz^2 = 1,$$

   where two of A, B, C are negative and the other is positive

4. The *cone*, the graph of

$$Ax^2 + By^2 + Cz^2 = 0,$$

   where two of A, B, C are positive and the other is negative

5. The *paraboloid*, the graph of

$$z = Ax^2 + By^2,$$

   where A, B are both positive or both negative

6. The *hyperbolic paraboloid*, the graph of

$$z = Ax^2 + By^2,$$

   where one of A, B is positive and the other is negative

The *trace* of a surface S in a plane Π is defined to be the intersection of S with Π. Graphs of many surfaces, including all quadrics, can be visualized by looking at their traces in various planes. We illustrate this technique with a typical example of each type of quadric.

D.1 *EXAMPLE.*

The ellipsoid $3x^2 + y^2 + 2z^2 = 6$.

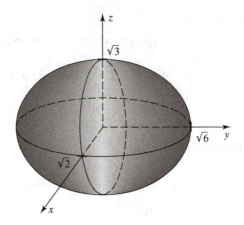

FIGURE D.1

SOLUTION. The trace of this surface in the xy-plane is the ellipse $3x^2 + y^2 = 6$. The trace of this surface in the yz-plane is the ellipse $y^2 + 2z^2 = 6$. And the trace of this surface in the xz-plane is the ellipse $3x^2 + 2z^2 = 6$. This surface is sketched in Figure D.1. ∎

D.2 EXAMPLE.

The hyperboloid of one sheet $x^2 + y^2 - z^2 = 1$.

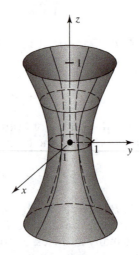

FIGURE D.2

SOLUTION. The trace of this surface in the plane $z = a$ is the circle $x^2 + y^2 = 1 + a^2$. The trace of this surface in $x = 0$ is the hyperbola $y^2 - z^2 = 1$. This surface is sketched in Figure D.2. ∎

D.3 *EXAMPLE.*

The hyperboloid of two sheets $x^2 - y^2 - z^2 = 1$.

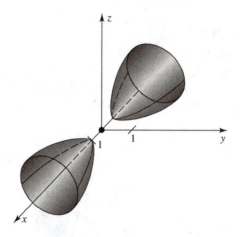

FIGURE D.3

SOLUTION.   The trace of this surface in the plane $z = 0$ is the hyperbola $x^2 - y^2 = 1$. The trace of this surface in $y = 0$ is the hyperbola $x^2 - z^2 = 1$. This surface has no trace in $x = 0$. This surface is sketched in Figure D.3.   ■

D.4 *EXAMPLE.*

The cone $z^2 = x^2 + y^2$.

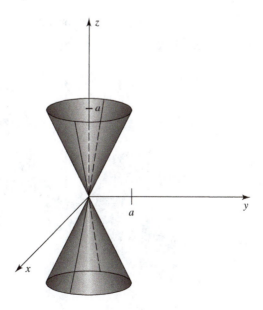

FIGURE D.4

SOLUTION. The trace of this surface in the plane $z = a$ is the circle $x^2 + y^2 = a^2$. The trace of this surface in $y = 0$ is a pair of lines $z = \pm x$. This surface is sketched in Figure D.4. ∎

D.5 *EXAMPLE.*

The paraboloid $z = x^2 + y^2$.

SOLUTION. If $a > 0$, the trace of this surface in the plane $z = a$ is the circle $x^2 + y^2 = a$. The trace of this surface in $y = 0$ is the parabola $z = x^2$. This surface is sketched in Figure D.5. ∎

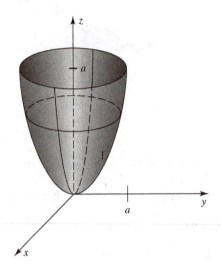

FIGURE D.5

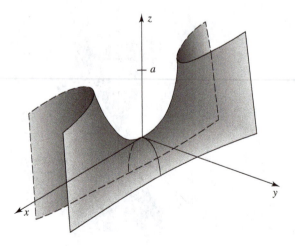

FIGURE D.6

D.6 EXAMPLE.

The hyperbolic paraboloid $z = x^2 - y^2$.

SOLUTION. The trace of this surface in the plane $z = a$ is the hyperbola $a = x^2 - y^2$. (It opens up around the xz-plane when $a > 0$, and around the yz-plane when $a < 0$.) The trace of this surface in the plane $y = 0$ is the parabola $z = x^2$. This surface is sketched in Figure D.6. (*Note*: The scale along the x-axis has been exaggerated to enhance perspective, so the hyperbolas below the $z = 0$ plane are barely discernible.) ∎

E. VECTOR CALCULUS AND PHYSICS

Throughout this appendix $C = (\varphi, I)$ is a smooth arc in $\mathbf{R}^2$, $S = (\psi, E)$ is a smooth surface in $\mathbf{R}^3$, $\{t_0, \ldots, t_N\}$ is a partition of I, and $\{R_1, \ldots, R_N\}$ is a grid on E.

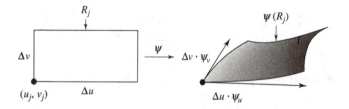

FIGURE E.1

E.1 Remark. *The integral*

$$\iint_S d\sigma = \int_E \|N_\psi(u, v)\| \, d(u, v) \tag{11}$$

can be interpreted as the surface area of S.

Let (u_j, v_j) be the lower left-hand corner of R_j and suppose that R_j has sides Δu, Δv (see Figure E.1). If R_j is small enough, the trace of each piece $S_j = (\psi, R_j)$ is approximately equal to the parallelogram determined by the vectors $\Delta u \, \psi_u$ and $\Delta v \, \psi_v$. Hence, by Exercise 8.2.7,

$$A(S_j) \approx \|(\Delta u \, \psi_u(u_j, v_j)) \times (\Delta v \, \psi_v(u_j, v_j))\|$$
$$= \|N_\psi(u_j, v_j)\| \, \Delta u \Delta v = \|N_\psi(u_j, v_j)\| \, |R_j|.$$

Summing over j, we obtain

$$A(S) \approx \sum_{j=1}^{N} \|N_\psi(u_j, v_j)\| \, |R_j|,$$

which is a Riemann sum of the integral (10).

E.2 Remark. *If w is a thin wire lying along C whose density (mass per unit length) at a point* (x, y) *is given by* $g(x, y)$, *then*

$$\int_C g \, ds$$

can be interpreted as the mass of ω.

Since mass is the product of density and length, an approximation to the mass of the piece of w lying along $C_k = (\phi, [t_{k-1}, t_k])$ is given by

$$g(t_k) \cdot L(C_k) = \int_{t_{k-1}}^{t_k} g(t_k) \|\phi'(t)\| \, dt$$

(see Definition 13.9). Summing over k, an approximation to the mass of w is

$$\sum_{j=1}^{N} \int_{t_{k-1}}^{t_k} g(t_k) \|\phi'(t)\| \, dt,$$

which is nearly a Riemann sum of the integral

$$\int_I g(\phi(t)) \|\phi'(t)\| \, dt = \int_C g \, ds.$$

The following remark has a similar justification.

E.3 Remark. *If S is a thin sheet of metal whose density at a point* (x, y, z) *is given by* $g(x, y, z)$, *then*

$$\iint_S g \, d\sigma$$

can be interpreted as the mass of S.

Work done by a force F acting on an object as it moves a distance d is defined to be $W = Fd$. There are many situations where the force changes from point to point. Examples include the force of gravity (which is weaker at higher altitudes), the velocity of a fluid flowing through a constricted tube (which gets faster at places where the tube narrows), the force on an electron moving through an electric field, and the force on a copper coil moving through a magnetic field.

E.4 Remark. *If an object acted on by a force* $F : \mathbf{R}^3 \to \mathbf{R}$ *moves along the curve* $C = (\phi, I)$, *then the unoriented line integral*

$$\int_C F \, ds$$

can be interpreted as the work done by F along C.

An approximation to the work done along $C_k = (\phi, [t_{k-1}, t_k])$ is

$$W_k \approx F(\phi(t_k))\|\phi(t_k) - \phi(t_{k-1})\| = \int_{t_{k-1}}^{t_k} F(\phi(t_k))\|\phi'(t)\|\,dt.$$

Summing over k, we find that an approximation to the total work along C is given by

$$\sum_{k=1}^{N} \int_{t_{k-1}}^{t_k} F(\phi(t_k))\|\phi'(t)\|\,dt,$$

which is nearly a Riemann sum of the integral

$$\int_I F(\phi(t))\|\phi'(t)\|\,dt = \int_C F\,ds.$$

The following remark explains why $F \cdot T$ is called the *tangential* component of F and $F \cdot \mathbf{n}$ is called the *normal* component of F.

E.5 Remark. *Let $\mathbf{u}$ be a unit vector in $\mathbf{R}^2$ (respectively, $\mathbf{R}^3$) and F be a function whose range is a subset of $\mathbf{R}^2$ (respectively, $\mathbf{R}^3$). If ℓ is the line in the direction $\mathbf{u}$ passing through the origin, then $|F \cdot \mathbf{u}|$ is the length of the projection of F onto ℓ (see Figure E.2a).*

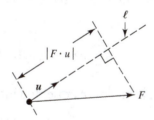

FIGURE E.2a

Let θ represent the angle between $\mathbf{u}$ and F. By (3) in Section 8.1,

$$|F \cdot \mathbf{u}| = \cos\theta\|F\|\,\|\mathbf{u}\| = \cos\theta\|F\|.$$

Hence, by trigonometry, $|F \cdot \mathbf{u}|$ is the length of the projection of F onto ℓ. Notice that $F \cdot \mathbf{u}$ is positive when θ is acute and negative when θ is obtuse.

Combining Remarks E.4 and E.5, we see that $\int_C F \cdot T\,ds$ represents the work done by the tangential component of a force field $F : \mathbf{R}^3 \to \mathbf{R}^3$ along C.

E.6 Remark. *If $S = (\psi, E)$ is a thin membrane submerged in an incompressible fluid which passes through S, and $F(x, y, z)$ represents the velocity vector of the flow of that fluid at the point (x, y, z), then the oriented integral of $F \cdot \mathbf{n}$ can be interpreted as the volume of fluid flowing through S in unit time.*

Let $\{E_j\}$ be a grid which covers E, and let h be the length of the line segment obtained by projecting F onto the normal line to S at a point $(x_j, y_j, z_j) \in \psi(E_j)$ (see Figure E.2b). If E_j is so small that F is essentially constant on the trace of $S_j = (\psi, E_j)$, then an approximation to the volume of fluid passing through S_j per unit time is given by

$$V_j = A(S_j) \cdot h = A(S_j) \cdot F(x_j, y_j, z_j) \cdot \mathbf{n}$$
$$= A(S_j) F(\psi(u_j, v_j)) \cdot N_\psi(u_j, v_j)/\|N_\psi(u_j, v_j)\|.$$

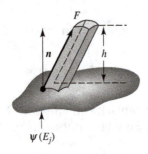

FIGURE E.2b

Summing over j and replacing $A(S_j)$ by $\|N_\psi\| \, |E_j|$ (see Remark E.1 above), we see that an approximation to the volume V of fluid passing through S per unit time is given by

$$\sum_{j=1}^{N} F(\psi(u_j, v_j)) \cdot N_\psi(u_j, v_j)|E_j|.$$

This is a Riemann sum of the oriented integral

$$\iint_E F(\psi(u, v)) \cdot N_\psi(u, v) \, dA = \iint_S F \cdot \mathbf{n} \, d\sigma.$$

F. EQUIVALENCE RELATIONS

A *partition* of a set X is a family of nonempty sets $\{E_\alpha\}_{\alpha \in A}$ such that

$$X = \bigcup_{\alpha \in A} E_\alpha \quad \text{and} \quad E_\alpha \cap E_\beta = \emptyset$$

for $\alpha \neq \beta$. A *binary relation* $\sim$ on X is a subset of $X \times X$. If (x, y) belongs to $\sim$, we shall write $x \sim y$. Examples of binary relations include $=$ on $\mathbf{R}$, $\leq$ on $\mathbf{R}$, and "parallel to" on the class of straight lines in $\mathbf{R}^2$.

A binary relation is called an *equivalence relation* if it satisfies three additional properties.

[THE REFLEXIVE PROPERTY] For every $x \in X, x \sim x$.
[THE SYMMETRIC PROPERTY] If $x \sim y$, then $y \sim x$.
[THE TRANSITIVE PROPERTY] If $x \sim y$ and $y \sim z$, then $x \sim z$.

Notice that $=$ is an equivalence relation on $\mathbf{R}$, "parallel to" is an equivalence relation on the class of straight lines in $\mathbf{R}^2$, but $\leq$ is not an equivalence relation on $\mathbf{R}$ (it fails to satisfy the Symmetric Property).

If $\sim$ is an equivalence relation on a set X, then

$$\overline{x}:=\{y \in X : y \sim x\}$$

is called the *equivalence class* of X which contains x.

F.1 Theorem. *If $\sim$ is an equivalence relation on a set X, then the set of equivalence classes $\{\overline{x} : x \in X\}$ forms a partition of X.*

Proof. Since $\sim$ is reflexive, each equivalence class $\overline{x}$ contains x (i.e., $\overline{x}$ is nonempty). Suppose that $\overline{x} \cap \overline{y} \neq \emptyset$ (i.e., that some $z \in X$ belongs to both these equivalence classes). Then $z \sim x$ and $z \sim y$. By the Symmetric Property and the Transitive Property, we have $x \sim y$ (i.e., $y \in \overline{x}$). By the Transitive Property, it follows that $\overline{y} \subseteq \overline{x}$. Reversing the roles of x and y, we also have $\overline{x} \subseteq \overline{y}$. Thus $\overline{x} = \overline{y}$. ∎

References

1. APOSTOL, TOM M., *Mathematical Analysis*. Reading, Massachusetts: Addison-Wesley Publishing Co., 1974.
2. BOAS, RALPH P., JR., *Primer of Real Functions*, Carus Monograph 13. New York: Mathematical Association of America and John Wiley and Sons, Inc., 1960.
3. COLLEY, SUSAN J., *Vector Calculus*, 3rd ed. Upper Saddle River, N.J.: Prentice Hall, Inc., 2006.
4. HOCKING, JOHN G. AND GAIL S. YOUNG, *Topology*. Reading, Massachusetts: Addison-Wesley Publishing Co., 1961.
5. KLINE, MORRIS, *Mathematical Thought from Ancient to Modern Times*. New York: Oxford University Press, 1972.
6. LOOMIS, LYNN H. AND SHLOMO STERNBERG, *Advanced Calculus*. Reading, Massachusetts: Addison-Wesley Publishing Co., 1968.
7. MARSDEN, JERROLD E., *Elementary Classical Analysis*. New York: W. H. Freeman & Co. Publishers, 1990.
8. MUNKRES, JAMES R., *Elementary Differential Topology*, Annals of Mathematical Studies. Princeton: Princeton University Press, 1963.
9. NOBLE, BEN AND JAMES W. DANIEL, *Applied Linear Algebra*, 2nd ed. Upper Saddle River, N.J.: Prentice Hall, Inc., 1977.
10. PRICE, G. BALEY, *Multivariable Analysis*. New York: Springer-Verlag New York, Inc., 1984.
11. RUDIN, WALTER, *Principles of Mathematical Analysis*, 3rd ed. New York: McGraw-Hill Book Co., 1976.
12. SPIVAK, MICHAEL, *Calculus on Manifolds*. New York: W. A. Benjamin, Inc., 1965.
13. TAYLOR, ANGUS E., *Advanced Calculus*. Boston: Ginn and Company, 1955.
14. WIDDER, DAVID V. *Advanced Calculus*, 2nd ed. Upper Saddle River, N.J.: Prentice Hall, Inc., 1961.
15. ZYGMUND, ANTONI, *Trigonometric Series*, Vol. I, 2nd ed. London and New York: Cambridge University Press, 1968.

Answers and Hints to Selected Exercises

CHAPTER 1

1.2 Ordered Field Axioms

0. b) False. d) True. It's vacuous.
1. b) Consider the cases $c = 0$ and $c \neq 0$.
2. To prove (7), multiply the first inequality in (7) by c and the second inequality in (7) by b. Prove (8) and (9) by contradiction.
4. a) $(-4, 3)$. b) $(0, 4)$. c) $(1/3, (\sqrt{3} - 1)/2) \cup (1, \infty)$. d) $(-\infty, 1)$.
   e) $(-1/2, 1/2)$.
5. a) Apply (6) to $a - 1$. b) Apply (6) to $a - 2$. c) Apply (6) to $1 - a$.
   d) Apply (6) to $a - 2$.
6. Observe that $(\sqrt{a} - \sqrt{b})^2 \geq 0$.
7. Factor. Then use Remark 1.5 and/or the triangle inequality.
8. a) $n > 99$. b) $n \geq 20$. c) $n \geq 23$.
9. a) Use uniqueness of multiplicative inverses to prove that $(nq)^{-1} = n^{-1}q^{-1}$. b) Use part a). c) Use proof by contradiction for the sum. Use a similar argument for the product, and identify all rationals q such that $xq \in \mathbf{Q}$ for a given $x \in \mathbf{R} \setminus \mathbf{Q}$. d) Use the Multiplicative Properties.
10. Show first that the given inequality is equivalent to $2abcd \leq b^2c^2 + a^2d^2$.
11. a) The Trichotomy Property implies i); the Additive and Multiplicative Properties imply ii).

1.3 The Completeness Axiom

0. b) True. d) False.
1. a) $\inf E = -3$, $\sup E = 1$. b) $\inf E = 0$, $\sup E = 3/2$. c) $\inf E = 0$, $\sup E = \sqrt{5}$. d) $\inf E = 0$, $\sup E = 3/2$. e) $\inf E = -1$, $\sup E = 3/2$. f) $\inf E = 7/4$, $\sup E = 3$.
3. Notice that $a - \sqrt{2} < b - \sqrt{2}$, and use Exercise 1.2.9c.
5. See Theorem 1.15.
6. b) Apply Theorem 1.14 to $-E$.
8. After showing that $\sup A$ and $\sup B$ exist, prove that $\max\{\sup A, \sup B\} \leq \sup E$.
9. Use the proof of Theorem 1.18 as a model.
10. $t_1 \leq t_2 \leq \ldots$.

1.4 Mathematical Induction

0. c) True.
1. Compare with Exercise 1.2.5.

3. c) You may use part a).
8. Use the Binomial Formula.
9. b) Show that $n^2 + 3n$ cannot be the square of an integer when $n > 1$.
   c) The expression is rational if and only if $n = 9$.
10. This recursion, discovered by P. W. Wade, generates all Pythagorean triples a, b, c which satisfy $c - b = 1$.

1.5 Inverse Functions and Images

0. a) False. It's $\pi - \arcsin(x)$. d) False.
1. α) $f^{-1}(x) = (x+7)/3$. β) $f^{-1}(x) = 1/\log x$. γ) $f^{-1}(x) = \arctan(x - \pi)$.
   δ) $f^{-1}(x) = -1 - \sqrt{6+x}$. ϵ) $f^{-1}(x) = (x + |x - 2| - |x - 4|)/3$.
   ζ) $f^{-1}(x) = (1 - \sqrt{1 - 4x^2})/2x$ when $x \neq 0$, and $f^{-1}(0) = 0$.
2. a) $f(E) = (-4, 5)$, $f^{-1}(E) = (0, 1)$. b) $f(E) = [1, 5]$, $f^{-1}(E) = [-1, 1]$.
   c) $f(E) = [-8, 1]$, $f^{-1}(E) = [1 - \sqrt{3}, 1 + \sqrt{3}]$. d) $f(E) = (0, \log 5]$,
   $f^{-1}(E) = (1 - \sqrt{e^3 - 1}, 1) \cup (1, 1 + \sqrt{e^3 - 1})$. e) $f(E) = [-1, 1]$,
   $f^{-1}(E) = \cup_{k \in \mathbf{Z}}[(4k - 1)\pi/2, (4k + 1)\pi/2]$.
3. a) $[-2, 2]$. b) $(0, 1]$. c) $\{0\}$. d) $[-1, 0]$. e) $(-\infty, 1)$. f) $\{1\}$.
4. First prove that $f(A) \backslash f(B) \subseteq f(A \backslash B)$ and $A \subseteq f^{-1}(f(A))$ hold whether f is 1–1 or not.

1.6 Countable and Uncountable Sets

0. b) False. d) False.
4. If ϕ is 1–1 from a set E into A, is $\psi(x) := f(\phi(x))$ 1–1 from E into B?
6. a) Prove it by induction on n. b) Use Exercise 1.6.5.

CHAPTER 2

2.1 Limits of Sequences

0. b) False. d) False.
2. a) Apply Definition 2.1 with $\varepsilon/2$ in place of ε. b) First prove that $1/x_n < 2$ for n large.
5. Definition 2.1 works for any positive ε including ε/C when $C > 0$.

2.2 Limit Theorems

0. b) True. c) False. d) You may wish to use derivatives to show that $2^x - x$ is increasing on $[2, \infty)$.
1. c) You may use Exercise 2.2.5. d) First prove that $n^2 \leq 2^n$ for $n = 4, 5, \ldots$.
3. a) $-4/3$. b) $1/2$. c) ∞. d) $1/2$.
5. You may wish to prove that $\sqrt{x_n} - \sqrt{x} = (x_n - x)/(\sqrt{x_n} + \sqrt{x})$.
6. Use Theorem 1.18.
8. If $x = \lim_{n \to \infty} x_n$ exists, what is $\lim_{n \to \infty} x_{n+1}$?

2.3 The Bolzano–Weierstrass Theorem

0. b) You only need to prove that $\{x_n\}$ *has* a convergent subsequence, not actually find it. d) False.

4. Prove that $\{x_n\}$ is monotone.

8. a) See Exercise 1.2.6.

2.4 Cauchy Sequences

0. a) False. c) True.

1. You may use Theorem 2.29.

5. Use Exercise 2.4.4.

6. Is it Cauchy? (Compare with Example 2.30 and see Exercise 1.4.4c.)

8. a) Use the Bolzano–Weierstrass Theorem.

2.5 Limits Supremum and Infimum

1. a) 2, 4. b) -1, 1. c) -1, 1. d) 1/2, 1/2. e) 0, 0. f) 0, ∞. g) ∞, ∞.

4. a) First prove that $\inf_{k \geq n} x_k + \inf_{k \geq n} y_k \leq \inf_{k \geq n}(x_k + y_k)$. c) By b), the first and final inequalities can only be strict if neither $\{x_n\}$ nor $\{y_n\}$ converges.

7. Let $s = \inf_{n \in \mathbb{N}}(\sup_{k \geq n} x_k)$ and consider the cases $s = \infty$, $s = -\infty$, and $s \in \mathbf{R}$.

8. Let $s = \liminf_{n \to \infty} x_n$ and consider the cases $s = \infty$, $s = 0$, and $0 < s < \infty$.

CHAPTER 3

3.1 Two-Sided Limits

0. b) False. c) False.

1. See Example 3.3.

2. b) The limit is zero. c) Does it exist as an extended real number? Why not?

3. a) 2. b) n. c) 0. d) 1/2. [Recall that $2\sin^2 x = 1 - \cos(2x)$.] e) 0.

7. b) Use Exercise 3.1.6.

8. b) Use Exercise 3.1.6.

3.2 One-Sided Limits and Limits at Infinity

0. a) False. c) True.

1. d) You may wish to multiply by $-1/-1$ to make the numerator positive as $x \to 1+$.

2. a) 2. b) 5/3. c) 1. d) ∞. e) $-\infty$. f) $\sqrt{2}/2$.

3. b) Use Theorem 3.8.

5. Prove that if $f(x)$ does not converge to L as $x \to \infty$, then there is a sequence $\{x_n\}$ such that $x_n \to \infty$ but $f(x_n)$ does not converge to L as $n \to \infty$.

6. See Exercise 2.2.6.

3.3 Continuity

0. a) True. Show that $J = [\alpha, \beta]$, where $\alpha = \inf\{f(x) : x \in [a, b]\}$ and $\beta = \sup\{f(x) : x \in [a, b]\}$. c) True.

1. a) and d) Recall that $\sqrt{x}$ is continuous on $[0, \infty)$.

2. c) Recall that $2^x = e^{x \log 2}$.

8. b) Use part a) to show that $f(x) \equiv f(mx/m) = mf(x/m)$ first. d) If the statement is true, then m must equal $f(1)$.
9. a) Begin by showing that $f(0) = 1$.

3.4 Uniform Continuity

0. d) True. (See Exercise 3.4.5.)
1. c) Recall that $\sin 2x - \sin 2a = 2\sin(x-a)\cos(x+a)$ and that $|\sin\theta| \le |\theta|$.
5. c) and e) Prove that $f(x) = x$ and $g(x) = x^2$ are both uniformly continuous on $(0, 1)$ but only one of them is uniformly continuous on $[0, \infty)$.
8. a) This is a function analogue of the Monotone Convergence Theorem.
9. You may wish to prove that if $P(x) = a_n x^n + \cdots + a_0$ is a polynomial of degree $n \ge 1$ whose leading coefficient satisfies $a_n > 0$, then $P(x) \to \infty$ as $x \to \infty$.

CHAPTER 4

4.1 The Derivative

0. a) False. d) True.
4. Use Definition 4.1.
5. a) $a = (2k+1)\pi/2$ and $b = a + (-1)^k$ for $k \in \mathbf{Z}$. b) $(1, 5)$ and $(-3, 29)$.

4.2 Differentiability Theorems

0. b) True.
1. a) $3a + c$. b) $(2b - d)/8$. c) bc. d) bc.
2. a) $8e + 3$. b) π. c) $(1 - 3\pi)/2$.
8. No, f is not differentiable at 0.
9. a) Use assumptions ii) and vi) to prove that $\sin x \to 0$ as $x \to 0$. Use assumption iii) to prove that $\cos x \to 1$ as $x \to 0$. b) First prove that $\sin x = \sin(x - x_0)\cos x_0 + \cos(x - x_0)\sin x_0$ for any $x, x_0 \in \mathbf{R}$. c) Assumption vi) and $0 \le 1 - \cos x \le 1 - \cos^2 x$ play a prominent role here. d) Use assumption iv) and part c).

4.3 The Mean Value Theorem

0. a) True. (But you cannot use Theorem 4.17i.) b) False.
10. a) Compare with Exercise 4.1.8.
11. This exercise has nothing to do with the Mean Value Theorem.
12. Use Darboux's Theorem and Theorem 4.18.

4.4 Taylor's Formula and l'Hôpital's Rule

0. a) False.
1. c) $n = 5$.
2. c) $n = 1999$.
5. a) 25. b) $-\infty$. c) $e^{1/6}$. d) 1. e) $-1/\pi$. f) 1. g) $-2\sqrt{2}$. h) $3/2$ (Rationalize the numerator and denominator. L'Hôpital's Rule only makes matters worse.)

6. a) Use one-variable calculus techniques to maximize $f(x) = \log x / x^\alpha$ for $x \in [1, \infty)$.

7. b) First prove that if $g(x) = e^{-1/x^2}/x^k$ for some $k \in \mathbf{N}$, then $g(x) \to 0$ as $x \to 0$. Next, prove that given $n \in \mathbf{N}$, there are integers $N = N(n) \in \mathbf{N}$ and $a_k = a_k^{(n)} \in \mathbf{Z}$ such that

$$f^{(n)}(x) = \begin{cases} \sum_{k=0}^{N}(a_k/x^k)e^{-1/x^2} & x \neq 0 \\ 0 & x = 0. \end{cases}$$

(*Note*: Although for each $n \in \mathbf{N}$ many of the a_k's are zero, this fact is not needed in this exercise.)

4.5 Inverse Function Theorems

0. a) False. c) False.
1. a) $1/\pi$. b) $1/e$. c) $1/\pi$.
2. b) $1/4e$.
3. Observe that if $x = \sin y$, then $\cos y = \sqrt{1 - x^2}$.
9. $f(x) = \pm\sqrt{\alpha}x + c$ for some $c \in \mathbf{R}$.
11. Use Darboux's Theorem.

CHAPTER 5

5.1 The Riemann Integral

0. a) False.
1. a) $L(f, P) = 17/16$. b) $L(f, P) = 11/8$. c) $U(f, P) \approx 0.5386697$.
4. a) Look at the proof of Lemma 3.28.
5. First show that $\int_I f(x)\,dx = 0$ for all subintervals I of $[a, b]$.
8. Notice that $|x_j - x_{j-1}| \leq \|P\|$ for each $j = 1, 2, \ldots, n$.
9. Prove that there is an absolute constant $C > 0$ such that $M_j(\sqrt{f}) - m_j(\sqrt{f}) \leq C(M_j(f) - m_j(f))$.

5.2 Riemann Sums

0. b) True. c) True.
1. a) 5. b) 9. c) $\pi a^2/2$. d) $10 + \pi/2$.
3. a) See Exercise 4.4.4 and use three nonzero terms for the upper bound. b) Use the Taylor polynomial with four nonzero terms.
5. Do not forget that f is bounded.
6. b) You may use the fact that $\int x^n\,dx = x^{n+1}/(n+1)$.
8. a) If $|f(x_0)| > M - \varepsilon/2$ for some $x_0 \in [a, b]$, can you choose a nondegenerate interval I such that $|f(x)| > M - \varepsilon$ for all $x \in I$? b) See Example 2.21.

5.3 The Fundamental Theorem of Calculus

0. b) True. c) True.
1. a) $-2xf(x^2)$. b) $3x^2 f(x^3) - 2xf(x^2)$. c) $(x\cos^2 x - x^2 \sin x \cos x)$ $f(x\cos x)$. d) $f(-x)$.

3. a) $-1/2$. b) -4.

6. Use the Fundamental Theorem of Calculus.

11. The hard part is showing that f^q is integrable on $[a, b]$. You may use Exercise 4.2.7a.

12. First show that $\log(a_n)$ is a Riemann sum.

5.4 Improper Riemann Integration

0. c) True.

1. a) $3/2$. b) $1/3$. c) $3/2$. d) -1.

2. a) $p > 1$. b) $p < 1$. c) $p > 1$. d) $p > 1$. e) $p > 1$.

3. Compare with Example 5.44.

4. a) Diverges. b) Diverges. c) Converges. d) Converges. e) Converges.

9. Integrate by parts first.

10. You might begin by verifying $\sin x \geq \sqrt{2}/2$ for $x \in [\pi/4, \pi/2]$ and $\sin x \geq 2x/\pi$ for $x \in [0, \pi/4]$.

5.5 Functions of Bounded Variation

1. c) Find a partition where the jumps are large.

4. Combine Theorems 4.18 and 3.39.

9. For the bounded case, prove that $(L) \int_a^b |f'(x)|\, dx \leq \text{Var } f \leq (U) \int_a^b |f'(x)|\, dx$.

5.6 Convex Functions

5. Use Remark 5.60.

CHAPTER 6

6.1 Introduction

0. a) False. d) True.

1. a) $e/(e + 1)$. b) $-\pi^2/(\pi^2 + 1)$. c) $64/5$. d) $13/35$.

2. a) 1. b) 4. c) $\log(2/3)$. d) $\sin 2$.

5. $|x| \leq 1$.

6. b) Consider the Geometric Series.

8. a) Is $na_{2n} \leq \sum_{k=n}^{\infty} a_k$?

9. c) Notice that if the partial sums of $\sum_{k=1}^{\infty} b_k$ are bounded, then $b = 0$.

10. b) See Exercise 6.1.9b. d) First prove that if $a_k \geq 0$ and $\sum_{k=0}^{\infty} a_k$ diverges, then $\sum_{k=0}^{\infty} a_k = \infty$.

11. Use Corollary 6.9.

6.2 Series with Nonnegative Terms

0. b) True. c) True.

1. c) If $p > 1$, are there constants $C > 0$ and $q > 1$ such that $\log k/k^p \leq Ck^{-q}$?

2. b) No, you cannot apply the p-Series Test to $k^{1-1/k}$ because the exponent $p := (1-1/k)$ is NOT constant but depends on k. d) Try the Integral Test.

4. It converges when $p > 1$ and diverges when $0 \leq p \leq 1$.

8. It diverges when $0 < q \leq 1$ and converges when $q > 1$.
10. Recall that $a^b := e^{b \log a}$ and observe that $\log(\log(\log k)) \to \infty$ as $k \to \infty$.

6.3 Absolute Convergence

0. b) False. c) False.
2. a) Convergent. b) Convergent. c) Convergent. d) Convergent. e) Convergent. f) Convergent. g) Divergent.
3. a) $(1, \infty)$. b) $\emptyset$. c) $(-\infty, -1) \cup (1, \infty)$. d) $(1/2, \infty)$. e) $(1, \infty)$.
   f) $(-\infty, \log_2(e))$ (use Stirling's Formula when $p = \log_2(e)$).
9. $\pi^2/8$.
10. See Definition 2.32.

6.4 Alternating Series

0. a) True. c) False.
1. b) See Example 6.34.
2. a) $[-1, 1)$. b) $(-\sqrt[3]{2}, \sqrt[3]{2})$. c) $(-1, 1]$. d) $[-3, -1]$.
3. a) Absolutely convergent. b) Absolutely convergent. c) Absolutely convergent. d) Conditionally convergent. e) Absolutely convergent.
6. Is it Cauchy?
7. Let $c_k = \sum_{j=k}^{\infty} a_j b_j$ and apply Abel's Formula to

$$\sum_{k=n}^{m} a_k \equiv \sum_{k=n}^{m} \frac{c_k - c_{k+1}}{b_k}.$$

8. See Example 6.34.
9. See Example 6.34.

6.5 Estimation of Series

1. a) At most 100 terms. b) At most 15 terms. c) At most 10 terms. (To prove that $\{a_k\}$ is monotone, verify that $a_{k+1}/a_k < 1$.)
2. a) $p > 1$.
3. a) $n = 8$. b) $n = 11$. c) $n = 14$. d) $n = 11$.

6.6 Additional Tests

1. a) Divergent. b) Absolutely convergent. c) Divergent. d) Absolutely convergent.
2. a) Absolutely convergent for $p > 0$ and divergent for $p \leq 0$. b) Absolutely convergent for $p > 0$ and divergent for $p \leq 0$. c) Absolutely convergent for $|p| < 1/e$, conditionally convergent for $p = -1/e$, and divergent otherwise. (Use Stirling's Formula when $p = \pm 1/e$.)
4. It actually converges absolutely.

CHAPTER 7

7.1 Uniform Convergence of Sequences

2. a) $(3^{34} - 1)/34$. b) 2. c) 14/3.
6. This is different from Theorem 7.9 because when E is not compact, uniform continuity is not the same as continuity.
8. Modify the proof of Example 4.22 to show that $(1+x/n)^n \uparrow e^x$ as $n \to \infty$. To prove that this is a uniform limit, choose N so large that $[a, b] \subset [-N, N]$ and find the maximum of $e^x - (1+x/N)^N$ on $[a, b]$.
9. a) Use Exercise 7.1.5c.

7.2 Uniform Convergence of Series

1. a) Recall that $|\sin \theta| \le |\theta|$ for all $\theta \in \mathbf{R}$.
5. Dominate, then telescope.
6. Is there a connection between $\sum_{k=1}^{\infty} k^{-1} \sin(xk^{-1})$ and $\sum_{k=1}^{\infty} \cos(xk^{-1})$?
7. Use Abel's Formula.
9. See Example 6.34.

7.3 Power Series

1. a) $R = 1$. b) $R = 1/3$. c) $R = 1/3$. d) $R = 1$.
2. a) $(-2, 2)$. b) $(3/4, 5/4)$. c) $[-1, 1)$. d) $[-1/\sqrt{2}, 1/\sqrt{2}]$. (Use Raabe's Test for the endpoints.)
7. a) $f(x) = 3x^2/(1 - x^3)$ for $x \in (-1, 1)$. b) $f(x) = (2 - x)/(1 - x)^2$ for $x \in (-1, 1)$. c) $f(x) = 2(1/x - 1 + \log x)/(1 - x)$ for $x \in (0, 2)$, $x \neq 1$, and $f(1) = 0$.
   d) $f(x) = \log(1/(1 - x^3))/x^3$ for $x \in [-1, 1)$, $x \neq 0$, and $f(0) = 1$.
8. Use Exercise 2.5.8 to prove that if $\limsup |a_k/a_{k+1}| < R$, then there is an $r < R$ such that $\{|a_k r^k|\}$ is increasing for k large (i.e., that $\sum_{k=1}^{\infty} a_k r^k$ diverges for some $r < R$).
9. Use the method of Example 7.36 to estimate $|f'(x)|$.
10. First prove that the radius of convergence of $\sum_{k=0}^{\infty} a_k x^k$ is ≥ 1.
11. a) Use Theorem 6.35 to estimate $\log(n!) = \sum_{k=1}^{n} \log k$. b) $x \in (-1/e, 1/e)$.

7.4 Analytic Functions

1. a) $x^2 + \cos(2x) = 1 - x^2 + \sum_{k=2}^{\infty} (-4)^k x^{2k}/(2k)!$. b) $x^2 3^x = \sum_{k=2}^{\infty} x^k \log^{k-2} 3/(k-2)!$. c) $\cos^2 x - \sin^2 x = \sum_{k=0}^{\infty} (-4)^k x^{2k}/(2k)!$.
   d) $(e^x - 1)/x = \sum_{k=0}^{\infty} x^k/(k+1)!$.
2. a) $x/(x^5 + 1) = \sum_{k=0}^{\infty} (-1)^k x^{5k+1}$.
   b) $e^x/(1 + x) = \sum_{k=0}^{\infty} (\sum_{j=0}^{k} (-1)^{k-j}/j!) x^k$.
   c) $\log(1/|x^2 - 1|) = \sum_{k=1}^{\infty} x^{2k}/k$.
   d) $\arcsin x = \sum_{k=0}^{\infty} \binom{-1/2}{k} (-1)^k x^{2k+1}/(2k + 1)$. (Use Theorem 7.52.)
3. a) $e^x = \sum_{k=0}^{\infty} e(x - 1)^k/k!$, valid for $x \in \mathbf{R}$.
   b) $\log_2 x^5 = 5 \sum_{k=1}^{\infty} (-1)^{k+1}(x - 1)^k/(k \log 2)$, valid for $x \in (0, 2]$.
   c) $x^3 - x + 5 = 5 + 2(x - 1) + 3(x - 1)^2 + (x - 1)^3$, valid for $x \in \mathbf{R}$.

d) Use Raabe's Test. $\sqrt{x} = 1 + (x-1)/2 + \sum_{k=2}^{\infty}(-1)^{k-1}1 \cdot 1 \cdot 3 \cdots$ $(2k-3)(x-1)^k/(2 \cdot 4 \cdot 6 \cdots (2k))$, valid on $[0, 2]$.

9. Let f_n be a Taylor polynomial of f and compute the integral of $f \cdot f_n$. Question: If $f_n \to f$ uniformly and f is bounded, does $f \cdot f_n \to f^2$ uniformly?

10. See Exercise 5.1.4 and use analytic continuation.

11. First use the Binomial Series to verify that $(1+x)^\beta \geq 1 + x^\beta$ for any $0 < x < 1$.

7.5 Applications

1. The first seven places of the only real root are given by -0.3176721.

6. Choose r_0 as in the proof of Theorem 7.58, define $\{x_n\}$ by (19), and find a δ so that $|f(x_0)| \leq \delta$ implies $|x_n - x_{n-1}| < r_0^{n+1}$.

CHAPTER 8

8.1 Algebraic Structure

4. Use (2).

6. b) $(a, a, a), a \neq 0$. c) $(a, (20-8a)/7, (8+a)/7), a \neq 0$.

7. $\arccos(1/\sqrt{n})$.

8. b) For part vi), write $\|\mathbf{x} \times \mathbf{y}\|^2 = (\mathbf{x} \times \mathbf{y}) \cdot (\mathbf{x} \times \mathbf{y})$ and use parts iv) and v).

8.2 Planes and Linear Transformations

2. a) $x - y + 2z + 5w = 1$. b) $x - y + w = 1$. c) Use Exercise 1.4.4a.

4. a)
$$\begin{bmatrix} 0 & 0 & 0 & 0 \\ 1 & 1 & 0 & 0 \\ 1 & 0 & -1 & 0 \\ 1 & 1 & 0 & 1 \end{bmatrix}.$$

b)
$$\begin{bmatrix} 1 & -1 & 1 \end{bmatrix}.$$

c)
$$\begin{bmatrix} 1 & 0 & \cdots & 0 & -1 \\ -1 & 0 & \cdots & 0 & 1 \end{bmatrix}.$$

5. Use the linear property to compute the $T(e_j)$'s.

a)
$$\begin{bmatrix} -1 & 4 \\ \pi & 0 \\ -1 & 1 \end{bmatrix}.$$

b)
$$\begin{bmatrix} 2 & e-2 & e-1 \\ 2 & \pi-2 & \pi-2 \end{bmatrix}.$$

c)
$$\begin{bmatrix} a & 4 & -1 & -\pi \\ b & 4 & 1 & -3 \end{bmatrix}.$$

for any choice of $a, b \in \mathbf{R}$.

d)
$$\begin{bmatrix} a & 5-a & 1 & -\pi \\ b & 4-b & 2 & -3 \\ c & 1-c & 0 & 1 \end{bmatrix}$$

for any choice of $a, b, c \in \mathbf{R}$.

6. Don't be an algebra zombie. Use $\mathbf{a}, \mathbf{b}, \mathbf{c}$ to produce a normal to Π.

8. If (x_0, y_0, z_0) does not lie on Π, let (x_2, y_2, z_2) be a point on Π different from (x_1, y_1, z_1), let θ represent the angle between $\mathbf{w} := (x_0 - x_2, y_0 - y_2, z_0 - z_2)$ and the normal (a, b, c), and compute $\cos \theta$ two different ways, once in terms of $\mathbf{w}$ and a second time in terms of the distance from (x_0, y_0, z_0) to Π.

10. a)
$$\begin{bmatrix} 2x \\ \cos x \end{bmatrix}.$$

8.3 Topology of $\mathbf{R}^n$

1. a) Open but not connected. b) Closed and connected. c) Neither open nor closed, but connected. d) Open but not connected. e) Closed and connected.
6. b) Try a proof by contradiction.
7. b) Try a proof by contradiction.
9. Notice that $\mathbf{a} \in E^c$ and E^c is open.
10. Diamonds and squares.

8.4 Interior, Closure, and Boundary

1. a) $E^o = \emptyset$, $\overline{E} = E \cup \{0\}$, $\partial E = \overline{E}$. b) $E^o = E$, $\overline{E} = [0, 1]$, $\partial E = \{1/n : n \in \mathbf{N}\} \cup \{0\}$. c) $E^o = \mathbf{R}$, $\overline{E} = \mathbf{R}$, $\partial E = \emptyset$. d) Use Theorem 1.18.
2. a) $E^o = \{(x, y) : x^2 + 4y^2 < 1\}$ and $\partial E = \{(x, y) : x^2 + 4y^2 = 1\}$. b) $E^o = \emptyset$ and $\partial E = E$. c) $E^o = \{(x, y) : y > x^2, 0 < y < 1\}$, $\overline{E} = \{(x, y) : y \geq x^2, 0 \leq y \leq 1\}$, and $\partial E = \{(x, y) : y = x^2, 0 \leq y \leq 1\} \cup \{(x, 1) : -1 \leq x \leq 1\}$. d) $\overline{E} = \{x^2 - y^2 \leq 1, -1 \leq y \leq 1\}$ and $\partial E = \{(x, y) : x^2 - y^2 = 1, -\sqrt{2} \leq x \leq \sqrt{2}\} \cup \{(x, 1) : -\sqrt{2} \leq x \leq \sqrt{2}\} \cup \{(x, -1) : -\sqrt{2} \leq x \leq \sqrt{2}\}$.
10. c) Use part b). You may assume that $\mathbf{R}^n$ is connected.

CHAPTER 9

9.1 Limits of Sequences

2. a) $(0, 2)$. b) $(1, 0, 1)$. c) $(-1/2, 1, 0)$.
6. b) Show that a set C is relatively closed in E if and only if $E \setminus C$ is relatively open in E.
8. b) Use the Bolzano–Weierstrass Theorem.

9.2 Heine–Borel Theorem

7. Use the Approximation Theorem for Infima.
8. You may use Exercise 4.4.7.

9.3 Limits of Functions

1. a) $\text{Dom} f = \{(x, y) : y \neq 1\}$ and the limit is $(0, 3)$. b) $\text{Dom} f = \{(x, y) : x \neq 0, y \neq 0, \text{ and } x/y \neq k\pi/2, k \text{ odd}\}$ and the limit is $(1, 0, 1)$. c) $\text{Dom} f = \{(x, y) : (x, y) \neq (0, 0)\}$ and the limit is $(0, 0)$. d) $\text{Dom} f = \{(x, y) : (x, y) \neq (1, 1)\}$ and the limit is $(0, 0)$.
2. a) $\lim_{y \to 0} \lim_{x \to 0} f(x, y) = \lim_{x \to 0} \lim_{y \to 0} f(x, y) = 0$, but $f(x, y)$ has no limit as $(x, y) \to (0, 0)$. b) $\lim_{y \to 0} \lim_{x \to 0} f(x, y) = 1/2$, $\lim_{x \to 0} \lim_{y \to 0} f(x, y) = 1$, so $f(x, y)$ has no limit as $(x, y) \to (0, 0)$. c) $\lim_{y \to 0} \lim_{x \to 0} f(x, y) = \lim_{x \to 0} \lim_{y \to 0} f(x, y) = 0$, and $f(x, y) \to 0$ as $(x, y) \to (0, 0)$.
7. Use the Mean Value Theorem.

9.4 Continuous Functions

4. For the converse, suppose not, and use the Sequential Characterization of Continuity.
5. See Exercise 8.3.8b.
8. See Theorem 3.40.
10. a) A polygonal path in E can be described as the image of a continuous function $f : [0, 1] \to E$. Use this to prove that every polygonal path is connected. c) Prove that if E is not polygonally connected, then there are nonempty open sets $U, V \subset E$ such that $U \cap V = \emptyset$ and $U \cup V = E$.

9.5 Compact Sets

1. a) Compact. b) Compact. c) Not compact. $H = E \cup \{(0, y) : -1 \leq y \leq 1\}$. d) Not compact. There is no compact set H which contains E.
5. Compare with Exercise 8.3.8.

9.6 Applications

2. See Exercise 7.2.7.
4. a) $\omega_f(t) = 1$ for all t. b) $\omega_f(t) = 0$ if $t \neq 0$ and $\omega_f(0) = 1$. c) $\omega_f(t) = 0$ if $t \neq 0$ and $\omega_f(0) = 2$.
7. a) $\sqrt{1/2}$. b) $f(0)/3$. c) $1/4$. d) $(e^4 - 1)/(2e^2)$.
9. d) You may wish to use Theorem 4.18.

CHAPTER 10

10.1 Introduction

8. c) Observe that $f_n(x) = x^n$ does not converge to a continuous function on $[0, 1]$.
10. a) Show that if E is not bounded, then there exist $x_n \in E$ and $a \in X$ such that $\rho(x_n, a) \to \infty$ as $n \to \infty$.

10.2 Limits of Functions

1. a) **R**. b) $[a, b]$. c) $\emptyset$. d) $\{x\}$ if E is infinite, $\emptyset$ if E is finite. e) $\emptyset$.
9. b) See the proof of Theorem 3.26.

10.3 Interior, Closure, and Boundary

1. a) $E^o = \emptyset$, $\overline{E} = E \cup \{0\}$, $\partial E = \overline{E}$. b) $E^o = E$, $\overline{E} = [0, 1]$, $\partial E = \{1/n : n \in \mathbf{N}\} \cup \{0\}$. c) $E^o = \mathbf{R}$, $\overline{E} = \mathbf{R}$, $\partial E = \emptyset$. d) See Theorem 1.18.
2. a) Closed. $E^o = \{(x, y) : x^2 + 4y^2 < 1\}$ and $\partial E = \{(x, y) : x^2 + 4y^2 = 1\}$. b) Closed. $E^o = \emptyset$ and $\partial E = E$. c) Neither open nor closed. $E^o = \{(x, y) : y > x^2, 0 < y < 1\}$, $\overline{E} = \{(x, y) : y \geq x^2, 0 \leq y \leq 1\}$, and $\partial E = \{(x, y) : y = x^2, 0 \leq y \leq 1\} \cup \{(x, 1) : -1 \leq x \leq 1\}$. d) Open. $\overline{E} = \{x^2 - y^2 \leq 1, -1 \leq y \leq 1\}$ and $\partial E = \{(x, y) : x^2 - y^2 = 1, -\sqrt{2} \leq x \leq \sqrt{2}\} \cup \{(x, 1) : -\sqrt{2} \leq x \leq \sqrt{2}\} \cup \{(x, -1) : -\sqrt{2} \leq x \leq \sqrt{2}\}$.
5. Notice that $a \in E^c$ and E^c is open.
8. See the description of relative balls following Definition 10.22.
9. To show that f is continuous at a, consider the open interval $I = (f(a) - \varepsilon, f(a) + \varepsilon)$.

10.4 Compact Sets

1. a) Compact. b) Compact. c) Not compact. $H = E \cup \{(0, y) : -1 \leq y \leq 1\}$. d) Not compact. There is no compact set H which contains E.
5. Compare with Exercise 10.3.10.
8. a) Notice that if $\cap H_k = \emptyset$, then $\{X \setminus H_k\}$ covers X.
10. a) Let $x_k \in E$. Does E contain a point a such that each $B_r(a)$, $r > 0$, contains x_k for infinitely many k's? b) See Exercise 10.1.10a.

10.5 Connected Sets

9. Use Exercise 10.5.8.
10. a) A polygonal path in E can be described as the image of a continuous function $f : [0, 1] \to E$. Use this to prove that every polygonal path is connected. c) Prove that if E is not polygonally connected, then there are nonempty open sets $U, V \subset E$ such that $U \cap V = \emptyset$ and $U \cup V = E$.
11. Try a proof by contradiction.

10.6 Continuous Functions

2. b) Note that $f^{-1}(-1, 1)$ is not open. Does this contradict Theorem 10.58?
4. You may wish to prove that A is relatively closed in E if and only if $E \setminus A$ is relatively open in E.
8. See Theorem 3.40.
9. See Remark 1.39.

10.7 Stone–Weierstrass Theorem

4. If P is a polynomial, what is $\int_a^b f(x) P(x)\, dx$?
7. If P is a trigonometric polynomial, what is $\int_0^{2\pi} f(x) P(x)\, dx$?

CHAPTER 11

11.1 Partial Derivatives and Partial Integrals

1. a) $f_{xy} = f_{yx} = e^y$. b) $f_{xy} = f_{yx} = -\sin(xy) - xy\cos(xy)$. c) $f_{xy} = f_{yx} = -2x/(x^2 + 1)^2$.

2. a) $f_x = (2x^5 + 4x^3y^2 - 2xy^4)/(x^2 + y^2)^2$ for $(x, y) \neq (0, 0)$ and $f_x(0, 0) = 0$. f_x is continuous on $\mathbf{R}^2$. b) $f_x = (2x/3) \cdot (2x^2 + 4y^2)/(x^2 + y^2)^{4/3}$ for $(x, y) \neq (0, 0)$ and $f_x(0, 0) = 0$. f_x is continuous on $\mathbf{R}^2$.

5. a) $e - 1$. b) $(e - 4)\cos(\pi - 1)$. c) $3 - \sqrt{3}$.

6. a) 2. b) 0. c) 3.

7. a) $9/10$. b) $e^{-\pi}/2$.

10. c) Choose $\delta > 0$ such that $|\phi(t)| < \varepsilon$ for $0 \leq t < \delta$ and break the integral in part b) into two pieces, one corresponding to $0 \leq t \leq \delta$ and the other to $\delta \leq t < \infty$. d) Combine part b) with Theorem 11.8.

11. a) $\mathcal{L}\{te^t\} = 1/(s - 1)^2$. b) $\mathcal{L}\{t \sin \pi t\} = 2s\pi/(s^2 + \pi^2)^2$. c) $\mathcal{L}\{t^2 \cos t\} = 2(s^3 - 3s)/(s^2 + 1)^3$.

11.2 The Definition of Differentiability

2. When $\mathbf{f}$ is differentiable at a and its domain is one dimensional, does $\|\mathbf{f}(a + h) - \mathbf{f}(a)\|/|h|$ have a limit as $h \to 0$?

9. b) The function f might not be differentiable when $\alpha = 1$.

11.3 Derivatives, Differentials, and Tangent Planes

1. a)
$$D(\mathbf{f} + \mathbf{g})(\mathbf{x}) = \begin{bmatrix} 2x + 1 & 2y - 1 \end{bmatrix}$$

and

$$D(\mathbf{f} \cdot \mathbf{g})(\mathbf{x}) = \begin{bmatrix} 3x^2 - 2xy + y^2 & 2xy - x^2 - 3y^2 \end{bmatrix}.$$

b)
$$D(\mathbf{f} + \mathbf{g})(\mathbf{x}) = \begin{bmatrix} x \cos x + \sin x + y & x + \sin y \end{bmatrix}$$

and

$$D(\mathbf{f} \cdot \mathbf{g})(\mathbf{x}) = \begin{bmatrix} x^2 y \cos x + 2xy \sin x - y \cos y & x^2 \sin x + xy \sin y - x \cos y \end{bmatrix}.$$

c)
$$D(\mathbf{f} + \mathbf{g})(\mathbf{x}) = \begin{bmatrix} -y\sin(xy) & 1 - x\sin(xy) \\ 1 + \log y & x/y \end{bmatrix}$$

and

$$D(\mathbf{f} \cdot \mathbf{g})(\mathbf{x}) = \begin{bmatrix} 2x \log y - y^2 \sin(xy) & \cos(xy) - xy \sin(xy) + x^2/y \end{bmatrix}.$$

d)
$$D(\mathbf{f} + \mathbf{g})(\mathbf{x}) = \begin{bmatrix} yz & xz + 1 & xy \\ 1 & 2y & -1 \end{bmatrix}$$

and

$$D(\mathbf{f} \cdot \mathbf{g})(\mathbf{x}) = \begin{bmatrix} y^2(z+1) & 2xyz + 2xy - 2yz & y^2(x-1) \end{bmatrix}.$$

2. a) $2x - 2y - z = 2$. b) $2x - 2y - z = 0$. c) $y - w = -1$.
3. $(-1/2, -1/2, 1/2)$, $2x + 2y + 2z = -1$.
4. a) $x - z = 0$. b) There are none.
8. a) $dz = 2x\,dx + 2y\,dy$. b) $dz = y\cos(xy)\,dx + x\cos(xy)\,dy$. c) $dz = (1 - x^2 + y^2)y/(1 + x^2 + y^2)^2\,dx + (1 + x^2 - y^2)x/(1 + x^2 + y^2)^2\,dy$.
9. $dw = .05$ and $\Delta w \approx 0.049798$.
10. L must be measured with no more than 3% error.

11.4 The Chain Rule

1.
$$\frac{\partial w}{\partial p} = \frac{\partial F}{\partial x}\frac{\partial x}{\partial p} + \frac{\partial F}{\partial y}\frac{\partial y}{\partial p} + \frac{\partial F}{\partial z}\frac{\partial z}{\partial p}, \qquad \frac{\partial w}{\partial q} = \frac{\partial F}{\partial x}\frac{\partial x}{\partial q} + \frac{\partial F}{\partial y}\frac{\partial y}{\partial q} + \frac{\partial F}{\partial z}\frac{\partial z}{\partial q},$$

$$\frac{\partial^2 w}{\partial p^2} = \frac{\partial}{\partial p}\left(\frac{\partial F}{\partial x}\right)\frac{\partial x}{\partial p} + \frac{\partial F}{\partial x}\frac{\partial^2 x}{\partial p^2} + \frac{\partial}{\partial p}\left(\frac{\partial F}{\partial y}\right)\frac{\partial y}{\partial p}$$

$$+ \frac{\partial F}{\partial y}\frac{\partial^2 y}{\partial p^2} + \frac{\partial}{\partial p}\left(\frac{\partial F}{\partial z}\right)\frac{\partial z}{\partial p} + \frac{\partial F}{\partial z}\frac{\partial^2 z}{\partial p^2}$$

$$= \frac{\partial F}{\partial x}\frac{\partial^2 x}{\partial p^2} + \frac{\partial F}{\partial y}\frac{\partial^2 y}{\partial p^2} + \frac{\partial F}{\partial z}\frac{\partial^2 z}{\partial p^2}$$

$$+ \frac{\partial^2 F}{\partial x^2}\left(\frac{\partial x}{\partial p}\right)^2 + \frac{\partial^2 F}{\partial y^2}\left(\frac{\partial y}{\partial p}\right)^2 + \frac{\partial^2 F}{\partial z^2}\left(\frac{\partial z}{\partial p}\right)^2$$

$$+ 2\frac{\partial^2 F}{\partial x \partial y}\left(\frac{\partial x}{\partial p}\right)\left(\frac{\partial y}{\partial p}\right) + 2\frac{\partial^2 F}{\partial x \partial z}\left(\frac{\partial x}{\partial p}\right)\left(\frac{\partial z}{\partial p}\right)$$

$$+ 2\frac{\partial^2 F}{\partial y \partial z}\left(\frac{\partial y}{\partial p}\right)\left(\frac{\partial z}{\partial p}\right).$$

6. Notice by Exercise 11.2.11 that this result still holds if "f is in C^2" is replaced by "the first-order partial derivatives of f are differentiable."
9. Take the derivative of $F(x, f(x)) = 0$ with respect to x.
10. Compute the derivative of $\mathbf{f} \cdot \mathbf{f}$ using the Dot Product Rule.

11.5 The Mean Value Theorem and Taylor's Formula

1. a) $f(x, y) = 1 - (x+1) + (y-1) + (x+1)^2 + (x+1)(y-1) + (y-1)^2$.
   b) $\sqrt{x} + \sqrt{y} = 3 + (x-1)/2 + (y-4)/4 - (x-1)^2/8 - (y-4)^2/64 + (x-1)^3/16\sqrt{c^5} + (y-4)^3/16\sqrt{d^5}$ for some $(c, d) \in L((x, y); (1, 4))$. c) $e^{xy} = 1 + xy + ((dx+cy)^4 + 12(dx+cy)^2 xy + 12x^2 y^2)e^{cd}/4!$ for some $(c, d) \in L((x, y); (0, 0))$.
2. Notice that by Exercise 11.2.11, this result still holds if "f is in C^p" is replaced by "the $(p-1)$-st order partial derivatives of f are differentiable."

10. Apply Taylor's Formula to $f(a + x, b + y)$ for $p = 3$, $x = r \cos\theta$, and $y = r \sin\theta$, and prove that

$$\frac{4}{\pi r^2} \int_0^{2\pi} f(a + r\cos\theta, b + r\sin\theta) \cos(2\theta)\, d\theta$$
$$= f_{xx}(a, b) - f_{yy}(a, b) + F(r),$$

where $F(r)$ is a function which converges to 0 as $r \to 0$.

11. c) Let (x_2, t_2) be the point identified in part b), and observe by one-dimensional theory that $u_t(x_2, t_2) = 0$. Use this observation and Taylor's Formula to obtain the contradiction $w_{xx}(x_2, t_2) - w_t(x_2, t_2) \geq 0$.

11.6 The Inverse Function Theorem

1. a) Since $\mathbf{f}(x, y) = (a, b)$ always has a solution by Cramer's Rule and $D\mathbf{f}$ is constant,

$$D\mathbf{f}^{-1}(a, b) = \begin{bmatrix} 5/17 & 1/17 \\ -2/17 & 3/17 \end{bmatrix} \quad \text{(see Theorem C.7)}.$$

b) Since $\mathbf{f}((4k + 1)\pi/2, -(4k + 1)\pi/2) = \mathbf{f}(2k\pi, -2k\pi) = (0, 1)$,

$$D\mathbf{f}^{-1}(0, 1) = \begin{bmatrix} 0 & 1 \\ 1 & -1 \end{bmatrix} \quad \text{or} \quad \begin{bmatrix} 1 & -1 \\ 0 & 1 \end{bmatrix} \quad \text{(see Theorem C.7)}$$

depending on which branch of $\mathbf{f}^{-1}$ you choose.

c) Since $\mathbf{f}(\pm 2, \pm 1) = \mathbf{f}(\pm 1, \pm 2) = (2, 5)$,

$$D\mathbf{f}^{-1}(2, 5) = \begin{bmatrix} \mp 1/3 & \pm 1/3 \\ \pm 2/3 & \mp 1/6 \end{bmatrix} \quad \text{or} \quad \begin{bmatrix} \pm 2/3 & \mp 1/6 \\ \mp 1/3 & \pm 1/3 \end{bmatrix}$$
(see Theorem C.7)

depending on which branch of $\mathbf{f}^{-1}$ you choose.

d) Since $\mathbf{f}(0, 1) = (-1, 0)$, one branch of $\mathbf{f}^{-1}$ satisfies

$$D(\mathbf{f}^{-1})(-1, 0) = \begin{bmatrix} -1/2 & 1 \\ -1/2 & 0 \end{bmatrix} \quad \text{(see Theorem C.7)}.$$

4. $F(x_0, y_0, u_0, v_0) = (0, 0)$, $x_0^2 \neq y_0^2$, and $u_0 \neq 0 \neq v_0$, where $F(x, y, u, v) = (xu^2 + yv^2 + xy - 9, xv^2 + yu^2 - xy - 7)$.

6. a) $\mathbf{f}^{-1}(s, t) = ((s + \sqrt{s^2 - 4t})/2, (s - \sqrt{s^2 - 4t})/2)$.

b) $D(\mathbf{f}^{-1})(\mathbf{f}(x, y)) = \begin{bmatrix} x/(x - y) & 1/(y - x) \\ y/(y - x) & 1/(x - y) \end{bmatrix}$ (see Theorem C.7).

7. Find the partial of $F(x_1, \ldots, g_{j+1}, \ldots, x_n)$ with respect to x_j.

9. Use the Implicit Function Theorem.

11. a) $ax + by = 1$, where $a^2 + b^2 = 1$. b) $x + y - z = \pm 1$.

11.7 Optimization

1. a) $f(1/3, 2/3) = -13/27$ is a local minimum and $(-1/4, -1/2)$ is a saddle point. b) Let $k, j \in \mathbf{Z}$. $f((2k+1)\pi/2, j\pi) = 2$ is a local maximum if k and j are even, $f((2k+1)\pi/2, j\pi) = -2$ is a local minimum if k and j are odd, and $((2k+1)\pi/2, j\pi)$ is a saddle point if $k + j$ is odd. c) This function has no local extrema. d) $f(0, 0) = 0$ is a local minimum if $a > 0$ and $b^2 - 4ac < 0$, a local maximum if $a < 0$ and $b^2 - 4ac < 0$, and $(0, 0)$ is a saddle point if $b^2 - 4ac > 0$.

2. a) $f(2, 0) = 8$ is the maximum and $f(-4/5, \pm\sqrt{21}/5) = -9/5$ is the minimum. b) $f(1, 2) = 17$ is the maximum and $f(1, 0) = 1$ is the minimum. c) $f(1, 1) = f(-1, -1) = 3$ is the maximum and $f(-1, 1) = -5$ is the minimum.

3. a) $f(-2, 0) = -2$ is the minimum and $f(1/2, \pm\sqrt{15}/2) = 17/4$ is the maximum. b) $f(\pm 2/\sqrt{5}, \pm 1/\sqrt{5}) = 0$ is the minimum and $f(\pm 1/\sqrt{5}, \mp 2/\sqrt{5}) = 5$ is the maximum. c) $\lambda = xy$, $3\mu = x + y$, $f(\pm 1/\sqrt{2}, \mp 1/\sqrt{2}, 0) = -1/2$ is the minimum and $f(\pm 1/\sqrt{6}, \pm 1/\sqrt{6}, \mp 2/\sqrt{6}) = 1/6$ is the maximum. d) $f(1, -2, 0, 1) = 2$ is the minimum and $f(1, 2, -1, -2) = 3$ is the maximum.

7. b) If $DE < 0$, then $ax + by + cz$ has no extremum subject to the constraint $z = Dx^2 + Ey^2$.

8. b) $f(2, 2, 4) = 48$ is the minimum. There is no maximum.

10. a) Use Cramer's Rule. b) Apply Theorem 11.59.

CHAPTER 12

12.1 Jordan Regions

1. α) $V(E; \mathcal{G}_1) = 3/4$, $V(E; \mathcal{G}_2) = 7/16$, $V(E; \mathcal{G}_3) = 15/64$; $v(E; \mathcal{G}_m) = 0$ for all m. β) $V(E; \mathcal{G}_1) = 1$, $v(E; \mathcal{G}_1) = 0$; $V(E; \mathcal{G}_2) = 13/16$, $v(E; \mathcal{G}_2) = 0$; $V(E; \mathcal{G}_3) = 43/64$, $v(E; \mathcal{G}_3) = 5/32$. γ) $V(E; \mathcal{G}_1) = 1$, $v(E; \mathcal{G}_1) = 0$; $V(E; \mathcal{G}_2) = 1$, $v(E; \mathcal{G}_2) = 1/4$; $V(E; \mathcal{G}_3) = 15/16$, $v(E; \mathcal{G}_3) = 1/2$.

2. c) First prove that E is a Jordan region if and only if there exist grids $\mathcal{G}_m$ such that $V(\partial E; \mathcal{G}_m) \to 0$ as $m \to \infty$.

5. a) See Theorem 8.15 or 10.39. b) You may use Exercise 12.1.6a.

6. d) Apply part c) to $E_1 = (E_1 \setminus E_2) \cup E_2$. e) Apply parts c) and d) to $(E_1 \cup E_2) = (E_1 \setminus (E_1 \cap E_2)) \cup (E_2 \setminus (E_1 \cap E_2)) \cup (E_1 \cap E_2)$.

7. Is it true for rectangles?

9. $\mathbf{a}$ is a cluster point of E if and only if $B_r(\mathbf{a}) \cap E$ contains infinitely many points for every $r > 0$ (see Exercise 2.4.8).

12.2 Riemann Integration on Jordan Regions

2. a) 5.

3. Show that the difference converges to zero as $r \to 0+$.

4. b) Area(E).

7. a) -1. b) $1/2$.

10. a) Let $\varepsilon > 0$ and choose δ by uniform continuity of ϕ. Choose a grid $\mathcal{G}$ such that $U(f, \mathcal{G}) - L(f, \mathcal{G}) < \delta^2$. Then break $U(\phi \circ f, \mathcal{G}) - L(\phi \circ f, \mathcal{G})$ into two pieces: those j which satisfy $M_j(\phi \circ f) - m_j(\phi \circ f) < \delta$, and those j which satisfy $M_j(\phi \circ f) - m_j(\phi \circ f) \geq \delta$. These two pieces are small for different reasons.

b) Use Example 3.34 and Theorem 12.29.

12.3 Iterated Integrals

1. a) 1. b) $28/9$. c) $(1 - \cos(\pi^2))/\pi$.

2. a) $E = \{(x, y) : 0 \leq x \leq 1, x \leq y \leq x^2 + 1\}$ and the integral of $x + 1$ over E is $5/4$. b) $E = \{(x, y) : 0 \leq x \leq 1, 0 \leq y \leq x\}$ and the integral of $\sin(x^2)$ over E is $(1 - \cos(1))/2$. c) $E = \{(x, y, z) : 0 \leq y \leq 1, \sqrt{y} \leq x \leq 1, 0 \leq z \leq x^2 + y^2\}$ and the volume of E is $26/105$. d) $E = \{(x, y, z) : 0 \leq x \leq 1, 0 \leq y \leq x^2, x^3 \leq z \leq 1\}$ and the integral of $\sqrt{x^3 + z}$ over E is $4(2\sqrt{2} - 1)/45$.

3. a) $(1 - \log 2)/2$. b) 1. c) $(e - 2)/2$. d) $1/8$.

4. a) 3π. b) $91/30$. c) $88/105$. d) $1/18$.

7. a) See Exercise 12.3.6.

12.4 Change of Variables

1. a) $\pi(1 - \cos 4)/4$. b) $3/10$. c) $(\sqrt{2} + \log(1 + \sqrt{2}))(b^3 - a^3)/6$. (Recall that the indefinite integral of $\sec \theta$ is $\int \sec \theta = \log|\sec \theta + \tan \theta| + C$.)

2. a) $(\pi\sqrt{3}/3) \sin 3$. b) $16^2/(3 \cdot 5 \cdot 7)$.

3. a) $(6\sqrt{6} - 7)4\pi/5$. b) $\pi(4e^3 - 1 - 2(\sqrt{8} - 1)e^{\sqrt{8}})$. c) $16\sqrt{2}/15$.

4. b) $\pi r^2 d/a$.

5. a) $4/27$. b) $9/112$. c) $3(e - 1)/e$. d) 5. (Use the change of variables $x = u + v, y = u - v$.)

6. See Exercise 12.2.3.

9. See Exercise 8.2.7.

10. d) $\pi^{n/2}$.

12.5 Partitions of Unity

3. See Theorem 7.56.

12.6 The Gamma Function and Volume

5. Let ψ_n represent the spherical change of variables in $\mathbf{R}^n$ and observe that the cofactor $|A_1|$ of $-\rho \sin \varphi_1 \sin \varphi_2 \ldots \sin \varphi_{n-2} \sin \theta$ in the matrix $D\psi_n$ is identical to $\Delta_{\psi_{n-1}}$ if, in $\Delta_{\psi_{n-1}}$, θ is replaced by φ_{n-2} and each entry in the last row of $D\psi_{n-1}$ is multiplied by $\sin \theta$.

8. $r^2 \operatorname{Vol}(B_r(0))/(n + 2)$.

CHAPTER 13

13.1 Curves

5. a) This curve spirals up the cone $x^2 + y^2 = z^2$ from $(0, 1, 1)$ to $(0, e^{2\pi}, e^{2\pi})$ and has arc length $\sqrt{3}(e^{2\pi} - 1)$. b) This curve coincides with the graph of $x = \pm y^{3/2}$, $0 \le y \le 1$ (looking like a stylized gull in flight) and has arc length $2(\sqrt{13^3} - 1)/27$. c) This curve is a straight line segment from $(0, 0, 0)$ to $(4, 4, 4)$ and has arc length $4\sqrt{3}$. d) The arc length of the astroid is 6.

6. a) 27. b) $ab(a^2 + ab + b^2)/(3(a + b))$. c) 12π. d) $(5 + 3\sqrt{5})/2$.

7. b) Use Dini's Theorem.

9. Analyze what happens to (x, y) and $dy/dx := (dy/dt)/(dx/dt)$ as $t \to -\infty$, $t \to -1-$, $t \to -t+$, $t \to 0$, and $t \to \infty$. For example, prove that, as $t \to -1-$, the trace of $\phi(t)$ lies in the fourth quadrant and is asymptotic to the line $y = -x$.

11. b) Take the derivative of $v' \cdot v'$ using the Dot Product Rule. d) Observe that $\phi(t) = v(\ell(t))$ and use the Chain Rule to compute $\phi'(t)$ and $\phi''(t)$. Then calculate $\phi' \times \phi''$ directly.

13.2 Oriented Curves

1. a) A spiral on the elliptic cylinder $y^2 + 9z^2 = 9$ oriented clockwise when viewed from far out the x-axis. b) A cubical parabola (it looks like a stylized gull in flight) on the plane $z = x$ oriented from left to right when viewed from far out the plane $y = x$. c) A sine wave on the parabolic cylinder $y = x^2$ oriented from right to left when viewed from far out the y-axis. d) An ellipse sliced by the plane $x = z$ out of the cylinder $y^2 + z^2 = 1$ oriented clockwise when viewed from far out the x-axis. e) A sine wave traced vertically on the plane $y = x$ oriented from below to above when viewed from far out the x-axis.

2. a) 128/3. b) $-\pi\sqrt{2}/2$. c) 0.

3. a) 5. b) $\pi(-1 + \sqrt{5})/2$. c) $|R|(2 - a - b)/2$. d) $-\sin(1) + 1/3$.

4. c) There exist functions ψ and τ on $[0, 1]$ which are C^1 on $(0, 1) \setminus \{j/N : j = 1, \ldots, N\}$ such that $\tau' > 0$ and $\psi = \phi_j \circ \tau$ on $((j - 1)/N, j/N)$ for each $j = 1, \ldots, N$.

7. c) If $\mathbf{F}$ is conservative, consider the case when C is smooth first. If (*) holds, use parts a) and b) to prove that $\mathbf{F}$ is conservative.

8. Use Jensen's Inequality.

13.3 Surfaces

1. a) $\sqrt{2}\pi(b^2 - a^2)$. b) $4\pi a^2$. c) $4\pi^2 ab$.

2. a) $\phi(u, v) = (u, v, u^2 - v^2)$, $E = \{(u, v) : -1 \le u \le 1, -|u| \le v \le |u|\}$, $\psi_1(t) = (1, t, 1 - t^2)$, $\psi_2(t) = (-1, t, 1 - t^2)$, $\psi_3(t) = (t, t, 0)$, $\psi_4(t) = (t, -t, 0)$, $I_1 = I_2 = I_3 = I_4 = [-1, 1]$, and $\iint_S g \, d\sigma = 22/3$. b) $\phi(u, v) = (u, u^3, v)$, $E = [0, 2] \times [0, 4]$, $\psi_1(t) = (t, t^3, 4)$, $\psi_2(t) = (t, t^3, 0)$, $\psi_3(t) = (0, 0, t)$, $\psi_4(t) = (2, 8, t)$, $I_1 = I_2 = [0, 2]$, $I_3 = I_4 = [0, 4]$, and $\iint_S g \, d\sigma = (4/27)(145^{3/2} - 1)$. c) $\phi(u, v) =$

$(3\cos u \cos v, 3\sin u \cos v, 3\sin v)$, $E = [0, 2\pi] \times [\pi/4, \pi/2]$, $\boldsymbol{\psi}_1(t) = ((3/\sqrt{2})\cos t, (3/\sqrt{2})\sin t, 3/\sqrt{2})$, $\boldsymbol{\psi}_2(t) = (3\cos t, 3\sin t, 0)$, $I = [0, 2\pi]$, and $\iint_S g\,d\sigma = 27\pi/2$.

5. b) Use Theorem 12.65.
6. If you got 52π, you gave up too much when you replaced $\|(x, y)\|$ by 3.

13.4 Oriented Surfaces

1. a) Since the x-axis lies to the left of the yz-plane when viewed from far out the positive y-axis, the boundary can be parametrized by $\boldsymbol{\phi}(t) = (3\sin t, 0, 3\cos t)$, $I = [0, 2\pi]$, and $\int_{\partial S} \mathbf{F} \cdot \mathbf{T}\,ds = -9\pi$. b) The boundary can be parametrized by $\boldsymbol{\phi}_1(t) = (0, -t, 1 + 2t)$, $I_1 = [-1/2, 0]$; $\boldsymbol{\phi}_2(t) = (t, 0, 1 - t)$, $I_2 = [0, 1]$; and $\boldsymbol{\phi}_3(t) = (-t, (1 + t)/2, 0)$, $I_3 = [-1, 0]$; and $\int_{\partial S} \mathbf{F} \cdot \mathbf{T}\,ds = -1/12$. c) The boundary can be parametrized by $\boldsymbol{\phi}_1(t) = (2\sin t, 2\cos t, 4)$, $I_1 = [0, 2\pi]$, and $\boldsymbol{\phi}_2(t) = (\cos t, \sin t, 1)$, $I_2 = [0, 2\pi]$, and $\int_{\partial S} \mathbf{F} \cdot \mathbf{T}\,ds = 3\pi$.
2. a) $\pi/2$. b) 16. c) $2\pi^2 ab^2$. d) $\pi/8$.
3. a) $-14/15$. b) $4\pi a^3/3$. c) $(3b^4 + 8a^3 - 8(a^2 - b^2)^{3/2})\pi/12$. d) $-2\pi/3$.
4. b) Use Theorem 12.65.

13.5 Theorems of Green and Gauss

1. a) 8/3. b) $3\log 3 + 2(1 - e^3)$. c) $-15\pi/4$.
2. a) $(b - a)(c - d)(c + d - 2)/2$. b) $-1/6$. c) 0.
3. a) $2(5 + e^3)$. b) π. c) 8. d) $\pi abc(a + b + c)/2$.
4. a) 224/3. b) $2(8\sqrt{2} - 7)/15$. c) 24π.
5. b) 3/2. c) $\text{Vol}(E) = (1/3)\int_{\partial E} x\,dy\,dz + y\,dz\,dx + z\,dx\,dy$. d) $2\pi^2 ab^2$.
9. c) Use Exercise 13.5.8 and Gauss's Theorem.
10. e) Use Green's Theorem and Exercise 12.2.3.

13.6 Stokes's Theorem

1. a) $-\pi/4$. b) $27\pi/4$.
2. a) 0. b) -3π. c) -10π. d) $-1/12$.
3. a) $\pi^2/5$. b) $-\pi/(8\sqrt{2})$. c) 28π (not -28π because $\mathbf{i} \times \mathbf{k} = -\mathbf{j}$). d) 32π. e) $-\pi$.
4. a) 18π. b) 8π. c) $3(1 - e) + 3\pi/2$. d) 0.
10. b) 2π.

CHAPTER 14

14.1 Introduction

1. a) $a_0(x^2) = 2\pi^2/3$, $a_k(x^2) = 4(-1)^k/k^2$, and $b_k(x^2) = 0$ for $k = 1, 2, \ldots$. b) All Fourier coefficients of $\cos^2 x$ are zero except $a_0(\cos^2 x) = 1/2$ and $a_2(\cos^2 x) = 1/2$.
6. a) $a_k(f) = 0$ for $k = 0, 1, \ldots$, $b_k(f) = 4/(k\pi)$ when k is odd and 0 when k is even.
   c) You may wish to use Theorem 9.40.

14.2 Summability of Fourier Series

5. b) See Exercise 10.7.6d. c) See Exercise 5.1.4b.
7. b) See Exercise 10.7.6d.
8. See Theorem 9.49.

14.3 Growth of Fourier Coefficients

4. See Exercise 14.2.4a and Theorem 7.12.

14.4 Convergence of Fourier Series

1. *Note*: It is not assumed that f is periodic.
2. c) $\pi^2/8$.
4. a) Use Abel's Formula. For the first identity, you must show that $\rho^N S_N \to 0$ as $N \to \infty$ for all $\rho \in (0, 1)$ if $\sum_{k=0}^{\infty} a_k r^k$ converges for all $r \in (0, 1)$.
5. a) Prove that for each fixed h, $a_k(f(x+h)) = a_k(f)\cos kh + b_k(f)\sin kh$.

Subject Index

Notation Index